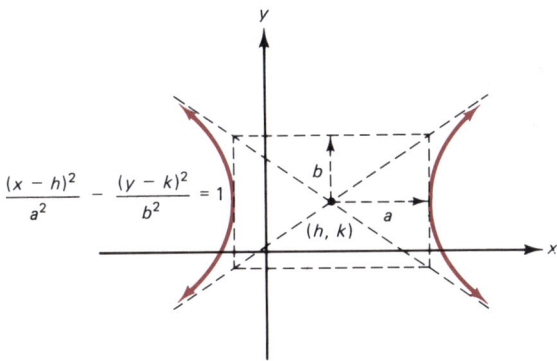

$$\frac{(x-h)^2}{a^2} - \frac{(y-k)^2}{b^2} = 1$$

Hyperbola

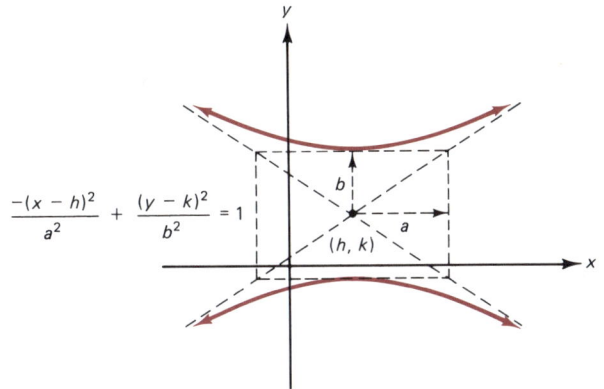

$$\frac{-(x-h)^2}{a^2} + \frac{(y-k)^2}{b^2} = 1$$

Hyperbola

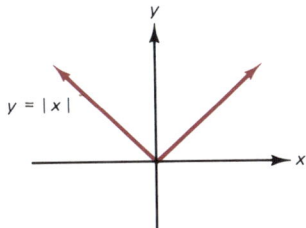

$y = |x|$

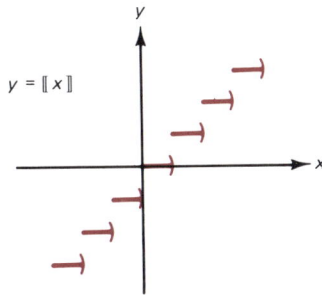

$y = [\![x]\!]$

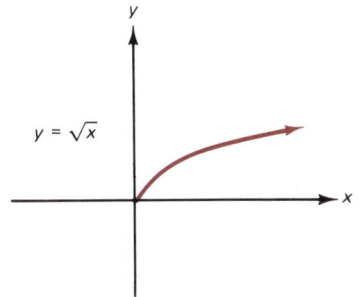

$y = \sqrt{x}$

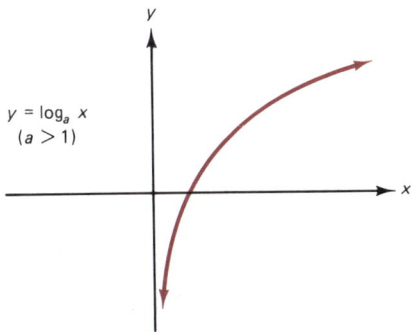

$y = \log_a x$
$(a > 1)$

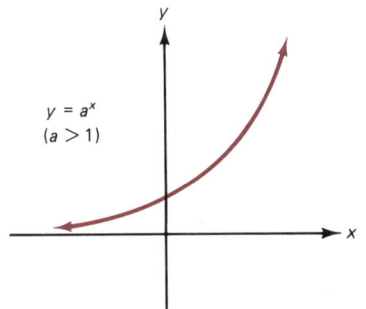

$y = a^x$
$(a > 1)$

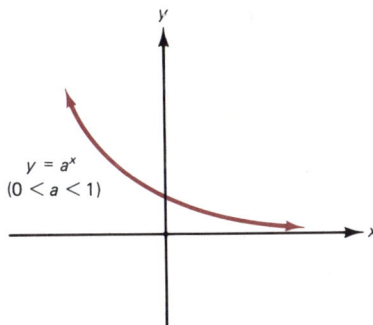

$y = a^x$
$(0 < a < 1)$

Algebra and Trigonometry

Algebra and Trigonometry

Howard A. Silver
Chicago State University

Prentice-Hall, Englewood Cliffs, New Jersey 07632

Library of Congress Cataloging-in-Publication Data

Silver, Howard A., (date)
  Algebra and trigonometry.

  Includes indexes.
  1. Algebra. 2. Trigonometry. I. Title.
QA154.2.S5187 1986   512'.13   85-25582
ISBN 0-13-021270-9

Editorial/production supervision: Kathleen M. Lafferty
Interior and cover designs: Jayne Conte
Editorial assistant: Susan Pintner
Manufacturing buyer: John B. Hall

Printed in the United States of America

10 9 8 7 6 5 4 3 2 1

ISBN 0-13-021270-9 01

Prentice-Hall International, Inc., *London*
Prentice-Hall of Australia Pty. Limited, *Sydney*
Editora Prentice-Hall do Brasil, Ltda., *Rio de Janeiro*
Prentice-Hall Canada Inc., *Toronto*
Prentice-Hall of India Private Limited, *New Delhi*
Prentice-Hall of Japan, Inc., *Tokyo*
Prentice-Hall of Southeast Asia Pte. Ltd., *Singapore*
Whitehall Books Limited, *Wellington, New Zealand*

To Becky and Lisa

CONTENTS

16 Sequences and Series 533

17 Further Topics in Algebra 555

Appendix A1

Answers to Selected Exercises A13

Index of Mathematical Terms A67

Index of Applications A73

PREFACE

Algebra and trigonometry is a critical course for students. It is the transition from the material learned in elementary and intermediate algebra to calculus and higher mathematics.

This text, *Algebra and Trigonometry*, fills this unique need. Our text reinforces and amplifies the concepts and skills learned in previous algebra courses. At the same time, the text's strong emphasis on graphing, functions, and advanced algebra topics provides an excellent preparation for calculus and other subsequent mathematics courses.

The material in this text is the set of topics standard to most college algebra courses. It is the presentation of these topics that is compelling and unique. The following special features are molded together to form a strong, cohesive text with which the students can learn and feel comfortable.

FORMAT

The distinctive, open, two-color layout invites students into the text and makes them feel comfortable.

READABILITY

The text is written in simple, down-to-earth English. Concepts are explained clearly and intuitively.

EXAMPLES

The key to learning mathematical skills is in the examples. Here, the examples are carefully chosen to illustrate the concepts. They are carefully explained with step-by-step flow charts beside the steps of the example.

PROBLEM SETS

There are more than 5700 exercises carefully structured to be an integral part of the learning experience of the text. The problem sets have the following three sections:

1. Every problem set begins with a few (8 to 18) warm-up or review exercises. These reflect the previously learned skills needed to work the new exercises. These warm-up exercises build the student's confidence (by working familiar material first) and integrate past and present material.
2. Following the warm-up exercises are carefully chosen exercises that are mixed to force the student to decide what procedure(s) to use.
3. At the end of almost every problem set are numerous applications of the material to business, engineering, physics, chemistry, health, life science, consumerism, more advanced mathematics, and so on. The applications both reinforce the concepts of the section as well as exhibit the relevance of the material.

YES/NO BOXES

Numerous YES/NO boxes warn the student of common errors. The correct procedure or statement is shown on the left under YES, while the common error is shown on the right under NO with a huge red X through it. Generally, a brief, reinforcing explanation is given below the boxes.

ILLUSTRATIONS

More than 760 illustrations appear throughout the text to enhance the subject matter.

PROCEDURES

All procedures, rules, theorems, and definitions are displayed on the page with a box for handy reference.

HAND CALCULATOR

Timely use of the scientific calculator is integrated into the text both to simplify the calculations and to reinforce the concepts themselves. Each calculator example shows the actual keys that are pressed, the 8-digit display as seen on the calculator, and a brief comment on each sequence of entries.

INTUITIVE

The trigonometric functions are developed from the intuitive right-triangle approach. The students are led carefully to the definitions for domains of real numbers.

REVIEW

Every chapter ends with a review that includes important definitions, theorems, and procedures. In addition, there is a chapter test on all the material in the chapter.

I gratefully acknowledge the help of the following reviewers during the various stages in the preparation of the manuscript: William Blair, Northern Illinois University; M. J. DeLeon, Florida Atlantic University; William Gordon, Bates College; Henry Green, Brookdale Community College; Curtis C. McKnight, University of Oklahoma; James Muhich, University of Wisconsin-Oshkosh; Paul Pontius, Pan American University; Kenneth D. Reeves, San Antonio College; Faye Thames, Lamar University; Andrew Wargo, Bucks County Community College; Carroll G. Wells, Western Kentucky University; Don Williams, Brazosport College; and Paul M. Young, Kansas State University.

I would like to thank the great Prentice-Hall staff for all their help and encouragement in the making of this text: my production editor, Kathleen Lafferty; my acquisitions editor, Bob Sickles; my field editors, Mike Ruhe and Karen Edwards; my designer, Jayne Conte; my marketing manager, Ed Moura; the other production and marketing people whose names I never learned; and to Jackie Blackmon and Richard Shields, who helped with the project. I would like to thank Jeni, Becky, and Lisa for being so patient with me while I wrote, rewrote, re-rewrote, and so on, this text.

I hope that this text proves successful and helpful to the students who use it. I would greatly appreciate any feedback by the student and faculty using this text. Good luck.

Howard Silver

The Number System

1.1

SETS OF NUMBERS AND THE REAL NUMBERS

In algebra we study numbers, such as 2 and $\sqrt{5}$, and symbols representing numbers, such as x and y. We study relations between numbers and operations on the numbers. We study individual numbers as well as sets of numbers.

We have two ways to express the elements of a **set** of numbers. The following shows how the same set can be written with the **roster method** or **set-builder notation**.

Roster: $A = \{1, 2, 3, 4, 5, 6, 7, 8\}$

Set-builder: $A = \{x \mid x$ is a whole number between 1 and 8, inclusive$\}$

        (This is read "The set of all x such that x is a whole number between 1 and 8, inclusive.")

We have special sets of numbers that we use over and over. First, we have the **natural numbers** (or **counting numbers**). This is the set

$$N = \{1, 2, 3, 4, \ldots\}$$

The natural numbers have their limitations. We cannot subtract $3 - 7$ or solve the equation $x + 7 = 3$. We therefore introduce negative numbers, such as -4, and zero. Now we have the set of **integers**

$$I = \{\ldots, -3, -2, -1, 0, 1, 2, 3, \ldots\}$$

The set I consists of the natural numbers, zero, and the negatives of the natural numbers.

The integers also have their limitations. We cannot divide $2 \div 5$ or solve the equation $5x = 2$. We therefore introduce fractions, such as 2/5. We now have the set of **rational numbers**

$$Q = \left\{ \frac{a}{b} \,\middle|\, a \text{ and } b \text{ are integers, } b \neq 0 \right\}$$

This set includes all fractions, such as $1/2$, $-5/7$, and $6 = 6/1$. All fractions can be written as either a terminating or a repeating decimal; for example, $9/8 = 1.125$ and $4/11 = 0.363636\ldots$.

Even the rational numbers are not complete. Mathematicians have found numbers such as π and $\sqrt{2}$ that are not rational. (We cannot write $\sqrt{2}$ as a fraction.) We have the set of **irrational numbers**

$$H = \{x \mid x \text{ is a nonterminating and nonrepeating decimal}\}$$

These are decimals, such as $5.18473\ldots$, that do not terminate or repeat. Examples include π, $\sqrt{2}$, $\sqrt{3}$, and $\sqrt[4]{5}$.

Finally, combining the rational and irrational numbers, we get the set of **real numbers**

$$R = \{x \mid x \text{ is any decimal}\}$$

Throughout most of this text, the numbers that we encounter are real.

Figure 1 shows the relations between these sets: N is contained in I, which is contained in Q, which is contained in R. The set H, which has nothing in common with Q, is also contained in R.

EXAMPLE 1   For each of the numbers listed in the table, we note all the sets in which the number is an element.

Number	Sets Containing Number
-8	I, Q, R
$\dfrac{2}{5}$	Q, R
0	I, Q, R
$\sqrt{7}$	H, R
15	N, I, Q, R
$0.777\ldots$	Q, R
$\pi = 3.1415\ldots$	H, R
0.359	Q, R

It is often convenient to picture the real numbers on a **number line**. There is a one-to-one match between all the real numbers and the points of the line. (For every real number, there is a point on the line; conversely, for every point on the line, there is a real number.)

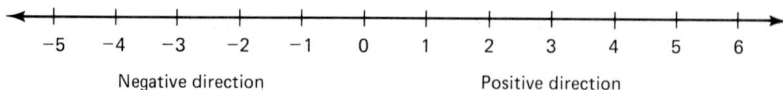

Negative direction   Positive direction

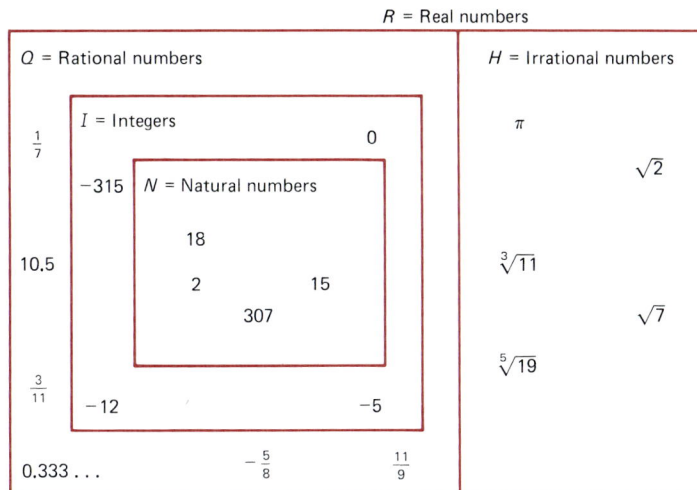

FIGURE 1

Just as there is no last real number, the number line extends without end in both directions. The real number associated with each point of the line is called its **coordinate**. The point 0 is called the **origin**. To the right of the origin is the **positive direction**; to the left of the origin is the **negative direction**.

EXAMPLE 2 On the following number line are examples of points and their coordinates. Note that we must sometimes estimate the coordinates and the locations.

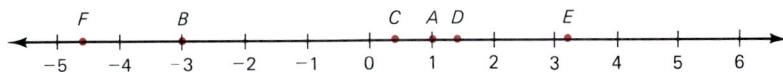

$$A = 1 \qquad B = -3 \qquad C = \frac{2}{5}$$

$$D = \sqrt{2} \qquad E = \pi \qquad F = -4.7$$

Two expressions are **equal** if they name the same number.

If two numbers are not equal, then one is larger than the other. We have **inequality symbols** to indicate when one number is larger (or smaller) than another. For instance, the "less than" symbol is defined as

$$a < b \quad \text{if} \quad b - a \text{ is positive}$$

On the number line the statement $a < b$ corresponds to "a is to the *left* of b."

EXAMPLE 3 Following are several examples of inequality symbols and their English equivalents.

Statement	English
(a) $6 < 9$	6 is less than 9
(b) $-7 \geq -15$	-7 is greater than or equal to -15
(c) $8 \leq 8$	8 is less than or equal to 8
(d) $x > -2$	x is greater than -2
(e) $-1 < a < 3$	a is between -1 and 3
(f) $t < 0$	t is less than 0 (t is negative)

EXAMPLE 4 Graph $\{x \mid x < -2\}$.

Solution $\{x \mid x < -2\}$ is the set of all real numbers less than (to the left of) -2. When we graph this, we use a right parenthesis ")" at -2 to show that it is *not* in the set.

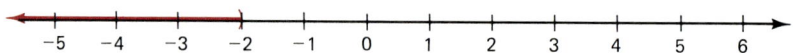

EXAMPLE 5 Graph $\{x \mid -3 \leq x \leq 4\}$.

Solution $\{x \mid -3 \leq x \leq 4\}$ is the set of all real numbers to the right of -3 and to the left of 4. Here we use square brackets "[ ]" at -3 and 4 to indicate that they are in the set.

**PROBLEM SET
1.1**

Perform the indicated operations.

*Warm-up
Exercises*

1. $35 + 87$

2. $402 - 189$

3. $\dfrac{1}{3} + \dfrac{1}{4}$

4. $7 - \dfrac{3}{5}$

5. $\left(\dfrac{2}{5}\right)\left(\dfrac{3}{7}\right)$

6. $\dfrac{3}{5} \div \dfrac{1}{7}$

7. $4.81 - 2.1$

8. $6.002 + 3.2 + 11.98$

9. $(0.002)(3.1)$

10. $4.8 \div 0.002$

Write the following sets using the roster method.

11. $\{x \mid x \text{ is a month that starts with "M"}\}$

12. $\{x \mid x \text{ is a day that starts with "S"}\}$

13. {x|x is a whole number between 2 and 10, inclusive}

14. {x|x is a multiple of 10 between 5 and 55}

Write the following using set-builder notation.

15. {1, 2, 3, 4, 5, 6}

16. {10, 11, 12, 13, 14}

17. {4, 6, 8, 10, 12, 14, 16}

18. {20, 40, 60}

For each of the following numbers, list all the sets (N, I, Q, H, or R) in which it is an element.

19. -10

20. $\dfrac{1}{2}$

21. 7.31

22. 5.12593...

23. 0

24. $\sqrt{5}$

25. π

26. 25

27. 0.414141...

28. 6.05

29. 17

30. $\dfrac{-3}{5}$

Answer the following statements as true or false.

31. All natural numbers are integers.

32. All fractions are rational numbers.

33. All real numbers are rational numbers.

34. All irrational numbers are real numbers.

35. All rational numbers are repeating decimals.

36. The set of numbers both rational and irrational is empty.

37. Some rational numbers are integers.

38. Some real numbers are rational numbers.

39. No real number is an integer.

40. No natural number is irrational.

Give (or estimate) the coordinates of the points shown on the following number line.

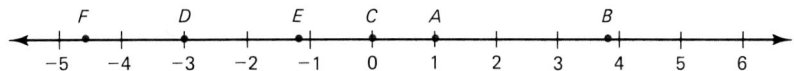

41. Point A

42. Point B

43. Point C

44. Point D

45. Point E

46. Point F

Translate the following sentences into mathematical symbols.

47. 5 is less than 8

48. 16 is greater than 10

49. -2 is greater than x

50. t is less than s

51. 10 is greater than or equal to r

52. m is less than or equal to k

53. y is between -2 and 3

54. z is between 5 and 10

Graph the following sets.

55. $\{x \mid x < 5\}$

56. $\{x \mid x > -2\}$

57. $\{x \mid x \geq -4\}$

58. $\{x \mid x \leq 3\}$

59. $\{x \mid -4 \leq x \leq 3\}$

60. $\{x \mid 0 \leq x \leq 10\}$

61. $\{x \mid -3 \leq x < 5\}$

62. $\{x \mid 2 < x \leq 12\}$

Technical Application

63. A bolt is to be 1.25 inches with a tolerance (the absolute value of the error) of 0.02 inch.

(a) What is the largest and smallest the bolt can be?

(b) Graph the set of lengths that the bolt can be.

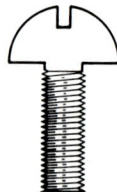

Life Science Application

64. A certain plant's temperature must remain between 60 and 80°F. Graph this set of temperatures.

Business Application

65. The Blowout Tire Company produces four lines of tires: two-ply, four-ply, snow, and radial. In one month it has 320,000 hours of available vulcanization time. The relation is

$$4t + 5f + 6s + 7r \leq 320,000$$

Determine if they can meet the following tire demands: $t = 8000$ two-ply, $f = 15,000$ four-ply, $s = 6000$ snow, and $r = 14,000$ radial.

1.2

PROPERTIES OF THE REAL NUMBERS

In the preceding section we saw that the real numbers consist of the set of all decimal numbers. Since we will be using real numbers throughout most of this text, it is important now to state some of the many properties that hold for real numbers. (When stating these properties, we use letters, such as a, b, and c, to represent real numbers.)

There are two basic operations on the real numbers: **addition** and **multiplication**. The addition of numbers a and b produces the **sum**, written $a + b$. The multiplication of a and b produces the **product**, written ab, $a \cdot b$, $a \times b$, $(a)b$, or $(a)(b)$.

Property 1

For real numbers a, b, and c:

		EXAMPLE
Closure Properties	$a + b$ is a real number. ab is a real number.	$6 + \pi$ is a real number. $2\sqrt{5}$ is a real number.
Commutative Properties	$a + b = b + a$ $ab = ba$	$2 + 5 = 5 + 2$ $3(5) = 5(3)$
Associative Properties	$a + (b + c) = (a + b) + c$ $a(bc) = (ab)c$	$2 + (3 + 4) = (2 + 3) + 4$ $2(3 \cdot 4) = (2 \cdot 3)4$
Identity Properties	$a + 0 = 0 + a = a$ $a \cdot 1 = 1 \cdot a = a$	$2 + 0 = 0 + 2 = 2$ $5 \cdot 1 = 1 \cdot 5 = 5$
Inverse Properties	For every real number a, there is a real number $(-a)$ such that $a + (-a) = 0$. For every *nonzero* real number a there is a real number $(1/a)$ such that $a \cdot (1/a) = 1$.	$8 + (-8) = 0$ $\dfrac{3}{5} \cdot \dfrac{5}{3} = 1$
Distributive Property	$a(b + c) = ab + ac$	$6(x + 2) = 6x + 12$

Two expressions are **equal** if they name the same number. For example, $8 + 1$, $14 - 5$, $18 \div 2$, and 3^2 all name the same number: *nine*. We have the following properties for equality.

Property 2

For real numbers a and b:

		EXAMPLE
Substitution Property	If $a - b$, then a can replace b (and b can replace a) in any statement or expression.	(i) If $x = 7$ and $y = 2x$, then $y = 2(7) = 14$. (ii) If $x = y$ and $y = a + b$, then $x = a + b$.

From Properties 1 and 2, we can derive other very important properties that will be used later when we solve equations for an unknown or unknowns.

Property 3

For real numbers a, b, and c:

EXAMPLE

Addition Property of Equality	If $a = b$, then $a + c = b + c$.	If $x = 7$, then $x + 3 = 10$.
Multiplication Property of Equality	If $a = b$, then $a \cdot c = b \cdot c$.	If $z = 4$, then $5z = 20$.
Zero-Multiplier Property	$a \cdot 0 = 0$	$9 \cdot 0 = 0$ or $(-3) \cdot 0 = 0$.
Zero-Product Property	If $a \cdot b = 0$, then either $a = 0$ or $b = 0$ (or both).	If $x(x - 5) = 0$, then either $x = 0$ or $x - 5 = 0$.

The inverse properties in Property 1 are very important. The number $-a$ is called the **additive inverse** (or **negative**) of a. For example, -6 is the additive inverse of 6, and 8 is the additive inverse of -8.

The number $1/a$ is called the **multiplicative inverse** (or **reciprocal**) of a (for $a \neq 0$). For example, $1/3$ is the reciprocal of 3, and $-2/5$ is the reciprocal of $-5/2$.

Real numbers and their negatives occur throughout this text. We cannot overstate their importance. We now state the following properties for additive inverses. Notice that **subtraction** $a - b$ is defined as the *addition of the additive inverse* $a + (-b)$.

Property 4

For real numbers a and b:

EXAMPLE

Double-Negative Property	$-(-a) = a$	$-(-9) = 9$
Addition Property	$(-a) + (-b) = -(a + b)$	$-10 + (-8) = -18$
Subtraction Definition	$a - b = a + (-b)$	$5 - (-8) = 5 + 8 = 13$
Multiplication Properties	$(-a)(-b) = ab$ $(-a)b = -(ab)$ $(-1)a = -a$	$(-7)(-10) = 70$ $(-12)5 = -60$ $(-1)(3) = -3$

The **division** of a and b produces the **quotient** a/b $\left(\text{or } \dfrac{a}{b} \text{ or } a \div b\right)$ defined as $a \cdot \dfrac{1}{b}$ (for $b \neq 0$). Basically, this states that division is the *multiplication by the reciprocal.* We now state some important properties for quotients. In the quotient (or fraction) a/b, a is called the **numerator** and b the **denominator**.

Property 5

For real numbers a, b, c, and d (all denominators are nonzero):

EXAMPLE

Division Definition	$\dfrac{a}{b} = a \cdot \dfrac{1}{b}$	$\dfrac{3}{2} = 3 \cdot \dfrac{1}{2}$
Division Properties	$\dfrac{-a}{b} = \dfrac{a}{-b} = -\dfrac{a}{b}$	$\dfrac{40}{-8} = -5 \quad \text{or} \quad \dfrac{-12}{3} = -4$
	$\dfrac{-a}{-b} = \dfrac{a}{b}$	$\dfrac{-20}{-2} = 10$
Equality Property	$\dfrac{a}{b} = \dfrac{c}{d} \quad \text{if} \quad ad = bc$	$\dfrac{9}{15} = \dfrac{12}{20}$ since $9(20) = 15(12)$
Cancellation Property	$\dfrac{a}{b} = \dfrac{ac}{bc}$	$\dfrac{4}{7} = \dfrac{4 \cdot 3}{7 \cdot 3} = \dfrac{12}{21}$
Addition of Quotients	$\dfrac{a}{c} + \dfrac{b}{c} = \dfrac{a+b}{c}$	$\dfrac{5}{7} + \dfrac{3}{7} = \dfrac{8}{7}$
Multiplication of Quotients	$\dfrac{a}{b} \cdot \dfrac{c}{d} = \dfrac{ac}{bd}$	$\dfrac{2}{7} \cdot \dfrac{3}{5} = \dfrac{6}{35}$
Division of Quotients	$\dfrac{a}{b} \div \dfrac{c}{d} = \dfrac{a}{b} \cdot \dfrac{d}{c}$	$\dfrac{2}{5} \div \dfrac{3}{7} = \dfrac{2}{5} \cdot \dfrac{7}{3} = \dfrac{14}{15}$

PROBLEM SET 1.2

For each of the following numbers, list all the sets (N, I, Q, H, or R) in which it is an element.

Warm-up Exercises

1. -46 **2.** 2.019 **3.** π

4. 34 **5.** $\dfrac{-4}{5}$ **6.** $\sqrt{7}$

For each of the following statements, state the real-number property that makes it true.

7. $6(7x) = (6 \cdot 7)x$ **8.** $\dfrac{2}{3} \cdot \dfrac{3}{2} = 1$

9. $7 + 0 = 7$ **10.** $\pi + 3$ is a real number.

11. $8(y + 3) = 8y + 24$

12. $7 \cdot a = a \cdot 7$

13. $5\sqrt{2}$ is a real number.

14. $7 + (10 + t) = (7 + 10) + t$

15. $6 + t = t + 6$

16. $x \cdot 1 = x$

17. $12 + (-12) = 0$

18. $4(a + 10) = 4a + 40$

19. If $x = 2$ and $y = 5x$, then $y = 5(2) = 10$.

20. If $a = b$ and $a = x^2 + 2$, then $b = x^2 + 2$.

21. If $x = 5$, then $3x = 15$.

22. If $(a + 2)(a - 3) = 0$, then either $a + 2 = 0$ or $a - 3 = 0$.

23. $(x^2 + x + 7) \cdot 0 = 0$

24. If $a = 7$, then $a + 2 = 9$.

25. If $x(x^2 - 1) = 0$, then either $x = 0$ or $x^2 - 1 = 0$.

26. $0 \cdot (4x) = 0$

27. If $x - 7 = 10$, then $x = 17$.

28. If $x/3 = 7$, then $x = 21$.

29. $4 - (-11) = 4 + 11 = 15$

30. $-(-3) = 3$

31. $(-3)(-6) = 18$

32. $(-80) + (-10) = -90$

33. $-[-(-8)] = -8$

34. $(-70)(10) = -700$

35. $(-10) + (-9) = -19$

36. $-4 - 20 = -4 + (-20) = -24$

37. $\dfrac{2}{9} + \dfrac{8}{9} = \dfrac{10}{9}$

38. $\dfrac{-10}{2} = -5$

39. $\dfrac{2}{7} \div \dfrac{8}{9} = \dfrac{2}{7} \cdot \dfrac{9}{8}$

40. $\dfrac{4}{11} \cdot \dfrac{2}{7} = \dfrac{8}{77}$

41. $\dfrac{4}{14} = \dfrac{6}{21}$ since $4 \cdot 21 = 14 \cdot 6 = 84$

42. $\dfrac{2}{9} + \dfrac{3}{5} = \dfrac{10}{45} + \dfrac{27}{45} = \dfrac{37}{45}$

43. $\dfrac{-28}{-7} = 4$

44. $\dfrac{-2}{5} \div \dfrac{3}{8} = \dfrac{-2}{5} \cdot \dfrac{8}{3} = \dfrac{-16}{15}$

Use the property given in parentheses to fill in the blank and make the statement true.

45. $-(-11) = $ _____ (double-negative property)

46. $6(a + 11) = $ _____ (distributive property)

47. $(7x^2 - 8x + 11) \cdot 0 = $ _____ (zero-multiplier property)

48. If $x = 10$, then $x + 2 = $ _____ (addition property)

49. $\dfrac{-21}{-7} = $ _____ (division property)

50. $0 + 13 = $ _____ (identity property)

51. $12 + \sqrt{7}$ is a _____ (closure property)

52. $t + 5 = 5 + $ _____ (commutative property)

53. $5\left(\dfrac{1}{5}\right) = $ _____ (inverse property)

54. $7 - (-5) = $ _____ (subtraction definition)

55. If $x = 4$ and $y = x^2$, then $y = $ _____ (substitution property)

56. $\dfrac{3}{7} \div \dfrac{4}{9} = $ _____ (division of quotients)

57. $\dfrac{2}{7} + \dfrac{4}{7} = $ _____ (addition of quotients)

58. $(-2)(-10) = $ _____ (multiplication property)

For each of the following proofs, provide the reason for each statement.

59. *Double-negative property:* $-(-a) = a$

(a) $-(-a) = -(-a) + 0$ _____

(b) $\quad\quad = -(-a) + (-a + a)$ _____

(c) $\quad\quad = [-(-a) + (-a)] + a$ _____

(d) $\quad\quad = 0 + a$ _____

(e) $\quad\quad = a$ _____

60. *Addition property:* $(-a) + (-b) = -(a + b)$

(a) $(-a) + (-b) = (-1)a + (-1)b$ _____

(b) $\quad\quad = (-1)(a + b)$ _____

(c) $\quad\quad = -(a + b)$ _____

1.3

**REAL-NUMBER
COMPUTATIONS**

In this section we review computations on the real numbers. We begin with the **absolute value** of a real number x; this is written $|x|$.

Definition

For any real number x,

$$|x| = \begin{cases} x & \text{if } x \geq 0 \\ -x & \text{if } x < 0 \end{cases}$$

In words, if x is positive, use it to compute $|x|$; if x is negative, use $-x$ (which is positive) to compute $|x|$; finally, if x is 0, then $|x| = 0$.

EXAMPLE 6   The following examples show how the absolute-value symbol works. Note that whatever is within the absolute-value bars becomes nonnegative.

(a) $|9| = 9$ (b) $|-8| = 8$

(c) $|0| = 0$ (d) $-|10| = -10$

EXAMPLE 7   The following are additions and subtractions of real numbers. (Recall that for subtraction, we add the additive inverse of the number to be subtracted.)

(a) $-3 + (-7) = -10$ (b) $-7 + 11 = 4$

(c) $-15 + 2 = -13$ (d) $-10 + (-2) = -12$

(e) $3 - 7 = 3 + (-7) = -4$ (f) $-5 - 8 = -5 + (-8) = -13$

(g) $2 - (-5) = 2 + 5 = 7$ (h) $-4 - (-6) = -4 + 6 = 2$

EXAMPLE 8   The following are examples of multiplication and division of real numbers.

(a) $(-3)(7) = -21$ (b) $(-8)(-5) = 40$

(c) $(-1)(-2)(-3) = -6$ (d) $(-1)(-2)(-2)(-5) = 20$

(e) $\dfrac{-28}{-7} = 4$

(f) $-10 \div 2 = -5$

(g) $\dfrac{(-1)(-10)}{(2)(-5)(1)} = -1$

(h) $\dfrac{(2)(-6)(-10)}{(-4)(5)} = \dfrac{120}{-20} = -6$

Exponents are a convenient way to express repeated factors.

Definition

For any real number b and natural number n,

$$b^n = \underbrace{b \cdot b \cdot b \cdot \ldots \cdot b}_{n \text{ factors}}$$

In the expression b^n, we call b the **base** and n the **exponent**. The expression b^n is called the **nth power** of b.

EXAMPLE 9 The following are examples of exponents.
(a) $5^2 = 5 \cdot 5 = 25$
(b) $2^5 = 2 \cdot 2 \cdot 2 \cdot 2 \cdot 2 = 32$
(c) $\left(\dfrac{1}{2}\right)^4 = \left(\dfrac{1}{2}\right)\left(\dfrac{1}{2}\right)\left(\dfrac{1}{2}\right)\left(\dfrac{1}{2}\right) = \dfrac{1}{16}$
(d) $(1.08)^3 = (1.08)(1.08)(1.08) = 1.259712$
(e) $(-3)^2 = (-3)(-3) = 9$
(f) $(a + b)^4 = (a + b)(a + b)(a + b)(a + b)$

Note that $5^2 \ne 2^5$. We discuss exponents in greater detail in the next chapter.

YES	NO
$(-3)^2 = (-3)(-3) = 9$	~~$(-3)^2 = -9$~~
$-3^2 = -9$	~~$-3^2 = 9$~~

A negative sign within the parentheses is a part of the base. When there are no parentheses, the negative sign is not a part of the base.

We discuss radicals in great detail later. Here let us simply recall that for $a \ge 0$, $\sqrt{a}$ is called **radical a** and answers the question: What nonnegative number times itself equals a? For example, $\sqrt{25} = 5$, $\sqrt{49} = 7$, and $\sqrt{100} = 10$.

In mathematics, we have a **standard order of operations** in which the operations are performed. In addition to algebra, this order is also followed in most major computer languages, such as BASIC, FORTRAN, Pascal, and C.

STANDARD ORDER OF OPERATIONS

1. All operations within parentheses (or other **grouping symbols**, such as braces, brackets, fraction bars, or radicals)
2. All exponents or radicals
3. All multiplications and divisions, left to right
4. All additions and subtractions, left to right

EXAMPLE 10   Simplify $3 \cdot (5 + 1)^2 + 7$.

Solution   We start with the parentheses.

Parentheses → Exponent → Multiplication → Addition

$$3 \cdot (5 + 1)^2 + 7 = 3 \cdot 6^2 + 7$$
$$= 3 \cdot 36 + 7$$
$$= 108 + 7$$
$$= 115$$

EXAMPLE 11   Simplify $-2[3^2 - 4(11 - 2^3)]$.

Solution   This expression contains **nested grouping symbols** (that is, parentheses within brackets). Here we work inside out: We first simplify within the inner parentheses and then the outer brackets.

Innermost parentheses → Outermost brackets

$$-2[3^2 - 4(11 - 2^3)] = -2[3^2 - 4(11 - 8)]$$
$$= -2[3^2 - 4 \cdot 3]$$
$$= -2[9 - 12]$$
$$= -2[-3] = 6$$

Throughout this text we use the **hand calculator** in the examples where it fits naturally. (Your instructor is the final judge of how much you can use the calculator for this course.) If you do not own a calculator or are in the market for a new one, here are some things to consider. (The examples in the text are based on calculators of this type.)

1. *Features.* The following keys will aid you in algebra, trigonometry, science, and business:

| sin | cos | tan | INV | $\sqrt{x}$ | log | ln x | 1/x | y^x | (|) |

This is often called a **scientific** (or slide rule) **calculator** and is made by Texas Instruments, Sharp, Casio, and others.

2. *Algebraic operating system (AOS).* This calculator has the features listed above and an $\boxed{=}$ key. It uses the standard order of operations (see page 13) used in algebra and most computer languages. The expression "$2 + 3 \cdot 4$" is evaluated as follows:

PRESS	DISPLAY	MEANING
$\boxed{2}\;\boxed{+}\;\boxed{3}\;\boxed{\times}\;\boxed{4}\;\boxed{=}$	14.	AOS logic

Let us redo two of the examples of this section on such a calculator.

	PRESS	DISPLAY	MEANING
Example 9(d):	$\boxed{1}\;\boxed{.}\;\boxed{0}\;\boxed{8}\;\boxed{y^x}\;\boxed{3}\;\boxed{=}$	1.259712	1.08^3
Example 10:	$\boxed{3}\;\boxed{\times}\;\boxed{(}\;\boxed{5}\;\boxed{+}\;\boxed{1}\;\boxed{)}\;\boxed{x^2}$	36.	$\left.\begin{array}{l} \\ \\ \end{array}\right\}\; 3 \cdot (5+1)^2 + 7$
	$\boxed{+}\;\boxed{7}\;\boxed{=}$	115.	

PROBLEM SET 1.3

Simplify the following.

Warm-up Exercises

1. $21 - 9$

2. $(0.8)(0.8)$

3. $\dfrac{0.48}{0.08}$

4. $\dfrac{40}{(2)(5)}$

5. $\left(\dfrac{1}{2}\right)\left(\dfrac{1}{2}\right)\left(\dfrac{1}{2}\right)$

6. $\dfrac{1}{4} + \dfrac{1}{5}$

7. $\dfrac{1}{3} - \dfrac{1}{4}$

8. $\sqrt{25}$

Simplify the following.

9. $-7 + 9$

10. $2 + 7$

11. $-6.1 + (-5.3)$

12. $10 + (-3)$

13. $-3 - 4$

14. $12 - 5$

15. $-10 - (-12)$

16. $3 - (-10)$

17. $(-40)(-2)$

18. $(-2)(-5)$

19. $(-5)(-6)(-1)(-2)$

20. $(-2)(-3)(4)$

21. $\dfrac{-100}{2}$

22. $\dfrac{25}{-5}$

23. $-40 \div (-8)$

24. $14 \div (-7)$

25. $\dfrac{(-9)(-2)(-1)}{(-6)(-3)}$

26. $\dfrac{(-8)(-4)(-1)}{2(-2)}$

27. $(-2)^4$

28. 6^2

29. $\left(\dfrac{-1}{2}\right)^3$

30. $\left(\dfrac{2}{5}\right)^2$

31. 0.7^3

32. 1.05^3

33. $-3 + 7(-2)$

34. $2 + 7 \cdot 3$

35. $(-5)4 - (-6)(-8)$

36. $-4 - 8 \cdot 2$

37. $2 \cdot 5^2$

38. $5^2 + 6^2$

39. $-10 + (-4)2^3$

40. $3 \cdot 4^3$

41. $(7 + 2)^2$

42. $2(5 + 7)$

43. $5(6 - 4)^2$

44. $(10 - 6)^3$

45. $2 + 5(3^2 + 1)^2$

46. $(2^2 + 1)^3$

47. $\sqrt{3 \cdot 8 + 1}$

48. $10(4^2 - 9) - 6^2$

49. $2[7 + 4(8 + 1)]$

50. $-3[12 - 2(5 + 2)]$

51. $[(1 + 2)^2 + (4 - 2)^3]^2$

52. $[(6 - 4)^4 - (5 - 2)^2]^2$

53. $\dfrac{4^2 + 2^2}{3^2 - 2^2}$

54. $\dfrac{4\sqrt{16} - 2\sqrt{25}}{2\sqrt{9}}$

55. $\sqrt{\dfrac{10 + 8}{3 - 1}}$

56. $\dfrac{\sqrt{5 \cdot 16 + 1}}{\sqrt{25} + \sqrt{9 + 7}}$

57. $\sqrt{-4 + 2^2}$

58. $\sqrt{16} + \sqrt{9}$

59. $\sqrt{16 + 9}$

60. $\sqrt{2^3 + 1}$

Physical Application

61. The fraction R of light reflected at a surface is given by

$$R = \frac{(n - 1)^2}{(n + 1)^2}$$

where n is the index of refraction. Find R when:
(a) $n = 1.5$ (glass)
(b) $n = 1.33$ (water)

Technical Application

62. The total resistance in 1000 feet of copper wire is approximated by

$$R = 2^{(g - 10)/3}$$

where g is the gauge of the wire. Find the resistance R when:
(a) $g = 13$
(b) $g = 19$
(c) $g = 40$

Consumer Application

63. If a person deposits P dollars in the bank *every* year for N years at interest rate r, the account will grow to

$$S = \frac{P[(1 + r)^N - 1]}{r}$$

Find S when:
(a) $P = \$1000$, $N = 5$, $r = 10\%$
(b) $P = \$2000$, $N = 10$, $r = 12\%$
(*Caution*: Be sure to write r as a decimal.)

Acoustical
Application

64. The lowest resonant frequency of a sound in a room is given by

$$f_0 = \frac{170}{\sqrt{d}} \sqrt{\frac{1}{M_1} + \frac{1}{M_2}}$$

where d is the separation of the walls and M_1 and M_2 are the surface densities. Find f_0 when:
(a) $d = 100$, $M_1 = 8$, $M_2 = 8$
(b) $d = 400$, $M_1 = 80$, $M_2 = 20$

CHAPTER 1 SUMMARY

Important Properties and Definitions

For real numbers a, b, c, and d (denominators cannot be zero):

$a + b$ is a real number.	ab is a real number.
$a + b = b + a$	$ab = ba$
$a + (b + c) = (a + b) + c$	$a(bc) = (ab)c$
$a + 0 = a$	$a \cdot 1 = a$
$a + (-a) = 0$	$a \cdot \dfrac{1}{a} = 1$
$a(b + c) = ab + ac$	$a \cdot 0 = 0$
$-(-a) = a$	$(-a) + (-b) = -(a + b)$
$a - b = a + (-b)$	$-a = (-1)a$
$(-a)(-b) = ab$	$(-a)(b) = -ab$
$\dfrac{-a}{b} = \dfrac{a}{-b} = -\dfrac{a}{b}$	$\dfrac{-a}{-b} = \dfrac{a}{b}$
$\dfrac{a}{b} = \dfrac{ac}{bc}$	$\dfrac{a}{c} + \dfrac{b}{c} = \dfrac{a + b}{c}$
$\dfrac{a}{b} \cdot \dfrac{c}{d} = \dfrac{ac}{bd}$	$\dfrac{a}{b} \div \dfrac{c}{d} = \dfrac{a}{b} \cdot \dfrac{d}{c} = \dfrac{ad}{bc}$
$|a| = \begin{cases} a & \text{if } a \geq 0 \\ -a & \text{if } a < 0 \end{cases}$	$b^n = b \cdot b \cdot \ldots \cdot b, \quad n \text{ factors}$

If $a = b$, then a can validly replace b in any statement or expression, or vice versa.
If $a = b$, then $a + c = b + c$ and $ac = bc$.
If $ab = 0$, then either $a = 0$, $b = 0$, or both.

Standard Order of Operations:
1. All operations within parentheses or other grouping symbols
2. All exponents or roots
3. All multiplications and divisions, left to right
4. All additions and subtractions, left to right

Review Exercises

1. Write $\{x | x$ is a multiple of 5 between 1 and 32$\}$ with the roster method.
2. Write $\{8, 10, 12, 14, 16, 18, 20\}$ with set-builder notation.
3. To what sets (N, I, Q, H, R) does -8 belong?

For Problems 4 to 9, state the real-number property that makes each statement true.

4. $9 \cdot x = x \cdot 9$

5. $-2(a - 5) = -2a + 10$

6. If $(x - 2)(x + 3) = 0$, then either $x - 2 = 0$ or $x + 3 = 0$.

7. If $x - 7 = 10$, then $x = 17$.

8. $-(-5) = 5$

9. $\dfrac{-2}{5} \div \dfrac{3}{8} = \dfrac{-2}{5} \cdot \dfrac{8}{3}$

10. Give the coordinates of points A and B.

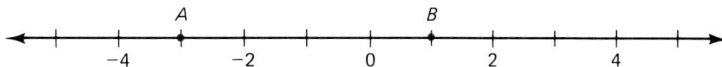

11. Translate "x is less than 5" into mathematical symbols.

12. Graph the set $\{x \mid x < 7\}$.

13. Graph the set $\{x \mid -2 \le x \le 5\}$.

Simplify as much as possible.

14. $|-3|$

15. $|-8| - |-9|$

16. $(-1)(-2)(-3)(-4)$

17. $\dfrac{(-8)(-3)(-4)}{(-1)(-6)}$

18. $-5 + 3 \cdot 4$

19. $7^2 - 2^3$

20. $6(5 - 2)^2$

21. $\sqrt{2 \cdot 5^2 - 1}$

22. $\dfrac{5^2 - 1^2}{7 + 1}$

23. $\sqrt{\dfrac{6^2 - 2^2}{1^2 + 1^2}}$

24. $\dfrac{6^2 - 1^2}{\sqrt{16} + \sqrt{9}}$

25. $\left(\dfrac{4 + 2}{1 + 1}\right)^2 + \sqrt{\dfrac{7^2 + 1^2}{2^2 - 1^2 - 1^2}}$

Polynomial Expressions

Recall from Chapter 1 that exponents are defined as follows:

$$b^n = \underbrace{b \cdot b \cdot \ldots \cdot b}_{n \text{ factors}}$$

where b is the **base** and n is the **exponent**. The expression b^n is called the **nth power** of b.

Let us now state some of the properties of exponents. (Although we do not prove these formally, in the examples we give some idea why they are true.)

Property
For real numbers a and b and natural numbers m and n (denominators are not zero):

E1 $a^m a^n = a^{m+n}$

E2 $(a^m)^n = a^{mn}$

E3 $(ab)^m = a^m b^m$

E4 $\dfrac{a^m}{a^n} = \begin{cases} a^{m-n} & \text{if } m > n \\ 1 & \text{if } m = n \\ \dfrac{1}{a^{n-m}} & \text{if } m < n \end{cases}$

E5 $\left(\dfrac{a}{b}\right)^n = \dfrac{a^n}{b^n}$

EXAMPLE 1   The following demonstrate Property E1.

(a) $2^4 \cdot 2^5 = (2 \cdot 2 \cdot 2 \cdot 2)(2 \cdot 2 \cdot 2 \cdot 2 \cdot 2)$
$$= 2 \cdot 2 \cdot 2 \cdot 2 \cdot 2 \cdot 2 \cdot 2 \cdot 2 \cdot 2 = 2^9 = 512$$

(b) $x^7 x^6 = x^{7+6} = x^{13}$

(c) $a^{2k} a^{3k} = a^{2k+3k} = a^{5k}$

Notice that the *base must be the same* for this property to hold.

EXAMPLE 2   The following demonstrate Property E2.

(a) $(7^2)^4 = (7^2)(7^2)(7^2)(7^2)$
$$= 7^{2+2+2+2} = 7^8 = 5{,}764{,}801$$

(b) $(t^4)^6 = t^{4 \cdot 6} = t^{24}$

(c) $(2^{4n})^3 = 2^{12n}$

EXAMPLE 3   The following demonstrate Property E3.

(a) $(6x)^4 = (6x)(6x)(6x)(6x)$
$$= (6 \cdot 6 \cdot 6 \cdot 6)(x \cdot x \cdot x \cdot x) = 6^4 x^4 = 1296x^4$$

(b) $(2ab^2)^5 = 2^5 a^5 (b^2)^5 = 32a^5 b^{10}$

(c) $(3p^a q^b)^c = 3^c p^{ac} q^{bc}$

EXAMPLE 4   The following demonstrate Property E4.

(a) $\dfrac{a^9}{a^4} = \dfrac{a \cdot a \cdot a \cdot a \cdot a \cdot \cancel{a} \cdot \cancel{a} \cdot \cancel{a} \cdot \cancel{a}}{\cancel{a} \cdot \cancel{a} \cdot \cancel{a} \cdot \cancel{a}}$

$$= a \cdot a \cdot a \cdot a \cdot a = a^5$$

(b) $\dfrac{t^2}{t^5} = \dfrac{\cancel{t} \cdot \cancel{t}}{\cancel{t} \cdot \cancel{t} \cdot t \cdot t \cdot t} = \dfrac{1}{t \cdot t \cdot t} = \dfrac{1}{t^3}$

(c) $\dfrac{8^5}{8^5} = 1$

(d) $\dfrac{7^{10}}{7^2} = 7^{10-2} = 7^8 = 5{,}764{,}801$

(e) $\dfrac{x^7}{x^{11}} = \dfrac{1}{x^{11-7}} = \dfrac{1}{x^4}$

As in Property E1, the *bases must be the same* for this property to hold.

EXAMPLE 5   The following demonstrate Property E5.

(a) $\left(\dfrac{x}{2}\right)^4 = \dfrac{x}{2} \cdot \dfrac{x}{2} \cdot \dfrac{x}{2} \cdot \dfrac{x}{2} = \dfrac{x \cdot x \cdot x \cdot x}{2 \cdot 2 \cdot 2 \cdot 2} = \dfrac{x^4}{2^4} = \dfrac{x^4}{16}$

(b) $\left(\dfrac{5a^4}{b}\right)^2 = \dfrac{5^2 (a^4)^2}{b^2} = \dfrac{25a^8}{b^2}$

EXAMPLE 6   Simplify $T = (-3a^2 b^5)(7a^4 bc^2)(2a^5 b^2 c^3 d^2)$.

Solution   We regroup by putting together factors with the same base.

$\boxed{\text{Given}}$	$T = (-3a^2b^5)(7a^4bc^2)(2a^5b^2c^3d^2)$
$\downarrow$	
$\boxed{\begin{array}{c}\text{Group terms}\\\text{with same base}\end{array}}$	$= (-3 \cdot 7 \cdot 2)(a^2a^4a^5)(b^5bb^2)(c^2c^3)(d^2)$
$\downarrow$	
$\boxed{\text{Simplify}}$	$= -42a^{11}b^8c^5d^2$

Property E4 suggests subtracting exponents when we divide terms with the same base. Now let us consider the expressions a^2/a^5 and a^4/a^4 worked out in two ways: using Property E4 and subtracting exponents.

Property E4	Subtracting Exponents
$\dfrac{a^2}{a^5} = \dfrac{1}{a^3}$	$\dfrac{a^2}{a^5} = a^{2-5} = a^{-3}$
$\dfrac{a^4}{a^4} = 1$	$\dfrac{a^4}{a^4} = a^{4-4} = a^0$

Since we want each method to produce the same answer, we make the following definitions.

Definition
For any real number a (except 0) and natural number n:

E6	**Zero exponent**	$a^0 = 1$
E7	**Negative exponent**	$a^{-n} = \dfrac{1}{a^n}$

EXAMPLE 7   The following examples demonstrate zero and negative exponents.

(a) $5^0 = 1$ (b) $t^0 = 1$

(c) $a^{-7} = \dfrac{1}{a^7}$ (d) $2^{-4} = \dfrac{1}{2^4} = \dfrac{1}{16}$

Using negative and zero exponents, we can now rewrite Property E4 simply as

$$\textbf{E4} \quad \frac{a^m}{a^n} = a^{m-n}$$

EXAMPLE 8   The following examples use Properties E1 to E7 for both positive and negative exponents.

(a) $x^{-7} \cdot x^{-4} = x^{-7+(-4)} = x^{-11} = \dfrac{1}{x^{11}}$

(b) $\dfrac{r^{-5}}{r^7} = r^{-5-7} = r^{-12} = \dfrac{1}{r^{12}}$

(c) $(x^3 y^{-2})^{-2} = (x^3)^{-2}(y^{-2})^{-2} = x^{-6} y^4 = \dfrac{y^4}{x^6}$

(d) $\left(\dfrac{y^{-3}}{y^{-5}}\right)^4 = \dfrac{y^{-12}}{y^{-20}} = y^{-12-(-20)} = y^8$

(e) $\dfrac{1}{m^{-2}} = \dfrac{m^0}{m^{-2}} = m^{0-(-2)} = m^2$

Notice that the last case suggests that $1/a^{-n} = a^n$.

EXAMPLE 9   Simplify $P = \left(\dfrac{x^{-2} y^4}{z^5}\right)^{-2} \left(\dfrac{x^{-1} z^2}{y^{-2}}\right)^3$.

Solution   As with many, many mathematics problems, we work the problem in small careful steps.

| Given | $P = \left(\dfrac{x^{-2} y^4}{z^5}\right)^{-2} \left(\dfrac{x^{-1} z^2}{y^{-2}}\right)^3$ |

$(a^m)^n = a^{mn}$

$$= \frac{x^4 y^{-8}}{z^{-10}} \cdot \frac{x^{-3} z^6}{y^{-6}}$$

Group terms with same base

$$= (x^4 x^{-3})\left(\frac{y^{-8}}{y^{-6}}\right)\left(\frac{z^6}{z^{-10}}\right)$$

$\dfrac{a^m}{a^n} = a^{m-n}$

$$= x^{4-3} y^{-8+6} z^{6+10}$$

Simplify

$$= x y^{-2} z^{16} = \frac{x z^{16}}{y^2}$$

Let us now note some common mistakes that students make with exponents, as well as the correct forms.

YES	NO
$a^2 \cdot a^5 = a^7$	$a^2 \cdot a^5 = a^{10}$
$\dfrac{a^{12}}{a^4} = a^8$	$\dfrac{a^{12}}{a^4} = a^3$
$(a^6)^2 = a^{12}$	$(a^6)^2 = a^{36}$
$3^5 \cdot 3^7 = 3^{12}$	$3^5 \cdot 3^7 = 9^{12}$
$5^{-2} = \dfrac{1}{5^2} = \dfrac{1}{25}$	$5^{-2} = -5^2 = -25$
$7^0 = 1$	$7^0 = 0$

Finally, let us discuss **scientific notation**, which is a notation used to express all numbers, especially very large and very small numbers, in a compact manner. Any positive real number r can be written

$$r = t \times 10^n$$

where $1 \le t < 10$ and n is an integer.

EXAMPLE 10   For numbers greater than 10, we move the decimal point n places to the *left* until the number is between 1 and 10. The power of 10 is 10^n.

(a) $7{,}500{,}000 = 7{,}500000 = 7.5 \times 1{,}000{,}000 = 7.5 \times 10^6$

(b) $138{,}400{,}000{,}000 = 1{,}38400000000 = 1.384 \times 10^{11}$

For numbers less than 1, we move the decimal point k places to the *right* until the number is between 1 and 10. The power of 10 is 10^{-k}.

(c) $0.00073 = 0.0007{,}3 = 7.3 \times 0.0001 = 7.3 \times 10^{-4}$

(d) $0.00000000123 = 0.000000001{,}23 = 1.23 \times 10^{-9}$

Scientific notation is important to know since most hand calculators automatically convert to scientific notation if an answer needs more places than the display has. For example, consider $234{,}845 \times 4{,}948{,}742$. The exact answer requires too many digits (13) for the display. On scientific calculators, the answer is displayed

DISPLAY	MEANING
1.1622 12	1.1622×10^{12}

EXAMPLE 11 When computing with numbers in scientific notation, we separate the digits from the powers of 10.

(a) $(2.7 \times 10^4) \times (1.9 \times 10^{-7}) = (2.7 \times 1.9) \times (10^4 \times 10^{-7})$
$$= 5.13 \times 10^{-3}$$

(b) $\dfrac{7.3 \times 10^8}{9.74 \times 10^{-3}} = \dfrac{7.3}{9.74} \times \dfrac{10^8}{10^{-3}}$
$$= 0.7495 \times 10^{11}$$
$$= 7.495 \times 10^{-1} \times 10^{11} = 7.495 \times 10^{10}$$

(c) $(1.03 \times 10^5)^{12} = 1.03^{12} \times (10^5)^{12} = 1.426 \times 10^{60}$

On a hand calculator such as a TI35, the multiplication of part (a) would be punched as follows:

PRESS	DISPLAY	MEANING
C	0.	Clear
2 . 7 EE 4	2.7 04	2.7×10^4
× 1 . 9 EE 7 +/−	1.9 − 07	Times 1.9×10^{-7}
=	5.13 − 03	Answer: 5.13×10^{-3}

Notice that the EE key replaces the phrase "times 10 to the ...."

PROBLEM SET 2.1

Simplify the following as much as possible. (Leave all exponents positive.)

Warm-up Exercises

1. $(2)(2)(2)$

2. $\left(\dfrac{1}{4}\right)\left(\dfrac{1}{4}\right)\left(\dfrac{1}{4}\right)$

3. $(-5)(-2)(4)$

4. $(-2)(3)(-4)$

5. 2^4

6. 5^3

7. $(-4)^3$

8. $(-1)^{10}$

9. 10^3

10. $\dfrac{1}{10^2}$

11. $\left(\dfrac{1}{2}\right)^5$

12. $\left(\dfrac{-1}{3}\right)^4$

Write the following numbers in scientific notation.

13. 520,000

14. 91,000,000

15. 1,300,000

16. 2,000,000,000

17. 0.005

18. 0.00073

19. 0.00000123

20. 0.0000000975

Write the following in the usual base-10 notation.

21. 7×10^4

22. 1.2×10^3

23. 6.31×10^7

24. 7.03×10^{10}

25. 8.2×10^{-2}

26. 4×10^{-4}

27. 6.5×10^{-1}

28. 5.7×10^{-5}

Simplify the following as much as possible. (Leave all exponents positive.)

29. $z^5 z^9 z^{12}$

30. $y^t y^{2t} y^{t+1}$

31. $[(a+b)^2]^3$

32. $(2^n)^3$

33. $(6x^4 y^3)^3$

34. $(4x^n y^m)^t$

35. $\dfrac{a^{10}}{a^2 a^4}$

36. $\dfrac{z^{6t}}{z^{2t}}$

37. $\left(\dfrac{x^2 y^3}{5}\right)^3$

38. $\left(\dfrac{a^n b^m}{c^k}\right)^p$

39. 5^{-2}

40. x^{-7}

41. $\dfrac{k^{-7}}{k^{-5}}$

42. $(a^{-5} b^3)^{-2}$

43. $(-ab^2 c^3)(4a^4 b^4)$

44. $(8ab^2 c^3)(-7a^2 b^4)(ac^5)$

45. $(3x^{2n} y^{3k})(-4x^{5n} y^k)$

46. $(-x^k y^m z^n)(2x^k y^{2m} z^{3n})$

47. $\left(\dfrac{2m^{-4} n^2}{p^{-4}}\right)^{-5} \left(\dfrac{4p^{-3} n^{-1}}{m^2}\right)^2$

48. $\left(\dfrac{y^{-2} z^{-3}}{x^4}\right)^{-4} \left(\dfrac{x^5 z^{-3}}{y}\right)^{-1}$

49. $x^4 x^5 x^6$

50. $(2^{10})^3$

51. $(2c^3)^4$

52. $\dfrac{t^4 t^6 t^7}{t^2 t^8}$

53. $\left(\dfrac{2r}{3}\right)^5$

54. $(2r^2 t)(-3r^3 t^5)$

55. $(a + b + c)^0$

56. $x^{-6} x^5 x^{-2}$

57. $(-3c^2)^{-2}$

58. $\left(\dfrac{2}{3}\right)^{-2}$

59. $(2ab^2 c^3 d^4)^5$

60. $(3 \times 10^4)^2$

61. $\dfrac{a^2 b^3 c^{-4}}{a^{-3} b^4 c^5}$

62. $\left(\dfrac{a^2 b^{-3}}{c^{-1} d^5}\right)^{-2}$

63. $(4 \times 10^5)(8 \times 10^{-3})$

64. $\left(\dfrac{a^x b^y}{c^z d^t}\right)^w$

65. $t(t^3)^4$

66. $\left(\dfrac{2^{-3}}{3^{-2}}\right)^{-1}$

67. $\dfrac{8.2 \times 10^{-4}}{4.1 \times 10^5}$

68. $(7.31 \times 10^{-13})^0$

69. $(2x^4 y^6)\left(\dfrac{4x^5 y^7}{x^8 y}\right)$

70. $(a^2 b^3 c^4)^3 (a^3 b^4 c^2)^{-1}$

71. $rr^2 r^3 r^4 r^5$

72. $s^{-2} s^{-4} s^{-6} s^{-8} s^{-10}$

73. $\dfrac{1}{2 \times 10^{-5}}$

74. $\left(\dfrac{u^2 v^{-3}}{s^{-1} t^7}\right)^{-2} \left(\dfrac{u^{-4} t^{-1}}{s^2 v^{-5}}\right)^3$

75. $\dfrac{(x - y)^7}{(x - y)^{-2}}$

76. t^{-5^2}

77. $\left(\dfrac{1}{x^{-5}}\right)^{-4}$

78. $\dfrac{(9 \times 10^2)(2 \times 10^{-3})^4}{6 \times 10^{-1}}$

79. $[(a^5)^2]^3$

80. $[(x^{-2})^{-3}]^{-1}$

81. $e^t e^{2t} e^{-3t}$

82. $(-2^3)^{-2} \left(\dfrac{-1}{2}\right)^{-5}$

83. $\dfrac{(8.2 \times 10^2)(7.39 \times 10^{-4})}{2.7 \times 10^{-7}}$

84. $\dfrac{7.25 \times 10^{11}}{(1.08 \times 10^{-3})(5.4 \times 10^7)}$

85. $(7.3 \times 10^4)^5$

86. $(1.734 \times 10^{-3})^8$

Life Science Application

87. If a man and a woman each have 2^{23} combinations on their 23 chromosomes, together they have $2^{23} \cdot 2^{23}$. Write this with a single exponent.

Business Application

88. The monthly mortgage payment M on a loan P for n months at monthly interest rate r is

$$M = \frac{Pr}{1 - (1 + r)^{-n}}$$

Find M when:
(a) $P = \$50,000$, $n = 300$ months, $r = 1.2\%$ (monthly rate)
(b) $P = \$4000$, $n = 48$ months, $r = 1.5\%$ (monthly rate)

Computer Application

89. A computer can perform a simple calculation in about 10^{-7} second. How long will it take to perform 10^4 (or 10,000) such calculations?

Miscellaneous Applications

The following problems involve the use of scientific notation from various fields of study. Write the numbers in the statements in scientific notation.

90. There are about 880,000,000 game pieces in a certain fast-food chain's contest.

91. The likelihood of getting the \$500,000 game card in the restaurant's contest is 0.00000000227.

92. The weight of the earth is 6,586,000,000,000,000,000,000 tons.

93. The wavelength of green light is 0.0000005 meter.

2.2 POLYNOMIALS

We have already seen letters used to represent numbers. Such a letter (or any other symbol) is called a **variable** (since it may be varied to represent different numbers). A number or symbol that cannot change is called a **constant**. For instance, if x can be any real number, it is a variable. The numbers 5, 1/4, and π are constant, since they do not change. Sometimes a letter, such as a, may be designated as a constant if it is not allowed to change.

A **term** is any product of a constant, variables, and powers of variables. The constant in the product is called the **coefficient**.

EXAMPLE 12 The following are algebraic terms.

Term	Coefficient
$3ab^2$	3
$-4x^2y^5z$	-4
pqr	1
$-m^2$	-1

A **polynomial** is a sum of terms, such as $7abc^2 - 8xy^5 + 3m^7n^5p^9$. We have special names for polynomials with a small number of terms. A **monomial** has one term, such as $6xy^5$. A **binomial** has two terms, such as $9a^2 - 16b^2$. A **trinomial** has three terms, such as $x^2 + 5x - 7$.

We pay special attention to **polynomials in one variable**. These have the form

$$a_n x^n + a_{n-1} x^{n-1} + \cdots + a_1 x + a_0 \qquad (a_n \neq 0)$$

where $a_n, a_{n-1}, \ldots, a_0$ are constant coefficients and x is the variable. The **degree** of a nonzero polynomial with only one variable is the highest power to which the variable with nonzero coefficient is raised. The polynomial 0 is called the **zero polynomial**. We do not define its degree.

EXAMPLE 13   The following are various polynomials, together with their degrees and special names.

Polynomial	Degree	Special Name
(a) $x^5 + 7x$	5	Binomial
(b) $8a^4$	4	Monomial
(c) $2t^7 - 4t^5 + 10t$	7	Trinomial
(d) $m^{10} - 9m^7 + 5m^4 - 2$	10	Polynomial
(e) 7	0	Constant polynomial
(f) 0	Not defined	Zero polynomial

Terms that are exactly the same except for the coefficient are called **like terms**. For example, $6x^2 y$, $10x^2 y$, and $-4x^2 y$ are like terms.

EXAMPLE 14   We use the distributive property to help us add (or combine) like terms.
(a) $6a^2 + 7a^2 = (6 + 7)a^2 = 13a^2$
(b) $8x^2 y^3 + 3x^2 y^3 - 4x^2 y^3 = (8 + 3 - 4)x^2 y^3 = 7x^2 y^3$

To add polynomials:

1. Group (or line up) like terms.
2. Add the coefficients.

EXAMPLE 15   Simplify

$$S = (5t^3 + 3t + 7) + (4t^2 - 8t - 1) + (9t^3 - 6t^2 - 4)$$

Solution   We work this problem horizontally by grouping like terms (using the commutative and associative properties).

Given	$S = (5t^3 + 3t + 7) + (4t^2 - 8t - 1) + (9t^3 - 6t^2 - 4)$
Group like terms	$= (5t^3 + 9t^3) + (4t^2 - 6t^2) + (3t - 8t) + (7 - 1 - 4)$
Simplify	$= 14t^3 - 2t^2 - 5t + 2$

Many students prefer to work problems like this vertically. (Note that we leave spaces for missing powers of t.)

$$
\begin{array}{r}
5t^3 \qquad\quad + 3t + 7 \\
4t^2 - 8t - 1 \\
9t^3 - 6t^2 \qquad\quad - 4 \\
\hline
14t^3 - 2t^2 - 5t + 2
\end{array}
$$

To subtract polynomials:

1. Replace the polynomial to be subtracted by its additive inverse; that is, change all the signs in the polynomial to be subtracted.
2. Add the resulting polynomials.

EXAMPLE 16 Simplify $D = (8t^3 - 2t^2 + 7t - 10) - (5t^3 - 9t^2 - 11t + 7)$.

Solution   We change all the signs in the second polynomial and add.

Change signs and add	$D = (8t^3 - 2t^2 + 7t - 10) + (-5t^3 + 9t^2 + 11t - 7)$
Group like terms	$= (8t^3 - 5t^3) + (-2t^2 + 9t^2) + (7t + 11t) + (-10 - 7)$
Simplify	$= 3t^3 + 7t^2 + 18t - 17$

To multiply polynomials:

1. Use the distributive law to multiply all the terms of one polynomial by the other.
2. Combine like terms.

EXAMPLE 17 We use the distributive law to multiply a monomial by a polynomial.

(a) $2a^3b^2(7ab^5 - 5a^2b^8) = 14a^4b^7 - 10a^5b^{10}$

(b) $-5xy^2z^3(6x - 7x^3y + 8y^2z) = -30x^2y^2z^3 + 35x^4y^3z^3 - 40xy^4z^4$

EXAMPLE 18 Multiply $M = (2x + 5)(x^2 - 7x - 3)$.

Solution We use the distributive law *twice* to multiply two polynomials.

Given	$M = (2x + 5)(x^2 - 7x - 3)$
Distributive law	$= (2x + 5)x^2 + (2x + 5)(-7x) + (2x + 5)(-3)$
Distributive law	$= 2x^3 + 5x^2 - 14x^2 - 35x - 6x - 15$
Combine terms	$= 2x^3 - 9x^2 - 41x - 15$

Many students prefer to multiply polynomials vertically, just as they multiply whole numbers.

$$\begin{array}{r} x^2 - 7x - 3 \\ 2x + 5 \\ \hline 5x^2 - 35x - 15 \\ 2x^3 - 14x^2 - 6x \\ \hline 2x^3 - 9x^2 - 41x - 15 \end{array}$$

Let us now state three very useful, special properties for multiplying binomials.

Property
For real numbers a, b, c, and d:

M1	**FOIL**	$(a + b)(c + d) = ac + ad + bc + bd$
M2	**Sum-and-difference property**	$(a + b)(a - b) = a^2 - b^2$
M3	**Square property**	$(a \pm b)^2 = a^2 \pm 2ab + b^2$

EXAMPLE 19 The following examples demonstrate the special-product properties. The memory device *FOIL* means *Firsts-Outers-Inners-Lasts.*

(a) $(2x + 7)(3x - 5) = 6x^2 - 10x + 21x - 35 = 6x^2 + 11x - 35$

(b) $(6a - 7b)(9a + 5b) = 54a^2 + 30ab - 63ab - 35b^2$
$$= 54a^2 - 33ab - 35b^2$$

(c) $(x - 7)(x + 7) = x^2 - 49$

(d) $(4a^2 - 5b)(4a^2 + 5b) = 16a^4 - 25b^2$

(e) $(2x + 5y)^2 = 4x^2 + 20xy + 25y^2$

(f) $(4a - 7b^3)^2 = 16a^2 - 56ab^3 + 49b^6$

Simplify the following.

1. $4 + (-7) + (-2)$

2. $7 - 3 - (-2) + (-1)$

3. $(-3)(-1)(4)$

4. $-5(-2 + 6)$

5. $6(x + 7)$

6. $-4(3t - 5)$

7. $z^2 z^3$

8. $t^3 t^4 t^5 t^6$

9. $(-2x^2 y^2)(-2x^2 y^2)$

10. $(-4a^2 b^3 c^4)(3ab^4)(b^5 c^6)$

For each of the following polynomials:

(a) Identify the coefficient of the first term.

(b) Give its degree.

(c) Give its special name (if any).

11. $4x^7 - 2x^3$

12. $40y^8 - 8y^3 + 2$

13. $5t^2 - 8t + 4$

14. $-8a^6$

15. $10a^4 - 9a^3 + 8a^2 - a$

16. $x^3 + 17$

17. $-z^5$

18. $5r^9 + 6r^7 - 3r^5 - r + 2$

Simplify the following as much as possible.

19. $8pq^2 - 10pq^2 - pq^2$

20. $-abc - 4abc + 13abc$

21. $(6t^3 - t - 3) + (7t^2 + 2t + 5) + (8t^3 - t^2 + 9)$

22. $(-x^3 - x + 4) + (2x^3 + 7x^2 - 7) + (5x^2 - x - 8)$

23. $(x^4 - x^3) + (x^4 - 3x^2) + (5x^3 + 3x) + (5x^2 + 7x + 1)$

24. $(p^2 - 4p) + (5p^3 - 2p^2) + (7p^4 - 8p^3) + (-3p^3 + p)$

25. $(5v^3 - 7v^2 - 4) - (2v^2 - 3v + 5)$

26. $(7m^3 - 8m + 1) - (10m^3 - m^2 + m)$

27. $(2x^2 + 5x) + (7x^2 - 5) - (8x^2 - x) - (4x + 5)$

28. $(3t^2 - t) - (5t + 1) - (10t^2 - 1) + (6t + 4)$

29. $(8y^3 - 2y + 1) + (4y^2 - 7y - 9) - (9y^3 - y - 10)$

30. $(12u^4 - 8u^2 + u) - (14u^2 - 7u + 1) - (3u^3 - 2u^2 + 10u)$

31. $-3a^2 b^3 (2ab^2 - 3a^2 b)$

32. $5rs^2 t^3 (2r^3 s - 4r^2 t^2 + 5st^5)$

33. $(4t + 1)(t^2 - t + 2)$

34. $(3r + 2)(5r^2 - r + 3)$

35. $(k - 2)(k^3 - 4k^2 + 2k + 1)$

36. $(2m + 1)(4m^3 - 3m^2 + 5m + 7)$

37. $(y^2 + 4y + 2)(2y^2 - 5y - 3)$

38. $(6a^2 - a + 5)(5a^2 + a - 4)$

39. $(2t + 9)(9t - 5)$

40. $(8u - 5)(4u + 3)$

41. $(5y - 2)(7y + 10)$

42. $(12a - 11)(5a + 6)$

43. $(g + 6)(g - 6)$

44. $(t - 5)(t + 5)$

45. $(2t - 7)(2t + 7)$

46. $(4a + 9)(4a - 9)$

47. $(4m^2 - 5n)(4m^2 + 5n)$

48. $(5u^3 - 3v^2)(5u^3 + 3v^2)$

49. $(x + 2)^2$

50. $(a - 7)^2$

51. $(2a - 1)^2$

52. $(4u - 3)^2$

53. $(4u^2 - 5v)^2$

54. $(10r^2 + 7s^5)^2$

55. $4t + 6t - 11t$

56. $-3x(4x^2 - x + 2)$

57. $(y^4 - y^3 + y - 5) - (5y^3 + 3y^2 - y + 4)$

58. $(2n - 3m) + (3n - 4m) - (n + 6m) - (5n - m)$

59. $(3x - 4)(5x + 1)$

60. $(2t + 3)^2$

61. $-a^2b(ab^4 - 3a^5b - b^7)$

62. $-x - 2x - 3x - 4x - 5x$

63. $(a - 3)(a^2 + 4a - 6)$

64. $(5t^3 - 3s^2)(5t^3 + 3s^2)$

65. $(2x + 1)(3x - 1) + (4x - 2)(3x + 2)$

66. $(4x + 3)(2x - 5) - (x + 9)(2x - 7) + x$

67. $(x^2 - 3)(2x^2 - 4x)$

68. $x - (x - x^2) - (x^2 - x^3) - (x^3 - x^4)$

69. $(1 + 3t^5s^3)(1 - t^5s^3)$

70. $(x + y)(x - 2y) + 7xy$

71. $(4r^2 + t)(3r - t) + (2r + t)(7r^3 - t) - 4rt + r^2$

72. $(1 + x^2 + x^4 + x^6)(x - x^3 + x^5 - x^7)$

73. $2x(x^2 + 7) + x^2(3x + 1)$ **74.** $a^3(2a^2 - 1) - 4a(3a^2 + a)$

75. $(2x + 7)(7x - 1) - 5x$ **76.** $7u^2 + (2u + 1)(3u - 5)$

77. $(2a + b - c)^2$ **78.** $(4x - 5y + z)^2$

79. $(2x + 1)(3x - 2)(4x + 5)$ **80.** $(3x - 1)(4x - 3)(5x + 2)$

81. $(x + 5)^3$ **82.** $(2x - 1)^3$

Business
Application

83. In business we have the relation

$$TP = TR - TC$$

where TP is the total profit, TR the total revenue, and TC the total cost. Complete the following table, where Q represents the quantity of an item produced.

TP	TR	TC
?	$8000Q$	$0.01Q^2 + 1200Q + 150{,}000$
?	$5200Q$	$0.006Q^2 + 2500Q + 74{,}000$
$-0.03Q^2 + 5Q - 2400$	$48Q$	?
$-0.007Q^2 + 13Q - 41{,}000$	$25Q$	?

Life Science
Applications

84. The population of a particular culture of bacteria is given by $5000(t + 3)^3$. Write this as a polynomial.

85. The total weight of a certain insect population is given by $W = n \cdot w$, where n is the total number and w is the average weight. Find W if $n = 2t^2 + 10t + 100$ and $w = t^2 + 15t + 1$.

Geometry
Applications

86. A rectangle a by b is increased to $(a + h)$ by $(b + k)$.
(a) What is the original area?
(b) What is the new area?
(c) What is the increase in area?

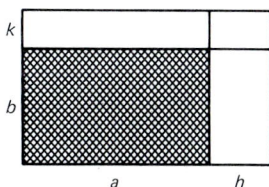

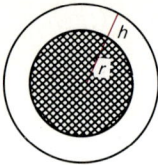

87. A circle radius r is increased to radius $(r + h)$.
  (a) What is the original area?
  (b) What is the new area?
  (c) What is the increase in area?

Physical Application

88. The distance d (in meters) that an objects falls in t seconds is given by

$$d = 4.9t^2$$

  (a) In $t + h$ seconds, it falls $4.9(t + h)^2$. Expand this polynomial.
  (b) What is the difference in distances between $t + h$ seconds and t seconds?

2.3
DIVISION OF POLYNOMIALS

In this section we discuss the division of polynomials (we considered addition, subtraction, and multiplication in the preceding section).

Let us first consider a polynomial divided by a monomial. Consider $(A + B + C) \div M$, where A, B, C, and M are monomials ($M \neq 0$). Thus we have

Division by M means multiplication by $1/M$

$$\frac{A + B + C}{M} = \frac{1}{M}(A + B + C)$$

Distributive law

$$= \frac{A}{M} + \frac{B}{M} + \frac{C}{M}$$

Here we used a polynomial with three terms (A, B, and C); however, the result holds for a polynomial with any number of terms: A, B, C, ..., and $M \neq 0$:

$$\frac{A + B + C + \cdots}{M} = \frac{A}{M} + \frac{B}{M} + \frac{C}{M} + \cdots$$

EXAMPLE 20 Divide $Q = (18x^6y^3 - 12x^5y^5 + 30x^4y^7) \div 6xy^2$.

Solution We rewrite this division as three fractions.

Rewrite as three fractions

$$Q = \frac{18x^6y^3 - 12x^5y^5 + 30x^4y^7}{6xy^2}$$

$$= \frac{18x^6y^3}{6xy^2} - \frac{12x^5y^5}{6xy^2} + \frac{30x^4y^7}{6xy^2}$$

Simplify

$$= 3x^5y - 2x^4y^3 + 5x^3y^5$$

EXAMPLE 21   Division of one polynomial by another mirrors the division of whole numbers. Consider the following two divisions side by side.

$$
\begin{array}{r}
12 \\
413\overline{)4987} \\
413 \\
\hline
857 \\
826 \\
\hline
31
\end{array}
\qquad
\begin{array}{r}
x + 2 \\
4x^2 + x + 3\overline{)4x^3 + 9x^2 + 8x + 7} \\
4x^3 + x^2 + 3x \\
\hline
8x^2 + 5x + 7 \\
8x^2 + 2x + 6 \\
\hline
3x + 1
\end{array}
$$

The similarity is very clear: Each power of x has its own column just as each power of 10 does in whole-number division. (Polynomial division, of course, may also involve negative coefficients.) Here $x + 2$ is the **quotient** and $3x + 1$ is the **remainder** (as we have in whole-number division).

The relation of the quotient and remainder to the divisor and dividend is the same for polynomials as it is for whole numbers:

$$4987 = (413)(12) + 31$$

$$\text{dividend} = (\text{divisor}) \cdot (\text{quotient}) + \text{remainder}$$

$$4x^3 + 9x^2 + 8x + 7 = (4x^2 + x + 3)(x + 2) + (3x + 1)$$

EXAMPLE 22   Divide $(x^4 + 5) \div (x + 1)$.

Solution   We insert each missing power of x with a zero coefficient. These zeros act as placeholders. At each step we divide the highest power of the divisor into the highest power of the dividend. After multiplying the new quotient term by the divisor, we subtract (be careful subtracting—add additive inverse).

$$
\begin{array}{r}
x^3 - x^2 + x - 1 \\
x + 1\overline{)x^4 + 0x^3 + 0x^2 + 0x + 5} \\
x^4 + x^3 \\
\hline
-x^3 + 0x^2 \\
x^3 x^2 \\
\hline
x^2 + 0x \\
x^2 + x \\
\hline
-x + 5 \\
-x - 1 \\
\hline
6
\end{array}
$$

The quotient is $x^3 - x^2 + x - 1$ with a remainder of 6. Thus,

$$x^4 + 5 = (x + 1)(x^3 - x^2 + x - 1) + 6$$

Perform the following long divisions, showing each quotient and remainder.

1. $81 \div 6$ **2.** $100 \div 7$

3. $7401 \div 27$ **4.** $40{,}003 \div 605$

Perform the indicated operations.

5. $(7x^2y^8)(-3x^4y^2)$ **6.** $8ab^2(2a^2b^3 - 5a^3b + 1)$

7. $(7a - 4b) - (9a - 5b)$ **8.** $3x^4 - (3x^4 - 6x^3)$

9. $(2x + 5)(x - 5) + 3$ **10.** $(x^2 - 1)(x^2 - x + 1) + (x - 3)$

Divide the following polynomials, showing quotient and remainder.

11. $(8x^4 - 6x^3 + 12x^2) \div 2x$

12. $(15a^3 - 20a^5 - 25a) \div 5a$

13. $(14m^4n^3 - 7m^5n^2 + 21m^9n^7) \div 7m^2n^2$

14. $(4a^2b^5 - 6a^5b^3 + 8a^7b^9) \div 2a^2b^3$

15. $(a^3 - 8a^2 + 13a + 6) \div (a - 3)$

16. $(6r^3 - 11r^2 - 7r + 15) \div (2r - 3)$

17. $(5b^3 - 7b^2 - 3b + 2) \div (b - 3)$

18. $(t^4 - 6t^3 - 4t^2 - 5t - 6) \div (t + 2)$

19. $(x^3 - 1) \div (x - 1)$

20. $(t^3 + 1) \div (t + 1)$

21. $(u^6 + u^3 + 2) \div (u^3 - 1)$

22. $p^6 \div (p^2 - 1)$

23. $(h^4 - 3h^3 + 4h^2 - 7h - 15) \div (h^2 - h + 5)$

24. $(m^4 - 3m^3 + 5m^2 - 10m + 6) \div (m^2 - 2m - 3)$

25. $(6p^6q^3 - 5p^3q^4 + 3p^2q^{10}) \div p^2q^3$

26. $(2m^3 + 3m^2 - 24m - 16) \div (m + 4)$

27. $(2z^4 - 3z^3 + 5z^2 - z - 1) \div (z - 3)$

28. $(2s^4 + 5s^3 - 11s^2 - 10s + 8) \div (s^2 + 3s - 2)$

29. $(t^2 + 12t + 35) \div (t + 5)$

30. $(10x^5 - 15x^3 - 5x) \div 5x^2$

31. $(4y^3 - 31y + 15) \div (2y - 5)$

32. $(27t^5 - 15t^3 - 6t^2) \div 3t^2$

33. $(b^6 + 1) \div (b + 1)$

34. $(12r^4 - 8r^3 + 11r^2 - 15) \div (3r^2 - 2r + 5)$

When multiplying $p = a \cdot b$, we call p the **product** and a and b the **factors**. **Factoring** is the process that we use to rewrite a number or polynomial as a product of factors. When multiplying we have two factors and wish to find the product:

$$\text{Multiplication: } (2x - 7)(3x + 1) = \boxed{ ? }$$

When factoring we have the product and wish to find its factors:

$$Factoring: 2x^2 - x - 15 = \boxed{?} \cdot \boxed{?}$$

Just as we factor natural numbers into primes, such as $60 = 2 \cdot 2 \cdot 3 \cdot 5$, we factor polynomials as much as possible.

The first type of factoring uses the distributive law to factor out the **greatest common factor (GCF)**, which is the largest expression that divides evenly into all the terms of the polynomial. For instance, $3a^2$ is the GCF of $3a^4$, $9a^3$, and $27a^2$. Note that the GCF of these terms is the product of the GCF of the coefficients (3) and the smallest power of each variable that is present (a^2).

EXAMPLE 23 We use the distributive law to factor the GCF out of the following polynomials.

| Factor out GCF, 6 | $6x - 12y + 18z = 6(x - 2y + 3z)$ |

| Factor out GCF, $7a$ | $7a^2b + 21ab^2 - 42a^2b^3 = 7ab(a + 3b - 6ab^2)$ |

| Factor out GCF, $(x + 5)$ | $6(x + 5) - x(x + 5) = (6 - x)(x + 5)$ |

YES	NO
$12x^2 + 6x = 6x(2x + 1)$	~~$12x^2 + 6x = 6x(2x)$~~
We factor the GCF from *all* the terms.	

Let us now look at some **special binomial factoring**. These involve binomials that can be factored very quickly.

Property
For real numbers a and b:

Difference of squares	$a^2 - b^2 = (a + b)(a - b)$
Difference of cubes	$a^3 - h^3 = (a - b)(a^2 + ab + b^2)$
Sum of cubes	$a^3 + b^3 = (a + b)(a^2 - ab + b^2)$

The *sum* of squares does *not* factor with real numbers.

EXAMPLE 24 The following are binomials that can be factored using the difference-of-squares property.
(a) $x^2 - 25 = (x + 5)(x - 5)$
(b) $64t^2 - 49s^2 = (8t + 7s)(8t - 7s)$
(c) $4 - (a + b)^2 = (2 + a + b)(2 - a - b)$

YES	NO
$x^2 - 25 = (x - 5)(x + 5)$	~~$x^2 + 25 = (x + 5)(x + 5)$~~

The sum of squares does not factor with real numbers.

EXAMPLE 25   The following use the special properties for the sum and difference of cubes.

(a) $x^3 - 8 = x^3 - 2^3 = (x - 2)(x^2 + 2x + 4)$
(b) $1 - 125t^3 = (1 - 5t)(1 + 5t + 25t^2)$
(c) $64k^3 + 27 = (4k + 3)(16k^2 - 12k + 9)$
(d) $1000m^3 + t^3 = (10m + t)(100t^2 - 10mt + t^2)$

We are now ready to factor second-degree trinomials, such as $12x^2 - 11x - 15$. These are generally factored by trial and error. When we multiply binomials, recall that the middle term of the product is a sum of two terms; thus, when we factor, we try to reverse this procedure and find out how the middle term came about.

EXAMPLE 26   The following trinomials have their leading coefficient equal to 1. To factor these trinomials, we search the factors of the last (constant) term for a pair whose sum is the coefficient of the middle term. (Be careful with negative signs.)

(a) $x^2 + 8x + 12 = (x + 2)(x + 6)$
$$\text{[since } (6)(2) = 12 \text{ and } 6 + 2 = 8]$$
(b) $t^2 - 3t - 18 = (t - 6)(t + 3)$
$$\text{[since } (-6)(3) = -18 \text{ and } -6 + 3 = -3]$$
(c) $a^2 - 8a + 15 = (a - 5)(a - 3)$
$$\text{[since } (-5)(-3) = 15 \text{ and } -5 + (-3) = -8]$$

EXAMPLE 27   Factor $3x^2 + x - 10$.

Solution   We use trial and error: We factor the first and last terms; then we look for the combination that produces the middle term, x.

| Factor first and last terms | $\overbrace{(3x\quad)(x\quad)}^{3x^2}$ | $\overbrace{\begin{array}{l}(\quad +1)(\quad -10)\\(\quad +2)(\quad -5)\end{array}}^{-10}$ |

COMBINATIONS	MIDDLE TERM
$(3x + 1)(x - 10)$	$-30x + x = -29x$ NO
$(3x - 10)(x + 1)$	$3x - 10x = -7x$ NO
$(3x + 2)(x - 5)$	$-15x + 2x = -13x$ NO
$(3x - 5)(x + 2)\sqrt{}$	$6x - 5x = x$ YES!

$$3x^2 + x - 10 = (3x - 5)(x + 2)$$

Try all combinations to get middle term, x

Choose correct combination

EXAMPLE 28 Factor $12x^2 - 4x - 5$.

Solution As in Example 27, we factor the first and last terms and look for a combination that produces the middle term, $-4x$.

Factor first and last terms	

$$\overbrace{\qquad 12x^2 \qquad}$$ $$\overbrace{\qquad -5 \qquad}$$

$(x\quad)(12x\quad)$
$(2x\quad)(6x\quad)$ $(\quad + 1)(\quad - 5)$
$(3x\quad)(4x\quad)$

Try all combinations to get middle term, $-4x$

COMBINATIONS	MIDDLE TERM
$(x + 1)(12x - 5)$	$-5x + 12x = 7x$ NO
$(x - 5)(12x + 1)$	$x - 60x = -59x$ NO
$(2x + 1)(6x - 5)\checkmark$	$-10x + 6x = -4x$ YES!

Choose correct combination

$$12x^2 - 4x - 5 = (2x + 1)(6x - 5)$$

Note that we do not check every combination: We stop when we find the correct one.

EXAMPLE 29 Factor $2m^2 + mn - 10m - 5n$.

Solution Sometimes we cannot factor a polynomial by dividing out a GCF, as a special product, or as a trinomial. Another method that we can try is **factoring by grouping**. Here we first factor in pairs and then factor the results.

Factor first and last pair separately

$$2m^2 + mn - 10m - 5n = m(2m + n) - 5(2m + n)$$

Factor GCF, $(2m + n)$

$$= (m - 5)(2m + n)$$

Let us combine all these rules and techniques into one general rule for factoring polynomials.

When factoring a polynomial:

1. Factor out the greatest common factor (GCF).
2. If there are two terms, try a special factoring.
3. If there are three terms, try factoring it as a trinomial.
4. If there are four terms, try grouping.

Continue this procedure until each factor can be factored no further.

EXAMPLE 30 The following polynomials need two or more steps to be factored completely.

(a) $x^5y^3 - xy^7 = xy^3(x^4 - y^4)$ [Factor out GCF]
$$= xy^3(x^2 + y^2)(x^2 - y^2) \quad \text{[Difference of squares]}$$
$$= xy^3(x^2 + y^2)(x - y)(x + y) \quad \text{[Difference of squares]}$$

(b) $4x^6 + 14x^5 - 30x^4 = 2x^4(2x^2 + 7x - 15)$ [Factor out GCF]
$$= 2x^4(2x - 3)(x + 5) \quad \text{[Factor trinomial]}$$

(c) $p^6 - q^6 = (p^3 + q^3)(p^3 - q^3)$ [Difference of squares]
$$= (p + q)(p^2 - pq + q^2)(p - q)(p^2 + pq + q^2)$$
$$\text{[Difference and sum of cubes]}$$

Just as there are prime numbers, such as 3 and 11, there are **prime** (or **irreducible**) **polynomials** that cannot be factored with integer coefficients. For example, $x^2 + 5x + 1$ is a prime polynomial.

PROBLEM SET 2.4

Multiply the following polynomials.

Warm-up Exercises

1. $2x(4x - 7)$

2. $4ab^2(a^2b + 3a^2b^5 + 1)$

3. $(3x - 2)(3x + 2)$

4. $\left(5p + \dfrac{1}{2}\right)\left(5p - \dfrac{1}{2}\right)$

5. $(3r - 5s)^2$

6. $(a - 2)(a^2 + 2a + 4)$

7. $(a - 4)(a + 5)$

8. $(a + 2)(a - b)$

9. $(4r - 3s)(5r + 7s)$

10. $x^2(x - 1)(x + 1)$

Factor the following polynomials completely.

11. $2x^4 - 3x^3 - 13x^2$

12. $12t^{12} - 9t^{10} - 6t^7 - 3t^4$

13. $21x^2yz^4 - 14x^3y^2z$

14. $15a^2bc^4 - 18a^2c^2 - 24a^4b^2c$

15. $7(ax - 1) - y(ax - 1)$

16. $5(2a - x) - x(2a - x)$

17. $16m^2 - 25n^2$

18. $t^4 - 16s^2$

19. $\dfrac{1}{9}u^2v^2 - a^2b^4$

20. $\dfrac{4}{25}x^4 - 9y^2$

21. $(m + n)^2 - (a + b)^2$

22. $(2p - q)^2 - (3u - v)^2$

23. $a^3 - 125b^3$

24. $t^3 + 8$

25. $r^3 - \dfrac{1}{64}$

26. $1 + \dfrac{z^3}{27}$

27. $a^2 + 4a - 32$

28. $k^2 + 13k + 30$

29. $r^2 - 11rs - 60s^2$

30. $p^2 + pq - 72q^2$

31. $u^2 - 3uv - 54v^2$

32. $x^2 - 5xy - 21y^2$

33. $x^4 + 7x^2 + 6$

34. $y^4 + 5y^2 - 14$

35. $10p^2 + 23pq - 5q^2$

36. $12a^2 + 29ab - 8b^2$

37. $8t^2 + 22ts - 21s^2$

38. $2m^2 + 17mn + 36n^2$

39. $8r^4 + 18r^2 + 7$

40. $20p^4 + 7p^2q^2 - 3q^4$

41. $ac - 5a + 2bc - 10b$

42. $2m^2 + mn - 10m - 5n$

43. $24xy + 54x - 4y - 9$

44. $12a^2 - 21a - 20ab + 35b$

45. $2rs - 3ra - 4ab + 6bc$

46. $2ax - 4cx - 5a + 10c$

47. $p^5 + p^2$

48. $k^{10} - k^2$

49. $3a^7b^2 - 3ab^8$

50. $a - 64a^7b^6$

51. $m^6 - n^6$

52. $h^9 + k^9$

53. $3a^4 + 6a^3 - 45a^2$

54. $5t^5 - 35t^4 + 30t^3$

55. $u^2w + u^2t - uvw - uvt$

56. $40x^3y^3 - 16x^3y^2 + 30x^2y^3 - 12x^2y^2$

57. $t^2 - tu - 12u^2$

58. $24m^2 + 7m - 5$

59. $15a^2 - 26ab + 8b^2$

60. $rs - rt - s^2 + st$

61. $8u^3 - 1000v^3$

62. $m^3 - m^2$

63. $4r^5 - 4r^4 - 80r^3$

64. $24a^2b - 42ab + 9b$

65. $2k - 10k^7$

66. $2a^7b - 128ab^7$

67. $x^2 - 8x + 15$

68. $9p^4 - \dfrac{1}{4}$

69. $1 - 125t^3$

70. $4x^4 - 20x^2y + 25y^2$

71. $a^2 - 3a - ab + 3b$

72. $75x^5 - 12x^3$

73. $x - x^3 + x^7 - x^{23}$

74. $12a^2 + 5ab - 28b^2$

75. $3x^3 + 12x^2 - 21x$

76. $a^9 + b^9$

77. $m^{12} - n^{12}$

78. $10t^4 + 47a^2 - 3$

79. $x^3 + x^2 - 2x - 10$

80. $r^9 - r$

81. $35y^2 - 17yz + 2z^2$

82. $10,000s^4 - 16t^8$

83. $1000a^3 + 27b^3$

84. $a^2b^2 - ac + 2ab^2 - 2c$

Engineering Application

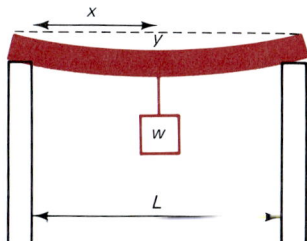

85. The maximum deflection (or sag) of a beam with a weight w on it is given by

$$y = \frac{-wx^4}{24EI} + \frac{wLx^3}{12EI} + \frac{-wL^3x}{24EI}$$

where x is the distance from the support, L the length of the beam, and E and I are constants. Factor this expression completely.

Physical Application

86. According to the Bohr theory, the frequency of light emitted when an electron jumps from orbit n_1 to orbit n_2 is given by

$$f = \frac{2\pi^2me^4}{h^3n_2^2} - \frac{2\pi^2me^4}{h^3n_1^2}$$

where m is the mass of an electron, e is the charge of an electron, and h is Planck's constant. Factor this expression completely.

87. If P dollars is invested at interest rate r, the account grows yearly as follows:

Year 1: $P + Pr$
Year 2: $P(1 + r) + P(1 + r)r$
Year 3: $P(1 + r)^2 + P(1 + r)^2 r$
Year n: $P(1 + r)^{n-1} + P(1 + r)^{n-1} r$

Factor each expression as much as possible.

Health
Application

88. An infection is spreading through a community at a rate given by $r = 2500y - 0.005y^2$, where y is the number of people with the disease. Factor this expression (first factor out -0.005).

2.5
RATIONAL EXPRESSIONS

The ratio of any two polynomials is called a **rational expression** (or **algebraic fraction**). For example,

$$\frac{2}{3}, \frac{-10}{t}, \frac{x^2 + 1}{y + 2}, \text{ and } \frac{a^2 + b^2}{c^2 - b^2}$$

are rational expressions (or fractions). As with rational numbers, the denominators cannot be zero.

On page 9, we stated some of the properties for quotients of real numbers. We can now restate these properties for rational expressions.

Property
For any polynomials A, B, C, and D (denominators are not zero):

RE1	**Equality property**	$\dfrac{A}{B} = \dfrac{C}{D}$ if and only if $AD = BC$
RE2	**Cancellation property**	$\dfrac{A}{B} = \dfrac{AC}{BC}$
RE3	**Multiplication property**	$\dfrac{A}{B} \cdot \dfrac{C}{D} = \dfrac{AC}{BD}$
RE4	**Division property**	$\dfrac{A}{B} \div \dfrac{C}{D} = \dfrac{A}{B} \cdot \dfrac{D}{C} = \dfrac{AD}{BC}$
RE5	**Addition property**	$\dfrac{A}{C} + \dfrac{B}{C} = \dfrac{A + B}{C}$
RE6	**Subtraction property**	$\dfrac{A}{C} - \dfrac{B}{C} = \dfrac{A + (-B)}{C}$

We say that a rational expression is **expressed in lowest terms** if the numerator and denominator have no common factors. The cancellation property (RE2) allows us to cancel common factors by dividing.

EXAMPLE 31   Express $\dfrac{m^2 - 36}{m^2 + 3m - 18}$ in lowest terms.

Solution

> *To express a rational expression in lowest terms:*
>
> 1. Factor the numerator and denominator completely.
> 2. Divide out common factors (Property RE2).

Factor completely

↓

Divide out common factor: $(m + 6)$

$$\frac{m^2 - 36}{m^2 + 3m - 18} = \frac{\cancel{(m + 6)}(m - 6)}{\cancel{(m + 6)}(m - 3)}$$

$$= \frac{m - 6}{m - 3}$$

YES	NO
$\dfrac{a^2 - 4}{a^2 + 2a} = \dfrac{(a - 2)(a + 2)}{a(a + 2)} = \dfrac{a - 2}{a}$	$\dfrac{\cancel{a^2} - 4}{\cancel{a^2} + 2a} = \dfrac{-4}{2a}$
We can only divide (or cancel) *factors*.	

YES	NO
$\dfrac{c}{bc} = \dfrac{1}{b}$	$\dfrac{c}{bc} = b$
When we cancel we are dividing the numerator and denominator by the same number. In the numerator here, $c \div c = 1$; in the denominator, $bc \div c = b$.	

EXAMPLE 32   Multiply: $P = \dfrac{a^2 + 3a + 2}{a^2 - 1} \cdot \dfrac{a^2 + 4a - 5}{a^4 + 5a^3}$.

Solution

To multiply rational expressions:

1. Factor all polynomials.
2. Divide any common factor out of the numerators and denominators.
3. Multiply (Property RE3).

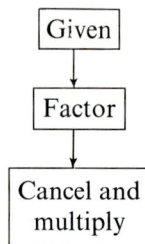

| Given |

$$P = \frac{a^2 + 3a + 2}{a^2 - 1} \cdot \frac{a^2 + 4a - 5}{a^4 + 5a^3}$$

| Factor |

$$= \frac{(a + 2)\cancel{(a + 1)}}{\cancel{(a + 1)}\cancel{(a - 1)}} \cdot \frac{(a + 5)\cancel{(a - 1)}}{a^3(a + 5)}$$

| Cancel and multiply |

$$= \frac{a + 2}{a^3}$$

EXAMPLE 33 Divide: $Q = \dfrac{a^2 + 16a + 60}{a^2 - 100} \div \dfrac{a^2 - a - 42}{a^2 - 3a - 28}$.

Solution

To divide rational expressions:

1. Invert the divisor (second fraction).
2. Multiply the resulting fractions (Property RE4).

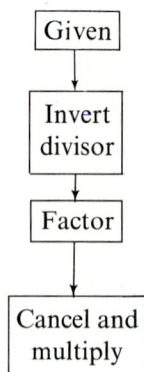

| Given |

$$Q = \frac{a^2 + 16a + 60}{a^2 - 100} \div \frac{a^2 - a - 42}{a^2 - 3a - 28}$$

| Invert divisor |

$$= \frac{a^2 + 16a + 60}{a^2 - 100} \cdot \frac{a^2 - 3a - 28}{a^2 - a - 42}$$

| Factor |

$$= \frac{\cancel{(a + 6)}\cancel{(a + 10)}}{(a - 10)\cancel{(a + 10)}} \cdot \frac{\cancel{(a - 7)}(a + 4)}{\cancel{(a - 7)}\cancel{(a + 6)}}$$

| Cancel and multiply |

$$= \frac{a + 4}{a - 10}$$

EXAMPLE 34 Add: $S = \dfrac{3}{x^2 - 4} + \dfrac{x}{x^2 + 3x + 2}$.

Solution

> *To add rational expressions*:
>
> 1. Find the lowest common denominator (LCD).
> 2. Rewrite the fractions with this LCD.
> 3. Add the numerators and divide by the LCD.

We can often find the LCD by first factoring the denominators to see what other factors are needed.

| Factor |

$$S = \frac{3}{(x+2)(x-2)} + \frac{x}{(x+2)(x+1)}$$

| Rewrite each fraction with LCD |

$$= \frac{3}{(x+2)(x-2)}\boxed{\frac{x+1}{x+1}} + \frac{x}{(x+2)(x+1)}\boxed{\frac{x-2}{x-2}}$$

| Simplify |

$$= \frac{3x+3}{(x+2)(x-2)(x+1)} + \frac{x^2-2x}{(x+2)(x+1)(x-2)}$$

| Add numerators; divide by LCD |

$$= \frac{x^2+x+3}{(x+2)(x-2)(x+1)}$$

EXAMPLE 35   Simplify $S = \dfrac{6}{a^2b^5} - \dfrac{7}{a^3b} - \dfrac{2}{ab^2}$.

Solution   *To subtract rational expressions*, we add the additive inverse of the fractions to be subtracted (Property RE6). Here the LCD is a^3b^5.

| Change signs and add |

$$S = \frac{6}{a^2b^5} + \frac{-7}{a^3b} + \frac{-2}{ab^2}$$

| Rewrite each fraction with LCD = a^3b^5 |

$$= \frac{6}{a^2b^5}\boxed{\frac{a}{a}} + \frac{-7}{a^3b}\boxed{\frac{b^4}{b^4}} + \frac{-2}{ab^2}\boxed{\frac{a^2b^3}{a^2b^3}}$$

| Simplify |

$$= \frac{6a}{a^3b^5} + \frac{-7b^4}{a^3b^5} + \frac{-2a^2b^3}{a^3b^5}$$

| Add numerators; divide by LCD |

$$= \frac{6a - 7b^4 - 2a^2b^3}{a^3b^5}$$

A **complex fraction** is a fraction that has other fractions within its numerator or denominator, or both. For example,

$$\frac{x + \dfrac{1}{x}}{x^2 + \dfrac{1}{x}}, \quad \frac{\dfrac{2}{3y}}{\dfrac{5}{5a}}, \quad \text{and} \quad \frac{\dfrac{5}{3x^2}}{x}$$

are complex fractions that we want to rewrite as single fractions.

EXAMPLE 36 Simplify $\dfrac{\dfrac{1}{10x^2} - \dfrac{2}{15x}}{\dfrac{6}{x} + \dfrac{1}{20}}$.

Solution

> *To simplify a complex fraction:*
>
> 1. Find the LCD of all the denominators.
>
> 2. Multiply the fraction by $1 = \dfrac{\text{LCD}}{\text{LCD}}$.
>
> 3. Simplify.

(There are other methods, but we give only this one.) Here the LCD is $60x^2$, so we multiply by $60x^2/60x^2$.

Multiply by

$$1 = \frac{60x^2}{60x^2}$$

$\downarrow$

Distributive law

$\downarrow$

Simplify

$$\frac{\dfrac{1}{10x^2} - \dfrac{2}{15x}}{\dfrac{6}{x} + \dfrac{1}{20}} = \frac{60x^2\left(\dfrac{1}{10x^2} - \dfrac{2}{15x}\right)}{60x^2\left(\dfrac{6}{x} + \dfrac{1}{20}\right)}$$

$$= \frac{\dfrac{60x^2}{10x^2} - \dfrac{120x^2}{15x}}{\dfrac{360x^2}{x} + \dfrac{60x^2}{20}}$$

$$= \frac{6 - 8x}{360x + 3x^2}$$

EXAMPLE 37 Simplify $\dfrac{x^{-1} + x^{-2}}{x^{-3} - x^{-4}}$.

Solution   Recall from Definition E7 (page 21) that $a^{-n} = 1/a^n$. Before we can simplify this expression, we must rewrite it with positive exponents. Then we can treat it as a complex fraction.

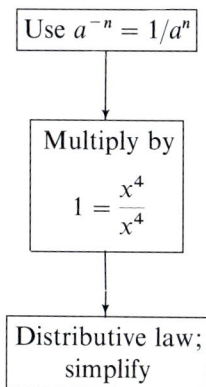

$$\frac{x^{-1} + x^{-2}}{x^{-3} - x^{-4}} = \frac{\dfrac{1}{x} + \dfrac{1}{x^2}}{\dfrac{1}{x^3} - \dfrac{1}{x^4}}$$

Use $a^{-n} = 1/a^n$

Multiply by
$$1 = \frac{x^4}{x^4}$$

$$= \frac{x^4\left(\dfrac{1}{x} + \dfrac{1}{x^2}\right)}{x^4\left(\dfrac{1}{x^3} - \dfrac{1}{x^4}\right)}$$

Distributive law; simplify

$$= \frac{\dfrac{x^4}{x} + \dfrac{x^4}{x^2}}{\dfrac{x^4}{x^3} - \dfrac{x^4}{x^4}} = \frac{x^3 + x^2}{x - 1}$$

PROBLEM SET 2.5

Perform the indicated operations and express the answers in lowest terms.

Warm-up Exercises

1. $\dfrac{7}{12} + \dfrac{1}{4} + \dfrac{5}{8}$ **2.** $\dfrac{1}{10} - \dfrac{2}{5} + \dfrac{11}{15}$

3. $\dfrac{7}{12} \cdot \dfrac{2}{3}$ **4.** $\dfrac{2}{5} \cdot \dfrac{15}{14} \cdot \dfrac{7}{6}$

5. $\dfrac{2}{5} \div \dfrac{3}{7}$ **6.** $\dfrac{4}{9} \div \dfrac{8}{3}$

Factor the following polynomials.

Warm-up Exercises

7. $x^2 - 9$ **8.** $a^2 + 3a - 10$

9. $6t^2 + t - 2$ **10.** $m^3 + 8$

Simplify the following as much as possible.

11. $\dfrac{20a^2 b^7 c^8}{24a^3 b^3 c^5}$ **12.** $\dfrac{45m^2 n^3 k^7}{-40m^5 p^8}$

13. $\dfrac{8a - 8b}{ab - b^2}$ **14.** $\dfrac{ax + ab}{ax - ab}$

15. $\dfrac{x^2 - y^2}{xy + y^2}$ **16.** $\dfrac{x^2 + 2x - 35}{x^2 - 25}$

17. $\dfrac{16x^8 y^9 z^{10}}{28yz^9 t^6} \cdot \dfrac{7z^2 tu^5}{25xyz^2}$ **18.** $\dfrac{-16ab^2}{9bc^2} \cdot \dfrac{15c^3 d^2}{14bx^2 y}$

19. $\dfrac{a^5}{a^3 + 7a^2} \cdot \dfrac{a^2 - 49}{a}$ **20.** $\dfrac{x^2 + 5x}{x^4} \cdot \dfrac{x^2}{x^2 - 25}$

21. $\dfrac{a^6 - 6a^5}{a^2 - a - 30} \cdot \dfrac{a^2 - 25}{a^4 - 36a^2}$ **22.** $\dfrac{x^2 - x - 42}{x^2 + 11x + 30} \cdot \dfrac{x^2 + 13x + 40}{x^2 + 4x - 32}$

23. $\dfrac{4xy^2z^6}{5u^7v^7w^5} \div \dfrac{-8w^3x^6y^4}{15u^5v^4z^5}$

24. $\dfrac{4a^3bc^4d}{3wx^2yz^4} \div \dfrac{8abc^4}{9x^4y^5z^9}$

25. $\dfrac{a-2}{(a+6)^3} \div \dfrac{a^2-a-2}{a^2-36}$

26. $\dfrac{x-4}{x+5} \div \dfrac{x^2-16}{x^2-25}$

27. $\dfrac{x^2-4x-12}{x^2-3x-18} \div \dfrac{x^2-6x-16}{x^2-4x-21}$

28. $\dfrac{t^2-49}{t^2-2t-35} \div \dfrac{t^2+2t-8}{t^2+7t+12}$

29. $\dfrac{1}{x^2} + \dfrac{1}{x^7} + \dfrac{1}{x^{10}}$

30. $\dfrac{7}{p} - \dfrac{4}{p^2} + \dfrac{6}{p^5}$

31. $\dfrac{a-b^2}{a^2b} - \dfrac{a^2-b}{ab^2}$

32. $\dfrac{4}{abc^2} + \dfrac{3}{a^2bc} - \dfrac{7}{ab^2c}$

33. $\dfrac{a}{a-5} + \dfrac{8}{2a-5}$

34. $\dfrac{t}{2t-1} - \dfrac{5}{t+1}$

35. $\dfrac{8}{m^2-12m+35} - \dfrac{5}{m^2-25}$

36. $\dfrac{k}{k^2-3k-28} + \dfrac{k-5}{k^2-7k}$

37. $3 - \dfrac{8}{4x} + \dfrac{5}{2x^2}$

38. $x - \dfrac{6}{x} + \dfrac{5}{x+4}$

39. $\dfrac{\dfrac{5}{12xy}}{\dfrac{7}{18yz}}$

40. $\dfrac{\dfrac{4}{9ab^2}}{\dfrac{7}{15a^2b}}$

41. $\dfrac{9 - \dfrac{1}{t^2}}{6 - \dfrac{1}{t}}$

42. $\dfrac{x^2 - \dfrac{3}{x^2}}{x + \dfrac{1}{x}}$

43. $\dfrac{\dfrac{7}{abc^2} - \dfrac{4}{a^2bc} + \dfrac{2}{ab^2c}}{\dfrac{1}{ab} - \dfrac{5}{ac} + \dfrac{3}{bc}}$

44. $\dfrac{\dfrac{7}{xy^2} - \dfrac{3}{xz^2} - \dfrac{1}{x^2y}}{\dfrac{2}{xyz} - \dfrac{5}{x^2yz} + \dfrac{3}{xyz^2}}$

45. $2^{-2} - 2^{-3}$

46. $5^{-1} + 5^{-2}$

47. $x^{-1} + y^{-1}$

48. $a^{-3} - a^{-2}$

49. $\dfrac{t^{-1} + t^{-4}}{t^{-2} + t^{-3}}$

50. $\dfrac{a^{-1} - b^{-2}}{a^{-3} + b^{-2}}$

51. $\dfrac{1}{xy}\left(\dfrac{x+y}{x^3y} + \dfrac{x-y}{x^2y^2}\right)$

52. $x^{-1}(x^{-2} + x^{-3})$

53. $\dfrac{1}{h}\left(\dfrac{1}{x+h} - \dfrac{1}{x}\right)$

54. $\dfrac{1}{a^{-2}} + \dfrac{1}{a^2}$

55. $\dfrac{1}{h}[(x+h)^{-2} - x^{-2}]$

56. $\dfrac{4}{x^2y^3z^4} + \dfrac{5}{x^3y^7} - \dfrac{1}{y^5z^8}$

57. $\left(\dfrac{4}{xy} - \dfrac{1}{y}\right) \div \left(\dfrac{1}{xy} + \dfrac{1}{x}\right)$

58. $\dfrac{5 - \dfrac{2}{x+4}}{\dfrac{3}{x-2} + \dfrac{7}{x^2+2x-8}}$

59. $\dfrac{x^2 + 4x - 32}{x^3 + 27} \cdot \dfrac{x^2 - 9}{x^2 + 13x + 40} \div \dfrac{x^3 - 4x^2}{x^5 - 3x^4 + 9x^3}$

60. $\dfrac{t^2 - 6t - 16}{t^2 - 4} \cdot \dfrac{t^3 - 8}{t^2 - 3t - 40} \div \dfrac{t^2 + 2t - 63}{t^2 - 2t - 35}$

61. $\dfrac{5}{x^2 - 11x + 30} + \dfrac{7}{x^2 - 10x + 24} - \dfrac{3}{x^2 - 9x + 20}$

62. $\dfrac{a}{2a^2 + 3a + 1} - \dfrac{a - 1}{2a^2 + 15a + 7} - \dfrac{a - 2}{a^2 + 8a + 7}$

63. $\dfrac{x^2 + 6x - 7}{x^3} \cdot \dfrac{7}{x^2 - 6x + 8} \div \dfrac{x^2 + 2x - 3}{x^2 - x - 12}$

64. $\dfrac{2}{x^2 - 4x - 5} - \dfrac{7}{x^2 + 5x + 4} + \dfrac{1}{x^2 - x - 20}$

65. $\dfrac{a^2 - 25}{a^2 + 12a + 35} \cdot \dfrac{a^2 - 49}{a^2 - 16a + 63} + \dfrac{a}{a^2 - 81}$

66. $\dfrac{p^2 + 9p + 20}{p^2 + 4p - 21} \div \dfrac{p^2 + 7p + 10}{p^2 + 9p + 14} \cdot \dfrac{p + 1}{p_2 - 9}$

Physical Application

67. In 2 seconds an object falls 64 feet; in t seconds it falls $16t^2$ feet. The average speed between 2 and t seconds is given by

$$\text{Average speed} = \dfrac{16t^2 - 64}{t - 2}$$

Simplify this fraction.

Optical Application

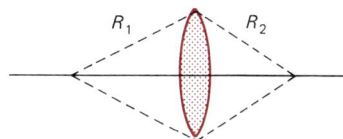

68. The *lensmaker's equation* is

$$\dfrac{1}{f} = (n - 1)\left(\dfrac{1}{R_1} - \dfrac{1}{R_2}\right)$$

where f is the focal length of the lens, n the index of refraction of glass, and R_1 and R_2 the radii of curvature, as shown. Write the expression on the right as a single fraction.

Business Application

69. The cost C to produce x units of a certain car is given by $C = -x^3 - 200x^2 + 3000x$. The average cost per car is given by

$$\text{Average cost} = \dfrac{C}{x} = \dfrac{-x^3 - 200x^2 + 3000x}{x}$$

Simplify this fraction.

Life Science Application

70. The weight W of a certain bacterial culture is given by

$$W = \dfrac{6400}{2 - 3^{-t}}$$

where t is the time in hours.

(a) Determine the weight when $t = 0$; $t = 1$; $t = 2$.
(b) Rewrite the equation as a simple fraction with only positive exponents.

71. The following formula is recommended in the construction of steel columns:

$$\frac{P}{A} = \frac{18{,}000}{1 + \dfrac{1}{18{,}000}\left(\dfrac{L}{k}\right)^2}$$

where P is the load, A the cross-sectional area, L the length, and k a constant. Simplify the expression on the right.

2.6
RADICALS

Just as factoring is the reverse of multiplication, **roots** are the reverse of exponents. Consider the following:

$$\text{Power: } 2^3 = \boxed{?} \qquad \text{Power is unknown.}$$

$$\text{Root: } \boxed{?}^3 = 8 \qquad \text{Base is unknown.}$$

Since $2^3 = 8$, we say that 2 is the *third root* of 8. We write this $2 = \sqrt[3]{8}$. In general, we write roots as follows.

Definition
For any real numbers r and a and positive integer n,

$$r = \sqrt[n]{a} \quad \text{means} \quad r^n = a$$

Here we call r the **nth root of a**, and $r = \sqrt[n]{a}$ answers the question: What number to the nth power gives a? The number n is called the **index**, a the **radicand**, and $\sqrt[n]{a}$ the **radical**. (Usually, $\sqrt[2]{a}$ is written simply $\sqrt{a}$.)

We have to be especially careful when n is even. For instance, $(-5)^2 = 25$ and $(5)^2 = 25$, but we say that $\sqrt{25} = 5$. (We insist that the symbol $\sqrt{25}$ represent *one and only one* number: 5.) We call this positive root the **principal root**. Also, $\sqrt{-36}$ is not a real number since there is no real number whose square is -36. (We discuss this situation in the chapter on complex numbers.)

1. If n is odd, there is always a unique root, $r = \sqrt[n]{a}$.
2. If n is even:
    (a) If $a \geq 0$, then $r = \sqrt[n]{a}$ is the nonnegative root.
    (b) If $a < 0$, then there is *no* real number for $\sqrt[n]{a}$.

EXAMPLE 38   The following are examples of radicals. (We assume that all letters represent positive numbers.)

(a) $\sqrt{49} = 7$ since $7^2 = 49$. (This is the *square root*.)

(b) $\sqrt[3]{-64} = -4$ since $(-4)^3 = -64$. (This is the *cube root*.)

(c) $\sqrt[5]{32} = 2$ since $2^5 = 32$. (This is the *fifth root*.)

(d) $\sqrt{\dfrac{64}{25}} = \dfrac{8}{5}$ since $\left(\dfrac{8}{5}\right)^2 = \dfrac{64}{25}$.

(e) $\sqrt[4]{0.0001} = 0.1$ since $(0.1)^4 = 0.0001$.

(f) $\sqrt{-100}$ is not a real number.

(g) $\sqrt{a^8} = a^4$ since $(a^4)^2 = a^8$.

(h) $\sqrt[5]{\dfrac{1}{x^{10}}} = \dfrac{1}{x^2}$ since $\left(\dfrac{1}{x^2}\right)^5 = \dfrac{1}{x^{10}}$.

Property

The following properties hold (assuming that the radicals exist as real numbers).

R1 $\sqrt[n]{ab} = \sqrt[n]{a}\sqrt[n]{b}$

R2 $\sqrt[n]{\dfrac{a}{b}} = \dfrac{\sqrt[n]{a}}{\sqrt[n]{b}} \qquad (b \neq 0)$

R3 $(\sqrt[n]{a})^n = a$

R4 $\sqrt[n]{a^n} = a$

R5 $\sqrt[m]{\sqrt[n]{a}} = \sqrt[mn]{a}$

Caution: When n is even, $a \geq 0$ and $b \geq 0$.

EXAMPLE 39 The following examples use Properties R1 to R5.

(a) $\sqrt{50} = \sqrt{25 \cdot 2} = \sqrt{25}\sqrt{2} = 5\sqrt{2}$ (R1)

(b) $\sqrt[3]{54} = \sqrt[3]{27 \cdot 2} = \sqrt[3]{27}\sqrt[3]{2} = 3\sqrt[3]{2}$ (R1)

(c) $\sqrt[4]{\dfrac{3}{16}} = \dfrac{\sqrt[4]{3}}{\sqrt[4]{16}} = \dfrac{\sqrt[4]{3}}{2}$ (R2)

(d) $(\sqrt{5})^2 = 5$ (R3)

(e) $\sqrt[3]{\sqrt[5]{7}} = \sqrt[15]{7}$ (R5)

(f) $\sqrt[7]{9^7} = 9$ (R4)

YES	NO
$\sqrt{4 \cdot 9} = \sqrt{4} \cdot \sqrt{9}$	$\sqrt{4+9} = \sqrt{4} + \sqrt{9}$
$\sqrt{36} = 2 \cdot 3 = 6$	$\sqrt{13} = 2 + 3 = 5$
$\sqrt{3a} = \sqrt{3}\sqrt{a}$	$\sqrt{3+a} = \sqrt{3} + \sqrt{a}$
Property R1 holds only for *products*.	

We have the following guidelines for keeping radicals as simplified as possible.

---

1. All possible powers are factored out of the radical.
2. No radicals contain denominators.
3. No denominators contain radicals.

Our next examples show how to meet these requirements.

EXAMPLE 40   Simplify $\sqrt[5]{64k^7m^{14}}$.

Solution   We simplify this by factoring out the largest fifth-power factor within the radicand; here it is $32k^5m^{10}$ $[=(2km^2)^5]$.

Factor largest fifth power	$\sqrt[5]{64k^7m^{14}} = \sqrt[5]{(32k^5m^{10})(2k^2m^4)}$
$\sqrt[n]{ab} = \sqrt[n]{a}\sqrt[n]{b}$	$= \sqrt[5]{32k^5m^{10}}\sqrt[5]{2k^2m^4}$
Simplify	$= 2km^2\sqrt[5]{2k^2m^4}$

EXAMPLE 41   Simplify $\sqrt[3]{\dfrac{5}{4x}}$.

Solution   We need to make the denominator a perfect cube; in this case, $8x^3$ is the easiest to achieve. To do this, we multiply by $\sqrt[3]{2x^2}$ since $\sqrt[3]{4x}\sqrt[3]{2x^2} = \sqrt[3]{8x^3} = 2x$.

$1 = \sqrt[3]{\dfrac{2x^2}{2x^2}}$	$\sqrt[3]{\dfrac{5}{4x}} = \sqrt[3]{\dfrac{5}{4x}}\sqrt[3]{\dfrac{2x^2}{2x^2}}$
$\sqrt[n]{ab} = \sqrt[n]{a}\sqrt[n]{b}$	$= \sqrt[3]{\dfrac{10x^2}{8x^3}}$
$\sqrt[n]{\dfrac{a}{b}} = \dfrac{\sqrt[n]{a}}{\sqrt[n]{b}}$	$= \dfrac{\sqrt[3]{10x^2}}{\sqrt[3]{8x^3}}$
Simplify	$= \dfrac{\sqrt[3]{10x^2}}{2x}$

Often we wish to move radical terms from the denominator to the numerator, or vice versa. We use pairs such as $\sqrt{7} + \sqrt{5}$ and $\sqrt{7} - \sqrt{5}$, called **conjugates**. The product of conjugates is always without a radical: $(\sqrt{7} + \sqrt{5})(\sqrt{7} - \sqrt{5}) = \sqrt{49} - \sqrt{25} = 2$.

EXAMPLE 42 Simplify $\dfrac{3}{\sqrt{7} + 2}$.

Solution We remove the radical from the denominator by **rationalizing**.

> *To rationalize a radical expression:*
>
> 1. Determine the conjugate of the expression.
>
> 2. Multiply by $1 = \dfrac{\text{conjugate}}{\text{conjugate}}$.
>
> 3. Simplify.

Here the conjugate of the denominator is $\sqrt{7} - 2$.

$\boxed{\text{Multiply by } 1 = \dfrac{\text{conjugate}}{\text{conjugate}}}$

$\dfrac{3}{\sqrt{7} + 2} = \dfrac{3}{\sqrt{7} + 2} \cdot \boxed{\dfrac{\sqrt{7} - 2}{\sqrt{7} - 2}}$

$\boxed{\begin{array}{c}\text{Distributive} \\ \text{law}\end{array}}$

$= \dfrac{3\sqrt{7} - 6}{\sqrt{49} - 4}$

$\boxed{\text{Simplify}}$

$= \dfrac{3\sqrt{7} - 6}{7 - 4} = \dfrac{3\sqrt{7} - 6}{3}$

EXAMPLE 43 Rationalize the numerator of $\dfrac{\sqrt{x + h} - \sqrt{x}}{h}$ (for $x > 0$ and $h \neq 0$).

Solution Generally, we rationalize the denominator, but there are circumstances (in calculus, for instance) for which it is convenient to rationalize the numerator. Here we multiply the fraction, top and bottom, by the conjugate of the numerator: $\sqrt{x + h} + \sqrt{x}$.

$$\dfrac{\sqrt{x + h} - \sqrt{x}}{h} = \dfrac{\sqrt{x + h} - \sqrt{x}}{h} \cdot \boxed{\dfrac{\sqrt{x + h} + \sqrt{x}}{\sqrt{x + h} + \sqrt{x}}}$$

$$= \dfrac{\sqrt{(x + h)^2} - \sqrt{x^2}}{h(\sqrt{x + h} + \sqrt{x})}$$

$$= \dfrac{(x + h) - x}{h(\sqrt{x + h} + \sqrt{x})}$$

$$= \dfrac{h}{h(\sqrt{x + h} + \sqrt{x})} = \dfrac{1}{\sqrt{x + h} + \sqrt{x}}$$

EXAMPLE 44 On a hand calculator with a $\boxed{\sqrt{x}}$ key, square roots are easy to compute. Consider $\sqrt{7.316}$.

PRESS	DISPLAY	MEANING
$\boxed{C}$	$\boxed{0.}$	Clear
$\boxed{7}\,\boxed{\cdot}\,\boxed{3}\,\boxed{1}\,\boxed{6}$	$\boxed{7.316}$	Enter radicand
$\boxed{\sqrt{x}}$	$\boxed{2.7048105}$	$\sqrt{7.316}$

For higher roots, we use a calculator with $\boxed{\text{INV}}$ and $\boxed{y^x}$ keys (or their equivalents). Consider $\sqrt[5]{416.028}$.

PRESS	DISPLAY	MEANING
$\boxed{C}$	$\boxed{0.}$	Clear
$\boxed{4}\,\boxed{1}\,\boxed{6}\,\boxed{\cdot}\,\boxed{0}\,\boxed{2}\,\boxed{8}$	$\boxed{416.028}$	Enter radicand
$\boxed{\text{INV}}\,\boxed{y^x}\,\boxed{5}$	$\boxed{5.}$	Fifth root
$\boxed{=}$	$\boxed{3.3406}$	$\sqrt[5]{416.028}$

PROBLEM SET 2.6

Simplify the following expressions as much as possible. (Assume that all letters are positive numbers.)

Warm-up Exercises

1. $(x + 3)(x - 3)$ **2.** $(2t - 5)(2t + 5)$

3. $(3ab^2c^3)^2$ **4.** $\left(\dfrac{2x^2y^3}{z^4}\right)^3$

5. $\sqrt{9}$ **6.** $\sqrt{16}$

7. $\sqrt{49}$ **8.** $\sqrt{100}$

9. $\sqrt{\dfrac{1}{4}}$ **10.** $\sqrt{\dfrac{1}{25}}$

Simplify the following expressions as much as possible. (Assume that all variables are positive.)

11. $\sqrt[4]{81/625}$ **12.** $\sqrt{0.04}$

13. $\sqrt[3]{0.001}$ **14.** $\sqrt{x^{10}}$

15. $\sqrt[4]{m^8 n^{12}}$ **16.** $\sqrt[3]{27y^3z^9}$

17. $\sqrt[5]{\dfrac{1}{x^{15}}}$ **18.** $\sqrt{\dfrac{25}{t^4}}$

19. $\sqrt[3]{64m^3/k^6n^{12}}$ **20.** $\sqrt{\dfrac{24}{25}}$

21. $\sqrt[3]{\dfrac{81}{125}}$ **22.** $(\sqrt{2})^2$

23. $\sqrt[11]{3^{11}}$

24. $\sqrt[4]{\sqrt[5]{3}}$

25. $\sqrt[2]{\sqrt[3]{\sqrt[4]{2}}}$

26. $\sqrt[3]{10,000s^5t^7}$

27. $\sqrt[5]{320a^7b^{12}c^{17}}$

28. $\sqrt[10]{2048p^7q^{19}r^{23}}$

29. $\dfrac{6}{\sqrt[3]{25x}}$

30. $\dfrac{8t}{\sqrt[5]{4t^4w^3}}$

31. $\sqrt[4]{\dfrac{5x}{27y^2}}$

32. $\dfrac{2}{\sqrt{11}+\sqrt{2}}$

33. $\dfrac{\sqrt{7}}{\sqrt{13}-\sqrt{t}}$

34. $\dfrac{1}{\sqrt{x}+\sqrt{3}}$

35. $\dfrac{7}{\sqrt{5}+2}$

36. $\dfrac{\sqrt{3}}{\sqrt{7}+\sqrt{5}}$

37. $\dfrac{\sqrt{11}}{4-\sqrt{8}}$

38. $\dfrac{\sqrt{3}+\sqrt{2}}{4}$

39. $\dfrac{5-\sqrt{11}}{2}$

40. $\dfrac{\sqrt{5+h}-\sqrt{5}}{h}$

41. $\dfrac{\sqrt{x}-\sqrt{y}}{x-y}$

42. $\dfrac{\sqrt{x}+\sqrt{5}}{x-5}$

43. $\dfrac{\sqrt{2x+t+1}-\sqrt{2x+1}}{t}$

44. $\sqrt[4]{\dfrac{3a}{2b^3}}$

45. $\sqrt{3}+\sqrt{12}$

46. $\dfrac{\sqrt{7}}{\sqrt{5}+\sqrt{3}}$

47. $\dfrac{3}{\sqrt[4]{8}}$

48. $\sqrt[3]{81a^2b^4c^6}$

49. $\sqrt{2}+\sqrt{8}+\sqrt{18}$

50. $\sqrt[3]{10}+\sqrt[3]{80}+\sqrt[3]{270}$

51. $4\sqrt[3]{81x}+6\sqrt[3]{24x^4}$

52. $10\sqrt[3]{54a}-3\sqrt[3]{250a^7}$

53. $\sqrt{2x^2y^3}\sqrt{8xy^7}$

54. $\sqrt[3]{6a^4b}\sqrt[3]{9ab^5}$

55. $\sqrt[4]{\dfrac{2x}{y^3z^5}}\sqrt[4]{\dfrac{32x^3y}{z^2}}$

56. $\sqrt{\dfrac{18a^3b^5}{25c^4d^9}}\sqrt{\dfrac{2ac^3}{b^7d^2}}$

Simplify the following with a hand calculator.

57. $\sqrt{20.17}$

58. $\sqrt{85.012}$

59. $\sqrt[3]{2181.972}$

60. $\sqrt[4]{0.0715}$

61. $\sqrt[10]{0.583}$

62. $\sqrt[20]{3.017}$

63. $\sqrt[10]{10}$

64. $\sqrt[100]{100}$

65. $\sqrt[1000]{1000}$

66. Without a calculator, determine which number in each of the following pairs is larger. (*Hint*: Raise both numbers to a common-multiple power.)

    (a) $\sqrt{2}$ or $\sqrt[5]{6}$ (b) $\sqrt{5}$ or $\sqrt[3]{11}$ (c) $\sqrt[3]{3}$ or $\sqrt[4]{4}$

67. Under a certain depreciation scheme, the yearly depreciation rate d is given by

$$d = 1 - \sqrt[n]{\frac{S}{C}}$$

where S is the salvage value, C the original cost, and n the number of useful years. Write this as a fraction without a radical in the denominator.

68. In 4 seconds, an object moves 20 meters; in t seconds, it moves $10\sqrt{t}$ meters. The average velocity in the time between 4 and t seconds is given by

$$\text{Average velocity} = \frac{10\sqrt{t} - 20}{t - 4}$$

Rationalize the numerator and simplify.

69. In an alternating-current circuit, the effective voltage V_{eff} and the maximum voltage V_{max} satisfy

$$V_{\text{eff}} = \frac{V_{\text{max}}}{\sqrt{2}}$$

Simplify this.

70. In a guitar, let the length from the neck to the bridge be L. Then the length from the bridge to the first fret is $L\sqrt[12]{1/2}$. Simplify this radical.

71. The fundamental frequency of a plucked, hammered, or bowed string is given by

$$f = \frac{1}{2L} \sqrt{\frac{T}{m}}$$

where L is the length of the string, T the tension, and m the mass per unit length. Simplify this radical.

72. For a salt (NaCl) crystal the distance between ions (or grating spacing) is given by

$$d = \sqrt[3]{\frac{M}{2pN_0}}$$

where M, p, and N_0 are constants. Simplify this radical.

2.7
RATIONAL
EXPONENTS

In addition to radicals, we have another, often more convenient way to express roots. We use **rational exponents**. As with zero and negative exponents, we want rational exponents to satisfy the basic exponent properties, E1 to E5 (see page 19). Consider $7^{1/3}$ and $\sqrt[3]{7}$:

Property E2: $(a^r)^s = a^{rs}$	Definition of radical
$(7^{1/3})^3 = 7^{3/3} = 7$	$(\sqrt[3]{7})^3 = 7$

Since $7^{1/3}$ and $\sqrt[3]{7}$ both produce 7 when cubed, it is reasonable to define $7^{1/3} = \sqrt[3]{7}$. In general, we have the following definition.

Definition

For real number a and natural number n, if $\sqrt[n]{a}$ exists,

$$\textbf{E8} \quad a^{1/n} = \sqrt[n]{a}$$

EXAMPLE 45   The following examples demonstrate this definition. (*Think*: Denominator of exponent = index of radical.)

(a) $9^{1/2} = \sqrt{9} = 3$ (b) $625^{1/4} = \sqrt[4]{625} = 5$

(c) $(-1000)^{1/3} = \sqrt[3]{-1000} = -10$ (d) $\sqrt[7]{x} = x^{1/7}$

Now consider $a^{m/n}$. We can rewrite this in two ways:

$$a^{m/n} = (a^{1/n})^m = (\sqrt[n]{a})^m$$
$$\text{or} \quad a^{m/n} = (a^m)^{1/n} = \sqrt[n]{a^m}$$

Definition

For real number a and rational number m/n, if $\sqrt[n]{a}$ exists,

$$\textbf{E9} \quad a^{m/n} = (\sqrt[n]{a})^m = \sqrt[n]{a^m}$$

EXAMPLE 46   The following examples use rational exponents. (*Think*: Denominator = index, numerator = power.) Generally, we compute the root first, then the power.

(a) $25^{3/2} = (\sqrt{25})^3 = 5^3 = 125$

(b) $(-1000)^{2/3} = (\sqrt[3]{-1000})^2 = (-10)^2 = 100$

(c) $\sqrt[9]{t^2} = t^{2/9}$

(d) $9^{-5/2} = (\sqrt{9})^{-5} = 3^{-5} = \dfrac{1}{3^5} = \dfrac{1}{243}$

(e) $\sqrt[7]{\dfrac{1}{x^3}} = \sqrt[7]{x^{-3}} = x^{-3/7}$

Although rational exponents, as in $9^{-5/2}$, do not appeal to our common sense, they are convenient since all the exponent properties now hold for rational exponents. We now restate the exponent properties.

Property

For real numbers a and b and rational numbers r and s, the following hold true whenever the expressions exist.

E1 $\quad a^r a^s = a^{r+s}$

E2 $\quad (a^r)^s = a^{rs}$

E3 $\quad (ab)^r = a^r b^r$

E4 $\quad \dfrac{a^r}{a^s} = a^{r-s}$

E5 $\quad \left(\dfrac{a}{b}\right)^r = \dfrac{a^r}{b^r}$

EXAMPLE 47 Simplify $x^{1/2}(x^{1/6} + x^{-1/6})$.

Solution We treat this as we would $x^5(x^2 + x^{-2}) = x^7 + x^3$: We use the distributive law and add the exponents.

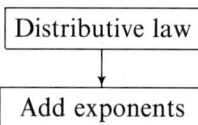

| Distributive law | $x^{1/2}(x^{1/6} + x^{-1/6}) = x^{1/2}x^{1/6} + x^{1/2}x^{-1/6}$ |

| Add exponents | $= x^{2/3} + x^{1/3}$ |

EXAMPLE 48 Simplify $P = \left(\dfrac{a^{1/3}b^{-2/5}}{c^{3/4}}\right)^{1/2} \cdot \left(\dfrac{a^{-1/6}}{b^{-1/10}c^{1/8}}\right)^{-1/2}.$

Solution This example is very much like Example 9 (page 22), except that we now have fractional exponents. We work this out *very slowly and carefully*, using the different exponent properties.

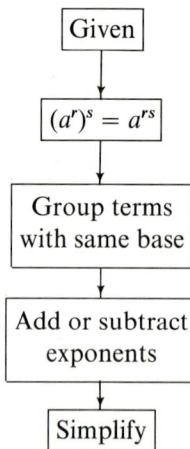

| Given | $P = \left(\dfrac{a^{1/3}b^{-2/5}}{c^{3/4}}\right)^{1/2} \cdot \left(\dfrac{a^{-1/6}}{b^{-1/10}c^{1/8}}\right)^{-1/2}$ |

| $(a^r)^s = a^{rs}$ | $= \dfrac{a^{1/6}b^{-1/5}}{c^{3/8}} \cdot \dfrac{a^{1/12}}{b^{1/20}c^{-1/16}}$ |

| Group terms with same base | $= a^{1/6}a^{1/12} \cdot \dfrac{b^{-1/5}}{b^{1/20}} \cdot \dfrac{1}{c^{3/8}c^{-1/16}}$ |

| Add or subtract exponents | $= \dfrac{a^{1/6+1/12} \cdot b^{-1/5-1/20}}{c^{3/8+(-1/16)}}$ |

| Simplify | $= \dfrac{a^{1/4}b^{-1/4}}{c^{5/16}} = \dfrac{a^{1/4}}{b^{1/4}c^{5/16}}$ |

EXAMPLE 49 We can also calculate fractional exponents with a hand calculator. Consider $9.08^{2/7}$.

PRESS	DISPLAY	MEANING
9 . 0 8	9.08	Enter base, 9.08
y^x (2 ÷ 7)	0.2857143	To 2/7 power
=	1.87819	Answer

Note that we use the parentheses keys for the exponent 2/7. Without the parentheses, the calculator gives top priority to $\boxed{y^x}$ and calculates $9.08^2/7$, which is *not* what we want.

Simplify the following (with or without a hand calculator).

1. $\sqrt{100}$ **2.** $\sqrt{200}$ **3.** $\sqrt[3]{8/27}$

4. $\sqrt{1.73}$ **5.** $\sqrt[3]{10}$ **6.** $\sqrt{16x^2y^4}$

7. 3^{-2} **8.** $\left(\dfrac{1}{2}\right)^{-1}$ **9.** $x^2x^3x^4$

10. $(a^2b^3)^4$ **11.** $\left(\dfrac{u^2}{v^{-3}}\right)^{-1}$ **12.** $x^2(x^3 - x^7)$

Evaluate the following expressions.

13. $125^{1/3}$ **14.** $10,000^{1/4}$ **15.** $\left(\dfrac{1}{16}\right)^{1/2}$

16. $\left(\dfrac{8}{27}\right)^{1/3}$ **17.** $8^{2/3}$ **18.** $81^{3/4}$

19. $125^{4/3}$ **20.** $\left(\dfrac{16}{625}\right)^{5/4}$ **21.** $9^{-1/2}$

22. $100,000^{-3/5}$ **23.** $\left(\dfrac{8}{125}\right)^{-2/3}$ **24.** $\left(\dfrac{9}{4}\right)^{-5/2}$

Write the following rational-exponent expressions as radicals (Problems 25 to 30); write the radicals with rational exponents (Problems 31 to 36).

25. $x^{1/9}$ **26.** $(a - b)^{1/3}$ **27.** $k^{3/2}$

28. $(1 - 3y)^{10/7}$ **29.** $(x + y)^{-1/6}$ **30.** $(1 + 3z)^{-5/3}$

31. $\sqrt{m + n}$ **32.** $\sqrt[7]{r - s}$ **33.** $\sqrt[3]{u^5}$

34. $(\sqrt[6]{t - u})^5$ **35.** $\dfrac{1}{\sqrt[3]{z^4}}$ **36.** $\sqrt{\dfrac{1}{1 + x}}$

Simplify the following expressions, leaving all exponents positive.

37. $a^{1/2}a^{2/3}$ **38.** $x^{1/4}x^{1/3}x^{1/2}$

39. $k^{1/2}k^{-1/4}$ **40.** $m^{1/3}m^{-1/2}$

41. $x^{1/3}(x^{1/2} + x^{4/5})$

42. $y^{1/7}(y^{1/3} - y^{2/3})$

43. $t^{-1/2}(t^{2/3} + t^{-1/3})$

44. $z^{1/6}(z^{-1/2} - z^{1/3})$

45. $(a^{1/2}b^{2/3})^{1/4}$

46. $(x^{1/7}y^{2/3})^{1/2}$

47. $\left(\dfrac{u^{-1/2}v^{1/3}}{w^{1/4}}\right)^{-1/4}$

48. $\left(\dfrac{a^{1/2}b^{-1/3}}{c^{-2/5}}\right)^{-1/3}$

49. $\left(\dfrac{a^{1/2}}{b^{-1/3}}\right)^{1/3}\left(\dfrac{a^{-1/3}}{b^{1/4}}\right)^{-1/2}$

50. $\left(\dfrac{x^{-1/5}v^{1/2}}{z^{1/3}t^{-2/7}}\right)^{-1/3}\left(\dfrac{x^{1/2}z^{-1/3}}{y^{-1/3}}\right)^{1/2}$

Write the following expressions as simply as possible (without radicals or negative exponents).

51. $9^{1/2}$

52. $\sqrt{a^5}$

53. $\sqrt[3]{\dfrac{1}{t}}$

54. $r^{1/3}r^{1/4}r^{1/5}$

55. $27^{-1/3}$

56. $\sqrt[3]{k}\sqrt[4]{k}$

57. $t^{-1/3}(t^{1/2} - t^{-1/5})$

58. $16^{5/4}$

59. $(a^{1/2}b^{-1/3})^{-1/4}$

60. $\dfrac{m^{1/2}}{m^{-1/3}}$

61. $\left(\dfrac{4}{25}\right)^{-3/2}$

62. $\sqrt[5]{\sqrt[4]{x^{1/3}}}$

63. $\sqrt[5]{x^{2/3}}$

64. $[x^{0.4}]^{-1}$

65. $(\sqrt{k})^3\sqrt[4]{k^5}$

66. $(9^{-1/2} + 8^{-1/3})^{-1}$

67. $\dfrac{(x + h)^{1/2} - x^{1/2}}{h}$

68. $\dfrac{1}{\sqrt{\dfrac{1}{x^{1/3}}}}$

69. $[(1000^{-4/3})(8^{-2/3})]^{-1/2}$

70. $\dfrac{(x + h)^{-1/2} - x^{-1/2}}{h}$

71. $a^{2/5}a^{-1/6}$

72. $x^{-1/2}(x^{1/3} - x^{-1/4})$

73. $\left(\dfrac{a^{1/5}b^{-1/4}}{c^{-1/3}}\right)^{-1/2}$

74. $\sqrt[3]{a^{1/2}b^{2/3}}$

75. $\dfrac{(m^{1/3}n^{-1/4})^{3/2}}{(m^{2/5}n^{1/2})^{-1/3}}$

76. $\sqrt[5]{\dfrac{1}{x^{-1/2}y^{2/3}}}$

Perform the following operations with (or without) a hand calculator.

77. $10^{5/3}$

78. $\sqrt{14^5}$

79. $1.031^{4/9}$

80. $\sqrt[7]{5 \cdot 2^3}$

81. $\sqrt[8]{0.031^5}$

82. $2.73^{11/13}$

83. $\sqrt{1.39 \times 10^7}$

84. $\sqrt[3]{4.07 \times 10^{-2}}$

Technical Application

85. The resistance in 1000 feet of copper wire can be approximated by

$$R = 2^{(g-10)/3}$$

where g is the gauge of the wire. Write this with radicals as simplified as possible.

86. The surface area of a human being is given by $S = 2W^{0.4}H^{0.7}$, where W is the weight in kilograms and H is the height in meters.

(a) Find S when $W = 74$ kilograms and $H = 1.8$ meters.

(b) Find S when $W = 60$ kilograms and $H = 1.7$ meters.

(c) Rewrite this formula using rational exponents.

(d) Rewrite this formula using radicals.

87. The sales S of a given product depend on the price p, the number of salespeople n, and the advertising budget a, as follows:

$$S = (40,000 - 2000p)n^{2/3}a^{1/2}$$

Find S when:

(a) $p = 5$, $n = 8$, $a = 2500$

(b) $p = 6$, $n = 1000$, $a = 10,000$

88. The recommended relation between stage volume V and the number of instruments on stage N is given by

$$N = 0.0021V^{0.855}$$

Find N (to the nearest whole number) when:

(a) $V = 9000$

(b) $V = 20,000$

(c) $V = 100,000$

CHAPTER 2 SUMMARY

Important Properties and Definitions

For real numbers a, b, c, d, and r, and integers m and n:

$b^n = b \cdot b \cdot b \cdot \ldots \cdot b$ (n factors)

$a^m a^n = a^{m+n}$

$(a^m)^n = a^{mn}$

$(ab)^n = a^n b^n$

$\dfrac{a^m}{a^n} = a^{m-n}$ $(a \neq 0)$

$\left(\dfrac{a}{b}\right)^n = \dfrac{a^n}{b^n}$ $(b \neq 0)$

$a^0 = 1$

$a^{-n} = \dfrac{1}{a^n}$ $(a \neq 0)$

$(a + b)(c + d) = ac + ad + bc + bd$

$(a + b)(a - b) = a^2 - b^2$

$(a + b)^2 = a^2 + 2ab + b^2$

$a^2 - b^2 = (a + b)(a - b)$

$a^3 + b^3 = (a + b)(a^2 - ab + b^2)$

$a^3 - b^3 = (a - b)(a^2 + ab + b^2)$

$\dfrac{a}{b} = \dfrac{c}{d}$ if $ad = bc$ $(b, d \neq 0)$

$\dfrac{a}{b} = \dfrac{ac}{bc}$ $(b, c \neq 0)$

$\dfrac{a}{b} \cdot \dfrac{c}{d} = \dfrac{ac}{bd}$ $(b, d \neq 0)$

$\dfrac{a}{b} \div \dfrac{c}{d} = \dfrac{a}{b} \cdot \dfrac{d}{c} = \dfrac{ad}{bc}$ $(b, c, d \neq 0)$

$\dfrac{a}{c} + \dfrac{b}{c} = \dfrac{a + b}{c}$ $(c \neq 0)$

$\dfrac{a}{c} - \dfrac{b}{c} = \dfrac{a - b}{c}$ $(c \neq 0)$

$r = \sqrt[n]{a}$ means $r^n = a$

$\sqrt[n]{ab} = \sqrt[n]{a}\sqrt[n]{b}$ $(a, b \geq 0)$

$\sqrt[n]{\dfrac{a}{b}} = \dfrac{\sqrt[n]{a}}{\sqrt[n]{b}}$ $(a \geq 0, b > 0)$

$(\sqrt[n]{a})^n = a$

$\sqrt[n]{a^n} = a$ (for $a \geq 0$)

$\sqrt[m]{\sqrt[n]{a}} = \sqrt[mn]{a}$

$a^{1/n} = \sqrt[n]{a}$

$a^{m/n} = \sqrt[n]{a^m} = (\sqrt[n]{a})^m$

Any positive real number r can be written $r = a \times 10^n$ (*for* $1 \le a < 10$). For monomials $A, B, C, \ldots$, and M ($\neq 0$), we have

$$\frac{A + B + C + \cdots}{M} = \frac{A}{M} + \frac{B}{M} + \frac{C}{M} + \cdots$$

Review Exercises

Simplify the following as much as possible. (Leave all expressions with positive exponents, and assume that all variables are positive.)

1. $\left(\dfrac{1}{10}\right)^3$　　　　　　　　　　　　　**2.** $x^5 x^7 x^9$

3. $(2^n)^3$　　　　　　　　　　　　　　**4.** $(7a^2 b^3 c^k)^2$

5. $\dfrac{a^5}{a^2 a^7}$　　　　　　　　　　　　　**6.** $\left(\dfrac{ab^3}{2}\right)^4$

7. $(8p^4 q^5)(5p^7 q^3)$　　　　　　　　**8.** 3^{-3}

9. $(a^{-1} b^{-5})^{-2}$　　　　　　　　**10.** $\left(\dfrac{2x^{-2}}{y^3 z^{-3}}\right)^{-1}\left(\dfrac{3x^2}{y^{-1} z^{-2}}\right)^{-2}$

11. $(7.3 \times 10^{-2}) \times (1.2 \times 10^5)$

12. Write 0.000782 in scientific notation.

13. $6ab^2 - 7ab^2 - 10ab^2$

14. $(8x^2 - 7x - 3) + (9x^2 - 10x + 2)$

15. $(7k^2 - 3k - 5) - (8k^2 - k + 1)$

16. $5a^2 b^3(6ab^2 c - 7a^2 bc^3)$

17. $(2k + 7)(3k - 5)$

18. $(3p^2 + p - 2)(2p^2 - p + 3)$

19. $(4x^2 y^7 + 12x^5 y^8 - 2xy^3) \div 2xy^3$

20. $(2x^3 - 3x^2 - 7x + 7) \div (x - 2)$

21. $(t^5 + 1) \div (t + 1)$

Factor the polynomials in Problems 22 to 28 completely.

22. $6x^9 y^2 - 12x^5 y^5 + 18x^3 y^4$　　　　**23.** $16x^2 - \dfrac{36}{49} y^2$

24. $a^3 - 125b^3$　　　　　　　　　　**25.** $z^2 + 5z - 14$

26. $6t^2 - 31t + 40$　　　　　　　　**27.** $6x^2 - 10xy + 3x - 5y$

28. $3x^5 + 6x^4 - 45x^3$

Simplify the following as much as possible.

29. Express $\dfrac{a^2 - 4}{a^3 - 8}$ in lowest terms.　　**30.** $\dfrac{8x^2 y^3 t^4}{5y^2 z t^6} \cdot \dfrac{15z^5 tu^4}{16t^4 uv^2}$

31. $\dfrac{t^2 + 3t + 2}{t^2 - 4} \cdot \dfrac{t^2 - 5t + 6}{t^2 + 2t - 15}$　　**32.** $\dfrac{a^2 - 6a + 9}{a^2 + 8a + 15} \div \dfrac{a^2 - 5a + 6}{a^2 + a - 6}$

33. $\dfrac{6}{a} + \dfrac{8}{a^2} - \dfrac{5}{a^3}$　　　　　　　**34.** $\dfrac{5}{y^2 - 4y - 5} - \dfrac{3}{y^2 + 5y + 4}$

35. $\dfrac{\dfrac{1}{9k} - \dfrac{2}{3k^2}}{\dfrac{7}{27} + \dfrac{1}{k}}$

36. $\dfrac{k^{-1} + k^{-2}}{k^{-3} - k^{-4}}$

37. $a^{-1}(a^{-2} - a^{-3})$

38. $\sqrt{\dfrac{16m^4}{n^8 k^{12}}}$

39. $\sqrt{\dfrac{48}{49}}$

40. $\sqrt{300ab^2 c^3 d^4}$

41. $\sqrt[4]{\dfrac{13}{2x^3}}$

42. $\dfrac{1}{\sqrt{x} - \sqrt{3}}$

43. $\sqrt{\dfrac{3a^3 b}{2c}} \sqrt{\dfrac{27ab^2}{8c}}$

44. $\left(\dfrac{8}{125}\right)^{-1/3}$

45. $z^{1/2}(z^{1/3} - z^{1/4})$

46. $\left(\dfrac{a^{1/2} b^{-1/3}}{c^{1/4}}\right)^{-1/2}$

47. $\sqrt{\dfrac{1}{x^{1/2} y^{-1/3}}}$

Equations and Inequalities

3.1

LINEAR EQUATIONS

In mathematics, some statements are true, some are false, and some may be true or false. Consider the following statements:

$4 + 5 = 9$ This is a *true* statement.

$2 + 3 = 8$ This is a *false* statement.

$x + 2 = 7$ Is this true or false? That depends on x: If $x = 5$, it is true; otherwise, it is false.

An equation such as $x + 2 = 7$ that can be true or false is called an **open** (or **conditional**) **equation**. Any number that makes the equation true is called a **solution**. We **solve** the equation by finding all the solutions.

Let us consider some of the situations that arise when we solve equations.

Equation	Solution	Discussion
(a) $x + 5 = 7$	$x = 2$	Some equations have *only one* solution.
(b) $x^2 = 9$	$x = 3$ or -3	Some equations have *more than one* solution.
(c) $x = x + 1$	No solution	Some equations have *no* solution.
(d) $x^2 - 4 = (x + 2)(x - 2)$	All real numbers	Some equations are *always* true.

An equation, such as $x^2 - 4 = (x + 2)(x - 2)$, that is always true is called an **identity**. Identities are proved, whereas open equations are solved. In this section we study a type of open equation called the **linear equation**, such as

$$2x + 5 = 9 \qquad \text{or} \qquad 6(3p - 1) - p = 8 - 2(5 - 2p)$$

Both of these can be put into the form $ax + b = c$.

To help us solve these equations, we recall the **addition and multiplication properties of equality** from Chapter 1 (see page 8). We now restate these properties.

PROPERTIES OF EQUALITY

1. If the same quantity is added to (or subtracted from) both sides of an equation, the new equation has the same solutions. In symbols:

$$\text{If } a = b, \text{ then } a + c = b + c.$$

2. If both sides of an equation are multiplied (or divided) by the same nonzero number, the new equation has the same solutions. In symbols:

$$\text{If } a = b, \text{ then } ac = bc.$$

EXAMPLE 1   Solve $2x + 7 = 5x - 8$.

Solution   This is a fairly simple linear equation. We use the properties of equalities to isolate x.

Given	$2x + 7 = 5x - 8$
Subtract $2x$	$7 = 3x - 8$
Add 8	$15 = 3x$
Divide by 3	$5 = x$

It is usually a good practice to check the solution by substituting $x = 5$ into the original equation:

$$\text{Left side: } 2(5) + 7 = 10 + 7 = 17$$

$$\text{Right side: } 5(5) - 8 = 25 - 8 = 17 \qquad Checks.$$

Let us now give a general procedure for solving linear equations.

To solve linear equations:

1. If there are fractions, multiply both sides by the LCD.
2. If there are parentheses, use the distributive law to remove them.
3. Combine like terms on each side of the equation.
4. By addition or subtraction, move all terms with the unknown to one side of the equation.
5. By addition or subtraction, move all other terms to the other side.
6. Multiply or divide to isolate the unknown.
7. Check (or verify) your result in the original equation.

If there is a solution, this procedure will find it; however, it may also produce other nonsolutions. Thus, step 7 (checking) is essential.

EXAMPLE 2 Solve $6(3p - 1) - p = 8 - 2(5 - 2p)$.

Solution We start by removing the parentheses with the distributive law.

Distributive law	$18p - 6 - p = 8 - 10 + 4p$
↓	
Combine terms	$17p - 6 = 4p - 2$
↓	
Add 6; subtract $4p$	$13p = 4$
↓	
Divide by 13	$p = \dfrac{4}{13}$

Thus, the solution is $p = 4/13$. The reader should check this.

EXAMPLE 3 Solve $\dfrac{x + 1}{3x - 9} - \dfrac{1}{2} = \dfrac{4}{3x - 9}$.

Solution We can make this a linear equation if we first multiply both sides by the LCD $= 2(3x - 9)$.

Multiply by LCD	$2(3x - 9)\left(\dfrac{x + 1}{3x - 9} - \dfrac{1}{2}\right) = 2(3x - 9)\left(\dfrac{4}{3x - 9}\right)$
↓	
Distributive law	$2(x + 1) - (3x - 9) = 2(4)$
↓	
Distributive law	$2x + 2 - 3x + 9 = 8$
↓	
Combine terms	$-x + 11 = 8$
↓	
Add x; subtract 8	$3 = x$

Let us check this result, $x = 3$:

$$\text{Left side: } \frac{3 + 1}{3(3) - 9} - \frac{1}{2} = \frac{4}{0} - \frac{1}{2} \quad ??$$

$$\text{Right side: } \frac{4}{3(3) - 9} = \frac{4}{0} \quad ??$$

What happened? We get the meaningless expression 4/0. Since our procedure was correct, we conclude that $x = 3$ is not really a solution. It is called an **extraneous solution** (but it is *not* really a solution). In this case, there is *no solution.*

Looking back at the original equation, we see that the $3x - 9$ in the denominator should have tipped us off that $x \neq 3$ (since this makes the denominator zero).

EXAMPLE 4 Solve $\dfrac{3x-1}{x^2} - \dfrac{2}{x} = \dfrac{3}{x^2}$.

Solution We can avoid the problem that we ran into in Example 3 by first investigating the denominator: Here $x \neq 0$.

Multiply by LCD, x^2	$x^2\left(\dfrac{3x-1}{x^2} - \dfrac{2}{x}\right) = x^2\left(\dfrac{3}{x^2}\right)$
Distributive law	$3x - 1 - 2x = 3$
Combine terms	$x - 1 = 3$
Add 1	$x = 4$

Let us check this solution, $x = 4$:

$$\text{Left side:} \frac{3(4)-1}{4^2} - \frac{2}{4} = \frac{11}{16} - \frac{2}{4} = \frac{3}{16}$$

$$\text{Right side:} \frac{3}{4^2} = \frac{3}{16} \qquad \textit{Checks.}$$

EXAMPLE 5 Solve $a = \dfrac{bx}{1-cx}$ for x.

Solution Here we have several letters (a, b, and c) in addition to the unknown (x). This is called a **literal equation**. We use the same procedure given on page 64: We move all the x-terms to one side, and all the other terms (without x) to the other side. (We circle all the x-terms so that we do not lose them.) Also, $1 \neq cx$ (or, $x \neq 1/c$). Why?

Circle unknown, x	$a = \dfrac{b\,ⓧ}{1-c\,ⓧ}$
Multiply by LCD	$(1-c\,ⓧ)a = (1-c\,ⓧ)\dfrac{b\,ⓧ}{1-c\,ⓧ}$
Distributive law	$a - ac\,ⓧ = b\,ⓧ$
Add acx	$a = ac\,ⓧ + b\,ⓧ$
Factor	$a = (ac + b)\,ⓧ$
Divide by $(ac + b)$	$\dfrac{a}{ac+b} = x$

Notice that circling the x helps us keep track of the letter that we are solving for.

Solve the following equations.

1. $x + 7 = 10$ **2.** $x + 3 = -5$

3. $x - 3 = 11$ **4.** $t - 7 = -2$

5. $2x = 10$ **6.** $-3s = 15$

7. $\dfrac{r}{-3} = 2$ **8.** $\dfrac{u}{-10} = -8$

Solve the following equations.

9. $-2x - 5 = -1$ **10.** $-5x - 8 = 2$

11. $2x - 7 = 4x - 9$ **12.** $3a + 7 = 7a - 2$

13. $4k + 9 = -2k - 5$ **14.** $-3m + 5 = -8m - 9$

15. $10x - 8 + 4x = 7 + 3x + 8x$ **16.** $4k - 10 - 6k = k - 8 + 1$

17. $8(2x + 4) - 3x = 7(x + 8) + 12$

18. $5(4a - 1) - 6a = 7(a + 2) - 1$

19. $3(u - 5) - u = 5(6 - u) - 4$

20. $2(3m - 5) - m = 6(7 - 3m) + 8$

21. $5(z - 2) - 4(2z + 3) = 7 - 2(3 - z)$

22. $8(2x - 1) + 3(x + 5) = x - 3(5 - x)$

23. $\dfrac{2}{3} - \dfrac{5}{t - 7} = \dfrac{1}{2}$

24. $\dfrac{m}{m - 5} + \dfrac{1}{2} = \dfrac{1}{2m - 10}$

25. $\dfrac{5}{k^2} + \dfrac{2}{k} = \dfrac{11}{k^2}$

26. $\dfrac{5}{3a^2} - \dfrac{1}{3a} = \dfrac{a + 3}{9a^2}$

27. $\dfrac{5}{x + 3} + \dfrac{1}{x} = \dfrac{7}{x + 3}$

28. $\dfrac{3}{m + 1} + \dfrac{3}{m} = \dfrac{3}{4m}$

29. $\dfrac{1}{u + 2} - \dfrac{2}{u - 1} = \dfrac{u - 5}{u^2 + u - 2}$

30. $\dfrac{3}{r + 5} + \dfrac{4}{r - 3} = \dfrac{9r - 1}{r^2 + 2r - 15}$

Solve the following equations for x.

31. $ax = bx + c$ **32.** $xyz - a = b - uvx$

33. $5(y + x) = 3(b - x) + a$ **34.** $a(x + z) + b(x + y) = c(x - t)$

35. $\dfrac{1}{a} + \dfrac{1}{x} = \dfrac{1}{b}$ **36.** $\dfrac{1}{x} - \dfrac{1}{y} = \dfrac{1}{z}$

37. $t = \dfrac{a - bx}{c + dx}$ **38.** $m = \dfrac{24x}{a(x + 1)}$

39. $2x - 7 = 5x - 10$

40. $\dfrac{x}{2} + \dfrac{3}{5} = \dfrac{11}{10}$

41. $5x + a = 3x + b$

42. $7(2x + 1) - 3 = 8(1 - 3x) - 4$

43. $\dfrac{1}{x} + \dfrac{1}{2} = \dfrac{5}{6}$

44. $\dfrac{15}{x} + \dfrac{100}{x^2} = \dfrac{35}{x}$

45. $6(x - 5) - 4x = 12(x - 10)$

46. $\dfrac{1}{x} - \dfrac{1}{a} - \dfrac{1}{b} = \dfrac{1}{c}$

47. $\dfrac{6}{x} - \dfrac{9}{x^2} = \dfrac{3}{x}$

48. $\dfrac{x + 20}{x - 20} = 9$

49. $\dfrac{1}{x} + \dfrac{1}{y} + \dfrac{1}{3} = \dfrac{1}{t}$

50. $5x - x + 6x - 300 = 3x - 50$

51. $4000 = \dfrac{2000x}{x - 1}$

52. $\dfrac{2}{x} + \dfrac{5}{x - 1} = \dfrac{61}{x^2 - x}$

53. $x + 2x + 3x - 100 = x - 50$

54. $\dfrac{u(vx - w)}{p(qx - r)} = 5$

55. $\dfrac{1}{x + 1} + \dfrac{1}{x - 1} = \dfrac{8}{x^2 - 1}$

56. $8(x - 2) + 2(x + 7) = 9(x - 1) + 5$

57. $a(bx + c) = d(ex + f)$

58. $-4x - 11 = -7x - 19$

59. $4(x + 3) - 5(x - 3) = 6(x - 9) + 1$

60. $-2x - u = px - v$

61. $5(3x + 7) = 2x + 10$

62. $\dfrac{4}{x} - \dfrac{5}{x^2} = \dfrac{6}{x}$

63. $a(x + b) + c = d(x + e)$

64. $\dfrac{1}{x + a} + \dfrac{1}{x} = \dfrac{b}{x + a}$

65. $\dfrac{1}{x + 1} + \dfrac{3}{x} = \dfrac{23}{x^2 + x}$

66. $\dfrac{1}{x} + \dfrac{1}{a} + \dfrac{1}{b} = \dfrac{1}{c}$

67. $2.5x - 3.73 = 8.18$

68. $\dfrac{2.81}{x} + \dfrac{1}{4.45} = 2.03$

69. $3.81(7.9x + 25.9) = 159$

70. $56.3(x + 82.9) = 7.85(16.2 - x)$

Life Science Application

71. The population P of a certain bacterial culture is approximated by

$$P = \dfrac{20,000t}{10 + t}$$

where t is the time in hours. Find t that gives $P = 15,000$.

72. For a lens we have the formula

$$\frac{1}{f} = \frac{1}{u} + \frac{1}{v}$$

where f is the focal length, u the distance to the object, and v the distance to the image. Solve for u in terms f and v.

Physical
Application

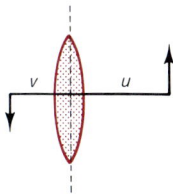

73. The *Doppler effect* predicts how the frequency f_s of a sound source moving at velocity v will sound to an observer f_0:

$$f_0 = \frac{331 f_s}{331 \pm v}$$

(The " $+$ " is for a sound source moving away, and the " $-$ " is for a sound source moving closer.)
(a) At what approaching speed v will middle C ($f_s = 262$) sound like G ($f_0 = 392$)?
(b) At what separating speed v will A ($f_s = 440$) sound like F # ($f_0 = 370$)?
(c) Solve for v in both the $+$ and $-$ cases.

Population
Application

74. The death rate at age x is given by

$$m_x = \frac{q_x}{1 - \frac{1}{2}q_x}$$

where q_x is the likelihood of dying at age x. Solve this for q_x.

Business
Application

75. A 33-year-old man buys an insurance policy that will pay $20,000 as well as return all his paid premiums when he dies. The annual premium P is given in the formula

$$P \cdot \frac{N_{33}}{D_{33}} = 20{,}000 \cdot \frac{M_{33}}{D_{33}} + P \cdot \frac{R_{33}}{D_{33}}$$

(a) Solve for P if $N_{33} = 640{,}776$, $D_{33} = 31{,}398$, $M_{33} = 12{,}734$, and $R_{33} = 346{,}564$.
(b) Solve for P in terms of the other letters.

Electrical
Application

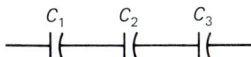

76. If capacitances C_1, C_2, and C_3 are put in series, the equivalent capacitance C satisfies

$$\frac{1}{C} = \frac{1}{C_1} + \frac{1}{C_2} + \frac{1}{C_3}$$

(a) Find C if $C_1 = 20$, $C_2 = 30$, and $C_3 = 60$.
(b) Solve for C_3 in terms of the other letters.

Health
Application

77. Young's formula for a child's dosage of medicine is

$$D_c = \frac{A}{A + 12} \cdot D_a$$

where A is the child's age, D_c the child's dosage, and D_a the adult's dosage.
(a) Find D_a if $A = 6$ and $D_c = 300$.
(b) Solve for A in terms of D_a and D_c.

Often in real life we are not directly given the equation to be solved; rather, we must derive the equation from some given verbal information. In this section we look at a few types of these problems.

To solve verbal problems:

1. Read and reread the problem. Understand the situation.
2. Identify the unknown (or unknowns) by a letter (or letters).
3. Draw a picture or make a table of the given information.
4. Translate the information into an equation.
5. Solve the equation.
6. Check the solution. (Ask yourself if the solution seems reasonable.)

Translation Problems

These problems involve only a direct translation from English into mathematics.

EXAMPLE 6   What number must be added to both the numerator and denominator of 9/19 to give 2/3?

Solution   We translate this directly from English to mathematical symbols.

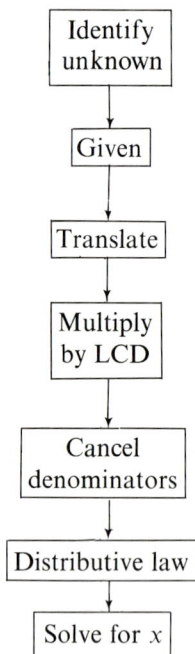

Identify unknown	Let x = number to be added

Add x to top and bottom of $\dfrac{9}{19}$ to get $\dfrac{2}{3}$

| Given | |

$$\frac{9 + x}{19 + x} = \frac{2}{3}$$

| Translate | |

| Multiply by LCD | $3(19 + x)\left(\dfrac{9 + x}{19 + x}\right) = 3(19 + x)\left(\dfrac{2}{3}\right)$ |

| Cancel denominators | $3(9 + x) = (19 + x)2$ |

| Distributive law | $27 + 3x = 38 + 2x$ |

| Solve for x | $x = 11$ |

We can check this solution, $x = 11$: $\dfrac{9 + 11}{19 + 11} = \dfrac{20}{30} = \dfrac{2}{3}$.

Interest Problems

These problems involve a sum of money invested at different interest rates. To work these problems, we recall the formula

$$I = P \cdot r$$

where I is the interest earned in 1 year, P is the principal invested, and r the rate of interest (expressed as a decimal).

EXAMPLE 7   Ms. Walker invested $8000 in two accounts: some at 12%, the rest at 7%. Her total interest for 1 year was $860. How much was invested at each rate?

Solution   Suppose that x dollars were invested at 12%. Since the total was $8000, then $8000 - x$ dollars were invested at 7%. It is helpful to make a $P - r - I$ table. (Recall that $I = P \cdot r$.)

	P	r	$I = Pr$
12% account	x	12%	$0.12x$
7% account	$8000 - x$	7%	$0.07(8000 - x)$
Total	8000	—	860

We use the fact that the sum of the interests is $860.

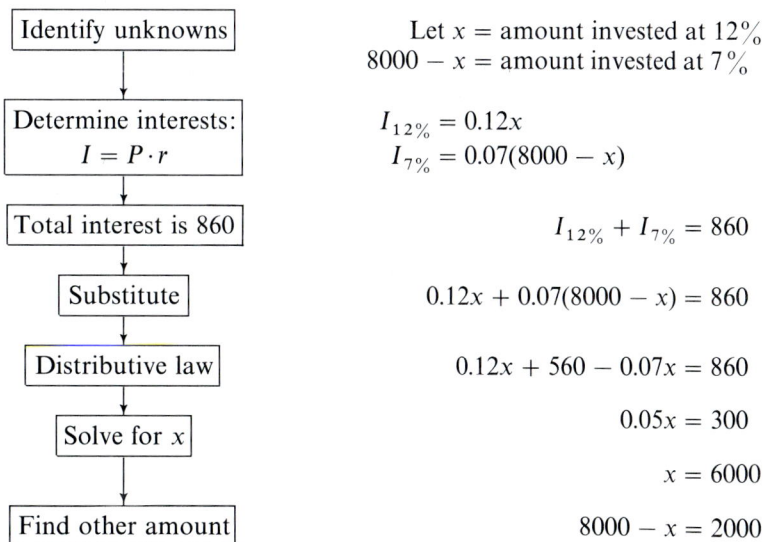

Identify unknowns	Let x = amount invested at 12% $8000 - x$ = amount invested at 7%
Determine interests: $I = P \cdot r$	$I_{12\%} = 0.12x$ $I_{7\%} = 0.07(8000 - x)$
Total interest is 860	$I_{12\%} + I_{7\%} = 860$
Substitute	$0.12x + 0.07(8000 - x) = 860$
Distributive law	$0.12x + 560 - 0.07x = 860$
Solve for x	$0.05x = 300$ $x = 6000$
Find other amount	$8000 - x = 2000$

Therefore, she invested $6000 at 12% ($=\$720$) and $2000 at 7% ($=\$140$). This yields a total of $860 interest.

Mixture Problems

Mixture problems are very much like interest problems. In these problems, we are blending various mixtures with different percents (strengths) of pure substance. Here we use the formula

$$A = P \cdot B$$

where A is the amount of pure substance (acid, iodine, antifreeze, and so on), P the percent, and B the base (total mixture). For instance, if $B = 10$ gallons of antifreeze solution and $P = 30\%$ strength, then $A = (0.30)(10) = 3$ gallons of pure antifreeze.

EXAMPLE 8   A nurse is mixing two iodine solutions: one 30% and the other 60%. How much of each must she mix to get 300 milliliters of a 50% solution?

Solution   Suppose that she uses x milliliters of the 60% solution. Since she needs 300 milliliters total, she uses $300 - x$ milliliters of 30% solution. As in Example 7, we use a table to help us sort out the information. [Since she wants 300 milliliters of 50% solution, this is $(0.50)(300) = 150$ milliliters of pure iodine.]

	Base, B	Percent, P	Pure Amount, $P \cdot B$
60% solution	x	60%	$0.60x$
30% solution	$300 - x$	30%	$0.30(300 - x)$
Total	300	50%	150

Identify unknowns	Let x = amount of 60% solution $300 - x$ = amount of 30% solution
Determine pure iodine amounts	$I_{60\%} = 0.60x$ $I_{30\%} = 0.30(300 - x)$
Total iodine = 150	$0.60x + 0.30(300 - x) = 150$
Distributive law	$0.60x + 90 - 0.30x = 150$
Solve for x	$0.30x = 60$ $x = 200$
Find other amount	$300 - x = 100$

Thus, the nurse needs 200 milliliters of 60% solution ($=120$ ml pure iodine) and 100 milliliters of 30% solution ($=30$ ml pure iodine). This yields 150 milliliters of iodine out of 300 total, or 50%.

Motion Problems

Motion problems usually involve two people making a similar trip, or one person making two trips. In all motion problems we use the formulas

$$d = r \cdot t \qquad r = \frac{d}{t} \qquad t = \frac{d}{r}$$

where d is the distance, r the rate (speed), and t the time. Usually, one of these three will be given completely, a second will be the unknown, and the third can be written in terms of the first two (using the formulas above).

EXAMPLE 9   Dave made a 3000-mile trip in 35 hours. He traveled part of it in a 50-mile-per-hour car and part of it in a 100-mile-per-hour train. What distances did he travel at each speed?

Solution   The distances are unknown and the rates known. If he traveled x miles in the car, then he traveled $3000 - x$ miles in the train. As in Examples 7 and 8, we use a table to help us, and we use the formula $t = d/r$.

	d	r	$t = d/r$
Car	x	50	$\dfrac{x}{50}$
Train	$3000 - x$	100	$\dfrac{3000 - x}{100}$
Total	3000	—	35

We use the fact that the total time is 35 hours.

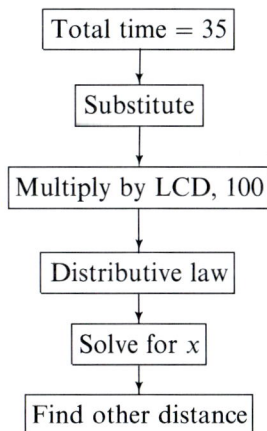

Total time = 35	Car time + train time = 35
Substitute	$\dfrac{x}{50} + \dfrac{3000 - x}{100} = 35$
Multiply by LCD, 100	$100\left(\dfrac{x}{50} + \dfrac{3000 - x}{100}\right) = 100(35)$
Distributive law	$2x + 3000 - x = 3500$
Solve for x	$x = 500$
Find other distance	$3000 - x = 2500$

Thus, Dave traveled 500 miles in the car (for $500/50 = 10$ hours) and 2500 miles in the train (for $2500/100 = 25$ hours). This is a total of 35 hours.

Work Problems

Work problems usually involve two people (or machines) working together at different rates. We want their combined rates.

We look at the fraction of the job that can be done in one period of time. For example, if Leon can paint a room in 3 hours, then in 1 hour he can paint 1/3 of a room. In general, we have the formula

$$\frac{1}{T_1} + \frac{1}{T_2} = \frac{1}{x}$$

where T_1 and T_2 are the times that it takes persons 1 and 2 to do the job and x is the time for their combined effort. [We subtract if the people (machines, or pipes) are working against each other.]

EXAMPLE 10   Mary can grade a stack of quizzes in 20 minutes; Ralph can grade the same stack in 30 minutes. If they work together, how long will it take them to grade the stack?

Solution   In 1 minute, Mary can grade 1/20 of the stack while Ralph can grade 1/30 of the stack. If their combined time is x, then in one minute they can grade $1/x$ of the stack.

$1/20 =$ amount Mary does in 1 minute

| Amount they do in 1 minute |

$1/30 =$ amount Ralph does in 1 minute

$1/x =$ amount both do in 1 minute

| In one minute: Mary + Ralph = both |

$$\frac{1}{20} + \frac{1}{30} = \frac{1}{x}$$

| Multiply by LCD, $60x$ |

$$60x\left(\frac{1}{20} + \frac{1}{30}\right) = 60x\left(\frac{1}{x}\right)$$

| Distributive law |

$$3x + 2x = 60$$

| Solve for x |

$$x = 12$$

Together, they can grade the stack in 12 minutes.

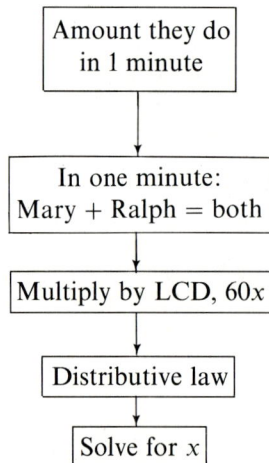

Translate the following phrases into algebraic symbols.

1. Five more than twice a number.

2. Eighty percent of a number.

3. One number is 6 less than three times another number.

4. Five is added to numerator and denominator of a fraction.

5. A number plus its reciprocal.

6. Three consecutive odd integers.

Solve the following equations.

7. $2x - 3 = 5x - 12$

8. $4(2x - 2) = -3(5x + 11) + 2$

9. $\dfrac{x}{40} + \dfrac{500 - x}{50} = 11$

10. $\dfrac{1}{5} + \dfrac{1}{7} = \dfrac{1}{x}$

11. $\dfrac{3 + x}{19 + x} = \dfrac{2}{5}$

12. $0.10x + 0.12(10,000 - x) = 1160$

Solve the following verbal problems.

13. Two hundred is 4% of what number?

14. Five thousand is 2% of what number?

15. The sum of three consecutive integers is 87. Find the integers.

16. The sum of three consecutive odd integers is 105. Find the integers.

17. What number must be added to both the numerator and denominator of 5/11 to get 3/4?

18. What number must be added to the denominator of 3/13 to get 1/10?

19. The sum of the reciprocals of two consecutive integers is 15 divided by the product of the integers. Find the integers.

20. If the reciprocal of a number is added to 1/2, the sum is 11 divided by twice the number. Find the number.

21. The Rosens invested some money at 11% and $6000 more at 12%. In 1 year their interest was $3480. How much was invested at each rate?

22. The Garcias invested some money at 10% and $2000 less at 14%. In 1 year their interest was $1400. How much was invested at each amount?

23. The Changs invest $4000 at interest rate r and $6000 at interest rate $r + 1$%. Their total interest is $1060. What are the interest rates?

24. High Bluff College invests $500,000 at interest rate r, $200,000 at interest rate $r + 1$%, and $100,000 at interest rate $r + 3$%. Their total interest is $77,000. What are the interest rates?

25. How many liters of a 20% iodine solution and a 40% iodine solution must be mixed to produce 80 liters of a 35% iodine solution?

26. How many quarts of a 50% antifreeze solution and an 80% antifreeze solution must be mixed to produce 30 quarts of a 60% antifreeze solution?

27. A goldsmith has 3 ounces of a 50% gold alloy. How many ounces of a 90% alloy must be added to produce a 60% alloy?

28. A pharmacist has 600 milliliters of a 20% solution of a certain drug. How many milliliters of a 50% solution must be added to produce a 30% solution?

29. Janet made a 2000-mile trip in 19 hours. She traveled part of it in a 50-mile-per-hour car and the rest of it in a 120-mile-per-hour train. What distance did she travel at each speed?

30. The LaRues made a 1000-mile trip in 3 hours: part at 40 miles per hour (driving to the airport) and the rest at 480 miles per hour (in a jet). What distance did they travel at each speed?

31. Al drove 1000 miles in 22 hours: part at 40 miles per hour and the rest at 50 miles per hour. How much time was spent at each speed?

32. Nancy and Louise make a 1560-mile trip in 32 hours. Nancy drove at 55 miles per hour and Louise drove at 45 miles per hour. How much time did each of them drive?

33. Donna, Ellen, and Fred are cleaning a hall after a dance. Alone, Donna would take 4 hours, Ellen 6 hours, and Fred 5 hours. How long will it take them working together?

34. Angela, Bob, Chuck, and Denise are campaigning door to door for a local candidate. Alone, Angela could canvass the precinct in 5 hours, Bob in 6 hours, Chuck in 4 hours, and Denise in 7 hours. How long will it take them working together?

35. Pipe A can fill a pool in 5 hours and pipe B can empty it in 15 hours. If both pipes are open, how long will it take to fill an empty pool?

36. It would take Steve 40 years to earn 1 million dollars (after taxes), and 50 years to spend 1 million dollars. At this rate, how long will it take him to save 1 million dollars?

37. Find three consecutive integers such that the sum of the first two is 20 more than the third.

38. Arthur inherits a large sum of money. He invests 1/4 of it at 10% and 3/4 of it at 12%. If 1 year's interest is $9200, what was the original sum of money?

39. The length of a rectangle is 20 more than twice the width. If the perimeter is 640, find the length and width.

40. A chemist has 100 milliliters of a 30% acid solution. How much pure acid (100%) must she add to bring the solution to 40% strength?

41. Amy has $1.30 (130 cents) in dimes and nickels. If there are 19 coins, how many of each coin does she have?

42. The Squirrel Junction Theater Company sells 650 tickets: some at $3 and the rest at $5. If their revenue is $2450, how many of each ticket were sold?

43. Train A pulls out of Foxville going 80 miles per hour. Train B leaves Foxville 1 hour later going 100 miles per hour in the same direction. How long will it take train B to catch train A? (*Hint*: When B catches A, what can you say about the distance both have traveled?)

44. Roger has scored 71, 81, 75, and 78 on four tests. What must he score on the fifth test to have an average of 80?

45. An endowment fund invests $100,000 at interest rate r, $200,000 at rate $r + 1\%$, and $300,000 at rate $r + 2\%$. Interest for 1 year is $62,000. What are the interest rates?

46. Lisa has $5.70 (570 cents) in dimes and 50-cent pieces. If she has 21 coins, how many of each does she have?

47. It takes Ed 2 weeks to earn $1000 and it takes Ethel 3 weeks to earn $1000. It takes them 1.5 weeks to spend $1000. How long will it take them to save $10,000?

48. Find three consecutive odd integers such that the sum of the first two is 7 more than the third.

49. How much 10% alcohol solution must be added to 4000 milliliters of pure alcohol to produce a 70% alcohol solution?

50. Mr. Pacini has $12,000 to invest. He will split it between an 8% account and an 11% account. How should he split the investments so that his $12,000 effectively earns 10% interest?

51. Professor X can write a book in 14 months and professor Y can write a book in 21 months. If they coauthor a book, how long will it take them?

52. Eighty is 5% of what number?

53. Mr. Hadley leaves $600,000 to his children, Nigel, Clarence, and Kent, in the ratio 5 to 4 to 3, respectively. How much does each child get?

54. Slugger O'Toole has batted 200 times and has a 0.230 batting average (he has gotten a hit in 23% of his at-bats). How many straight hits does he need to have a 0.300 average?

3.3 QUADRATIC EQUATIONS

In this section we study equations with terms of degree two. These are called **quadratic equations** and have the standard form

$$ax^2 + bx + c = 0 \qquad (a \neq 0)$$

For instance,

$$x^2 - 9x + 5 = 0 \qquad \text{and} \qquad 3x^2 - 4x - 11 = 0$$

are quadratic equations in **standard form**: All the nonzero terms are on one side (usually the left), and zero is on the other. We now consider four different methods for solving quadratic equations:

1. Factoring method
2. Square-root method
3. Completing the square
4. Quadratic formula

Factoring Method

The **factoring method** uses the **zero-product property**, which we discussed in Chapter 1 (page 8). Let us restate this property now.

Theorem 1 (*Zero-Product Property*)
For any real numbers a and b:

> If $a \cdot b = 0$, then either $a = 0$, $b = 0$, or both.

In words, if a product of two real numbers is zero, then one (or both) of the factors must be zero. We use this property as follows.

> To solve quadratic equations by factoring:
>
> 1. Put the equation into standard form, $ax^2 + bx + c = 0$.
> 2. Factor the polynomial.
> 3. Set each factor equal to zero.
> 4. Solve each of the resulting equations.

EXAMPLE 11   Solve $6x^2 - 11x - 10 = 0$.

Solution   We first factor this trinomial (it is already in standard form). Since the product is zero, one of the factors must be zero.

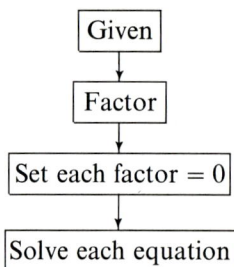

Given		$6x^2 - 11x - 10 = 0$
Factor		$(2x - 5)(3x + 2) = 0$
Set each factor $= 0$		$2x - 5 = 0$ or $3x + 2 = 0$
Solve each equation		$x = \dfrac{5}{2}$ or $x = \dfrac{-2}{3}$

Thus, the solution set has *two* solutions: $\left\{ \dfrac{5}{2}, \dfrac{-2}{3} \right\}$.

YES	NO
$(m + 3)(m - 2) = 0$	$(m + 3)(m - 2) = 14$
$m + 3 = 0$ or $m - 2 = 0$	$m + 3 = 14$ or $m - 2 = 14$
The zero-product property holds only when the product is *zero*.	

EXAMPLE 12   Solve $(m + 3)(m - 2) = 14$.

Solution We cannot use the zero-product property until we have the equation in standard form, with a *zero* on one side of the equation.

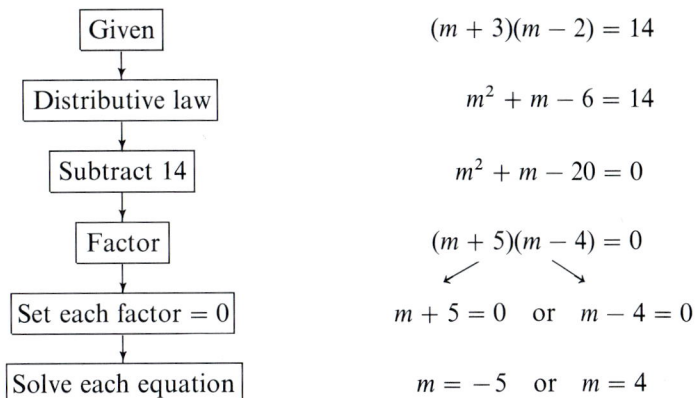

Given	$(m + 3)(m - 2) = 14$
↓	
Distributive law	$m^2 + m - 6 = 14$
↓	
Subtract 14	$m^2 + m - 20 = 0$
↓	
Factor	$(m + 5)(m - 4) = 0$
↓	↙　　↘
Set each factor $= 0$	$m + 5 = 0$　or　$m - 4 = 0$
↓	
Solve each equation	$m = -5$　or　$m = 4$

The solutions are -5 and 4. (Verify these solutions in the original equation.)

Square-Root Method

Sometimes, a quadratic equation can best be solved by finding the square roots of both sides. Consider the equation $x^2 = a$, for $a \geq 0$. We rewrite this in standard form $x^2 - a = 0$, and factor:

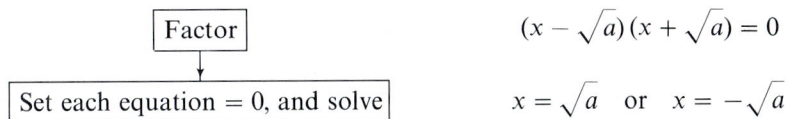

Factor	$(x - \sqrt{a})(x + \sqrt{a}) = 0$
↓	
Set each equation $= 0$, and solve	$x = \sqrt{a}$　or　$x = -\sqrt{a}$

Theorem 2
For any real number $a \geq 0$:

$$\text{If } x^2 = a, \text{ then } x = \pm\sqrt{a}.$$

We use the symbol $\pm m$ to denote the two numbers m and $-m$.

EXAMPLE 13 The following use Theorem 2.
    (a) $x^2 = 64$ has solutions 8 and -8 (or ± 8).
    (b) $x^2 = 10$ has solutions $\sqrt{10}$ and $-\sqrt{10}$ (or $\pm\sqrt{10}$).
    (c) $x^2 = -9$ has *no* real solutions.

EXAMPLE 14 Solve $(k - 7)^2 = 13$.

Solution We first use Theorem 2 by taking the square root of both sides. (This method works only when the unknown expression is a perfect square.)

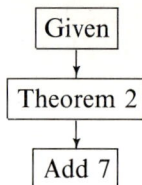

Given	$(k - 7)^2 = 13$

$$\downarrow$$

Theorem 2	$k - 7 = \sqrt{13}$ or $k - 7 = -\sqrt{13}$

$$\downarrow$$

Add 7	$k = 7 + \sqrt{13}$ or $k = 7 - \sqrt{13}$

Thus, the solutions are $7 \pm \sqrt{13}$.

Completing the Square

Example 14 suggests another method for solving quadratic equations. If we can make the left side of the equation a perfect square, we can then use Theorem 2. We call this method **completing the square**.

EXAMPLE 15   Solve $x^2 + 6x + 1 = 0$.

Solution   To complete the square we first rewrite the equation and compare it to the perfect square $(x + a)^2 = x^2 + 2ax + a^2$.

$$\textit{Perfect square:} \quad x^2 + \boxed{2a}\, x + a^2$$
$$\textit{Our expression:} \quad x^2 + \boxed{6}\, x \qquad = -1$$

We match $2a$ with 6, or $a = 3$. Thus, we need $a^2 = 3^2$ to make the left side a perfect square.

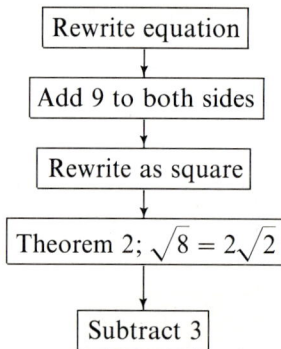

Rewrite equation	$x^2 + 6x \qquad = -1$

$$\downarrow$$

Add 9 to both sides	$x^2 + 6x + 9 = 8$

$$\downarrow$$

Rewrite as square	$(x + 3)^2 = 8$

$$\downarrow$$

Theorem 2; $\sqrt{8} = 2\sqrt{2}$	$x + 3 = 2\sqrt{2}$ or $x + 3 = -2\sqrt{2}$

$$\downarrow$$

Subtract 3	$x = -3 + 2\sqrt{2}$ or $x = -3 - 2\sqrt{2}$

Thus, the solutions are $-3 + 2\sqrt{2}$ and $-3 - 2\sqrt{2}$.

To solve a quadratic equation by completing the square:

1. Rewrite the equation in the form $x^2 + mx \qquad = n$.
2. Compare the left-hand side to $x^2 + 2ax + a^2$ to find $a = m/2$.
3. Add $a^2 = (m/2)^2$ to both sides of the equation.
4. Solve, using Theorem 2 since the left-hand side is now a perfect square.

Quadratic Formula

Sometimes a quadratic equation cannot be factored easily, and it does not lend itself easily to the square-root method or completing the square. Our last resort is the ultimate method: the **quadratic formula**. We start with the form $ax^2 + bx + c = 0$ (with $a \neq 0$) and develop the formula by completing the square.

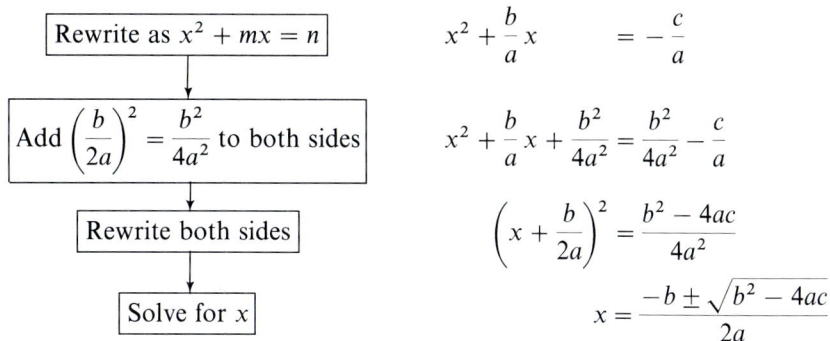

$\boxed{\text{Rewrite as } x^2 + mx = n}$ $\qquad x^2 + \dfrac{b}{a}x \qquad = -\dfrac{c}{a}$

$\boxed{\text{Add } \left(\dfrac{b}{2a}\right)^2 = \dfrac{b^2}{4a^2} \text{ to both sides}}$ $\qquad x^2 + \dfrac{b}{a}x + \dfrac{b^2}{4a^2} = \dfrac{b^2}{4a^2} - \dfrac{c}{a}$

$\boxed{\text{Rewrite both sides}}$ $\qquad \left(x + \dfrac{b}{2a}\right)^2 = \dfrac{b^2 - 4ac}{4a^2}$

$\boxed{\text{Solve for } x}$ $\qquad x = \dfrac{-b \pm \sqrt{b^2 - 4ac}}{2a}$

Theorem 3 (*Quadratic Formula*)

The solutions to $ax^2 + bx + c = 0$ $(a \neq 0)$ are given by

$$x = \frac{-b \pm \sqrt{b^2 - 4ac}}{2a}$$

EXAMPLE 16 Solve $3x^2 - 5x = 1$.

Solution We first rewrite this in standard form, $3x^2 - 5x - 1 = 0$. Since this cannot be factored with integers, we use the quadratic formula. We start by identifying the terms a, b, and c.

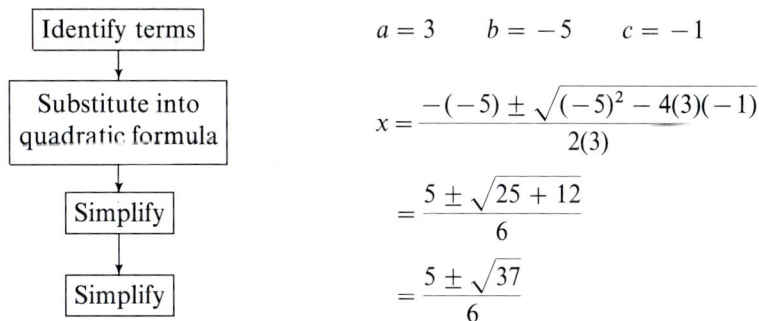

$\boxed{\text{Identify terms}}$ $\qquad a = 3 \qquad b = -5 \qquad c = -1$

$\boxed{\begin{array}{c}\text{Substitute into}\\\text{quadratic formula}\end{array}}$ $\qquad x = \dfrac{-(-5) \pm \sqrt{(-5)^2 - 4(3)(-1)}}{2(3)}$

$\boxed{\text{Simplify}}$ $\qquad = \dfrac{5 \pm \sqrt{25 + 12}}{6}$

$\boxed{\text{Simplify}}$ $\qquad = \dfrac{5 \pm \sqrt{37}}{6}$

The solution set can be written $\left\{ \dfrac{5 + \sqrt{37}}{6}, \dfrac{5 - \sqrt{37}}{6} \right\}$.

We can write the solutions to Example 16 as three-place decimals. We use a hand calculator with a $\boxed{\sqrt{x}}$ key.

PRESS	DISPLAY	MEANING
$\boxed{\text{C}}$	0.	Clear
$\boxed{5}$	5.	5
$\boxed{+}\ \boxed{3}\ \boxed{7}\ \boxed{\sqrt{x}}$	6.0827625	Plus $\sqrt{37}$
$\boxed{=}$	11.082763	$5 + \sqrt{37}$
$\boxed{\div}\ \boxed{6}$	6.	Divided by 6
$\boxed{=}$	1.8471271	The " + " solution

Similarly, we find the " $-$ " solution as -0.1804604. Thus, the set of approximate solutions is $\{1.847, -0.180\}$.

We can also use the calculator to verify the solution. We take the solution, as just computed, and we press $\boxed{\text{STO}}$ to put it into the memory. Whenever x appears in the expression, we press $\boxed{\text{RCL}}$ to recall its value.

PRESS	DISPLAY	MEANING
$\boxed{\text{STO}}$	1.8471271	Store " + " solution
$\boxed{3}\ \boxed{\times}\ \boxed{\text{RCL}}\ \boxed{x^2}$	3.4118784	$3x^2$
$\boxed{-}\ \boxed{5}\ \boxed{\times}\ \boxed{\text{RCL}}$	1.8471271	Minus $5x$
$\boxed{=}$	0.9999999	Close enough to 1

Because of roundoff within the calculator, we do not get exactly 1; however, we are satisfied with 0.9999999. We conclude that we have the correct answer. (On some calculators, the memory-store key is $\boxed{\text{M}+}$, and the memory-recall key is $\boxed{\text{MR}}$.)

EXAMPLE 17   Solve $x^2 - 3x + 11 = 0$.

Solution   We use the quadratic formula since we cannot factor the trinomial.

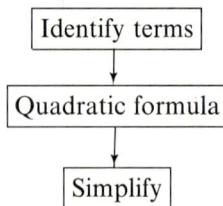

$\boxed{\text{Identify terms}}$ $a = 1 \qquad b = -3 \qquad c = 11$

$\boxed{\text{Quadratic formula}}$ $x = \dfrac{-(-3) \pm \sqrt{(-3)^2 - 4(1)(11)}}{2(1)}$

$\boxed{\text{Simplify}}$ $= \dfrac{3 \pm \sqrt{-35}}{2}$

There are *no* real solutions. We discuss this situation later in the text.

The number $b^2 - 4ac$ within the radical of the quadratic formula is called the **discriminant**, **D**, and it gives us the nature of the solutions to $ax^2 + bx + c = 0$.

---

1. If $D > 0$, there are two real solutions.
2. If $D = 0$, there is only one real solution.
3. If $D < 0$, there are no real solutions.

EXAMPLE 18 The following table shows how the discriminant of a quadratic equation can be used to determine the nature of the solutions.

Equation	$D = b^2 - 4ac$	Sign	Type of solution(s)
$2x^2 - 5x - 7 = 0$	$(-5)^2 - 4(2)(-7) = 81$	$+$	Two real solutions
$x^2 + 6x + 9 = 0$	$6^2 - 4(1)(9) = 0$	0	One real solution
$x^2 + 2x + 8 = 0$	$2^2 - 4(1)(8) = -28$	$-$	No real solutions

Applications

Many verbal problems lead to quadratic equations. As noted on page 70, the key is translating the information carefully into a mathematical equation; then we solve the equation.

EXAMPLE 19 Becky made a 140-mile trip: 40 miles at one speed and 100 miles at 10 miles per hour faster. The entire trip took 3 hours. Find the two speeds.

Solution Recall from Example 9 that we use a *d-r-t* table to help us solve motion problems. Here we let one rate be x and the other $x + 10$ (since it is 10 miles per hour faster). We find the times using $t = d/r$.

	d	r	$t = d/r$
Speed 1	40	x	$\dfrac{40}{x}$
Speed 2	100	$x + 10$	$\dfrac{100}{x + 10}$
Total trip	140	—	3

We use the fact that the total time is 3 hours.

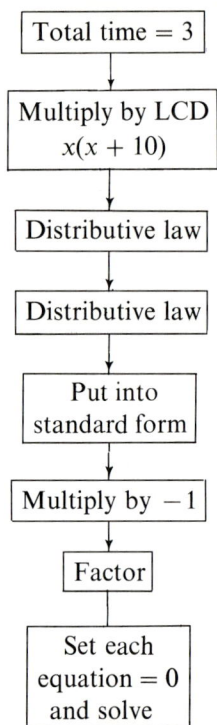

Total time = 3	$$\frac{40}{x} + \frac{100}{x + 10} = 3$$
Multiply by LCD $x(x + 10)$	$$x(x + 10)\left(\frac{40}{x} + \frac{100}{x + 10}\right) = 3x(x + 10)$$
Distributive law	$$40(x + 10) + 100x = 3x(x + 10)$$
Distributive law	$$40x + 400 + 100x = 3x^2 + 30x$$
Put into standard form	$$-3x^2 + 110x + 400 = 0$$
Multiply by -1	$$3x^2 - 110x - 400 = 0$$
Factor	$$(3x + 10)(x - 40) = 0$$
Set each equation = 0 and solve	$$x = \frac{-10}{3} \quad \text{or} \quad 40$$

Since a speed of $-10/3$ makes no sense here, the solution is 40 miles per hour (for 40 miles) and 50 miles per hour (for 100 miles).

**PROBLEM SET
3.3**

Solve the following equations.

*Warm-up
Exercises*

1. $x + 2 = 0$

2. $2x - 5 = 0$

3. $2x - 7 = 4x - 9$

4. $5(t - 2) = 2t + 1$

5. $\dfrac{1}{x} + \dfrac{1}{2} = \dfrac{5}{6}$

6. $\dfrac{15}{x} + \dfrac{100}{x^2} = \dfrac{35}{x}$

Factor the following polynomials.

*Warm-up
Exercises*

7. $x^2 - 4$

8. $x^2 - 100$

9. $x^2 - 8x + 16$

10. $x^2 + 2x + 1$

11. $x^2 - 7x + 10$

12. $6x^2 + 7x - 5$

Solve the following quadratic equations by the factoring method.

13. $(x - 7)(x + 10) = 0$

14. $(2x - 7)(5x + 1) = 0$

15. $t^2 - 8t + 7 = 0$

16. $u^2 + 18u + 17 = 0$

Solve the following equations using the square-root method.

17. $a^2 = 100$

18. $x^2 = 169$

19. $(10u - 3)^2 = 5$

20. $(7k - 5)^2 = 11$

Solve the following equations by completing the square.

21. $x^2 + 4x = 21$ **22.** $t^2 - 8t = 20$

Solve the following using the quadratic formula.

23. $x^2 + 5x + 2 = 0$ **24.** $a^2 - 3a - 9 = 0$

25. $y^2 - 10y + 3 = 0$ **26.** $2k^2 - 5k + 1 = 0$

Solve the following quadratic equations using any method.

27. $x^2 - 13x + 30 = 0$ **28.** $t^2 + 22 = 13t$

29. $r^2 - 36 = 0$ **30.** $m^2 + 11m = -18$

31. $(t - 3)(t - 4) = 6$ **32.** $(y + 2)(y + 5) = 18$

33. $(a - 2)(a + 4) = 5a + 2$ **34.** $(2z + 1)(z - 3) = 2z - 3$

35. $\dfrac{x^2}{9} + \dfrac{2x}{3} = 3$ **36.** $\dfrac{36}{r^2} + 1 = \dfrac{13}{r}$

37. $\dfrac{-3}{k} + k = \dfrac{11}{2}$ **38.** $\dfrac{4}{y - 2} = 1 + \dfrac{10}{y - 3}$

39. $y^2 = 5$ **40.** $t^2 = -1$

41. $(x + 1)^2 = 25$ **42.** $(k - 2)^2 = 36$

43. $(u - 7)^2 = 15$ **44.** $(m + 5)^2 = 2$

45. $x^2 - 2x = 4$ **46.** $z^2 + 10x = 7$

47. $p^2 + p = 5$ **48.** $k^2 - 3k = 8$

49. $2x^2 + 4x = 9$ **50.** $3w^2 - 5w = 11$

51. $6m^2 = 17 - 3m$ **52.** $5y^2 = 13 + 2y$

53. $(2n + 1)(n - 3) = 7$ **54.** $(3x - 2)(x + 5) = 8$

55. $(t - 2)(t + 5) = t + 3$ **56.** $(u + 6)(2u - 1) = 3u - 7$

57. $\dfrac{x^2}{2} + \dfrac{x}{3} = \dfrac{5}{6}$ **58.** $\dfrac{a^2}{3} - \dfrac{a}{6} = \dfrac{7}{9}$

59. $\dfrac{4}{3a^2} - \dfrac{5}{6a} = \dfrac{10}{9}$ **60.** $\dfrac{5}{2x^2} + \dfrac{7}{4x} + \dfrac{1}{8} = 0$

61. $x^2 - 3x = 11$ **62.** $7u^2 - 9u = 2$

63. $x^2 + 2xy - y^2 = 0$, for x. (Assume $y \geq 0$.)

64. $zx^2 - 2yx - yz = 0$, for x. (Assume $y \geq z > 0$.)

For each of the following equations:

(a) Use the discriminant to determine the nature of the solutions.

(b) For equations with real solutions, write the solutions as three-place decimals.

65. $x^2 + 2x - 5 = 0$ **66.** $w^2 + 9w + 7 = 0$

67. $u^2 - 7u = 2$ **68.** $2m^2 - 3m = 10$

69. $16z^2 = -1 + 8z$ **70.** $5r^2 - 2 = 6r$

71. The sum of a number and its reciprocal is $41/20$. Find the number.

72. A number plus 4 times its reciprocal is $17/2$. Find the number.

73. The length of a rectangle is 8 more than its width. If the area is 48, find the dimensions.

74. The perimeter of a rectangle is 46 and the area is 60. Find the dimensions. (*Hint*: Write the length in terms of the width and the perimeter.)

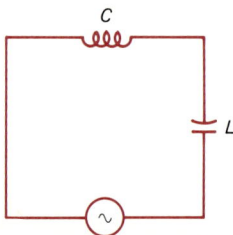

75. A rectangle 10 by 20 is increased all the way around by a uniform border of width x. If the area of the border is 64, find the width of the border. (*Hint:* Area of border = new area − old area.)

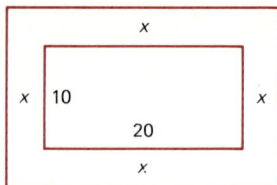

76. A rectangular mall is 20 meters by 30 meters. There is a pathway of uniform width around it. If the total area is 816 square meters, find the width of the border.

Business Application

77. The supply S and demand D for a certain cassette tape are given by

$$S = 200{,}000p - 700{,}000 \qquad D = \frac{1{,}500{,}000}{p}$$

where p is the price. Set $S = D$ and solve for the equilibrium price p.

Health Application

78. The rate at which a disease is spreading through a community is given by

$$R = \frac{y(5000 - y)}{50{,}000}$$

where y is the number of people with the disease. Find y if $R = 45$.

Optical Application

79. The length L from a lens to an image is given by

$$L = \left(M + \frac{1}{M} + 2 \right) f + n$$

where M is the magnifaction, f the focal length of the lens, and n the separation of nodal points. Find M if $L = 57$, $f = 5$, and $n = 2$.

Life Science Application

80. The probability p that a certain organism will die after x hours is given by

$$p = 1 - \frac{3}{x^2 + x + 3}$$

At what time x is there a two-thirds chance that the organism will be dead? (If $p = 2/3$, find x.)

Electrical Application

81. In an alternating-current circuit, the resonant frequency occurs when

$$2\pi f L = \frac{1}{2\pi f C}$$

where L in the inductance and C the capacitance. Solve for f (which is positive).

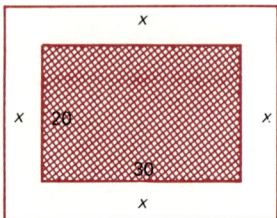

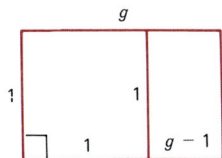

82. The *golden rectangle* shown at the left has dimensions 1 by g. If a 1-by-1 square is cut out, the remaining rectangle has dimensions $(g - 1)$ by 1, which is related to the original rectangle by

$$\frac{\text{Length}}{\text{Width}} = \frac{g}{1} = \frac{1}{g - 1}$$

Solve for g. (Consider only the positive solution.)

3.4
OTHER TYPES
OF EQUATIONS

This section deals with equations that do not fit neatly into any one class. Some of these equations involve radicals; some involve fractional exponents; and some can be reduced to quadratic equations.

Radical Equations

A **radical equation** is an equation that contains at least one radical. For example, $\sqrt{2x + 5} = \sqrt{x + 3}$ is a radical equation.

> To solve radical equations that involve an *n*th root:
>
> 1. Get one of the radicals alone on one side of the equation.
> 2. Raise both sides to the *n*th power. [Recall that $(\sqrt[n]{a})^n = a$, for positive a.]
> 3. If there are still radicals, go back to step 1.
> 4. When there are no more radicals, solve for the unknown.
> 5. *Important*: Check to see that each result is really a solution to the original equation. (A hand calculator may be helpful here—see page 82.)

Step 5 is very important. This procedure gives all the solutions, but it may also give results that are not solutions (called **extraneous solutions**).

EXAMPLE 20 Solve $x - 7 = \sqrt{x + 5}$.

Solution We square both sides of the equation.

Square both sides	$(x - 7)^2 = (\sqrt{x + 5})^2$
Simplify	$x^2 - 14x + 49 = x + 5$
Put in standard form	$x^2 - 15x + 44 = 0$
Factor	$(x - 4)(x - 11) = 0$
Solve for x	$x = 4$ or 11

We must check both $x = 4$ and $x = 11$.

$\boxed{\text{Check } x = 4}$
Left side: $4 - 7 = -3$
Right side: $\sqrt{4 + 5} = \sqrt{9} = 3$ Does *not* check.

$\boxed{\text{Check } x = 11}$
Left side: $11 - 7 = 4$
Right side: $\sqrt{11 + 5} = \sqrt{16} = 4$ *Checks.*

Thus, the only solution is 11. Again we see the importance of checking the solutions.

EXAMPLE 21   Solve $\sqrt{3x + 4} = \sqrt{x + 2} + 2$.

Solution   Since one radical is alone on one side of the equation, we can begin by squaring both sides.

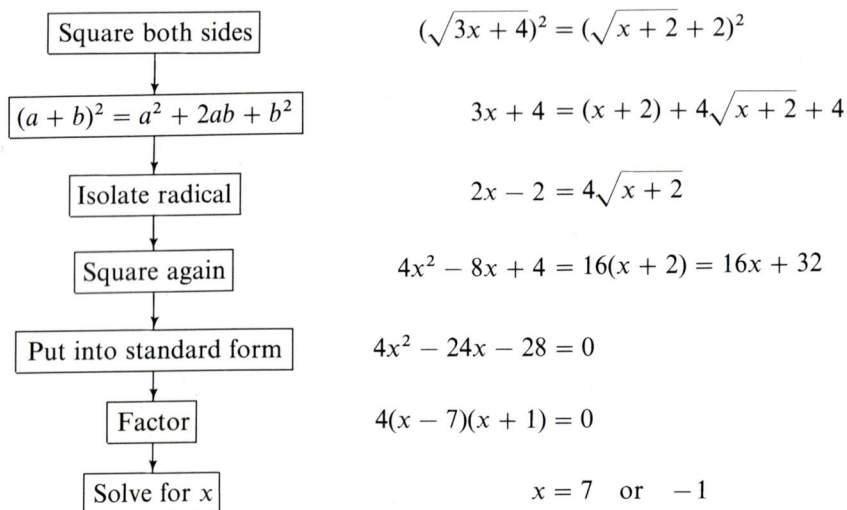

$\boxed{\text{Square both sides}}$ $(\sqrt{3x + 4})^2 = (\sqrt{x + 2} + 2)^2$

$\boxed{(a + b)^2 = a^2 + 2ab + b^2}$ $3x + 4 = (x + 2) + 4\sqrt{x + 2} + 4$

$\boxed{\text{Isolate radical}}$ $2x - 2 = 4\sqrt{x + 2}$

$\boxed{\text{Square again}}$ $4x^2 - 8x + 4 = 16(x + 2) = 16x + 32$

$\boxed{\text{Put into standard form}}$ $4x^2 - 24x - 28 = 0$

$\boxed{\text{Factor}}$ $4(x - 7)(x + 1) = 0$

$\boxed{\text{Solve for } x}$ $x = 7$ or -1

We now check $x = 7$ and $x = -1$.

$\boxed{\text{Check } x = 7}$
Left side: $\sqrt{3(7) + 4} = \sqrt{25} = 5$
Right side: $\sqrt{7 + 2} + 2 = 3 + 2 = 5$ *Checks.*

$\boxed{\text{Check } x = -1}$
Left side: $\sqrt{-3 + 4} = \sqrt{1} = 1$
Right side: $\sqrt{-1 + 2} + 2 = 1 + 2 = 3$ Does *not* check.

Thus, the only solution is 7.

EXAMPLE 22   Solve $(3x - 7)^{1/5} = 2$.

Solution   This equation might also be written as the radical equation

$$\sqrt[5]{3x - 7} = 2$$

and would be solved the same way: Raise both sides to the fifth power.

Raise to fifth power	$[(3x-7)^{1/5}]^5 = 2^5$
↓	
$(a^{1/n})^n = a$	$3x - 7 = 32$
↓	
Solve for x	$x = 13$

Let us check $x = 13$.

$$\text{Left side: } (3 \cdot 13 - 7)^{1/5} = 32^{1/5} = 2$$

$$\text{Right side: } 2 \quad \textit{Checks.}$$

Thus, the solution is 13.

Substitution Method

Often a tricky-looking equation can be reduced to a quadratic equation by a clever substitution. (We know how to solve quadratic equations—see page 77.) We call this the **substitution method**.

EXAMPLE 23   Solve $x^4 - 6x^2 + 5 = 0$.

Solution   This is a fourth-degree equation; however, by letting $m = x^2$ (and $m^2 = x^4$), it becomes a nice, solvable equation.

Substitute: $m = x^2$; $m^2 = x^4$	$m^2 - 6m + 5 = 0$
↓	
Factor	$(m - 5)(m - 1) = 0$
↓	
Solve for m	$m = 1 \quad \text{or} \quad 5$

Now that we know m, we can solve for x.

$$m = 1 \quad \text{or} \quad 5$$

Since $x^2 = m$	$x^2 = 1 \quad \text{or} \quad x^2 = 5$
↓	
Theorem 2	$x = \pm 1 \quad \text{or} \quad x = \pm\sqrt{5}$

There are *four* solutions: $1, -1, \sqrt{5}$, and $-\sqrt{5}$. We check two of these and leave the other two to the reader.

Check $x = -1$	$(-1)^4 - 6(-1)^2 + 5$
	$= 1 - 6 + 5 = 0 \quad \textit{Checks.}$

Check $x = \sqrt{5}$	$(\sqrt{5})^4 - 6(\sqrt{5})^2 + 5$
	$= 25 - 30 + 5 = 0 \quad \textit{Checks.}$

EXAMPLE 24   Solve $6x^{1/2} - 7x^{1/4} - 3 = 0$.

Solution   As in Example 23, this awkward equation can be reduced to a nice quadratic equation by substituting $m = x^{1/4}$ (and $m^2 = x^{1/2}$).

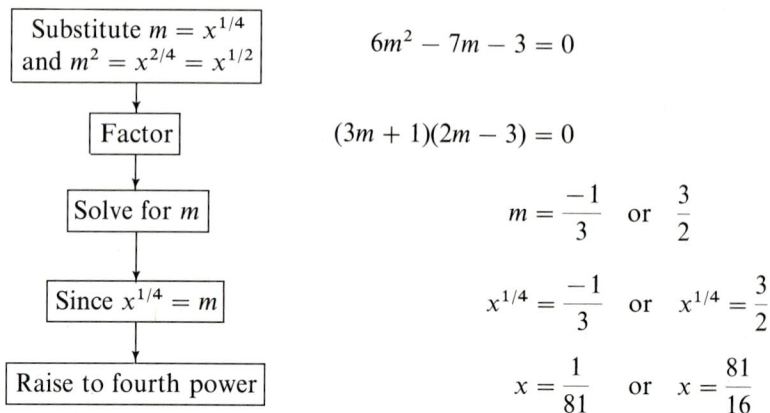

| Substitute $m = x^{1/4}$ and $m^2 = x^{2/4} = x^{1/2}$ | $6m^2 - 7m - 3 = 0$ |

Factor $\qquad (3m + 1)(2m - 3) = 0$

Solve for m $\qquad m = \dfrac{-1}{3}$ or $\dfrac{3}{2}$

Since $x^{1/4} = m$ $\qquad x^{1/4} = \dfrac{-1}{3}$ or $x^{1/4} = \dfrac{3}{2}$

Raise to fourth power $\qquad x = \dfrac{1}{81}$ or $x = \dfrac{81}{16}$

Let us check these solutions.

$\boxed{x = \dfrac{1}{81}}$ $6\left(\dfrac{1}{81}\right)^{1/2} - 7\left(\dfrac{1}{81}\right)^{1/4} - 3 = \dfrac{6}{9} - \dfrac{7}{3} - 3 = \dfrac{-14}{3}$ Does *not* check.

$\boxed{x = \dfrac{81}{16}}$ $6\left(\dfrac{81}{16}\right)^{1/2} - 7\left(\dfrac{81}{16}\right)^{1/4} - 3 = \dfrac{54}{4} - \dfrac{21}{2} - 3 = 0$ *Checks.*

Thus, the only solution is 81/16.

PROBLEM SET 3.4

Simplify the following.

Warm-up Exercises

1. $(x + 4)^2$ **2.** $(2x - 1)^2$

3. $(\sqrt{x + 5})^2$ **4.** $(-\sqrt{4x - 7})^2$

5. $(3 + \sqrt{x + 2})^2$ **6.** $(4 - \sqrt{x + 5})^2$

Solve the following equations. (Be sure to check for extraneous solutions.)

Warm-up Exercises

7. $2x + 5 = 4x + 11$ **8.** $5x - 2 = 3(x + 2)$

9. $p^2 = 49$ **10.** $(x - 3)^2 = 25$

11. $k^2 - 8k + 15 = 0$ **12.** $8z^2 = 6z + 9$

Solve the following equations. (Be sure to check for extraneous solutions.)

13. $\sqrt{2x + 5} = 7$ **14.** $\sqrt{5u - 3} - 2 = 8$

15. $\sqrt{x - 1} = -5$ **16.** $\sqrt{x - 10} + 5 = 1$

17. $\sqrt{4x - 7} = \sqrt{2x + 1}$

18. $\sqrt{10u - 3} = \sqrt{7u + 6}$

19. $\sqrt{2t - 1} - \sqrt{t} = 0$

20. $\sqrt{5a - 2} - \sqrt{6a + 1} = 0$

21. $\sqrt{2x + 1} = \sqrt{6x + 1} - 2$

22. $\sqrt{3u + 6} = \sqrt{u - 1} + 3$

23. $\sqrt{2a + 5} + \sqrt{a + 1} = 7$

24. $\sqrt{v} + \sqrt{5v - 2} = 6$

25. $\sqrt{7u + 1} = \sqrt{u - 5} + 3$

26. $\sqrt{r + 2} + 5 = \sqrt{5r - 3}$

27. $(x^2 - 1)^{1/3} = 2$

28. $(t^3 + 1)^{1/2} = 3$

29. $(3x + 5)^{1/5} = (5x - 1)^{1/5}$

30. $(4u - 1)^{1/3} = (x + 2)^{1/3}$

31. $(4x - 15)^{1/4} = 3^{1/2}$

32. $(2x - 5)^{1/3} = 3^{2/3}$

33. $2y^6 - 13y^3 + 6 = 0$

34. $6a^6 - 5a^3 - 4 = 0$

35. $z^8 = 16z^4 - 63$

36. $r^5 = 15 - 6r^{10}$

37. $2x^4 - x^2 - 15 = 0$

38. $y^4 - 3y^2 = 10$

39. $3x^{-2} + 13x^{-1} - 10 = 0$

40. $y^{-4} + 2y^{-2} = 35$

41. $x - 7\sqrt{x} + 12 = 0$

42. $a + 7a^{1/2} + 10 = 0$

43. $(t + 1)^{1/2} - (t + 1)^{1/4} - 6 = 0$

44. $(2u - 3)^{1/3} - 5(2u - 3)^{1/6} + 6 = 0$

45. $\sqrt[3]{x} = 4$

46. $a^{1/2} - 10a^{1/4} + 21 = 0$

47. $\sqrt{a + 7} = 5$

48. $(k^2 - k - 4)^{1/2} = 4$

49. $z^4 - 7z^2 + 10 = 0$

50. $r^{1/3} + 8r^{1/6} = 9$

51. $(t^2 - 9)^{1/3} = 3$

52. $\sqrt{4x + 1} + \sqrt{6x - 5} = 0$

53. $x^{-2} - x^{-1} = 20$

54. $(m - 6)^{1/2} = (5m + 6)^{1/4}$

55. $\sqrt[3]{14 - t} = \sqrt[3]{\dfrac{33}{t}}$

56. $1 = \sqrt[3]{\dfrac{13}{z^2} - \dfrac{36}{z^4}}$

57. $x = \sqrt{\dfrac{11}{2} + \dfrac{3}{x^2}}$

58. $x^3 = \sqrt{x^3 + 30}$

59. $t^4 = \sqrt{8t^4 + 9}$

60. $k^2 = \sqrt[3]{\dfrac{11k^3 + 7}{6}}$

Music Application

61. The frequency of the tone of a plucked, hammered, or bowed string is given by

$$f = \frac{1}{2L}\sqrt{\frac{T}{m}}$$

(see page 54). Solve for T in terms of the other letters.

Electrical Application

62. If a resistance R and a reactance X are in parallel, the equivalent impedance Z satisfies

$$Z = \frac{RX}{\sqrt{R^2 + X^2}}$$

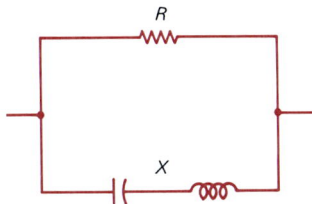

Solve this equation for X.

Acoustical Application

63. The recommended relation between the number of instruments N and the stage volume V is

$$N = 0.0021V^{0.855}$$

Solve for V in terms of N.

64. According to the theory of relativity, a rest mass m_0 grows to mass m when moving at velocity v, by the formula

$$m = \frac{m_0}{\sqrt{1 - \left(\dfrac{v}{c}\right)^2}}$$

where c is the speed of light. How fast must a 70-kilogram man travel to double his mass? (Find v if $m_0 = 70$, $m = 140$, and $c = 3 \times 10^8$.)

3.5
LINEAR INEQUALITIES

Just as we have open equations, inequalities such as

$$2x + 7 \leq 13 \qquad \text{and} \qquad x^2 - 3x + 2 > 0$$

are called **open inequalities**, since their truth depends on x. In general, the solutions to inequalities are not single numbers (such as $x = 3$) but rather, intervals on the real-number line. For example, one such solution might be $x < 5$, which would pictured as follows (see page 4):

We have a special **interval notation** to help describe such inequality solutions.

Symbol	Meaning
(	Left endpoint is *not* included.
)	Right endpoint is *not* included.
[	Left endpoint *is* included.
]	Right endpoint *is* included.
$(-\infty,$	No left endpoint.
$,\infty)$	No right endpoint.

The symbol ∞ is called **infinity**, but it is *not* a real number; rather, it is just a symbol to show that there is no end to the interval.

We have the following types of intervals.

Name	Form	Meaning
Open	(a, b)	*Neither* endpoint is included.
Closed	$[a, b]$	*Both* endpoints are included.
Half-open	$(a, b]$ or $[a, b)$	*Only one* endpoint is included.
Infinite	(a, ∞) or $[a, \infty)$	Interval continues without end in one direction.
	$(-\infty, b)$ or $(-\infty, b]$	

EXAMPLE 25 In the following table we show an interval with inequality symbols and interval notation; then we sketch it and give its type.

Inequality	Interval	Graph	Type
(a) $-2 < x < 5$	$(-2, 5)$		Open interval
(b) $1 \leq x \leq 4$	$[1, 4]$		Closed interval
(c) $2 < x \leq 6$	$(2, 6]$		Half-open interval
(d) $3 < x$	$(3, \infty)$		Infinite interval
(e) $x \leq 4$	$(-\infty, 4]$		Infinite interval
(f) $x \geq -2$	$[-2, \infty)$		Infinite interval

A **linear inequality** can be put in one of the following forms:

$$ax + b < 0 \qquad ax + b > 0$$
$$ax + b \leq 0 \qquad ax + b \geq 0$$

Note that the linear inequality has the same appearance as the linear equation, except for the inequality symbol. For this reason we expect to have properties for solving linear inequalities that are similar to those for linear equations (see page 64). In fact, we do have very similar properties, with one exception.

Theorem 4

Let a, b, and c be real numbers:

EXAMPLE

	If $a < b$, then:	If $x < 5$, then:
Addition Property	1. $a + c < b + c$	1. $x + 3 < 8$
Multiplication Property	2a. $ac < bc$ (for positive c)	2a. $2x < 10$
	2b. $ac > bc$ (for negative c)	2b. $-3x > -15$

These relations also hold true for $\leq$, $>$, and $\geq$. Note that multiplying (or dividing) by a *negative* number *reverses* the inequality symbol.

EXAMPLE 26   Solve $-5x + 2 < 12$ and graph the solution.

Solution   We isolate x as we would with an equation.

Given	$-5x + 2 < 12$
↓	
Subtract 2	$-5x < 10$
↓	
Divide by -5; reverse inequality	$x > -2$

This solution $x > -2$ can be written as the interval $(-2, \infty)$ and has the graph

EXAMPLE 27   Solve $3x + 7 \geq 5x + 1$ and graph the solution.

Solution   We solve this by getting x alone.

Subtract $3x$	$7 \geq 2x + 1$
↓	
Subtract 1	$6 \geq 2x$
↓	
Divide by 2	$3 \geq x$

We can write this as an interval $(-\infty, 3]$ and graph as follows:

EXAMPLE 28   Solve $-2 < \dfrac{x - 3}{5} < 1$.

Solution   In some courses, especially calculus and statistics, we often see a double inequality such as this one. It is really an abbreviation for *two* inequalities:

$$-2 < \frac{x - 3}{5} \quad and \quad \frac{x - 3}{5} < 1$$

We can use the same rules as before and solve both inequalities together.

Multiply by 5	$-10 < x - 3 < 5$
↓	
Add 3	$-7 < \quad x \quad < 8$

This solution can be written as the open interval $(-7, 8)$ and is graphed as follows:

EXAMPLE 29 Solve $2x - 1 \geq 3$ *and* $4x + 1 \leq 17$ together.

Solution This is called a **compound inequality**. We solve each inequality separately and then take the **intersection** (or overlap) of the two solution sets.

Solve $2x - 1 \geq 3$

$$2x - 1 \geq 3$$
$$2x \geq 4$$
$$x \geq 2$$

Solve $4x + 1 \leq 17$

$$4x + 1 \leq 17$$
$$4x \leq 16$$
$$x \leq 4$$

Graph solutions; take intersection

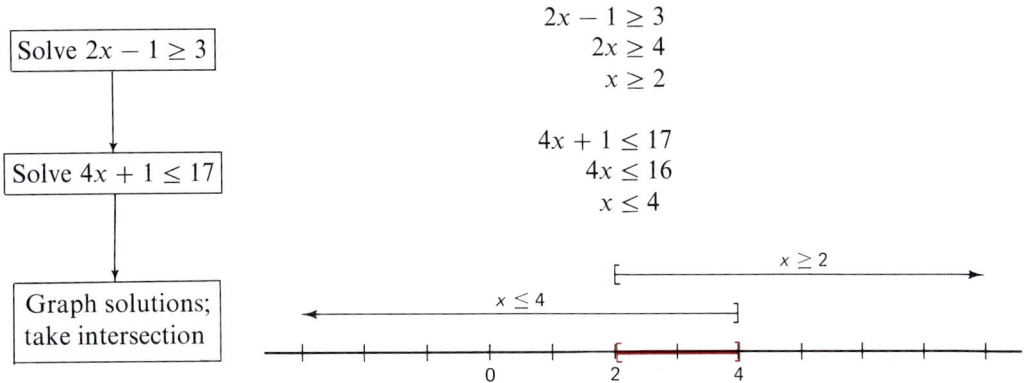

We can see that the solution is $2 \leq x \leq 4$, or the closed interval $[2, 4]$.

EXAMPLE 30 Solve $3x - 1 > 8$ *or* $2x + 5 < 3$ together.

Solution Like Example 29, this is a compound inequality. However, the "or" indicates the union of the individual solution sets. (The **union** of sets A and B, written $A \cup B$, is the set of elements in A, or B, or both.)

Solve $3x - 1 > 8$

$$3x - 1 > 8$$
$$3x > 9$$
$$x > 3$$

Solve $2x + 5 < 3$

$$2x + 5 < 3$$
$$2x < -2$$
$$x < -1$$

Graph solutions; take union

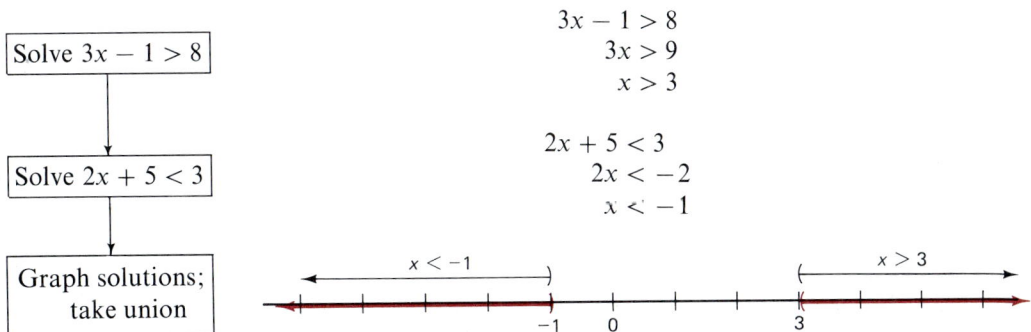

We see that the solution is the union of the two nonoverlapping intervals $(-\infty, -1)$ and $(3, \infty)$. We write the solution

$$(-\infty, -1) \cup (3, \infty)$$

Solve the following equations.

1. $2x + 1 = 9$

2. $4x - 7 = 7x + 2$

3. $2(t + 5) + 4 = 5(t - 2) + 3$

4. $\dfrac{x - 4}{2} = 5$

Graph the following sets.

5. $\{x \mid x \le 5\}$

6. $\{x \mid x > -2\}$

7. $\{x \mid -2 < x < 4\}$

8. $\{x \mid -3 \le x \le 3\}$

Write the following inequalities in interval notation.

9. $2 < x < 8$

10. $3 \le x \le 5$

11. $x > 6$

12. $x \le -3$

13. $x < 4$

14. $x \ge -1$

15. $-2 < x \le -1$

16. $6 \le x < 11$

Write the following intervals as inequalities.

17. $(4, 7)$

18. $[-1, 5]$

19. $(2, 8]$

20. $[6, 9)$

21. $(-\infty, 3)$

22. $(5, \infty)$

23. $[-2, \infty)$

24. $(-\infty, 4]$

Solve the following inequalities and express the solutions as intervals.

25. $10p \le -20$

26. $\dfrac{w}{-3} \ge -4$

27. $4x - 2 \ge 18$

28. $3y + 5 < 26$

29. $-5r - 2 < -7$

30. $-7u + 2 \le -3$

31. $6t + 5 \ge 9t - 4$

32. $10x + 4 < 3x + 39$

33. $15q + 1 \le 7q - 4$

34. $3y - 10 \ge -3y + 11$

35. $5(x + 1) \ge 8(x - 1) + 1$

36. $4(n + 5) > 5(2n + 7) - n$

37. $3 < 2x + 1 \le 7$

38. $-2 \le 3z - 5 < 1$

39. $-1 < \dfrac{n - 1}{5} < 1$

40. $-2 \le \dfrac{u + 5}{3} \le 3$

41. $4 \le \dfrac{2x + 7}{5} < 7$

42. $-1.5 < \dfrac{3x - 2}{10} \le 1.5$

43. $k \ge 5 \text{ and } k \le 11$

44. $r < 12 \text{ and } r > 7$

45. $x - 2 > 4 \text{ and } x + 1 < 11$

46. $t + 3 < 7 \text{ and } t - 2 > -3$

47. $2m + 1 < 15 \text{ and } 3m - 1 \ge 2$

48. $5r - 1 \le 14 \text{ and } 2r + 1 > -5$

49. $t \le 1 \text{ or } t \ge 5$

50. $2m > 10 \text{ or } 5m < -10$

51. $2x + 3 < 7 \text{ or } 5x - 1 > 19$

52. $4t - 2 \ge 14 \text{ or } 3m + 10 \le 1$

53. $2(r - 5) > 5(r + 1)$

54. $x < 2 \text{ and } 3x > 3$

55. $4 < x + 2 < 7$

56. $-4 \le \dfrac{4x + 8}{2} \le 4$

57. $x \geq 1$ *and* $2x \leq 8$

58. $-5t - 3 < -3t - 7$

59. $-1 \leq \dfrac{2z + 5}{3} \leq 1$

60. $4m + 1 \geq 5$ *or* $2m - 3 \leq 11$

61. $-4x - 1 > -3x + 7$

62. $-3 \leq \dfrac{x + 5}{-1} \leq 3$

63. $2t + 1 \leq 7$ *or* $3t - 1 \geq 2$

64. $-2t - 5 \geq -11$

65. $-2 \leq \dfrac{x - 7}{-4} \leq 1$

66. $\dfrac{2x - 6}{4} > \dfrac{2x + 1}{5}$

67. $\dfrac{2x + 5}{-3} < \dfrac{2x + 10}{4}$

68. $-7t - 5 > -6t - 9$

*Population
Application*

69. The population P of a bacterial culture is given by

$$P = \frac{20{,}000t}{10 + t}$$

where t is the time in hours. For what times is the population greater than 15,000? (Solve $P \geq 15{,}000$ for $t \geq 0$.)

*Business
Applications*

70. The value of a depreciating machine is given by

$$V = 100{,}000 - 6000t$$

where t is the age in years. For what years t is $V \geq 40{,}000$?

71. If a certain state income tax for a family of four is given by

$$T = 0.025(I - 4000)$$

where I is the family income, for what incomes I is $T \leq 1000$?

*Life Science
Application*

72. The relation between temperature Celsius C and Fahrenheit F is

$$C = \frac{5}{9}(F - 32)$$

A certain plant must be kept between 15° and 25°C. To what temperatures Fahrenheit does this correspond? [Solve $15 \leq \frac{5}{9}(F - 32) \leq 25$.]

*Health
Application*

73. A dietitian determines that a patient should have between 56 and 72 grams of protein a day. Using a diet of milk, hamburger, beans, and potatoes, the dietitian has the relation

$$56 \leq 8m + 7h + 15b + 2p \leq 72$$

Find m if $h = 3$ ounces, $b = 1$ cup, and $p = 2$ potatoes.

**3.6
QUADRATIC
INEQUALITIES**

A **quadratic inequality** is an inequality that can be put into one of the standard forms (notice the similarity to quadratic equations, with $a \neq 0$):

$ax^2 + bx + c < 0$	$ax^2 + bx + c > 0$
$ax^2 + bx + c \leq 0$	$ax^2 + bx + c \geq 0$

In this section we solve quadratic inequalities and other inequalities that lead to quadratic inequalities. We solve these inequalities by factoring (as we did with quadratic equations). We then look at the signs of the product, recalling the facts:

$$(+)(+) = (+)$$
$$(-)(-) = (+)$$
$$(+)(-) = (-)$$

EXAMPLE 31 Solve $x^2 + 3x - 10 > 0$.

Solution We begin by factoring; then we find where each factor is positive and where each is negative. Finally, we look at the combinations of $+$ and $-$ signs that give a positive product.

$$(x + 5)(x - 2) > 0$$

Factor

$(x + 5)$ is $\begin{cases} > 0 & \text{if } x > -5 \\ = 0 & \text{if } x = -5 \\ < 0 & \text{if } x < -5 \end{cases}$

Analyze $(x + 5)$

$(x - 2)$ is $\begin{cases} > 0 & \text{if } x > 2 \\ = 0 & \text{if } x = 2 \\ < 0 & \text{if } x < 2 \end{cases}$

Analyze $(x - 2)$

Graph the sign of each factor

$(x + 5)$: $- \ - \ - \ -$ O $+ \ + \ + \ + \ + \ + \ + \ + \ + \ + \ + \ + \ + \ + \ + \ + \ +$

Determine the sign of the product

$(x - 2)$: $- \ - \ - \ - \ - \ - \ - \ - \ - \ - \ - \ - \ - \ - \ - \ - \ -$ O $+ \ + \ + \ +$

Product: $+ \ + \ + \ +$ O $- \ - \ - \ - \ - \ - \ - \ - \ - \ - \ - \ - \ - \ -$ O $+ \ + \ + \ +$

Solution is where product is $+$

Solution set $= \{x \mid x < -5 \text{ or } x > 2\}$

$$= (-\infty, -5) \cup (2, \infty)$$

Note that there are two regions where the product is positive (both factors positive or both factors negative). We can write this either in terms of two inequalities or as the union of two infinite intervals.

We can use this same sign method to solve inequalities involving fractions, by recalling the following facts:

$$\frac{(+)}{(+)} = (+) \qquad \frac{(-)}{(-)} = (+) \qquad \frac{(+)}{(-)} = (-) \qquad \frac{(-)}{(+)} = (-)$$

EXAMPLE 32 Solve $\dfrac{x-1}{x+4} \leq 0$.

Solution Here we look at the signs of the numerator and denominator. Then we look for a combination whose quotient is zero or negative. (We must be careful since the denominator cannot be zero.)

Analyze $(x - 1)$

$(x - 1)$ is $\begin{cases} > 0 & \text{if } x > 1 \\ = 0 & \text{if } x = 1 \\ < 0 & \text{if } x < 1 \end{cases}$

Analyze $(x + 4)$

$(x + 4)$ is $\begin{cases} > 0 & \text{if } x > -4 \\ = 0 & \text{if } x = -4 \\ < 0 & \text{if } x < -4 \end{cases}$

Graph the signs of top and bottom

$(x - 1)$: $- \ - \ - \ - \ - \ - \ - \ - \ - \ - \ - \ - \ - \ - \ O \ + + + + + + +$

$(x + 4)$: $- \ - \ - \ - \ - \ O \ + + + + + + + + + + + + + + + + + +$

Determine sign of quotient

Quotient: $+ + + + + \ \square \ - \ - \ - \ - \ - \ - \ - \ - \ - \ O \ + + + + + + +$

Solution is where quotient is 0 or $-$

$$\text{Solution set} = \{x \mid -4 < x \leq 1\}$$
$$= (-4, 1]$$

Note that in the quotient we avoided the point $x = -4$. Why? This point makes the denominator zero, so we cannot include it. Thus, we get the half-open interval $(-4, 1]$.

EXAMPLE 33 Solve $\dfrac{9x - 2}{3x + 2} \geq 2$.

Solution We would like to solve this as we did Example 32. But the 2 on the right is a problem, since we must have a 0 on the right. Thus, we subtract 2 from both sides and rewrite the fraction. (Why can we *not* simply multiply both sides by $3x + 2$?)

Subtract 2	$$\dfrac{9x-2}{3x+2} - 2 \geq 0$$

↓

Rewrite $-2 = \dfrac{-2(3x+2)}{(3x+2)}$	$$\dfrac{9x-2}{3x+2} + \dfrac{-6x-4}{3x+2} \geq 0$$

↓

Add fractions	$$\dfrac{3x-6}{3x+2} \geq 0$$

↓

Analyze $(3x-6)$	$(3x-6)$ is $\begin{cases} >0 & \text{if } x > 2 \\ =0 & \text{if } x = 2 \\ <0 & \text{if } x < 2 \end{cases}$

↓

Analyze $(3x+2)$	$(3x+2)$ is $\begin{cases} >0 & \text{if } x > \dfrac{-2}{3} \\ =0 & \text{if } x = \dfrac{-2}{3} \\ <0 & \text{if } x < \dfrac{-2}{3} \end{cases}$

↓

Graph the signs of top and bottom

$(3x-6)$: $-\ -\ -\ -\ -\ -\ -\ -\ -\ -\ -\ \bigcirc\ +\ +\ +\ +$

↓

Determine the sign of the quotient

$(3x+2)$: $-\ -\ -\ -\ -\ -\ \bigcirc\ +\ +\ +\ +\ +\ +\ +\ +\ +\ +$

Quotient: $+\ +\ +\ +\ +\ +\ \square\ -\ -\ -\ -\ -\ \bigcirc\ +\ +\ +\ +$

↓

Solution is where quotient is 0 or $+$

Number line showing values $-\frac{2}{3}$, 0, 2.

$$\text{Solution set} = \left\{ x \,\middle|\, x < \frac{-2}{3} \text{ or } x \geq 2 \right\}$$

$$= \left(-\infty, \frac{-2}{3} \right) \cup [2, \infty)$$

Here we avoid the point $x = -2/3$, since it causes the denominator to be zero.

PROBLEM SET

3.6   *Factor the following polynomials.*

Warm-up **1.** $x^2 - 4$ | **2.** $r^2 - 49$

Exercises **3.** $k^2 + 5k - 14$ | **4.** $t^2 + 5t - 36$

Simplify the following.

Warm-up Exercises

5. $\dfrac{1}{x} + \dfrac{1}{y}$

6. $\dfrac{7}{x} + 5$

7. $\dfrac{2}{u+3} - 1$

8. $\dfrac{3}{t-5} - \dfrac{1}{2}$

Solve the following inequalities and express the solutions as intervals.

Warm-up Exercises

9. $x + 1 > 0$

10. $x + 1 < 0$

11. $-5t - 10 < 0$

12. $-5t - 10 > 0$

Solve the following inequalities and express the solutions as intervals.

13. $2y^2 - 13y + 6 < 0$

14. $3a^2 + 13a - 10 > 0$

15. $x^2 + x \geq 12$

16. $r \leq 15 - 6r^2$

17. $\dfrac{x^2}{9} + \dfrac{2x}{3} < 3$

18. $(t-3)(t-4) < 6$

19. $\dfrac{k}{2k-6} < 0$

20. $\dfrac{3-t}{5+t} \geq 0$

21. $\dfrac{2r-3}{3r+5} \leq 0$

22. $\dfrac{5n+6}{3n-5} < 0$

23. $\dfrac{7k-1}{k} \leq 4$

24. $\dfrac{3u+4}{u} \geq -5$

25. $\dfrac{2x+1}{x-7} < \dfrac{x-1}{x-7}$

26. $\dfrac{5t-2}{3t-1} > \dfrac{t+6}{3t-1}$

27. $\dfrac{a+2}{a+3} \geq \dfrac{1}{2}$

28. $\dfrac{m-2}{m+5} \leq \dfrac{1}{3}$

29. $(t+2)(t-3) \leq 14$

30. $\dfrac{4x+3}{x} > \dfrac{1}{5}$

31. $\dfrac{k+5}{k+4} > 0$

32. $\dfrac{x^2}{18} + 1 < \dfrac{1}{2}$

33. $\dfrac{2u-5}{u} < \dfrac{1}{2}$

34. $\dfrac{k-5}{k-2} \geq \dfrac{1}{4}$

35. $\dfrac{x^2}{8} + \dfrac{x}{4} \geq 1$

36. $12p^2 + 5 \leq 23p$

37. $\dfrac{r+3}{r} \leq \dfrac{1}{3}$

38. $(m-5)(m-3) > 8$

39. $6x^2 + x > 35$

40. $\dfrac{(a+2)(a+3)}{(a+4)(a+5)} < 0$

41. $2(x+5) > 5(x-3)$

42. $x(x+1) < 6x + 14$

43. $x(x+1)(x+2) \leq 0$

44. $x^3 - x \geq 0$

45. $\dfrac{x+1}{x+2} < 0$

46. $\dfrac{2x-1}{x+5} > \dfrac{1}{2}$

47. $\dfrac{x+3}{(x+1)(x+2)} \geq 0$

48. $\dfrac{x^2-1}{x-5} \leq 0$

49. The population of a community is given by

$$P = 20{,}000 + 3000t - 50t^2$$

where t is the number of years after 1980. For what times t will $P \geq 45{,}000$?

50. The amount of sugar in the blood after eating a certain food is given by $S = 4.0 + 0.5t - 0.1t^2$, where t is the time in hours. Find the times at which the sugar level $S \geq 4.4$.

51. The probability of a certain organism dying after t hours is given by

$$p = 1 - \frac{3}{t^2 + 4t + 3}$$

Find the positive times at which $p \geq 7/8$.

52. The rate at which a disease is spreading through a community is given by

$$R = \frac{(5000 - y)y}{50{,}000}$$

where y is the number of people who already have the disease. For what numbers y is $R \geq 80$?

3.7
ABSOLUTE-VALUE EQUATIONS AND INEQUALITIES

Recall from Chapter 1 that the **absolute value** is given by

$$|x| = \begin{cases} x & \text{if } x \geq 0 \\ -x & \text{if } x < 0 \end{cases}$$

For example, $|5| = 5$ and $|-6| = 6$.

Note that $|x|$ might be x or $-x$. We use this fact in solving **absolute-value equations and inequalities**. These are equations and inequalities with an unknown within the absolute-value bars, such as $|x + 2| = 7$ or $|a - 8| < 3$.

We observe that $|x|$ can be viewed as the distance from x to 0 on the number line.

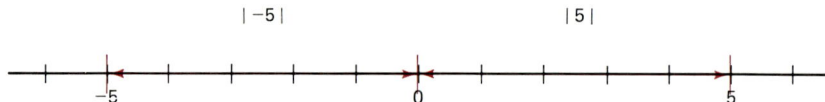

EXAMPLE 34   Find the following values.

(a) $|x| = 4$ (b) $|x| < 4$ (c) $|x| > 4$

Solution   (a) $|x| = 4$ means that x is 4 units from 0; hence, x is 4 or -4.

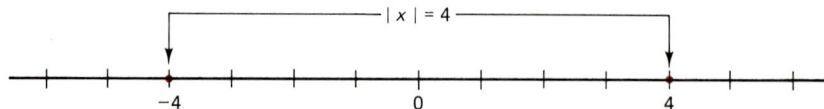

(b) $|x| < 4$ means that x is less than 4 units from 0; hence, x is between -4 and 4, or $-4 < x < 4$.

(c) $|x| > 4$ means that x is more than 4 units from 0; hence, x is greater than 4 or less than -4. Thus, we have two regions, $x > 4$ or $x < -4$.

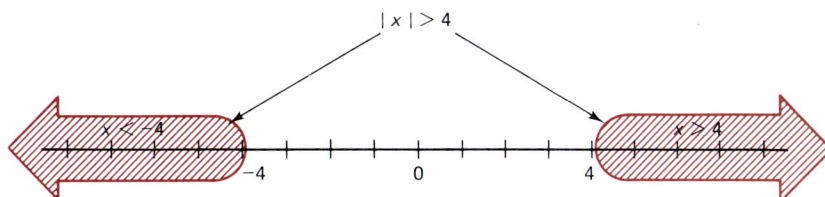

We can use this example to obtain a general rule.

Property
For $c > 0$ and any expression P:

EXAMPLE

$|P| = c$ means $P = c$ or $P = -c$	$|P| = 7$ means $P = 7$ or $P = -7$
$|P| < c$ means $-c < P < c$	$|P| < 7$ means $-7 < P < 7$
$|P| > c$ means $P < -c$ or $P > c$	$|P| > 7$ means $P < -7$ or $P > 7$

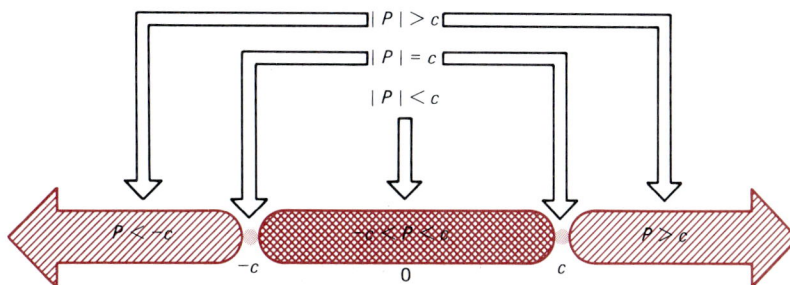

To solve an absolute-value equation of the form $|P| = c$:

1. Split this into two equations: $P = c$ or $P = -c$.
2. Solve $P = c$ and $P = -c$ separately.

Note: If c is negative, $|P| = c$ has no solution, since $|P|$ is always positive or zero.

EXAMPLE 35 Solve $|x - 3| = 8$.

Solution We split this into two equations.

$$|x - 3| = 8$$

Split into two equations

$$x - 3 = 8 \quad \text{or} \quad x - 3 = -8$$

Solve for x

$$x = 11 \quad \text{or} \quad x = -5$$

There are two solutions, 11 and -5. Written as a set, this is $\{11, -5\}$. The reader should check that both 11 and -5 satisfy the original equation.

EXAMPLE 36 Solve $|2x - 5| = 6$.

Solution We again split this into two equations.

$$|2x - 5| = 6$$

Split into two equations

$$2x - 5 = 6 \quad \text{or} \quad 2x - 5 = -6$$

$$2x = 11 \quad \text{or} \quad 2x = -1$$

Solve for x

$$x = \frac{11}{2} \quad \text{or} \quad x = \frac{-1}{2}$$

The solution set is $\left\{ \dfrac{11}{2}, \dfrac{-1}{2} \right\}$.

To solve inequalities of the form $|P| < c$:

1. Rewrite $|P| < c$ as $-c < P < c$.
2. Solve this inequality.

Note: This procedure also holds for $|P| \leq c$. Also, c must be positive.

EXAMPLE 37 Solve $|x - 4| < 5$.

Solution We rewrite this as a double inequality.

$$|x - 4| < 5$$

Rewrite as $-c < P < c$

$$-5 < x - 4 < 5$$

Solve for x

$$-1 < x < 9$$

The solution is the interval $(-1, 9)$ and can be graphed as follows:

EXAMPLE 38 Solve $|2x - 1| \le 7$.

Solution We rewrite this as a double inequality:

Rewrite as $-c \le P \le c$	$-7 \le 2x - 1 \le 7$
↓	$-6 \le\ \ 2x\ \ \le 8$
Solve for x	$-3 \le\ \ \ x\ \ \ \le 4$

The solution is the interval $[-3, 4]$ and can be graphed as follows:

To solve inequalities of the form $|P| > c$:

1. Split the inequality into two inequalities: $P > c$ or $P < -c$.
2. Solve each of these inequalities.

Note: This procedure also holds for $|P| \ge c$. Also, c must be positive.

YES	NO
$|P| > 7$	$|P| > 7$
$P > 7$ or $P < -7$	$P > 7$ or $P > -7$

The inequality $|P| > c$ splits into $P > c$ or $P < -c$.

EXAMPLE 39 Solve $|3x + 1| \ge 5$.

Solution We split this into two inequalities.

Split into $P > c$ or $P < -c$	$|3x + 1| \ge 5$
↓	$3x + 1 \ge 5$ or $3x + 1 \le -5$
Solve for x	$3x \ge 4$ or $3x \le -6$
	$x \ge \dfrac{4}{3}$ or $x \le -2$

The solution is $(-\infty, -2] \cup [\frac{4}{3}, \infty)$ and has the following graph:

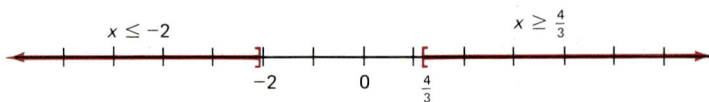

$x \leq -2$ $x \geq \frac{4}{3}$

-2 0 $\frac{4}{3}$

EXAMPLE 40   Solve $|2x + 5| > a$ for x (assume that $a > 0$).

Solution   This is a **literal inequality** (review Example 5, page 66). We solve for x in the usual way.

$$|2x + 5| > a$$

| Split into |
| $P > c$ or $P < -c$ |

$2x + 5 > a$ or $2x + 5 < -a$

$2x > a - 5$ or $2x < -a - 5$

| Solve for x |

$x > \dfrac{a - 5}{2}$ or $x < \dfrac{-a - 5}{2}$

We leave our solution in terms of the letter a (since we do not know its value). We write the solution in interval notation as

$$\left(-\infty, \frac{-a - 5}{2}\right) \cup \left(\frac{a - 5}{2}, \infty\right)$$

PROBLEM SET 3.7

Warm-up Exercises

Simplify the following.

1. $|3|$ **2.** $|-5|$ **3.** $|0|$

4. $-|-7|$ **5.** $|-|-4||$ **6.** $-|-|-8||$

Solve the following.

Warm-up Exercises

7. $x - 7 = 3$ **8.** $x - 7 = -3$

9. $2x + 5 = -9$ **10.** $2x + 5 = 9$

11. $x + 5 < 4$ **12.** $x + 5 > -4$

13. $2x - 7 < -3$ **14.** $2x - 7 > 3$

15. $5x + 1 > a$ **16.** $5x + 1 < -a$

Solve the following and express the solutions to the inequalities in interval notation.

17. $|x + 5| = 3$ **18.** $|t - 2| = 7$

19. $|2m - 1| = 5$ **20.** $|3k + 2| = 4$

21. $|3r - 5| = 0$ **22.** $|5n + 3| = \dfrac{1}{2}$

23. $|y - 3| \leq 4$ **24.** $|z + 10| < 1$

25. $|6g - 5| < 7$ **26.** $|7t + 2| \leq 5$

27. $|2x + 3| \leq 0$

28. $|3k - 1| < \dfrac{1}{3}$

29. $|x + 4| \geq 2$

30. $|z - 3| > 5$

31. $|7r - 1| > 3$

32. $|6m + 4| \geq 3$

33. $|3t + 2| \geq 0$

34. $|2k - 5| > \dfrac{1}{4}$

Solve the following for x. (Assume that all other letters represent positive numbers.)

35. $|x + 1| < a$

36. $|2x + 1| \leq r$

37. $|x + 5| \geq b$

38. $|3x + 5| > z$

39. $|x + 5| < 3$

40. $|x + 5| \geq 2$

41. $|2x + 3| = 7$

42. $|2x + 3| < z$

43. $|x - 4| \geq 8$

44. $|4x - 1| = k$

45. $|2x - 1| > a$

46. $|3x + 7| \geq 5t$

47. $|3x + 4| = 2x$

48. $|7x - 3| = x + 1$

49. $|5x - 1| > 3x + 1$

50. $|2x + 3| < 7x - 1$

51. $|x + 2| \leq x + 3$

52. $|3x - 8| \geq 5x$

53. $|ax| = b$

54. $|mx + n| = p$

55. $|x + s| \leq t$

56. $|x - y| > z$

Technical
Application
57. The width of a certain bolt has a tolerance of 0.002 and satisfies $|w - 1.755| \leq$ 0.002. Solve this equation for the range of w.

CHAPTER 3
SUMMARY   Important Properties

Let a, b, c, and P be real numbers.

If $a = b$, then:

   1. $a + c = b + c$

   2. $ac = bc$

If $ab = 0$, then either $a = 0$, $b = 0$, or both.

If $x^2 = a$ $(a > 0)$, then $x = \pm\sqrt{a}$.

The solutions to $ax^2 + bx + c = 0$ $(a \neq 0)$ are

$$x = \frac{-b \pm \sqrt{b^2 - 4ac}}{2a}$$

If $a < b$, then:

   1. $a + c < b + c$

  2a. $ac < bc$ (if c is positive)

  2b. $ac > bc$ (if c is negative)

$|P| = c$ means $P = c$ or $P = -c$ (for $c > 0$)

$|P| < c$ means $-c < P < c$ (for $c > 0$)

$|P| > c$ means $P < -c$ or $P > c$ (for $c > 0$)

CHAPTER 3 Summary   **107**

Review Exercises

Solve the following problems. (Express the solutions to inequalities in interval notation.)

1. $5x - 9 = 7x + 15$

2. $2(5m + 3) - m = 4(2m - 1)$

3. $\dfrac{x + 1}{2x^2} - \dfrac{3}{4x} = \dfrac{3}{4x^2}$

4. $\dfrac{6}{x - 2} - \dfrac{3}{x + 4} = \dfrac{6}{x^2 + 2x - 8}$

5. $\dfrac{1}{x} + \dfrac{1}{y} = \dfrac{1}{z}$ (for x)

6. The sum of three consecutive even integers is 72. Find the numbers.

7. What number must be added to both the numerator and denominator of 2/7 to give 1/2?

8. The Grozas invest \$5000: some at 10% and the rest at 9%. If their total interest is \$460, how much is invested at each rate?

9. How many liters of a 20% solution must be mixed with a 50% solution to produce 300 liters of a 40% solution?

10. Roy made a 900-mile trip in 20 hours. He traveled part of it at 40 miles per hour and the rest at 50 miles per hour. What distances were traveled at each speed?

11. Sue can paint a room in 4 hours and Jo can paint it in 3 hours. How long will it take them working together?

12. $m^2 + 3m = 10$

13. $x + \dfrac{35}{x} = 12$

14. $(x + 2)^2 = 49$

15. $t^2 - 2t - 7 = 0$

16. $(k + 2)(k + 3) = 11k$

17. $x - \dfrac{3}{2} = \dfrac{6}{x}$

18. The sum of a number and its reciprocal is 29/10. Find the number.

19. $\sqrt{5a - 1} = \sqrt{3a + 7}$

20. $\sqrt{3t - 2} = \sqrt{t + 2}$

21. $\sqrt[3]{x + 5} = 2$

22. $x^8 + 6x^4 = 7$

23. $a^{1/2} + 30 = 11a^{1/4}$

24. $x = \sqrt{\dfrac{x^4 + 2}{3}}$

25. $-6t - 1 \geq 11$

26. $3(x + 1) < 2(2x + 5) - 1$

27. $-2 < \dfrac{x + 5}{7} < 2$

28. $x + 3 > 1$ *and* $2x - 1 < 15$

29. $2r^2 + 11r \geq 6$

30. $\dfrac{x - 3}{2x - 1} < 0$

31. $\dfrac{3x - 1}{x + 3} > \dfrac{1}{2}$

32. $x - 1 < 5$ *or* $2x + 1 > 13$

33. $|x + 5| = 3$

34. $|2x - 1| \leq 5$

35. $|3u - 2| > 8$

36. $|x + 2| < t$ (for x with $t > 0$)

Graphing

4.1

THE CARTESIAN COORDINATE SYSTEM

In Chapter 3 we saw statements whose truth depended on the value of a certain variable. Now consider the following.

$2 + 2 = 4$ This is a true statement.

$2 + x = 6$ The truth of this depends on x.

$x + y = 7$ The truth of this depends on *both* x and y.

In this chapter we deal with equations in *two* variables. The solutions to these equations are **ordered pairs** of numbers, (x, y). There are many (infinitely many) solutions to $x + y = 7$: $(-1, 8)$, $(2, 5)$, $(3, 4)$ $(6, 1)$, $(7, 0)$, $(10, -3)$, and so on.

Recall that there is a one-to-one correspondence between the real numbers and the points of the number line. Similarly, there is a one-to-one correspondence between ordered pairs of real numbers and the points of the plane. This connection between algebra (the ordered pairs) and geometry (the plane) is very valuable since it allows us to visualize an equation such as $x + y = 7$ or $y = x^2$.

We label the points of the plane by their corresponding ordered pairs, and we call this the **Cartesian** (or **rectangular**) **coordinate system** (named for its inventor, René Descartes). This system is formed by two perpendicular number lines, called **coordinate axes**.

From Figure 1 (on the next page) we can see the following properties about the Cartesian coordinate system.

1. The positive directions are up and to the right.
2. The negative directions are down and to the left.
3. The horizontal axis is called the **x-axis**.
4. The vertical axis is called the **y-axis**.
5. The plane is divided into four **quadrants**, labeled I, II, III, and IV (counterclockwise) as in Figure 1.
6. The point where the axes cross is called the **origin**. Its coordinates are $(0, 0)$.

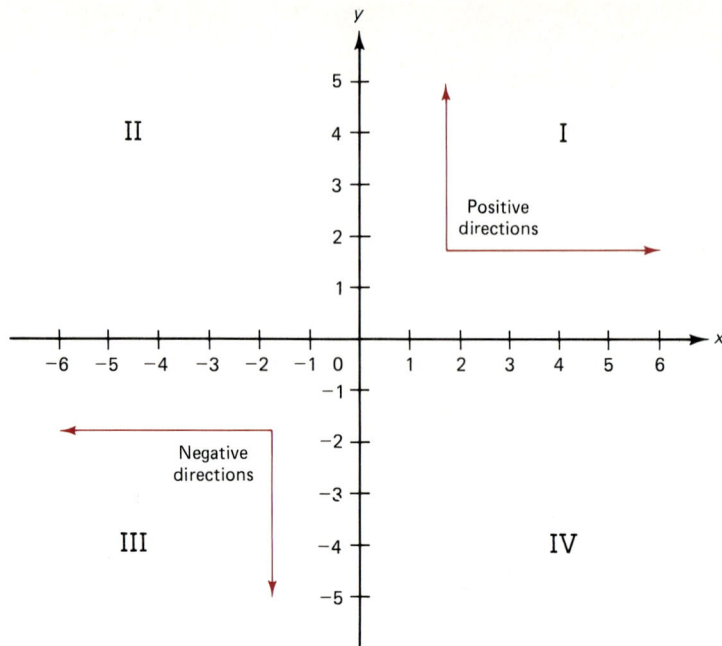

FIGURE 1

We note that the *x*- and *y*-axes need not be scaled the same; however, unless otherwise indicated, the axes are understood to be 1, 2, 3, and so on.

For the ordered pair (x, y):

1. *x* means horizontal movement from the origin (right for positive, left for negative). This is called the **x-coordinate**.
2. *y* means vertical movement from the origin (up for positive, down for negative). This is called the **y-coordinate**.

EXAMPLE 1   Locate the following points on the coordinate plane.

$$A = (5, 2) \qquad B = (2, 5) \qquad C = (-3, 1)$$
$$D = (1, -4) \qquad E = (0, -1) \qquad F = (-2, -6\tfrac{1}{2})$$

Solution   We first determine what each pair means:

$$A = (5, 2) \text{ means 5 to the right, 2 up.}$$

$$B = (2, 5) \text{ means 2 to the right, 5 up.}$$

$$C = (-3, 1) \text{ means 3 to the left, 1 up.}$$

$D = (1, -4)$ means 1 to the right, 4 down.

$E = (0, -1)$ means no units to the right or left, 1 down.

$F = (-2, -6\frac{1}{2})$ means 2 to the left, $6\frac{1}{2}$ down.

We plot these pairs in Figure 2. Note that $(5, 2)$ is *not* the same as $(2, 5)$. That is why (x, y) is called an *ordered pair*: The order *is* important.

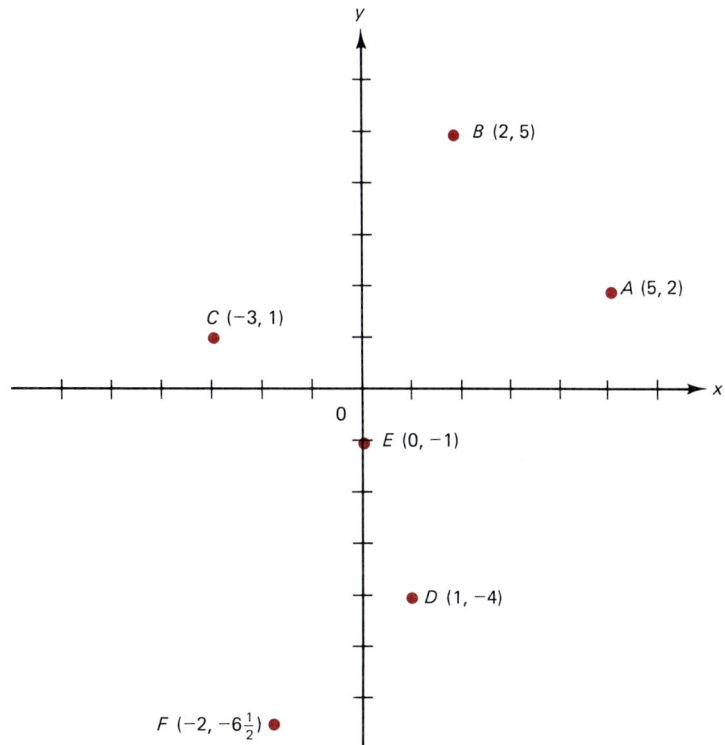

FIGURE 2

EXAMPLE 2 Give the coordinates of the points in Figure 3 (page 112).

Solution We start at the origin and "walk" to the points (horizontally first, vertically second).

G is 3 to the right, 1 up. Thus, $G = (3, 1)$.

H is 0 to the right or left, 4 up. Thus, $H = (0, 4)$.

I is 5 to the left, 1 up. Thus, $I = (-5, 1)$.

J is 2 to the left, 0 up or down. Thus, $J = (-2, 0)$.

K is 3 to the left, $3\frac{1}{3}$ down. Thus, $K = (-3, -3\frac{1}{3})$.

L is 3.4 to the right, 2 down. Thus, $L = (3.4, -2)$.

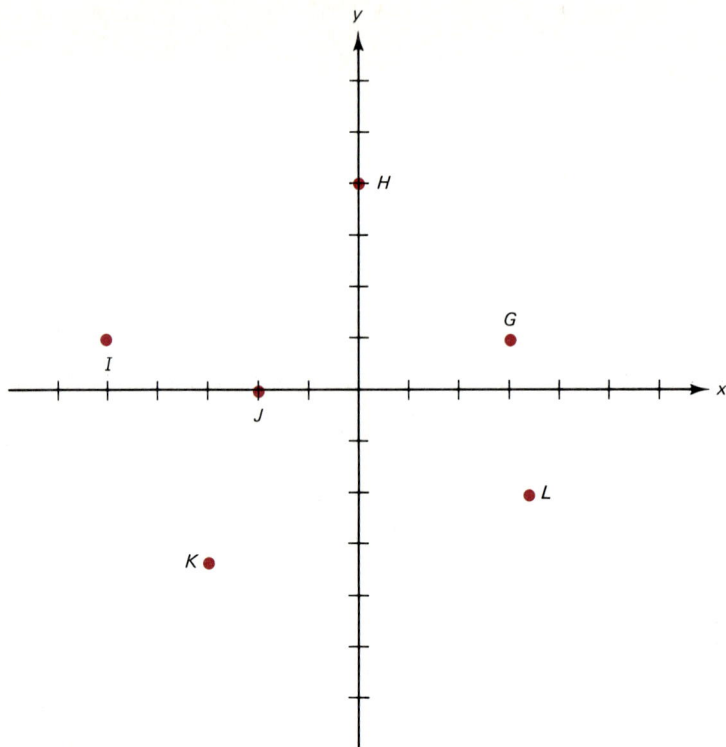

FIGURE 3

Distance Formula

The distance between two points on the number line is given by $d = |b - a|$, where b and a are the points' coordinates. We now wish to find the **distance** between two points in the plane.

To find the distance between two points, we use the **Pythagorean theorem**: For right triangles with legs a and b and hypotenuse c, we have $c^2 = a^2 + b^2$, or $c = \sqrt{a^2 + b^2}$ (see Figure 4).

FIGURE 4

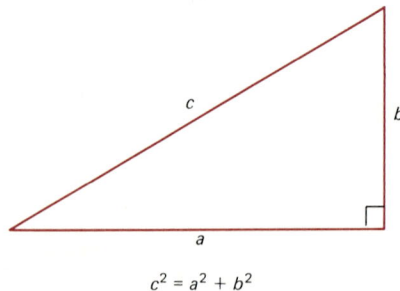

$$c^2 = a^2 + b^2$$

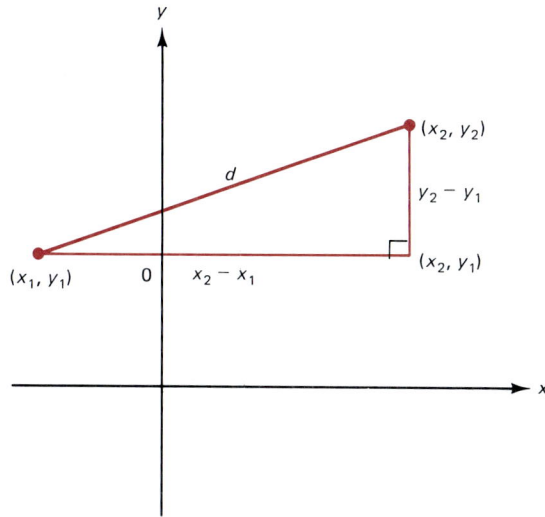

FIGURE 5

Suppose now that we have two points (x_1, y_1) and (x_2, y_2) (see Figure 5). [It does not matter which point is labeled "1" and which "2"; however, we generally call the leftmost point (x_1, y_1).] Comparing the two right triangles (Figures 4 and 5), we let $a = x_2 - x_1$ (the change in the x-coordinate), $b = y_2 - y_1$ (the change in the y-coordinate), and $c = d$ (the distance between the points).

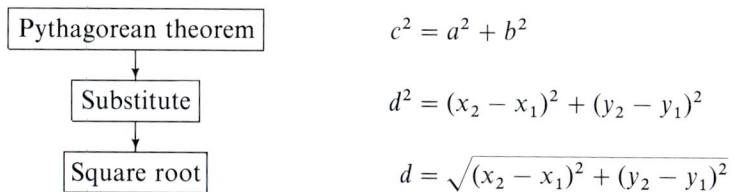

Pythagorean theorem	$c^2 = a^2 + b^2$
Substitute	$d^2 = (x_2 - x_1)^2 + (y_2 - y_1)^2$
Square root	$d = \sqrt{(x_2 - x_1)^2 + (y_2 - y_1)^2}$

This is our **distance formula**.

Theorem 1

The distance d between any two points (x_1, y_1) and (x_2, y_2) in the plane is given by

$$d = \sqrt{(x_2 - x_1)^2 + (y_2 - y_1)^2}$$

EXAMPLE 3 Find the distance between $(-2, 1)$ and $(5, 4)$.

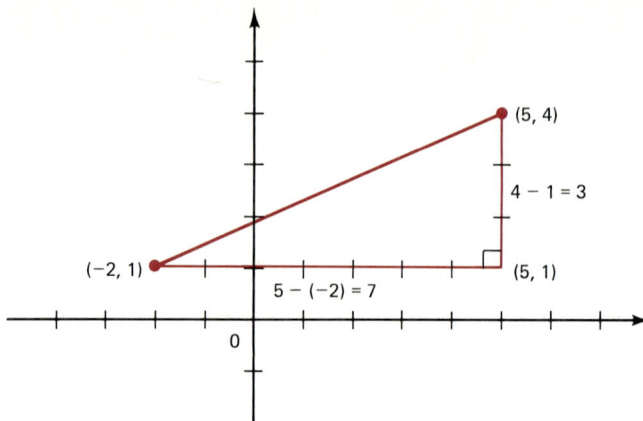

FIGURE 6

Solution Figure 6 shows these points and the right triangle they form. To use the distance formula, we let $(x_1, y_1) = (-2, 1)$ and $(x_2, y_2) = (5, 4)$.

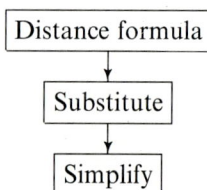

Distance formula	$d = \sqrt{(x_2 - x_1)^2 + (y_2 - y_1)^2}$
Substitute	$= \sqrt{[5 - (-2)]^2 + (4 - 1)^2}$
Simplify	$= \sqrt{7^2 + 3^2} = \sqrt{49 + 9} = \sqrt{58}$

Thus, the distance is $\sqrt{58}$.

EXAMPLE 4   Find the distance between $(-3, -4)$ and $(-1, 7)$.

Solution   We substitute into the distance formula and *subtract carefully*.

$$d = \sqrt{[-1 - (-3)]^2 + [7 - (-4)]^2}$$
$$= \sqrt{2^2 + 11^2} = \sqrt{125} = 5\sqrt{5}$$

Midpoint Formula

What is the midpoint M between the points 5 and 9 on the number line?

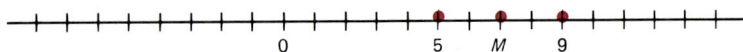

We see that the midpoint is just the average, $M = \dfrac{5 + 9}{2} = 7$.

Suppose now that we have two points (x_1, y_1) and (x_2, y_2) in the plane. The **midpoint** is given by the average of the coordinates, as follows.

Theorem 2

The midpoint M between (x_1, y_1) and (x_2, y_2) is given by

$$M = \left(\frac{x_1 + x_2}{2}, \frac{y_1 + y_2}{2}\right)$$

EXAMPLE 5   Find the midpoint of the points $(7, 1)$ and $(-5, 3)$, as shown in Figure 7.

Solution   To find the midpoint, we average the x-coordinates and average the y-coordinates.

| Midpoint formula | $M = \left(\dfrac{x_1 + x_2}{2}, \dfrac{y_1 + y_2}{2}\right)$ |

$$\downarrow$$

| Substitute | $= \left(\dfrac{7 + (-5)}{2}, \dfrac{1 + 3}{2}\right)$ |

$$\downarrow$$

| Simplify | $= (1, 2)$ |

Thus, the midpoint is $(1, 2)$, as seen in Figure 7.

FIGURE 7

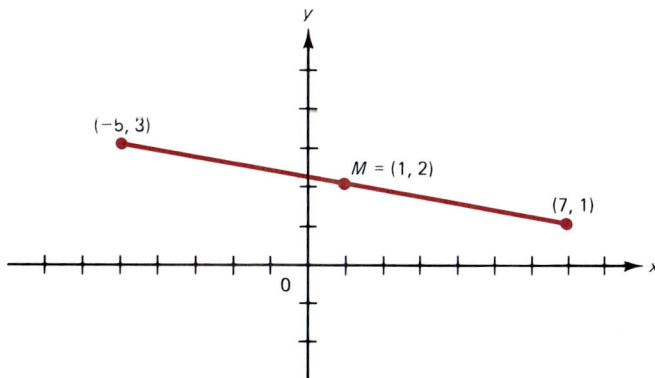

Identify the coordinates of the points on the following number line.

1. *A*   **2.** *B*   **3.** *C*

4. *D*   **5.** *E*   **6.** *F*

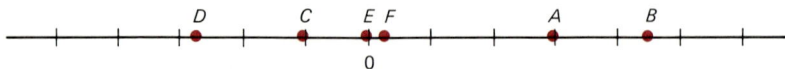

Simplify the following expressions.

7. $-2 - 3$ **8.** $2 - (-8)$

9. $[4 - (-7)]^2 + (-2 - 5)^2$ **10.** $\sqrt{(-2 + 9)^2 + (1 - 6)^2}$

Locate the following points on a rectangular coordinate system.

11. $A = (6, 2)$ **12.** $B = (3, 7)$ **13.** $C = (0, 5\frac{1}{2})$

14. $D = (-1, 5)$ **15.** $E = (-3, 1)$ **16.** $F = (-4, 0)$

17. $G = (-5, -1)$ **18.** $H = (-2, -3\frac{1}{2})$ **19.** $I = (0, -2)$

20. $J = (4, -4)$ **21.** $K = (2\frac{1}{4}, -6\frac{1}{2})$ **22.** $L = (4, 0)$

Identify the coordinates of the points shown in Figure 8.

23. *A*   **24.** *B*   **25.** *C*

26. *D*   **27.** *E*   **28.** *F*

29. *G*   **30.** *H*   **31.** *I*

32. *J*   **33.** *K*   **34.** *L*

FIGURE 8

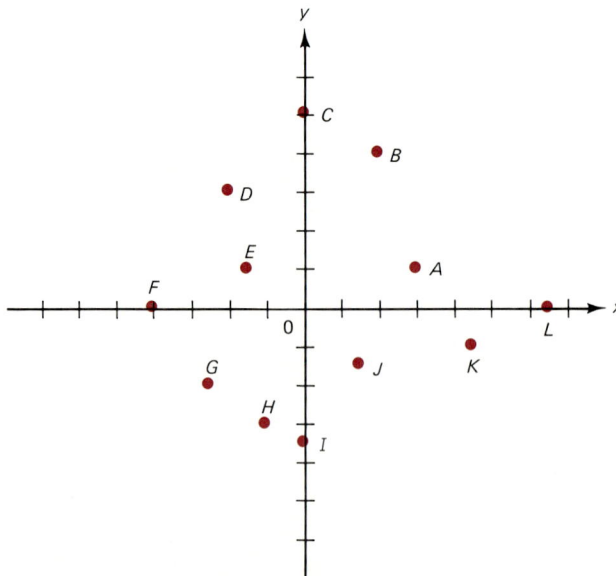

For each of the following pairs of points, find:
(a) The distance between the points.
(b) The midpoint between the points.

35. (1, 2) and (9, 8)   **36.** (2, 5) and (4, 7)
37. (0, 4) and (5, 6)   **38.** (2, 0) and (5, 10)
39. (−3, 1) and (1, 7)   **40.** (−2, −6) and (8, 4)
41. (−2, 4) and (4, −6)   **42.** (−1, −9) and (3, −5)

For each of the following pairs of points, as shown in Figure 9:
(a) Give the coordinates of each point.
(b) Find the distance between the points.
(c) Find the midpoint between the points.

43. *A* and *B*   **44.** *C* and *D*   **45.** *D* and *E*
46. *A* and *D*   **47.** *B* and *C*   **48.** *A* and *C*

For each of the following equations:
(a) Find the missing coordinate in each of the ordered pairs. (Recall that x is the first coordinate and y is the second.)
(b) Graph the ordered pairs.
(c) Draw a smooth curve or a line through the points.

49. $x + y = 5$ (3,); (0,); (, 0); (, 1)
50. $x - y = 2$ (5,); (3,): (, 0); (, 4)
51. $y = x^2$ (0,); (1,); (2,); (3,)
52. $y = \sqrt{x}$ (0,); (1,); (4,); (9,)

FIGURE 9

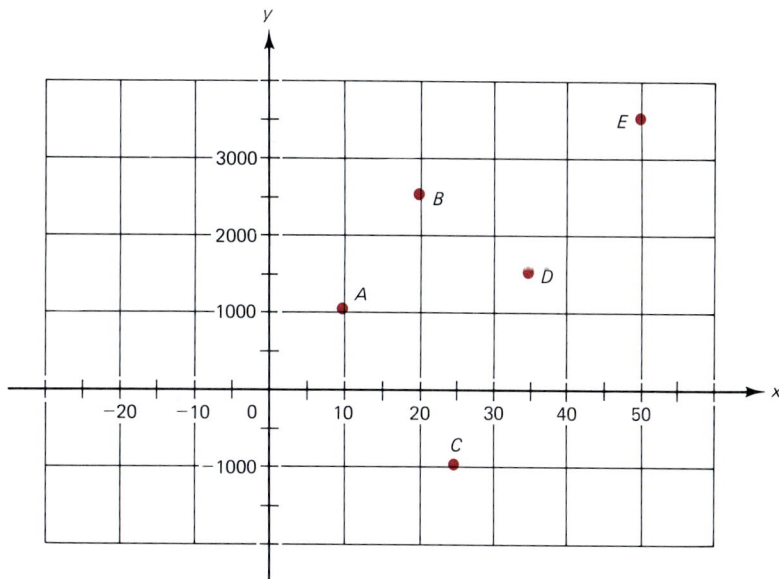

53. The following table (from an insurance mortality table) shows how many of 1,000,000 births are still alive at various ages through 100. Graph these data.

x (age)	y (number living)
0	1,000,000
10	971,804
20	951,483
30	924,609
40	883,342
50	810,900
60	677,771
70	454,548
80	181,765
90	21,577
100	0

(It is a mathematical convenience that zero people are assumed to be alive at age 100. This is, of course, not really accurate.)

54. Shown is a segment of an electrocardiogram of a heart beating. Each little horizontal division denotes 0.04 second, and each little vertical division 0.1 millivolt. A normal heartbeat is characterized by five waves: P, Q, R, S, and T.
(a) Find the coordinates of the five peaks shown (P, Q, R, S, and T).
(b) How much time occurs between the first R peak and the next R peak?
(c) Compute the heartbeat rate: 60 ÷ (time between peaks).

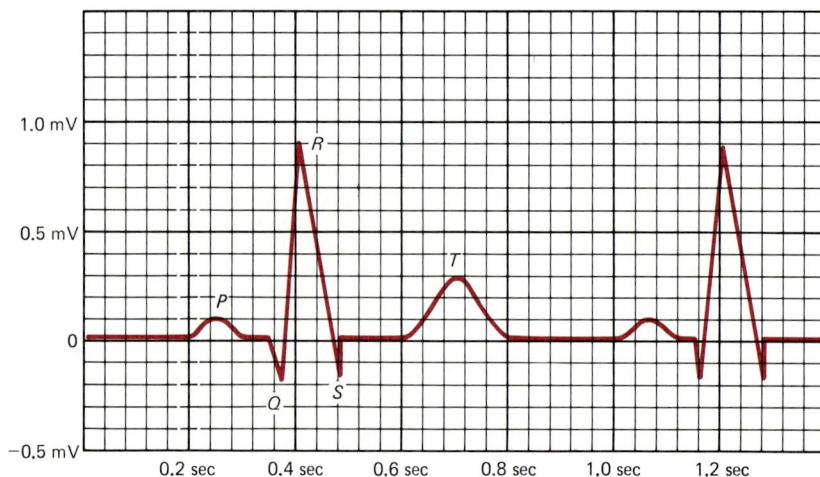

55. The Manik-Deprez Yo-Yo Company has had its ups and downs lately, as seen in the accompanying graph of its sales (on the next page). Give the coordinates of points *A, B, C, D,* and *E.*

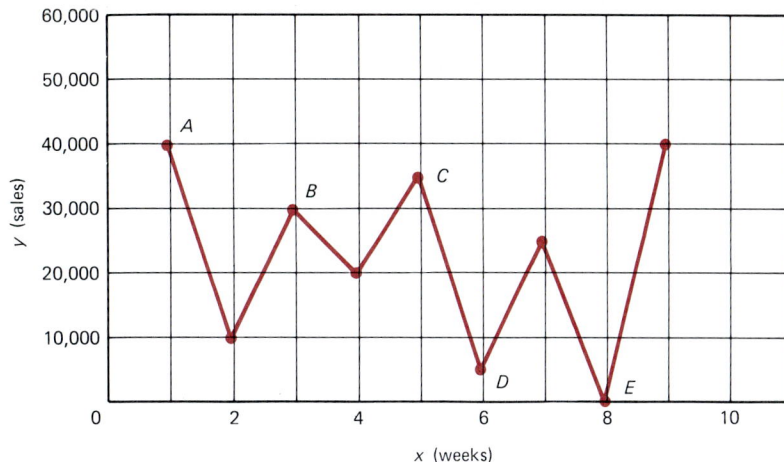

4.2 GRAPHING EQUATIONS

The truth of equations such as $x + y = 7$ depends on both x and y. The solutions to such equations in two variables are **ordered pairs**. As discussed in the preceding section, we picture ordered pairs on a Cartesian coordinate system. We call this illustration of the equation its **graph**.

In graphing an equation, we cannot plot all the solutions (since there are too many); rather, we try to plot enough sample points so that we can fill in the sketch.

Let us restate and emphasize the following.

The graph of an equation with two variables (usually x and y) is an illustration (in the plane) of the set of *all* ordered pairs and that satisfy the equation.

EXAMPLE 6   Graph the equation $x + y = 7$.

Solution   This is an easy equation to start with since we can easily see many solutions. We make an x-y table for the solutions.

Make a table of (x, y) pairs

x	y
0	7
1	6
3	4
6	1
8	−1

Plot points

Draw a curve through points

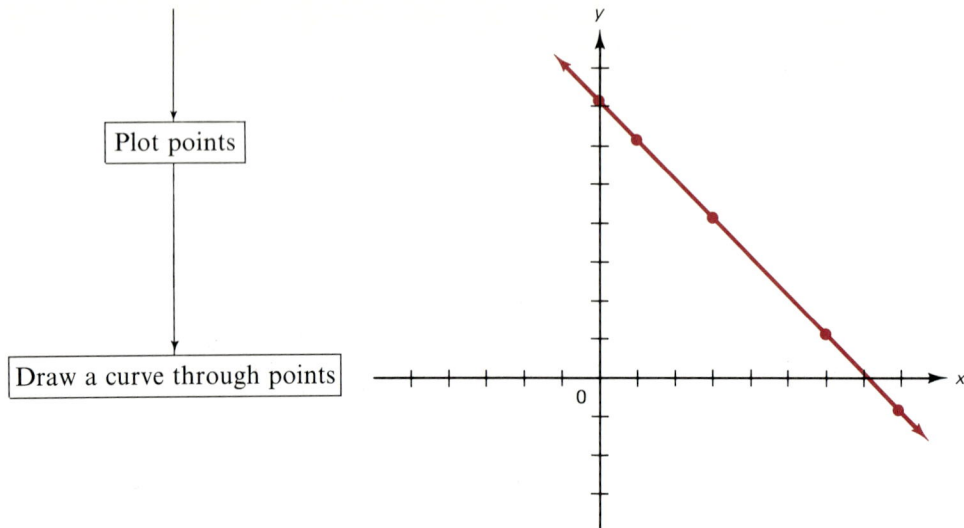

Not all equations are this simple to graph: The solutions may be harder to find, the axes may have to be scaled to fit the points, and the graph may not be a straight line. In general, here is a standard procedure that we use.

To graph an equation with two variables:

1. Find several solutions to the equation. (An x-y table is useful in displaying these solutions.)
2. Scale the x- and y-axes to fit all the data.
3. Plot the points.
4. Draw a smooth curve through the points.

EXAMPLE 7   Graph $y = x^2 + 2$.

Solution   We substitute some x-values (positive and negative) and find the y-values.

Make a table of (x, y) pairs

x	y
-4	18
-2	6
0	2
1	3
3	11
5	27

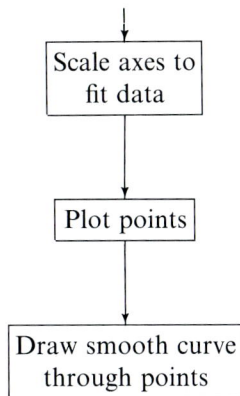

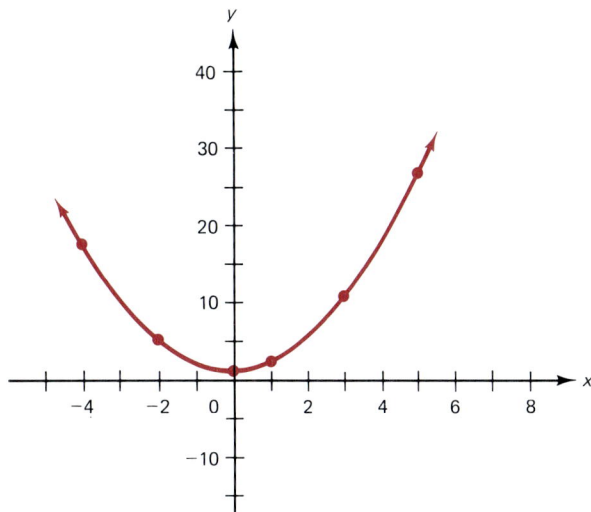

Note that the x- and y-axes are scaled differently since the range of the y-values is much larger than the range of the x-values.

EXAMPLE 8   Graph $x = y^2$.

Solution   Here it is more convenient to pick the y-values first and then compute x.

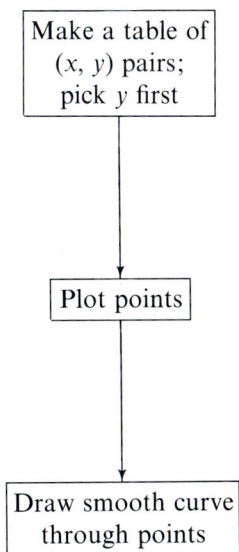

x	y
9	-3
4	-2
1	-1
0	0
1	1
4	2
9	3

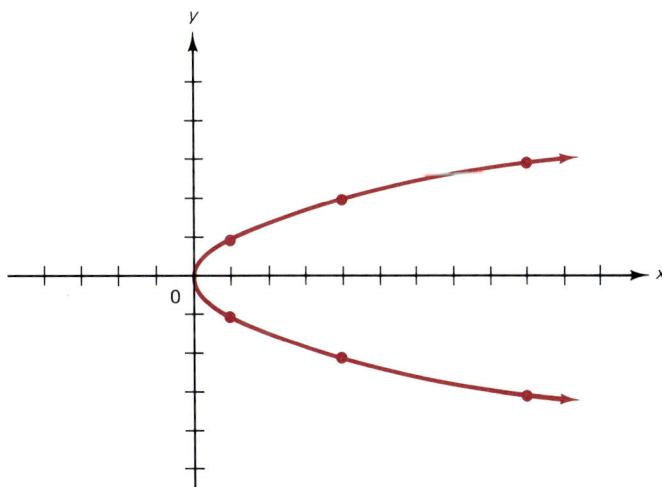

EXAMPLE 9   Graph $y = x^3 - 5x$.

Solution   Here we pick the x-values first since y is given in terms of x. Because this curve has several turns, we find some noninteger x-values. (In general, the higher the degree of the polynomial, the more turns the graph has.) We use a hand calculator to help.

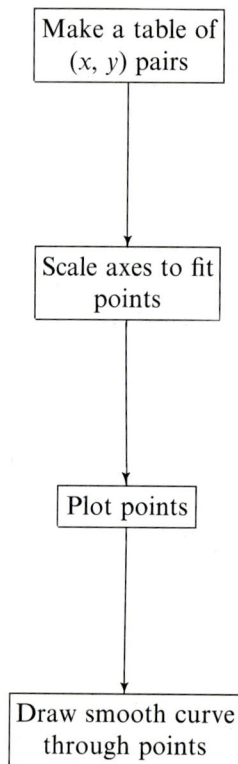

x	y
-3	-12.000
-2.5	-3.125
-2	2.000
-1.5	4.125
-1	4.000
0	0.000
1	-4.000
1.5	-4.125
2	-2.000
2.5	3.125
3	12.000

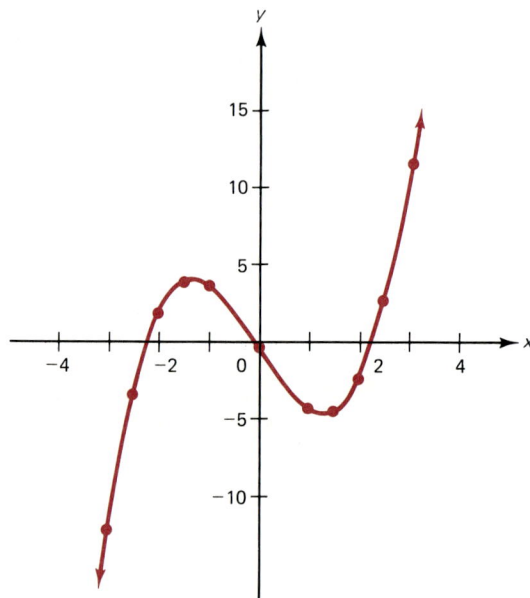

Make a table of (x, y) pairs

Scale axes to fit points

Plot points

Draw smooth curve through points

EXAMPLE 10   Graph $y = \sqrt{x + 1}$.

Solution We must be careful when we pick our x-values. Recall that we cannot have a negative number under the radical. Any x that produces a negative under the radical must be discarded.

x	y
-3	—
-2	—
-1	$\sqrt{0} = 0$
0	$\sqrt{1} = 1$
1	$\sqrt{2} \approx 1.41$
2	$\sqrt{3} \approx 1.73$
3	$\sqrt{4} = 2$
4	$\sqrt{5} \approx 2.24$

Make table of (x, y) pairs; discard x-values that produce negative radicands

↓

Plot points

↓

Draw smooth curve through points

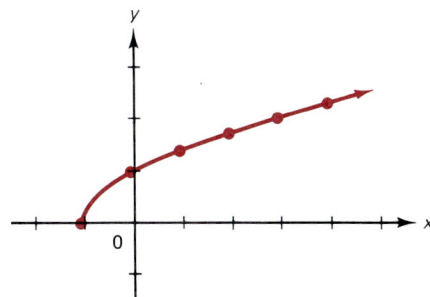

Some graphs have a **symmetry**; that is, one half of the graph looks like a mirror image of the other half. We have three major types of symmetry (see Table 1 on the next page).

1. **Symmetry about the *x*-axis.** Here the x-axis acts as a mirror. (See Example 8.)
2. **Symmetry about the *y*-axis.** Here the y-axis acts as a mirror. (See Example 7.)
3. **Symmetry about the origin.** Here the points occur in pairs on opposite sides of the origin. (See Example 9.)

Note: A graph may not have any of these symmetries. (See Example 10.)

TABLE 1

Symmetry Type	Test	Sketch
x-axis	Both (x, y) and $(x, -y)$ are on the graph.	
y-axis	Both (x, y) and $(-x, y)$ are on the graph.	
Origin	Both (x, y) and $(-x, -y)$ are on the graph.	

PROBLEM SET 4.2

Locate the following in the coordinate plane.

Warm-up Exercises

1. $(0, 3)$ **2.** $(2, 3)$ **3.** $(5, 0)$

4. $(-1, 6)$ **5.** $(0, -4)$ **6.** $(-2, -5)$

Evaluate the following.

Warm-up Exercises

7. $3^2 - 2(3) + 1$ **8.** $2(-1)^2 + 3(-1) - 5$

9. $\sqrt{2^3 + 1}$ **10.** $\sqrt[3]{3^2 - 1}$

For the following equations, find the missing coordinate in each of the ordered pairs.

Warm-up Exercises

11. $2x + y = 12$ $(0, \ )$; $(\ , 0)$; $(4, \ )$; $(\ , 2)$

12. $y = x^2 + x$ $(0, \ )$; $(1, \ )$; $(2, \ )$; $(-1, \ )$

Graph the following equations and identify any symmetries.

13. $x - y = 3$ **14.** $y - x = 3$

15. $2x - y = 8$ **16.** $x + 3y = 6$

17. $y = 2x + 3$ **18.** $y = 5 - x$

19. $x = 6 - y$

20. $x = 8 - 2y$

21. $y = x^2 + x + 1$

22. $y = 2x^2 - x - 2$

23. $x = y^2 + 3$

24. $x = 4 - y^2$

25. $y = x^3$

26. $y = x^3 + 1$

27. $y = x^3 + x^2$

28. $y = x^4 + x^3$

29. $y = \sqrt[3]{3 + x}$

30. $y = \sqrt[3]{4 - x}$

31. $x = \sqrt{y}$

32. $x = \sqrt[3]{y}$

33. $x = \sqrt{y + 5}$

34. $x = \sqrt[3]{1 + y}$

35. $y = 2x + 5$

36. $x = \sqrt{y^2 + y}$

37. $2x + y = 10$

38. $y = x^2 - x$

Physical Application

39. The period T of a small swing of a pendulum depends only on the length L, as follows:

$$T = 2\pi \sqrt{\frac{L}{32}}$$

Graph this equation with L on the x-axis and $L = 2, 4, 6, 8,$ and 10 feet.

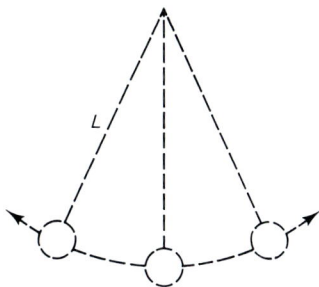

Population Application

40. The population P of a certain town is related to the time t (years after 1986) by the formula

$$P = 20{,}000 + 800t$$

Graph this equation with t on the x-axis and $t = 0, 2, 4, 6, 8,$ and 10 years.

Business Application

41. The supply S and demand D for a certain car wax are given by

$$S = 20{,}000p - 15{,}000 \qquad D = \frac{100{,}000}{p}$$

where p is the price. Graph both of the equations on the same plane with p on the x-axis (S and D each on the y-axis) and $p = 1, 2, 3, 4,$ and 5 dollars.

Health Application

42. The amount of sugar S in the blood t hours after eating a certain food is given by $S = 4 + 0.5t - 0.1t^2$. Graph this equation with t on the x-axis and $t = 0, 1, 2, 3, 4,$ and 5 hours.

4.3
THE STRAIGHT LINE

In the preceding section we saw that different equations give different graphs. In this section we focus on the **straight line**. It can be shown that the equation of the straight line (called a **linear equation**) has the following **general form**:

$$\boxed{Ax + By = C}$$

where A and B are constants not both zero.

Linear equations in standard form can easily be graphed by finding the intercepts:

1. The **x-intercept** is the x-value where the line crosses the x-axis (where $y = 0$).
2. The **y-intercept** is the y-value where the line crosses the y-axis (where $x = 0$).

EXAMPLE 11   Graph $2x - 5y = 10$.

Solution   We first find the intercepts.

Set $y = 0$; solve for x

$$2x - 5(0) = 10$$
$$2x = 10$$
$$x = 5$$

Set $x = 0$; solve for y

$$2(0) - 5y = 10$$
$$-5y = 10$$
$$y = -2$$

Now we can graph the line. [We need only two points to graph a line; however, a third point would be a good check. For instance, $(2.5, -1)$ also satisfies the equation.]

Find intercepts and third point

x	y
5	0
0	-2
2.5	-1

Plot points

Draw line through points

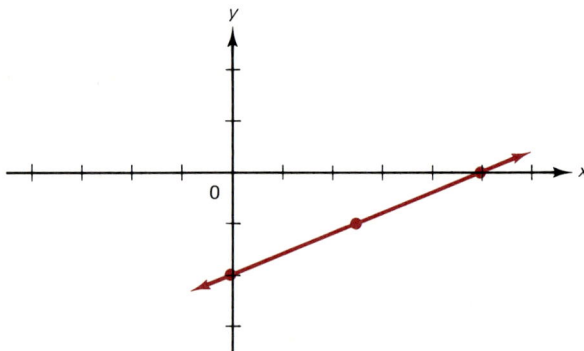

EXAMPLE 12   Graph $5x + 30y = 7500$.

Solution   We first find the intercepts. Here we have to scale the axes carefully to fit the intercepts.

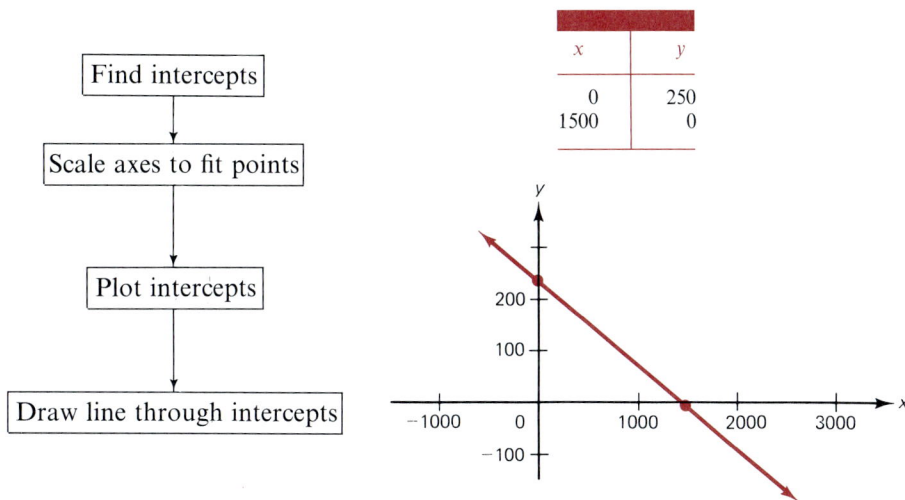

x	y
0	250
1500	0

Find intercepts

Scale axes to fit points

Plot intercepts

Draw line through intercepts

EXAMPLE 13   Graph the lines:
(a) $y = 2$
(b) $x = -3$

Solution   (a) The equation $y = 2$ must seem incomplete without an x-term. Here this means that y is *always* 2, no matter what x is. Thus, $(1, 2)$, $(-4, 2)$, $(5, 2)$, and so on, are all solutions (the second number of the pair is always 2). As we see in Figure 10, this is a horizontal line. In fact, all lines of the form $y = k$ are horizontal.
(b) Similarly, $x = -3$ means that x is always -3, no matter what y is. As we see in Figure 10, this produces a vertical line. In fact, all lines of the form $x = k$ are vertical.

FIGURE 10

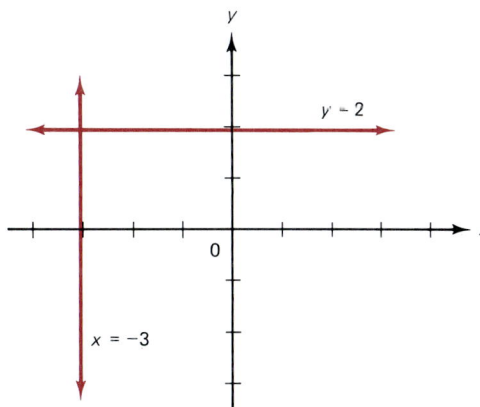

A linear equation represents a relation between two variables. Often we wish to know how rapidly the y-value is increasing (or decreasing) as the x-value increases. On the graph this corresponds to the steepness, or slope, of the line.

Definition 1
The **slope** m of the line through points (x_1, y_1) and (x_2, y_2) is given by

$$m = \frac{\text{rise}}{\text{run}} = \frac{\text{change in } y}{\text{change in } x} = \frac{y_2 - y_1}{x_2 - x_1}$$

EXAMPLE 14 Find the slope of the lines shown in Figure 11:
(a) Through $(1, -2)$ and $(5, 4)$.
(b) Through $(-3, 4)$ and $(-1, -2)$.

Solution For each line we divide the **rise** (difference in y-coordinates) by the **run** (difference in x-coordinates).

(a) Line 1: $m = \dfrac{\text{change in } y}{\text{change in } x} = \dfrac{4 - (-2)}{5 - 1} = \dfrac{6}{4} = \dfrac{3}{2}$

(b) Line 2: $m = \dfrac{\text{change in } y}{\text{change in } x} = \dfrac{-2 - 4}{-1 - (-3)} = \dfrac{-6}{2} = -3$

FIGURE 11

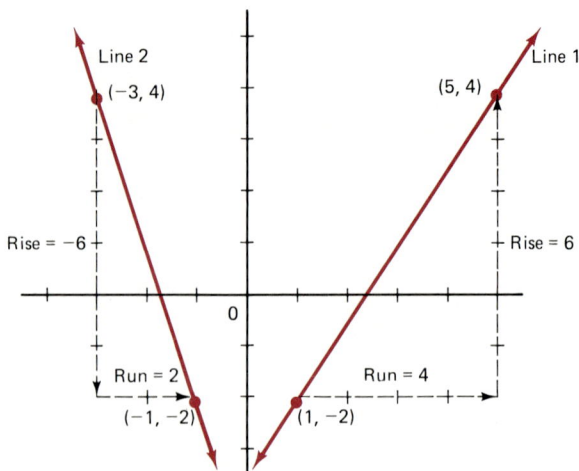

The following table summarizes how slopes and lines are related. (Note that a vertical line has *no* slope.)

Line	Slope	Sketch
Rising (left to right)	Positive	$m > 0$
Falling (left to right)	Negative	$m < 0$
Horizontal	Zero	$m = 0$
Vertical	Undefined	m is undefined

We can use the slopes of two lines to determine if they are parallel or perpendicular.

1. Two lines are **parallel** if and only if their slopes are equal: $m_1 = m_2$.
2. Two lines are **perpendicular** if and only if their slopes are the negative reciprocals of each other: $m_2 = -1/m_1$ (or $m_1 m_2 = -1$).

EXAMPLE 15 Figure 12 illustrates these relations. In Figure 12(a) the parallel lines all have the same slope: 3/2. In Figure 12(b) the perpendiculars have slopes that are negative reciprocals of each other: 7/4 and $-4/7$.

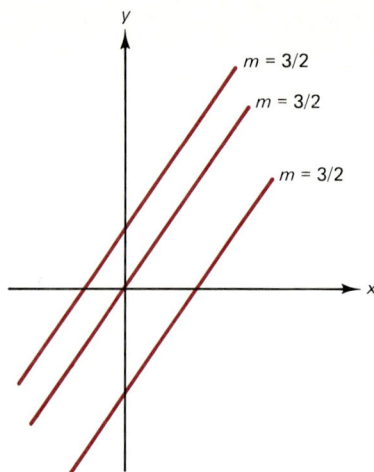

(a) Parallel lines
(b) Perpendicular lines

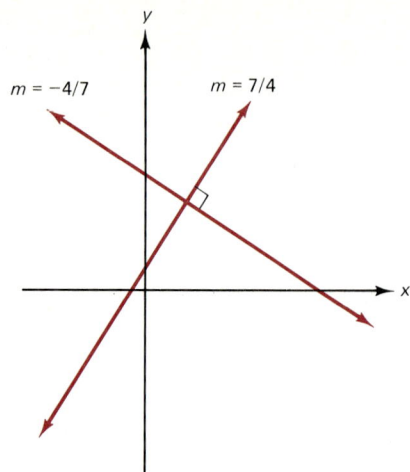

FIGURE 12

Evaluate the following.

Warm-up
Exercises

1. $\dfrac{12 - 4}{4 - 2}$

2. $\dfrac{10 - 2}{2 - 4}$

3. $\dfrac{-2 - 10}{-1 - (-3)}$

4. $\dfrac{-5 - (-10)}{1 - (-4)}$

Solve the following equations.

Warm-up
Exercises

5. $2x + 7 = 3$

6. $-10 + 7y = 25$

7. $\dfrac{7}{6} m = 1$

8. $\dfrac{-4}{5} m = 1$

Graph the following linear equations.

9. $x + y = 5$

10. $x - y = 3$

11. $2x + y = 8$

12. $x - 5y = 10$

13. $2x - 7y = 14$

14. $3x + 4y = 12$

15. $4x + 5y = 20$

16. $7y - 5x = 35$

17. $\dfrac{x}{4} - \dfrac{y}{5} = 1$

18. $\dfrac{x}{6} + \dfrac{y}{8} = 1$

19. $x = 5$

20. $y = 4$

21. $y = -1$

22. $x = -2$

23. $x = 0$

24. $y = 0$

Find the slope of the line through each of the following pairs of points.

25. $(1, 3)$ and $(3, 9)$

26. $(2, 5)$ and $(3, 11)$

27. $(-1, 3)$ and $(2, -7)$

28. $(-2, -7)$ and $(5, -2)$

29. $(2, -3)$ and $(6, -3)$

30. $(-3, 8)$ and $(1, -2)$

31. $(5, -2)$ and $(5, 6)$ **32.** $(4, -2)$ and $(5, -7)$

33. $(4.5, 12)$ and $(8.2, -2.3)$ **34.** $(-2.01, 8.2)$ and $(1.9, -45.1)$

35. $(10, 2000)$ and $(20, 4500)$ **36.** $(500, 0.01)$ and $(700, 0.02)$

For each of the following slopes, determine the slopes of the lines that are parallel and perpendicular.

37. $m = 2$ **38.** $m = -5$ **39.** $m = \dfrac{1}{3}$

40. $m = \dfrac{3}{5}$ **41.** $m = \dfrac{-9}{4}$ **42.** $m = \dfrac{-1}{5}$

For each set of four points in the following problems, determine if any pair of lines through A, B, C, and D are parallel or perpendicular. (Hint: Sketch the points first.)

43. $A = (3, 9)$; $B = (6, 1)$; $C = (8, 0)$; $D = (16, 3)$

44. $A = (2, 8)$; $B = (-6, 3)$; $C = (4, -2)$; $D = (-8, -3)$

45. $A = (6, -4)$: $B = (0, 3)$; $C = (1, -2)$; $D = (-7, 5)$

46. $A = (1, 5)$; $B = (3, 7)$; $C = (6, 2)$; $D = (9, 5)$

Complete the following table.

	Change in x	Change in y	Slope, m
47.	-30	3	?
48.	18	-9	?
49.	$-3 - 7$	$5 - 3$	?
50.	$-1 - (-14)$	$1 - 6$	?
51.	2	?	-5
52.	-3	?	4
53.	?	8	4
54.	?	-10	5

Engineering Application

55. The coefficient of expansibility α for a 1-meter bar is

$$\alpha = \frac{L_2 - L_1}{T_2 - T_1}$$

where L is the length and T is the temperature (°C). Find α for a hard-rubber bar having the following (T, L) data: $(0°, 1.0000)$ and $(70°, 1.0056)$.

Business Application

56. The *marginal cost* (average cost to produce one extra item) is given as the slope

$$MC = \frac{C_2 - C_1}{x_2 - x_1}$$

where C is the cost to produce x items. Find the MC for the following (x, C) pairs:

(a) $(150, 5000)$ and $(250, 8000)$

(b) $(2000, 400)$ and $(3000, 500)$

Health Application

57. A certain patient needs 64 grams of protein. She can get this by satisfying $8m + 2b = 64$, where m is glasses of milk and b is slices of bread. Graph this equation with m on the x-axis.

58. The average velocity of a moving object is given by

$$v = \frac{s_2 - s_1}{t_2 - t_1}$$

where s is the distance the object travels in time t. Calculate the average velocity for the following (t, s) pairs.
(a) (5 seconds, 100 meters) and (15 seconds, 150 meters)
(b) (2 hours, 105 miles) and (3.5 hours, 180 miles)
(c) (3 days, 700 kilometers) and (5 days, 950 kilometers)

59. Jack computes that in 1 year it costs him $1440 to drive 8000 miles and $1760 to drive 12,000. Find the slope through the points (8000, 1440) and (12,000, 1760). This slope is the cost per mile to drive the car.

4.4
THE EQUATION OF THE STRAIGHT LINE

In previous sections we were given an equation of a line and asked to find a few ordered pairs that satisfy the equation and then to graph it. Now we have the reverse problem: We are given a few specific facts about a line (such as the slope and a point or two points) and we want to find its equation. (The equation of a line tells us how each x- and y-coordinate pair is related.) Suppose that we are given the slope m and a point (x_1, y_1) of a line. Can we find an equation for the coordinates of the typical point (x, y)? Yes; see Figure 13. We start with the definition of the slope.

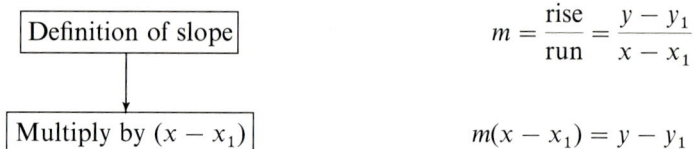

$$\boxed{\text{Definition of slope}} \qquad m = \frac{\text{rise}}{\text{run}} = \frac{y - y_1}{x - x_1}$$

$$\boxed{\text{Multiply by } (x - x_1)} \qquad m(x - x_1) = y - y_1$$

Theorem 3
An equation of the line with slope m through point (x_1, y_1) is given by

$$\boxed{y - y_1 = m(x - x_1)}$$

FIGURE 13

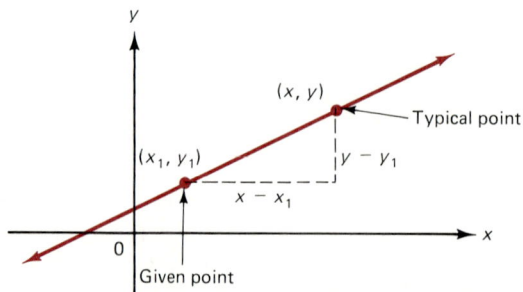

This form is called the **point-slope form**, since a point and slope are given. [Recall that a vertical line has no slope and has the form $x = k$. The vertical line through (x_1, y_1) is $x = x_1$.]

EXAMPLE 16   Find an equation of the line with slope 5 and through the point $(-3, 2)$.

Solution   We substitute these values into the point-slope form: $m = 5$, $x_1 = -3$, $y_1 = 2$. (Be careful with the subtraction.)

$$\boxed{y - y_1 = m(x - x_1)}$$

$$y - 2 = 5[x - (-3)]$$

$$\boxed{\text{Distributive law}}$$

$$y - 2 = 5x + 15$$

$$\boxed{\text{Add 2}}$$

$$y = 5x + 17$$

EXAMPLE 17   Find an equation of the line through $(2, 7)$ and $(4, 1)$.

Solution   We need a point and the slope. We have a point (two, in fact), but we must find the slope.

$$\boxed{\text{Find slope}}$$

$$m = \frac{1 - 7}{4 - 2} = \frac{-6}{2} = -3$$

$$\boxed{\text{Use slope} = -3; \text{point} = (2, 7)}$$

$$y - 7 = -3(x - 2)$$

$$\boxed{\text{Distributive law}}$$

$$y - 7 = -3x + 6$$

$$\boxed{\text{Add 7}}$$

$$y = -3x + 13$$

Here we used $(2, 7)$ as the point in the formula. Could we have used $(4, 1)$ instead? Yes, and we would get the same answer. (Try it.)

Note that this equation has a slightly different form than $Ax + By = C$. In general for nonvertical lines,

$$\boxed{y = mx + b}$$

is called the **slope-intercept form**. This is because m (the coefficient of x) is the slope and b is the y-intercept [since $(0, b)$ satisfies the equation]. See Figure 14 on the next page.

We can use the slope-intercept form, $y = mx + b$, to help find the slope and y-intercept of a line: m is the slope and b is the y-intercept.

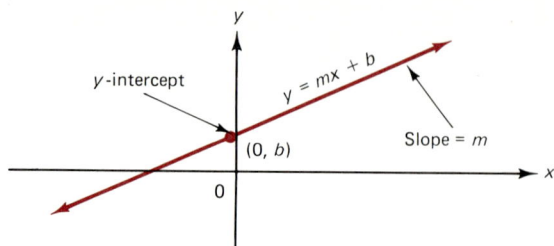

FIGURE 14

EXAMPLE 18   We can read the slope and y-intercept from the equation of a line (in slope-intercept form).
(a) For $y = 5x - 2$, the slope is 5; the y-intercept is -2.
(b) For $y = 4x$, the slope is 4; the y-intercept is 0.
(c) For $y = 8$, the slope is 0; the y-intercept is 8.
(d) For $2x + 3y = 7$, we must first solve for y:

$$2x + 3y = 7$$

$$3y = -2x + 7$$

$$y = \frac{-2}{3}x + \frac{7}{3}$$

Now we can see that the slope is $-2/3$, and the y-intercept is $7/3$.

EXAMPLE 19   Find an equation of the line through $(2, -5)$ and perpendicular to $2x - 3y = 8$.

Solution   Whenever we see the phrase "Find an equation of the line...," we should immediately look for a point and a slope. We have a point $(2, -5)$, so we need to look for the slope.
　　　　We know that the line that we want is perpendicular to $2x - 3y = 8$, so that the slope we want is the negative reciprocal of the slope of $2x - 3y = 8$.

$$2x - 3y = 8$$

Solve for y
to find slope

$$-3y = -2x + 8$$

$$y = \frac{2}{3}x - \frac{8}{3}$$

Thus, the slope of the given line is $2/3$. Therefore, the slope of the desired perpendicular is the negative reciprocal, $-3/2$. Now we can use the point-slope form, since we have a point and the slope.

$$\boxed{\begin{array}{c}\text{Use slope} = -3/2; \\ \text{point} = (2, -5)\end{array}} \qquad\qquad y - (-5) = \frac{-3}{2}(x - 2)$$

$$\downarrow$$

$$\boxed{\text{Simplify}} \qquad\qquad\qquad\qquad y + 5 = \frac{-3}{2}x + 3$$

$$\downarrow$$

$$\boxed{\text{Subtract 5}} \qquad\qquad\qquad\qquad y = \frac{-3}{2}x - 2$$

If we multiply both sides by 2 and rearrange terms, we get

$$3x + 2y = -4$$

PROBLEM SET
4.4

Graph the following equations.

1. $y = x + 2$ **2.** $y = 2x - 3$

3. $2x - y = 8$ **4.** $3x + 4y = 12$

For each of the following pairs of points:
(a) Find the slope of the line through them.
(b) Find the slope of the perpendicular to this line.

5. (2, 5) and (4, 11)   **6.** (2, 9) and (5, −1)

7. (5, −2) and (7, −8)   **8.** (−3, 7) and (2, −10)

For each of the following equations, identify the slope and y-intercept.

9. $y = 5x - 3$ **10.** $y = 6x + 7$

11. $y = \frac{2}{3}x - 3$ **12.** $y = \frac{-4}{5}x - \frac{5}{7}$

13. $2x + y = 5$ **14.** $4x - 3y = 7$

15. $3x + 5y + 1 = 0$ **16.** $8x - 3y - 11 = 0$

Find the equations of the lines with the following conditions.

17. Through (−2, −4) with slope 2.

18. Through (1, −5) with slope −3.

19. Through (2, −4) with slope $\frac{-1}{3}$.

20. Through (−2, −5) and (1, 10).

21. Through (1, −4) and (5, 11).

22. Through (−5, 10) and (−2, −6).

23. Perpendicular to $y = 3x - 1$, through (2, −1).

24. Perpendicular to $y = \frac{-2}{5}x + 1$, through (−2, 1).

25. Parallel to $3x - 5y = 7$, through (1, −3).

26. Perpendicular to $2x + 7y = 1$, through (−3, 5).

27. Through (1, −3) and (3, 5).

28. Parallel to $y = 2x + 3$, through $(-2, 1)$.

29. Through $(1, 8)$ with slope 2.

30. Perpendicular to $y = \dfrac{x}{4} + 5$, through $(1, 5)$.

31. Parallel to $y = 5x - 1$, through $(4, -3)$.

32. Through $(-1, 7)$ with slope 10.

33. Perpendicular to $y = -3x + 2$, through $(2, -5)$.

34. Parallel to $x - y = 9$, through $(1, -8)$.

35. With slope 2 and y-intercept 4.

36. With slope -3 and x-intercept 1.

37. With x-intercept 3 and y-intercept 4.

38. With x-intercept -7 and y-intercept 3.

39. Through $(-1, 2)$ and perpendicular to the line through $(1, 2)$ and $(3, 8)$.

40. Through $(0, 7)$ and parallel to the line through $(-1, 6)$ and $(3, 4)$.

41. Vertical, through $(2, -5)$.

42. Horizontal, through $(-3, -7)$.

Business Applications
43. A new machine is worth $80,000. Nine years later its salvage value is $3500. Find an equation of the line through $(0, \$80{,}000)$ and $(9, \$3500)$. This is called *straight-line depreciation*. What does the slope mean?

44. A company finds that it can rent 300 cars at $25 per day and 400 cars at $20 per day. Find an equation of the line through $(300, 25)$ and $(400, 20)$. This is called the *demand line*.

45. A manufacturer finds that it costs a fixed $200,000 plus $200 for each stereo set that it produces. Write this as an equation for the cost (y) in terms of the number of stereo sets produced (x).

Physical Application
46. In a certain experiment, at $0°C$ a gas had a volume of 300, and at $20°C$ it had a volume of 322. Find an equation of the line through $(0, 300)$ and $(20, 322)$. At what temperature is the volume zero? (This is called *absolute zero*.)

Consumer Application
47. A certain store charges $2.96 to develop and print 12 photographs, and $4.64 for 20. Find an equation of the line between $(12, 2.96)$ and $(20, 4.64)$. What does the slope mean? What does the y-intercept mean?

Health Application
48. The daily calorie needs of a person depend on his or her desired weight. Find the equations of the following lines through the given points (for people aged 18 to 35):
(a) Women: (110 pounds, 1850 calories) and (165 pounds, 2550 calories).
(b) Men: (110 pounds, 2200 calories) and (176 pounds, 3250 calories).

Mathematics Application
49. When using tables (such as square-root, logarithm, or trigonometry tables) we often need the value of a number between two given numbers. For instance, the square-root table gives the following square roots for 21 and 22:

x	$\sqrt{x}$
21	4.583
22	4.690

(a) Find an equation of the line through $(21, 4.583)$ and $(22, 4.690)$. This is the *linear interpolation* and approximates the square roots between 21 and 22.

(b) Use the equation to approximate $\sqrt{21.3}$ and check its accuracy on a hand calculator.

50. In Chapter 7 we will study logarithms. Consider the following portion of a logarithm table.

x	$\log x$
3.68	0.5658
3.69	0.5670

(a) Find an equation of the line through (3.68, 0.5658) and (3.69, 0.5670).
(b) Use this equation to approximate the value of log 3.687 and check its accuracy on a hand calculator (press $\boxed{3.687}$ $\boxed{\log}$).

CHAPTER 4 SUMMARY

Important Properties and Definitions

The *distance d* between (x_1, y_1) and (x_2, y_2) is given by

$$d = \sqrt{(x_2 - x_1)^2 + (y_2 - y_1)^2}$$

The *midpoint M* of points (x_1, y_1) and (x_2, y_2) is given by

$$M = \left(\frac{x_1 + x_2}{2}, \frac{y_1 + y_2}{2} \right)$$

The *slope m* of the line through (x_1, y_1) and (x_2, y_2) is given by

$$m = \frac{\text{rise}}{\text{run}} = \frac{y_2 - y_1}{x_2 - x_1}$$

Two lines are *parallel* if $m_1 = m_2$ (equal slopes).

Two lines are *perpendicular* if $m_1 = -1/m_2$ (negative reciprocals).

An equation of the line through (x_1, y_1) with slope m is given by

$$y - y_1 = m(x - x_1)$$

The general form of a straight line is

$$Ax + By = C \qquad (A \text{ and } B \text{ not both } 0)$$

For a nonvertical line written $y = mx + b$, m is the slope and b is the y-intercept.

Review Exercises

For Problems 1 to 5, let $A = (1, -3)$, $B = (3, 5)$, and $C = (-2, 7)$.

1. Plot the points A, B, and C.
2. Find the distance between A and B.

3. Find the distance between A and C.

4. Find the midpoint between B and C.

5. Find the midpoint between A and B.

Graph the following equations and identify any symmetries.

6. $x = y + 10$　　　　　　　　　**7.** $y = x^3 + 1$

8. $x = \sqrt{y - 2}$　　　　　　　　**9.** $y = -x^2$

10. $7x - 3y = 21$　　　　　　　　**11.** $x = -2$

For Problems 12 to 18, let $A = (1, -3)$, $B = (3, 5)$, and $C = (-2, 7)$.

12. Find the slope of the line through A and B.

13. Find the slope of the line through B and C.

14. Is the line through A and B perpendicular to the line through B and C?

15. Find the equation of the line through A with slope -3.

16. Find the equation of the line through C, perpendicular to the line through A and B.

17. Find the equation of the line through B, parallel to $2x - 3y = 5$.

18. Find the equation of the line through B and C.

19. Identify the slope and y-intercept of $4x + 5y = 7$.

Functions

5.1
FUNCTIONS

In science, health, industry, and business we frequently study how one quantity affects or predicts another. For example:

1. How does the *price* of an object predict its *sales*?
2. How does the *velocity* of an object affect its *energy*?
3. How does the *age* of a child determine his or her medicine *dosage*?
4. How does the *current* in a circuit affect its *power*?
5. How does the *diameter* of a beam predict the *load* it can support?

In answering these types of questions, we generally look for a rule or formula that can connect the first quantity to the second. Also, we expect the prediction to give *unique* answers; that is, we do not want a price to predict two different sales or a child's age to produce two different medicine dosages. We expect (and demand) one and only one answer.

At the gateway to almost all higher mathematics is the idea of the *function*. The function concept is so important that we now introduce it in several different ways.

> A **function** is a rule that assigns every number x in a set to exactly one number y in another set.

The set of numbers from which x is chosen is called the **domain**. The set of numbers in which y belongs is called the **range**.

EXAMPLE 1 The following are rules that assign each x of the domain to a number of the range.

Domain	Range	Domain	Range
⋮	⋮	⋮	⋮
-3 ⟶	9	-3 ⟶	-1
-2 ⟶	4	-2 ⟶	1
-1 ⟶	1	-1 ⟶	3
0 ⟶	0	0 ⟶	5
1 ⟶	1	1 ⟶	7
2 ⟶	4	2 ⟶	9
3 ⟶	9	3 ⟶	11
4 ⟶	16	4 ⟶	13
⋮	⋮	⋮	⋮
x ⟶	x^2	x ⟶	$2x + 5$

In the example on the left we square the first number to get the second. In the example on the right we double the first number and add 5 to get the second. In both cases we used a rule or formula.

YES	NO

Two different numbers can be assigned to the same number, but one number *cannot* be assigned to two different numbers.

At a more intuitive level we can view a function as a machine that transforms *x*-values from the domain into *y*-values in the range. Figure 1 illustrates this idea. The machine on the lower left is the "square machine," which takes 10 in and puts out 100. The machine on the lower right is the "double-and-add-5 machine," which takes 6 in and puts out 17.

Our last approach to functions is somewhat more formal, but very important for graphing and more advanced ideas.

A **function** is a set of ordered pairs (x, y) such that each *x*-value is paired with only one *y*-value.

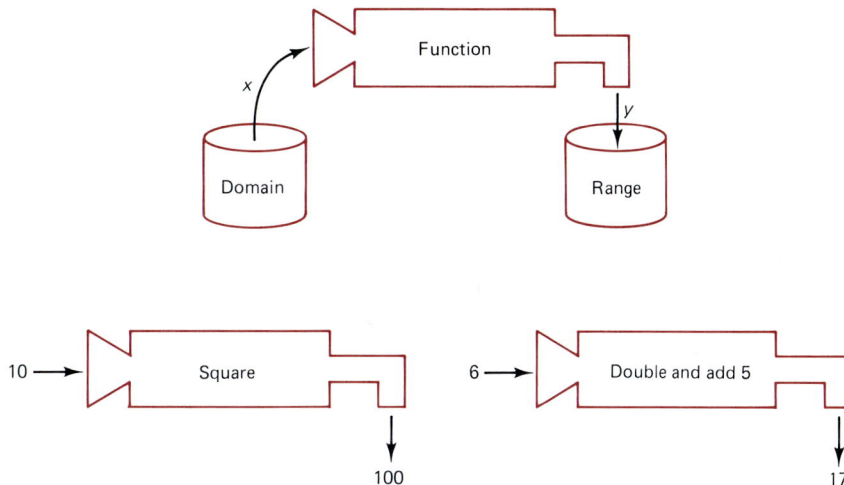

FIGURE 1

The connection between this definition and the "rule" definition is to consider the ordered pair (x, y) as the assignment $x \rightarrow y$. For example,

$$\{(1, 1), (2, 4), (3, 9), (4, 16)\} \quad \text{and} \quad \begin{array}{l} 1 \rightarrow 1 \\ 2 \rightarrow 4 \\ 3 \rightarrow 9 \\ 4 \rightarrow 16 \end{array}$$

represent the same function (the squaring function). On the other hand,

$$\{(1, 2), (1, 4), (1, -5)\} \quad \text{and} \quad \begin{array}{l} 1 \rightarrow 2 \\ 1 \rightarrow 4 \\ 1 \rightarrow -5 \end{array}$$

are *not* functions, since the same x-value (1) is paired with (or assigned to) different y-values (2, 4, and -5).

We usually represent functions with letters such as f, g, h, F, or G. For x in the domain, we write

$$\boxed{y = f(x)}$$

(read "f of x") as the number to which x is assigned. This notation, $f(x)$, is *not* multiplication but rather, the result of a rule f acting on x.

EXAMPLE 2 Let us show how this function notation works. Note that the rule f acts on whatever number is within the parentheses. Let $f(x) = 3x^2 + 2$ and $G(x) = \dfrac{x}{x+1}$.

$$f(1) = 3(1)^2 + 2 = 5 \qquad\qquad G(0) = \frac{0}{0+1} = 0$$

$$f(4) = 3(4)^2 + 2 = 50 \qquad\qquad G(1) = \frac{1}{1+1} = \frac{1}{2}$$

$$f(10) = 3(10)^2 + 2 = 302 \qquad\qquad G(4) = \frac{4}{4+1} = \frac{4}{5}$$

$$f(-2) = 3(-2)^2 + 2 = 14 \qquad\qquad G(70) = \frac{70}{70+1} = \frac{70}{71}$$

$$f(0) = 3(0)^2 + 2 = 2 \qquad\qquad G(k) = \frac{k}{k+1}$$

$$f(t) = 3t^2 + 2 \qquad\qquad G(m+n) = \frac{m+n}{m+n+1}$$

$$f(a+b) = 3(a+b)^2 + 2$$
$$= 3a^2 + 6ab + 3b^2 + 2 \qquad G(-1) \text{ is undefined. Why?}$$

EXAMPLE 3 For $f(x) = 2x^2 - 5x + 3$, compute:

(a) $f(a+h)$ (b) $f(a+h) - f(a)$ (c) $\dfrac{f(a+h) - f(a)}{h}$

Solution This is an important calculation for future studies in areas such as calculus. We simply compute each expression by carefully substituting into the formula for f.

(a)
> Substitute
> Distributive law

$$f(a+h) = 2(a+h)^2 - 5(a+h) + 3$$
$$= 2a^2 + 4ah + 2h^2$$
$$- 5a - 5h + 3$$

(b)
> Substitute
> Subtract polynomials

$$f(a+h) - f(a) = 2a^2 + 4ah + 2h^2$$
$$- 5a - 5h + 3$$
$$- (2a^2 - 5a + 3)$$
$$= 4ah + 2h^2 - 5h$$

(c)
> $\dfrac{\text{Substitute}}{\text{Divide}}$

$$\frac{f(a+h) - f(a)}{h} = \frac{4ah + 2h^2 - 5h}{h}$$
$$= 4a + 2h - 5$$

EXAMPLE 4 For $f(x) = 1/x$, compute:

(a) $f(a + h)$ (b) $f(a + h) - f(a)$ (c) $\dfrac{f(a + h) - f(a)}{h}$

Solution   This example is similar to Example 3, except that it involves fractions and an LCD.

(a) | Substitute | $f(a + h) = \dfrac{1}{a + h}$

(b) | Substitute | $f(a + h) - f(a) = \dfrac{1}{a + h} - \dfrac{1}{a}$

| Rewrite with LCD | $= \dfrac{a}{a(a + h)} + \dfrac{-(a + h)}{a(a + h)}$

| Simplify | $= \dfrac{-h}{a(a + h)}$

(c) | Divide | $\dfrac{f(a + h) - f(a)}{h} = \dfrac{-1}{a(a + h)}$

Recall that the **domain** is the set of numbers from which x is chosen. We now assume that the *domain of a function is the largest allowable set of real numbers*. (Recall that R is the set of real numbers.) We can take smaller domains, if we wish.

The domain of a function is the set of all real numbers R, except:

1. We exclude an x-value that makes a denominator zero.
2. We exclude an x-value that causes a radicand with even index to be negative.

EXAMPLE 5   Determine the domains of the following functions:

(a) $f(x) = x^4 - 2x + 1$ (b) $f(x) = \dfrac{x + 7}{x^2 - 9}$ (c) $f(x) = \sqrt{2x - 5}$

Solution   The domains are the entire set R, except for two possible trouble spots: a zero in the denominator or a negative number under a radical of even index.

(a) $f(x) = x^4 - 2x + 1$ has no denominator or radical to worry about; hence, the domain is all R.

(b) $f(x) = \dfrac{x + 7}{x^2 - 9}$ has a denominator, so we must check for zeros. We

see that if $x = \pm 3$, then $x^2 - 9 = 0$. Thus, the domain is all real numbers except ± 3.

(c) $f(x) = \sqrt{2x - 5}$ has a radical. The expression within the radical cannot be negative; thus,

$$\boxed{\begin{array}{c} \text{Determine where} \\ 2x - 5 \geq 0 \end{array}}$$

$$2x - 5 \geq 0$$
$$2x \geq 5$$
$$x \geq \frac{5}{2}$$

Therefore, the domain is the set of real numbers $x \geq 5/2$.

PROBLEM SET
5.1

Simplify the following.

Warm-up
Exercises

1. $2(3)^2 - 5(3) + 4$

2. $3\left(\frac{1}{2}\right)^2 - 5\left(\frac{1}{2}\right) + 1$

3. $\dfrac{(x + 3)^2 - x^2}{3}$

4. $\dfrac{\sqrt{x + 2} - \sqrt{x}}{2} \cdot \dfrac{\sqrt{x + 2} + \sqrt{x}}{\sqrt{x + 2} + \sqrt{x}}$

Solve the following.

Warm-up
Exercises

5. $3x - 5 = 0$

6. $x^2 - 16 = 0$

7. $5x + 2 \geq 0$

8. $x^2 - 3x + 2 \geq 0$

Locate the following ordered pairs on the coordinate plane.

Warm-up
Exercises

9. $(1, 7)$

10. $(-2, 3)$

11. $(0, -1)$

12. $(-1, -2)$

For each of the following rules, complete the assignments.

13.　$x \to x^3$
　　　$1 \to 1$
　　　$2 \to 8$
　　　$3 \to$
　　　$4 \to$
　　　$-5 \to$
　　　$1/2 \to$

14.　$x \to \sqrt{x}$
　　　$1 \to 1$
　　　$4 \to 2$
　　　$9 \to$
　　　$49 \to$
　　　$100 \to$
　　　$900 \to$

15.　$x \to |x|$
　　　$1 \to 1$
　　　$-2 \to 2$
　　　$3 \to$
　　　$-4 \to$
　　　$5 \to$
　　　$0 \to$

16.　$x \to 1/x$
　　　$1 \to 1$
　　　$2 \to 1/2$
　　　$3 \to$
　　　$-4 \to$
　　　$1/5 \to$
　　　$7/3 \to$

For the following functions, find:

(a) $f(0)$ (b) $f(-1)$ (c) $f(2)$
(d) $f(20)$ (e) $f(t)$ (f) $f(h + k)$

17. $f(x) = x + 3$ **18.** $f(x) = 3x - 2$

19. $f(x) = x^2 + 1$ **20.** $f(x) = x^3 - x$

21. $f(x) = \dfrac{1}{x - 1}$ **22.** $f(x) = \dfrac{1}{x^2 + 1}$

23. $f(x) = \sqrt{x + 1}$ **24.** $f(x) = \sqrt{x}$

25. $f(x) = |x|$ **26.** $f(x) = |2x - 3|$

For the following functions, find:

(a) $f(a + h)$ (b) $f(a + h) - f(a)$ (c) $\dfrac{f(a + h) - f(a)}{h}$

27. $f(x) = 3x$ **28.** $f(x) = -2x$

29. $f(x) = 5x - 3$ **30.** $f(x) = 3 - x$

31. $f(x) = x^2 + 6$ **32.** $f(x) = 2x^2 + x - 5$

33. $f(x) = 2x^3$ **34.** $f(x) = x^3 - x + 1$

35. $f(x) = \dfrac{3}{x}$ **36.** $f(x) = \dfrac{1}{x^2}$

37. $f(x) = \sqrt{x}$ **38.** $f(x) = \sqrt{x + 1}$

(Hint: For Problems 37 and 38, rationalize the radical—see page 51.)

Find the domain for each of the following functions.

39. $f(x) = 3x^2 + 5$ **40.** $f(x) = 1 - x - x^2 - x^3$

41. $F(x) = \dfrac{1}{x}$ **42.** $g(x) = \dfrac{x + 2}{21}$

43. $f(x) = \dfrac{1}{x - 2}$ **44.** $G(x) = \dfrac{7}{2x + 5}$

45. $f(x) = \sqrt{x}$ **46.** $F(x) = \sqrt{x - 3}$

47. $g(x) = \dfrac{1}{x^2 - 25}$ **48.** $h(x) = \dfrac{1}{x^2 - 3x + 2}$

49. $F(x) = \sqrt{x^2 - 1}$ **50.** $G(x) = \sqrt{x^2 - 5x + 4}$

51. $f(x) = \sqrt{x} + \dfrac{1}{x - 3}$ **52.** $F(x) = \dfrac{1}{\sqrt{x}} - \dfrac{1}{x - 5}$

In this section we have seen different (yet equivalent) approaches to the function concept. The following table and exercises summarize their connection. For each form of the function, fill in the missing values and then rewrite the function in the other forms.

	Assignment	Ordered Pairs	Function Notation
Example:	$x \to x^2$ $1 \to 1$ $2 \to 4$ $-3 \to 9$	(x, x^2) $(1, 1)$ $(2, 4)$ $(-3, 9)$	$f(x) = x^2$ $f(1) = 1$ $f(2) = 4$ $f(-3) = 9$
53.	$x \to 2x + 7$ $1 \to$ $2 \to$ $-3 \to$		
54.		$\left(x, \dfrac{1}{x}\right)$ $(1, \quad)$ $(2, \quad)$ $(-3, \quad)$	
55.			$f(x) = 5x^2 - 1$ $f(1) =$ $f(2) =$ $f(-3) =$
56.	$x \to \dfrac{x + 1}{x}$ $1 \to$ $2 \to$ $-3 \to$		

For the following functions, complete the indicated ordered pairs. Then graph the ordered pairs and draw a smooth curve through the points. Also, state the domain of the function.

57. $f(x) = x^2 + 1$ $(-2, \quad); (-1, \quad); (0, \quad); (1, \quad); (2, \quad)$

58. $f(x) = 5 - x$ $(-2, \quad); (-1, \quad); (0, \quad); (1, \quad); (2, \quad)$

59. $f(x) = \dfrac{1}{x}$ $\left(\dfrac{1}{4}, \quad\right); \left(\dfrac{1}{2}, \quad\right); (1, \quad); (2, \quad); (5, \quad)$

60. $f(x) = \sqrt{x}$ $(0, \quad); (1, \quad); (4, \quad); (9, \quad); (16, \quad)$

Business Applications   **61.** The demand for an item is given by

$$D(x) = \frac{100{,}000}{x}$$

where x is the price.

(a) Find $D(50)$, $D(100)$, and $D(1000)$.

(b) Find $D(a + h)$, $D(a + h) - D(a)$, and $\dfrac{D(a + h) - D(a)}{h}$.

62. The sales that a company makes depend on its advertising budget according to the function

$$S(x) = 10,000\sqrt{x}$$

where x is the advertising budget (in thousands of dollars).
(a) Find $S(100)$, $S(900)$, and $S(2500)$.

(b) Find $S(a + h)$, $S(a + h) - S(a)$, and $\dfrac{S(a + h) - S(a)}{h}$.

Chemical Application **63.** The increase in temperature due to a catalyst of concentration u is given by

$$T(u) = \frac{10u}{\sqrt{u + 1}}$$

Find $T(1)$, $T(4)$, and $T(25)$.

Health Application **64.** The amount of sugar in the blood t hours after ingesting a certain food is given by $S(t) = 4 + 0.5t - 0.1t^2$.
(a) Find $S(0)$, $S(1)$, $S(2)$, and $S(3)$.

(b) Find $S(a + h)$, $S(a + h) - S(a)$, and $\dfrac{S(a + h) - S(a)}{h}$.

Physical Applications **65.** The distance that an object rises above the ground in t seconds is given by

$$y(t) = 80t - 16t^2$$

(a) Find $y(0)$, $y(1)$, and $y(2)$.
(b) For what times t is $y(t) = 0$?

(c) Find $y(a + h)$, $y(a + h) - y(a)$, and $\dfrac{y(a + h) - y(a)}{h}$.

66. The gravitational potential energy that an object has is given by

$$E(r) = \frac{10^{15}}{r}$$

where r is the distance from the center of the earth (in meters).
(a) Find $E(10^7)$, $E(10^8)$, and $E(10^{10})$.

(b) Find $E(a + h)$, $E(a + h) - E(a)$, and $\dfrac{E(a + h) - E(a)}{h}$.

Computer Applications **67.** Not all functions assign numbers to numbers. In setting up computer files, computer scientists often use a function h, which assigns letters to integers. For instance,

$$h: \quad A \rightarrow 1$$
$$B \rightarrow 2$$
$$\vdots$$
$$Z \rightarrow 26$$

(a) Find $h(\text{L})$.   (b) Find $h(\text{I})$.
(c) Find $h(\text{S})$.   (d) Find $h(\text{A})$.

68. The function h of Problem 67 can be used to assign a name to an integer. We add the numbers assigned to each letter of the name. In symbols, for a name with letters $L_1 L_2 L_3 \cdots$, we say

$$H(L_1 L_2 L_3 \cdots) = h(L_1) + h(L_2) + h(L_3) + \cdots$$

For example, $H(\text{JOE}) = h(\text{J}) + h(\text{O}) + h(\text{E}) = 10 + 15 + 5 = 30$.
(a) Find $H(\text{LISA})$.   (b) Find $H(\text{BECKY})$.
(c) Find $H(\text{AMY})$.   (d) Find $H(\text{REGINALD})$.

5.2
GRAPHS OF FUNCTIONS

In the last chapter we saw how to graph equations. In the preceding section we saw that functions are frequently rules that are defined by equations. We now put these ideas together to produce the **graph of a function**.

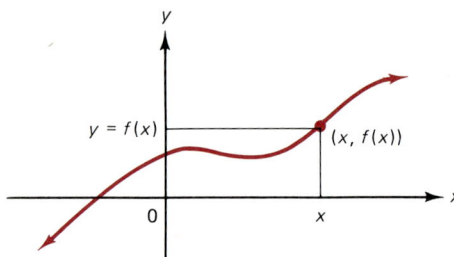

FIGURE 2

To graph a function we graph the equation $y = f(x)$; that is, we graph the ordered pairs $(x, f(x))$. Figure 2 illustrates the graph of a function with the y-coordinate being the value of the function $f(x)$.

Since a function assigns each x to only one y, each vertical line through a given x-value crosses the graph only once.

> **The Vertical-Line Test:** A curve is the graph of a function if it is crossed at most once by any vertical line.

EXAMPLE 6 In Figure 3 curves (a) and (b) are functions, but (c) and (d) are *not* functions since some vertical line crosses them more than once.

> To graph a function:
>
> 1. Determine the domain for the x-values.
> 2. Make a table of (x, y) pairs, where $y = f(x)$.
> 3. Plot the points and draw a curve through the points.

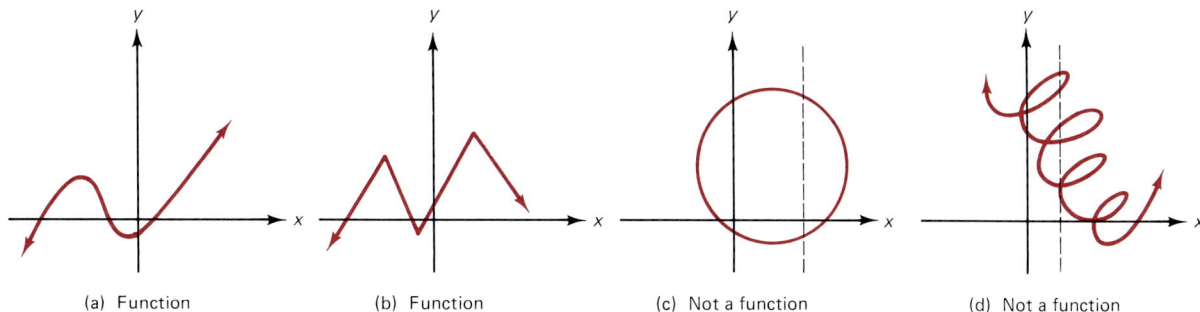

(a) Function   (b) Function   (c) Not a function   (d) Not a function

FIGURE 3

Later in the chapter we will consider the graphs of polynomial functions and those of rational functions (quotients of polynomial functions). In this section we consider some of the special functions that occur frequently in mathematics and its applications.

EXAMPLE 7   Graph $f(x) = 3$.

Solution   This is called a **constant function**, since all x-values are assigned to 3. We graph the equation $y = 3$, which is a horizontal line (all constant functions have horizontal graphs). The domain is all the real numbers.

x	$y = f(x)$
-2	3
0	3
5	3

Make a table of $(x, f(x))$ pairs

Plot points

Draw line through points

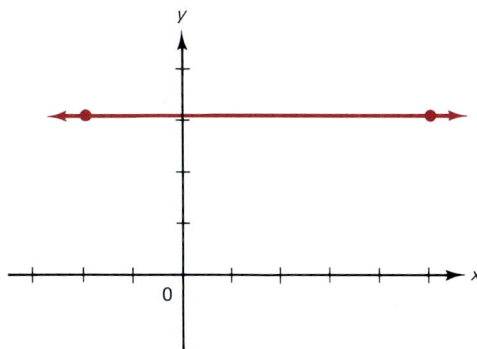

EXAMPLE 8   Graph $f(x) = 2x - 1$.

Solution   This is a **linear function**, as we studied on page 125. Its domain is all R, and its graph is of the equation $y = 2x - 1$. (Recall also that the slope is 2.)

x	$y = 2x - 1$
-1	-3
0	-1
1	1
2	3

Make a table of $(x, f(x))$ pairs

Plot points

Draw line through points

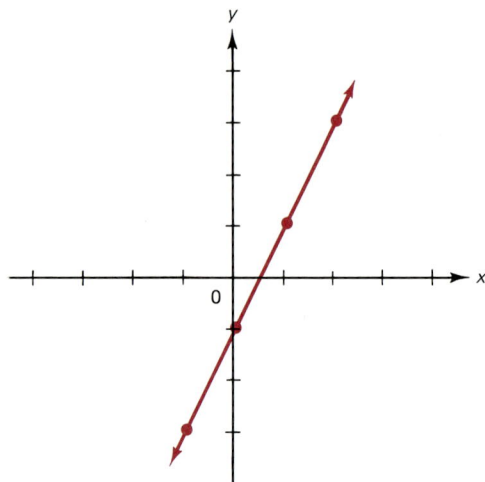

EXAMPLE 9 Graph $f(x) = |x|$.

Solution This is an **absolute-value function**. Since all x-values are allowable, the domain is R. Note that the graph is symmetric about the y-axis, since x and $-x$ always produce the same y.

x	$y = |x|$
-2	2
-1	1
0	0
1	1
2	2

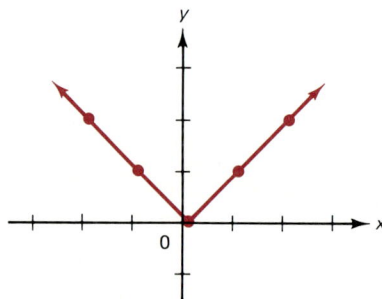

The graph is two rays that are connected at the origin.

EXAMPLE 10 Graph $f(x) = \sqrt{2 - x}$.

Solution Since this function involves a radical, we must first determine the domain: We must have $2 - x \geq 0$, or $2 \geq x$. Now we make our table for $2 \geq x$ (and using a hand calculator).

x	$y = \sqrt{2-x}$
2	$\sqrt{0} = 0$
1	$\sqrt{1} = 1$
0	$\sqrt{2} \approx 1.4$
-1	$\sqrt{3} \approx 1.7$
-2	$\sqrt{4} = 2$
-3	$\sqrt{5} \approx 2.2$
-4	$\sqrt{6} \approx 2.4$

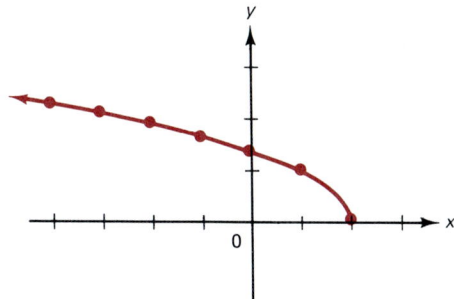

EXAMPLE 11   Graph $f(x) = \sqrt{x^2 - 4}$.

Solution   As with Example 10, we must first find where the term under the radical is nonnegative. We factor $x^2 - 4 = (x + 2)(x - 2)$ to examine the signs. (Review quadratic inequalities, page 97.)

Examine signs to
determine domain:
$\{x \leq -2 \text{ or } x \geq 2\}$

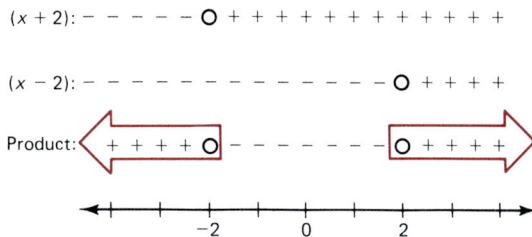

x	$y = \sqrt{x^2 - 4}$
± 2	$\sqrt{0} = 0$
± 3	$\sqrt{5} \approx 2.2$
± 4	$\sqrt{12} \approx 3.5$
± 5	$\sqrt{21} \approx 4.6$
± 6	$\sqrt{32} \approx 5.7$

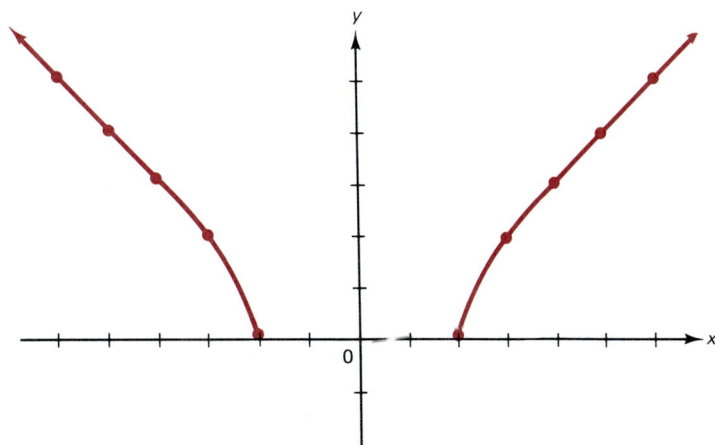

EXAMPLE 12   Graph $f(x) = \begin{cases} x & \text{for } x < 0 \\ x^2 & \text{for } 0 \leq x \leq 2 \\ 2x - 5 & \text{for } x > 2. \end{cases}$

Solution This type of graph is called a **piecewise function** since it is defined differently on different pieces of the domain. We graph this type of function in pieces: We make a separate table for each region.

	{x < 0}		{0 ≤ x ≤ 2}		{x > 2}
x	$y = x$	x	$y = x^2$	x	$y = 2x - 5$
-3	-3	0	0	2	-1
-2	-2	1	1	3	1
-1	-1	2	4	4	3
0	0			5	5

Make (x, y) table for each region

Plot points and draw curves for each region

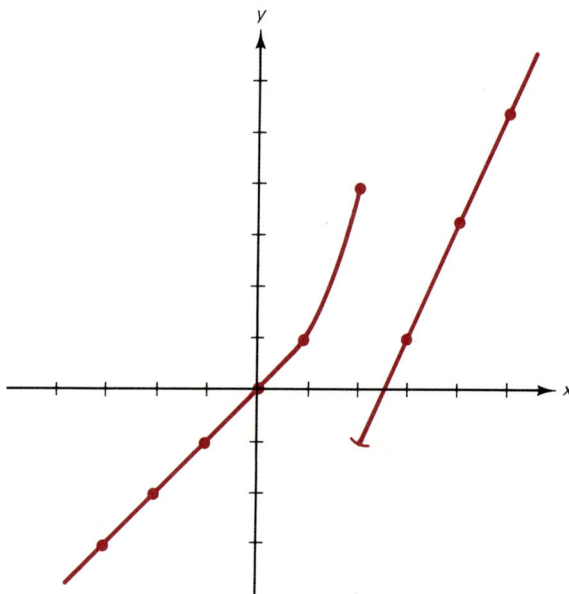

Note that we compute and plot the endpoints for each interval (even if an endpoint is not in the region) to help us complete the graph. We can use the open-interval notation for a point not on the curve, such as $(2, -1)$ in this case.

EXAMPLE 13 Graph $f(x) = [\![x]\!]$.

Solution This is the **greatest integer function**, where $[\![x]\!]$ is the greatest integer less than or equal to x. Examples are $[\![4.3]\!] = 4$, $[\![2/3]\!] = 0$, $[\![6.9]\!] = 6$, $[\![7]\!] = 7$, $[\![-2.1]\!] = -3$, and $[\![-1/5]\!] = -1$. Note that we always *round down* to the next lowest integer, unless x is itself an integer.

[Compare this to the truncation function in many computer languages: for instance, INT(X) in BASIC or TRUNC(X) in Pascal.]

x	$y = [\![x]\!]$
−1.3	−2
−1	−1
−0.4	−1
0	0
0.4	0
0.9	0
1	1
1.2	1
1.9	1
2	2
2.3	2
2.8	2
3	3

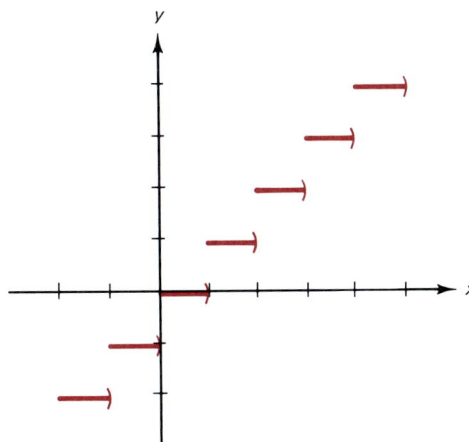

This graph is often called a **step function** because of its shape. Note that it jumps at each integer; for example, $[\![1.999]\!] = 1$, but $[\![2]\!] = 2$.

EXAMPLE 14   A weekday long-distance phone call from Beaver Creek to Buffalo Grove is $2.50 for the first 3 minutes, plus $0.35 for each additional minute. Write this as a piecewise function and graph.

Solution   We can look at this as follows:
(a) For minutes 0 to 3, the cost is a constant $2.50.
(b) For minutes 3 to 4, the cost is $2.50 + $0.35 = $2.85.
(c) For minutes 4 to 5, the cost is $2.85 + $0.35 = $3.20.

And so on. If we let x be the number of minutes and $C(x)$ be the cost, we can write this as a piecewise function.

$$C(x) = \begin{cases} 2.50 & \text{for } 0 < x \leq 3 \\ 2.85 & \text{for } 3 < x \leq 4 \\ 3.20 & \text{for } 4 < x \leq 5 \\ 3.55 & \text{for } 5 < x \leq 6 \\ \text{and so on} \end{cases}$$

Figure 4 on the next page shows the graph of this function.

Generally, we are given a rule or equation, and we want to draw the graph. Sometimes, however, the opposite is true: We have the graph and want to determine some of the assignments.

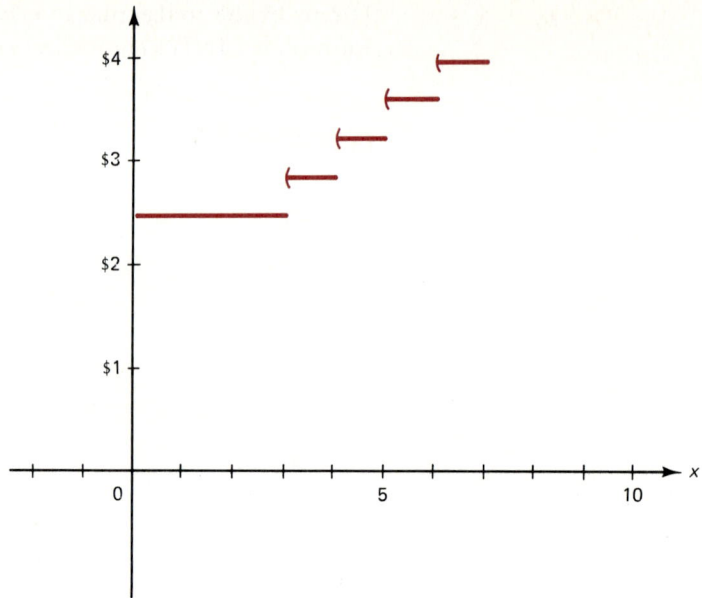

FIGURE 4

EXAMPLE 15 Consider Figure 5. Here we are given the graph of some function. We use the graph to help evaluate the function. (As with most graphical methods, the values will be approximate.) For instance, to approximate $f(10)$, we start at $x = 10$ on the x-axis, move up vertically to the curve, and then over horizontally to the y-axis at $y = 80$. Thus, $(10, 80)$ is on the curve and $f(10) = 80$. Similarly, we can use the graph to see that $f(20) = 180$ and $f(30) = 250$.

FIGURE 5

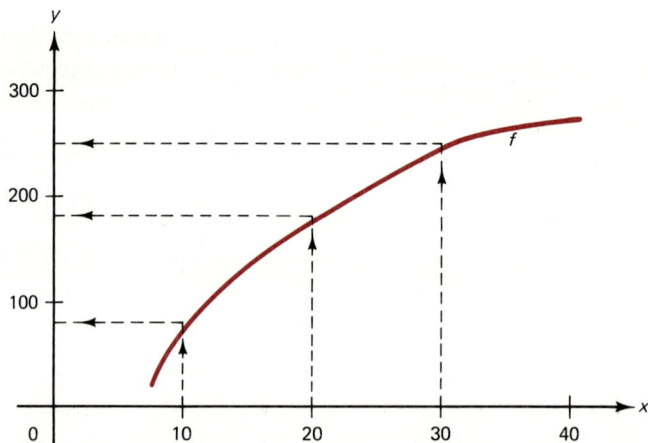

Locate the following points on the coordinate plane.

1. $(-2, 1)$ **2.** $(0, 5)$ **3.** $(4, 0)$

4. $(-3, -2)$ **5.** $(4, -1)$ **6.** $(3, 1)$

Graph the following.

7. $y = 2x + 3$ **8.** $y = x^2 + 1$

9. $y = \sqrt{x}$ **10.** $y = \dfrac{1}{x}$

For each of the following functions, complete the indicated ordered pairs.

11. $f(x) = 5 - x$ $(0, \quad); (1, \quad); (2, \quad); (-1, \quad)$

12. $f(x) = x^2 + 1$ $(-1, \quad); (0, \quad); (1, \quad); (2, \quad)$

13. $f(x) = \sqrt{x}$ $(0, \quad); (1, \quad); (4, \quad); (9, \quad)$

14. $f(x) = 6$ $(-2, \quad); (0, \quad); (3, \quad); (7, \quad)$

State whether each of the following sketches represents a function.

15.

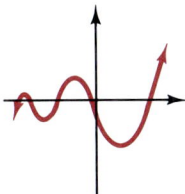

16.

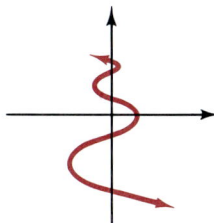

17.

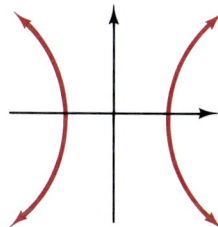

18.

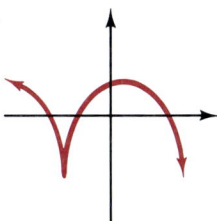

19.

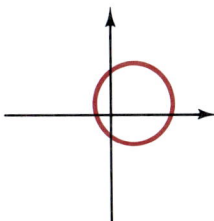

20.

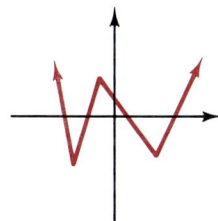

Graph the following functions.

21. $f(x) = 3x - 4$ **22.** $f(x) = 5x + 1$

23. $f(x) = 1 + 4x$ **24.** $f(x) = 4 - 2x$

25. $f(x) = |x + 1|$ **26.** $f(x) = |x - 5|$

27. $f(x) = |2x - 3|$ **28.** $f(x) = |3x + 1|$

29. $f(x) = \sqrt{4 - x}$ **30.** $f(x) = \sqrt{3 + 2x}$

31. $f(x) = \sqrt{9 - x^2}$ **32.** $f(x) = \sqrt{25 - x^2}$

33. $f(x) = \begin{cases} 2x - 5 & \text{if } x < 3 \\ -2x + 7 & \text{if } x \geq 3 \end{cases}$ **34.** $f(x) = \begin{cases} x^2 & \text{if } x < 1 \\ 1 & \text{if } x \geq 1 \end{cases}$

35. $f(x) = \begin{cases} 0 & \text{if } x < -2 \\ x + 2 & \text{if } -2 \leq x \leq 2 \\ 4 & \text{if } x > 2 \end{cases}$ **36.** $f(x) = \begin{cases} -x & \text{if } x < 0 \\ 3x - x^2 & \text{if } 0 \leq x \leq 3 \\ x - 3 & \text{if } x > 3 \end{cases}$

37. $f(x) = \left[\!\!\left[\dfrac{x}{2}\right]\!\!\right]$

38. $f(x) = [\![3x + 1]\!]$

39. $f(x) = -1$

40. $f(x) = \begin{cases} -1 & \text{if } x < 0 \\ 1 & \text{if } x \geq 0 \end{cases}$

41. $f(x) = \sqrt{x - 1}$

42. $f(x) = |x + 5|$

43. $f(x) = \begin{cases} 0 & \text{if } x \leq 0 \\ 2x & \text{if } x > 0 \end{cases}$

44. $f(x) = [\![3x]\!]$

45. $f(x) = |2x - 1|$

46. $f(x) = \sqrt{x^2 - 36}$

47. $f(x) = [\![x - 1]\!]$

48. $f(x) = |5 + x|$

49. $f(x) = \sqrt{16 - x^2}$

50. $f(x) = 0$

51. The Metro Taxi Company charges $2 for the first 1/2 mile and 50 cents for each 1/2 mile thereafter. Express this charge as a piecewise function.

52. The Jones Rent-a-Car Company charges $30 per day, plus 20 cents per mile. Express this charge as a linear function of x, the number of miles. Then graph.

53. As this text is written, the cost of first-class mail is 22 cents for the first ounce and 17 cents for each fraction of an ounce thereafter. Express the cost as a piecewise function of x, the weight of the letter or package.

54. A telephone call from town A to town B is 7 units for the first 3 minutes and 2 units for each fraction of a minute thereafter. Express this charge using the greatest integer function of x, the number of minutes. Then graph.

Shown on the next page are the graphs of two functions, f and g. Use these curves to approximate the following.

55. $f(100)$

56. $g(100)$

57. $g(300)$

58. $f(300)$

59. $f(400)$

60. $g(400)$

61. $g(150)$

62. $f(150)$

63. $f(450)$

64. $g(450)$

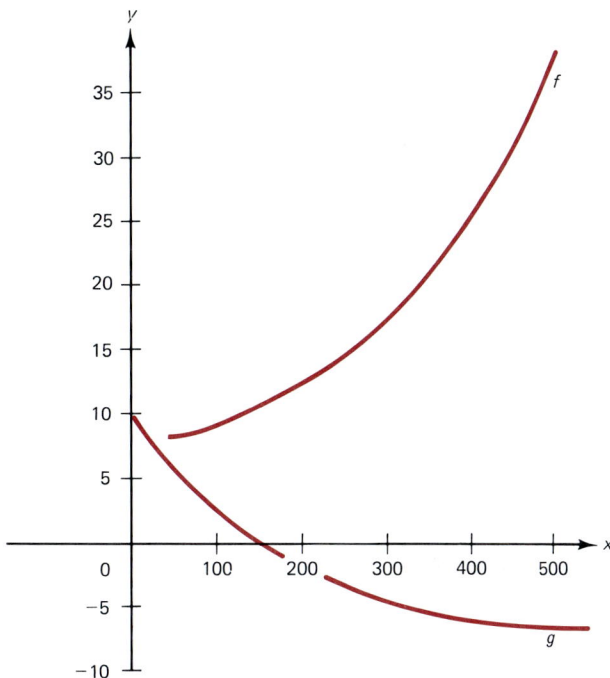

*Physical
Application*

65. A 1-gram piece of ice (at $-30°C$) is heated. The temperature depends on the heat added (in calories) x by

$$T(x) = \begin{cases} 0.5x - 30 & \text{if } 0 \le x < 60 \\ 0 & \text{if } 60 \le x < 140 \\ x - 140 & \text{if } 140 \le x < 240 \\ 100 & \text{if } 240 \le x < 780 \\ 0.4x - 212 & \text{if } x \ge 780 \end{cases}$$

Graph this function.

*Engineering
Application*

66. The Brinell hardness number N depends on the depression d made by a test object as follows:

$$N(d) = \frac{1910 + 191\sqrt{100 - d^2}}{d}$$

Find the domain (d must be positive) and graph.

67. The voltage coming out of a triangular-wave generator has the form

$$V(t) = \begin{cases} t - 1 & \text{if } 0 \le t < 2 \\ 3 - t & \text{if } 2 \le t < 4 \\ t - 5 & \text{if } 4 \le t < 6 \\ 7 - t & \text{if } 6 \le t < 8 \\ t - 9 & \text{if } 8 \le t < 10 \\ 11 - t & \text{if } 10 \le t < 12 \end{cases}$$

Graph this function.

68. A 35-year-old person buys a certain $10,000 decreasing term insurance policy. (The payoff upon death decreases every year.) The payoff (or *face value*) is given by $F(x) = 10{,}000 - 500 \cdot [\![x - 35]\!]$, where x is the age at death. Graph this function for $35 \le x \le 55$.

5.3
POLYNOMIAL
FUNCTIONS

We discussed polynomials on page 26. We now discuss **polynomial functions**, which have the general form

$$f(x) = a_n x^n + a_{n-1} x^{n-1} + \cdots + a_2 x^2 + a_1 x + a_0$$

where the a's are real-number coefficients ($a_n \ne 0$). In this section we look at polynomial functions of various degrees and study their graphs.

Polynomial Functions of Degree Two or Less

For degree two or less, polynomial functions have special names and forms, as shown in the following table. (In each case, $a \ne 0$.)

Degree	Name	Form
0	Constant	$f(x) = a$
1	Linear	$f(x) = ax + b$
2	Quadratic	$f(x) = ax^2 + bx + c$

We have already discussed the constant and linear functions (see Examples 7 and 8, page 149), so we turn our attention to the **quadratic** (or **second-degree**) **functions**. We have already graphed several quadratic equations (see Examples 7 and 8, page 120). Here we consider equations of the form

$$y = ax^2 + bx + c \qquad (a \ne 0)$$

These have a cuplike shape and open either up or down. (Why can they *not* open right or left?) Also, they have a turning point, or **vertex**, which is either a high point (**maximum**) or a low point (**minimum**). We find these by completing the square.

$$f(x) = ax^2 + bx + c$$

$$= a\left(x^2 + \frac{b}{a}x \qquad\right) + c$$

$$= a\left(x^2 + \frac{b}{a}x + \frac{b^2}{4a^2}\right) + c - \frac{b^2}{4a}$$

$$= a\left(x + \frac{b}{2a}\right)^2 + \left(c - \frac{b^2}{4a}\right)$$

Let us examine these two terms. The rightmost term, $c - b^2/4a$, is a constant (since there is no x-term). In the other term, we have $(x + b/2a)^2 \geq 0$. (Why?) The two cases depend on the sign of a:

1. If $a > 0$, then $a(x + b/2a)^2 \geq 0$, and is zero only for $x = -b/2a$. Thus, $f(-b/2a) = c - b^2/4a$, and all other values are *greater* than this. Therefore, $(-b/2a, f(-b/2a))$ is a vertex or a low point (minimum).

2. Similarly, if $a < 0$, $f(-b/2a) = c - b^2/4a$, and all other values are *less* than this. Therefore, $(-b/2a, f(-b/2a))$ is a vertex or a high point (maximum).

Figure 6 shows these cases.

Theorem 1

A quadratic function $f(x) = ax^2 + bx + c$ $(a \neq 0)$ satisfies:

(a) Vertex $= \left(\dfrac{-b}{2a}, c - \dfrac{b^2}{4a}\right) = \left(\dfrac{-b}{2a}, f\left(\dfrac{-b}{2a}\right)\right)$.

(b) If $a > 0$, the graph opens up and the vertex is a **valley** (minimum).

(c) If $a < 0$, the graph opens down and the vertex is a **peak** (maximum).

FIGURE 6

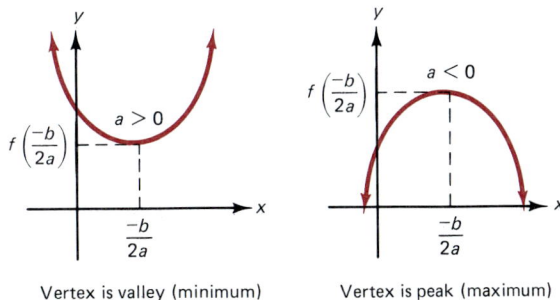

Vertex is valley (minimum)　　　Vertex is peak (maximum)

EXAMPLE 16   We use Theorem 1 to compute the vertex of each of the following quadratic functions.

Equation	Vertex	Type	Sketch
$f(x) = ax^2 + bx + c$	$\left(\dfrac{-b}{2a}, f\left(\dfrac{-b}{2a}\right)\right)$	$a > 0$: valley $a < 0$: peak	
$f(x) = x^2 - 4x + 7$	$(2, 3)$	Valley (minimum)	
$f(x) = -x^2 + 3x - 1$	$\left(\dfrac{3}{2}, \dfrac{5}{4}\right)$	Peak (maximum)	
$f(x) = 2x^2 + x - 2$	$\left(\dfrac{-1}{4}, \dfrac{-17}{8}\right)$	Valley (minimum)	

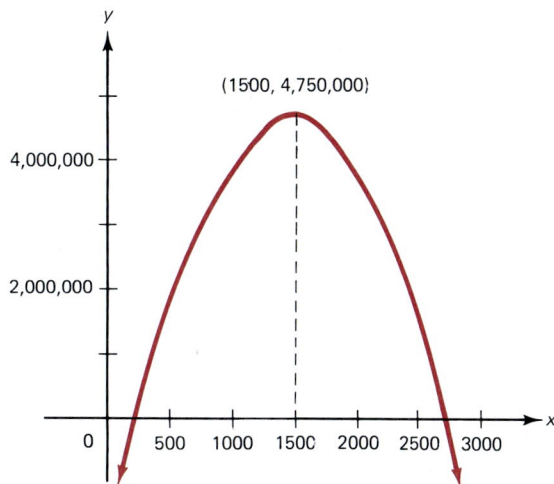

FIGURE 7

EXAMPLE 17   The ABC Truck Company has determined that the profit P that it can make by producing x trucks is given by

$$P(x) = -3x^2 + 9000x - 2,000,000$$

What number of trucks will maximize their profit, and what is the maximum profit?

Solution   This is a quadratic function. We use Theorem 1 to find that the vertex has x-coordinate

$$x = \frac{-b}{2a} = \frac{-9000}{2(-3)} = 1500 \text{ trucks}$$

This means that they should produce 1500 trucks to make a maximal profit of $P(1500) = 4,750,000$. Figure 7 shows this situation.

Polynomial Functions of Degree Three or More

Let us start by looking at the simple functions $f(x) = x^n$. Figure 8 shows the graphs of $y = x^2$, $y = x^4$, and $y = x^6$ with one set of axes, and $y = x$, $y = x^3$, and $y = x^5$ with another set of axes.

FIGURE 8

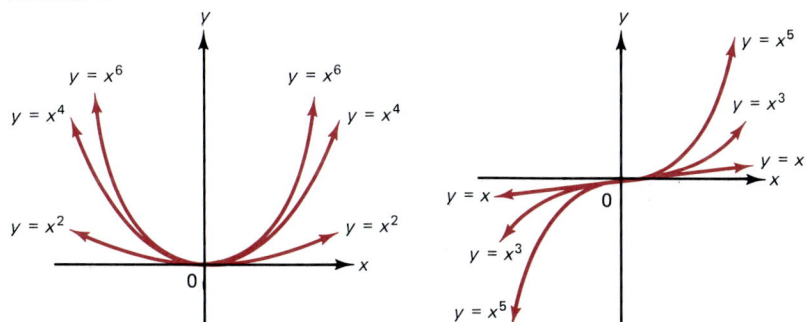

We have purposely separated the functions like this to suggest the definitions of the following classes of functions:

1. A function is **even** if $f(-x) = f(x)$, for all x in the domain. That is, both (x, y) and $(-x, y)$ are on the curve. (Recall that this means that the curve is symmetric about the y-axis.) Polynomials containing only even exponents $(1, x^2, x^4,$ and so on) are even functions. (This is where the name came from, although there are other even functions.)

2. A function is **odd** if $f(-x) = -f(x)$, for all x in the domain. That is, both (x, y) and $(-x, -y)$ are on the curve. (Thus, the curve is symmetric about the origin.) Polynomials containing only odd exponents $(x, x^3, x^5,$ and so on) are odd functions.

Note: A function with both odd and even exponents, such as $f(x) = x^3 + x^2 + x + 1$, is neither odd nor even. Also, the coefficients do *not* affect the function's being odd or even. For instance, $f(x) = 5x^2$ is even.

EXAMPLE 18 The following are examples of odd and even polynomial functions.

Function	Odd/Even?	Sketch
$f(x) = 1 - x^2 + x^4$	Even	
$f(x) = 2x^3 - 5x$	Odd	

$f(x) = 5x^3 - 12x^2 + 7x + 1$	Neither	

With higher-degree polynomials, we usually have to plot many points to find the graph. However, we can sometimes factor the polynomial and analyze the signs of the factors.

To graph a polynomial function by factoring:

1. If possible, factor the polynomial.
2. Determine intervals where the polynomial is positive, negative, or zero.
3. Where positive, the graph is above the x-axis.
4. Where negative, the graph is below the x-axis.
5. Plot a few points (perhaps, one per interval).

EXAMPLE 19 Graph $f(x) = x^4 - x^2$.

Solution We factor and analyze the signs. Also, this is an even function.

Factor

$$x^4 - x^2 = x^2(x^2 - 1)$$
$$= x^2(x - 1)(x + 1)$$

Analyze signs of factors and product

x^2: +   +   +   +   +   +   +   O   +   +   +   +   +   +

$(x - 1)$:   –   –   –   –   –   –   –   O   +   +   +   +

$(x + 1)$: –   –   –   –   –   O   +   +   +   +   +   +   +   +

Product: +   +   +   +   +   O   –   O   –   O   +   +   +   +

$$-3 \quad -2 \quad -1 \quad 0 \quad 1 \quad 2 \quad 3$$

| Determine a few points (one per interval) |

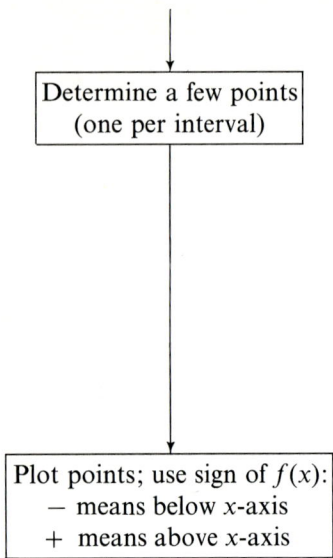

x	y
$\pm\dfrac{1}{2}$	$\dfrac{-3}{16}$
± 2	12

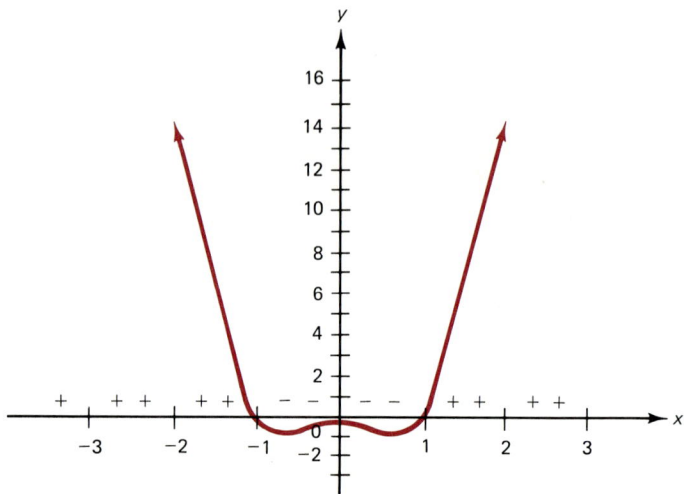

| Plot points; use sign of $f(x)$:
 − means below x-axis
 + means above x-axis |

Since this is an even function, note that we can pick our x-values in pairs, such as $\pm 1/2$ or ± 2.

EXAMPLE 20 Graph $f(x) = x^5 - 6x^4 + 8x^3$.

Solution We factor this polynomial and examine the signs; then we plot a few points to help. This function is neither odd nor even.

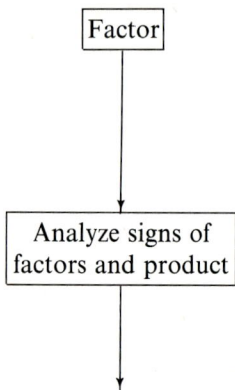

| Factor |

$$f(x) = x^3(x^2 - 6x + 8)$$
$$= x^3(x - 2)(x - 4)$$

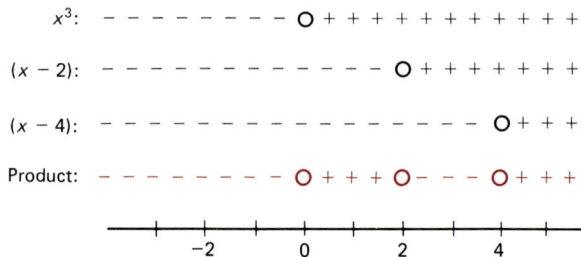

| Analyze signs of factors and product |

x^3: − − − − − − − − O + + + + + + + + + + +

$(x - 2)$: − − − − − − − − − − − O + + + + + + +

$(x - 4)$: − − − − − − − − − − − − − − O + + +

Product: − − − − − − − O + + + O − − − O + + +

x	y
-1	-15
1	3
3	-27
5	375

Find a few
in-between points

Plot points and zeros; use signs:
$-$ means below x-axis
$+$ means above x-axis

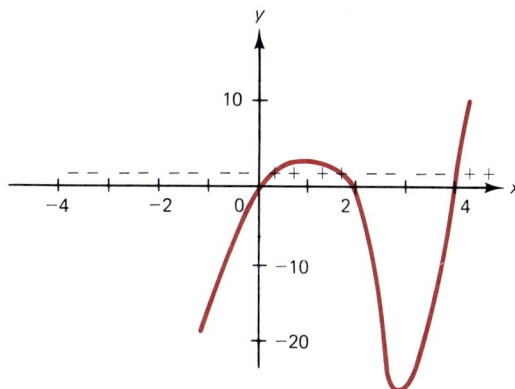

PROBLEM SET
5.3

Factor the following polynomials completely.

Warm-up
Exercises

1. $x^2 - 16$

2. $x^2 + x - 12$

3. $x^5 - 25x^3$

4. $x^3 - 2x^2 + 2x - 4$

Graph the following functions.

Warm-up
Exercises

5. $f(x) = 3$

6. $f(x) = 2x + 1$

7. $f(x) = \sqrt{x + 1}$

8. $f(x) = x^2 + 1$

State whether each of the following functions is odd, even, or neither.

9. $f(x) = 4x^3$

10. $f(x) = 7x^2$

11. $f(x) = x^2 - 5x$

12. $f(x) = x^5 - 7x^3 + x$

13. $g(x) = 9x^4 - x^2 + 2$

14. $G(x) = 4x^4 - 3x^3 + 2x^2$

15. $f(x) = 10x^7 - 8x^3$

16. $F(x) = -8x^{10} - 5x^4 + 7$

Graph the following polynomial functions.

17. $f(x) = x^3$

18. $f(x) = x^4$

19. $f(x) = (x - 1)(x - 2)$

20. $F(x) = (x - 3)(x + 1)$

21. $f(x) = x(x - 1)(x - 5)$

22. $G(x) = (x + 2)(x - 2)(x - 5)$

23. $F(x) = x^3 - 4x$

24. $g(x) = x^4 - 9x^2$

For each of the following functions:
(a) *State whether the function is odd, even, or neither.*
(b) *Find (or estimate) the vertices (if any).*
(c) *Graph the function.*

25. $f(x) = 2x + 5$	**26.** $F(x) = 7$
27. $f(x) = 4 - x^2$	**28.** $f(x) = x^2 - 3x + 4$
29. $g(x) = -2$	**30.** $f(x) = x^6 - x^4$
31. $f(x) = x^2 - 7x + 6$	**32.** $G(x) = x^3 + x^2 - 4x - 4$
33. $F(x) = x^3 - x^2$	**34.** $f(x) = x^5 - x$
35. $f(x) = 1 + x + x^2 + x^3$	**36.** $F(x) = \sqrt{x^2 - 49}$
37. $h(x) = x - x^3$	**38.** $g(x) = |x^2 - 25|$
39. $f(x) = \sqrt{x^2 - 6x + 5}$	**40.** $F(x) = \sqrt{x^3 - 2x^2 - x + 2}$
41. $g(x) = |x^2 + x - 6|$	**42.** $f(x) = |3x - 1|$

Geometry Application

43. A farmer has 100 meters of fencing to enclose a rectangular field.
(a) Show that the sides can be represented as x and $50 - x$.
(b) Write the area as a function of x.
(c) For what x is the area greatest? What is the area?
(d) Graph this area function.

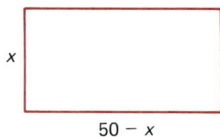

x

$50 - x$

Health Application

44. The sugar in the blood t hours after ingesting a certain food is given by $S(t) = 4 + 0.5t - 0.1t^2$.
(a) At what time is the amount of sugar greatest?
(b) Graph this function for $0 \le t \le 6$.

Physical Application

45. The height that a ball rises after being thrown upward at 96 feet per second is given by $y(t) = -16t^2 + 96t$, where t is the time in seconds.
(a) At what time is the height greatest? What is this height?
(b) At what times is the ball on the ground ($y = 0$)?

Electrical Application

46. The current through an inductance at time t is given by

$$i(t) = -t^3 + 15t^2 - 56t$$

Graph this function.

Business Application

47. If a company produces x lamps, its profit is given by

$$P(x) = -0.005x^2 + 15x - 3500$$

What number of lamps gives the greatest profit? What is this profit?

5.4 RATIONAL FUNCTIONS

On page 40, we studied rational expressions, which are the quotients of polynomials. Here we study **rational functions**, which are quotients of polynomial functions; the standard form is

$$f(x) = \frac{P(x)}{Q(x)}$$

where P and Q are polynomial functions like the ones seen in the preceding section. [As always, we cannot divide by zero, so we restrict our domain to the real numbers where $Q(x) \neq 0$.]

Let us begin with a simple set of rational functions.

EXAMPLE 21 Graph the functions $f(x) = 1/x^n$, for $n = 1, 2, 3, 4, 5,$ and 6.

Solution We make tables for the functions $y = 1/x$ and $y = 1/x^2$; the others are similar and are left as an exercise. Note that $x = 0$ is not in the domains of these functions.

<table>
<tr><td rowspan="7">Make table for $y = 1/x$ and $y = 1/x^2$</td><td>x</td><td>$y = \dfrac{1}{x}$</td><td>x</td><td>$y = \dfrac{1}{x^2}$</td></tr>
<tr><td>1/10</td><td>10</td><td>± 2</td><td>1/4</td></tr>
<tr><td>1/2</td><td>2</td><td>± 1</td><td>1</td></tr>
<tr><td>1</td><td>1</td><td>± 5</td><td>1/25</td></tr>
<tr><td>3</td><td>1/3</td><td>$\pm 1/3$</td><td>9</td></tr>
<tr><td>$-1/7$</td><td>-7</td><td>$\pm 1/10$</td><td>100</td></tr>
<tr><td>-2</td><td>$-1/2$</td><td></td><td></td></tr>
</table>

In Figure 9, we show the graphs of $y = 1/x$, $y = 1/x^3$, and $y = 1/x^5$ together, and the graphs of $y = 1/x^2$, $y = 1/x^4$, and $y = 1/x^6$ together. (Compare this to the graphs of $y = x^n$ in Figure 8, page 161.)

FIGURE 9

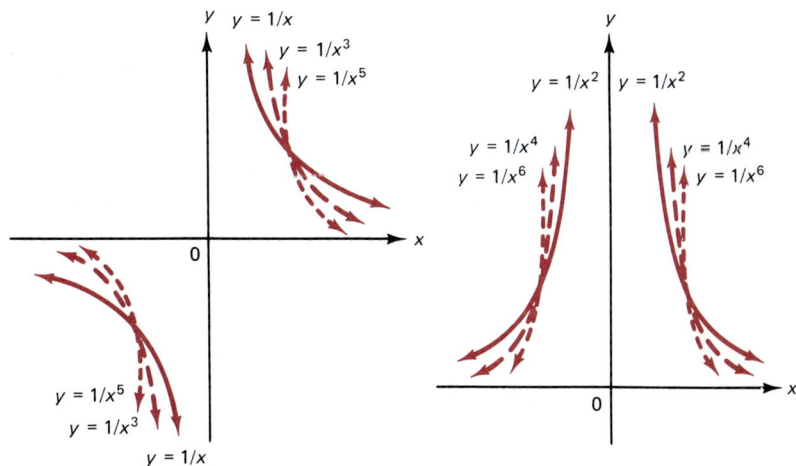

We can see the following facts from Figure 9:

1. The functions (such as $y = 1/x^2$ or $y = 1/x^4$) with *even* exponents are *even* functions (symmetric about the y-axis).
2. The functions (such as $y = 1/x$ or $y = 1/x^3$) with *odd* exponents are *odd* functions (symmetric about the origin).
3. As x becomes closer and closer to zero, y becomes very large in absolute value, and the curve approaches the y-axis as an asymptote.
4. As x increases or decreases without bound, y approaches zero, and the curve approaches the x-axis as an asymptote.

An **asymptote** is a line that a curve approaches but never crosses. We have two special types of asymptotes used in graphing rational functions (see Figure 10):

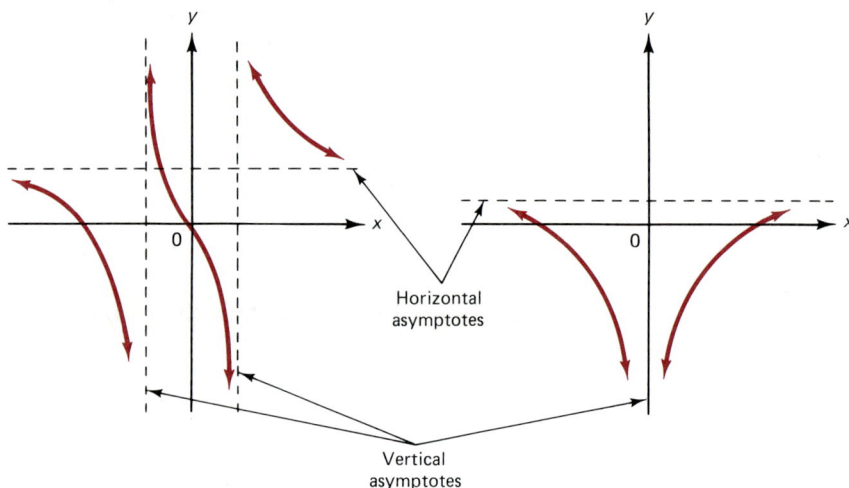

FIGURE 10

---

1. **Vertical Asymptotes**. These are vertical lines of the form $x = c$ that the curve approaches but never crosses. They are usually found where the *denominator is zero* (and the numerator is not zero).
2. **Horizontal Asymptotes**. These are horizontal lines of the form $y = k$ that the curve approaches. We find them by examining the fraction for very large values of x, as follows:
   (a) Divide every term (top and bottom) by the highest power of x that occurs.
   (b) As x gets large, all terms with an x in the denominator (such as $1/x$ or $1/x^2$) go to zero.
   (c) What remains is the horizontal asymptote. (A zero in the denominator means that there is no horizontal asymptote.)

EXAMPLE 22 Find the vertical and horizontal asymptotes of

$$f(x) = \frac{2x^2 - x - 7}{x^2 - 9}$$

Solution *Vertical Asymptotes.* Where is the denominator zero? At $x = \pm 3$. Thus, $x = 3$ and $x = -3$ are the vertical asymptotes.

Horizontal Asymptotes. The highest power of x here is x^2; thus, we divide every term by x^2 and get

$$y = \frac{2x^2 - x - 7}{x^2 - 9} = \frac{2 - \dfrac{1}{x} - \dfrac{7}{x^2}}{1 - \dfrac{9}{x^2}} \rightarrow \frac{2 - 0 - 0}{1 - 0} = 2$$

since the terms $1/x$, $7/x^2$, and $9/x^2$ go to zero as x gets large. Thus, $y = 2$ is the horizontal asymptote.

Let us now give a general procedure that can be used to graph rational functions.

To graph rational functions:

1. Find the vertical asymptotes (where the denominator is zero).
2. Find the horizontal asymptotes (divide by the highest power of x).
3. Find the x-intercepts (where the numerator is zero).
4. Find the y-intercept (where x is zero).
5. Analyze signs of factors to find where the graph is above or below the x-axis.
6. If necessary, plot a few extra points to help complete the sketch.

EXAMPLE 23 Graph $f(x) = \dfrac{2x + 5}{x - 1}$.

Solution We first find the asymptotes and intercepts. We then look at the signs of this quotient to find where the graph is above or below the x-axis. Finally, we find a few extra points to fill in the graph.

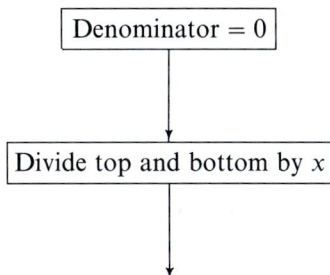

| Denominator $= 0$ | $x = 1$ is vertical asymptote. |

| Divide top and bottom by x | $y = \dfrac{2 + \dfrac{5}{x}}{1 - \dfrac{1}{x}} \rightarrow \dfrac{2 + 0}{1 - 0} = 2$ is horizontal asymptote. |

| Numerator = 0 |

$x = \dfrac{-5}{2}$ is x-intercept.

| $x = 0$ |

$y = \dfrac{0 + 5}{0 - 1} = -5$ is y-intercept.

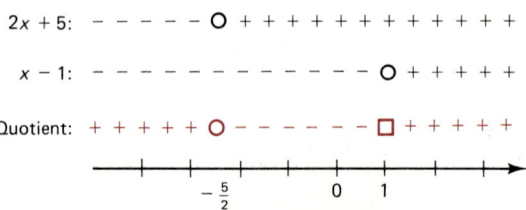

$2x + 5$: $-$ $-$ $-$ $-$ $-$ O $+$ $+$ $+$ $+$ $+$ $+$ $+$ $+$ $+$ $+$ $+$ $+$

$x - 1$: $-$ $-$ $-$ $-$ $-$ $-$ $-$ $-$ $-$ $-$ $-$ O $+$ $+$ $+$ $+$ $+$

Quotient: $+$ $+$ $+$ $+$ O $-$ $-$ $-$ $-$ $-$ $-$ □ $+$ $+$ $+$ $+$

| Analyze signs of quotient |

$-\dfrac{5}{2}$ 0   1

x	y
-4	$3/5$
-1	$-3/2$
3	$11/2$
4	$13/3$

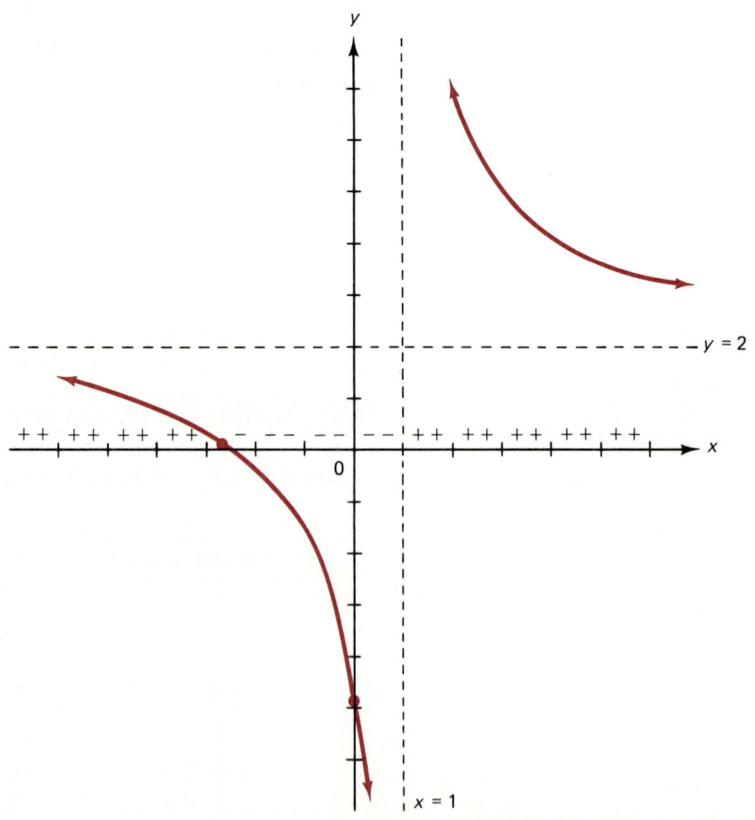

| Find a few extra points |

$y = 2$

| Fill in asymptotes, intercepts, and extra points; then sketch |

$x = 1$

EXAMPLE 24 Graph $f(x) = \dfrac{x-3}{x^2-4}$.

Solution We first find the asymptotes and intercepts. Then we look at the signs and find a few extra points.

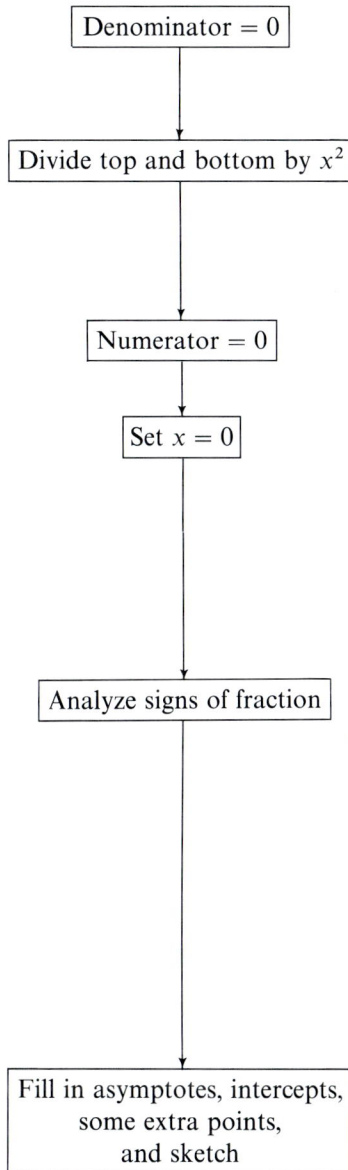

| Denominator = 0 | $x = \pm 2$ are vertical asymptotes. |

| Divide top and bottom by x^2 | $y = \dfrac{\dfrac{1}{x} - \dfrac{1}{x^2}}{1 - \dfrac{4}{x^2}} \to \dfrac{0-0}{1-0} = 0$ |

is horizontal asymptote.

| Numerator = 0 | $x = 3$ is x-intercept. |

| Set $x = 0$ | $y = \dfrac{0-3}{0-4} = \dfrac{3}{4}$ is y-intercept. |

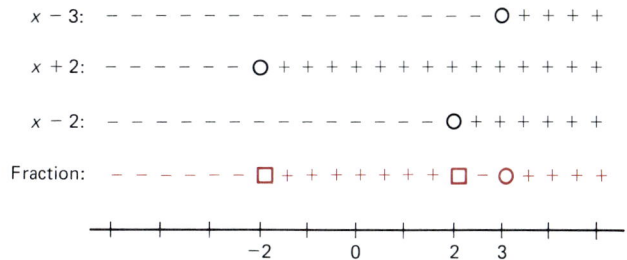

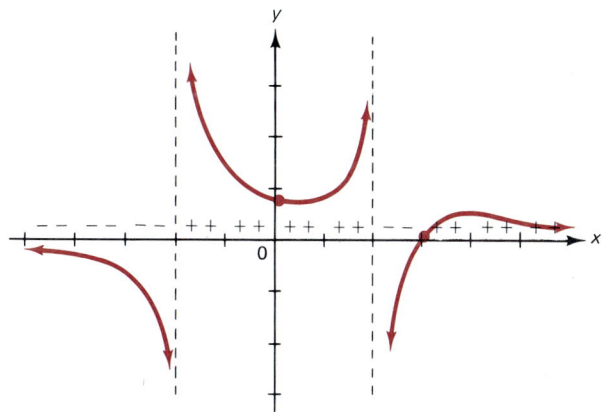

$x - 3$: − − − − − − − − − − − − − − ◯ + + + +

$x + 2$: − − − − − − ◯ + + + + + + + + + + + + +

$x - 2$: − − − − − − − − − − − − − − ◯ + + + + + +

| Analyze signs of fraction |

Fraction: − − − − − − ☐ + + + + + + + ☐ − ◯ + + + +

                    −2   0   2 3

| Fill in asymptotes, intercepts, some extra points, and sketch |

EXAMPLE 25 Graph $f(x) = \dfrac{x^4}{x^4 + 2}$.

Solution We start with the asymptotes and the intercepts.

Denominator $= 0$

$x^4 + 2 > 0$ means there is *no* vertical asymptote.

Divide top and bottom by x^4

$y = \dfrac{1}{1 + \dfrac{2}{x^4}} \rightarrow \dfrac{1}{1 + 0} = 1$

is horizontal asymptote.

Numerator $= 0$

$x = 0$ is x-intercept.

Set $x = 0$

$y = \dfrac{0}{0 + 2} = 0$ is y-intercept.

x^4: $+ + + + + + \, O \, + + + + +$

$x^4 + 2$: $+ + + + + + + + + + + +$

Analyze signs

Fraction: $+ + + + + \, O \, + + + + +$

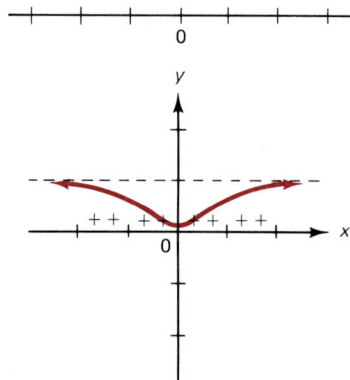

Fill in asymptote, intercept, a few extra points, and sketch

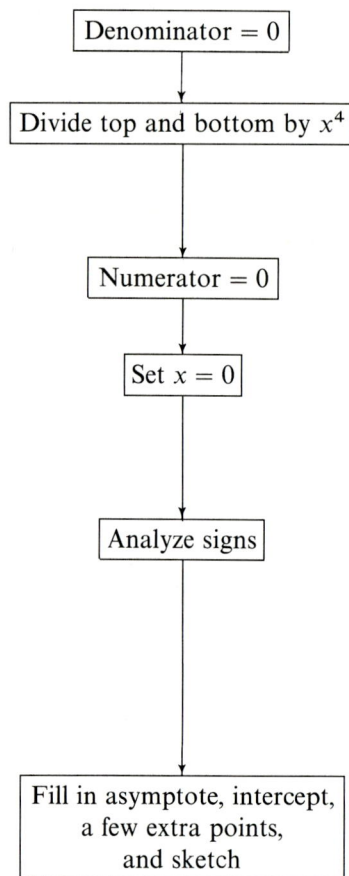

PROBLEM SET
5.4 *Solve the following.*

Warm-up
Exercises

1. $x - 7 = 0$ **2.** $2x + 7 = 0$

3. $5x^2 - 20 = 0$ **4.** $x^2 - 7x + 12 = 0$

5. $x + 4 < 0$ **6.** $3x - 4 > 0$

7. $x^2 + 9 \geq 0$ **8.** $x^2 - 16 \leq 0$

Graph the following functions.

Warm-up
Exercises

9. $f(x) = 3$ **10.** $f(x) = 5 - 3x$

11. $f(x) = x^2 + 1$ **12.** $F(x) = x^3 - x$

CHAPTER 5 Functions

Simplify the following, using only positive exponents.

13. x^{-3}

14. 5^{-2}

15. $x^{-1} + x^{-2}$

16. $(x^{-2} - x^{-3})^{-1}$

Give the vertical and horizontal asymptotes (if any) of the following rational functions.

17. $f(x) = \dfrac{x-1}{x+2}$

18. $f(x) = \dfrac{2x+7}{x-2}$

19. $F(x) = \dfrac{x-1}{x^2-9}$

20. $g(x) = \dfrac{2x+1}{x^2-2x-8}$

21. $f(x) = \dfrac{2x^2+1}{3x^2-3}$

22. $F(x) = \dfrac{x^2}{x^2+1}$

Graph the following rational functions.

23. $F(x) = \dfrac{1}{x^6}$

24. $f(x) = \dfrac{1}{x^5}$

25. $f(x) = \dfrac{2x-4}{3x+4}$

26. $G(x) = \dfrac{4x+3}{5x-7}$

27. $f(x) = \dfrac{x^2-1}{x^2-9}$

28. $F(x) = \dfrac{x^2-4}{x^2-2x-3}$

29. $f(x) = \dfrac{x^4-1}{2x^4+3}$

30. $F(x) = \dfrac{x^4-3x^2+2}{x^4+1}$

31. $h(x) = \dfrac{1}{x-2}$

32. $f(x) = \dfrac{x}{x+5}$

33. $F(x) = \dfrac{x}{x^2+4}$

34. $g(x) = \dfrac{x^2}{x^2-1}$

35. $F(x) = \dfrac{1}{x^2-3x-10}$

36. $K(x) = \dfrac{x^2-2x-15}{2x^2+7x-4}$

37. $f(x) = 1 + \dfrac{4}{x-2}$

38. $G(x) = 2 - \dfrac{3}{x+5}$

39. $f(x) = \dfrac{(x-2)(x-3)}{(x-4)(x-5)}$

40. $f(x) = \dfrac{(x+1)(x+3)(x+5)}{(x-2)(x-4)}$

41. The percent of a market familiar with a product after t weeks of an advertising campaign might be given by

$$P(t) = \frac{100t}{4+t}$$

Graph this function for $t \geq 0$.

42. The gain of a certain amplifier is given by

$$G(R) = \frac{100R}{5000+R}$$

where R is the load resistance. Graph this function for $R \geq 0$.

43. The coefficient of performance of a refrigerator operating between $300°K$ $(27°C)$ and a lower temperature T is given (ideally) by

$$K(T) = \frac{T}{300 - T}$$

Graph this function for $0 \leq T < 300$.

44. If the adult dosage of a medicine is 300 milligrams, the dosage for a child at age a might be given by

$$D(a) = \frac{300a}{12 + a}$$

Graph this function for $0 \leq a \leq 18$.

5.5
OPERATIONS ON FUNCTIONS

If we have two functions f and g, we can combine them algebraically; that is, we can form new functions $f + g$, $f - g$, $f \cdot g$, f/g, and g/f.

EXAMPLE 26   Let $f(x) = \sqrt{x}$ and $g(x) = x - 5$. Below are possible algebraic combinations of f and g. (Notice that the domain D in each case is the intersection of the domains of f and g less the x-values that make the denominator zero.)

(a) $(f + g)(x) = f(x) + g(x) = \sqrt{x} + x - 5$ $\qquad D = \{x \geq 0\}$

(b) $(f - g)(x) = f(x) - g(x) = \sqrt{x} - x + 5$ $\qquad D = \{x \geq 0\}$

(c) $(g - f)(x) = g(x) - f(x) = x - 5 - \sqrt{x}$ $\qquad D = \{x \geq 0\}$

(d) $(f \cdot g)(x) = f(x) \cdot g(x) = \sqrt{x}(x - 5)$ $\qquad D = \{x \geq 0\}$

(e) $\left(\dfrac{g}{f}\right)(x) = \dfrac{g(x)}{f(x)} = \dfrac{x - 5}{\sqrt{x}}$ $\qquad D = \{x \geq 0; x \neq 0\}$

(f) $\left(\dfrac{f}{g}\right)(x) = \dfrac{f(x)}{g(x)} = \dfrac{\sqrt{x}}{x - 5}$ $\qquad D = \{x \geq 0; x \neq 5\}$

In the last two cases we removed from the domain those x-values that make the denominator zero. We can illustrate $f + g$ above by use of our function-machine idea. In Figure 11, we put $x = 9$ into the machine, where it is acted on by both f ("square root") and g ("subtract 5"). The results $f(9) = 3$ and $g(9) = 4$ are then fed into the " + " box, and this gives the final result, 7.

FIGURE 11

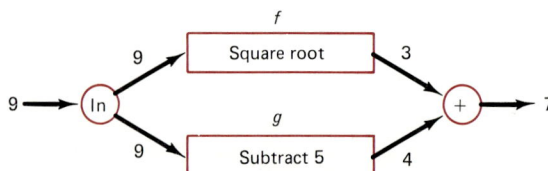

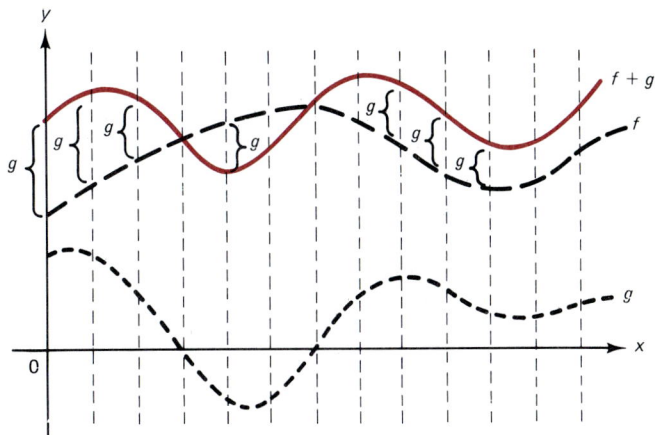

FIGURE 12

We can also show $f + g$ graphically. (The others, $f - g$, $f \cdot g$, and f/g, can also be graphed, but they are harder.) We graph both f and g on the same coordinate system, and then add the y-coordinates along various vertical lines.

In Figure 12, we add the heights $g(x)$ to the heights $f(x)$ for various x-values. These produce the new heights for $f + g$.

EXAMPLE 27   Let $f(x) = x$ and $g(x) = 1/x$. Show $f + g$ graphically.

Solution   We graph f and g separately. For convenience, we add the height of the shorter curve to the height of the taller.

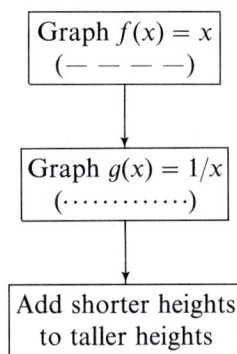

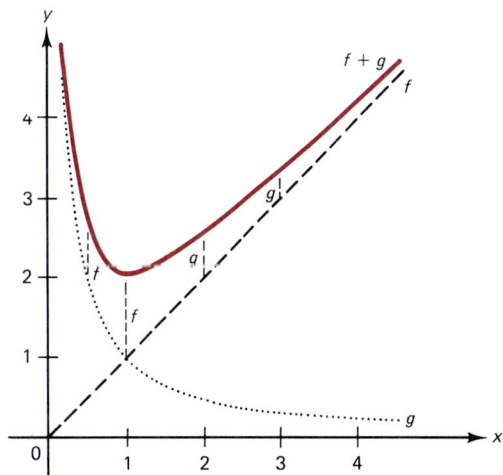

Another very important operation is the composition of two functions. Here we apply one function after the other.

Definition

If f and g are functions (with the range of g in the domain of f), then

$$(f \circ g)(x) = f[g(x)]$$

is the **composite** of f and g: First, we find $g(x)$; then we apply f to $g(x)$ to get $f[g(x)]$. Figure 13 illustrates this.

FIGURE 13

EXAMPLE 28 Let $f(x) = \sqrt{x}$, $g(x) = x^3 + 5$, and $h(x) = 1/x$. The following are various possible composites of these functions. Note that $f \circ g \neq g \circ f$.

(a) $(f \circ g)(x) = f[g(x)] = f[x^3 + 5] = \sqrt{x^3 + 5}$ [First g, then f]

(b) $(g \circ f)(x) = g[f(x)] = g[\sqrt{x}] = (\sqrt{x})^3 + 5$ [First f, then g]

(c) $(h \circ g)(x) = h[g(x)] = h[x^3 + 5] = \dfrac{1}{x^3 + 5}$ [First g, then h]

(d) $(f \circ h)(x) = f[h(x)] = f[1/x] = \sqrt{1/x}$ [First h, then f]

(e) $(f \circ g \circ h)(x) = f[g[h(x)]] = f[g[1/x]]$

$$= f\left[\left(\frac{1}{x}\right)^3 + 5\right]$$

$$= \sqrt{\left(\frac{1}{x}\right)^3 + 5}$$ [First h, then g, then f]

Figure 14 illustrates the case $(f \circ h)(4)$. First, we put 4 into h and out comes its reciprocal, 1/4. We then put 1/4 into f and out comes its square root, 1/2. This is exactly how the composite $f \circ h$ works.

FIGURE 14

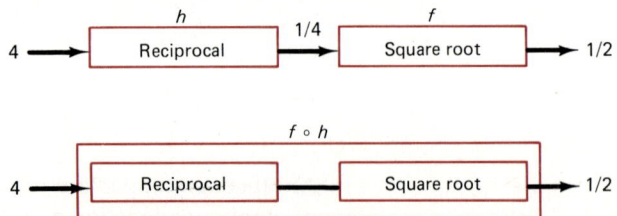

For the following functions, find:

(a) $f(4)$ (b) $f(a + h)$ (c) $f(a + h) - f(a)$ (d) $\dfrac{f(a + h) - f(a)}{h}$

*Warm-up
Exercises*

1. $f(x) = 2x + 7$ **2.** $f(x) = 2x^2 - 3x + 5$

3. $f(x) = x^3$ **4.** $f(x) = \dfrac{1}{x}$

Graph the following functions.

*Warm-up
Exercises*

5. $f(x) = 3x - 1$ **6.** $f(x) = 4 - x^2$

7. $f(x) = \dfrac{1}{x - 2}$ **8.** $f(x) = \sqrt{x + 3}$

Let $f(x) = 2x^2 + 5x - 7$, $g(x) = \sqrt[3]{x}$, and $h(x) = 2x + 3$. Find the indicated functions and their new domains.

9. $(f + g)(x)$ **10.** $(g - h)(x)$

11. $(f \cdot h)(x)$ **12.** $\left(\dfrac{f}{g}\right)(x)$

13. $\left(\dfrac{g}{h}\right)(x)$ **14.** $(f + g - h)(x)$

15. $(g \cdot h)(x)$ **16.** $\left(\dfrac{h}{f}\right)(x)$

17. $\left(\dfrac{fg}{h}\right)(x)$ **18.** $(f \cdot g \cdot h)(x)$

For each of the following pairs of functions, f and g:
(a) Graph f and g on the coordinate plane.
(b) Graph f + g on the same plane. (Be careful with negative y-coordinates.)

19. $f(x) = 2x + 1$; $g(x) = x - 1$ **20.** $f(x) = x^2$; $g(x) = 1 - x$

21. $f(x) = \sqrt{x}$; $g(x) = \dfrac{1}{x}$ **22.** $f(x) = 4 - x^2$; $g(x) = x - 2$

23. $f(x) = 4$; $g(x) = x^2$ **24.** $f(x) = x^2$; $g(x) = \dfrac{1}{x}$

Let $f(x) = x^2 - 1$, $g(x) = \sqrt[3]{x}$, $h(x) = 1/x$, and $k(x) = 5x + 2$. Find the indicated composites.

25. $(f \circ g)(x)$ **26.** $(g \circ f)(x)$

27. $(h \circ f)(x)$ **28.** $(h \circ k)(x)$

29. $(k \circ g)(x)$ **30.** $(k \circ f)(x)$

31. $(f \circ h)(x)$ **32.** $(g \circ k)(x)$

33. $(f \circ g \circ h)(x)$ **34.** $(g \circ h \circ k)(x)$

For the following pairs of functions, f and g, verify that $(f \circ g)(x) = x$; that is, show that g followed by f returns the original x.

35. $f(x) = 5x$; $g(x) = \dfrac{x}{5}$ **36.** $f(x) = x + 5$; $g(x) = x - 5$

37. $f(x) = 2x - 1$; $g(x) = \dfrac{x+1}{2}$ **38.** $f(x) = x^3 - 5$; $g(x) = \sqrt[3]{x+5}$

39. $f(x) = \dfrac{1}{x}$; $g(x) = \dfrac{1}{x}$ **40.** $f(x) = \dfrac{1}{x^2}$; $g(x) = \dfrac{1}{\sqrt{x}}$

Let $f(x) = 1/x^2$, $g(x) = x^5 + x + 1$, $h(x) = \sqrt[6]{x}$, and $k(x) = x^4$. Find the indicated functions.

41. $(f + g)(x)$ **42.** $(f \circ g)(x)$

43. $(h \cdot k)(x)$ **44.** $\left(\dfrac{g}{f}\right)(x)$

45. $(h \circ k)(x)$ **46.** $(g - k)(x)$

47. $(k \circ g)(x)$ **48.** $(f + g + h)(x)$

49. $(h + k \cdot g)(x)$ **50.** $(f - h \cdot k)(x)$

51. $(f \circ g \circ h)(x)$ **52.** $(g \circ f \circ k)(x)$

53. $(f \circ f)(x)$ **54.** $(h \circ h)(x)$

55. $(k \circ k \circ k)(x)$ **56.** $(f \circ f \circ f \circ f)(x)$

Electrical Application

57. It is possible to add the signals of different waves to produce another signal. Shown are three waves whose sum approximates the dashed *square wave*. Graphically add the three waves shown.

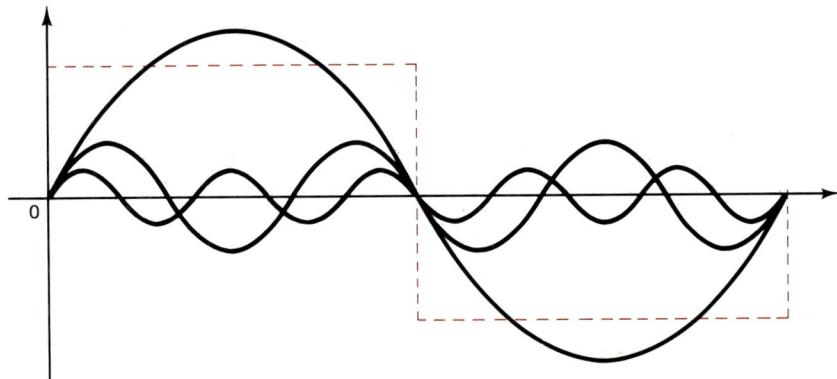

Business Application

58. A company is deciding how many valves q it should order to replenish its inventory. It has two costs to consider:

$$\text{Storage costs:} \quad f(q) = \frac{q}{10}$$

$$\text{Ordering costs:} \quad g(q) = \frac{40{,}000}{q}$$

Graph the functions f and g for $0 \le q \le 1000$. Then show graphically the total cost function $C(q) = f(q) + g(q)$. At what q is the cost a minimum?

5.6
INVERSE FUNCTIONS

We now look at the possibility of reversing the action of a function. One problem is that some functions assign two (or more) x-values to the same y. For example, $f(x) = x^2$ assigns both 4 and -4 to 16. If we tried to reverse the square function, which number would we send 16 back to?

The solution is to look at only those functions that send different x-values to different y-values. These are called **one-to-one functions**; that is, if $a \neq b$, then $f(a) \neq f(b)$. Figure 15 shows a function that is one-to-one and another that is not.

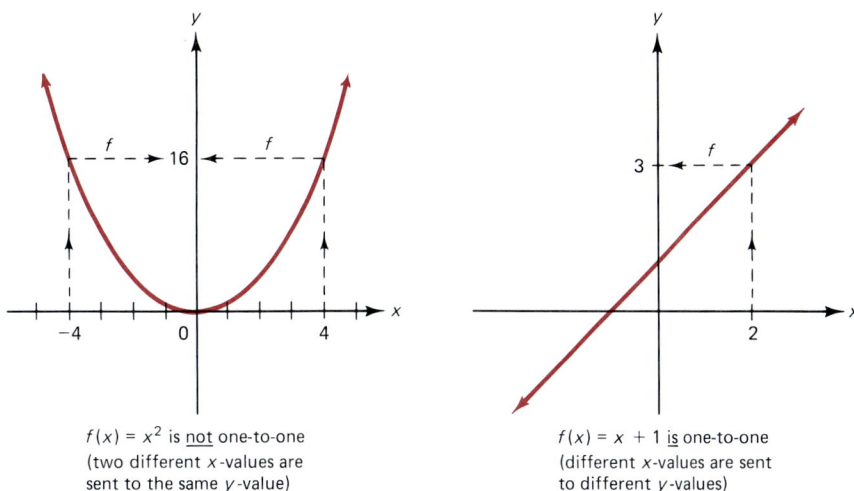

$f(x) = x^2$ is **not** one-to-one (two different x-values are sent to the same y-value)

$f(x) = x + 1$ **is** one-to-one (different x-values are sent to different y-values)

FIGURE 15

To test a function's being one-to-one, we can use the **horizontal-line test** on its graph: If all horizontal lines cross the curve at most once, the function is one-to-one; if some horizontal line crosses the curve at least twice, it is not one-to-one. (Compare this to the vertical-line test, page 148.)

EXAMPLE 29 In Figure 16, curves (a) and (b) are one-to-one, while (c) and (d) are not one-to-one since some horizontal line crosses each of them more than once.

FIGURE 16

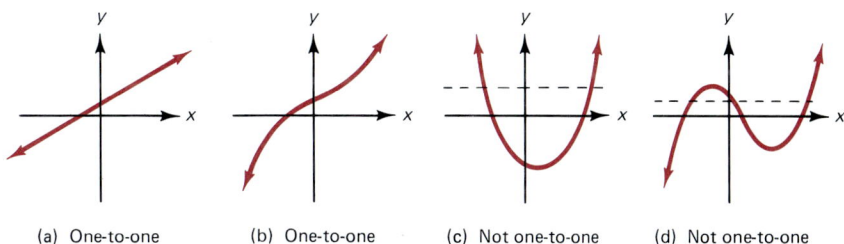

(a) One-to-one   (b) One-to-one   (c) Not one-to-one   (d) Not one-to-one

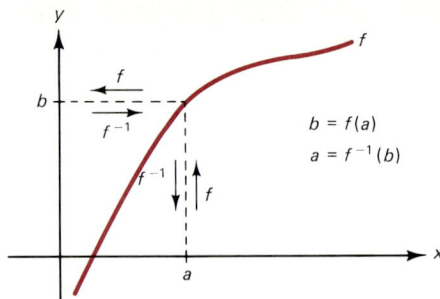

FIGURE 17

If f is one-to-one, then each y-value in the range comes from only one x-value. Thus, if $b = f(a)$, we can reverse the assignment from $a \to b$ to $b \to a$, as shown in Figure 17. If $b = f(a)$, then we say $a = f^{-1}(b)$, where f^{-1} is the notation for the inverse of a one-to-one function f.

Definition
If f is a one-to-one function, then the **inverse function** f^{-1} (or f-**inverse**) is the function that satisfies

$$(f^{-1} \circ f)(x) = x \qquad \text{and} \qquad (f \circ f^{-1})(x) = x$$

In words, if we apply f and then f^{-1}, we get back the original x; or f^{-1} "undoes" whatever f does. Figure 18 illustrates this. Note that the domain of f = the range of f^{-1} and that the range of f = the domain of f^{-1}.

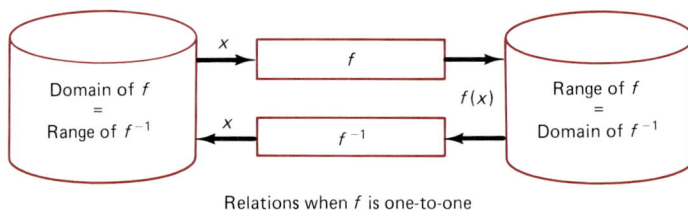

Relations when f is one-to-one

FIGURE 18

YES	NO
f^{-1} is the inverse of f	$f^{-1}(x) = \dfrac{1}{f(x)}$
The -1 is *not* an exponent here, but a symbol for the inverse.	

EXAMPLE 30   The following are simple examples of functions and their inverses. In each case, note how the inverse "undoes" the action of the function.

(a) $f(x) = 4x; f^{-1}(x) = x/4$ [Multiply by 4; divide by 4]
(b) $g(x) = x + 7; g^{-1}(x) = x - 7$ [Add 7; subtract 7]
(c) $h(x) = x^3; h^{-1}(x) = \sqrt[3]{x}$ [Cube; take cube root]

Since a function can be viewed as a set of ordered pairs, we can view the inverse as the interchange of the x- and y-coordinates of all the pairs. For instance,

$$f = \{(1, 1), (2, 4), (3, 9)\} \quad \text{and} \quad f^{-1} = \{(1, 1), (4, 2), (9, 3)\}$$

are inverses, since the assignments, such as $3 \to 9$, are reversed to $9 \to 3$. In general, if (a, b) belongs to f, then (b, a) belongs to f^{-1} (see Figure 19). This means that the graphs of f and f^{-1} will be symmetric about the line $y = x$.

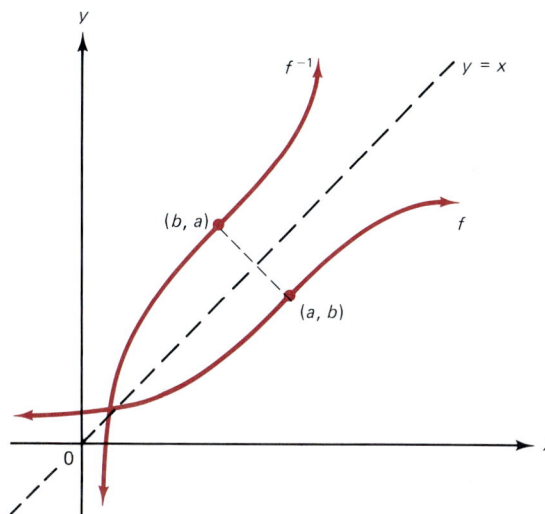

FIGURE 19

If the function f is given as an equation involving x and y, we can find f^{-1} by interchanging x and y in the equation. For instance, if f is defined by $y = 3x + 4$, then f^{-1} is defined by $x = 3y + 4$. However, since it is customary to write y in terms of x, we would solve this equation for y: $y = \dfrac{x - 4}{3}$. This is the equation for f^{-1}.

To find and graph the inverse of a one-to-one function (defined by an equation):

1. Interchange x and y in the equation.
2. Solve this new equation for y. This is the equation for f^{-1}.
3. Graph each function. (Make a table for f; reverse columns for f^{-1}.)
4. As a check, the graphs of f and f^{-1} are symmetric about the line $y = x$.

EXAMPLE 31 Find and graph the inverse of $f(x) = 2x - 5$.

Solution We write the function $y = 2x - 5$, interchange x and y, and solve for y.

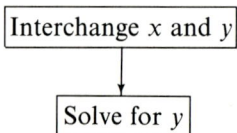

| Interchange x and y |

| Solve for y |

$$f: \quad y = 2x - 5$$
$$f^{-1}: \quad x = 2y - 5$$
$$x + 5 = 2y$$
$$\frac{x + 5}{2} = y$$

Thus, $f^{-1}(x) = \dfrac{x + 5}{2}$. To graph, we make a table for f and reverse the columns for f^{-1}.

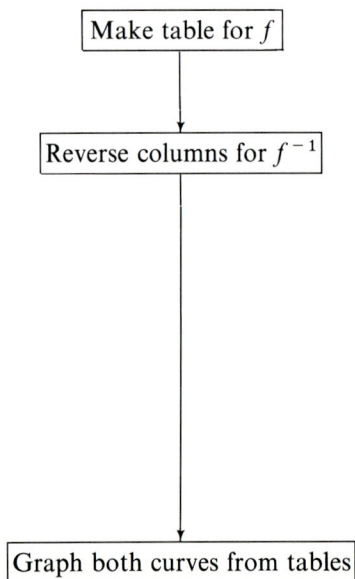

| Make table for f |

| Reverse columns for f^{-1} |

| Graph both curves from tables |

f		f^{-1}	
x	y	x	y
0	-5	-5	0
1	-3	-3	1
2	-1	-1	2
3	1	1	3

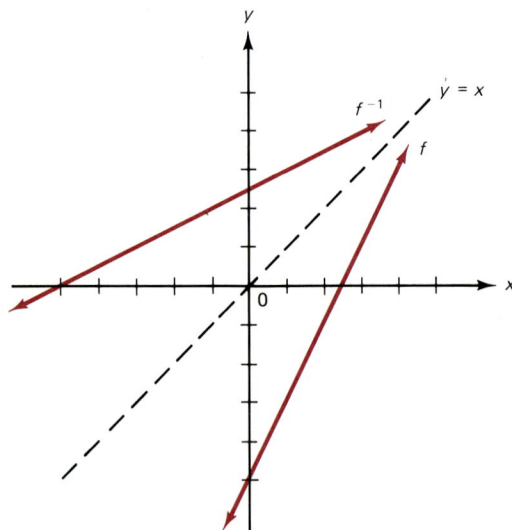

Note how the lines are symmetric about the line $y = x$ (45° line).

EXAMPLE 32 Find and graph the inverse of $f(x) = x^3 - 4$.

Solution We first find the inverse.

| Interchange x and y |

$$f: \quad y = x^3 - 4$$
$$f^{-1}: \quad x = y^3 - 4$$
$$x + 4 = y^3$$
$$\sqrt[3]{x + 4} = y$$

↓

| Solve for y |

Thus, $f^{-1}(x) = \sqrt[3]{x + 4}$. Now we graph f and f^{-1}.

| Make table for f |

↓

| Reverse columns for f^{-1} |

f			f^{-1}	
x	y		x	y
-2	-12		-12	-2
-1	-5		-5	-1
0	-4		-4	0
1	-3		-3	1
2	4		4	2

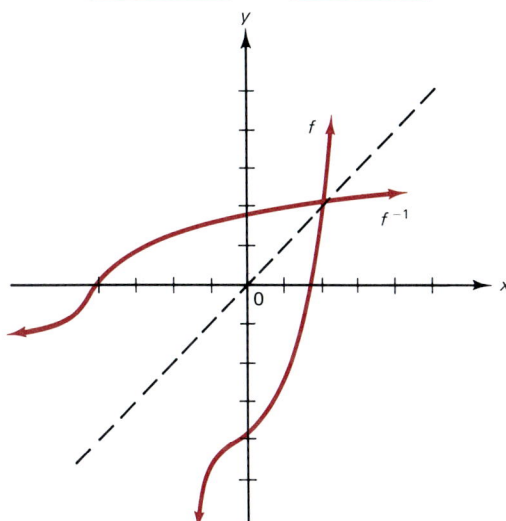

↓

| Graph f and f^{-1} from table (note symmetry) |

EXAMPLE 33 Find and graph the inverse of $f(x) = x^2$ for the domain $[0, \infty)$.

Solution Why do we restrict our domain? We know that $y = x^2$ is *not* one-to-one if the domain is all real numbers; however, by looking only at $x \geq 0$, the function $y = x^2$ becomes one-to-one, as shown below.

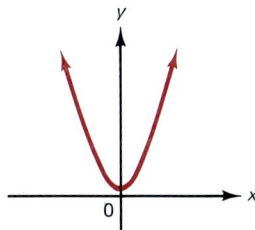

If domain = R,
$y = x^2$ is not one-to-one

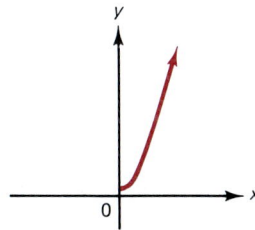

If domain = $[0, \infty]$,
$y = x^2$ is one-to-one

Now we find and graph the inverse.

$$\boxed{\text{Interchange } x \text{ and } y \\ \text{and solve for } y}$$

$$f: \quad y = x^2$$
$$f^{-1}: \quad x = y^2$$
$$\sqrt{x} = y$$

$$\boxed{\text{Graph } f(x) = x^2 \\ \text{and } f^{-1}(x) = \sqrt{x}}$$

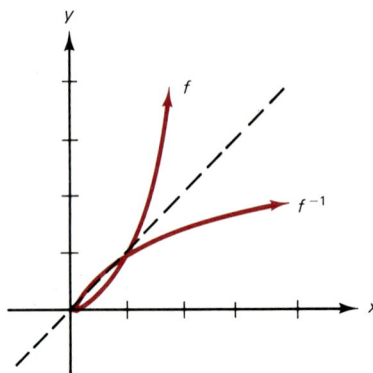

Find the composition $f \circ g$ in the following cases.

Warm-up
Exercises

1. $f(x) = 2x + 1$; $g(x) = x^2 - 7$

2. $f(x) = x^2 + x$; $g(x) = 3x - 2$

3. $f(x) = 5x + 1$; $g(x) = \dfrac{x - 1}{5}$

4. $f(x) = \dfrac{1}{x + 1}$; $g(x) = \dfrac{1 - x}{x}$

Solve the following equations for y (in terms of x).

Warm-up
Exercises

5. $x = y + 1$ **6.** $x = 2y - 7$

7. $x = 2y^3 - 3$ **8.** $x = \sqrt[4]{y + 2}$

9. $x = \dfrac{2}{y - 3}$ **10.** $x = \dfrac{y - 1}{y - 2}$

For each of the following graphs of functions, state whether it is one-to-one or not. If not, restrict the domain to make the function one-to-one.

11.

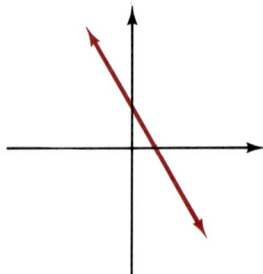

12.

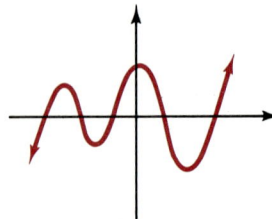

13.

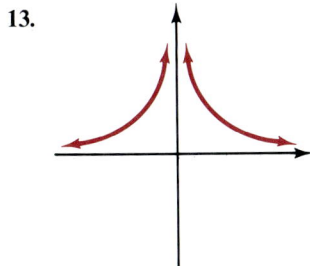

14.

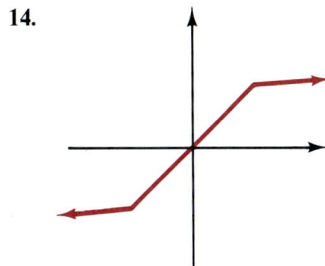

For each of the following functions:
(a) State the action of f in words.
(b) State the action of f^{-1} in words.
(c) Write $f^{-1}(x)$ as an equation.

15. $f(x) = 2x$ **16.** $f(x) = x + 9$ **17.** $f(x) = x - 3$

18. $f(x) = x/5$ **19.** $f(x) = \sqrt[3]{x}$ **20.** $f(x) = x^5$

For the following pairs of functions, verify that f and f^{-1} are inverses by finding both
$f \circ f^{-1}$ and $f^{-1} \circ f$.

21. $f(x) = x^5 + 2; f^{-1}(x) = \sqrt[5]{x - 2}$

22. $f(x) = \dfrac{x - 3}{4}; f^{-1}(x) = 4x + 3$

23. $f(x) = -x; f^{-1}(x) = -x$

24. $f(x) = 5\sqrt[3]{x}; f^{-1}(x) = \left(\dfrac{x}{5}\right)^3$

For each of the following functions, f:
(a) Graph f.
(b) Using the graph of f and the horizontal-line test, restrict the domain of f (if
*   necessary) so that f is one-to-one.*
(c) Find f^{-1}.
(d) Find $(f^{-1} \circ f)(x)$.
(e) Graph f^{-1} on the same plane as f.

25. $f(x) = \dfrac{1}{x}$ **26.** $f(x) = x^3 + 1$

27. $f(x) = -x$ **28.** $f(x) = \dfrac{1}{x - 2}$

29. $f(x) = \dfrac{x}{x + 1}$ **30.** $f(x) = \dfrac{x - 1}{x}$

31. $f(x) = x^2 - 1$ on $[0, \infty)$ **32.** $f(x) = \dfrac{1}{x^2}$ on $[0, \infty)$

33. $f(x) = 3x$ **34.** $f(x) = 4x - 8$

35. $f(x) = x^2 + 2$ **36.** $f(x) = \dfrac{2}{x + 3}$

37. $f(x) = \sqrt{x - 1}$ **38.** $f(x) = 2x^2 - 5$

39. $f(x) = \dfrac{4}{x}$ **40.** $f(x) = x^5 - 9$

41. The temperature Fahrenheit F can be written as a function of the temperature Celsius: $F = \frac{9}{5}C + 32$. Find the inverse of this function by writing C as a function of F.

42. The annual cost C to operate a certain car is a function of the number of miles driven: $C = 1500 + 0.09m$. Find the inverse by writing m (miles) as a function of C (cost).

43. A company's sales S are a function of its advertising budget a (in thousands of dollars): $S = 10,000\sqrt{a}$. Find the inverse by writing a as a function of S.

44. The demand D for an item is a function of its price: $D = 20,000 - 1000p$. Find the inverse by writing p as a function of D.

45. The horizontal distance d (in miles) that an antenna can project a signal is a function of its height (in feet): $d = 1.22\sqrt{h}$. Find the inverse by writing h as a function of d.

**CHAPTER 5
SUMMARY**

Important Definitions and Theorems

A quadratic function $f(x) = ax^2 + bx + c$ $(a \neq 0)$ satisfies the following:

1. The vertex is at $\left(\dfrac{-b}{2a}, f\left(\dfrac{-b}{2a} \right) \right)$.

2. If $a > 0$, the curve opens up and the vertex is a minimum.

3. If $a < 0$, the curve opens down and the vertex is a maximum.

$$(f \circ g)(x) = f[g(x)] \qquad (f^{-1} \circ f)(x) = x$$

Review Exercises

Let $f(x) = 1/x$.

1. Find $f(2)$.

2. Find $f\left(\dfrac{2}{5} \right)$.

3. Find $f(-3)$

4. Find $f(a + h)$.

5. Find $f(a + h) - f(a)$.

6. Find $\dfrac{f(a + h) - f(a)}{h}$.

7. Give the domain of f.

State whether each of the following sketches represents a function.

8.

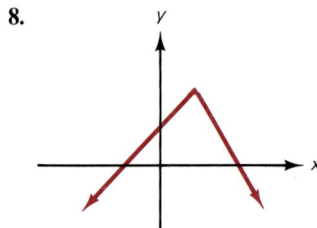

9.

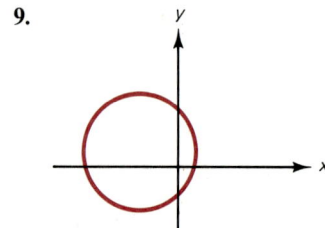

Graph the following functions.

10. $f(x) = 1 - 4x$

11. $f(x) = |x - 2|$

12. $f(x) = \sqrt{x - 2}$

13. $f(x) = \sqrt{x^2 - 4}$

14. $f(x) = [\![2x]\!]$

15. $f(x) = \begin{cases} 0 & \text{if } x < 0 \\ 2x & \text{if } 0 \le x \le 3 \\ 9 - x & \text{if } x > 3 \end{cases}$

16. $f(x) = -x^2 + 4x + 2$ (Also give the vertex.)

17. $f(x) = (x + 3)(x - 1)$

18. $f(x) = x^3 - x$

19. $f(x) = \dfrac{1}{x^2}$

20. $f(x) = \dfrac{x - 1}{2x - 5}$

21. $f(x) = \dfrac{x - 5}{x^2 - 4}$

Let $f(x) = 1/x$, $g(x) = x^2 + 5$, $h(x) = \sqrt[5]{x}$. Find the indicated functions.

22. $(f + g)(x)$

23. $(g - h)(x)$

24. $(f \cdot h)(x)$

25. $\left(\dfrac{f}{g}\right)(x)$

26. $(f \circ g)(x)$

27. $(g \circ h)(x)$

State whether each of the following functions is one-to-one.

28.

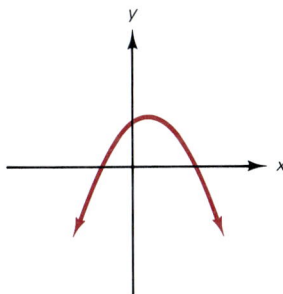

29.

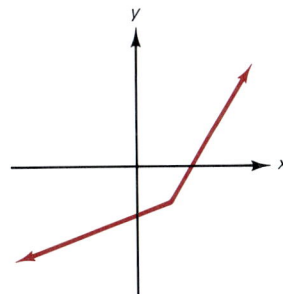

Find f^{-1} and graph f and f^{-1}.

30. $f(x) = 3x - 2$

31. $f(x) = \dfrac{x - 2}{x}$

More Graphs and Functions

TRANSLATION   Often two different equations produce almost the same graph, except that one is shifted, right or left, up or down, from the other. Figure 1 illustrates this idea.

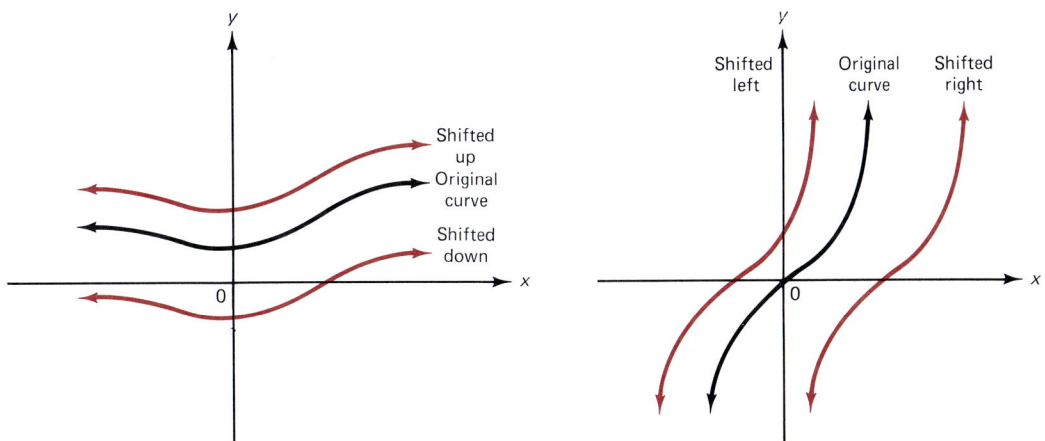

FIGURE 1

EXAMPLE 1   Figure 2 shows four related equations and their graphs.
   (a) Figure 2(a) shows the familiar graph of the equation $y = x^2$, with its vertex at $(0, 0)$.
   (b) Figure 2(b) shows the graph of the equation $y = (x - 3)^2$. Its shape is the same as that of $y = x^2$, except shifted 3 units *right*. Its vertex is at $(3, 0)$.
   (c) Figure 2(c) shows the graph of the equation $(y - 2) = x^2$. Its shape is the same as that of $y = x^2$, except shifted 2 units *up*. Its vertex is at $(0, 2)$.
   (d) Figure 2(d) shows the graph of the equation $(y - 2) = (x - 3)^2$. Its shape is the same as that of $y = x^2$, except shifted both 3 units right and 2 units up. Its vertex is at $(3, 2)$.

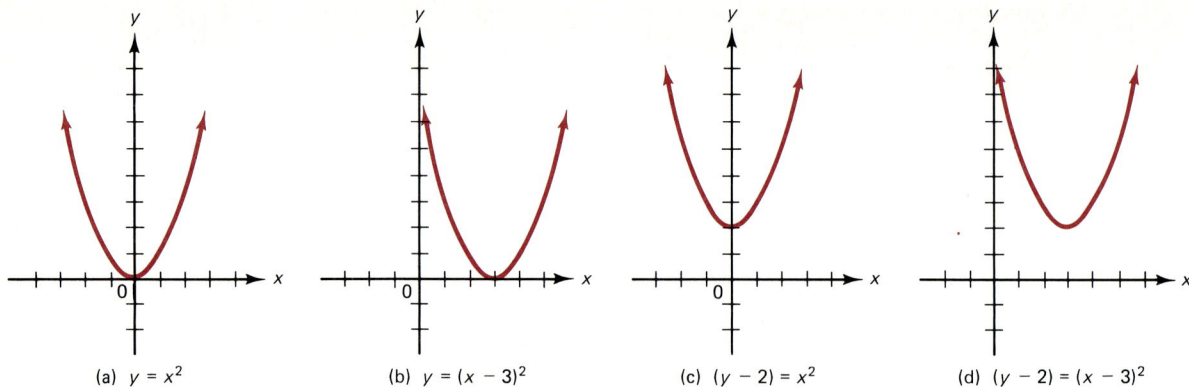

(a) $y = x^2$ (b) $y = (x - 3)^2$ (c) $(y - 2) = x^2$ (d) $(y - 2) = (x - 3)^2$

FIGURE 2

We observe that simply replacing x by $(x - 3)$ or y by $(y - 2)$ has resulted in a shift of the graph right or up. This shift (horizontally, vertically, or both) is also called a **translation**. Let us summarize the rules for the translation of a graph.

Replacing y by $y - a$ results in a **vertical translation** of the graph:
(a) If $a > 0$, the translation is $|a|$ units *up*.
(b) If $a < 0$, the translation is $|a|$ units *down*.

Replacing x by $x - a$ results in a **horizontal translation** of the graph:
(a) If $a > 0$, the translation is $|a|$ units *right*.
(b) If $a < 0$, the translation is $|a|$ units *left*.

YES	NO
$y = (x + 3)^2 = (x - (-3))^2$ is $y = x^2$ translated 3 units *left*.	~~$y = (x + 3)^2$ is $y = x^2$ translated 3 units right.~~

When we make the replacement of $x + a$ or $y + a$, the sign of a and the direction of the shift are *opposite*: $a > 0$ shifts left or down, while $a < 0$ shifts right or up.

EXAMPLE 2   Graph the equation $y = \dfrac{1}{x + 2}$.

Solution   We consider the graph of the equation $y = 1/x$ shifted 2 units to the *left* [since we have $x + 2 = x - (-2)$].

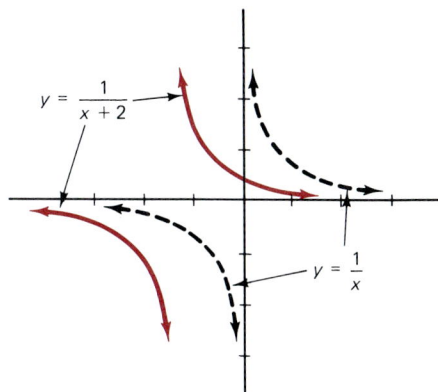

Graph $y = 1/x$

Then shift curve 2 units left

$y = \dfrac{1}{x+2}$

$y = \dfrac{1}{x}$

EXAMPLE 3 Graph the equation $y + 4 = |x - 3|$.

Solution We consider the graph of $y = |x|$, shifted 4 units *down* (since we have $y - (-4)$) and 3 units *right* (since we have $x - 3$).

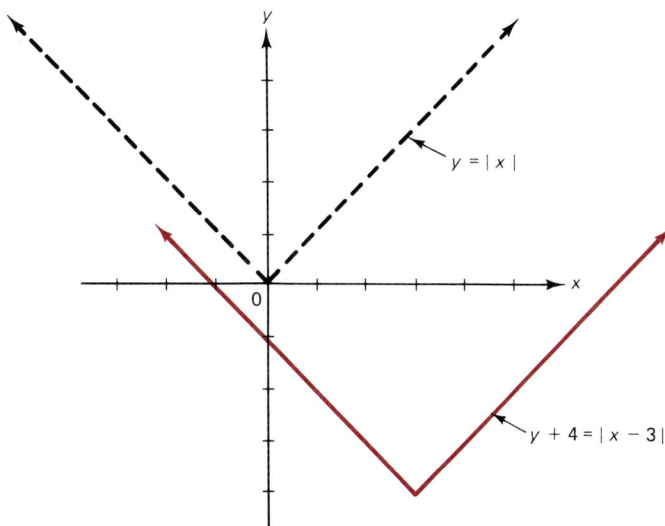

Graph $y = |x|$

Then shift curve 3 units right, 4 units down

$y = |x|$

$y + 4 = |x - 3|$

EXAMPLE 4 Consider the equation $x^2 + 4y^2 = 10$. If we translate the graph of this equation 6 units left and 3 units up, what is the equation of the new graph?

Solution This is the reverse problem. Here we have the translations and want the new equation. A shift of 6 to the left means replacing x by $x + 6$; a shift of 3 up means replacing y by $y - 3$. Thus, for the new graph, we get

$$(x + 6)^2 + 4(y - 3)^2 = 10$$

Graph the following equations.

1. $y = 2x - 1$ **2.** $y = |x|$

3. $x = y^2$ **4.** $y = \sqrt{x}$

Solve the following equations by completing the square.

5. $x^2 - 6x + 9 = 49$ **6.** $x^2 - 10x + 6 = 17$

Use the translation concept to graph the following equations.

7. $y = (x - 2)^3$ **8.** $y = 2(x + 5) + 1$

9. $y = (x + 3)^2$ **10.** $y = \dfrac{1}{x - 1}$

11. $x = (y - 1)^2$ **12.** $x = \dfrac{1}{y + 2}$

13. $x - 1 = y^2$ **14.** $x + 3 = y^2$

15. $y + 3 = (x + 2)^2$ **16.** $y - 1 = (x + 5)^2$

17. $y - 2 = \sqrt{x + 1}$ **18.** $y + 3 = \sqrt{x - 3}$

Rewrite the following equations with the indicated translations. (*Do not graph.*)

19. $y = x^2$, translated 2 units left.

20. $y = x^3$, translated 3 units right.

21. $y = \sqrt{x}$, translated 5 units up.

22. $y = \dfrac{1}{x}$, translated 3 units down.

23. $y = x^2 + x$, translated 3 units left, 2 down.

24. $y = \dfrac{1}{x^2}$, translated 2 units right, 1 down.

25. $x = 4y^2$, translated 4 units left, 3 up.

26. $x = 5y^3 - y$, translated 6 units right, 4 up.

27. $x^2 + y^2 = 25$, translated 4 units right, 2 up.

28. $4x^2 + 9y^2 = 36$, translated 2 units left, 6 up.

29. $\dfrac{x^2}{4} - \dfrac{y^2}{25} = 1$, translated 1 unit right, 8 down.

30. $\dfrac{-x^2}{100} + \dfrac{y^2}{49} = 1$, translated 2 units left, 3 down.

For each of the following equations, perform the necessary algebraic procedure (such as adding a number to both sides or completing the square) to rewrite it as the translation of a simpler equation.

31. $y = x^2 - 1$ **32.** $y = (x - 1)^2 + 4$

33. $y = x^2 - 4x + 4$ **34.** $x = y^2 + 10y + 25$

35. $x^2 + 2x + y^2 = 3$ **36.** $x^2 + y^2 - 6y = 7$

37. $x^2 + 8x - y^2 - 8y = 4$ **38.** $4x^2 + 8x - y^2 + 6y = 3$

39. Show in two ways that $y = f(x) + a$ is a translation of $y = f(x)$ a units up (for $a > 0$):

    (a) Using the ideas of this section.

    (b) By considering the sum of the functions $y = f(x)$ and $y = a$.

40. Show that $y = f(x - a)$ is a translation of $y = f(x)$ a units right (for $a > 0$).

6.2
CONIC SECTIONS I (PARABOLAS AND CIRCLES)

Over 2000 years ago the Greeks were enchanted by a special set of curves, called **conic sections**. The curves got this name since they can be cut from cones, as shown in Figure 3. In this section we discuss the parabola and the circle. In the next section we discuss the ellipse and the hyperbola.

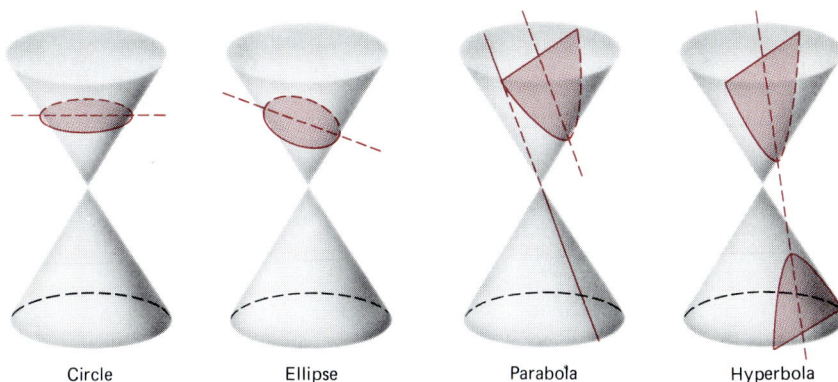

Circle   Ellipse   Parabola   Hyperbola

FIGURE 3

The Parabola

The **parabola** can be seen as the path of a thrown object, the cross section of a radar scanner, or the cross section of a flashlight reflector (as shown in Figure 4).

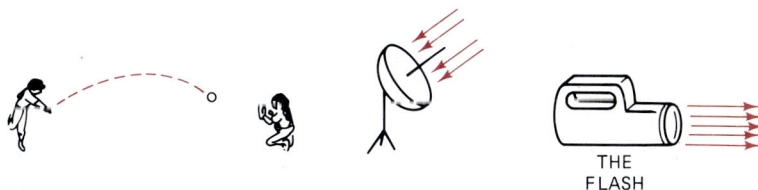

THE FLASH

FIGURE 4

Formally, the definition of a **parabola** is the set of all points equidistant from a point (called the **focus**) and a line (called the **directrix**). Figure 5 illustrates this definition for a parabola in a standard position [focus at $(0, b)$, directrix of $y = -b$, and vertex at the origin].

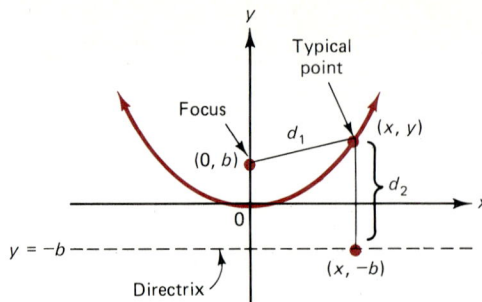

FIGURE 5

The typical point (x, y) on the parabola satisfies $d_1 = d_2$, or more conveniently $d_1^2 = d_2^2$. We use the distance formula to find an equation involving x and y.

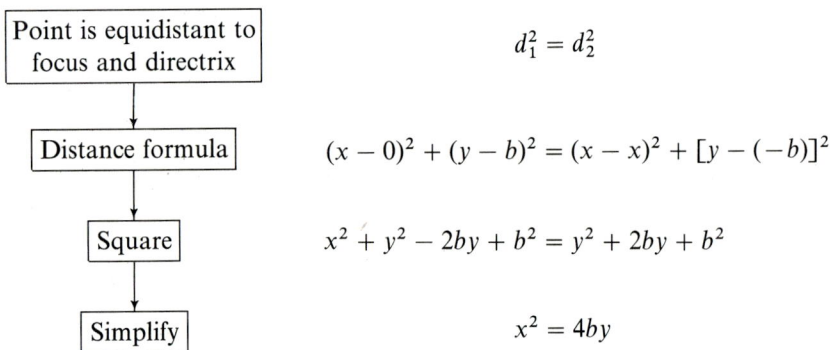

| Point is equidistant to focus and directrix | $d_1^2 = d_2^2$ |

↓

| Distance formula | $(x - 0)^2 + (y - b)^2 = (x - x)^2 + [y - (-b)]^2$ |

↓

| Square | $x^2 + y^2 - 2by + b^2 = y^2 + 2by + b^2$ |

↓

| Simplify | $x^2 = 4by$ |

In a similar way we can find the equation of a parabola opening down, left, or right. As we saw in the preceding section, replacing x by $x - h$ and y by $y - k$ translates a graph right or left and up or down, depending on the signs of h and k. We now look at the equation of a parabola translated so that its vertex is at (h, k).

For $a \neq 0$ the equations of the parabolas with vertices at (h, k) are

$$y - k = a(x - h)^2 \quad \begin{cases} \text{opens up if } a > 0 \\ \text{opens down if } a < 0 \end{cases}$$

$$x - h = a(y - k)^2 \quad \begin{cases} \text{opens right if } a > 0 \\ \text{opens left if } a < 0 \end{cases}$$

Table 1 summarizes these facts.

TABLE 1

Equation Form	$a > 0$	$a < 0$
$y - k = a(x - h)^2$	Opens up Symmetric about $x = h$	Opens down Symmetric about $x = h$
$x - h = a(y - k)^2$	Opens right Symmetric about $y = k$	Opens left Symmetric about $y = k$

EXAMPLE 5 Graph $y = x^2 - 6x + 7$.

Solution We want $y - k = a(x - h)^2$. Recall (see page 29) that $(x - h)^2 = x^2 - 2hx + h^2$. We compare this with our expression:

$$\textit{Perfect square:} \quad x^2 - 2h\,x + h^2$$

$$\textit{Our expression:} \quad x^2 - 6\,x + 7$$

We match -6 with $-2h$ or $h = 3$. Thus, we need $h^2 = 3^2 = 9$ to produce the perfect square $x^2 - 6x + 9 = (x - 3)^2$.

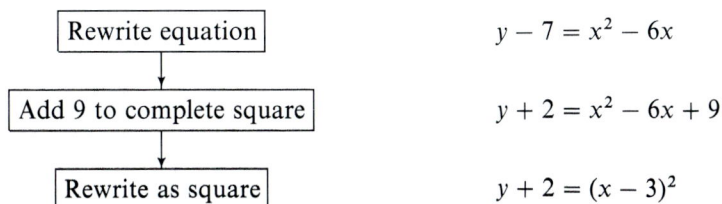

Rewrite equation	$y - 7 = x^2 - 6x$
Add 9 to complete square	$y + 2 = x^2 - 6x + 9$
Rewrite as square	$y + 2 = (x - 3)^2$

This procedure is called **completing the square**, which we saw on page 80. Notice that we add 9 to both sides.

Let us compare this equation to the standard form:

$$\textit{Standard form:} \quad y \boxed{-k} = a(x \boxed{-h})^2$$

$$\textit{Our equation:} \quad y \boxed{+2} = (x \boxed{-3})^2$$

Thus, $a = 1$ and the curve opens up. The vertex $(h, k) = (3, -2)$. Also, the y-intercept is 7.

Vertex = $(3, -2)$

$\downarrow$

y-intercept = 7

$\downarrow$

Curve opens up

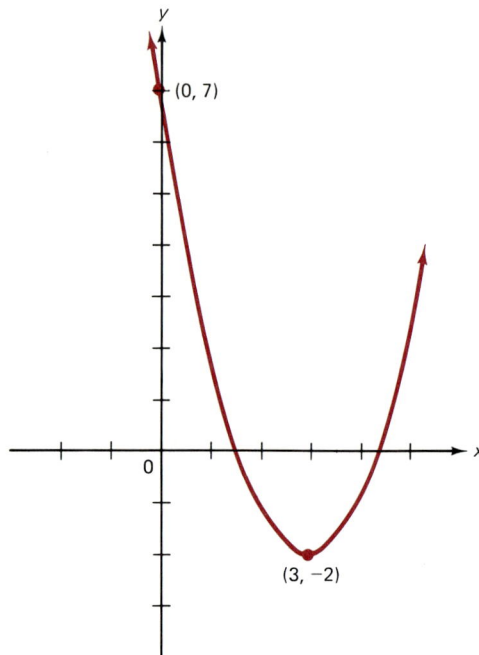

The Circle

We recall from geometry that a **circle** is a set of points the same distance from a center. We use the distance formula to translate the definition of a circle into an equation. As shown in Figure 6, the distance from the center (h, k) to a typical point (x, y) on the circle is always r. Thus, we get the following.

Theorem
The equation of the circle with radius r and center (h, k) is given by

$$(x - h)^2 + (y - k)^2 = r^2$$

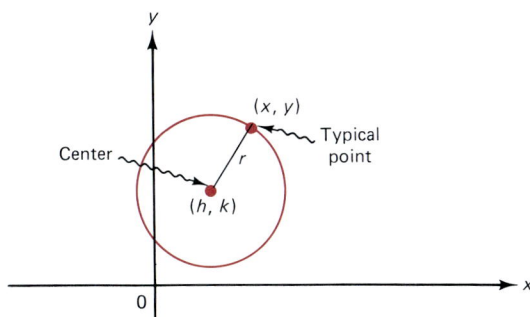

FIGURE 6

EXAMPLE 6   Find the equations of the circles:
(a) With radius 6 and center $(1, -4)$.
(b) With center $(2, 3)$ and through $(1, 6)$.

Solution   (a) We substitute $r = 6$ and $(h, k) = (1, -4)$.

$$\boxed{\text{Substitute}} \qquad (x - 1)^2 + (y - (-4))^2 = 6^2$$

$$\boxed{\text{Simplify}} \qquad (x - 1)^2 + (y + 4)^2 = 36$$

(b) Here we have the center, but no radius. Since $(1, 6)$ is on the circle, the radius is the distance between center $(2, 3)$ and point $(1, 6)$. This distance is

$$d = \sqrt{(1 - 2)^2 + (6 - 3)^2} = \sqrt{1 + 9} = \sqrt{10}$$

We now substitute center $(2, 3)$ and radius $\sqrt{10}$ to get

$$(x - 2)^2 + (y - 3)^2 = 10$$

The graphs of these equations are shown in Figure 7.

FIGURE 7

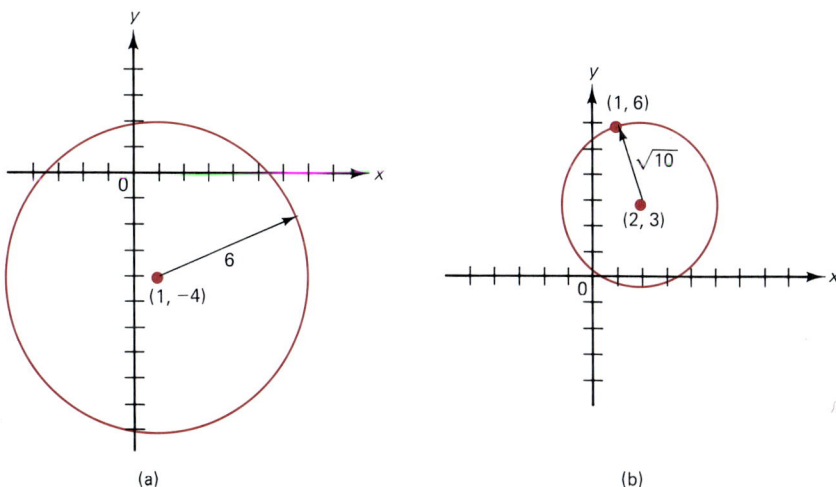

(a)

(b)

EXAMPLE 7   Graph the circle $x^2 + y^2 + 10x - 6y = 2$.

Solution   As in Example 5, we complete the square. We start by comparing our equation to the standard form: $(x - h)^2 + (y - k)^2 = r^2$.

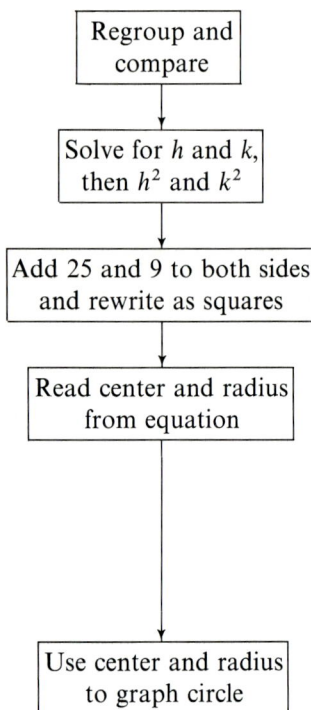

Regroup and compare

Given: $(x^2 \boxed{+ 10}\, x \quad\;) + (y^2 \boxed{- 6}\, y \quad\;) = 2$

Standard: $(x^2 \boxed{- 2h}\, x + h^2) + (y^2 \boxed{- 2k}\, y + k^2) = r^2$

Solve for h and k, then h^2 and k^2

$$h = -5 \qquad\qquad k = 3$$
$$h^2 = 25 \qquad\qquad k^2 = 9$$

Add 25 and 9 to both sides and rewrite as squares

$$(x^2 + 10x + \boxed{25}) + (y^2 - 6y \boxed{+9}) = 2 \boxed{+ 25 + 9}$$
$$(x + 5)^2 \quad + \quad (y - 3)^2 \quad = 36 = 6^2$$

Read center and radius from equation

$$\text{Center} = (h, k) = (-5, 3); \text{ radius} = 6$$

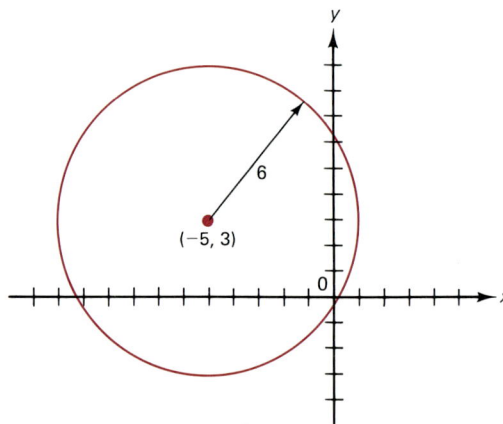

Use center and radius to graph circle

Multiply the following.

Warm-up Exercises

1. $(x - 3)^2$ **2.** $(y - 4)^2$

3. $\left(y + \dfrac{1}{4}\right)^2$ **4.** $\left(x + \dfrac{1}{2}\right)^2$

Solve the following equations by completing the square.

Warm-up Exercises

5. $x^2 - 2x + 1 = 9$ **6.** $x^2 + 6x + 9 = 25$

7. $x^2 - 4x = 5$ **8.** $x^2 + 8x + 1 = 3$

Graph the following quadratic equations by substituting various values for one of the variables.

Warm-up Exercises

9. $y = x^2 + x$ **10.** $x = y^2 - 3$

Graph the following by completing the square and identifying:
(a) *The vertex.*
(b) *The symmetry line.*
(c) *The easiest-to-find intercept.*
(d) *The direction in which the graph opens.*

11. $y = (x - 1)^2 + 2$ 12. $x = -(y + 3)^2 + 4$

13. $x = 3(y - 1)^2 - 5$ 14. $y = -2(x - 5)^2 + 1$

15. $y = x^2 - 2x + 1$ 16. $x = 3y^2 - 6y + 1$

17. $x = y^2 + 4y + 6$ 18. $y = x^2 - 8x - 1$

19. $y = -x^2 + 4x + 1$ 20. $x = -y^2 - 8x + 3$

Find the equation for each of the following circles.

21. Radius $= 2$; center at $(1, 5)$.

22. Radius $= 3$; center at $(-3, 5)$.

23. Radius $= 5$; center at $(0, -4)$.

24. Radius $= 1$; center at $(0, 0)$.

25. Through $(1, 5)$; center at $(3, 2)$.

26. Through $(-2, 1)$; center at $(5, -1)$.

27. Tangent to x-axis; center at $(2, 4)$. (*Hint:* Sketch first.)

28. Tangent to y-axis; center at $(-1, 3)$. (*Hint:* Sketch first.)

Graph the following circles.

29. $(x - 1)^2 + (y - 3)^2 = 25$ 30. $(x + 1)^2 + (y - 7)^2 = 4$

31. $x^2 + (y - 2)^2 = 1$ 32. $x^2 + y^2 = 9$

33. $x^2 + 2x + y^2 = 3$ 34. $x^2 + y^2 - 4y = 5$

35. $x^2 + 4x + y^2 - 6y = 3$ 36. $x^2 + y^2 - 10x + 12y = 7$

37. $x^2 + y^2 + 3x - 9y = 1$ 38. $x^2 + y^2 - 5x - 11y = \dfrac{-1}{2}$

Graph the following.

39. $y = x^2 - 3$ 40. $x = 1 - y^2$

41. $x^2 + y^2 = 64$ 42. $x^2 + 2x + y^2 = 3$

43. $y^2 = 25 - x^2$ 44. $y - x^2 = 5$

45. $x - y + y^2 - 3 = 0$ 46. $x^2 - 10x = 6y - y^2$

47. $x^2 + y^2 - 4x - 2y = 11$ 48. $x^2 - 2x + 5 = y$

49. $x = y^2 + 6y - 1$ 50. $x^2 + 6x = 10y - y^2 + 2$

Find the equations of the following.

51. Circle: radius $= 4$; center at $(2, 4)$.

52. Circle: radius $= 2$; center at $(-2, 1)$.

53. Circle: tangent to $x = 2$; center at $(3, 5)$.

54. Circle: tangent to $y = -3$; center at $(1, 3)$.

55. Parabola: vertex at $(2, 4)$; through $(0, 0)$ and opens down.

56. Parabola: vertex at $(-1, 3)$; through $(2, 0)$ and opens right.

57. The path of a thrown ball is given by

$$y = \frac{-x^2}{100} + x + 5$$

(a) Graph this equation.
(b) How far out in the x-direction does the ball hit the ground ($y = 0$)?
(c) How high does the ball rise?

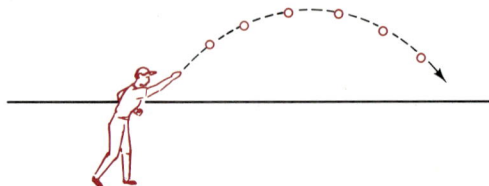

58. The parabolic arch of a concrete bridge is given by

$$y = \frac{(x - 100)^2}{-120} + 70$$

(a) Graph the arch.
(b) How high and how long is the arch?

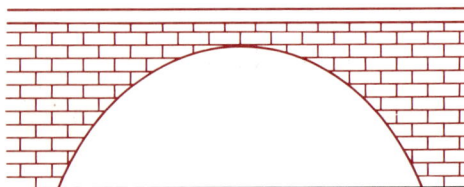

6.3
CONIC SECTIONS II (ELLIPSES AND HYPERBOLAS)

We have just studied the parabola and the circle. We now continue our study of the conic sections and look at the ellipse and the hyperbola.

The Ellipse

The **ellipse** is an oval-like figure that is the path that a planet travels about the sun or the shape of the floor of a whispering gallery. (The ellipse has two focal points, and a person standing on one focal point can hear a whispering person standing on the other.) Figure 8 shows this.

FIGURE 8

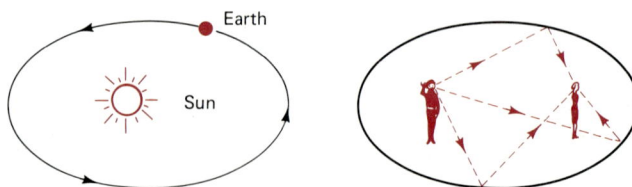

Earth

Sun

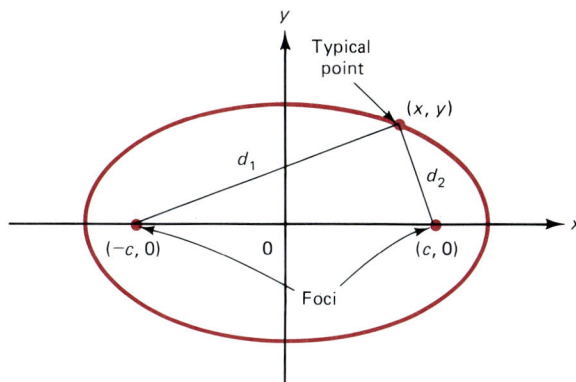

FIGURE 9   For an ellipse, $d_1 + d_2 = $ constant.

Formally, the definition of an ellipse is the set of points the sum of whose distance to two fixed points (called **foci**—the plural of **focus**) is constant. Figure 9 shows an ellipse in a standard position (foci on the x- or y-axis, symmetric about the origin).

Consider the typical point (x, y) on the ellipse. The foci are at $(-c, 0)$ and $(c, 0)$. For convenience, we call the constant total distance $2a$. We start and finish the derivation of the equation of this ellipse; we leave the middle details as an exercise—see Problem 35, page 209.

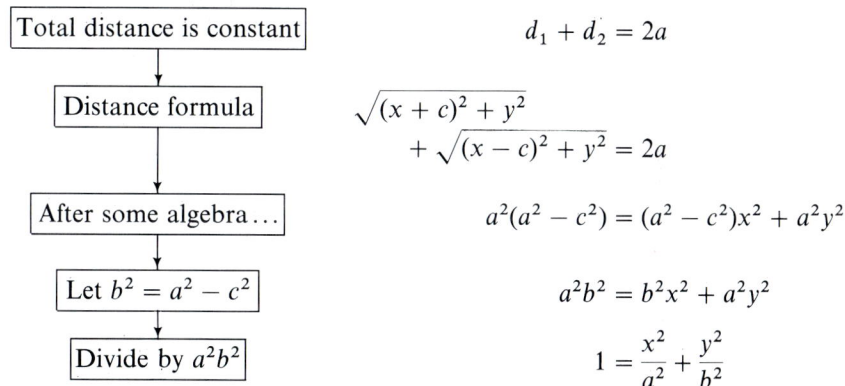

Total distance is constant	$d_1 + d_2 = 2a$
Distance formula	$\sqrt{(x + c)^2 + y^2}$ $+ \sqrt{(x - c)^2 + y^2} = 2a$
After some algebra ...	$a^2(a^2 - c^2) = (a^2 - c^2)x^2 + a^2y^2$
Let $b^2 = a^2 - c^2$	$a^2b^2 = b^2x^2 + a^2y^2$
Divide by a^2b^2	$1 = \dfrac{x^2}{a^2} + \dfrac{y^2}{b^2}$

Therefore, the equation of the ellipse centered at the origin has the standard form

$$\frac{x^2}{a^2} + \frac{y^2}{b^2} = 1$$

where $\pm a$ are the x-intercepts and $\pm b$ are the y-intercepts. We call the four points $(-a, 0)$, $(a, 0)$, $(0, b)$, and $(0, -b)$ the **vertices** of the ellipse.

If $a > b$, the foci are on the x-axis, and the x-axis is called the **major axis** and the y-axis the **minor axis**. If $a < b$, the foci are on the y-axis, and the y-axis is called the major axis and the x-axis the minor axis.

EXAMPLE 8   Graph $9x^2 + 4y^2 = 36$.

Solution   We divide both sides by 36 to put this equation into standard form.

| Divide by 36 |

$$\frac{x^2}{4} + \frac{y^2}{9} = 1$$

| Read intercepts from equation |

$$a^2 = 4 \qquad b^2 = 9$$
$a = \pm 2$ are x-intercepts
$b = \pm 3$ are y-intercepts
The vertices are $(-2, 0)$, $(2, 0)$,
$(0, -3)$, and $(0, 3)$.

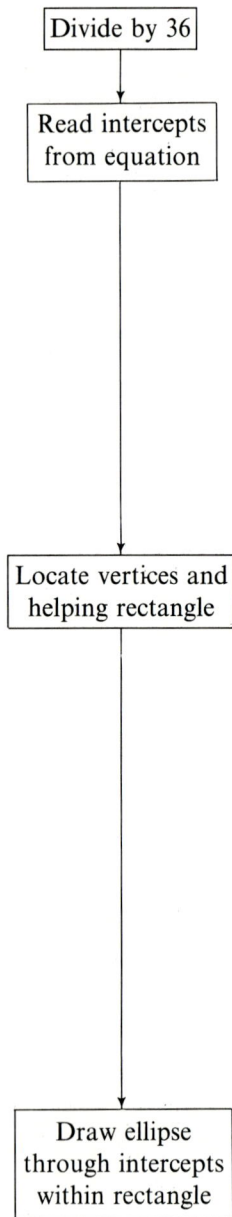

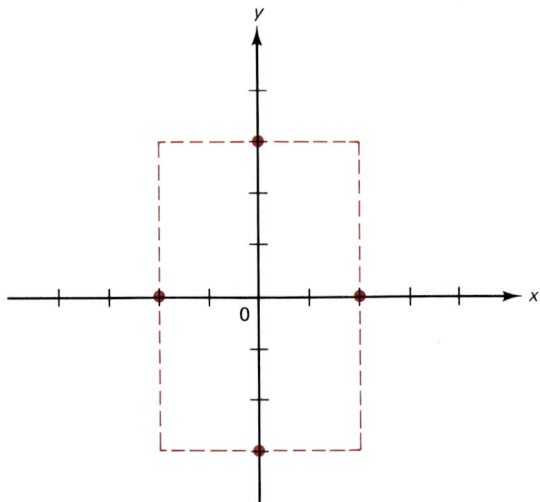

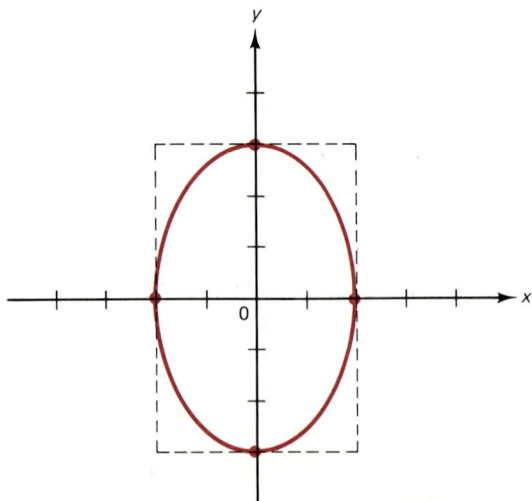

| Locate vertices and helping rectangle |

| Draw ellipse through intercepts within rectangle |

By translating, we see that the standard form of an ellipse centered at (h, k) is given by

$$\frac{(x - h)^2}{a^2} + \frac{(y - k)^2}{b^2} = 1$$

where a and b give the helping rectangle. The four vertices are now $(h + a, k)$, $(h - a, k)$, $(h, k + b)$, and $(h, k - b)$.

EXAMPLE 9 Graph $\dfrac{(x - 1)^2}{25} + \dfrac{(y + 3)^2}{16} = 1$.

Solution This is an ellipse centered at $(1, -3)$. The helping rectangle extends ± 5 in the x-direction and ± 4 in the y-direction. The vertices are $(6, -3)$, $(-4, -3)$, $(1, -7)$, and $(1, 1)$.

Locate center $= (1, -3)$ and vertices; put helping rectangle around center

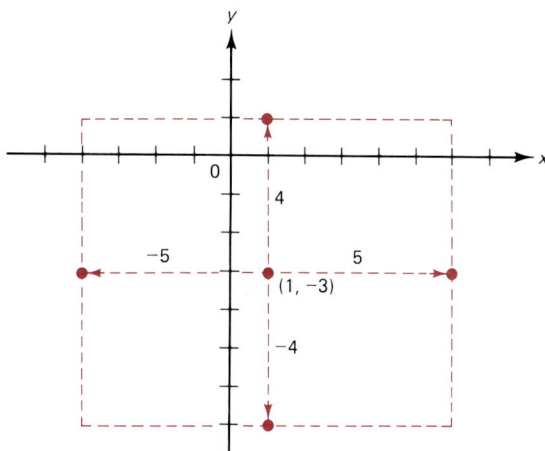

Draw ellipse within rectangle

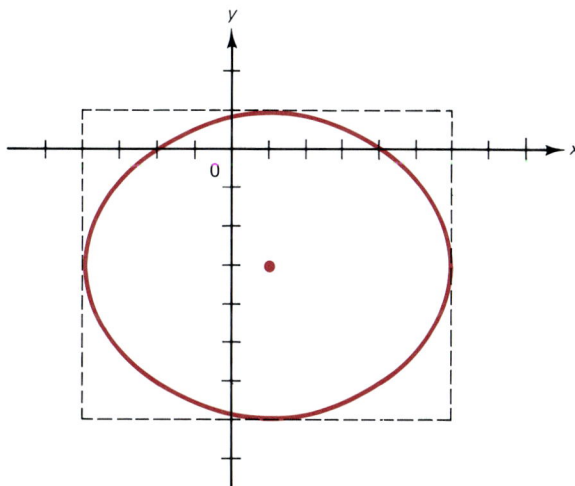

The Hyperbola

The **hyperbola** is a curve that is the path of a "scattered" atomic particle or the edge of the shadow of a lamp shade (see Figure 10).

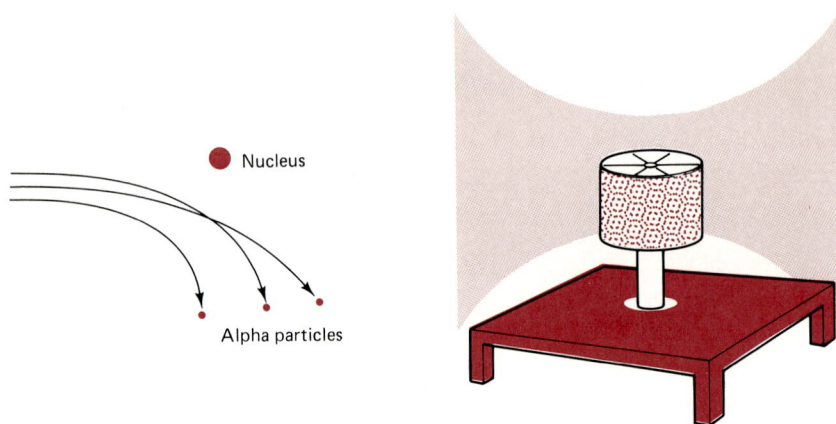

Nucleus

Alpha particles

Formally, the definition of a hyperbola is the set the difference of whose distance to two fixed points (called foci) is constant (see Figure 11). Note that this definition involves "difference," while the definition of the ellipse involves "sum."

FIGURE 11   For a hyperbola, $d_1 - d_2 =$ constant, or $d_1 - d_2 =$ constant.

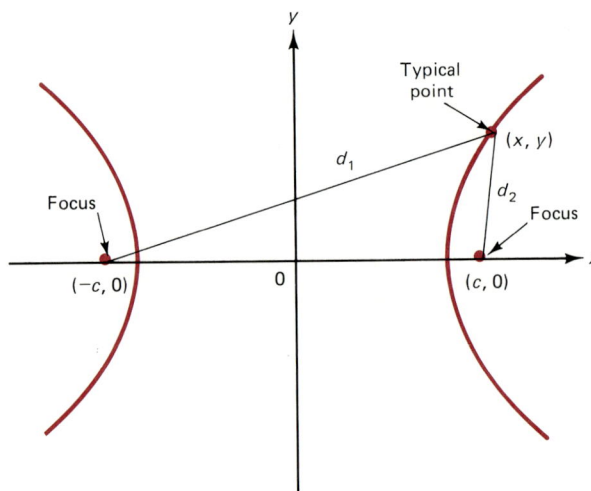

Typical
point

(x, y)

d_1

d_2

Focus

Focus

$(-c, 0)$

0

$(c, 0)$

The derivation of the equation of the hyperbola follows that for the ellipse very closely, except that we start with $d_1 - d_2 = 2a$ and make the replacement $b^2 = c^2 - a^2$. Ultimately, we find that the equations of a hyperbola centered at the origin are given by

$$\frac{x^2}{a^2} - \frac{y^2}{b^2} = 1 \qquad \text{and} \qquad -\frac{x^2}{a^2} + \frac{y^2}{b^2} = 1$$

Note that one term is positive and the other is negative. Unlike the ellipse, the hyperbola has only one pair of vertices: on the *x*-axis or the *y*-axis, but *not* both.

However, like the ellipse, the hyperbola has a helping rectangle with dimension 2a by 2b (see Figure 12). The diagonals through the corners are called **asymptotes**. These are lines that the hyperbola approaches, but never crosses (see page 168). The asymptotes are *not* part of the graph, just helpers. If the hyperbola opens left and right, the vertices are $(-a, 0)$ and $(a, 0)$; if it opens up and down, the vertices are $(0, b)$ and $(0, -b)$.

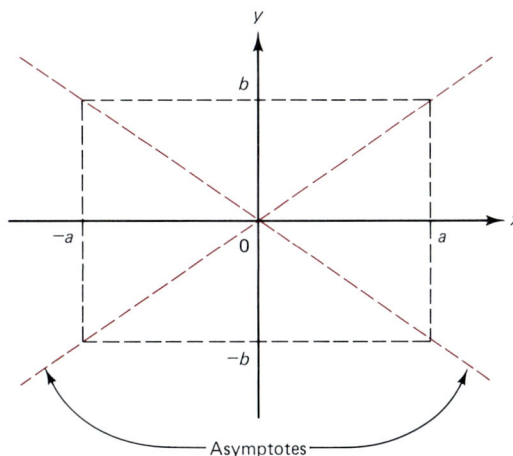

FIGURE 12

EXAMPLE 10 Graph $\dfrac{x^2}{25} - \dfrac{y^2}{4} = 1$.

Solution If we set $y = 0$, we find that the *x*-intercepts are at $x = \pm 5$. However, if we set $x = 0$, we get $y^2 = -4$, which has no real solution. Although there are no *y*-intercepts, we use the values ± 2 to construct the helping rectangle.

Locate vertices;
sketch helping rectangle
and asymptotes

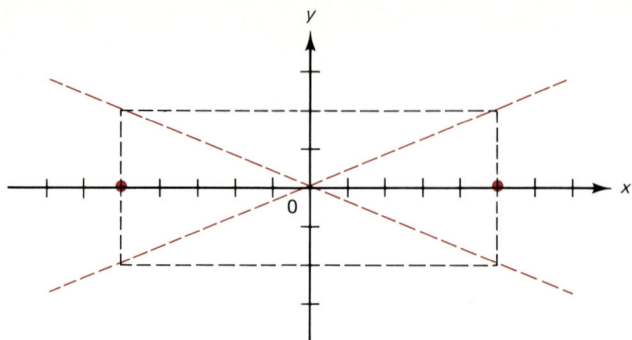

Draw hyperbola
through vertices
approaching asymptotes

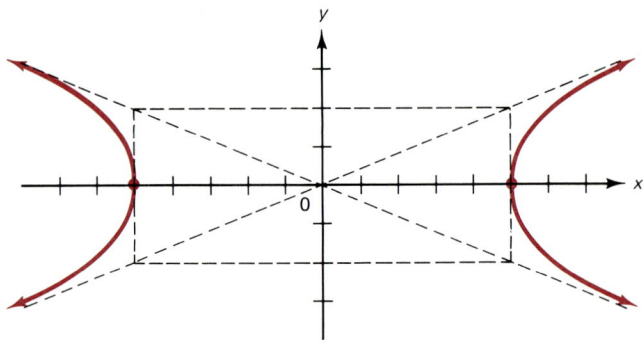

By translating, we can center a hyperbola at (h, k) and get the standard forms

$$\frac{(x-h)^2}{a^2} - \frac{(y-k)^2}{b^2} = 1 \quad \text{and} \quad -\frac{(x-h)^2}{a^2} + \frac{(y-k)^2}{b^2} = 1$$

In the first case, the vertices are $(h + a, k)$ and $(h - a, k)$; in the second case, the vertices are $(h, k - b)$ and $(h, k + b)$.

EXAMPLE 11 Graph $-\dfrac{(x-1)^2}{9} + \dfrac{(y+2)^2}{16} = 1$.

Solution We can look at the equation and read off the following information:

1. The center is at $(1, -2)$.
2. The helping rectangle extends ± 3 in the x-direction and ± 4 in the y-direction. The vertices are $(1, 2)$ and $(1, -6)$.
3. Since the y-term is positive, the hyperbola opens up and down.

Locate center $(1, -2)$ and vertices; put helping rectangle around center

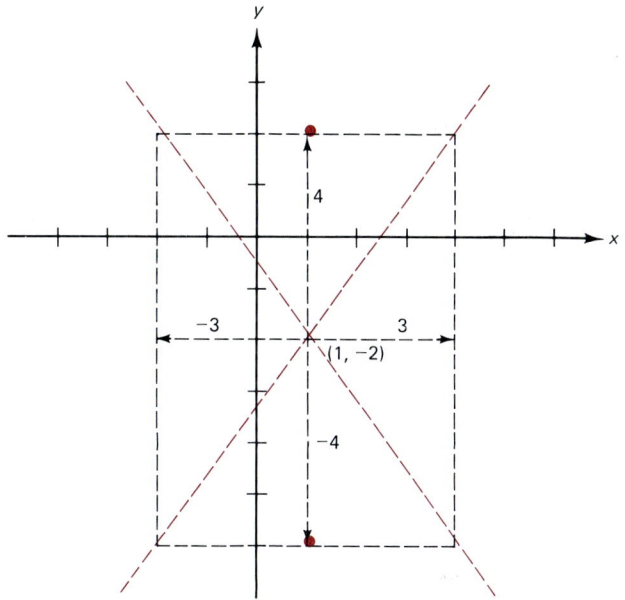

Draw hyperbola opening up and down through vertices approaching asymptotes

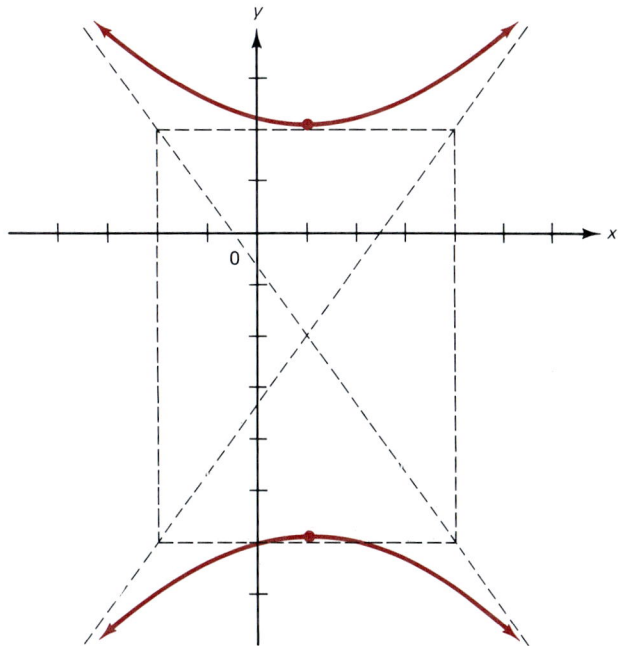

We summarize our discussion of ellipses and hyperbolas in Table 2.

TABLE 2

Equation	Curve	Sketch
$\dfrac{(x-h)^2}{a^2} + \dfrac{(y-k)^2}{b^2} = 1$	Ellipse Vertices: $(h+a, k)$, $(h-a, k)$, $(h, k+b)$, $(h, k-b)$	
$\dfrac{(x-h)^2}{a^2} - \dfrac{(y-k)^2}{b^2} = 1$	Hyperbola, opening right and left Vertices: $(h+a, k)$, $(h-a, k)$	
$-\dfrac{(x-h)^2}{a^2} + \dfrac{(y-k)^2}{b^2} = 1$	Hyperbola, opening up and down Vertices: $(h, k+b)$, $(h, k-b)$	

Graph the following equations.

1. $x + y = 5$

2. $2x - y = 8$

3. $x^2 - 2x + y^2 + 4y = 20$

4. $y = (x - 3)^2 + 4$

Identify and graph the following conic sections.

5. $\dfrac{x^2}{1} + \dfrac{y^2}{16} = 1$

6. $\dfrac{x^2}{100} + \dfrac{y^2}{49} = 1$

7. $4x^2 + y^2 = 16$

8. $9x^2 + 16y^2 = 144$

9. $\dfrac{(x + 1)^2}{16} + \dfrac{(y + 2)^2}{49} = 1$

10. $\dfrac{(x + 5)^2}{25} + \dfrac{(y - 1)^2}{4} = 1$

11. $(x - 4)^2 + 4(y + 1)^2 = 4$

12. $25(x + 1)^2 + 4(y - 3)^2 = 100$

13. $\dfrac{-x^2}{100} + \dfrac{y^2}{9} = 1$

14. $\dfrac{x^2}{81} - \dfrac{y^2}{25} = 1$

15. $4x^2 - 9y^2 = 36$

16. $-25x^2 + 4y^2 = 100$

17. $\dfrac{-(x + 5)^2}{49} + \dfrac{(y + 1)^2}{16} = 1$

18. $\dfrac{(x - 4)^2}{9} - \dfrac{(y - 2)^2}{4} = 1$

19. $\dfrac{y^2}{64} - \dfrac{x^2}{25} = 1$

20. $\dfrac{(x - 1)^2}{16} - \dfrac{(y - 2)^2}{25} = 1$

21. $25x^2 + 4y^2 = 100$

22. $x^2 + 16y^2 = 64$

23. $4x^2 + 8x + y^2 - 2y = 11$

24. $x^2 - 6x - y^2 - 2y = 17$

25. $x = y^2 - 2y$

26. $x^2 + 10x + 9y^2 + 36y = 20$

27. $x^2 + 4x - 4y^2 - 8y = 16$

28. $x^2 - 4x - y = 2$

A circle, an ellipse, a hyperbola, and often a parabola cannot be the graph of a function.
(Why?) For each of the following equations:
(a) Graph the equation.
(b) Restrict the range so that the curve becomes the graph of a function. Then express y
as a function of x.
(c) Restrict the domain so that the function is also one-to-one (see page 179). Then find
and graph the inverse function.

29. $x^2 + y^2 = 4$

30. $x^2 + 4y^2 = 16$

31. $x^2 - 4y^2 = 4$

32. $-9x^2 + 4y^2 = 36$

33. $25x^2 + 4y^2 = 100$

34. $x = y^2$

35. Complete the derivation of the equation of an ellipse in standard position.

36. Complete the derivation of the equation of a hyperbola in standard position.

37. Often people need to draw an ellipse for a garden or to make a piece of furniture. One method is to tack down a string at the two focal points. A pencil

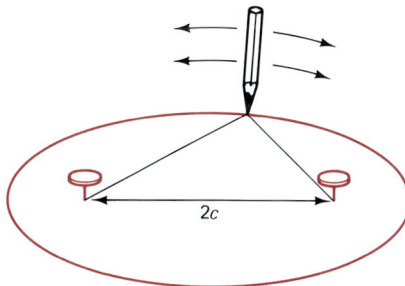

or pen is then moved around with the string held taut, as shown at the left. If the desired dimensions are $2a$ by $2b$ (assume that $a > b$), then the string length is $2a$ (why?). The tacks are placed at a distance of $2c$, where $c^2 = a^2 - b^2$ (why?).

(a) Answer the "why" questions above.

(b) Use this technique to draw ellipses of total dimension 3 by 2 and 6 by 7.

Physical Application   **38.** The earth has a radius of about 4000 miles. Suppose that a satellite is launched that has an elliptical orbit centered at the center of the earth. It is 6000 miles from the earth at its farthest, and 2000 miles at its closest. Find the equation of this ellipse.

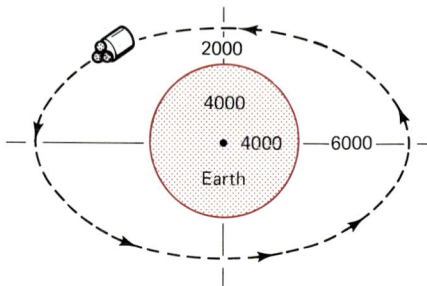

6.4
GRAPHING
INEQUALITIES

On pages 92 to 100 we studied inequalities in one variable, such as $x + 2 < 7$ or $x^2 - 4 \geq 5$. Recall that the solutions were intervals on the number line rather than an isolated point or two.

In this section we look at **inequalities in two variables**, such as

$$x + 2y < 4 \qquad x^2 + y^2 \geq 9 \qquad \text{or} \qquad \frac{x^2}{4} - \frac{y^2}{9} \leq 1$$

These inequalities, in general, separate the plane into regions:

1. The YES region, where the points all satisfy the inequality. This is the solution set; so we want the YES region.

2. The NO region, where none of the points satisfies the inequality. Thus, the solution set (the YES region) is, in general, the *other* region.

Figure 13 illustrates this. The key to these inequalities is first graphing the corresponding equation.

FIGURE 13

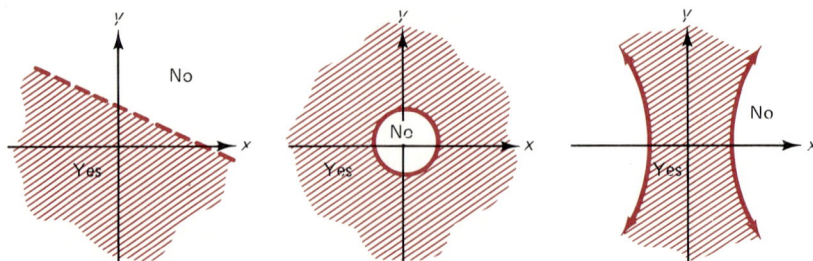

210   CHAPTER 6 More Graphs and Functions

To graph inequalities in two variables:

1. Graph the corresponding equation with a light dashed line or curve, dividing the plane into regions.
2. Choose a test point P (not on the curve) and determine if it satisfies the inequality.
3a. If P *does satisfy* the inequality, it is in the YES region. Shade the region containing P.
3b. If P does *not* satisfy the inequality, it is in the NO region. In general, the *other* region is the YES region to be shaded.
4a. If the inequality is $\leq$ or $\geq$, include the curve.
4b. If the inequality is $<$ or $>$, exclude the curve (use a dashed line).

EXAMPLE 12   Graph $2x - 5y \leq 10$.

Solution   We begin by graphing the equation $2x - 5y = 10$. This was done on page 126. We use $(0, 0)$ as the test point.

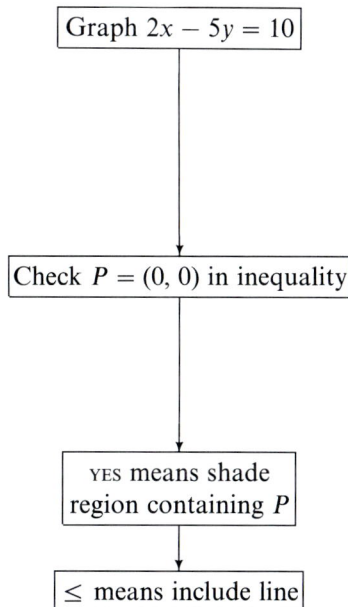

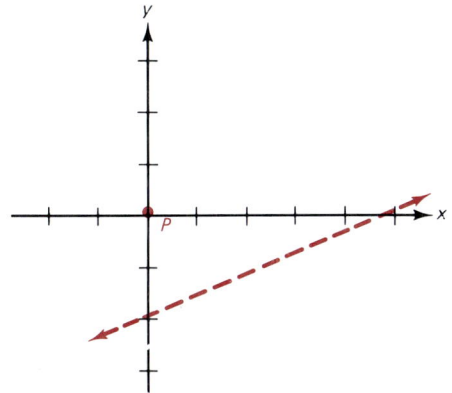

| Graph $2x - 5y = 10$ |

| Check $P = (0, 0)$ in inequality |

Is $2(0) - 5(0) \leq 10$?   YES

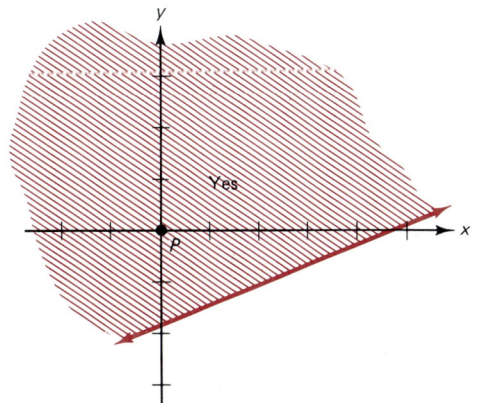

| YES means shade region containing P |

| $\leq$ means include line |

EXAMPLE 13   Graph $y > x^2 - 6x + 7$.

Solution   We first graph the equation $y = x^2 - 6x + 7$. This was done in Example 5 (see page 195 for details). We again use $(0, 0)$ as a test point.

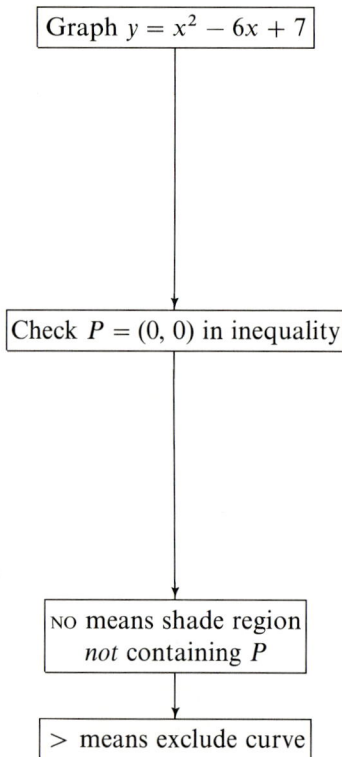

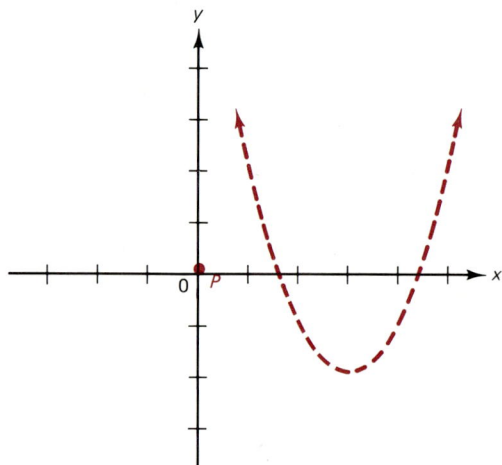

| Graph $y = x^2 - 6x + 7$ |

| Check $P = (0, 0)$ in inequality | Is $0 > 0^2 - 6(0) + 7$?   NO

| NO means shade region *not* containing P |

| > means exclude curve |

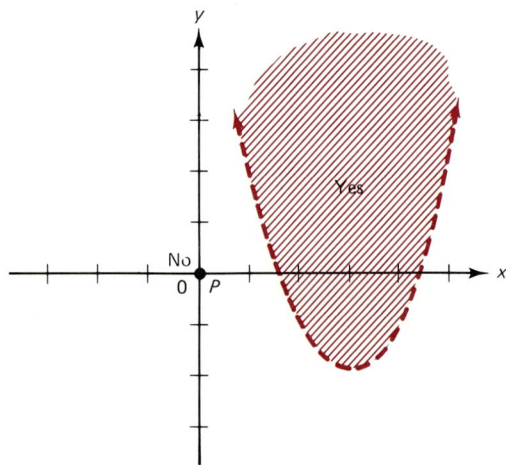

EXAMPLE 14   Graph $\dfrac{(x - 1)^2}{25} + \dfrac{(y + 3)^2}{16} < 1$.

Solution   We begin by graphing the corresponding equation, which was done in Example 9 (see page 203 for details). Here we use the center of the ellipse $(1, -3)$ as the test point.

Graph

$$\frac{(x-1)^2}{25} + \frac{(y+3)^2}{16} = 1$$

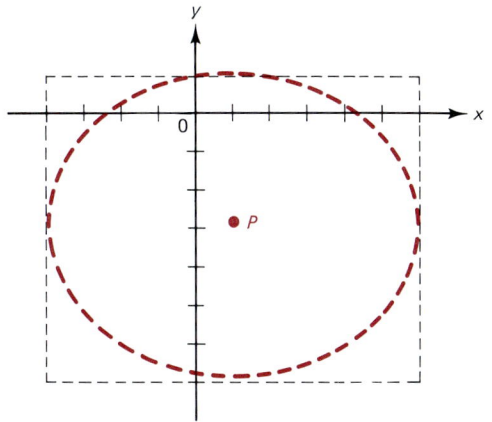

Check $P = (1, -3)$ in inequality

Is $\dfrac{(1-1)^2}{25} + \dfrac{(-3+3)^2}{16} < 1?$ YES

YES means shade region containing P

< means exclude curve

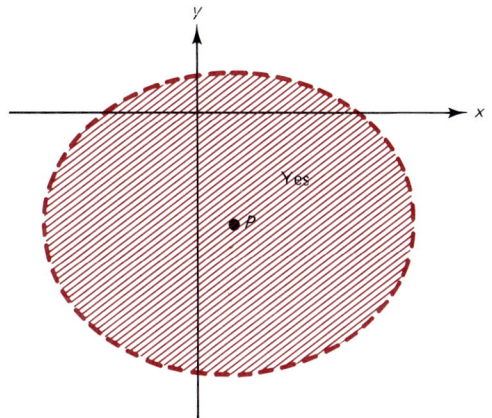

EXAMPLE 15   Graph $\dfrac{x^2}{25} - \dfrac{y^2}{4} \geq 1$.

Solution   The corresponding equation was graphed in Example 10 (see page 205 for details). We use $(0, 0)$ as a test point.

Graph $\dfrac{x^2}{25} - \dfrac{y^2}{4} = 1$

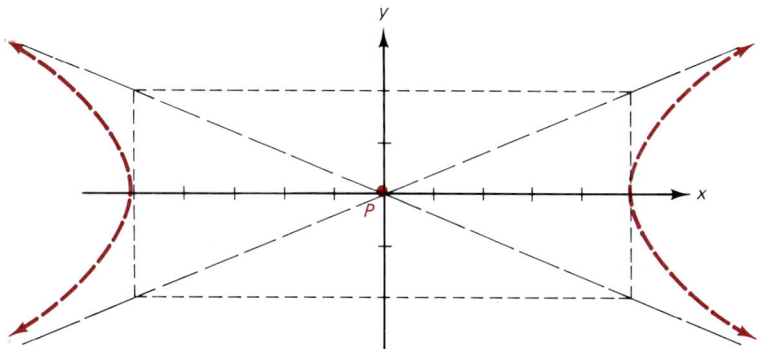

Check $P = (0, 0)$ in inequality

Is $\dfrac{0^2}{25} - \dfrac{0^2}{4} \geq 1?$ NO

NO means shade regions *not* containing P

$\geq$ means include curve

Yes

No

P

Yes

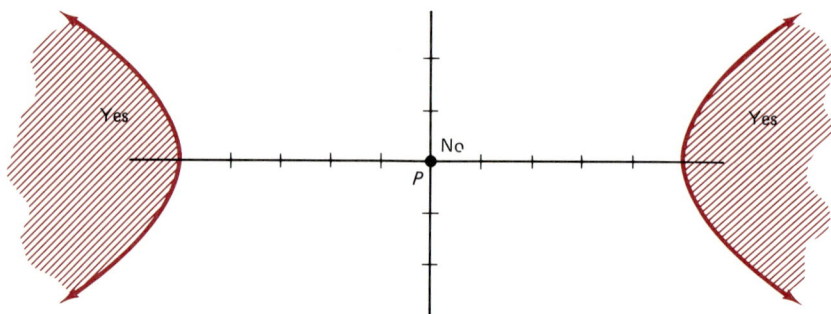

PROBLEM SET
6.4

Graph the following equations.

Warm-up
Exercises

1. $y = 5x + 2$

2. $y = 3x^2 - 5$

3. $x = y^2 - y + 4$

4. $4x^2 + y^2 = 36$

5. $(x - 3)^2 + (y - 2)^2 = 1$

6. $\dfrac{-x^2}{4} + \dfrac{y^2}{16} = 1$

Graph the following inequalities.

7. $2x - 3y \geq 6$

8. $4x + 3y > 12$

9. $y > 2x - 1$

10. $y \leq -3x + 5$

11. $x \geq 2y^2 + 1$

12. $y > 3x^2 - 5$

13. $y > -x^2 - 2x + 1$

14. $x \leq 2y^2 + y - 5$

15. $9x^2 + 4y^2 \geq 36$

16. $4x^2 + 36y^2 > 36$

17. $\dfrac{(x + 1)^2}{16} + \dfrac{(y - 1)^2}{4} > 1$

18. $\dfrac{(x - 2)^2}{9} + \dfrac{(y - 3)^2}{9} \leq 1$

19. $x^2 - 9y^2 \geq 36$

20. $-4x^2 + y^2 > 16$

21. $\dfrac{(x - 1)^2}{4} - \dfrac{(y + 1)^2}{9} \leq 1$

22. $\dfrac{-(x + 1)^2}{16} + \dfrac{(y + 2)^2}{4} < 1$

23. $y \geq 2x + 5$

24. $x > 4$

25. $36x^2 + 9y^2 > 36$

26. $x^2 + 2x + y^2 \leq 3$

27. $x^2 - 9y^2 \leq 9$

28. $x^2 + 4x + 4y^2 < 5$

29. $y < x^3$

30. $x \geq y^3$

Health
Application

31. A dietitian must see that a patient has at least 70 grams of protein. Because of cholesterol restriction, the dietitian can use only peanut butter and wheat bread. Graph the relation $2b + 5p \geq 70$, where b (on the x-axis) is the number of slices of bread and p (on the y-axis) is the number of tablespoons of peanut butter.

Business
Application

32. An automobile manufacturer makes two cars: Foxes and Beavers. The company has only 450 tons of steel. Graph the relation $1.5F + 1.0B \leq 450$, where F is the number of Foxes produced and B the number of Beavers.

Technical
Application

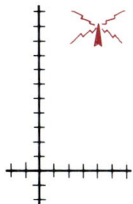

33. On a coordinate plane, a transmitter is set up at (4, 10). It can transmit to all points within a radius of 7.
   (a) What inequality do the points within the listening area satisfy?
   (b) Graph this inequality.

Engineering
Application

34. A certain column will support a weight if

$$\frac{P/A}{20{,}000} + \frac{M/Z}{15{,}000} \le 1$$

Graph this inequality with P/A on the x-axis and M/Z on the y-axis.

6.5
VARIATION

In science, health, industry, and business, we see two types of variation between quantities: **direct** and **inverse**. The following table shows how these two variations are defined.

Variation Type	Definition	Equation
Direct variation	x increases as T increases	$x = kT$
Inverse variation	x decreases as T increases	$x = \dfrac{k}{T}$

In both definitions k is called the **constant of variation**.

EXAMPLE 16   The following are examples of direct and inverse variation.

Statement	Equation
x varies directly with the square root of z	$x = k\sqrt{z}$
a varies directly with the cube of b	$a = kb^3$
F varies inversely with the square of r	$F = \dfrac{k}{r^2}$
t varies inversely with the cube root of s	$t = \dfrac{k}{\sqrt[3]{s}}$

To solve variation problems:

1. Label the quantities of the problem with letters.
2. Translate the statement into an algebraic equation.
3. Substitute the given quantities and solve for the constant k.
4. Substitute k and other information into the equation and simplify.

EXAMPLE 17 A company's sales vary directly with the square root of its advertising budget. If the company had 1000 sales with a $10,000 budget, how many sales will it have with a $90,000 budget?

Solution Let S = sales and B = budget. We first translate our statement; then we use $S = 1000$ and $B = 10,000$ to find k; finally, we substitute $B = 90,000$ to find S.

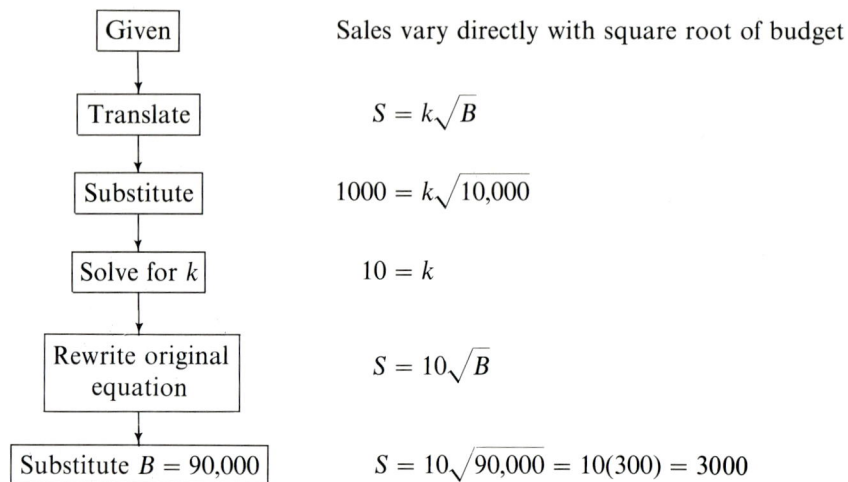

Given	Sales vary directly with square root of budget
Translate	$S = k\sqrt{B}$
Substitute	$1000 = k\sqrt{10,000}$
Solve for k	$10 = k$
Rewrite original equation	$S = 10\sqrt{B}$
Substitute $B = 90,000$	$S = 10\sqrt{90,000} = 10(300) = 3000$

Thus, their sales will be 3000.

EXAMPLE 18 The intensity of sound varies inversely with the square of the distance from the source. If the intensity is 60 at a distance of 3, what is the intensity at a distance of 4?

Solution Let I = intensity and d = distance. We translate and solve for k.

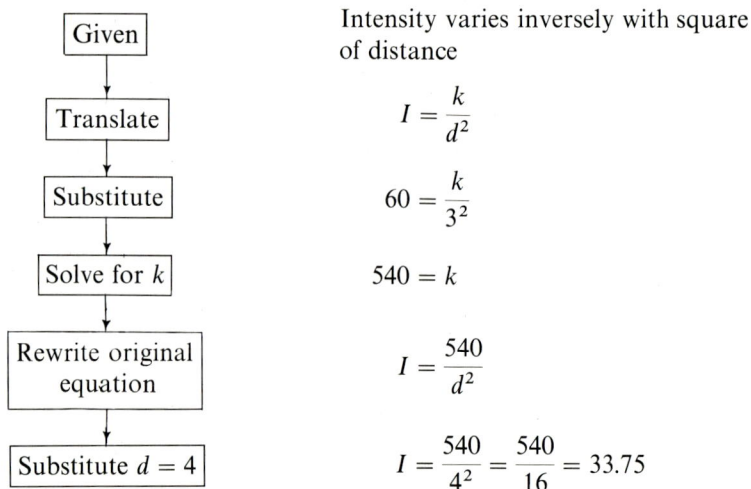

Given	Intensity varies inversely with square of distance
Translate	$I = \dfrac{k}{d^2}$
Substitute	$60 = \dfrac{k}{3^2}$
Solve for k	$540 = k$
Rewrite original equation	$I = \dfrac{540}{d^2}$
Substitute $d = 4$	$I = \dfrac{540}{4^2} = \dfrac{540}{16} = 33.75$

When a quantity varies with two or more variables, we say that the first quantity **varies jointly** with the others.

EXAMPLE 19   The frequency of the pitch of a guitar string varies directly with the square root of the tension and inversely with the length. One string with a tension of 121 and length of 26 has a frequency of 220. What is the frequency if the tension is 144 and the length is 24?

Solution   Let F = frequency, T = tension, and L = length. As in Examples 17 and 18, we first translate and solve for k.

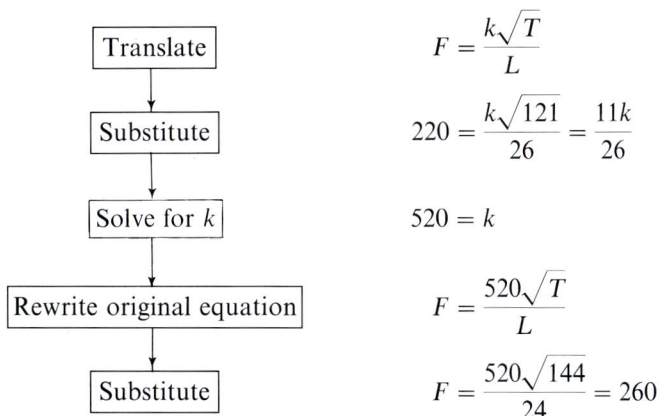

Translate	$F = \dfrac{k\sqrt{T}}{L}$
↓	
Substitute	$220 = \dfrac{k\sqrt{121}}{26} = \dfrac{11k}{26}$
↓	
Solve for k	$520 = k$
↓	
Rewrite original equation	$F = \dfrac{520\sqrt{T}}{L}$
↓	
Substitute	$F = \dfrac{520\sqrt{144}}{24} = 260$

Thus, the frequency is 260.

PROBLEM SET 6.5

Solve the following equations for x.

Warm-up Exercises

1. $2x = 24$

2. $-3x = 30$

3. $\dfrac{x}{2} = 10$

4. $\dfrac{x}{-4} = 9$

5. $\dfrac{3x}{2} = \dfrac{6 \cdot 25}{10}$

6. $\dfrac{7x}{10} = \dfrac{2 \cdot 21}{3}$

7. $\dfrac{4x}{a} = \dfrac{3y}{10}$

8. $\dfrac{5sx}{t} = \dfrac{8d^2}{e}$

For each of the following problems:
(a) Translate the first statement into an algebraic equation.
(b) Use the given information to find the constant of variation.
(c) Find the indicated variable.

9. u varies inversely with the fourth power of v: $u = 4$ and $v = 2$. Find u if $v = 3$.

10. m varies inversely with the square root of n: $m = 3$ and $n = 25$. Find m if $n = 16$.

11. a varies directly with the cube of x and inversely with the square of y and fourth root of z: $a = 8$, $x = 2$, $y = 3$, and $z = 16$. Find a if $x = 3$, $y = 2$, and $z = 1$.

12. p varies directly with a and the square root of b and inversely with the sixth power of c: $p = 36$, $a = 3$, $b = 4$, and $c = 1$. Find p if $a = 5$, $b = 100$, and $c = 2$.

13. *a* varies inversely with the square root of *t*. When *t* = 100, then *a* = 10. Find *a* when *t* = 4.

14. *x* varies directly with *a* and inversely with the square of *b*. When *a* = 6 and *b* = 2, then *x* = 15. Find *x* when *a* = 40 and *b* = 5.

15. *p* varies directly with the square root of *u* and inversely with the cube of *v*. When *u* = 25 and *v* = 2, then *p* = 25. Find *p* when *u* = 36 and *v* = 1.

16. *z* varies directly with *a*, *b*, and *c*, and inversely with the cube root of *d*. When *a* = 2, *b* = 3, *c* = 4, and *d* = 8, then *z* = 48. Find *z* when *a* = 4, *b* = 6, *c* = 2, and *d* = 27.

17. When *x* = 2, *y* = 32; when *x* = 3, *y* = 72; when *x* = 10, *y* = 800. How does *y* vary with *x*?

18. When *b* = 1, *a* = 48; when *b* = 4, *a* = 24; when *b* = 16, *a* = 12. How does *a* vary with *b*?

Business Applications

19. The demand for a product varies inversely with its price. If it sells 100,000 at $6, how many will it sell at $5?

20. The number of sales of an item varies directly with the square root of the number of salespeople and the two-thirds power of the advertising budget and inversely with the price. If there are 25 salespeople, an $8000 advertising budget, and the price is $20, then the sales are 2000. How many sales will there be if there are 100 salespeople, an advertising budget of $1000, and a price of $10?

Photographic Application

21. With an electronic flash, the proper f/stop varies inversely with the distance to the subject. If a certain flash requires an f/stop of 8 when the subject is 9 feet away, what will the f/stop be when the subject is 18 feet away? (The constant of variation here is often called the *guide number*.)

Life Science Application

22. In a simple bacterial culture, the rate of growth varies directly with the population size. If the rate of growth is 40 when the population is 1000, what is the rate of growth when the population is 20,000?

Health Application

23. A man's weight varies with the cube of his height. If a 170-pound man is 6 feet tall, how heavy would a 10-foot-tall man be (if there were 10-foot-tall men)?

Electrical Application

24. The resistance in a wire varies directly with the length of the wire and inversely with the square of the diameter of the wire. A 1000-foot wire with a diameter of 81 mils has a resistance of 1.62 ohms. How much resistance is there in 20,000 feet of wire of diameter 40 mils?

Technical Applications

25. The lumber sizing of a log varies directly with the number of boards that can be cut, and their length, width, and thickness. One log can provide 50 boards, 2 by 4 by 24; its lumber sizing is 800. What is the lumber sizing of another log that provides 80 boards, 1 by 12 by 40?

26. The rate at which heat travels through a substance varies directly with area and the temperature difference and inversely with the thickness. The rate is 4 for a window of area 100, thickness 2, and temperature difference (outside − inside) of 10°. What is the rate for a window of area 200, thickness 1, and temperature difference of 25°?

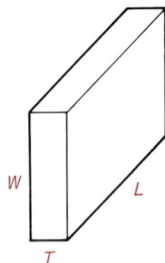

Physical Applications

27. The escape velocity from a planet varies directly with the square root of its mass and inversely with the square root of its radius. The mass of the earth is 5.97×10^{24}, its radius 6.37×10^{6}, and its escape velocity 1.12×10^{4} (meters per second). What is the escape velocity of the moon given that its mass = 7.33×10^{22} and its radius = 1.74×10^{6}?

28. The pressure on a gas varies directly with the number of moles and its absolute temperature and inversely with its volume. If there are 2 moles of a gas at $300°K$ with a volume of 10, then the pressure is 498. What is the pressure if there are 5 moles of gas at $350°K$ with a volume of 5?

Engineering Application

29. The critical load of a cylindrical column varies directly as the fourth power of its diameter and inversely with the square of its height. If a 300-centimeter-tall column with a 30-centimeter diameter can support 2700 kilograms, how much load can a column 40 centimeters in diameter and 200 centimeters high support?

Computer Application

30. The time to sort (alphabetically) a group of names in a computer varies with the square of the number of names. If 100 names take 5×10^{-4} second, how long will 1000 names take to sort?

CHAPTER 6 SUMMARY

Important Properties and Definitions

Replacing x by $x - a$ in an equation translates the graph:
(a) $|a|$ units right if $a > 0$.
(b) $|a|$ units left if $a < 0$.

Replacing y by $y - b$ in an equation translates the graph:
(a) $|b|$ units up if $b > 0$.
(b) $|b|$ units down if $b < 0$.

The equation of the parabola with vertex at (h, k) is given by

$$y - k = a(x - h)^2 \quad \text{or} \quad x - h = a(y - k)^2$$

The equation of the circle with center (h, k) and radius r is given by

$$(x - h)^2 + (y - k)^2 = r^2$$

The equation of the ellipse with center (h, k) is given by

$$\frac{(x - h)^2}{a^2} + \frac{(y - k)^2}{b^2} = 1$$

The equation of the hyperbola with center (h, k) is given by

$$\frac{(x - h)^2}{a^2} - \frac{(y - k)^2}{b^2} = 1 \quad \text{or} \quad \frac{-(x - h)^2}{a^2} + \frac{(y - k)^2}{b^2} = 1$$

Review Exercises

Rewrite the following equations with the indicated translations. (Do not graph.)

1. $y = x^4$, translated 5 units right.
2. $y^2 = x^3$, translated 3 units left, 2 units up.

Graph the following.

3. $y = (x - 1)^2 - 2$
4. $x = -3(y + 2)^2 + 3$
5. $x = y^2 - 4y + 1$
6. $y = 2x^2 - 4x - 5$

7. $x^2 + (y - 1)^2 = 4$

8. $x^2 + 4x + y^2 - 6y = 3$

9. $4x^2 + 9y^2 = 36$

10. $x^2 - 9y^2 = 36$

11. $\dfrac{x^2}{16} + \dfrac{y^2}{25} = 1$

12. $\dfrac{-x^2}{9} + \dfrac{y^2}{36} = 1$

13. $\dfrac{(x - 1)^2}{4} - \dfrac{(y + 1)^2}{16} = 1$

14. $\dfrac{(x + 2)^2}{9} + \dfrac{(y - 3)^2}{25} = 1$

15. $y < 2x - 1$

16. $x > y^2 - y + 1$

17. $4x^2 + 25y^2 \geq 100$

For the following problems:
(a) Translate the statement into an equation.
(b) Find the constant of variation.
(c) Solve for the indicated variable.

18. x varies directly with the square root of y. If $y = 16$, then $x = 20$. Find x if $y = 100$.

19. p varies directly with x and inversely with the cube of t. If $x = 10$ and $t = 2$, then $p = 80$. Find p if $x = 7$ and $t = 4$.

20. The weight of a fish varies directly with its length and the square of its girth. A fish that is 10 inches long and 4 inches across weighs 2 pounds. How much will a fish weigh that is 15 inches long and 6 inches across?

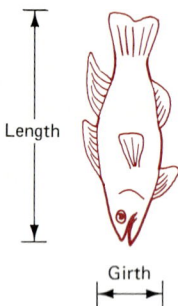

Length

Girth

CHAPTER 7

Exponential and Logarithmic Functions

7.1
EXPONENTIAL FUNCTIONS

In Chapter 2 we discussed various types of exponents:

Positive-integer exponent: $2^5 = 32$; $3^4 = 81$

Zero exponent: $7^0 = 1$; $10^0 = 1$

Negative-integer exponent: $2^{-3} = \dfrac{1}{8}$; $10^{-2} = \dfrac{1}{100}$

Rational exponent: $9^{1/2} = 3$; $16^{1.25} = 16^{5/4} = 32$

But what about irrational exponents, such as in 2^π? Here, we can sneak up on 2^π by taking rational-decimal exponents closer and closer to π:

$$2^3 = 8$$

$$2^{3.1} = 8.57419$$

$$2^{3.14} = 8.81524$$

$$2^{3.141} = 8.82135$$

$$2^{3.1415} = 8.82441$$

and so on

In more advanced mathematics courses it can be shown that a^x can be found for any real x ($a > 0$). We use this fact to define the following function.

Definition
For all real numbers x, the **exponential function** with base a ($a > 0$, $a \neq 1$) has the form

$$f(x) = a^x$$

Exponential functions have the properties shown in Table 1.

TABLE 1

Property	Meaning	Sketch
E1 Let $a > 1$. If $x < y$, then $a^x < a^y$.	For base $a > 1$, the function $f(x) = a^x$ *increases* as x increases.	
E2 Let $0 < a < 1$. If $x < y$, then $a^x > a^y$.	For base a between 0 and 1, the function $f(x) = a^x$ *decreases* as x increases.	

EXAMPLE 1   Graph $f(x) = 2^x$.

Solution   Since the base 2 is greater than 1, Property E1 tells us that this function increases as x increases. We can see this fact as we make a table of values and graph the points.

x	$y = 2^x$
-3	$2^{-3} = 1/8$
-2	$2^{-2} = 1/4$
-1	$2^{-1} = 1/2$
0	$2^0 = 1$
1	$2^1 = 2$
2	$2^2 = 4$
3	$2^3 = 8$

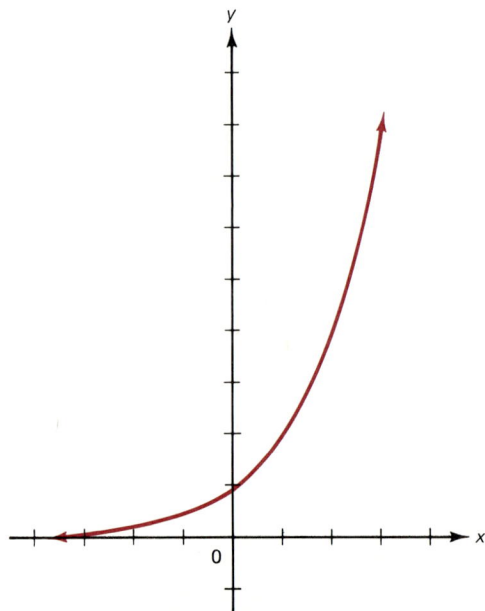

EXAMPLE 2   Graph $f(x) = 3^{-x}$.

Solution   We again make a table of (x, y) pairs. We can also write this function as

$$f(x) = 3^{-x} = \frac{1}{3^x} = \left(\frac{1}{3}\right)^x$$

This function decreases as x increases since the base $1/3$ is between 0 and 1.

x	$y = 3^{-x}$
-3	$3^{-(-3)} = 27$
-2	$3^{-(-2)} = 9$
-1	$3^{-(-1)} = 3$
0	$3^0 = 1$
1	$3^{-1} = 1/3$
2	$3^{-2} = 1/9$
3	$3^{-3} = 1/27$

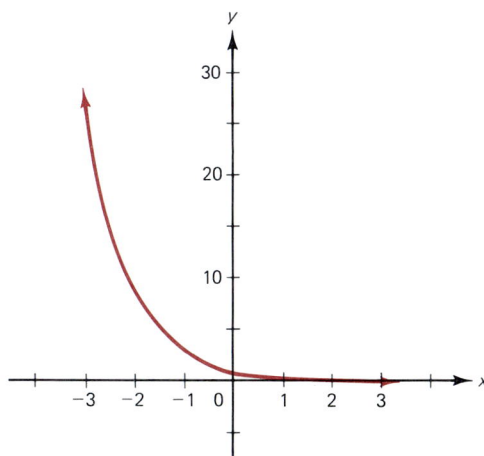

EXAMPLE 3   Graph $f(x) = e^x$, where $e \approx 2.7183$.

Solution   The number e is very important in both pure and applied mathematics. We discuss it later in this chapter and in the problem set. This function appears in statistics and physics. We use a calculator to help with the computations; for instance,

PRESS	DISPLAY	MEANING
2 · 7 1 8 3 y^x 2 =	7.38915	$e^2 \approx 2.7183^2$

x	$y = 2.7183^x$
-3	0.050
-2	0.135
-1	0.368
0	1.000
1	2.718
2	7.389
3	20.086

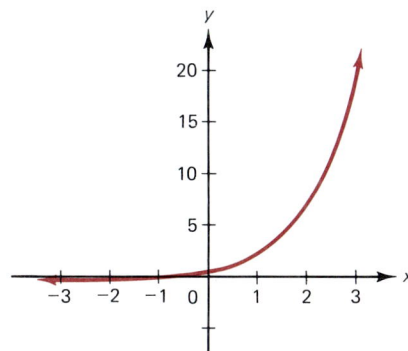

EXAMPLE 4 Graph $f(x) = e^{-x^2}$.

Solution As in Example 3, we encounter the number $e \approx 2.7183$. This function has several names (normal and Gaussian), and its graph is called the *bell-shaped curve*. It is very important in physics and statistics.

x	$y = 2.7183^{-x^2}$
-2	$2.7183^{-4} = 0.018$
-1	$2.7183^{-1} = 0.368$
-0.5	$2.7183^{-0.25} = 0.779$
0	$2.7183^{0} = 1$
0.5	$2.7183^{-0.25} = 0.779$
1	$2.7183^{-1} = 0.368$
2	$2.7183^{-4} = 0.018$

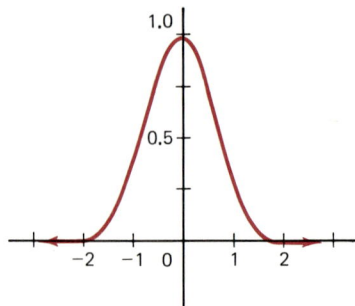

Exponential functions are used in situations where a quantity (money, population, radioactivity, and so on) is growing or shrinking at a rate proportional to its current size. These have the forms

$$f(x) = C \cdot a^x \qquad \text{and} \qquad f(x) = C \cdot a^{-x}$$

where C is a constant. This is shown in Table 2.

TABLE 2

Applications	Example	Sketch
Growth, Compound interest	$y = 1000(1.09)^x$	
Decay, Depreciation	$y = 200 \cdot e^{-2x}$ $y = 500(0.8)^x$	

EXAMPLE 5 If $1000 is invested at 9% interest, after t years the balance is given by

$$f(t) = 1000(1.09)^t$$

Find $f(0)$, $f(5)$, $f(10)$, $f(15)$, and $f(20)$. Then graph.

Solution   This is a *growth application* (compound interest). We use a hand
calculator to help us with the powers.

$$f(0) = 1000(1.09)^0 = 1000$$
$$f(5) = 1000(1.09)^5 = 1539$$
$$f(10) = 1000(1.09)^{10} = 2367$$
$$f(15) = 1000(1.09)^{15} = 3642$$
$$f(20) = 1000(1.09)^{20} = 5604$$

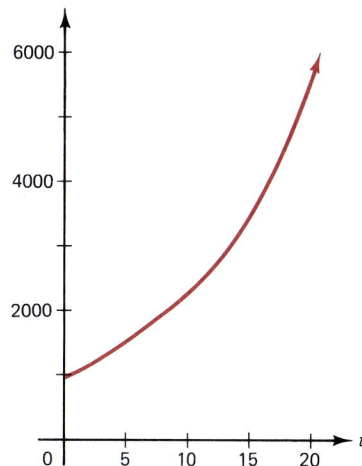

On many hand calculators, the last expression is computed as follows:

PRESS	DISPLAY	MEANING
1000 × 1.09 y^x 20 =	5604.4108	$1000(1.09)^{20}$

EXAMPLE 6   The radioactive count in a certain substance after t hours is
given by

$$f(t) = 4000 \cdot 2^{-t/10}$$

Find $f(0)$, $f(10)$, $f(20)$, $f(30)$, and $f(40)$. Then graph.

Solution   This is a *decay application*.

$$f(0) = 4000(2^0) = 4000$$
$$f(10) = 4000(2^{-1}) = 2000$$
$$f(20) = 4000(2^{-2}) = 1000$$
$$f(30) = 4000(2^{-3}) = 500$$
$$f(40) = 4000(2^{-4}) = 250$$

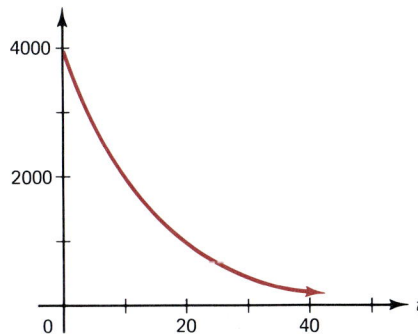

We have another exponent property.

Property
For real numbers x, y, and a ($a > 0$, $a \neq 1$),

E3 $a^x = a^y$ if and only if $x = y$

We use this property to solve simple equations involving exponents: *We rewrite the expressions to have the same base.*

EXAMPLE 7   In the examples below we solve for x: Once we have the bases equal, then the exponents must be equal.

(a) $3^x = 81$

$$3^x = 3^4 \quad \textit{(Rewrite with same base: 3)}$$

$$x = 4 \quad \textit{(Equate exponents)}$$

(b) $2^{3x} = 64$

$$2^{3x} = 2^6 \quad \textit{(Rewrite with same base: 2)}$$

$$3x = 6 \quad \textit{(Equate exponents)}$$

$$x = 2 \quad \textit{(Solve for x)}$$

(c) $\quad 4^x = \dfrac{1}{8}$

$$(2^2)^x = 2^{-3} \quad \textit{(Rewrite with same base: 2)}$$

$$2x = -3 \quad \textit{(Equate exponents)}$$

$$x = \dfrac{-3}{2} \quad \textit{(Solve for x)}$$

(d) $\quad 27^{1-x} = 9$

$$(3^3)^{1-x} = 3^2 \quad \textit{(Rewrite with same base: 3)}$$

$$3 - 3x = 2 \quad \textit{(Equate exponents)}$$

$$-3x = -1$$

$$x = \dfrac{1}{3} \quad \textit{(Solve for x)}$$

PROBLEM SET 7.1

Simplify the following.

Warm-up Exercises

1. 5^2

2. 2^5

3. 4^0

4. $\left(\dfrac{1}{5}\right)^0$

5. 2^{-3}

6. 3^{-2}

7. $9^{1/2}$

8. $25^{-3/2}$

9. $10 \cdot 2^3$

10. $100\left(\dfrac{1}{2}\right)^2$

11. 2.55^{10}

12. 1.085^{20}

Graph the following.

Warm-up Exercises

13. $f(x) = \dfrac{1}{x}$

14. $f(x) = \sqrt{x}$

Graph the following exponential functions. (Where needed, use $e = 2.7183$.)

15. $f(x) = 3^x$

16. $f(x) = 4^x$

17. $f(x) = 2^{-x}$

18. $f(x) = 5^{-x}$

19. $f(x) = e^x$ **20.** $f(x) = \pi^x$

21. $f(x) = 1.5^x$ **22.** $f(x) = (1.07)^x$

23. $f(x) = e^{-x}$ **24.** $f(x) = 2.5^{-x}$

25. $f(x) = \left(\dfrac{1}{4}\right)^x$ **26.** $f(x) = (0.8)^x$

27. $f(x) = 2^{-x^2}$ **28.** $f(x) = 4^{x/2}$

29. $f(x) = 500 \cdot 2^x$ for $x = 0, 1, 2, 3, 4$

30. $f(x) = 10{,}000(1.02)^x$ for $x = 0, 10, 20, 30, 40$

31. $f(x) = 2000(1.5)^{x/20}$ for $x = 0, 20, 40, 60, 80$

32. $f(x) = 4000(0.98)^x$ for $x = 0, 10, 20, 30, 40$

33. $f(x) = 1000(1.2)^{-x}$ **34.** $f(x) = 400(4.1)^{x/2}$

35. $f(x) = 2^x + 2^{-x}$ **36.** $f(x) = 2^x - 2^{-x}$

Solve the following equations for x.

37. $10^x = \dfrac{1}{100}$ **38.** $4^x = \dfrac{1}{64}$

39. $5^x = 1$ **40.** $7^x = 1$

41. $2^{-x} = 8$ **42.** $3^{-x} = 81$

43. $2^{3x+1} = \dfrac{1}{32}$ **44.** $5^{2x-1} = \dfrac{1}{125}$

45. $9^x = 27$ **46.** $25^x = \dfrac{1}{125}$

47. $10^{-x^2} = \dfrac{1}{10{,}000}$ **48.** $2^{-x^2} = \dfrac{1}{16}$

Consumer Application

49. If P dollars are invested at interest rate i for t years, the balance grows to

$$f(t) = P(1 + i)^t$$

For the following deposits P and interest rates i (expressed as a decimal), find $f(0)$, $f(5)$, $f(10)$, $f(15)$, and $f(20)$. Then graph.
(a) $P = 10{,}000$, $i = 10\%$ (b) $P = 500{,}000$, $i = 13\%$
(c) $P = 2000$, $i = 8\%$ (d) $P = 60{,}000$, $i = 15\%$

Business Application

50. A machine of original value V depreciates at a rate d every year. Its value after t years is given by the function

$$f(t) = V(1 - d)^t$$

For the following machine values V and depreciation rates d (expressed as a decimal), find $f(0)$, $f(2)$, $f(4)$, $f(6)$, $f(8)$, and $f(10)$. Then graph.
(a) $V = 30{,}000$, $d = 10\%$ (b) $V = 100{,}000$, $d = 20\%$
(c) $V = 10{,}000$, $d = 8\%$ (d) $V = 1{,}000{,}000$, $d = 15\%$

Physical Application

51. Carbon-14, which is used for dating ancient artifacts, has a half-life of about 5600 years. That is, carbon-14 loses half its radioactive content every 5600 years. The carbon content present after t years is given by

$$f(t) = A(\tfrac{1}{2})^{t/5600}$$

where A is the original amount. Using $A = 100$, find $f(0)$, $f(5600)$, $f(11{,}200)$, and $f(16{,}800)$. Then graph.

52. The population of a country in t years can be approximated by

$$f(t) = P(1 + r)^t$$

where P is the present population and r is the rate of increase per year (essentially, birth rate $-$ death rate). For each of the following countries, find $f(0)$, $f(5)$, $f(10)$, $f(15)$, and $f(20)$. Then graph.
(a) China: $P = 1,000,000,000$, $r = 1.6\%$
(b) Mexico: $P = 70,000,000$, $r = 3.3\%$
(c) Saudi Arabia: $P = 10,400,000$, $r = 5.6\%$
(d) United States: $P = 230,000,000$, $r = 0.7\%$

*Psychology
Application*

53. The number of tasks that a worker can perform after x days on a job is given by

$$f(x) = 400(1 - 2^{-x})$$

Find $f(0)$, $f(1)$, $f(2)$, $f(3)$, and $f(4)$.

*Mathematics
Application*

54. The number e appears in mathematics after studying numbers of the form

$$\left(1 + \frac{1}{n}\right)^n$$

as n becomes larger and larger. Compute the following:
(a) $(1.5)^2$ (b) $(1.25)^4$
(c) $(1.1)^{10}$ (d) $(1.01)^{100}$
(e) $(1.001)^{1000}$ (f) $(1.0001)^{10,000}$
(g) Continue this process on your calculator until the display stops changing. This is an approximation of e.

7.2
LOGARITHMS

A **logarithm** is an exponent. Consider the statement

$$10^{\boxed{?}} = 100$$

where we know that the base is 10 and the result is 100, but the exponent is unknown. We are asking: 10 to what power gives 100? Of course, the exponent $\boxed{?} = 2$. We write this as

$$\log_{10} 100 = 2 \qquad (\text{since } 10^2 = 100)$$

EXAMPLE 8   The following are simple examples of logarithms. Note how the logarithm answers the question: What is the exponent?
(a) $\log_{10} 10 = 1$ (since $10^1 = 10$)
(b) $\log_{10} 100 = 2$ (since $10^2 = 100$)
(c) $\log_{10} 1000 = 3$ (since $10^3 = 1000$)
(d) $\log_{10} 0.1 = -1$ (since $10^{-1} = 0.1$)
(e) $\log_{10} \dfrac{1}{1000} = -3$ $\left(\text{since } 10^{-3} = \dfrac{1}{1000}\right)$
(f) $\log_{10} 10^k = k$
(g) $\log_2 2 = 1$ (since $2^1 = 2$)
(h) $\log_2 8 = 3$ (since $2^3 = 8$)
(i) $\log_2 64 = 6$ (since $2^6 = 64$)
(j) $\log_2 \dfrac{1}{2} = -1$ $\left(\text{since } 2^{-1} = \dfrac{1}{2}\right)$

(k) $\log_2 \dfrac{1}{16} = -4$ $\left(\text{since } 2^{-4} = \dfrac{1}{16}\right)$

(l) $\log_2 2^k = k$

Definition

For real numbers y, $x > 0$, and $b > 0$ $(b \neq 1)$,

$$\log_b x = y \quad \text{means} \quad b^y = x$$

We call $b^y = x$ the **exponent form** and $\log_b x = y$ the **logarithm form**. We call b the **base** and y the **logarithm** or **exponent**. We read the expression $\log_b x = y$ as "log to the base b of x equals y."

YES	NO
$\log_{10} 1000 = 3$	~~$\log_{10} -1000 = -3$~~

We cannot find the logarithm of zero or a negative number, such as -1000.

EXAMPLE 9 Exponent and logarithm forms are interchangeable. The table below shows how each form is easily converted to the other.

Logarithm form		Exponent form
(a) $\log_5 125 = 3$	$\longleftrightarrow$	$5^3 = 125$
(b) $\log_3 \dfrac{1}{9} = -2$	$\longleftrightarrow$	$3^{-2} = \dfrac{1}{9}$
(c) $\log_{64} 8 = \dfrac{1}{2}$	$\longleftrightarrow$	$64^{1/2} = 8$
(d) $\log_{16} 8 = \dfrac{3}{4}$	$\longleftrightarrow$	$16^{3/4} = 8$
(e) $\log_{36} \dfrac{1}{6} = \dfrac{-1}{2}$	$\longleftrightarrow$	$36^{-1/2} = \dfrac{1}{6}$

EXAMPLE 10 We can solve simple equations involving logarithms by converting them to exponent form.

(a) $\log_5 \dfrac{1}{25} = x$ means $5^x = \dfrac{1}{25}$ (*Exponent form*)

$\qquad\qquad\qquad\qquad 5^x = 5^{-2}$ (*Same base:* 5)

$\qquad\qquad\qquad\qquad x = -2$ (*Equal exponents*)

(b) $\log_{16} 32 = x$ means $16^x = 32$ (*Exponent form*)

$\qquad\qquad\qquad (2^4)^x = 2^5$ (*Same base:* 2)

$\qquad\qquad\qquad\qquad 4x = 5$ (*Equal exponents*)

$\qquad\qquad\qquad\qquad x = \dfrac{5}{4}$ (*Solve for x*)

(c) $\log_x 8 = 3$ means $x^3 = 8$ (*Exponent form*)

$$x = \sqrt[3]{8} = 2 \quad (\textit{Cube root})$$

(d) $\log_5 x = -1$ means $5^{-1} = x$ (*Exponent form*)

$$\frac{1}{5} = x \quad (\textit{Evaluate})$$

Let us now consider the **logarithmic function**

$$f(x) = \log_b x$$

In the definition of $\log_b x$, we require that $x > 0$. Thus, we say that the *domain of the logarithmic function is the set of positive real numbers.*

EXAMPLE 11   Graph the functions

(a) $y = \log_{10} x$ (b) $y = \log_2 x$

Solution   We use some of the values found in Example 8 to help us plot these graphs.

(a)

x	$y = \log_{10} x$
1/100	-2
1/10	-1
1	0
10	1
100	2

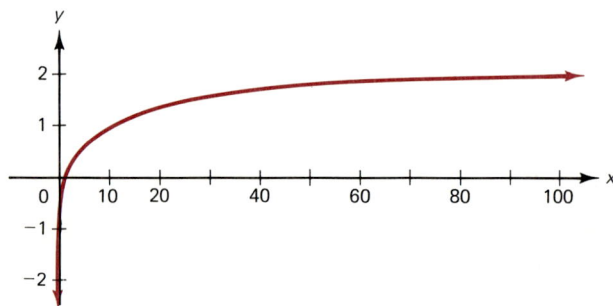

(b)

x	$y = \log_2 x$
1/8	-3
1/4	-2
1/2	-1
1	0
2	1
4	2
8	3

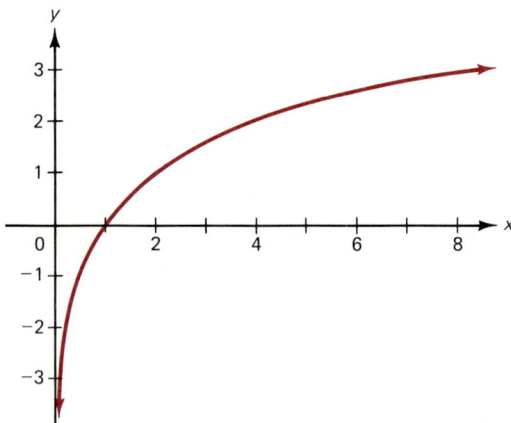

Despite the different bases (10 and 2) in these two functions, the similarities are obvious; in fact, the graph of $f(x) = \log_b x$ $(b > 1)$ always has the same characteristics:

1. It passes through $(1, 0)$.
2. It has a vertical asymptote at $x = 0$.
3. As x (always positive) increases, y increases (but very slowly).

The logarithm and exponent forms are interchangeable. This is an inverse relation (review pages 179 to 186). Figure 1 illustrates this idea with base 2. With the exponential function, 5 goes in and 32 comes out. With the logarithmic function, the 32 goes in and the 5 comes out again.

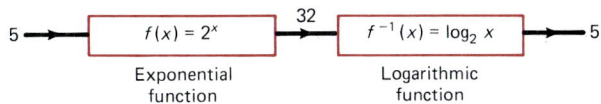

$$5 \longrightarrow \boxed{f(x) = 2^x} \xrightarrow{\;32\;} \boxed{f^{-1}(x) = \log_2 x} \longrightarrow 5$$

| Exponential function | Logarithmic function |

FIGURE 1

The following table summarizes the relationship between domain and range of the logarithmic and exponential functions.

Function	Domain	Range
$y = \log_b x$	$x > 0$	All real numbers y
$y = b^x$	All real numbers x	$y > 0$

EXAMPLE 12 Graph the inverse functions:
   (a) $f(x) = 2^x$
   (b) $f^{-1}(x) = \log_2 x$

Solution We make a table for both functions. Note that the table for $y = \log_2 x$ is the reverse of that for $y = 2^x$ (as it should be for inverse functions). Note the usual symmetry about the line $y = x$.

x	$y = 2^x$
-2	1/4
-1	1/2
0	1
1	2
2	4
3	8

x	$y = \log_2 x$
1/4	-2
1/2	-1
1	0
2	1
4	2
8	3

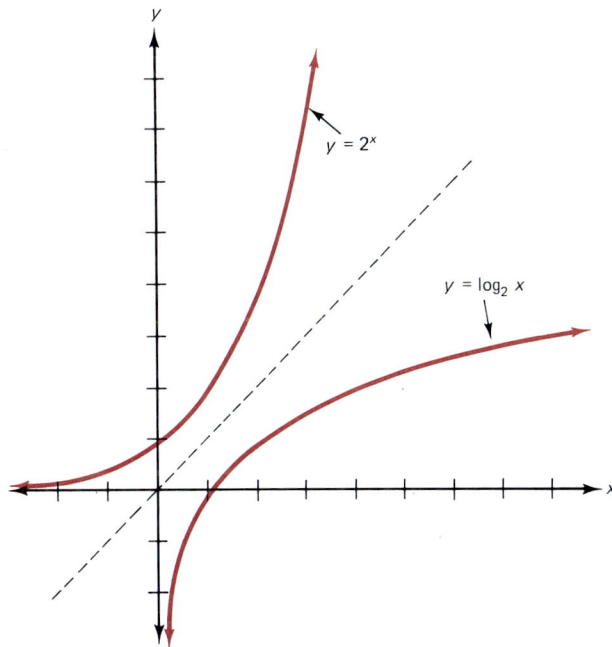

Let us now give some very important and useful properties for logarithms.

Property

For positive real numbers M, N, and b ($b \neq 1$), and real number r:

L1 $\log_b M \cdot N = \log_b M + \log_b N$

L2 $\log_b \dfrac{M}{N} = \log_b M - \log_b N$

L3 $\log_b N^r = r \log_b N$

L4 $\log_b b^M = M$

L5 $b^{\log_b M} = M$

L6 $\log_b 1 = 0$

Properties L1 to L3 are especially important since they allow us to rewrite one arithmetic operation in terms of a simpler one. For instance, L1 lets us write multiplication in terms of addition of logarithms; L2 lets us write division in terms of subtraction of logarithms; and L3 lets us write raising to a power as multiplication of a logarithm.

We sketch a proof for L1. (The others are similar and left as exercises.) Let $x = \log_b M$ and $y = \log_b N$. Then

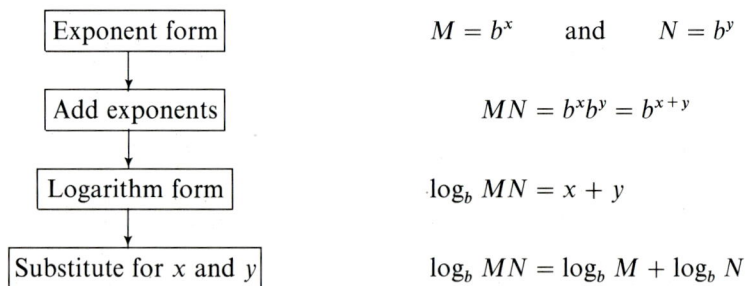

Exponent form	$M = b^x$ and $N = b^y$
Add exponents	$MN = b^x b^y = b^{x+y}$
Logarithm form	$\log_b MN = x + y$
Substitute for x and y	$\log_b MN = \log_b M + \log_b N$

EXAMPLE 13   The following examples use Properties L1 to L6.

(a) $\log_5 7ab = \log_5 7 + \log_5 a + \log_5 b$ (L1)

(b) $\log_2 \dfrac{x}{9} = \log_2 x - \log_2 9$ (L2)

(c) $\log_{10} m^3 = 3 \log_{10} m$ (L3)

(d) $\log_4 a^2 b^3 = \log_4 a^2 + \log_4 b^3$ (L1)

         $= 2 \log_4 a + 3 \log_4 b$ (L3)

(e) $\log_3 \dfrac{x^2 y^5}{\sqrt[4]{z}} = \log_3 x^2 y^5 - \log_3 z^{1/4}$ (L2)

$$= \log_3 x^2 + \log_3 y^5 - \log_3 z^{1/4} \quad \text{(L1)}$$

$$= 2\log_3 x + 5\log_3 y - \frac{1}{4}\log_3 z \quad \text{(L3)}$$

(f) $\log_{10} 10^3 = 3$ (L4)

(g) $4^{\log_4 7} = 7$ (L5)

(h) $\log_5 1 = \log_7 1 = \log_2 1 = 0$ (L6)

EXAMPLE 14 These examples are the opposite of Example 13. Here we wish to combine serveral logarithmic expressions as a single logarithmic expression.

(a) $\log_7 m + \log_7 n = \log_7 mn$ (L1)

(b) $\log_{10} 3 - \log_{10} t = \log_{10} \dfrac{3}{t}$ (L2)

(c) $\dfrac{1}{2} \log_8 z = \log_8 z^{1/2}$ or $\log_8 \sqrt{z}$ (L3)

(d) $2\log_5 u - 3\log_5 v = \log_5 u^2 - \log_5 v^3$ (L3)

$$= \log_5 \dfrac{u^2}{v^3} \quad \text{(L2)}$$

YES	NO
$\log_{10} 8x = \log_{10} 8 + \log_{10} x$	$\log_{10} 8x = \log_{10} 8 \cdot \log_{10} x$
$\log_3 2 + \log_3 5 = \log_3 10$	$\log_3 2 + \log_3 5 = \log_3 7$
$\log_6 \dfrac{a}{b} = \log_6 a - \log_6 b$	$\log_6 \dfrac{a}{b} = \dfrac{\log_6 a}{\log_6 b}$
$\log_4 35 - \log_4 7 = \log_4 5$	$\log_4 35 - \log_4 7 = \log_4 28$
$\log_5 x^3 = 3\log_5 x$	$\log_5 x^3 = (\log_5 x)^3$
$\log_{10} 1 = 0$	$\log_{10} 0 = 1$

PROBLEM SET 7.2

Simplify the following.

Warm-up Exercises

1. 5^2 **2.** 5^3 **3.** 5^4

4. 5^0 **5.** 5^{-1} **6.** 5^{-2}

7. $27^{1/3}$ **8.** $49^{1/2}$ **9.** $8^{4/3}$

10. $100^{5/2}$ **11.** $25^{-1/2}$ **12.** $27^{-2/3}$

Solve the following equations.

13. $2^x = 16$

14. $3^x = 81$

15. $8^x = 32$

16. $4^x = \dfrac{1}{16}$

17. $x^{1/2} = 7$

18. $x^{-1/2} = \dfrac{1}{5}$

Complete the following tables.

N	$\log_5 N$
5	1
25	2
19. 125	?
20. 625	?
21. 1/5	?
22. 1/25	?
23. 1/125	?
24. 1	?

N	$\log_3 N$
3	1
9	2
25. 27	?
26. 81	?
27. 243	?
28. 1/81	?
29. 1/3	?
30. 1	?

Write the following in logarithm form.

31. $2^6 = 64$

32. $10^4 = 10{,}000$

33. $81^{1/2} = 9$

34. $8^{2/3} = 4$

35. $4^{-2} = \dfrac{1}{16}$

37. $2^{-7} = \dfrac{1}{128}$

37. $25^{-1/2} = \dfrac{1}{5}$

38. $16^{-3/4} = \dfrac{1}{8}$

Write the following in exponent form.

39. $\log_6 36 = 2$

40. $\log_2 1024 = 10$

41. $\log_{16} 4 = \dfrac{1}{2}$

42. $\log_{27} 3 = \dfrac{1}{3}$

43. $\log_6 \dfrac{1}{6} = -1$

44. $\log_{10} \dfrac{1}{1000} = -3$

45. $\log_{1000} \dfrac{1}{100} = \dfrac{-2}{3}$

46. $\log_9 \dfrac{1}{3} = \dfrac{-1}{2}$

Solve the following equations for x.

47. $\log_{27} 9 = x$

48. $\log_{25} \dfrac{1}{125} = x$

49. $\log_{10} 10^7 = x$

50. $\log_7 7^{10} = x$

51. $\log_x 64 = 3$

52. $\log_x 10{,}000 = 2$

53. $\log_x \dfrac{1}{9} = -2$

54. $\log_x 5 = \dfrac{1}{2}$

55. $\log_2 x = -3$

56. $\log_4 x = \dfrac{-1}{2}$

Verify that each of the following pairs of functions are inverses and graph the pair on the same plane.

57. $f(x) = 4^x$ and $f^{-1}(x) = \log_4 x$

58. $f(x) = 5^x$ and $f^{-1}(x) = \log_5 x$

59. $f(x) = 3^x$ and $f^{-1}(x) = \log_3 x$

60. $f(x) = 10^x$ and $f^{-1}(x) = \log_{10} x$

Write the following in terms of simpler logarithmic terms.

61. $\log_{10} 5xy$

62. $\log_2 4mnk$

63. $\log_7 \dfrac{6}{p}$

64. $\log_9 \dfrac{x}{y}$

65. $\log_{10} x^4$

66. $\log_2 p^2$

67. $\log_7 \sqrt{z}$

68. $\log_8 \sqrt{t}$

69. $\log_2 \sqrt[3]{\dfrac{p^2 q^3}{r}}$

70. $\log_3 \sqrt[4]{\dfrac{a^2 b}{c^4}}$

Write each of the following in terms of a single logarithm with coefficient 1.

71. $\log_3 a + \log_3 b$

72. $\log_7 8 + \log_7 x + \log_7 y$

73. $\log_2 7 - \log_2 x$

74. $\log_{10} y - \log_{10} z$

75. $p \log_2 5$

76. $5 \log_2 p$

77. $2 \log_5 x + 3 \log_5 y$

78. $6 \log_2 a - 4 \log_2 b - 2 \log_2 c$

79. $\dfrac{1}{2} \log_{10} k - \dfrac{1}{3} \log_{10} m$

80. $\dfrac{2}{3} \log_6 p - \dfrac{3}{4} \log_6 q - \dfrac{2}{5} \log_6 r$

It is known that $\log_7 2 = 0.3562$ and $\log_7 3 = 0.5646$. Use these facts and Properties L1 to L3 to find the following. (Hint: Write each number as products and powers of 2 and 3. For example, since $6 = 2 \cdot 3$, $\log_7 6 = \log_7 2 + \log_7 3 = 0.9208$.)

81. $\log_7 9$

82. $\log_7 4$

83. $\log_7 16$

84. $\log_7 27$

85. $\log_7 12$

86. $\log_7 18$

For each of the following functions, f:

(a) *Give the domain for f. (Recall that we cannot find the logarithm of zero or a negative number.)*

(b) *Find f^{-1}. (Review the procedure for finding an inverse—page 181.)*

87. $f(x) = \log_2 x$

88. $f(x) = \log_7 x$

89. $f(x) = \log_3 (x + 1)$

90. $f(x) = \log_9 (2x + 1)$

91. $f(x) = \log_{10}(x^3 - 1)$

92. $f(x) = \log_e(x^3 + 1)$

93. $f(x) = 4^x$

94. $f(x) = 10^x$

95. $f(x) = 2^{x-1}$

96. $f(x) = 5^{2x+3}$

Acoustical Application

97. Sound intensity I is measured in *decibels* using the formula

$$I = 10 \log_{10} R$$

where R is the ratio of the pressure of the given sound to the pressure of the least audible sound. Find I in the following cases.
(a) Whisper: $R = 1000$
(b) Ordinary conversation: $R = 1,000,000$
(c) Jackhammer (at 20 feet): $R = 100,000,000$
(d) Airplane engine (at 15 feet): $100,000,000,000$
(e) Rock concert: $R = 1,000,000,000,000$

Photographic Applications

98. The *transmission T* of a film negative is the fraction of light that passes through the negative. The *density D* of the negative is given by

$$D = \log_{10}(1/T)$$

Find the densities in the following cases:
(a) $T = 1/10$ (b) $T = 1/100$ (c) $T = 1/1000$

99. Two numbers are used to measure the speed of film: ASA and DIN. The relation between these numbers is

$$\text{DIN} = 15 + 3 \log_2(\text{ASA}/25)$$

Find the DIN for the following ASA values:
(a) ASA = 50   (b) ASA = 100   (c) ASA = 400

Electrical Application

100. The relation between the *gauge G* of a copper wire and its diameter d (in mils$-1/1000$ inch) can be approximated by

$$G = 50 - 20 \log_{10} d$$

Find the gauge of each of the following wires:
(a) $d = 100$ mils   (b) $d = 10$ mils   (c) $d = 1$ mil

Information Application

101. Logarithms to the base 2 are used to measure the *information I* in a message. Suppose that a code contains four symbols (say a, b, c, and d) which occur with frequency p_a, p_b, p_c, and p_d. Its information is given by

$$I = -(p_a \cdot \log_2 p_a + p_b \cdot \log_2 p_b + p_c \cdot \log_2 p_c + p_d \cdot \log_2 p_d)$$

Find I if $p_a = 1/2$, $p_b = 1/4$, $p_c = 1/8$, and $p_d = 1/8$.

7.3 COMMON LOGARITHMS

Although we have discussed many possible logarithmic bases, there are only two that are used with great frequency. The first type uses a base of $e = 2.718\ldots$. These are called *natural logarithms* and occur in the study of calculus. We discuss them later in the chapter.

The other type of logarithm uses a base of 10. These are called **common logarithms** and fit nicely into our base-10 system of numeration. For simplicity, we write log x for $\log_{10} x$. (The base of 10 is understood.)

EXAMPLE 15   The powers of 10 have simple common logarithms.

(a) $\log 10 = 1$ (b) $\log 100 = 2$

(c) $\log 1000 = 3$ (d) $\log 10{,}000 = 4$

(e) $\log 10^9 = 9$ (f) $\log 1 = 0$

(g) $\log 0.1 = -1$ (h) $\log 0.01 = -2$

(i) $\log 0.001 = -3$ (j) $\log 10^{-6} = -6$

In general, we have (from Property L4)

$$\log 10^k = k$$

What about the common logarithms of all the other numbers? For numbers between 1 and 10, we use a **table of common logarithms** (see the Appendix). For example, to find log 6.78 in the table, we find the 6.7 row and the 8 column. These meet at .8312; thus, log 6.78 = 0.8312. We see this in the following cutaway table.

N	7	8	9
6.6	.8241	.8248	.8254
6.7	.8306	.8312	.8319
6.8	.8370	.8376	.8382

Recall that with scientific notation (review pages 23 to 24) we can write any positive number u as the product of a power of 10 and a number between 1 and 10. We use scientific notation and Property L1 to find the common logarithm of any number. For example, consider log 7900.

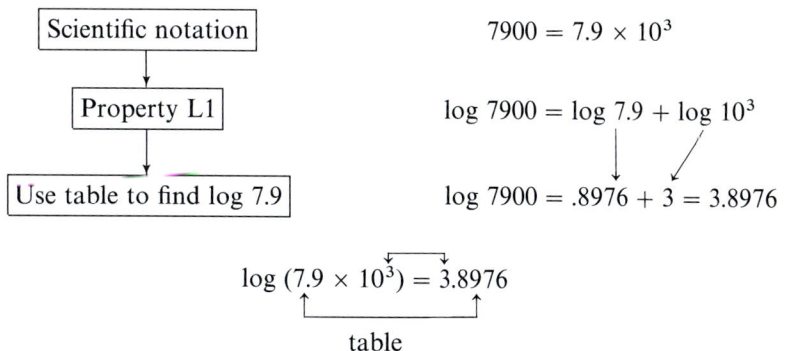

Scientific notation	$7900 = 7.9 \times 10^3$
↓	
Property L1	$\log 7900 = \log 7.9 + \log 10^3$
↓	
Use table to find log 7.9	$\log 7900 = .8976 + 3 = 3.8976$

$$\log (7.9 \times 10^3) = 3.8976$$

table

For any number $u = d \times 10^n$, log u has two parts: The integer part n, called the **characteristic**, and the decimal part log d, called the **mantissa** (found in the table).

EXAMPLE 16 For numbers written $u = d \times 10^n$, we have $\log u = \log d + n$, where we find $\log d$ in the table.

u	Scientific notation	$\log u$
(a) 5100	5.1×10^3	3.7076
(b) 200,000	2.0×10^5	5.3010
(c) 83,500	8.35×10^4	4.9217
(d) 6.08	6.08×10^0	0.7839

EXAMPLE 17 For numbers less than 1, the characteristic is negative. In these cases we usually rewrite it as the difference of nonnegative numbers: for example, $-3 = 0 - 3$, or $7 - 10$, or $17 - 20$.

u	Scientific notation	$\log u$
(a) 0.0054	5.4×10^{-3}	$0.7324 - 3$ $(-3 = 0 - 3)$
(b) 0.000792	7.92×10^{-4}	$6.8987 - 10$ $(-4 = 6 - 10)$
(c) 0.153	1.53×10^{-1}	$19.1847 - 20$ $(-1 = 19 - 20)$

YES	NO
$\log 0.153 = 9.1847 - 10$	$\log 0.153 = -1.1847$

Only the characteristic -1 is negative. Writing -1.1847 would *incorrectly* make 0.1847 negative, also. We can, however, subtract $9.1847 - 10 = -0.8153$.

We can also compute common logarithms on a calculator with a $\boxed{\log}$ key. Let us redo $\log 5100$ and $\log 0.153$ with a calculator.

PRESS	DISPLAY	MEANING
$\boxed{5}\,\boxed{1}\,\boxed{0}\,\boxed{0}\,\boxed{\log}$	3.7075702	$\log 5100$
$\boxed{.}\,\boxed{1}\,\boxed{5}\,\boxed{3}\,\boxed{\log}$	-0.8153086	$\log 0.153$
$\boxed{5}\,\boxed{+/-}\,\boxed{\log}$	E 0.	Error; $\log(-5)$ is undefined.

Note that the calculator automatically writes $\log 0.153 = -0.8153086$. Also, we get an error when we attempt to compute $\log(-5)$, since the calculator is wired for the fact that the domain of the logarithmic function is the set of positive real numbers.

We can reverse the process. Suppose that we are given $\log x = 5.9047$, and we want to find x. We are looking for the **antilogarithm**, written antilog 5.9047. We know that the 5 corresponds to 10^5. We then find .9047 in the body of the table and see that it is in the 8.0 row and the 3 column:

Hence, antilog $5.9047 = 8.03 \times 10^5 = 803{,}000$.

EXAMPLE 18 The following are examples of antilogarithms. (Here if we cannot find the mantissa in the table exactly, we take the closest value.)

x	antilog x
(a) 3.7513	$5.64 \times 10^3 = 5640$
(b) 8.0215 − 10	$1.05 \times 10^{-2} = 0.0105$

Since 10^x and antilog x are inverses of the same function ($\log x$), it follows that for any real number x

$$10^x = \text{antilog } x$$

EXAMPLE 19 We can compute 10^x by using antilogarithms.
(a) $10^{1.5}$ = antilog $1.5000 = 3.16 \times 10^1 = 31.6$
(b) $10^{2.79}$ = antilog $2.7900 = 6.17 \times 10^2 = 617$

We can compute antilog x or 10^x on a calculator with $\boxed{\text{INV}}$ and $\boxed{\log}$ keys ($\boxed{\text{INV}}$ stands for inverse). Consider the following.

PRESS	DISPLAY	MEANING
$\boxed{3}\boxed{\cdot}\boxed{7}\boxed{5}\boxed{1}\boxed{3}\boxed{\text{INV}}\boxed{\log}$	5640.2713	antilog 3.7513
$\boxed{2}\boxed{\cdot}\boxed{7}\boxed{9}\boxed{\text{INV}}\boxed{\log}$	616.595	$10^{2.79}$

Write the following numbers in scientific notation.

1. 60,000   **2.** 720,000   **3.** 52.1

4. 0.0072   **5.** 0.0004   **6.** 0.000000053

Write the following in terms of simpler logarithmic terms. (*Do not evaluate.*)

7. $\log_{10} 3xy$ **8.** $\log (4.1)(5.7)$

9. $\log \dfrac{6}{x}$ **10.** $\log_{10} \dfrac{(8.1)(98)}{53.1}$

Find the following common logarithms.

11. $\log 10$ **12.** $\log 100{,}000$ **13.** $\log 10^{12}$

14. $\log 0.01$ **15.** $\log 0.00001$ **16.** $\log 10^{-7}$

17. $\log 7.81$ **18.** $\log 4.01$ **19.** $\log 70.2$

20. $\log 8000$ **21.** $\log 500{,}000$ **22.** $\log 420$

23. $\log 0.0034$ **24.** $\log 0.99$ **25.** $\log 0.00022$

26. $\log 0.0000004$ **27.** $\log 0.0777$ **28.** $\log 0.111$

Find (*or approximate*) the following antilogarithms.

29. antilog 1.5966   **30.** antilog 3.6920

31. antilog 5.7481   **32.** antilog 2.5432

33. antilog $(7.8102 - 10)$ **34.** antilog $(4.0374 - 10)$

35. antilog $(1.5123 - 5)$ **36.** antilog $(2.7811 - 4)$

37. $10^{2.8716}$ **38.** $10^{3.5490}$

39. $10^{1.2}$ **40.** $10^{0.9}$

Evaluate the following.

41. $\log 7100$ **42.** $\log 0.00001$

43. $\log 0.03$ **44.** $\log 57{,}000$

45. $\log 100{,}000$ **46.** $\log 0.00073$

47. antilog 2.0789   **48.** $10^{5.77}$

49. $10^{4.23}$ **50.** $10^{\log 82}$

51. $\log 10^{2.5}$ **52.** antilog 3.1893

53. $10^{\log 41}$ **54.** antilog $(\log 250)$

55. $\log (\text{antilog } 1.5)$ **56.** $10^{\log 530}$

57. antilog $(\log 7500)$ **58.** $\log 10^{3.37}$

59. The pH of a solution is given by

$$\text{pH} = -\log_{10} [\text{H}^+]$$

where $[\text{H}^+]$ is the molar concentration of hydrogen ions. Find the pH in each of the following cases. [*Hint*: Simplify the logarithmic expression to a single number; for example, for soda pop, $[\text{H}^+] = 2.0 \times 10^{-3}$. Then pH $= -\log [\text{H}^+] = -(0.3010 - 3) = 2.699 \approx 2.7$.]
(a) Nonalkaline shampoo: $[\text{H}^+] = 1.6 \times 10^{-6}$
(b) Fruit juice: $[\text{H}^+] = 8.0 \times 10^{-5}$
(c) Wine: $[\text{H}^+] = 4.0 \times 10^{-4}$
(d) Lye: $[\text{H}^+] = 3.2 \times 10^{-14}$
(e) Milk: $[\text{H}^+] = 4.0 \times 10^{-7}$

60. The density of a film negative is given by

$$D = \log \frac{1}{T}$$

(see page 236). Find the density in the following cases:

(a) $T = \dfrac{1}{2}$ (b) $T = \dfrac{1}{16}$ (c) $T = \dfrac{1}{50}$

61. The intensity of sound (in decibels) is given by

$$I = 10 \log R$$

(see page 236). Compute the intensity in each of the following cases:
(a) One person talking: $R = 1{,}000{,}000$
(b) Two people talking: $R = 2{,}000{,}000$
(c) Three people talking: $R = 3{,}000{,}000$
(d) Ten people talking: $R = 10{,}000{,}000$
(e) 100 people talking: $R = 100{,}000{,}000$

62. The power gain PG in an amplifier is measured by

$$\text{PG} = 10 \log \frac{P_{\text{out}}}{P_{\text{in}}}$$

where P_{out} is the power out and P_{in} is the power in. Find PG in the following cases:

(a) $P_{\text{in}} = 5,\ P_{\text{out}} = 600$ (b) $P_{\text{in}} = 2,\ P_{\text{out}} = 400$

63. The *Richter scale* gives the magnitude M of an earthquake as follows:

$$M = \log \frac{I}{I_0}$$

where I is the intensity of the quake and I_0 is a minimum intensity used as a standard. Compute M for each of the following cases:
(a) I is $50{,}000{,}000$ times I_0.
(b) I is $200{,}000{,}000$ times I_0.
(c) I is $500{,}000{,}000$ times I_0.

7.4 LINEAR INTERPOLATION AND COMPUTATIONS WITH LOGARITHMS

In the preceding section we saw how to evaluate the logarithmic function using the tables; for instance, log 721 = 2.8579. However, the table is limited: We can only evaluate the logarithm of a number with at most three nonzero digits. What about log 8.358?

To evaluate expressions such as log 8.358, we use a technique called **linear interpolation**. This method helps us find the logarithmic values *between* the given values in the table. This technique is also very important for use with other types of tables that are not programmed into a calculator. (We show several of these in the problem set.)

EXAMPLE 20 Consider log 8.358. Linear interpolation involves fitting a line segment between the two known points to help approximate the in-between points. Figure 2 shows this situation: We know log 8.35 = 0.9217 and log 8.36 = 0.9222, but we cannot find log 8.358. We can see that it is between 0.9217 and 0.9222. But how much?

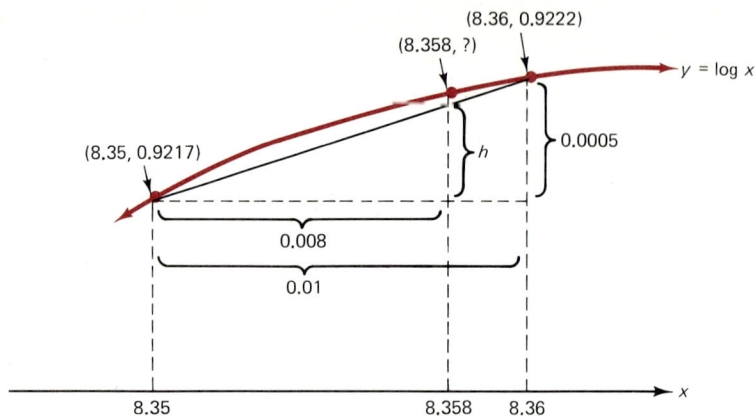

(8.36, 0.9222)

(8.358, ?)

y = log x

(8.35, 0.9217)

0.0005

h

0.008

0.01

8.35   8.358   8.36   x

FIGURE 2

We draw the line between the known points and solve for h (which is *approximately* what we have to add to log 8.35 to get log 8.358). Since we have similar right triangles, we have the following proportion:

$$\boxed{\frac{\text{height 1}}{\text{height 2}} = \frac{\text{base 1}}{\text{base 2}}}$$

$$\frac{h}{0.0005} = \frac{0.008}{0.01}$$

$$\boxed{\text{Solve for } h}$$

$$h = 0.0005\left(\frac{0.008}{0.01}\right) = 0.0004$$

We add this to log 8.35 to approximate log 8.358:

$$\log 8.358 \approx 0.9217 + 0.0004 = 0.9221$$

EXAMPLE 21 Find log 5632.

Solution Using the logarithmic tables, we see that log 5630 = 3.7505 and log 5640 = 3.7513. To approximate log 5632, we could sketch a portion of the graph as we did in Figure 2. However, it is usually more convenient to use a table such as the following (the differences between values correspond to the sides of the triangles in Figure 2):

	x	$\log x$	
2 [	5630	3.7505	] h
	5632	?	
10	5640	3.7513	0.0008

Again, we want h and use proportionality to find it:

$$\frac{h}{0.0008} = \frac{2}{10}$$

$$h = 0.00016$$

This is what we add to log 5630 to approximate log 5632:

$$\log 5632 \approx 3.7505 + 0.00016 = 3.75066$$

EXAMPLE 22   Find x such that $\log x = 0.7138$ (or equivalently, find antilog 0.7138).

Solution   We scan the tables for 0.7138, but the closest values that we find are log $5.17 = 0.7135$ and log $5.18 = 0.7143$. Thus, we know that x is between 5.17 and 5.18. We use linear interpolation (as in Example 21) to find k (what we must add to 5.17 to approximate x).

We now solve the following proportion for k:

$$\frac{k}{0.01} = \frac{0.0003}{0.0008}$$

$$k = 0.00375 \approx 0.004$$

Now we can find x:

$$x = \text{antilog } 0.7138 \approx 5.17 + 0.004 = 5.174$$

Computations with Common Logarithms

Not many years ago, common logarithms (and the slide rule based on them) were among the main tools for complicated calculations. The inexpensive hand calculator has drastically reduced the usage of common logarithms in calculations; however, there are still instances in which they are used. We now give a brief discussion of calculations with common logarithms. (Also, calculations with logarithms serve as a good reinforcement of the logarithmic properties, which are *still* important.)

Let us recall the logarithmic properties.

L1	$\log MN = \log M + \log N$
L2	$\log \dfrac{M}{N} = \log M - \log N$
L3	$\log N^r = r \log N$

EXAMPLE 23   Compute $P = (7800)(520)(0.0293)$.

Solution

> To multiply numbers using logarithms:
>
> 1. Find the logarithm of each number.
> 2. Add the logarithms.
> 3. Find the antilogarithm of the sum.

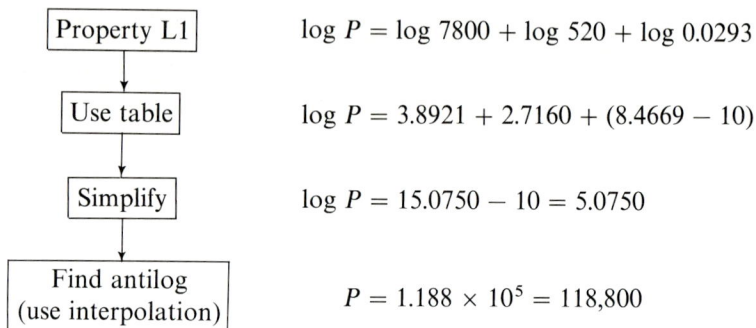

| Property L1 | $\log P = \log 7800 + \log 520 + \log 0.0293$ |

| Use table | $\log P = 3.8921 + 2.7160 + (8.4669 - 10)$ |

| Simplify | $\log P = 15.0750 - 10 = 5.0750$ |

| Find antilog (use interpolation) | $P = 1.188 \times 10^5 = 118{,}800$ |

Note: This is merely an approximation: The true answer is 118,840.8.

EXAMPLE 24   Compute $Q = \dfrac{(21{,}000)(0.0314)}{0.593}$.

Solution   We now add and subtract logarithms.

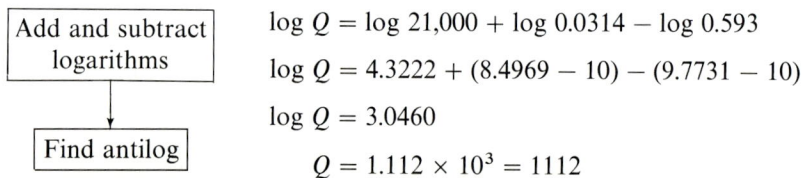

Add and subtract logarithms	$\log Q = \log 21{,}000 + \log 0.0314 - \log 0.593$
	$\log Q = 4.3222 + (8.4969 - 10) - (9.7731 - 10)$
	$\log Q = 3.0460$

| Find antilog | $Q = 1.112 \times 10^3 = 1112$ |

EXAMPLE 25   Compute $C = \sqrt[4]{0.00731}$.

Solution   We rewrite this as $(0.00731)^{1/4}$ and use Property L3. The characteristic is -3, which we first write as $7 - 10$. Since 10 is not divisible by 4, we rewrite this as $17 - 20$.

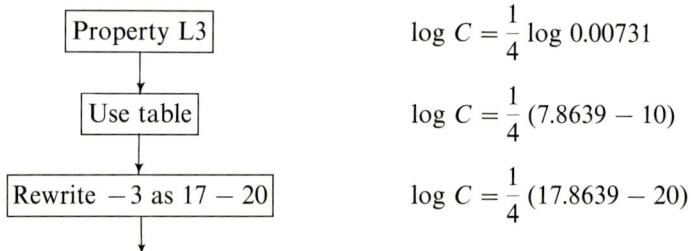

| Property L3 | $\log C = \dfrac{1}{4} \log 0.00731$ |

| Use table | $\log C = \dfrac{1}{4}(7.8639 - 10)$ |

| Rewrite -3 as $17 - 20$ | $\log C = \dfrac{1}{4}(17.8639 - 20)$ |

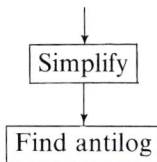

$$\boxed{\text{Simplify}}$$

$\log C = 4.4660 - 5$

$$\boxed{\text{Find antilog}}$$

$C = 2.924 \times 10^{-1} = 0.2924$

PROBLEM SET
7.4

Evaluate the following using the logarithm tables.

Warm-up
Exercises

1. log 7.2   **2.** log 481,000   **3.** log 0.00026

4. antilog 0.8149   **5.** antilog 3.4567   **6.** $10^{2.301}$

Simplify the following.

Warm-up
Exercises

7. $\log_7 abc$ **8.** $\log_2 \dfrac{x^2}{y^3}$

9. $\log \sqrt[3]{t^4}$ **10.** $\log \left(\dfrac{x^2 y}{zt^4}\right)^{1/5}$

Evaluate the following using the logarithm tables.

11. log 1234   **12.** log 82,880

13. log 795,300   **14.** log 1,575,000

15. log 0.001237   **16.** log 0.7573

17. log 0.0005555   **18.** log 0.06008

19. antilog 7.1245   **20.** antilog 0.3141

21. antilog 3.0605   **22.** antilog 2.1111

23. antilog $(8.6565 - 10)$ **24.** antilog $(4.8181 - 10)$

25. antilog $(13.4456 - 20)$ **26.** antilog $(3.9317 - 4)$

27. $10^{1.2347}$ **28.** $10^{4.9874}$

29. $10^{-2.9753}$ **30.** $10^{-1.9673}$

Use logarithms to calculate the following.

31. $(6.41)(12.3)$ **32.** $(540)(6000)$

33. $(810)(720)(0.48)$ **34.** $(4700)(0.55)(0.0731)$

35. $(1900)(404)(0.52)$ **36.** $(24,500)(0.0613)(0.0021)$

37. $\dfrac{25,000}{625}$ **38.** $\dfrac{400,000}{8000}$

39. $\dfrac{(21.2)(18.9)}{12.4}$ **40.** $\dfrac{(604)(1900)}{20.7}$

41. $\dfrac{17.6}{(4.03)(24.7)}$ **42.** $\dfrac{0.031}{(0.72)(0.0045)}$

43. $2.5^{1.3}$ **44.** $4.8^{2.1}$

45. 1.09^{20} **46.** 1.15^{10}

47. $\sqrt{171}$ **48.** $\sqrt[3]{4.92}$

49. $\sqrt[3]{0.0211}$ **50.** $\sqrt[5]{0.000659}$

51. $7000(0.8)^{10}$ **52.** $\dfrac{(1.73)^8}{(2.78)^5}$

53. One of the most important functions in statistics is the *normal distribution*

$$f(z) = \frac{1}{\sqrt{2\pi}} e^{-z^2/2}$$

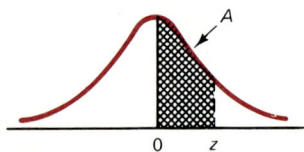

FIGURE 3

as sketched in Figure 3. Tables are given to provide the area A under the curve from 0 to z. A portion of this table is shown below. Suppose that $z = 0.73$; we then find the intersection of the 0.7 row and .03 column, which gives $A = 0.2673$. Use linear interpolation to find A for the following z-values:

(a) $z = 0.534$ (b) $z = 1.038$ (c) $z = 0.333$

z	.00	.01	.02	.03	.04	.05	.06	.07	.08	.09
0.0	.0000	.0040	.0080	.0120	.0160	.0199	.0239	.0279	.0319	.0359
0.1	.0398	.0438	.0478	.0517	.0557	.0596	.0636	.0675	.0714	.0753
0.2	.0793	.0832	.0871	.0910	.0948	.0987	.1026	.1064	.1103	.1141
0.3	.1179	.1217	.1255	.1293	.1331	.1368	.1406	.1443	.1480	.1517
0.4	.1554	.1591	.1628	.1664	.1700	.1736	.1772	.1808	.1844	.1879
0.5	.1915	.1950	.1985	.2019	.2054	.2088	.2123	.2157	.2190	.2224
0.6	.2257	.2291	.2324	.2357	.2389	.2422	.2454	.2486	.2517	.2549
0.7	.2580	.2611	.2642	.2673	.2704	.2734	.2764	.2794	.2823	.2852
0.8	.2881	.2910	.2939	.2967	.2995	.3023	.3051	.3078	.3106	.3133
0.9	.3159	.3186	.3212	.3238	.3264	.3289	.3315	.3340	.3365	.3389
1.0	.3413	.3438	.3461	.3485	.3508	.3531	.3554	.3577	.3599	.3621
1.1	.3643	.3665	.3686	.3708	.3729	.3749	.3770	.3790	.3810	.3830
1.2	.3849	.3869	.3888	.3907	.3925	.3944	.3962	.3980	.3997	.4015
1.3	.4032	.4049	.4066	.4082	.4099	.4115	.4131	.4147	.4162	.4177

54. Below is a table relating the brightness of a straight tungsten wire in a vacuum (as in a light bulb) to the absolute temperature ($°K$). Use linear interpolation to find the brightness for the following temperatures:

(a) $1300°K$ (b) $2250°K$ (c) $2930°K$

Temperature	Brightness
1000	966
1200	1149
1400	1330
1600	1509
1800	1684
2000	1857
2200	2026
2400	2192
2600	2356
2800	2516
3000	2673
3200	2827
3400	2978

In this section we study equations that involve exponential or logarithmic functions. For example,

$$2^x = 10 \qquad \text{and} \qquad \log_2 x + \log_2 (x - 2) = 3$$

are an **exponential** and a **logarithmic equation**.

Exponential Equations

In solving many exponential equations, we use the following property.

> **Property**
> Let M, N, and b be positive real numbers ($b \neq 1$); then

> **L7**   If $M = N$, then $\log_b M = \log_b N$.

In words, *we can take the logarithm of both sides of an equation, and the results are still equal.* (We usually use the common or natural logarithm.)

EXAMPLE 26   Solve $2^x = 15$.

Solution   We solve this exponential equation by *taking the common logarithm of both sides.*

log of both sides	$\log 2^x = \log 15$
L3: $\log N^r = r \log N$	$x \log 2 = \log 15$
Divide by log 2	$x = \dfrac{\log 15}{\log 2}$
Simplify	$x = 3.90689$

We can get this answer with a hand calculator:

PRESS	DISPLAY	MEANING
1 5 log ÷ 2 log =	3.9068907	$\dfrac{\log 15}{\log 2}$

We can also check this answer on a calculator:

PRESS	DISPLAY	MEANING
2 y^x 3 . 9 0 6 8 9 =	15.	$2^{3.90689} = 15$

EXAMPLE 27   Solve $8^{3x-1} = 5^{x+4}$.

Solution   Again we take the logarithm of both sides and isolate x.

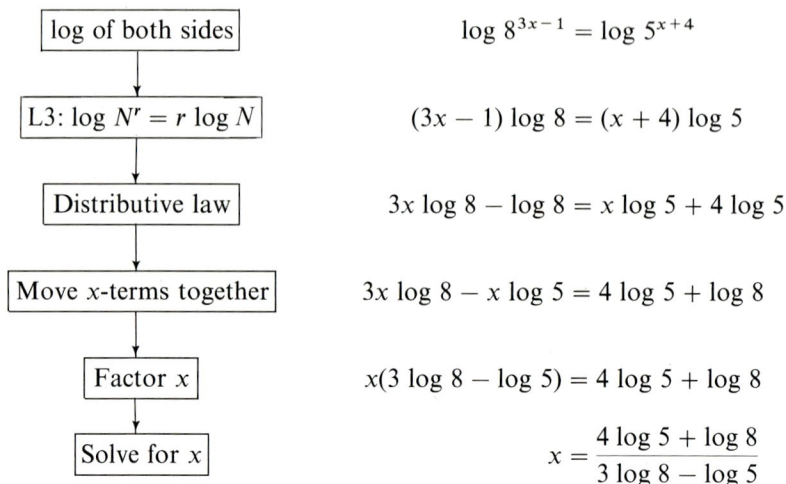

log of both sides	$\log 8^{3x-1} = \log 5^{x+4}$
L3: $\log N^r = r \log N$	$(3x - 1) \log 8 = (x + 4) \log 5$
Distributive law	$3x \log 8 - \log 8 = x \log 5 + 4 \log 5$
Move x-terms together	$3x \log 8 - x \log 5 = 4 \log 5 + \log 8$
Factor x	$x(3 \log 8 - \log 5) = 4 \log 5 + \log 8$
Solve for x	$x = \dfrac{4 \log 5 + \log 8}{3 \log 8 - \log 5}$

We can simplify this on a hand calculator as follows (notice how we use the parentheses keys):

PRESS	DISPLAY	MEANING
(4 × 5 log + 8 log)	3.69897	Numerator
÷ (3 × 8 log − 5 log)	2.0103	Denominator
=	1.840009	Answer

EXAMPLE 28   Solve $x^7 = 500$.

Solution   Here the base is the unknown, but again we take the common logarithm of both sides.

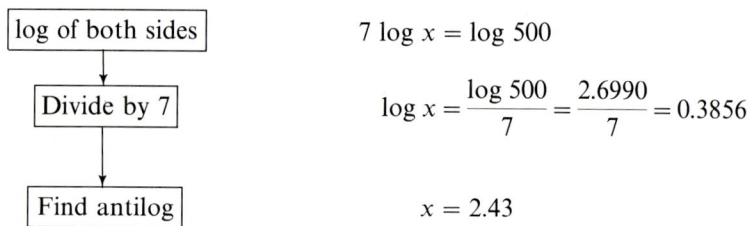

log of both sides	$7 \log x = \log 500$
Divide by 7	$\log x = \dfrac{\log 500}{7} = \dfrac{2.6990}{7} = 0.3856$
Find antilog	$x = 2.43$

We could also have solved this problem by taking the seventh root of both sides to get $x = \sqrt[7]{500} = 2.42978$.

Logarithmic Equations

Logarithmic equations usually have two or more logarithmic expressions in them. We use the logarithmic properties L1 to L3 to help solve these equations.

EXAMPLE 29   Solve $\log_5 2 + \log_5 10 + \log_5 x = 3$.

Solution   We use Property L1 to rewrite this with only one logarithmic expression. We then put it into exponent form and solve.

$$\boxed{\text{L1: } \log M + \log N = \log MN} \qquad\qquad \log_5 (2 \cdot 10 \cdot x) = 3$$

$$\downarrow$$

$$\boxed{\text{Exponent form}} \qquad\qquad 20x = 5^3 = 125$$

$$\downarrow$$

$$\boxed{\text{Divide by 20}} \qquad\qquad x = \frac{125}{20} = 6.25$$

EXAMPLE 30   Solve $\log_2 x + \log_2 (x - 2) = 3$.

Solution   Again, we first combine the terms into a single logarithmic expression. Since we can take the logarithm of a positive number only, we must be careful that $x - 2 > 0$, or $x > 2$.

$$\boxed{\log M + \log N = \log MN} \qquad\qquad \log_2 x(x - 2) = 3$$

$$\downarrow$$

$$\boxed{\text{Exponent form}} \qquad\qquad x(x - 2) = 2^3 = 8$$

$$\downarrow$$

$$\boxed{\text{Standard quadratic form}} \qquad\qquad x^2 - 2x - 8 = 0$$

$$\downarrow$$

$$\boxed{\text{Factor}} \qquad\qquad (x - 4)(x + 2) = 0$$

$$\downarrow$$

$$\boxed{\text{Solve for } x} \qquad\qquad x = 4, \ -2$$

Since $\log_2 \ -2$ is not allowed, the only solution is $x = 4$.

**PROBLEM SET
7.5**

Solve the following equations.

*Warm-up
Exercises*

1. $2x + 3 = 17$ 　　　　　　　　 **2.** $4x - 7 = 17$

3. $2(x - 3) + 7 = -2(x + 2) + 21$ 　　 **4.** $4(2x - 1) + x = 5(x + 1) + 23$

5. $x^2 - 9x + 14 = 0$ 　　　　　　 **6.** $x^2 + 5x = 24$

Use the logarithmic properties L1 to L3 to rewrite the following expressions in a different form.

Warm-up
Exercises

7. $\log 3^x$

8. $\log 2^{x-1}$

9. $\log x^3$

10. $\log 5^{-x}$

11. $\log_2 5 + \log_2 x$

12. $\log_7 (x - 1) + \log_7 (x + 1)$

Solve the following equations.

13. $4^x = 13$

14. $1.7^x = 3.2$

15. $1.09^x = 2$

16. $1.15^x = 3$

17. $12^{x+1} = 15^x$

18. $7^{2x+1} = 11^x$

19. $5^{2x-1} = 6^{x+3}$

20. $15^{x-4} = 9^{2x+1}$

21. $x^{10} = 2$

22. $x^4 = 5.3$

23. $x^{1.9} = 2.3$

24. $x^{4.3} = 8.42$

25. $\log_{10} 5 + \log_{10} x = 4$

26. $\log_6 x - \log_6 2 = 1$

27. $\log_7 x^2 - \log_7 x = 2$

28. $\log_4 x^3 - \log_4 x^2 = \dfrac{1}{2}$

29. $\log_3 x + \log_3 (x - 6) = 3$

30. $\log_2 x + \log_2 (x + 4) = 5$

31. $\log_6 (x - 3) + \log_6 (x + 2) = 2$

32. $\log_4 (x + 6) + \log_4 (x - 6) = 3$

33. $6^x = 7$

34. $x^{2.3} = 100$

35. $x^{1.5} = 5.2$

36. $2^{x+4} = 7^{x-1}$

37. $\log_4 2 + \log_4 x = \dfrac{5}{2}$

38. $\log_6 x^4 - \log_6 x^2 = 2$

39. $3^{x+1} = 5^x$

40. $1.08^x = 3$

41. $a^x = b$ (for x)

42. $x^c = d$ (for x)

43. $\log_2 (\log_2 x) = 3$

44. $(\log_3 x)^2 = 4$

45. $\log_x 4 + \log_x 9 = 2$

46. $\log_x 2 + \log_x 32 = 3$

47. $\log_{x^2} 250 - \log_{x^2} 2 = \dfrac{3}{2}$

48. $\log_{x^2} 54 - \log_{x^2} 6 = 1$

Business
Applications

49. The time t for an investment to double in value at 9% interest is the solution to the equation

$$(1.09)^t = 2$$

Solve the equation for t, the *doubling time*.

50. The rate of interest r needed to double an investment in 7 years is the solution to

$$(1 + r)^7 = 2$$

Solve for r. (*Hint*: First solve for $1 + r$; then find r.)

Consumer
Applications

51. Maria buys a $6400 car. Five years later, she sells it for $2000. Find the rate of depreciation by solving

$$6400(1 - r)^5 = 2000$$

52. A woman has a choice between two jobs: One starts at $18,000 with 10% annual raises, while the other starts at $24,000 with 7% annual raises. The time t at which the salaries will be equal is given by

$$18,000(1.10)^t = 24,000(1.07)^t$$

Solve this for t.

Population
Application

53. The population of India is about 700,000,000. Its population is increasing at the rate of 1.9% per year. Find the number of years y that it will take India to reach 1,000,000,000 people by solving

$$700{,}000{,}000(1.019)^y = 1{,}000{,}000{,}000$$

Physical
Application

54. The carbon-14 content in a certain object is given by

$$C = 100\left(\frac{1}{2}\right)^{t/5600}$$

where t is the age in years (see page 227). In how many years t will the content be 90? Also, solve the equation for t in terms of C.

Electrical
Application

55. The power gain in an amplifier is measured by

$$PG = 10 \log \frac{P_{out}}{P_{in}}$$

(see page 241). Solve this equation for P_{out} in terms of PG and P_{in}.

Photographic
Application

56. The relation between ASA and DIN (two measures of film speed) is

$$DIN = 15 + 3 \log_2 \frac{ASA}{25}$$

Solve this equation for ASA in terms of DIN.

7.6
NATURAL LOGARITHMS AND OTHER BASES

Sometimes we wish to convert from one logarithm base to another. Suppose that we want $\log_b N$, but we only know facts about logarithms in base a. Here is how we convert. Let $x = \log_b N$ and solve for x.

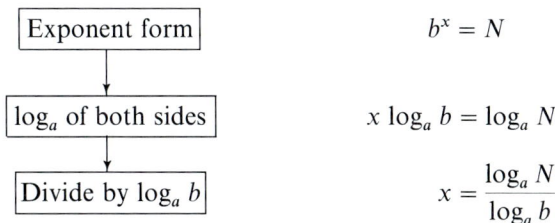

Exponent form	$b^x = N$
$\log_a$ of both sides	$x \log_a b = \log_a N$
Divide by $\log_a b$	$x = \dfrac{\log_a N}{\log_a b}$

This becomes our **change-of-base property**.

Property
For bases a and b and $N > 0$,

$$\textbf{L8} \quad \log_b N = \frac{\log_a N}{\log_a b}$$

Usually, the known base a will be 10 or e, since $\log_{10} b$ or $\log_e b$ is easily found.

EXAMPLE 31   Find $\log_2 20$.

Solution   We use Property L8 with $a = 10$.

$$\boxed{\text{L8: } \log_b N = \frac{\log N}{\log b}}$$

$$\log_2 20 = \frac{\log 20}{\log 2}$$

$$\downarrow$$

$$\boxed{\text{Simplify}}$$

$$= 4.3219$$

EXAMPLE 32   Find $\log_e 6$.

Solution   As mentioned earlier, logarithms to the base $e = 2.7183\ldots$ are called **natural logarithms**. We use L8 with $a = 10$ and the fact that $\log e = 0.4343$.

$$\boxed{\text{L8: } \log_b N = \frac{\log N}{\log b}}$$

$$\log_e 6 = \frac{\log 6}{\log e}$$

$$\downarrow$$

$$\boxed{\text{Find logs and simplify}}$$

$$= \frac{0.7782}{0.4343} = 1.7918$$

The natural-logarithm function $\log_e x$ is also written ln x. Although e must seem like a strange base to use, the reasons for its use become clear in calculus. Just as there is a table of common logarithms, there is a table of natural logarithms (see the Appendix).

On many calculators, we compute the natural logarithm with an $\boxed{\text{ln } x}$ key. Also, e^x is computed using $\boxed{\text{INV}}$ and $\boxed{\text{ln } x}$ together (since e^x is the inverse of ln x). For example, consider ln 6 and $e^{1.5}$:

PRESS	DISPLAY	MEANING
$\boxed{6}$ $\boxed{\text{ln } x}$	$\boxed{1.7917595}$	ln 6 (or $\log_e 6$)
$\boxed{1}$ $\boxed{\cdot}$ $\boxed{5}$ $\boxed{\text{INV}}$ $\boxed{\text{ln } x}$	$\boxed{4.4816891}$	$e^{1.5}$

Just as we have the property $\log 10^k = k$, we have a similar property for natural logarithms:

$$\boxed{\ln e^k = k}$$

For example, $\ln e^4 = 4$. We can use this fact in solving exponential equations involving a base of e.

EXAMPLE 33   The population of a certain bacterial culture is given by

$$P = 4000e^{0.1t}$$

where t is the time in hours. Find the time at which $P = 10{,}000$.

Solution   We must solve the exponential equation $4000e^{0.1t} = 10{,}000$. Since this equation involves e, we take the natural logarithm of both sides.

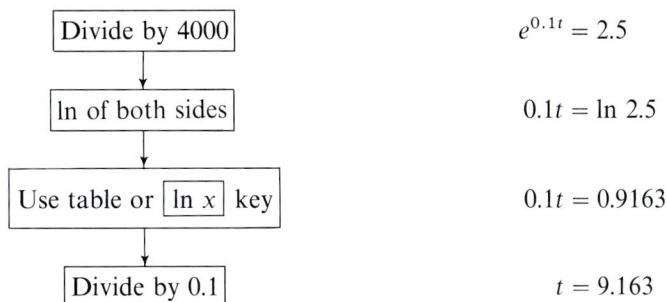

Divide by 4000	$e^{0.1t} = 2.5$
ln of both sides	$0.1t = \ln 2.5$
Use table or $\boxed{\ln x}$ key	$0.1t = 0.9163$
Divide by 0.1	$t = 9.163$

Thus, it takes about 9.2 hours.

EXAMPLE 34   The radioactivity R in a certain substance starts at 1000. After t years it is given by

$$R = 1000e^{-kt}$$

where k is a constant.
(a) If after 100 years $R = 960$, find k.
(b) Find the *half-life*; that is, at what time is the radioactivity down to half its original size (or $R = 500$)?

Solution   (a) We substitute $t = 100$ and $R = 960$ and solve the equation for k.

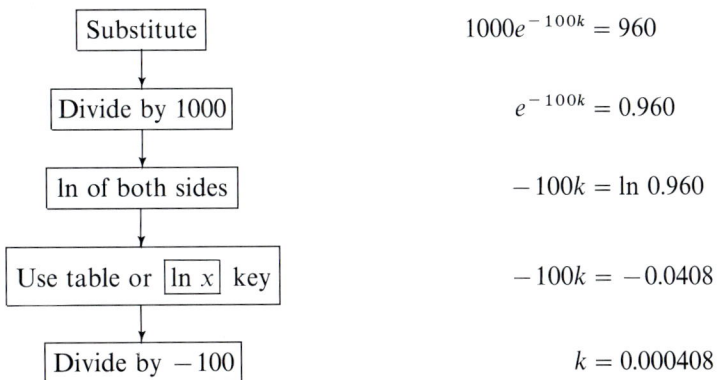

Substitute	$1000e^{-100k} = 960$
Divide by 1000	$e^{-100k} = 0.960$
ln of both sides	$-100k = \ln 0.960$
Use table or $\boxed{\ln x}$ key	$-100k = -0.0408$
Divide by -100	$k = 0.000408$

Thus, $R = 1000e^{-0.000408t}$.

(b) To find the half-life, we substitute $k = 0.000408$ and $R = 500$ (when only half the original 1000 is left) and solve for t.

Substitute	$500 = 1000e^{-0.000408t}$
Divide by 1000	$0.5 = e^{-0.000408t}$
$\ln 0.5 = -.6931$	$-0.6931 = -0.000408t$
Solve for t	$1699 = t$

Thus, the half-life is approximately 1700 years.

PROBLEM SET
7.6

Find the following logarithms.

Warm-up Exercises

1. log 100

2. log 0.0001

3. log 47000

4. log 17

5. log 0.02

6. log 0.00023

7. $\log_2 16$

8. $\log_5 \dfrac{1}{25}$

Solve the following equations.

Warm-up Exercises

9. $2(x + 4) = 3x + 1$

10. $(x - 2)(x + 3) = 24$

11. $2^x = 7$

12. $3^x = 2$

13. $5^{-x} = 4$

14. $7^{-2x} = 3$

Evaluate the following.

15. $\log_4 8$

16. $\log_{25} 125$

17. $\log_9 5$

18. $\log_3 11$

19. $\log_6 0.3$

20. $\log_{12} 0.75$

21. $\ln 2$

22. $\log_e 1.06$

23. $\ln 0.4$

24. $\ln 0.78$

25. $\ln e^A$

26. $\log_e e^{-xy}$

Find x in each of the following equations.

27. $e^{2x} = 4$

28. $e^{3x} = 10$

29. $2e^{0.1x} = 5$

30. $10e^{0.5x} = 30$

31. $e^{-x} = 0.5$

32. $e^{-x} = 0.35$

33. $50e^{-0.5x} = 20$

34. $200e^{-0.2} = 20$

35. $x = \log_5 7$ **36.** $e^x = 4.2$

37. $x = \ln 2.9$ **38.** $x = \log_3 8.5$

39. $e^x = 15$ **40.** $x = \log_7 7$

41. $e^{-x} = 0.7$ **42.** $x = \ln 2.1$

43. $x = \log_{12} 12$ **44.** $500e^{2x} = 1500$

45. $2^{-x} = 0.15$ **46.** $x = \log_9 0.01$

47. $e^{x^2} = 15$ **48.** $e^{-x^2} = 0.3$

49. $e^{-(x-1)^2/2} = 0.8$ **50.** $e^{(x+1)^2/4} = 10$

Electrical Application

51. The current i in a certain direct-current circuit is given by

$$i = 2(1 - e^{-50t})$$

(a) At what time t will the current be 1?
(b) Solve for t in terms of i.

Information Application

52. A code contains two symbols (a and b) that are used with frequencies 30% and 70%. Find the information I, which is given by

$$I = -(0.3 \log_2 0.3 + 0.7 \log_2 0.7)$$

Life Science Application

53. The population of a certain bacterial culture is given by

$$P = 10{,}000e^{0.2t}$$

where t is the time in hours.
(a) Find the time t at which $P = 20{,}000$.
(b) Find the time t at which $P = 100{,}000$.
(c) Solve for t in terms of P.

Physical Application

54. The radioactivity R in a certain substance starts at 200. After t years it is given by

$$R = 200e^{-kt}$$

where k is a constant.
(a) After 500 years, $R = 188$; find k.
(b) Find the *half-life*; that is, at what time is the radioactivity down to half its original size (or $R = 100$)?
(c) Solve for t in terms of R.
(d) Suppose the substance is found with $R = 2.5$. How long has the substance been decaying?

55. Newton's law of cooling states that the rate of decrease in temperature in an object is proportional to the difference in temperature. For instance, if a 60°C object is dropped into a vat of 20°C water, then the temperature T of the object is given by

$$T = 20 + 40e^{-0.06t}$$

where t is the time in seconds.
(a) Graph this for $t = 0, 10, 20, 30,$ and 40.
(b) At what time will the temperature be 30°C?
(c) Solve for t in terms of T.

Important Definitions and Properties

For all real numbers $a, b, M, N > 0$ $(a \neq 1, b \neq 1)$:

If $a > 1$, then $f(x) = a^x$ is increasing.

If $0 < a < 1$, then $f(x) = a^x$ is decreasing.

$a^x = a^y$ if and only if $x = y$.

$\log_b N = x$ means $b^x = N$.

$\log_b MN = \log_b M + \log_b N$

$\log_b \dfrac{M}{N} = \log_b M - \log_b N$

$\log_b N^r = r \log_b N$

$\log_b b^M = M \qquad \log_b 1 = 0$

$b^{\log_b M} = M \qquad \log_b b = 1$

$10^x = \text{antilog } x$

If $M = N$, then $\log_b M = \log_b N$.

$\log_b N = \dfrac{\log_a N}{\log_a b}$

$\log_{10} N = \log N \qquad \log_e N = \ln N$

$\log 10^k = k \qquad \ln e^k = k$

$10^{\log k} = k \qquad e^{\ln k} = k$

Review Exercises

Graph the following functions.

1. $f(x) = 2^x$

2. $f(x) = 0.9^x$

3. $f(x) = 1000 \cdot 3^x$

4. $f(x) = \log_2 x$

Solve the following equations.

5. $5^x = \dfrac{1}{25}$

6. $2^{-x} = 16$

7. $3^{-x^2} = \dfrac{1}{81}$

Write the following in logarithm form.

8. $8^{5/3} = 32$

9. $25^{-3/2} = \dfrac{1}{125}$

Write the following in exponent form.

10. $\log_{10} 10{,}000 = 4$

11. $\log_8 \dfrac{1}{4} = \dfrac{-2}{3}$

Solve the following equations.

12. $\log_{16} 4 = x$

13. $\log_x 64 = 3$

14. $\log_5 x = -3$

Write the following in terms of logarithm terms.

15. $\log_7 6ab^2$

16. $\log_2 \dfrac{t^4}{s^3}$

17. $\log_{10} \sqrt[7]{\dfrac{x^2 y^5}{z^6}}$

Write the following in terms of a single logarithm (with coefficient 1).

18. $\log_7 a + \log_7 b - \log_7 c$

19. $5 \log_2 x - \dfrac{2}{3} \log_2 y - \dfrac{3}{2} \log_2 z$

Find the following common logarithms and antilogarithms.

20. $\log 10$

21. $\log 1500$

22. $\log 0.007214$

23. antilog 3.0981

24. antilog $(8.7817 - 10)$

25. $10^{2.7127}$

Use logarithms to calculate the following.

26. $(49.8)(5100)(0.7)$

27. $\dfrac{(610)(802)}{720{,}000}$

28. $\sqrt{40{,}200}$

29. $\sqrt[3]{\dfrac{2500}{407}}$

Solve each of the following equations.

30. $1.8^x = 3$

31. $15^{x+1} = 9^{2x}$

32. $x^{1.7} = 2.6$

33. $\log_3 36 - \log_3 x = 2$

34. $\log_4 (x + 4) + \log_4 (x - 4) = 2$

35. $5^{-x} = 3$

36. $1000 \cdot 2^{-x} = 500$

Find the following logarithms.

37. $\log_5 8$

38. $\log_9 0.8$

39. $\log_e 7$

40. $\ln 0.7$

Solve the following equations.

41. $e^{3x} = 10$

42. $7e^{0.5x} = 20$

43. $100e^{-0.7x} = 25$

Introduction to Trigonometry

Originally, **trigonometry** began as the study of functions of angles in triangles. The applications of these functions have been of great importance to physicists, astronomers, surveyors, aviators, civil engineers, and so on.

More recently, trigonometry has grown in scope. The trigonometric functions are defined for all real numbers (and not just acute angles). The applications now include many fields with a periodic nature: electricity, sound, business, medicine, meteorology, and so on.

Our ultimate goal is to study the more general trigonometric functions defined for all real numbers. First, however, we start with the trigonometric functions based on acute angles since this approach is both historic and more intuitive. Once comfortable with these functions, we then expand our definitions to all angles and finally to all real numbers.

8.1
ANGLES AND ARCS

We begin with one of the simplest concepts of geometry: the angle. An **angle** is a rotation from one ray (called the **initial side**) about a common endpoint (called the **vertex**) to another ray (called the **terminal side**). As Figure 1 shows, counterclockwise rotation is considered positive, and clockwise negative.

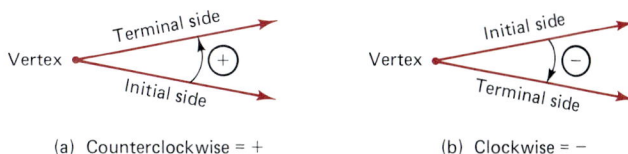

(a) Counterclockwise = +   (b) Clockwise = −

FIGURE 1

The most familiar way to measure angles is in degrees.

> 1 **degree** $= 1° = \dfrac{1}{360}$ of a complete revolution.
>
> A complete revolution is 360°.

EXAMPLE 1 Figure 2 shows some common angles measured in degrees. Note that if an angle is one-half of a revolution, it is 1/2 of 360° = 180°, and so on.

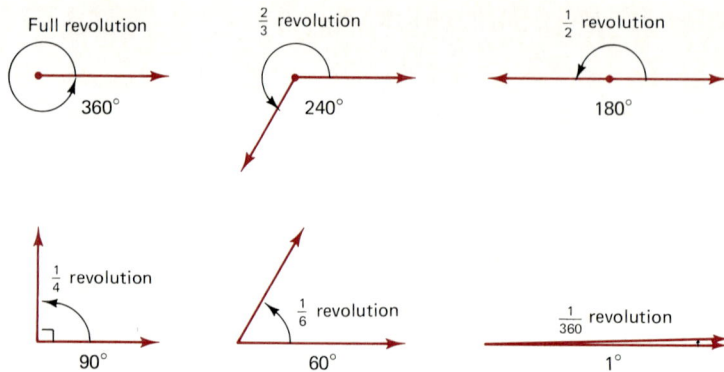

FIGURE 2

For angles less than 1°, we have minutes and seconds.

$$1 \text{ minute} = 1' = \frac{1}{60} \text{ of } 1° \qquad (60' = 1°)$$

$$1 \text{ second} = 1'' = \frac{1}{60} \text{ of } 1' = \frac{1}{3600} \text{ of } 1° \qquad (60'' = 1')$$

For example, $17\frac{1}{2}° = 17°30'$ (since $30'$ is 1/2 of a degree). With more and more people using calculators, fractions of degrees are more often written with decimals. For instance,

$$42°29' = 42° + \frac{29}{60}° = 42.48°$$

Angles between 0° and 90° are called **acute angles**, and a 90° angle is often called a **right angle**.

Often, two different angles have the same initial and terminal sides. We call such angles **equivalent** (or **coterminal**).

EXAMPLE 2 Figure 3 shows three equivalent (or coterminal) angles. Although they all have the same initial and terminal sides, they each have a *different* measure:
(a) This is 60°.
(b) Here, we rotate in a *negative* direction (clockwise) 300° from the initial side to the terminal side. This is −300°.
(c) Here, we go counterclockwise one complete revolution (360°), plus another 60°. This gives 420°.

Given an angle larger than 360° (or less than 0°), it is often convenient to find an equivalent angle between 0° and 360°. To do this, we subtract (or add) 360° until the angle is between 0° and 360°.

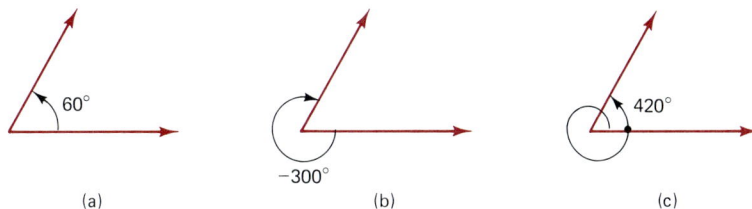

(a)   (b)   (c)

FIGURE 3

EXAMPLE 3   For each of the following angles, we find its equivalent between 0° and 360°.

Angle	Equivalent between 0° and 360°	Comment
(a) 140°	140°	Already between 0° and 360°
(b) 410°	50°	$410° - 360° = 50°$
(c) 800°	80°	$800° - 360° - 360° = 80°$
(d) $-70°$	290°	$-70° + 360° = 290°$

When dealing with many angles of different sizes, it is often convenient to put the angles into a **standard position**: the vertex at the origin and the initial side on the positive x-axis.

EXAMPLE 4   Figure 4 shows two angles, each in standard position: the vertex at the origin and the initial side on the positive x-axis.

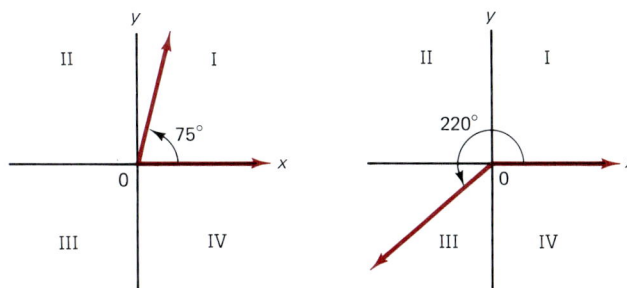

FIGURE 4

In Figure 4, note that the terminal side of an angle in standard position usually lies in any of the four quadrants. We say an angle is in quadrant II if its terminal side is in quadrant II, and so on.

Angles are very much related to the arcs of circles. In fact, an **arc** (a connected portion of a circle) is measured by the central angle that cuts it.

The **length of an arc** is proportional to the measure of the arc in degrees. For example, a 180° arc (1/2 circle) has a length of 1/2 the circumference, and a 90° arc (1/4 circle) has a length of 1/4 the circumference. Recall that the circumference of a circle of radius r is given by

$$C = 2\pi r$$

where $\pi \approx 3.14$. (Also, $C = \pi D$, where D is the diameter.) Thus, we have the following theorem.

Theorem
In a circle of radius r, the length L of an arc of A degrees satisfies the proportion

$$\frac{L}{2\pi r} = \frac{A}{360}$$

EXAMPLE 5   Find the length of the arc shown in Figure 5.

Solution   The arc is 110°; therefore, it is 110/360 of the whole circle. The circumference is $2\pi r = 2(3.14)(20) = 125.6$. Thus, the arc length is

$$\frac{L}{2\pi r} = \frac{A}{360}$$

Solve for L

$$\frac{L}{125.6} = \frac{110}{360}$$

$$L = 38.4$$

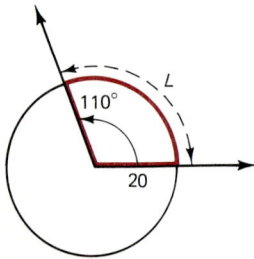

FIGURE 5

**PROBLEM SET
8.1**

Locate the following points on a rectangular coordinate plane.

*Warm-up
Exercises*

1. (2, 1)   **2.** (5, 3)   **3.** (−2, 4)
4. (−3, 0)   **5.** (0, −1)   **6.** (−1, −2)

Find the circumferences of each of the following circles.

*Warm-up
Exercises*

7. Radius = 100   **8.** Radius = 70
9. Diameter = 12   **10.** Diameter = 50

Complete the following table.

	Angle (degrees)	Fraction of a revolution
11.	?	1/5
12.	?	1/10
13.	?	3/4
14.	?	7/12
15.	15°	?
16.	20°	?
17.	300°	?
18.	210°	?

Problems 19 and 20 each show three equivalent angles. Use the first angle to find the measure of the other two as sketched.

19.

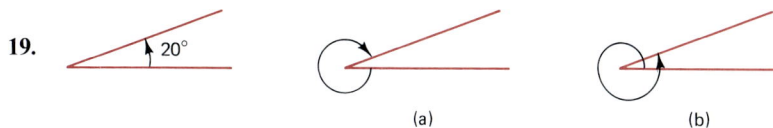

(a)   (b)

20.

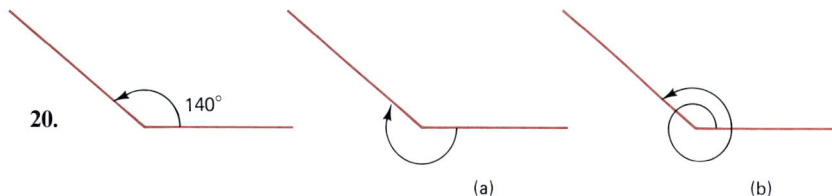

(a)   (b)

For each of the following angles, find an equivalent angle between 0° and 360°.

21. $-55°$	**22.** $-601°$	**23.** $205°$
24. $315°$	**25.** $385°$	**26.** $426°$
27. $557°$	**28.** $619°$	**29.** $1200°$
30. $1700°$		

Sketch each of the following angles in standard position.

31. $45°$	**32.** $150°$	**33.** $220°$
34. $290°$	**35.** $-60°$	**36.** $-110°$
37. $460°$	**38.** $510°$	

Complete the following table for central angles and their corresponding arcs in circles.

	Central angle	Measure of arc (degrees)
39.	$25°$	?
40.	$140°$	?
41.	?	$260°$
42.	?	$190°$

Complete the following table for central angles and arcs within a circle.

	Radius	Central angle (degrees)	Arc length
43.	10	$70°$	?
44.	6	$140°$	?
45.	50	$240°$	?
46.	100	$10°$	?
47.	12	?	4π
48.	4	?	3π
49.	1	?	$\pi/2$
50.	1	?	$\pi/6$

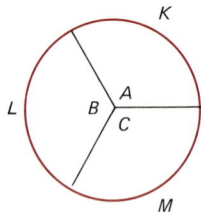

Consider the circle (sketched at the left, but not to scale) with angles A, B, and C, and corresponding arc lengths K, L, and M. Complete the following table.

	Radius	A	B	C	K	L	M	Circumference
51.	10	70°	80°	?	?	?	?	?
52.	40	100°	?	120°	?	?	?	?
53.	50	?	?	?	?	25π	50π	?
54.	12	?	?	?	4π	6π	?	?
55.	?	?	30°	?	4π	?	?	12π
56.	?	?	?	45°	?	?	20π	40π

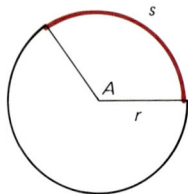

Another very important angle measure involves radians, which we will discuss in great detail in the next chapter. As shown at the left, if a central angle A in a circle radius r cuts an arc of length s, we define the radian measure as

$$A = \frac{s}{r} \text{ radians}$$

In words, the radian measure of a central angle is the ratio of the arc length to the radius of the circle. For example, if angle A cuts an arc of 7 in a circle radius 10, the angle A = 7/10 radian. Complete the following table, which finds the radian measure of some common angles.

	r	Angle (degrees)	Fraction of circle	Circumference of circle	s = arc length	Angle = s/r (radians)
57.	10	90°	?	?	?	?
58.	20	90°	?	?	?	?
59.	30	180°	?	?	?	?
60.	100	180°	?	?	?	?
61.	10	60°	?	?	?	?
62.	30	60°	?	?	?	?
63.	40	360°	?	?	?	?
64.	100	360°	?	?	?	?

Technical Application

65. A mechanic gives a bolt an eighth of a turn.
(a) How many degrees is this?
(b) If the wrench used is 20 centimeters long, how long is the arc of the end of the wrench?

Geographical Application

66. There are 24 time zones, and the radius of the earth is about 3960 miles (6370 kilometers).
(a) How many degrees are in each time zone (assuming uniform time zones)?
(b) How many miles (kilometers) are in a time zone?

Historical Application

67. More than 1700 years before Columbus "proved" that the world was round, a Greek mathematician Eratosthenes provided a brilliant, yet incredibly simple procedure for measuring the circumference of the earth. During the summer solstice, the sun was directly overhead in Syene, Egypt (0°). In Alexandria (500 miles away), the sun's rays made a 7.2° angle with a sundial (see page 265). Use this information to calculate the circumference of the earth. (Compare this to the accepted circumference by using r = 3960 miles.)

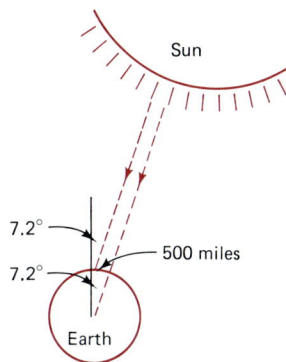

Sun

7.2°

7.2°

500 miles

Earth

68. Most single-lens-reflex (SLR) cameras have changeable lenses. The *angle of view* θ is related to the focal length f of the lens as shown in the table below. Graph the table (with f on the *x*-axis).

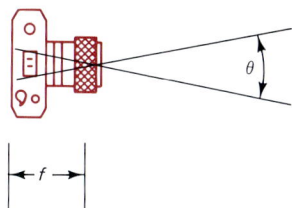

f (mm)	15	28	50	85	135	200	300	600	1200
θ	110°	75°	46°	28°	18°	12°	8°	4°	2°

8.2
RIGHT TRIANGLES

Historically, trigonometry began with the right triangle. In this section we look at several of the important properties of triangles that lead to the trigonometric functions. We recall that a **right triangle** is a triangle with a right angle (90°). The side opposite the right angle is called the **hypotenuse**, and each of the other two sides is called a **leg**. Figure 6 illustrates this.

One of the most remarkable of all theorems of plane geometry is the **Pythagorean theorem**, which relates the sides of a right triangle (as sketched in Figure 6).

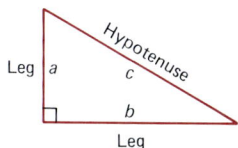

FIGURE 6

Pythagorean Theorem
In a right triangle with legs of length a and b and hypotenuse of length c, we have

$$a^2 + b^2 = c^2$$

EXAMPLE 6   We use the Pythagorean theorem to find the missing sides of the right triangles in Figure 7.

Solution   Since the triangles are all right triangles, the Pythagorean theorem applies: $a^2 + b^2 = c^2$. Triangles (b) and (c) are special triangles: the 45°–45°–90° triangle and the 30°–60°–90° triangle.

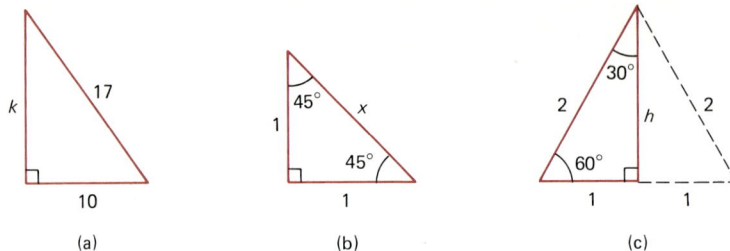

(a)

(b)

(c)

FIGURE 7

(a) $k^2 + 10^2 = 17^2$
$k^2 = 289 - 100$
$k = \sqrt{189} \approx 13.7$

(b) $1^2 + 1^2 = x^2$
$2 = x^2$
$\sqrt{2} = x$ (or $x \approx 1.41$)

(c) $1^2 + h^2 = 2^2$
$h^2 = 4 - 1$
$h = \sqrt{3} \approx 1.73$

Other important facts about triangles are the following.

> The sum of the angles in any triangle is 180°.

> The sum of the acute angles in any *right* triangle is 90°. Such angles are called **complementary**.

Often two triangles have the same shape, but one has longer sides than the other. More precisely, two triangles are **similar** if the three angles of one are equal to the three angles of the other and their sides are proportional.

Theorem

> Two right triangles are similar if an acute angle of one triangle is equal to an acute angle of the other.

Figure 8 shows two right triangles with $A = A'$; therefore, the triangles are similar. Thus, the corresponding ratios are equal. For instance,

$$\frac{a}{b} = \frac{a'}{b'} \qquad \frac{a}{c} = \frac{a'}{c'} \qquad \text{and so on}$$

In fact, any right triangle with an angle equal to A (or A') is similar to these triangles, and its sides have the same ratios. We have special names for all the possible ratios associated with angle A. These ratios are called

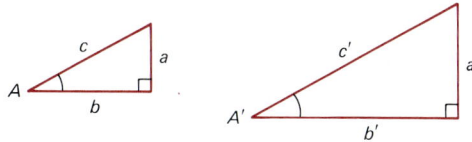

FIGURE 8

trigonometric functions. As you use them, they will become quite familiar to you. For now, however, you are advised to memorize them.

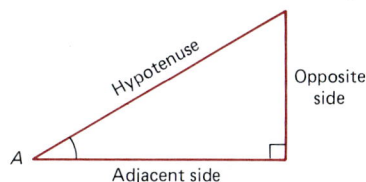

FIGURE 9

Figure 9 shows a general right triangle with an acute angle A. The three sides have special names:

1. The leg opposite angle A is called the **opposite side** (abbreviated **opp**).
2. The leg on one of the rays of angle A is called the **adjacent side** (abbreviated **adj**).
3. The remaining side (opposite the right angle) is called the **hypotenuse** (abbreviated **hyp**).

Definition
For a right triangle with acute angle A (as shown in Figure 9), we define the trigonometric functions as follows.

$$\textbf{sine } A = \sin A = \frac{\text{opposite side}}{\text{hypotenuse}} = \frac{\text{opp}}{\text{hyp}}$$

$$\textbf{cosine } A = \cos A = \frac{\text{adjacent side}}{\text{hypotenuse}} = \frac{\text{adj}}{\text{hyp}}$$

$$\textbf{tangent } A = \tan A = \frac{\text{opposite side}}{\text{adjacent side}} = \frac{\text{opp}}{\text{adj}}$$

$$\textbf{cotangent } A = \cot A = \frac{\text{adjacent side}}{\text{opposite side}} = \frac{\text{adj}}{\text{opp}}$$

$$\textbf{secant } A = \sec A = \frac{\text{hypotenuse}}{\text{adjacent side}} = \frac{\text{hyp}}{\text{adj}}$$

$$\textbf{cosecant } A = \csc A = \frac{\text{hypotenuse}}{\text{opposite side}} = \frac{\text{hyp}}{\text{opp}}$$

For now the six trigonometric functions are defined only for acute angles A. Ultimately, we will expand our definitions to angles of any size. Let us now look at some examples for acute angles.

EXAMPLE 7   Find the six trigonometric functions for angles A and B in the right triangles in Figure 10.

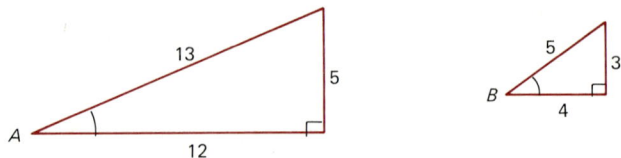

FIGURE 10

Solution   We evaluate the functions by substituting into the definitions. (Here, we use the abbreviations adj = adjacent, opp = opposite, and hyp = hypotenuse.)

$$\sin A = \frac{\text{opp}}{\text{hyp}} = \frac{5}{13} \qquad \sin B = \frac{\text{opp}}{\text{hyp}} = \frac{3}{5}$$

$$\cos A = \frac{\text{adj}}{\text{hyp}} = \frac{12}{13} \qquad \cos B = \frac{\text{adj}}{\text{hyp}} = \frac{4}{5}$$

$$\tan A = \frac{\text{opp}}{\text{adj}} = \frac{5}{12} \qquad \tan B = \frac{\text{opp}}{\text{adj}} = \frac{3}{4}$$

$$\cot A = \frac{\text{adj}}{\text{opp}} = \frac{12}{5} \qquad \cot B = \frac{\text{adj}}{\text{opp}} = \frac{4}{3}$$

$$\sec A = \frac{\text{hyp}}{\text{adj}} = \frac{13}{12} \qquad \sec B = \frac{\text{hyp}}{\text{adj}} = \frac{5}{4}$$

$$\csc A = \frac{\text{hyp}}{\text{opp}} = \frac{13}{5} \qquad \csc B = \frac{\text{hyp}}{\text{opp}} = \frac{5}{3}$$

**PROBLEM SET
8.2**

Solve the following equations for x (positive x only).

Warm-up
Exercises

1. $x + 50 + 70 = 180$

2. $x + 41 + 87 = 180$

3. $\dfrac{x}{10} = \dfrac{5}{9}$

4. $\dfrac{x}{12} = \dfrac{7}{20}$

5. $x^2 = 2^2 + 5^2$

6. $x^2 = 3^2 + 6^2$

7. $3^2 + x^2 = 7^2$

8. $x^2 + 2^2 = 6^2$

In the table below, angles A, B, and C are the angles of a triangle. Find the missing angles in each case.

	A	B	C
9.	60°	70°	?
10.	17.4°	38.9°	?
11.	42°	?	90°
12.	7°	?	90°
13.	?	88.2°	90°
14.	?	0.5°	90°
15.	x	2x	3x
16.	2x	3x	4x

For each of the following right triangles, evaluate the six trigonometric functions: sin A, cos A, tan A, cot A, sec A, and csc A.

17.

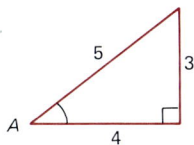

18.

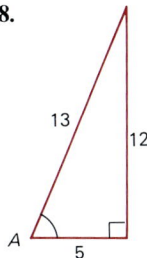

19.

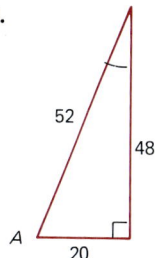

20.

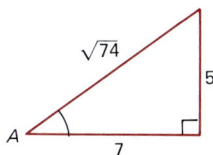

Consider the right triangle at the left with sides a, b, and c, and acute angles A and B. (The sketch is not to scale.) Complete the tables below. What patterns do you observe in the tables?

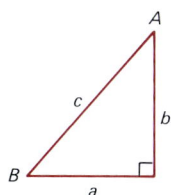

	a	b	c	sin A	sin B	cos A	cos B
21.	7	8	?	?	?	?	?
22.	6	?	9	?	?	?	?
23.	6	?	12	?	?	?	?
24.	12	?	15	?	?	?	?
25.	4	5	?	?	?	?	?
26.	3	10	?	?	?	?	?
27.	7	?	10	?	?	?	?
28.	?	4	8	?	?	?	?
29.	?	?	20	$\frac{1}{2}$	?	?	?
30.	?	?	10	?	?	?	$\frac{2}{5}$
31.	?	12	?	?	$\frac{2}{3}$	?	?
32.	8	?	?	?	?	$\frac{1}{3}$	?

	a	b	c	$\tan A$	$\cot A$	$\tan B$	$\cot B$
33.	4	?	9	?	?	?	?
34.	?	10	13	?	?	?	?
35.	?	20	?	?	?	1.5	?
36.	8	?	?	?	0.4	?	?

In the following problems, sketch a right triangle with an angle A that satisfies the given condition. Then use the sketch to evaluate the indicated function.

37. $\sin A = \dfrac{3}{5}$; find $\cos A$.

38. $\tan A = \dfrac{5}{7}$; find $\sin A$.

39. $\cos A = \dfrac{5}{13}$; find $\tan A$.

40. $\cot A = \dfrac{5}{2}$; find $\cos A$.

41. $\sec A = 2$; find $\sin A$.

42. $\csc A = 2.5$; find $\tan A$.

Engineering Application

43. An antenna h feet above ground transmits a signal to a point of tangency to the earth. It is $3960 + \dfrac{h}{5280}$ miles from the center of the earth. Use the sketch at the left to show that the distance d is approximately

$$d \approx 1.22\sqrt{h}$$

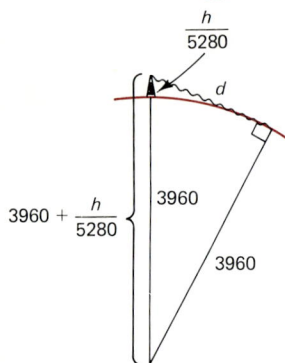

Physical Application

44. An object is thrown with a horizontal speed of 50 and vertical speed 70.
(a) What is the resulting speed v?
(b) What is $\tan A$, where A is the angle to the horizontal?

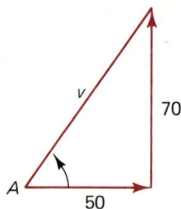

Sports Application

45. A pool player wishes to bounce a ball off the cushion and into the far corner. The angles indicated are equal. What can you say about lengths x and y?

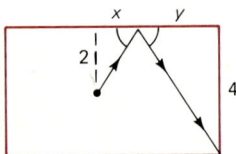

46. A solar panel is installed as the roof of the house shown at the left.
   (a) What is the length L?
   (b) What is tan A? (This is the *pitch* of the roof.)

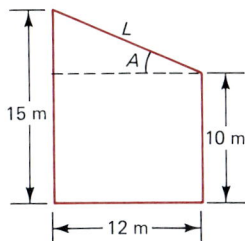

8.3
TRIGONOMETRIC
FUNCTIONS OF
ANY ANGLE

In the preceding section we introduced and defined the six trigonometric functions for acute angles. Recall that these definitions were based on the fact that an acute angle produces the same ratios no matter what size of right triangle it is in (since the triangles would all be similar).

We now want to extend our definitions to all angles. These definitions are based on the definitions for acute angles. Recall from page 261 that an angle is in standard position if its vertex is at the origin and its initial side is on the positive x-axis (see Figure 11). In naming angles, we now graduate from capital letters, such as A and B, to Greek letters, such as θ (theta) and ϕ (phi).

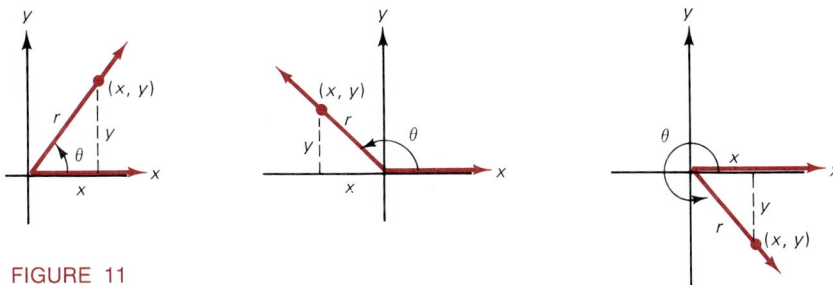

FIGURE 11

Definition

For an angle θ in standard position, let (x, y) be any point on the terminal side of θ with $r = \sqrt{x^2 + y^2}$ (see Figure 11). We then define the trigonometric functions as follows (where denominators are not zero):

$$\sin \theta = \frac{y}{r} \qquad \csc \theta = \frac{r}{y}$$

$$\cos \theta = \frac{x}{r} \qquad \sec \theta = \frac{r}{x}$$

$$\tan \theta = \frac{y}{x} \qquad \cot \theta = \frac{x}{y}$$

Before we work some examples, let us first make some observations.

1. For acute angles these ratios are the same as defined on page 267.
2. Unlike the acute-angle case, x or y (or both) may be positive or negative; thus, the ratios themselves may be positive or negative.
3. The distance r is always positive.

EXAMPLE 8 Find the six trigonometric ratios for the angles shown in Figure 12.

Solution We use the (x, y) coordinates in the definitions. Note that the distance r from the origin to (x, y) is always positive.

(a) $\sin \theta = \dfrac{y}{r} = \dfrac{4}{5} = 0.80$ (b) $\sin \phi = \dfrac{y}{r} = \dfrac{-5}{\sqrt{74}} \approx -0.58$

$\cos \theta = \dfrac{x}{r} = \dfrac{-3}{5} = -0.60$ $\cos \phi = \dfrac{x}{r} = \dfrac{-7}{\sqrt{74}} \approx -0.81$

$\tan \theta = \dfrac{y}{x} = \dfrac{4}{-3} \approx -1.33$ $\tan \phi = \dfrac{y}{x} = \dfrac{-5}{-7} \approx 0.71$

$\cot \theta = \dfrac{x}{y} = \dfrac{-3}{4} = -0.75$ $\cot \phi = \dfrac{x}{y} = \dfrac{-7}{-5} = 1.40$

$\sec \theta = \dfrac{r}{x} = \dfrac{5}{-3} \approx -1.67$ $\sec \phi = \dfrac{r}{x} = \dfrac{\sqrt{74}}{-7} \approx -1.23$

$\csc \theta = \dfrac{r}{y} = \dfrac{5}{4} = 1.25$ $\csc \phi = \dfrac{r}{y} = \dfrac{\sqrt{74}}{-5} \approx -1.72$

Note that many of these ratios are negative.

FIGURE 12

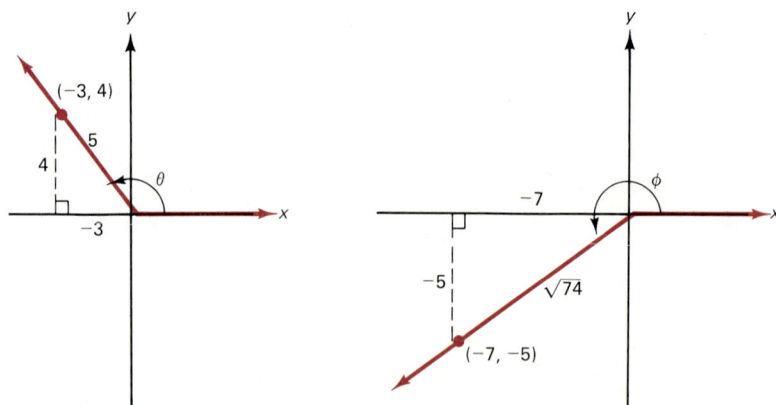

TABLE 1

Quadrant I	Quadrant II	Quadrant III	Quadrant IV
$x = +$ $y = +$ $r = +$	$x = -$ $y = +$ $r = +$	$x = -$ $y = -$ $r = +$	$x = +$ $y = -$ $r = +$
$\sin\theta = \dfrac{+}{+} = +$ $\cos\theta = \dfrac{+}{+} = +$ $\tan\theta = \dfrac{+}{+} = +$	$\sin\theta = \dfrac{+}{+} = +$ $\cos\theta = \dfrac{-}{+} = -$ $\tan\theta = \dfrac{+}{-} = -$	$\sin\theta = \dfrac{-}{+} = -$ $\cos\theta = \dfrac{-}{+} = -$ $\tan\theta = \dfrac{-}{-} = +$	$\sin\theta = \dfrac{-}{+} = -$ $\cos\theta = \dfrac{+}{+} = +$ $\tan\theta = \dfrac{-}{+} = -$
All are positive.	Only sine is positive.	Only tangent is positive.	Only cosine is positive.

Let us now look at the signs of the trigonometric functions sine, cosine, and tangent as they occur in each of the four different quadrants. The signs of the functions depend on the signs of x and y, which change from quadrant to quadrant. Table 1 summarizes the signs of x, y, and the functions in each of the four quadrants. The following memory device (all-sin-tan-cos) tells us the functions that are *positive* in each of the four quadrants. (In the next section we will see how to extend this rule to cotangent, secant, and cosecant.)

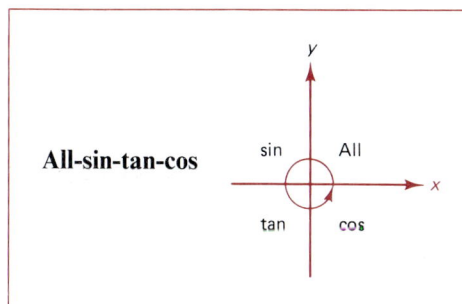

All-sin-tan-cos

On page 287 we will see how to evaluate functions such as $\sin 169°$ using trigonometric tables or a hand calculator. But now, we look at the trigonometric functions for special angles.

EXAMPLE 9 Let us look at the angles $0°$, $90°$, $180°$, and $270°$, as shown in Figure 13(a). We have chosen to use the points $(1, 0)$, $(0, 1)$, and so on, (since they all have $r = 1$) but we could have used other points, such as

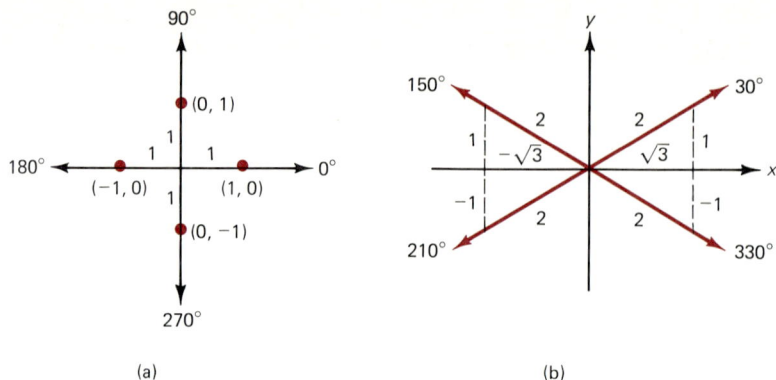

(a)

(b)

FIGURE 13

$(2, 0)$, $(0, 2)$, and so on. We evaluate a few of the functions and leave the rest as exercises.

$$\sin 0° = \frac{0}{1} = 0 \qquad\qquad \cos 0° = \frac{1}{1} = 1$$

$$\tan 90° \text{ is undefined} \qquad \sin 270° = \frac{-1}{1} = -1$$

$$\cos 180° = \frac{-1}{1} = -1 \qquad \tan 180° = \frac{0}{-1} = 0$$

Note that $\tan 90°$ is *not* defined since it involves a $1/0$ ratio. It is important to note that some of the trigonometric functions are not defined for certain angles (generally, where a denominator is zero).

EXAMPLE 10   Let us look at the angles $30°$, $150°$, $210°$, and $330°$, as shown in Figure 13(b). Note that all these angles are based on the 1–$\sqrt{3}$–2 right triangle [see also Figure 7(c)]. We have to make sure that we have the correct signs (positive or negative).

$$\sin 30° = \frac{1}{2} \qquad\qquad \tan 30° = \frac{1}{\sqrt{3}} = \frac{\sqrt{3}}{3}$$

$$\cos 150° = \frac{-\sqrt{3}}{2} \qquad\qquad \sin 210° = \frac{-1}{2}$$

$$\tan 210° = \frac{-1}{-\sqrt{3}} = \frac{\sqrt{3}}{3} \qquad \cos 330° = \frac{\sqrt{3}}{2}$$

The functions for angles based on a similar triangle (60°, 120°, 240°, and 300°) can also be similarly computed (see the problem set). The functions of 45°, 135°, 225°, 315°, and so on, are based upon the 1-1-$\sqrt{2}$ triangle that we saw in Example 6 and Figure 7(b). You will be asked to compute these in the exercises.

Let us make another observation. Since equivalent angles, such as 30° and 390°, have the same terminal side, they have the same values for the trigonometric functions (since these functions are based on points lying on the terminal side). For example, $\sin 30° = \sin 390° = \sin 750° = 1/2$. In general, for any of the six trigonometric functions f, we have

$$f(\theta) = f(\theta + 360°)$$

for any angle θ. We call such a function **periodic**.

Before we leave this section, let us show how we can use given trigonometric functions to find the other functions. We use the given information to draw a **reference right triangle**.

EXAMPLE 11 Suppose that $\sin \theta = -5/\sqrt{41}$ and $\cos \theta = 4/\sqrt{41}$. Sketch θ and the reference right triangle; then find the other functions.

Solution   Where is θ? The sine is negative (which happens in quadrants III and IV), and the cosine is positive (which happens in quadrants I and IV). Thus, θ is in quadrant IV (since this is where the sine is negative and the cosine is positive).

In quadrant IV, we can now construct a reference right triangle with sides -5, 4, $\sqrt{41}$ that satisfies $\sin \theta = -5/\sqrt{41}$ and $\cos \theta = 4/\sqrt{41}$ (see Figure 14). Using the sketch, we can find the other values:

$$\tan \theta = \frac{-5}{4} \qquad \cot \theta = \frac{-4}{5} \qquad \sec \theta = \frac{\sqrt{41}}{4} \qquad \csc \theta = \frac{-\sqrt{41}}{5}$$

FIGURE 14

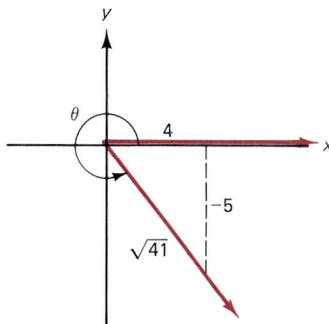

8.3

For each of the following angles, find an equivalent (or coterminal) angle in the ranges given.

	Angle between 0° and 360°	Angle between 360° and 720°	Angle between −360° and 0°
1.	51°	?	?
2.	218°	?	?
3.	?	502°	?
4.	?	719°	?
5.	?	?	−17°
6.	?	?	−189°

Warm-up Exercises (1, 2)

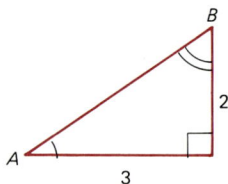

Warm-up Exercises (7, 8)

For the right triangle shown at the left, find the hypotenuse and complete the following table.

Angle	sin	cos	tan	cot	sec	csc
7. A	?	?	?	?	?	?
8. B	?	?	?	?	?	?

Each of the following points lies on the terminal side of an angle θ. For each θ, find (a) sin θ, (b) cos θ, and (c) tan θ.

9.

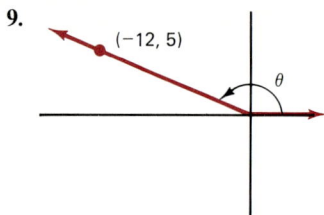

10.

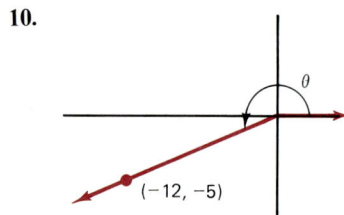

11. (1, 5) **12.** (−2, 3) **13.** (4, 0)

14. (0, −4) **15.** (−5, −2) **16.** (3, −1)

Use geometry facts to evaluate the following.

17. tan 45°   **18.** sin 60°

19. cos 90°   **20.** sin 90°

21. cos 135°   **22.** sin 150°

23. tan 210°   **24.** sin 225°

25. tan 270°   **26.** cos 300°

27. sin 390°   **28.** cos 405°

29. sin 480°   **30.** cos 540°

31. sin 810°   **32.** cos 720°

33. sin −90°   **34.** cos −45°

35. cot 0°   **36.** sec 45°

37. sec 90°   **38.** cot −30°

Sketch θ and the reference right triangle using the following information.

39. $\sin \theta = \dfrac{2}{3}$; quadrant II

40. $\cos \theta = \dfrac{-1}{5}$; quadrant II

41. $\tan \theta = \dfrac{3}{4}$; quadrant III

42. $\cot \theta = \dfrac{-2}{5}$; quadrant IV

43. $\sin \theta = \dfrac{2}{\sqrt{13}}$ and $\cos \theta = \dfrac{3}{\sqrt{13}}$

44. $\sin \theta = \dfrac{1}{2}$ and $\cos \theta = \dfrac{-\sqrt{3}}{2}$

45. $\sin \theta = \dfrac{-1}{4}$ and $\cos \theta = \dfrac{-\sqrt{15}}{4}$

46. $\sin \theta = \dfrac{-1}{3}$ and $\cos \theta = \dfrac{2\sqrt{2}}{3}$

In the following problems, the quadrant and one trigonometric function are given. With this information, use a sketch of θ to evaluate the other functions.

	Quadrant of θ	sin θ	cos θ	tan θ
47.	I	$\dfrac{1}{2}$	?	?
48.	I	?	$\dfrac{2}{5}$	?
49.	II	?	?	$\dfrac{-3}{4}$
50.	II	$\dfrac{4}{7}$	?	?
51.	III	?	$\dfrac{-\sqrt{2}}{2}$	?
52.	III	?	?	$\dfrac{5}{3}$
53.	IV	$\dfrac{-1}{\sqrt{5}}$	?	?
54.	IV	?	$\dfrac{\sqrt{10}}{10}$	?

Complete the following tables for the special angles.

55.

θ	sin θ	cos θ	tan θ	cot θ	sec θ	csc θ
0°						
90°						
180°						
270°						

56.

θ	sin θ	cos θ	tan θ	cot θ	sec θ	csc θ
45°						
135°						
225°						
315°						

57.

θ	$\sin \theta$	$\cos \theta$	$\tan \theta$	$\cot \theta$	$\sec \theta$	$\csc \theta$
30°						
150°						
210°						
330°						

58.

θ	$\sin \theta$	$\cos \theta$	$\tan \theta$	$\cot \theta$	$\sec \theta$	$\csc \theta$
60°						
120°						
240°						
300°						

Physical Application

59. The movement (or displacement) of a weight on a spring going up and down from its rest position can be viewed as the y-coordinate of an object traveling around a circle. Suppose that the radius of the circle is 3.

(a) Show that the displacement is given by $y = 3 \sin \theta$.

(b) What is the maximum displacement?

(c) Complete the following table for the displacement y as θ travels through one revolution.

θ	0°	45°	90°	135°	180°	225°	270°	315°	360°
y									

(d) Graph the table in part (c).

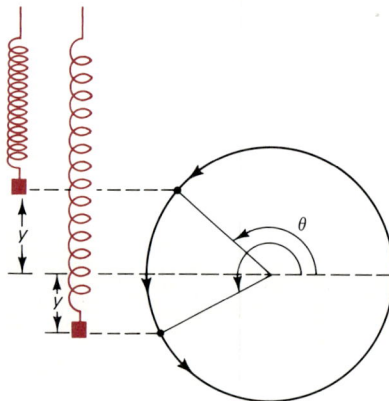

Electrical Application

60. The voltage from an alternating current (ac) might be given by $V = 110 \sin \theta$, where θ can be viewed as an angle that is continually rotating. Find V in the following table.

θ	0°	45°	90°	135°	180°	225°	270°	315°	360°
V									

Then graph the points (θ, V) above.

On page 267 we defined the six trigonometric functions for acute angles in right triangles. On page 271 we generalized our definitions of the functions to all angles. In working problems in those sections, perhaps you noticed that there were relations between the different functions. In this section we look at some of the important relations between the functions. These relations are important since we can often replace or substitute one function by another that we may know more about, or wish to know more about.

Let us begin by looking at the relation between complementary acute angles. (Recall that angles are complementary if their sum is $90°$; for example, $42°$ and $48°$ are complementary.)

EXAMPLE 12 Angles A and B shown in the right triangles of Figure 15 are complementary. Let us evaluate the six trigonometric functions for the right triangle in Figure 15(a). We then examine the table for relations.

Angle	sin	cos	tan	cot	sec	csc
A	$\dfrac{2}{\sqrt{29}}$	$\dfrac{5}{\sqrt{29}}$	$\dfrac{2}{5}$	$\dfrac{5}{2}$	$\dfrac{\sqrt{29}}{5}$	$\dfrac{\sqrt{29}}{2}$
B	$\dfrac{5}{\sqrt{29}}$	$\dfrac{2}{\sqrt{29}}$	$\dfrac{5}{2}$	$\dfrac{2}{5}$	$\dfrac{\sqrt{29}}{2}$	$\dfrac{\sqrt{29}}{5}$

The arrows point out that $\sin A = \cos B$, $\sin B = \cos A$, and so on.

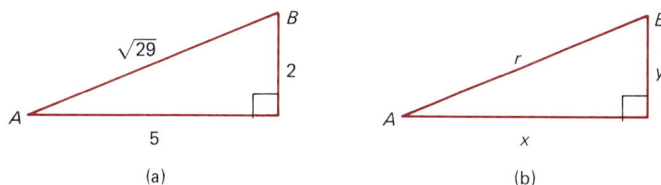

(a)   (b)

FIGURE 15

It appears from Example 12 that $\sin A = \cos(90° - A)$, and so on. But is this always true? Let us look at the general right triangle as pictured in Figure 15(b). We see that

$$\sin A = \frac{y}{r} \quad \text{and} \quad \cos B = \frac{y}{r} \quad \text{(Thus, } \sin A = \cos B\text{.)}$$

$$\cos A = \frac{x}{r} \quad \text{and} \quad \sin B = \frac{x}{r} \quad \text{(Thus, } \cos A = \sin B\text{.)}$$

$$\tan A = \frac{y}{x} \quad \text{and} \quad \cot B = \frac{y}{x} \quad \text{(Thus, } \tan A = \cot B\text{.)}$$

And so on. We can continue these and derive the following **complementary relations**. For any acute angle θ, we have

$$\sin \theta = \cos (90° - \theta)$$
$$\cos \theta = \sin (90° - \theta)$$
$$\tan \theta = \cot (90° - \theta)$$
$$\cot \theta = \tan (90° - \theta)$$
$$\sec \theta = \csc (90° - \theta)$$
$$\csc \theta = \sec (90° - \theta)$$

Why are these relations important? Consider the following example.

EXAMPLE 13 Suppose that we know the values of the trigonometric functions for 17°, as shown below.

Angle	sin	cos	tan	cot	sec	csc
17°	0.2924	0.9563	0.3057	3.2709	1.0457	3.4203

With this information and the complementary relations we can evaluate the functions for 73° (the complement of 17°).

Angle	sin	cos	tan	cot	sec	csc
17°	0.2924	0.9563	0.3057	3.2709	1.0457	3.4203
73°	0.9563	0.2924	3.2709	0.3057	3.4203	1.0457

Now let us look again at the first row of the table on page 279.

Angle	sin	cos	tan	cot	sec	csc
A	$\dfrac{2}{\sqrt{29}}$	$\dfrac{5}{\sqrt{29}}$	$\dfrac{2}{5}$	$\dfrac{5}{2}$	$\dfrac{\sqrt{29}}{5}$	$\dfrac{\sqrt{29}}{2}$

Note that the arrows point out values that are *reciprocals* of each other. This is true in general; for instance,

$$\sin A = \frac{y}{r} \quad \text{and} \quad \csc A = \frac{r}{y} \quad (y \neq 0)$$

imply that $\csc A = \dfrac{1}{\sin A}$. We now state the **reciprocal relations**. For any angle θ, we have the following (except where denominators are zero).

$$\csc \theta = \frac{1}{\sin \theta} \qquad \sec \theta = \frac{1}{\cos \theta} \qquad \cot \theta = \frac{1}{\tan \theta}$$

EXAMPLE 14 Use the reciprocal relations to complete the following table.

Angle	sin	cos	tan	cot	sec	csc
θ	$\dfrac{4}{\sqrt{41}}$	$\dfrac{-5}{\sqrt{41}}$	$\dfrac{-4}{5}$			

Solution Once we know $\sin \theta$, $\cos \theta$, and $\tan \theta$, we can find $\cot \theta$, $\sec \theta$, and $\csc \theta$ by taking the reciprocals.

Angle	sin	cos	tan	cot	sec	csc
θ	$\dfrac{4}{\sqrt{41}}$	$\dfrac{-5}{\sqrt{41}}$	$\dfrac{-4}{5}$	$\dfrac{-5}{4}$	$\dfrac{-\sqrt{41}}{5}$	$\dfrac{\sqrt{41}}{4}$

Another relation that is very important (but not as obvious as the ones before) involves the ratio of $\sin \theta$ to $\cos \theta$. Since they have the same denominator, their ratio is easy to simplify.

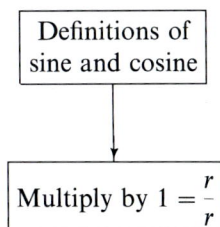

| Definitions of sine and cosine |

Multiply by $1 = \dfrac{r}{r}$

$$\frac{\sin \theta}{\cos \theta} = \frac{\dfrac{y}{r}}{\dfrac{x}{r}}$$

$$= \frac{\dfrac{y}{r} \cdot r}{\dfrac{x}{r} \cdot r} = \frac{y}{x} = \tan \theta$$

Similarly, if we divide $\cos \theta$ by $\sin \theta$, we get $\cot \theta$. Summarizing, we have the **quotient relations**.

$$\tan \theta = \frac{\sin \theta}{\cos \theta} \qquad \cot \theta = \frac{\cos \theta}{\sin \theta}$$

EXAMPLE 15   Suppose that we are given the sine and cosine of an angle. We can then find the other four functions.

Angle	sin	cos	tan	cot	sec	csc
θ	$\dfrac{-2}{\sqrt{29}}$	$\dfrac{5}{\sqrt{29}}$				

We can complete the table above by using our relations. First, we find the tangent using the quotient relation.

$$\tan \theta = \frac{\sin \theta}{\cos \theta} = \frac{\dfrac{-2}{\sqrt{29}}}{\dfrac{5}{\sqrt{29}}} = \frac{-2}{5}$$

Thus, $\tan \theta = -2/5$. We now use the reciprocal relations to complete the table.

Angle	sin	cos	tan	cot	sec	csc
θ	$\dfrac{-2}{\sqrt{29}}$	$\dfrac{5}{\sqrt{29}}$	$\dfrac{-2}{5}$	$\dfrac{-5}{2}$	$\dfrac{\sqrt{29}}{5}$	$-\dfrac{\sqrt{29}}{2}$

Example 15 shows us how we can obtain all six functions by just knowing the sine and cosine. We now have another important relation that connects the sine and cosine to each other. If (x, y) is any point on the terminal side of θ, then we have

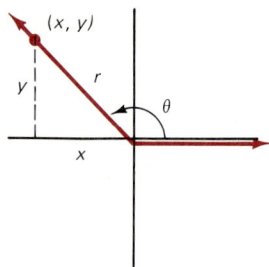

Distance formula	$x^2 + y^2 = r^2$
Divide by r^2	$\dfrac{x^2}{r^2} + \dfrac{y^2}{r^2} = 1$
$\cos \theta = \dfrac{x}{r}$	
$\sin \theta = \dfrac{y}{r}$	$(\cos \theta)^2 + (\sin \theta)^2 = 1$

Before we box in this equation, let us first note how we write powers of trigonometric functions. We write $(\sin \theta)^2 = \sin^2 \theta$, $(\sin \theta)^3 = \sin^3 \theta$, and so on. Since this equation is derived from the Pythagorean theorem, we call it a **Pythagorean relation**. For any angle θ,

$$\sin^2 \theta + \cos^2 \theta = 1$$

EXAMPLE 16 Suppose that $\sin \theta = 0.6$ and θ is in quadrant II. Find the other five functions.

Solution Why do we have to know that θ is in quadrant II? Watch what happens when we use the Pythagorean relation to find $\cos \theta$.

$$\boxed{\sin^2 \theta + \cos^2 \theta = 1} \qquad (0.6)^2 + \cos^2 \theta = 1$$

$$\downarrow \qquad \qquad \cos^2 \theta = 1 - 0.36 = 0.64$$

$$\boxed{\text{Find } \cos \theta} \qquad \cos \theta = \pm 0.8$$

Is it 0.8 or -0.8? We need to know the quadrant. Since θ is in quadrant II, the cosine is negative; thus, $\cos \theta = -0.8$. Now that we have $\sin \theta$ and $\cos \theta$, we find $\tan \theta$.

$$\boxed{\tan \theta = \frac{\sin \theta}{\cos \theta}} \qquad \tan \theta = \frac{0.6}{-0.8} = -0.75$$

Using the reciprocal relations, we find the remaining values.

$$\cot \theta = \frac{1}{-0.75} = -1.33 \qquad \sec \theta = \frac{1}{-0.8} = -1.25$$

$$\csc \theta = \frac{1}{0.6} = 1.67$$

Angle	sin	cos	tan	cot	sec	csc
θ	0.6	-0.8	-0.75	-1.33	-1.25	1.67

PROBLEM SET
8.4   *For each θ given in Problems 1 to 8, complete the following table.*

Angle	sin	cos	tan	cot	sec	csc
θ						

1.

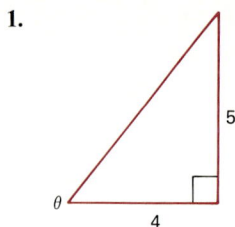

2.

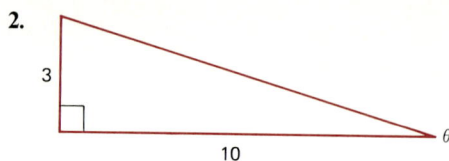

3.

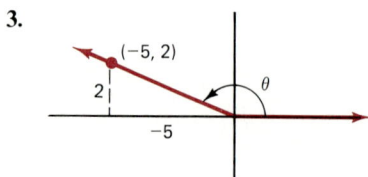

4.

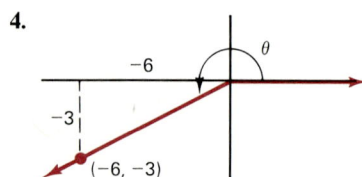

5. (3, 7) is on the terminal side of θ.

6. (-2, 6) is on the terminal side of θ.

7. (-5, -7) is on the terminal side of θ.

8. (4, -2) is on the terminal side of θ.

Evaluate the following using only the definitions and geometry facts.

9. $\sin 0°$ **10.** $\cos 30°$ **11.** $\tan 45°$

12. $\sin 120°$ **13.** $\cos 180°$ **14.** $\tan 300°$

Complete the following tables.

15.

Angle	sin	cos	tan	cot	sec	csc
θ	$\dfrac{4}{\sqrt{41}}$	$\dfrac{5}{\sqrt{41}}$	$\dfrac{4}{5}$	$\dfrac{5}{4}$	$\dfrac{\sqrt{41}}{5}$	$\dfrac{\sqrt{41}}{4}$
$90° - \theta$						

16.

Angle	sin	cos	tan	cot	sec	csc
θ	$\dfrac{5}{\sqrt{26}}$	$\dfrac{1}{\sqrt{26}}$	5	$\dfrac{1}{5}$	$\sqrt{26}$	$\dfrac{\sqrt{26}}{5}$
$90° - \theta$						

17.

Angle	sin	cos	tan	cot	sec	csc
42°	0.6691	0.7431			1.3456	1.4945
48°			1.1106	0.9004		

18.

Angle	sin	cos	tan	cot	sec	csc
7°			0.1228	8.1443		
83°	0.9925	0.1219			8.2055	1.0075

19.

Angle	sin	cos	tan	cot	sec	csc
40°	0.6428		0.8391		1.3054	
50°	0.7660		1.1918		1.5557	

20.

Angle	sin	cos	tan	cot	sec	csc
11°	0.1908	0.9816		5.1446		
79°				0.1944	5.2408	1.0187

For each of the following angles, find the cotangent, secant, and cosecant.

	Angle	sin	cos	tan	cot	sec	csc
21.	A	$\dfrac{5}{\sqrt{61}}$	$\dfrac{6}{\sqrt{61}}$	$\dfrac{5}{6}$			
22.	B	$\dfrac{-4}{\sqrt{17}}$	$\dfrac{1}{\sqrt{17}}$	-4			
23.	22°	0.3746	0.9272	0.4040			
24.	60°	0.8660	0.5000	1.7321			
25.	144°	0.5878	-0.8090	-0.7265			
26.	302°	-0.8480	0.5299	-1.6003			

For the following angles, find the tangent and cotangent.

	Angle	sin	cos	tan	cot
27.	A	$\dfrac{1}{\sqrt{26}}$	$\dfrac{5}{\sqrt{26}}$		
28.	B	$\dfrac{9}{\sqrt{85}}$	$\dfrac{2}{\sqrt{85}}$		
29.	C	$\dfrac{-7}{\sqrt{58}}$	$\dfrac{3}{\sqrt{58}}$		
30.	D	$\dfrac{-1}{\sqrt{37}}$	$\dfrac{6}{\sqrt{37}}$		
31.	29°	0.4848	0.8746		
32.	319°	-0.6561	0.7547		

For the following angles, find the missing values.

	Angle	Quadrant	sin	cos
33.	A	I	$\dfrac{1}{2}$	
34.	B	I		$\dfrac{1}{3}$
35.	C	II	$-\dfrac{1}{2}$	
36.	D	II		$\dfrac{-2}{5}$
37.	E	III		$\dfrac{-2}{\sqrt{29}}$
38.	F	IV	$\dfrac{-3}{\sqrt{79}}$	

For each of the following angles, fill in the missing values. (Hint: Sketch first.)

	Angle	Quadrant	sin	cos	tan	cot	sec	csc
39.	A	I	$\dfrac{3}{5}$	?	?	?	?	?
40.	B	I	?	$\dfrac{12}{13}$	?	?	?	?
41.	C	II	?	$\dfrac{-1}{2}$	?	?	?	?
42.	D	II	0.8	?	?	?	?	?
43.	E	III	$\dfrac{-1}{4}$	?	?	?	?	?
44.	F	III	?	$\dfrac{-2}{5}$	?	?	?	?
45.	G	IV	?	$\dfrac{2}{\sqrt{13}}$	?	?	?	?
46.	H	IV	$\dfrac{-1}{\sqrt{10}}$	?	?	?	?	?
47.	I	I	?	?	$\dfrac{4}{3}$	?	?	?
48.	J	II	?	?	?	$\dfrac{-12}{5}$	?	?
49.	K	?	$\dfrac{-1}{\sqrt{5}}$	?	$\dfrac{1}{2}$	?	?	?
50.	L	?	?	$\dfrac{-3}{\sqrt{34}}$	?	$\dfrac{-3}{5}$	?	?

The following are also Pythagorean relations (similar to $\sin^2\theta + \cos^2\theta = 1$). Prove them.

51. $\tan^2\theta + 1 = \sec^2\theta$

52. $\cot^2\theta + 1 = \csc^2\theta$

Physical Applications

53. An object of weight W is just about to slide down an incline at angle θ. At this instant, the downward force (gravity) $W\sin\theta$ is exactly equal to the upward force (friction) $\mu W\cos\theta$. Show that the coefficient of friction of the incline μ, satisfies $\mu = \tan\theta$.

54. A spring is traveling up and down with maximum displacement A (see Problem 59, page 278). The kinetic energy K and potential energy U are given by

$$K = \frac{1}{2}kA^2\cos^2\theta \qquad \text{and} \qquad U = \frac{1}{2}kA^2\sin^2\theta$$

Show that the total energy $E = K + U$ is given by $E = \frac{1}{2}kA^2$.

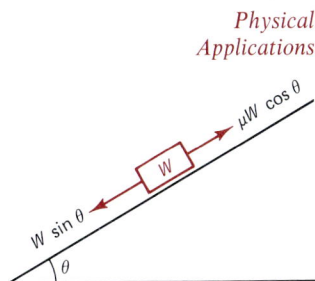

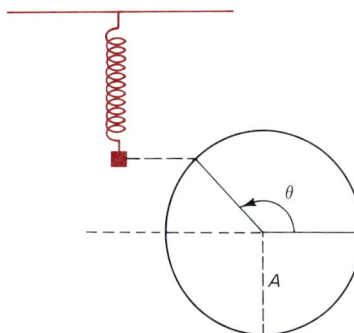

8.5 EVALUATING TRIGONOMETRIC FUNCTIONS

Although we have looked at a few special angles ($0°$, $30°$, $45°$, and so on), we have not discussed how we find values such as $\sin 138°$ or $\cos 138°$. One method, of course, would be to construct very carefully a reference right triangle for a $138°$ angle and then compute the appropriate ratios. Needless to say, this method would be very time consuming and fairly inaccurate. Fortunately, we do have two simple methods for evaluating these functions:

1. Trigonometric tables
2. Hand calculators

Trigonometric Tables

Long before the advent of the hand calculator in the 1970s, elaborate **trigonometric tables** were computed based on very sophisticated and complicated formulas (see Problem 87, page 294). Such a table appears in the Appendix (Table C). Table 2 shows a cutaway portion of the table. Let us make a few immediate observations about this table:

1. Most entries are rounded to four-decimal-place accuracy. (Therefore, almost all values are *approximate*.)

TABLE 2

Deg	Sin	Tan	Cot	Cos	
18.0	0.3090	0.3249	3.078	0.9511	72.0
.1	.3107	.3269	3.060	.9505	71.9
.2	.3123	.3288	3.042	.9500	.8
.3	.3140	.3307	3.024	.9494	.7
.4	.3156	.3327	3.006	.9489	.6
.5	.3173	.3346	2.989	.9483	.5
.6	.3190	.3365	2.971	.9478	.4
.7	.3206	.3385	2.954	.9472	.3
.8	.3223	.3404	2.937	.9466	.2
.9	.3239	.3424	2.921	.9461	71.1
19.0	0.3256	0.3443	2.904	0.9455	71.0
.1	.3272	.3463	2.888	.9449	70.9
.2	.3289	.3482	2.872	.9444	.8
	Cos	Cot	Tan	Sin	Deg

2. The angles are given in 0.1° intervals.

3. The degree column on the left goes from 0.0° *down* to 45.0°.

4. The degree column on the right goes from 45.0° *up* to 90.0°.

5. There is no secant or cosecant column. (We must use the reciprocal relations for these functions.)

EXAMPLE 17 Using the portion shown in Table 2, we can see that:
(a) sin 18.0° = 0.3090   (b) cos 18.3° = 0.9494

(c) tan 18.9° = 0.3424   (d) $\sec 18.7° = \dfrac{1}{\cos 18.7°} = 1.0557$

Note that we find sec 18.7° by taking the reciprocal of cos 18.7° (= 0.9472). (We again note that these values are *approximate*; for example, sin 18.0° ≈ 0.3090, but we write it with an equal sign.)

What about angles between 45 and 90°? Recall the *complementary relations*: For an acute angle A,

$$\sin A = \cos (90° - A) \qquad \tan A = \cot (90° - A) \qquad \text{and so on}$$

For example, sin 75° = cos 15°. These relations are built right into the tables. Note that the angles in the far-right column are the complements of the angles in the far-left column. (For example, 18.0° is opposite its complement 72°.) Note also that the functions at the bottom of the table are the cofunctions of the ones at the top. (For example, "Cos" is at the bottom of the "Sin" column.)

EXAMPLE 18 For angles between 45° and 90°, we use the angles at the *right* of the table and the functions at the *bottom* and read up. For instance,

(a) $\sin 71.0° = 0.9455$

(b) $\cos 71.6° = 0.3156$

(c) $\cot 71.9° = 0.3269$

(d) $\csc 72.0° = \dfrac{1}{\sin 72.0°} = 1.0515$

For angles whose accuracy is more than one decimal place, we can use *linear interpolation*, as explained on page 241.

As we can see, the table gives us the values of the trigonometric functions for acute angles. Now, what about all the other angles (greater than 90° or less than 0°)? Consider angles (such as those in Figure 16) in standard position. Note that the base acute angle of each right triangle formed has the same trigonometric ratios (except for sign) as the given angle. We call the *acute* angle formed by the terminal side of θ and the x-axis the **reference angle**.

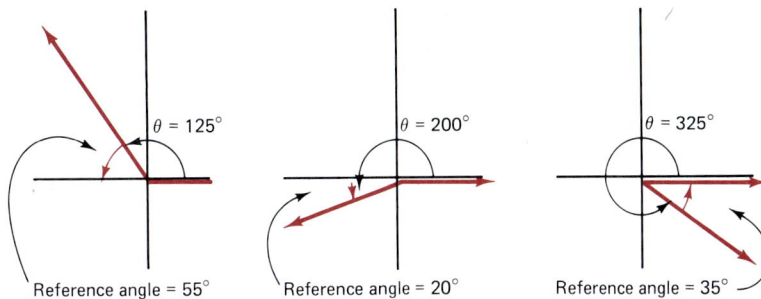

FIGURE 16

To compute the reference angle, we take the positive acute angle between the terminal side of θ and the *nearest* part of the x-axis.

We can use the following table for finding the reference angle for θ between 0° and 360°.

QUADRANT FOR θ	REFERENCE ANGLE
I	θ
II	$180° - \theta$
III	$\theta - 180°$
IV	$360° - \theta$

If the angle is greater than 360° or less than 0°, we first find an equivalent angle between 0° and 360°. Recall also that the sign of each trigonometric function depends on the quadrant of the angle. The memory device for the signs is: all-sin-tan-cos. This means that all functions are positive in quadrant I, only the sine is positive in quadrant II, and so on.

We can put all this information together to evaluate trigonometric functions for any angle.

> To evaluate a trigonometric function for any angle θ:
>
> 1. Make a sketch of θ.
> 2. If necessary, find an equivalent angle between $0°$ and $360°$.
> 3. Determine the quadrant that θ is in.
> 4. Find the reference angle.
> 5. Evaluate the trigonometric function for the reference angle.
> 6. Attach the proper sign (all-sin-tan-cos).

EXAMPLE 19   Find sin 125°.

Solution   Using Figure 16, we see that the reference angle is 55°. In quadrant II, the sine is positive; thus,

$$\sin 125° = \sin 55° = 0.8192$$

EXAMPLE 20   Find cos 200°.

Solution   Using Figure 16, we see that the reference angle is 20°. In quadrant III, the cosine is negative; thus,

$$\cos 200° = -\cos 20° = -0.9397$$

EXAMPLE 21   Find tan 685°.

Solution   First, we find an equivalent angle: $685° - 360° = 325°$. As we see in Figure 16, the reference angle for 325° is 35°. In quadrant IV, the tangent is negative; thus,

$$\tan 685° = \tan 325° = -\tan 35° = -0.7002$$

YES	NO
$\cos 200° = -0.9397$	~~$\cos 200° = 0.9397$~~
Do not forget to attach the proper sign.	

EXAMPLE 22   What acute angle has $\sin \theta = 0.32$?

Solution   This is the reverse problem: Here we have the value and we want the angle. In the tables, we go down the "Sin" column until we find the closest number to 0.32. We find the number 0.3206, which corresponds to the angle 18.7°. (Actually, there may be many other angles, such as 378.7°, that also have a sine of 0.3206; but for now, we want only the acute angle.) Thus, $\theta = 18.7°$. For a slightly more accurate answer, we can use linear interpolation, as explained on page 241.

Hand Calculators

More and more, people are turning to their **hand calculators** to evaluate trigonometric functions. Many calculators (even in the $10 to $20 range) have the keys

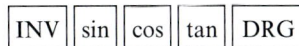

$$\boxed{\text{INV}}\ \boxed{\text{sin}}\ \boxed{\text{cos}}\ \boxed{\text{tan}}\ \boxed{\text{DRG}}$$

which are used for trigonometric functions. When you finish reading this subsection on calculators, you will probably never want to see a trigonometric table again. This is understandable, but somewhat unfortunate, since there are certain insights to be gained by using the tables.

EXAMPLE 23   We now rework some of the earlier examples of this section using a hand calculator. (Be sure that your calculator is in the degree mode—most calculators are when first turned on. The display will probably show a tiny DEG below the digits.)
(a) cos 18.3°   (b) sin 125°   (c) cos 200°
(d) tan 685°   (e) tan 90°

PRESS	DISPLAY	MEANING
(a) $\boxed{1}\boxed{8}\boxed{.}\boxed{3}$	18.3 DEG	Enter angle
$\boxed{\cos}$	0.9494255 DEG	cos 18.3°
(b) $\boxed{1}\boxed{2}\boxed{5}$	125. DEG	Enter angle
$\boxed{\sin}$	0.8191521 DEG	sin 125°
(c) $\boxed{2}\boxed{0}\boxed{0}$	200. DEG	Enter angle
$\boxed{\cos}$	−0.9396926 DEG	cos 200°
(d) $\boxed{6}\boxed{8}\boxed{5}$	685. DEG	Enter angle
$\boxed{\tan}$	−0.7002075 DEG	tan 685°
(e) $\boxed{9}\boxed{0}$	90. DEG	Enter angle
$\boxed{\tan}$	E 0. DEG	Error: tan 90° is undefined

Note what happened when we tried to sneak tan 90° past the calculator: We got an error, since tan 90° is not defined.

EXAMPLE 24   Most calculators do not have contangent, secant, or cosecant keys. Thus, we must use the other keys and the reciprocal key $\boxed{1/x}$. (Recall the reciprocal relations, page 281.) Consider the values:

(a) $\cot 79°$ (b) $\sec(-217.8°)$

PRESS	DISPLAY	MEANING
(a) $\boxed{7}\boxed{9}$	79. _{DEG}	Enter angle
$\boxed{\text{tan}}$	5.1445541 _{DEG}	$\tan 79°$
$\boxed{1/x}$	0.1943803 _{DEG}	$\cot 79° = \dfrac{1}{\tan 79°}$

- -

PRESS	DISPLAY	MEANING
(b) $\boxed{2}\boxed{1}\boxed{7}\boxed{\cdot}\boxed{8}\boxed{+/-}$	-217.8 _{DEG}	Enter angle
$\boxed{\text{cos}}$	-0.790155 _{DEG}	$\cos(-217.8°)$
$\boxed{1/x}$	-1.2655744 _{DEG}	$\sec(-217.8°) = \dfrac{1}{\cos(-217.8°)}$

EXAMPLE 25   Use a calculator to find acute angles A and B that satisfy:
(a) $\sin A = 0.2100$ (b) $\cot B = 1.500$

Solution   Like Example 22, this is an inverse problem: Given the value, find the angle. Here, we use the $\boxed{\text{INV}}$ key. With the cotangent, we must first use the $\boxed{1/x}$ key to convert to tangent.

PRESS	DISPLAY	MEANING
(a) $\boxed{\cdot}\boxed{2}\boxed{1}\boxed{0}\boxed{0}$	0.2100 _{DEG}	$\sin A = 0.2100$
$\boxed{\text{INV}}\boxed{\text{sin}}$	12.122352 _{DEG}	$A \approx 12.1°$

- -

PRESS	DISPLAY	MEANING
(b) $\boxed{1}\boxed{\cdot}\boxed{5}\boxed{0}\boxed{0}$	1.500 _{DEG}	$\cot B = 1.500$
$\boxed{1/x}$	0.6666667 _{DEG}	$\tan B = 0.6666667$
$\boxed{\text{INV}}\boxed{\text{tan}}$	33.690068 _{DEG}	$B \approx 33.7°$

Write each of the following in terms of an angle between 0° and 45°. (Do not evaluate.)

1. $\cos 71° = \sin$ _____

2. $\tan 83° = \cot$ _____

3. $\sin 56.7° = \cos$ _____

4. $\cot 89.9° = \tan$ _____

Write each of the following in terms of a sine, cosine, or tangent. (Do not evaluate.)

5. $\cot 51° =$ _____

6. $\sec 184° =$ _____

7. $\csc(-23°) =$ _____

8. $\cot 450° =$ _____

For each of the following angles, find an equivalent angle between 0° and 360°.

9. $400°$

10. $500°$

11. $605°$

12. $750°$

13. $1070°$

14. $4000°$

For each of the quadrants, state the sign of the sine, cosine, and tangent.

15. Quadrant I

16. Quadrant II

17. Quadrant III

18. Quadrant IV

For each of the following angles, find its reference angle.

19. $56°$

20. $165°$

21. $218.2°$

22. $298.8°$

23. $410.2°$

24. $900.1°$

25. $-102°$

26. $-276°$

Use the trigonometric tables to evaluate the following.

27. $\sin 17°$

28. $\cos 24.4°$

29. $\sin 0.1°$

30. $\sec 27°$

31. $\tan 55.1°$

32. $\cos 82°$

33. $\tan 89.9°$

34. $\sec 89.5°$

35. $\sin 159°$

36. $\sin 342°$

37. $\cos 401°$

38. $\tan 123°$

39. $\cot 357°$

40. $\sec 111.1°$

41. $(\cot -10°)$

42. $\sin (-23.5°)$

Use a hand calculator to evaluate the following.

43. $\sin 31°$

44. $\sin 123.5°$

45. $\cos 459.9°$

46. $\tan 56.89°$

47. cot 9°   **48.** cot 218.9°

49. sec 9000°   **50.** csc 6.001°

51. sin (−345°)   **52.** cos $\dfrac{1}{2}°$

53. cot (−179.99°)   **54.** sec (−270°)

For each of the following equations, find the acute angle A.

55. sin $A = 0.3100$ **56.** cos $A = 0.690$

57. tan $A = 1.25$ **58.** cot $A = 2.17$

59. sec $A = 2.50$ **60.** csc $A = 1.85$

61. sin $A = 0.0982$ **62.** tan $A = 5.789$

63. cos $A = 0.993$ **64.** sec $A = 3.21589$

65. cot $A = 10$ **66.** csc $A = 1.0001$

Evaluate each of the following pairs, X and Y. (Use the tables or a calculator.) Can you see any patterns?

67. $X = 2(\cos 50°)(\sin 50°)$; $Y = \sin 100°$

68. $X = \sin 150°$; $Y = 2(\sin 75°)(\cos 75°)$

69. $X = \cos^2 40° - \sin^2 40$; $Y = \cos 80°$

70. $X = 2(\cos^2 10°)$; $Y = \cos 20°$

71. $X = \cos 200°$; $Y = 2(\cos^2 100°) - 1$

72. $X = 1 - 2(\sin^2 80°)$; $Y = \cos 160°$

For each of the following trigonometric functions, use the table to determine whether the function increases or decreases as the angle increases from 0° to 90°.

73. sine   **74.** tangent   **75.** cosine

76. cotangent   **77.** secant   **78.** cosecant

For each of the following equations, find the angle θ in the indicated quadrant (for $0° \le \theta < 360°$).

79. sin $\theta = 0.2$ (quadrant I)

80. cos $\theta = 0.7$ (quadrant I)

81. sin $\theta = 0.8$ (quadrant II)

82. cos $\theta = 0.65$ (quadrant IV)

83. tan $\theta = 2.5$ (quadrant III)

84. sec $\theta = 1.8$ (quadrant IV)

85. cos $\theta = -0.33$ (quadrant II)

86. sin $\theta = -0.85$ (quadrant III)

One approximation for the sine function is as follows. If θ is an angle measured in degrees,

$$\sin \theta \approx \frac{\pi\theta}{180} - \frac{1}{3!}\left(\frac{\pi\theta}{180}\right)^3 + \frac{1}{5!}\left(\frac{\pi\theta}{180}\right)^5 - \frac{1}{7!}\left(\frac{\pi\theta}{180}\right)^7$$

where $3! = 1 \cdot 2 \cdot 3 = 6$, $5! = 1 \cdot 2 \cdot 3 \cdot 4 \cdot 5 = 120$, and $7! = 1 \cdot 2 \cdot 3 \cdot 4 \cdot 5 \cdot 6 \cdot 7 = 5040$. (*The smaller the angle, the better this approximation is.*) *Use this formula to approximate the following and then check their accuracy.*

θ	$\sin \theta$ (formula)	$\sin \theta$ (tables or calculator)
87. 5°		
88. 30°		
89. 60°		
90. 90°		

Physical Application

91. Light can appear to bend as it passes from air into another substance. This is called *refraction*. *Snell's law* gives the relation between the incident angle θ and the refracted angle θ_2:

$$\sin \theta_1 = n \sin \theta_2$$

where n is the index of refraction. Complete the following table.

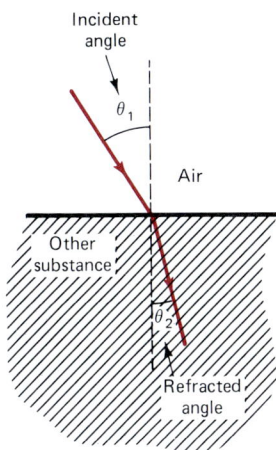

n (substance)	θ_1	θ_2
1.33 (water)	40°	?
1.33 (water)	?	43°
1.50 (glass)	38.2°	?
1.50 (glass)	?	32°
? (salt)	30°	19.1°
? (salt)	50°	30°

Optical Application

92. Often lenses (such as in sunglasses) are *polarized* in a certain direction. If a lens meets light oriented at an angle θ to its polarization, the fraction of light that gets through is $\cos^2 \theta$.

(a) Complete the following table.

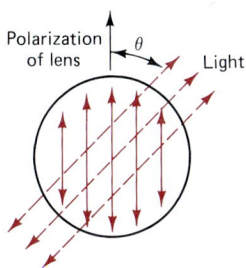

θ	0°	15°	30°	45°	60°	75°	90°
Fraction							

(b) Graph this table.

93. The period P of a pendulum is given by

$$P = 2\sqrt{\frac{L \cos \theta}{g}}$$

where L is the length (in meters), θ is the angle of the swing, and $g = 9.8$ (gravity constant). Find P in the following cases.

(a) $L = 2$; $\theta = 2°$ (b) $L = 2$; $\theta = 3°$

(c) $L = 2$; $\theta = 5°$ (d) $L = 2$; $\theta = 10°$

Mathematics Application

94. Recall that the slope m of a nonvertical line is given by $m = b/a$ (change-in-y/change-in-x). Further, the angle of inclination θ satisfies $\tan \theta = b/a$. Thus

$$\tan \theta = m$$

Use this relation to find the angle of inclination of the following lines.

(a) $y = 2x - 1$ (b) $y = \frac{1}{3}x + 7$

(c) $2x - 5y = 8$ (d) $3x + 5y = 1$

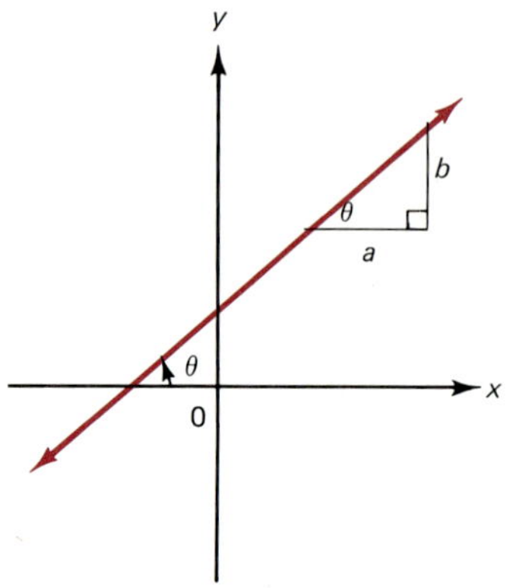

8.6 APPLICATIONS OF RIGHT-TRIANGLE TRIGONOMETRY

In this section we look at a few simple applications of the trigonometric functions for right triangles. Generally, we are given a side and an acute angle of the triangle, and we are to find one of the other sides. The solution lies in selecting one of the six trigonometric functions that relates the unknown side to the given side.

> To find an unknown side of a right triangle when another side and an angle are given:
>
> 1. Draw a reasonable picture; make a rough estimate based on the sketch.
> 2. Select a trigonometric function of the given angle that is the ratio of the unknown side to the given side.
> 3. Evaluate the function for the angle (by table or calculator).
> 4. Set this value equal to the ratio and solve for the unknown.

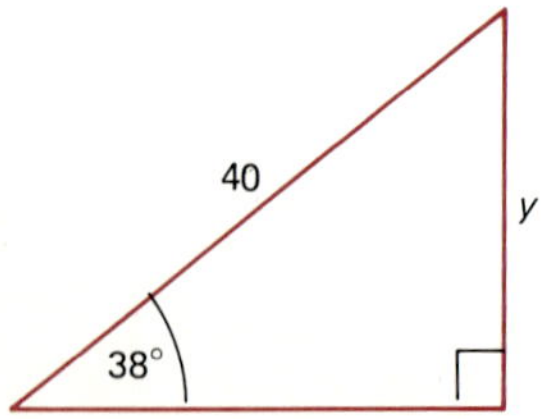

EXAMPLE 26 Find the side y in the right triangle shown at the left.

Solution The given angle is 38°. What function relates y (opposite 38°) to 40 (hypotenuse)? The sine. Thus, we find $\sin 38°$, set it equal to $y/40$, and solve. (*An estimate:* y is somewhat less than 40, say about 27.)

$$\boxed{\sin 38° = \frac{\text{opp}}{\text{hyp}}}$$

$$\sin 38° = \frac{y}{40}$$

$$\boxed{\begin{array}{c}\text{Evaluate} \\ \sin 38°\end{array}}$$

$$0.6157 = \frac{y}{40}$$

$$\boxed{\text{Multiply by 40}}$$

$$24.628 = y$$

Thus, $y \approx 25$.

One of the common applications of right-triangle trigonometry is in using the **angle of elevation** or **depression**, the angle formed by a horizontal line and a ray up or down from the horizontal line. Figure 17 shows an illustration of an angle of elevation and an angle of depression.

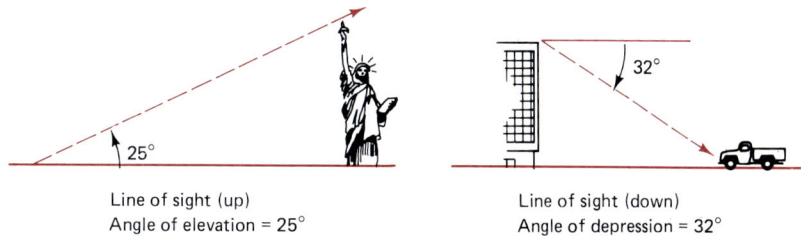

Line of sight (up)
Angle of elevation = 25°

Line of sight (down)
Angle of depression = 32°

FIGURE 17

EXAMPLE 27   A man is 250 meters from the foot of a building. The angle of elevation to the top of the building is 31°. How tall is the building?

Solution   We begin with a reasonable sketch, as in Figure 18. (*Our estimate: h* is somewhat less than 250, say 175.) We select the trigonometric ratio of the unknown height h to the known side 250. This is the tangent.

$$\boxed{\tan 31° = \frac{\text{opp}}{\text{adj}}}$$

$$\tan 31° = \frac{h}{250}$$

$$\boxed{\begin{array}{c}\text{Evaluate} \\ \tan 31°\end{array}}$$

$$0.6009 = \frac{h}{250}$$

$$\boxed{\text{Multiply by 250}}$$

$$150.2 = h$$

Thus, the building is about 150 meters tall. See page 298.

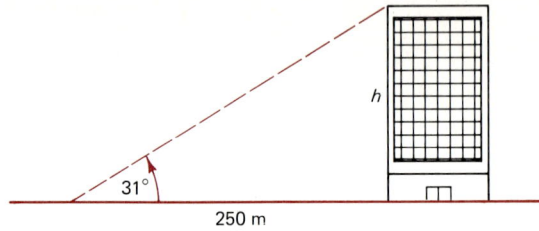

FIGURE 18

EXAMPLE 28 A man 15 feet above the water level is pulling a boat into the dock. At the moment, the angle of depression of the rope is 19°.
(a) How far is the boat from the base of the dock?
(b) How much rope is out?

Solution We begin with the sketch shown in Figure 19. Here we have two unknown distances: x (the distance to the foot of the dock) and r (the length of the rope). (*Our estimate:* Both x and r are quite a bit longer than 15, say about 40; also, r should be slightly longer than x. Why?)
(a) To find x, we use the tangent, which compares 15 to x.

$$\boxed{\tan 19° = \frac{\text{opp}}{\text{adj}}} \qquad\qquad \tan 19° = \frac{15}{x}$$

$$\downarrow$$

$$\boxed{\begin{array}{c}\text{Evaluate}\\\tan 19°\end{array}} \qquad\qquad 0.3443 = \frac{15}{x}$$

$$\downarrow$$

$$\boxed{\begin{array}{c}\text{Solve}\\\text{for } x\end{array}} \qquad\qquad x = \frac{15}{0.3443} = 43.6$$

Thus, the boat is about 44 feet from the foot of the dock.
(b) To find r, we use the sine, which compares 15 to r.

$$\boxed{\sin 19° = \frac{\text{opp}}{\text{hyp}}} \qquad\qquad \sin 19° = \frac{15}{r}$$

$$\downarrow$$

$$\boxed{\begin{array}{c}\text{Evaluate}\\\sin 19°\end{array}} \qquad\qquad 0.3256 = \frac{15}{r}$$

$$\downarrow$$

$$\boxed{\begin{array}{c}\text{Solve}\\\text{for } r\end{array}} \qquad\qquad r = \frac{15}{0.3256} = 46.1$$

Thus, about 46 feet of rope is out.

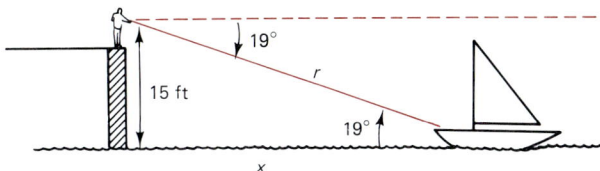

FIGURE 19

PROBLEM SET
8.6

Complete the following table for the right triangles below.

Angle	sin	cos	tan	cot	sec	csc
A						
B						

Warm-up Exercises

1.

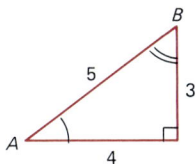

2.

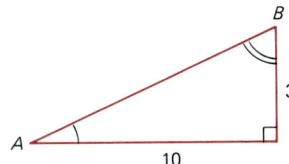

Evaluate the following.

Warm-up Exercises

3. tan 51°   **4.** sin 38°   **5.** cos 11.3°

6. cot 77.7°   **7.** sin 119°   **8.** cos 272°

Solve the following equations.

Warm-up Exercises

9. $\dfrac{x}{12} = 3$

10. $\dfrac{2x}{4} = 14$

11. $\dfrac{4}{x} = 0.8$

12. $\dfrac{5}{3x} = 0.25$

FIGURE 20

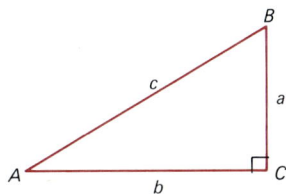

Figure 20 shows a general right triangle, with sides a, b, and c, and angles A, B, and C. In each of the following problems, use the given information to find the unknown.

13. $b = 145$, $A = 1°$; find a.

14. $b = 20$, $B = 55°$; find a.

15. $b = 7.2$, $B = 23°$; find c.

16. $b = 1900$, $A = 85°$; find c.

17. $c = 0.7$, $A = 14°$; find a.

18. $c = 651$, $B = 28°$; find a.

19. $c = 1500$, $B = 45°$; find b.

20. $c = 1.23$, $A = 61°$; find b.

21. $a = 5$, $c = 8$; find B.

22. $a = 12$, $B = 14°$; find c.

23. $a = 10$, $b = 15$; find A.

24. $a = 6$, $c = 10$; find B.

25. $c = 20$, $B = 51°$; find A.

26. $a = 5$, $c = 9$; find b.

27. $c = 100$, $A = 19°$; find b.

28. $a = 10$, $A = 72°$; find B.

29. A road rises at an incline of 5°. How far along the surface of the road must one drive to be elevated 20 feet?

30. A kite is flying at an elevation of 59°. If 350 meters of string are out, how high above the ground is the kite?

31. A woman on a 40-meter tower sees a car at an angle of depression of 15°. How far is the car from the base of the tower?

32. A rope is tied at an angle of depression of 55° from the top of a 100-foot building to the ground. How long must the rope be?

33. A 50-meter building casts a 19-meter shadow. What is the angle of elevation of the sun?

34. An airplane rises 1000 feet in its first 11,000 feet of travel (measured along the ground). What is the angle of elevation?

35. A man looks down from a 72-meter building and sees two cars approaching, one at an angle of depression of 17° and the other at 19°. How far apart are the cars?

36. A person is standing 1000 meters from a giant statue. The angles of elevation to the top and bottom of the face are 23° and 21°. How tall is the face?

Building Application

37. Shown is the sketch of the roof of a house.
    (a) Find the lengths, x and y.
    (b) Find the angles of *pitch*, θ and ϕ.

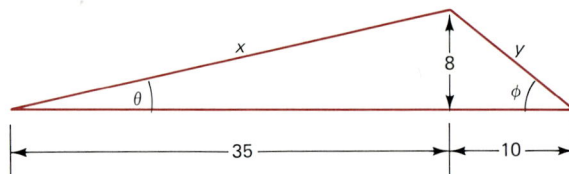

Photographic Application

38. Anne is photographing Bob at a distance of 8 feet, and 3 feet below the ceiling. She wants to bounce the flash off the ceiling to soften the light and shadow. (See Figure 21.)
    (a) What is the total distance the light travels, flash to face?
    (b) What is the angle of elevation of the flash?

FIGURE 21

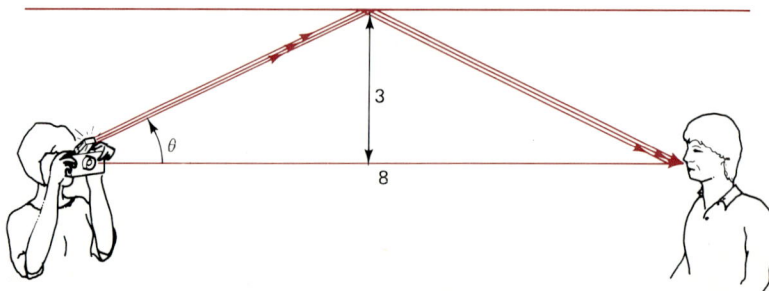

39. A force F is exerted by a muscle on a bone. The useful component U is perpendicular to the bone, and the wasted component W is parallel to the bone (see Figure 22).

(a) If $F = 100$, find U and W for $\theta = 10°$, $85°$, and $145°$.

(b) Using part (a), what phase of a chin-up is the easiest for an athlete?

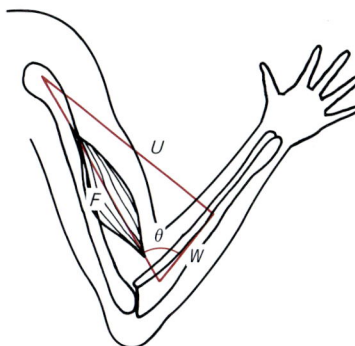

FIGURE 22

40. The slopes of the different cores in a dam are given as ratios, such as $2 : 1$, of the base to the height. For one such dam, these ratios are shown in Figure 23. Find angles A, B, C, D, E, and F.

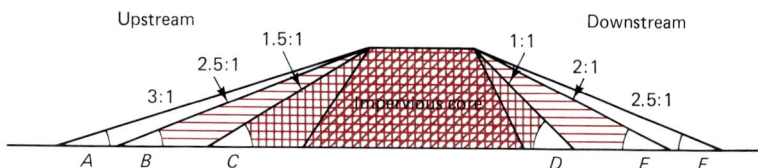

FIGURE 23

CHAPTER 8 SUMMARY

Important Theorems and Definitions

$1° = 1/360$ of a complete revolution.

$1' = 1/60$ of $1°$ and $1'' = 1/60$ of $1'$.

In a circle with radius r, the length of an arc of A degrees is given by

$$\frac{L}{2\pi r} = \frac{A}{360}$$

(Pythagorean theorem) In a right triangle with legs a and b and hypotenuse c, we have

$$a^2 + b^2 = c^2$$

The sum of the angles in any triangle is $180°$.

Two right triangles are similar if an acute angle of one equals an acute angle of the other.

In a right triangle,

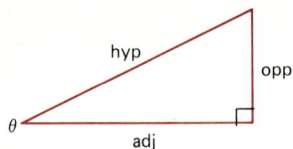

$$\sin \theta = \frac{\text{opp}}{\text{hyp}} \qquad \cos \theta = \frac{\text{adj}}{\text{hyp}}$$

$$\tan \theta = \frac{\text{opp}}{\text{adj}} \qquad \cot \theta = \frac{\text{adj}}{\text{opp}}$$

$$\sec \theta = \frac{\text{hyp}}{\text{adj}} \qquad \csc \theta = \frac{\text{hyp}}{\text{opp}}$$

For an angle θ in standard position,

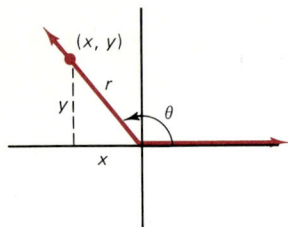

$$\sin \theta = \frac{y}{r} \qquad \cos \theta = \frac{x}{r}$$

$$\tan \theta = \frac{y}{x} \qquad \cot \theta = \frac{x}{y}$$

$$\sec \theta = \frac{r}{x} \qquad \csc \theta = \frac{r}{y}$$

(Functions are not defined for angles that cause a denominator to be zero.)

$$\sin \theta = \cos (90° - \theta)$$
$$\cos \theta = \sin (90° - \theta)$$
$$\tan \theta = \cot (90° - \theta)$$
$$\cot \theta = \tan (90° - \theta)$$
$$\sec \theta = \csc (90° - \theta)$$
$$\csc \theta = \sec (90° - \theta)$$

$$\csc \theta = \frac{1}{\sin \theta}$$

$$\sec \theta = \frac{1}{\cos \theta}$$

$$\cot \theta = \frac{1}{\tan \theta}$$

$$\tan \theta = \frac{\sin \theta}{\cos \theta} \qquad \cot \theta = \frac{\cos \theta}{\sin \theta}$$

$$\sin^2 \theta + \cos^2 \theta = 1$$

$$\tan^2 \theta + 1 = \sec^2 \theta \qquad \cot^2 \theta + 1 = \csc^2 \theta$$

Quadrant for θ	Reference Angle
I	θ
II	$180° - \theta$
III	$\theta - 180°$
IV	$360° - \theta$

Review Exercises

1. What fraction of a complete revolution is 150°?
2. How many degrees is 1/9 of a complete revolution?
3. The angles below are equivalent (or coterminal). Give the measures in degrees.

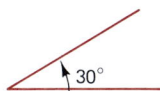   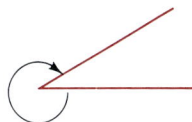   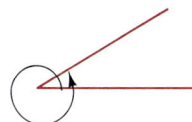

(a)   (b)

4. For each of the following angles, find an equivalent angle between 0° and 360°.
   (a) 600°   (b) 850°   (c) −35°
5. Draw each of the following angles in standard position.
   (a) 147°   (b) 305°
6. Give the length of the following arcs in a circle cut by central angle A.
   (a) radius $= 10$; $A = 75°$
   (b) radius $= 50$; $A = 220°$
7. Find the missing sides in the following right triangles (see Figure 20).
   (a) $a = 3$; $b = 7$; $c = ?$
   (b) $a = 2$; $b = ?$; $c = 10$
8. A triangle has angles 41° and 73.4°. What is the third angle?
9. Complete the following table for the triangle at the left.

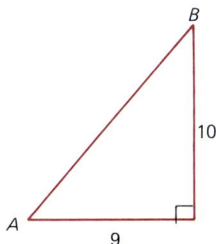

Angle	sin	cos	tan	cot	sec	csc
A						
B						

Find sin θ, cos θ, tan θ, and cot θ for each of the following angles.

10. $(-4, 3)$ is on the terminal side of θ.
11. $(-8, -3)$ is on the terminal side of θ.
12. Find sin 0°.   13. Find cos 30°.   14. Find tan 45°.
15. Find sin 120°.   16. Find cos 180°.   17. Find tan 270°.

Complete the following tables.

18.

Angle	sin	cos	tan	cot	sec	csc
θ	$\dfrac{7}{\sqrt{53}}$	$\dfrac{2}{\sqrt{53}}$			$\dfrac{\sqrt{53}}{2}$	$\dfrac{\sqrt{53}}{7}$
$90° - \theta$			$\dfrac{2}{7}$	$\dfrac{7}{2}$		

19.

Angle	sin	cos	tan	cot	sec	csc
A	$\dfrac{-3}{\sqrt{13}}$	$\dfrac{2}{\sqrt{13}}$	$\dfrac{-3}{2}$			

20.

Angle	sin	cos	tan	cot	sec	csc
B	-0.8	0.6				

21.

Angle	Quadrant	sin	cos	tan	cot	sec	csc
A	II	$\dfrac{1}{2}$					
B		0.6	-0.8				

Complete the following table.

Angle	Reference angle	sin	cos	tan	cot
22. $151°$					
23. $279°$					
24. $562°$					
25. $746°$					
26. $-19°$					

Find x in the following right triangles.

27.

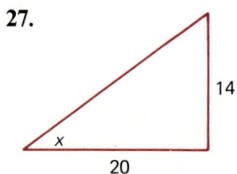

14
20
x

28.

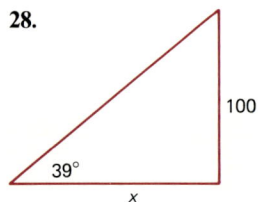

100
39°
x

29. The angle of elevation to the top of a tower is $36.5°$. If the person is 150 meters from the base, how tall is the tower?

Circular Functions

9.1
RADIAN
MEASURE

We are very familiar with degrees; we now introduce a new angle measure called **radians**, which are based on the angle's relation to the arcs of circles. (In fact, *radian* gets its name from *radi*us and *an*gle.) Radian measure is very important for studying the trigonometric functions in mathematics and science.

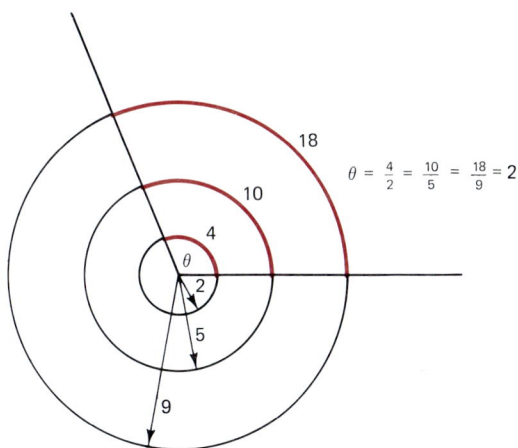

$$\theta = \frac{4}{2} = \frac{10}{5} = \frac{18}{9} = 2$$

FIGURE 1

FIGURE 2

$\theta = \frac{s}{r}$ radians

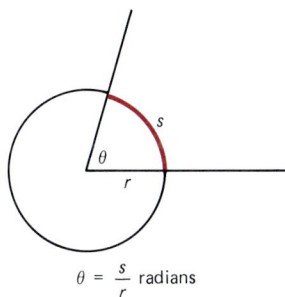

Figure 1 shows three circles with the same center. The angle θ cuts through each circle, and each arc length is twice the radius. Here we say that θ is 2 radians.

For any given angle θ, the arc-length-to-radius ratio is always the same. We use this fact to measure the angle in radians. Figure 2 shows a circle of radius r and an angle θ that cuts an arc length of s. Using Figure 2, we define the **radian measure** of θ:

$$\theta = \frac{s}{r}$$

In words, θ (in radians) is the ratio of the arc length to the radius. (For any θ, this ratio is always the same no matter what size circle we use, as we saw in Figure 1.)

Note that radians are defined as length-to-length ratios (such as centimeter/centimeter or inch/inch). Since these units cancel, we say that a radian is a *dimensionless quantity*. (The trigonometric functions, defined as length-to-length ratios, are also dimensionless.)

EXAMPLE 1   Figure 3 shows three examples of radian measure.
   (a) The arc length is 5, and the radius is 10. Thus, we measure the angle as $\theta = 5/10 = 1/2$ radian.

   (b) The arc length is r, and the radius is r. Thus, the angle is $\theta = \dfrac{r}{r} = 1$ radian.

   (c) The arc length is the whole circumference, $2\pi r$. Thus, the radian measure of the whole circle is $\theta = \dfrac{2\pi r}{r} = 2\pi$ radians.

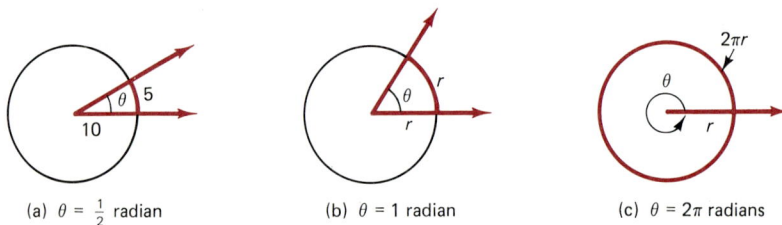

(a) $\theta = \frac{1}{2}$ radian   (b) $\theta = 1$ radian   (c) $\theta = 2\pi$ radians

FIGURE 3

Example 1(c) shows a very important relation between degree measure and radian measure. We see that a whole circle is 2π radians and of course $360°$. Thus, $360° = 2\pi$ radians. Dividing both sides by 2, we get the following **degree–radian conversion**.

$$180° = \pi \text{ radians}$$

EXAMPLE 2   We often use this formula to convert from radians to degrees by replacing π radians by $180°$.
   (a) $\dfrac{\pi}{2}$ radians $= \dfrac{180°}{2} = 90°$

   (b) $\dfrac{\pi}{3}$ radians $= \dfrac{180°}{3} = 60°$

   (c) $\dfrac{2\pi}{3}$ radians $= \dfrac{2(180°)}{3} = 120°$

   (d) 5π radians $= 5(180°) = 900°$

   (e) $\dfrac{-\pi}{4}$ radian $= \dfrac{-180°}{4} = -45°$

In general, we have the following proportion to convert between degrees and radians.

$$\frac{D \text{ degrees}}{180°} = \frac{R \text{ radians}}{\pi \text{ radians}}$$

To use this relation, we substitute the given measure (D or R) into the equation and solve for the other.

EXAMPLE 3 To express 1 radian in degrees, we let $R = 1$ and solve for D.

$$\boxed{\frac{D}{180°} = \frac{R}{\pi}}$$

$$\boxed{\text{Solve for } D}$$

$$\frac{D}{180°} = \frac{1}{\pi}$$

$$D = \frac{180°}{\pi} \approx 57.3°$$

Thus, 1 radian $\approx 57.3°$ [see Figure 3(b)].

EXAMPLE 4 To express $1°$ in radians, we let $D = 1°$.

$$\boxed{\frac{D}{180°} = \frac{R}{\pi}}$$

$$\boxed{\text{Solve for } R}$$

$$\frac{1°}{180°} = \frac{R}{\pi}$$

$$\frac{\pi}{180°} = R$$

Thus, $1° = \pi/180$ radian ≈ 0.0175 radian.

EXAMPLE 5 To express $75°$ in radians, we let $D = 75°$ and solve for R.

$$\boxed{\frac{D}{180°} = \frac{R}{\pi}}$$

$$\boxed{\text{Solve for } R}$$

$$\frac{75°}{180°} = \frac{R}{\pi}$$

$$\frac{75\pi}{180} = R$$

Thus $75° = 75\pi/180$ radians ≈ 1.31 radians.

Figures 4 and 5 (page 308) summarize the relations between radians and degrees. Just as equivalent angles can have different degree measures (such as $90°$ and $450°$), note that equivalent angles can also have different radian measures (such as $\pi/2$ and $5\pi/2$).

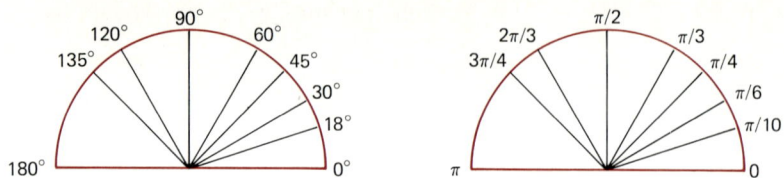

FIGURE 4

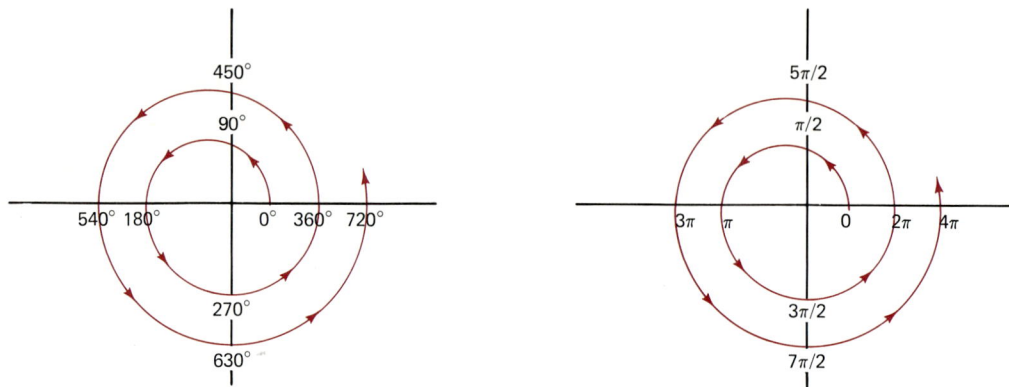

FIGURE 5

PROBLEM SET

9.1 *Complete the following table for arcs (in degrees) of circles.*

		Radius of circle	Arc (degrees)	Arc length
Warm-up	**1.**	10	90°	?
Exercises	**2.**	20	360°	?
	3.	100	60°	?
	4.	40	270°	?
	5.	8	?	8π
	6.	12	?	2π

Complete the following table for angles in degrees.

		Angle θ (degrees)	Fraction of a circle	$\sin\theta$	$\cos\theta$	$\tan\theta$
Warm-up	**7.**	30°	?	?	?	?
Exercises	**8.**	300°	?	?	?	?
	9.	20°	?	?	?	?
	10.	?	2/3	?	?	?
	11.	?	1/20	?	?	?
	12.	?	2/5	?	?	?

Express the central angles θ in radians in the following circles with radius r and arc length s.

13. $r = 5$; $s = 5$

14. $r = 7$; $s = 21$

15. $r = 10$; $s = 35$

16. $r = 30$; $s = 10$

Convert each of the following radian measures into degrees (to the nearest tenth of a degree).

17. π **18.** $\dfrac{\pi}{2}$

19. $\dfrac{\pi}{4}$ **20.** $\dfrac{\pi}{10}$

21. $\dfrac{2\pi}{5}$ **22.** $\dfrac{3\pi}{2}$

23. 4π **24.** $\dfrac{7\pi}{2}$

25. 2   **26.** 3

27. 0.1   **28.** 1.5

29. -1 **30.** -4

Convert each of the following degree measures into radians (in terms of π).

31. $90°$ **32.** $45°$

33. $180°$ **34.** $360°$

35. $20°$ **36.** $10°$

37. $270°$ **38.** $450°$

39. $1000°$ **40.** $950°$

41. $0.1°$ **42.** $-120°$

Complete the following table for a circle of radius r with a central angle θ.

	r	Circumference of circle	θ (degrees)	θ (radians)	Arc length
43.	100	?	$90°$	?	?
44.	24	?	$45°$	?	?
45.	20	?	?	?	5π
46.	12	?	?	?	8π
47.	30	?	?	π	?
48.	90	?	?	$\pi/6$	?
49.	?	?	?	$\pi/4$	π
50.	?	?	?	$\pi/3$	4π
51.	?	24π	$30°$	?	?
52.	?	100π	$135°$	?	?
53.	?	200π	?	?	50π
54.	?	100π	?	?	75π

Complete the following table (without tables or calculator).

	θ (radians)	θ (degrees)	$\sin \theta$	$\cos \theta$	$\tan \theta$
55.	$\pi/3$	?	?	?	?
56.	$\pi/4$	?	?	?	?
57.	π	?	?	?	?
58.	$3\pi/4$	?	?	?	?
59.	?	$90°$	?	?	?
60.	?	$30°$	?	?	?
61.	?	$120°$	?	?	?
62.	?	$360°$	?	?	?

For each angle (in radians), find an equivalent angle between 0 and 2π in terms of π.

63. 7π

64. $\dfrac{11\pi}{4}$

65. $\dfrac{9\pi}{4}$

66. $\dfrac{13\pi}{6}$

67. $\dfrac{-7\pi}{2}$

68. $\dfrac{-19\pi}{4}$

Calculator Application

69. Your calculator probably has a $\boxed{\text{DRG}}$ key for the mode of angle measure used: degrees, radians, or grads. Used somewhat in Europe, a *grad* is 1/100 of a right angle. (Hence, a complete revolution is 400 grads, written 400ᵍ.) Complete the following table.

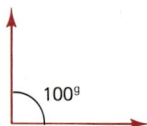
100ᵍ

	θ (degrees)	θ (radians)	θ (grads)
Example:	$90°$	$\pi/2$	100^g
(a)	$180°$	?	?
(b)	$45°$	?	?
(c)	?	$\pi/6$	?
(d)	?	$5\pi/2$	?
(e)	?	?	150^g
(f)	?	?	700^g

Physical Application

70. A ball of weight W is at the end of a pendulum. The weight has two components: T is equal and opposite to the tension in the string, and F is perpendicular to the string and restores the ball to the middle.
  (a) Find F and T in terms of W and θ.
  (b) Simplify the expression for F, assuming that θ is very small and measured in radians. (For small angles θ measured in radians, we have $\sin \theta \approx \theta$. This approximation is used quite often in science.)

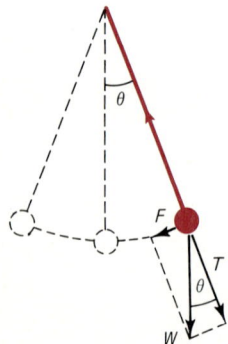

71. Most of the high-level computer languages (such as BASIC, FORTRAN, Pascal, and PL/1) have built-in trigonometric functions (sine, cosine, and tangent) for angles expressed *in radians.* For example, in BASIC

$$Y = SIN(1.75)$$

assigns to Y the sine of 1.75 radians. Rewrite the following assignments in a computer language. [Be careful with (e) and (f).]

(a) $y = 6 \cos x$

(b) $y = \dfrac{\tan x}{5x}$

(c) $y = 1 - 2 \sin^2 x$

(d) $y = t\sqrt{\sin(b - c)} + u$

(e) $y = \cot x$

(f) $y = \sec x$

9.2
CIRCULAR FUNCTIONS

In the preceding chapter we introduced the six trigonometric functions for angles expressed in degrees. We now want to consider the same trigonometric functions for angles expressed in radians.

We have a table for the trigonometric functions of angles in radians. This is Table D in the Appendix. It is used just as the table for angles in degrees is used. When necessary, we find the **reference angle**. Table 1 below provides a handy table for determining the quadrant that an angle lies in as well as its reference angle. As we did in Chapter 8, we find the reference angle as the acute angle formed by the terminal side and the nearest x-axis.

TABLE 1

Quadrant for θ	Interval in Radians (Exact)	Interval in Radians (Approximate)	Reference Angle
I	$0 < \theta < \dfrac{\pi}{2}$	$0 < \theta < 1.57$	θ
II	$\dfrac{\pi}{2} < \theta < \pi$	$1.57 < \theta < 3.14$	$\pi - \theta$
III	$\pi < \theta < \dfrac{3\pi}{2}$	$3.14 < \theta < 4.71$	$\theta - \pi$
IV	$\dfrac{3\pi}{2} < \theta < 2\pi$	$4.71 < \theta < 6.28$	$2\pi - \theta$

EXAMPLE 6 The following examples use Table D in the Appendix exactly as we do with degrees. We find the reference angle just as before (see Figure 6 on page 312).

(a) Find $\sin 0.31$. The angle 0.31 radian is in the first quadrant. We simply look down the sine column until we reach the 0.31 row. We see that $\sin 0.31 = 0.30506$.

(b) Find $\cos 2.78$. Since $1.57 \leq 2.78 \leq 3.14$, the angle is in quadrant II. Thus, its reference angle is $3.14 - 2.78 = 0.36$. (This is $\pi - \theta$.) Since the cosine is negative in quadrant II, $\cos 2.78 = -\cos 0.36 = -0.93590$.

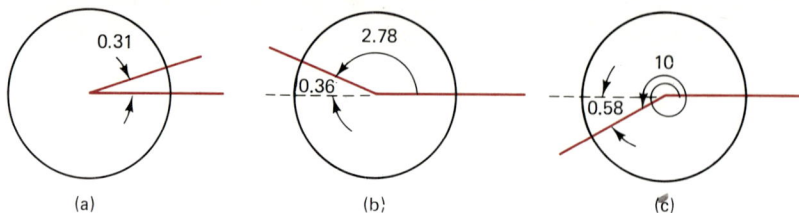

FIGURE 6

(c) Find tan 10. Since $2\pi \leq 10 \leq 4\pi$, $\theta = 10$ is more than one revolution. Therefore we first find an equivalent angle by subtracting 2π: $10 - 6.28 = 3.72$. This angle is in quadrant III, and its reference angle is $3.72 - 3.14 = 0.58$. Since the tangent is positive in quadrant III,

$$\tan 10 = \tan 3.72 = \tan 0.58 = 0.65517$$

We can also evaluate the trigonometric functions of angles expressed in radians using many hand calculators. Since the machine probably expects degrees, you may have to change the mode. (Check your operating manual. On many machines, there is a $\boxed{\text{DRG}}$ key: Turn the machine on and you are in the degree mode; press $\boxed{\text{DRG}}$ and you are in the radian mode. Most calculators use the abbreviation "RAD" on the display.)

PRESS	DISPLAY	MEANING
$\boxed{\text{ON}}$ $\boxed{\text{DRG}}$	0. RAD	Put into radian mode.
(a) $\boxed{\cdot}\boxed{3}\boxed{1}$	0.31 RAD	Enter angle (in radians)
$\boxed{\sin}$	0.3050586 RAD	sin 0.31
(b) $\boxed{2}\boxed{\cdot}\boxed{7}\boxed{8}$	2.78 RAD	Enter angle (in radians)
$\boxed{\cos}$	-0.9353348 RAD	cos 2.78
(c) $\boxed{1}\boxed{0}$	10. RAD	Enter angle (in radians)
$\boxed{\tan}$	0.648361 RAD	tan 10

Note that there are slight differences between the evaluation using the table and the evaluation using the calculator, since we had to round π to 3.14 to use the table. Thus, the calculator result is much more accurate.

EXAMPLE 7 For some special angles, we do not need tables or calculators to evaluate the trigonometric functions.

(a) $\sin \dfrac{\pi}{3} = \sin 60° = \dfrac{\sqrt{3}}{2} \approx 0.87$

(b) $\cos \dfrac{\pi}{2} = \cos 90° = 0$

(c) $\tan \dfrac{3\pi}{4} = \tan 135° = -\tan 45° = -1$

(d) $\cot \pi = \cot 180°$ is undefined.

(e) $\cos 2\pi = \cos 360° = \cos 0° = 1$

(f) $\sin \dfrac{3\pi}{2} = \sin 270° = -1$

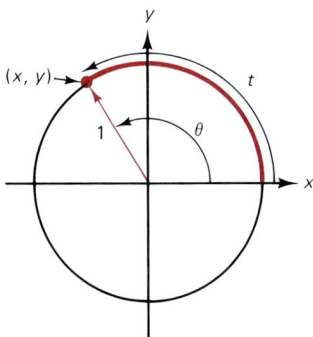

FIGURE 7 On a unit circle, $t = \theta$.

At this point, let us review what we have developed. We defined the trigonometric functions for acute angles. We then defined the functions for angles of any size. We have discussed the functions for angles expressed in radians. *We have now reached our ultimate goal: to define the trigonometric functions for domains of all real numbers (not just angles).*

We now consider the **unit circle** centered at the origin. This circle has radius 1 and satisfies $x^2 + y^2 = 1$ (see page 196). We can view any real number as starting at $(1, 0)$ and being wrapped around the unit circle, like a cloth tape measure, as shown in Figure 7. Thus, a real number t becomes an arc length t on the circle. Since $r = 1$, the arc length $t = r\theta = 1\theta = \theta$. Therefore, *on a unit circle, we can identify any real number t with a central angle t measured in radians.* (*Note:* The circle must be a unit circle, and the angle must be measured in radians; this is why unit circles and radians are so important.)

> We define each trigonometric function for a real number t as that same function for the angle t measured in radians.

For example, we define

$$\sin 2 = \sin (2 \text{ radians}) = 0.9093$$

$$\cos \frac{1}{4} = \cos \left(\frac{1}{4} \text{ radian} \right) = 0.9689$$

$$\tan (-5.1) = \tan (-5.1 \text{ radians}) = 2.4494$$

Since $\sin t = y/r$ and $r = 1$ on the unit circle, we have $\sin t = y$. Similarly, since $\cos t = x/r$ and $r = 1$ on the unit circle, we have $\cos t = x$. In like manner, we obtain the following property.

Property

Let (x, y) be a point on the unit circle, $x^2 + y^2 = 1$. If t is any real number (hence, central angle in radians) as shown in Figure 7, we have

$$\sin t = y \qquad \csc t = \frac{1}{y}$$

$$\cos t = x \qquad \sec t = \frac{1}{x}$$

$$\tan t = \frac{y}{x} \qquad \cot t = \frac{x}{y}$$

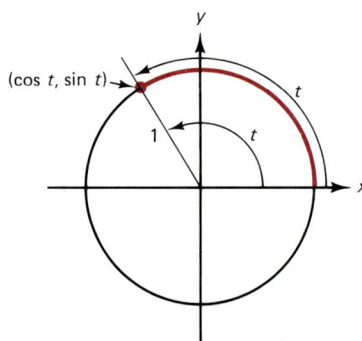

FIGURE 8

Figure 8 illustrates this property. The point on the unit circle at angle t radians has coordinates $(\cos t, \sin t)$.

EXAMPLE 8   If the real number 9 is wrapped as an arc length around a unit circle starting at $(1, 0)$, it makes a rotation of 9 radians and terminates at the point $(\cos 9, \sin 9)$ on the circle (see Figure 9).

FIGURE 9

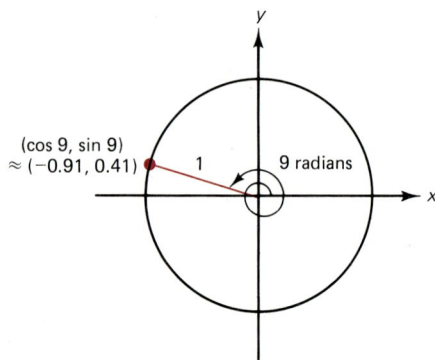

To evaluate cos 9, we first find an angle between 0 and 2π that is equivalent to 9 radians. This is $9 - 2\pi \approx 2.72$. Why? We now find the reference angle for 2.72. Since this angle is in quadrant II, its reference angle is $\pi - 2.72 \approx 0.42$. In quadrant II the sine is positive and the cosine is negative. Thus, we obtain

$$\cos 9 \approx -\cos 0.42 = -0.9131$$

Similarly,

$$\sin 9 \approx +\sin 0.42 = 0.4078$$

On a calculator, this appears as follows.

PRESS	DISPLAY	MEANING
DRG 9 cos	−0.9111302 RAD	cos 9 (remember radian mode)
9 sin	0.4121187 RAD	sin 9

If Examples 6 and 8 seem very similar, this is not an accident. By wrapping a real number around the unit circle, the real number is equal to the central angle expressed in radians. Therefore, we evaluate the trigonometric functions the same for the real numbers as for the angles in radians.

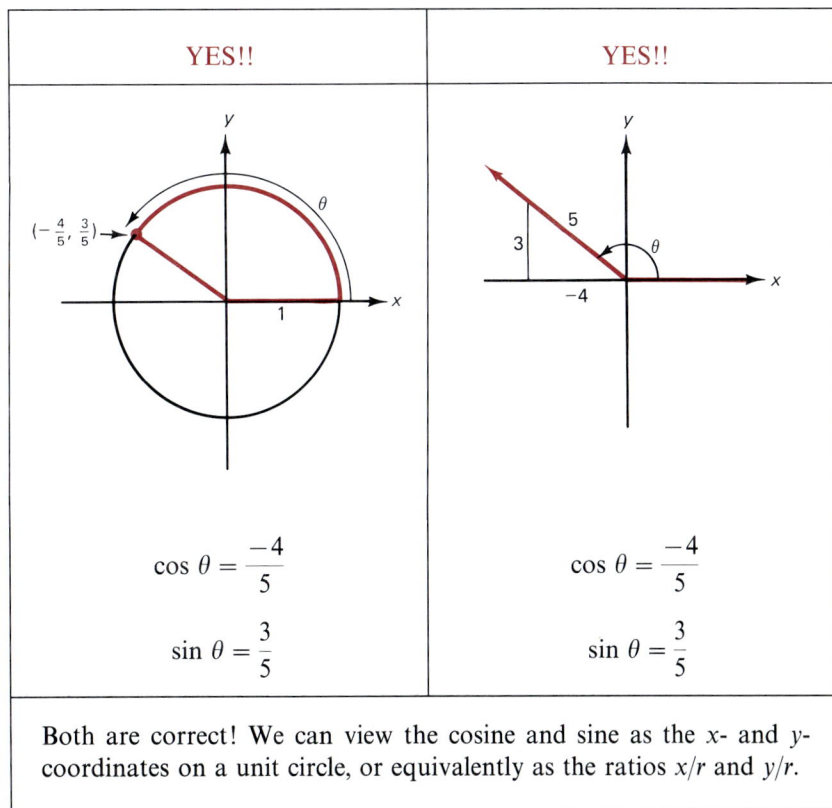

YES!! YES!!

$$\cos \theta = \frac{-4}{5}$$

$$\sin \theta = \frac{3}{5}$$

$$\cos \theta = \frac{-4}{5}$$

$$\sin \theta = \frac{3}{5}$$

Both are correct! We can view the cosine and sine as the x- and y-coordinates on a unit circle, or equivalently as the ratios x/r and y/r.

TABLE 2

Function	On Unit Circle	Domain
$f(t) = \sin t$	y	All real numbers t
$f(t) = \cos t$	x	All real numbers t
$f(t) = \tan t$	y/x	$t \neq \pm\dfrac{\pi}{2}, \pm\dfrac{3\pi}{2}, \pm\dfrac{5\pi}{2}$, and so on
$f(t) = \cot t$	x/y	$t \neq 0, \pm\pi, \pm 2\pi$, and so on
$f(t) = \sec t$	$1/x$	$t \neq \pm\dfrac{\pi}{2}, \pm\dfrac{3\pi}{2}, \pm\dfrac{5\pi}{2}$, and so on
$f(t) = \csc t$	$1/y$	$t \neq 0, \pm\pi, \pm 2\pi$, and so on

We have now achieved our goal of defining the trigonometric functions on the set of real numbers. Table 2 summarizes these functions, their relationships to the unit circle (as shown in Figure 7), and their domains (some t-values must be excluded).

PROBLEM SET

9.2

Complete the following table for θ (in degrees). (For Problems 5 to 8, do not use a calculator or tables.)

Warm-up Exercises

	θ (degrees)	Equivalent 0 to 360°	Quadrant	Reference angle	$\sin \theta$	$\cos \theta$	$\tan \theta$
1.	143°	?	?	?	?	?	?
2.	319°	?	?	?	?	?	?
3.	444°	?	?	?	?	?	?
4.	567°	?	?	?	?	?	?
5.	120°	?	?	?	?	?	?
6.	315°	?	?	?	?	?	?
7.	495°	?	?	?	?	?	?
8.	960°	?	?	?	?	?	?

Complete the following table for θ (in radians). (For Problems 21 to 28, do not use a calculator or tables.)

	θ (radians)	Quadrant (or axis)	Reference angle	$\sin \theta$	$\cos \theta$	$\tan \theta$
9.	0.10	?	?	?	?	?
10.	0.20	?	?	?	?	?
11.	0.56	?	?	?	?	?
12.	1.31	?	?	?	?	?
13.	1.50	?	?	?	?	?
14.	2.97	?	?	?	?	?
15.	3.74	?	?	?	?	?

16.	5.09	?	?	?	?	?
17.	6.03	?	?	?	?	?
18.	12.59	?	?	?	?	?
19.	−1.14	?	?	?	?	?
20.	−2.03	?	?	?	?	?
21.	$\pi/4$	?	?	?	?	?
22.	$\pi/3$	?	?	?	?	?
23.	$\pi/2$	?	?	?	?	?
24.	$2\pi/3$	?	?	?	?	?
25.	π	?	?	?	?	?
26.	$5\pi/4$	?	?	?	?	?
27.	$3\pi/2$	?	?	?	?	?
28.	2π	?	?	?	?	?

For each of the following equations, find a real number $0 \le t \le \pi/2$ that satisfies the equation. (Reminder: If you use a calculator, put the machine in the radian mode.)

29. $\sin t = 0.83$ **30.** $\cos t = 0.53$

31. $\tan t = 2.1$ **32.** $\sin t = 0.01$

33. $\cos t = 0.98$ **34.** $\tan t = 0.71$

35. $\sin t = 0.99$ **36.** $\cos t = 0.01$

37. $\tan t = 1$ **38.** $\sin t = 0.5$

If a real number t is wrapped around a unit circle, it terminates at (x, y), as shown in Figure 7. Complete the following table.

	x	y	t	Quadrant where (x, y) lies
39.	?	?	6.0	?
40.	?	?	−4.2	?
41.	0.5	?	?	I
42.	0.63	?	?	I
43.	−0.71	?	?	II
44.	−0.21	?	?	III
45.	?	−0.82	?	IV
46.	?	0.59	?	II
47.	0.60	−0.80	?	?
48.	−0.80	−0.60	?	?

Figure 10 shows a unit circle on a square grid (each small division is 0.1). The lengths 0.5, 1.0, 1.5, and so on are marked off along the circle [as they are wrapped counterclockwise from (1, 0)]. Use only the graph and the definitions to evaluate the following.

49. $\cos 1.5$ **50.** $\sin 2.0$

51. $\sin 3.0$ **52.** $\cos 4.5$

53. $\tan 5.5$ **54.** $\cot 6.0$

55. $\csc 2.5$ **56.** $\tan 3.5$

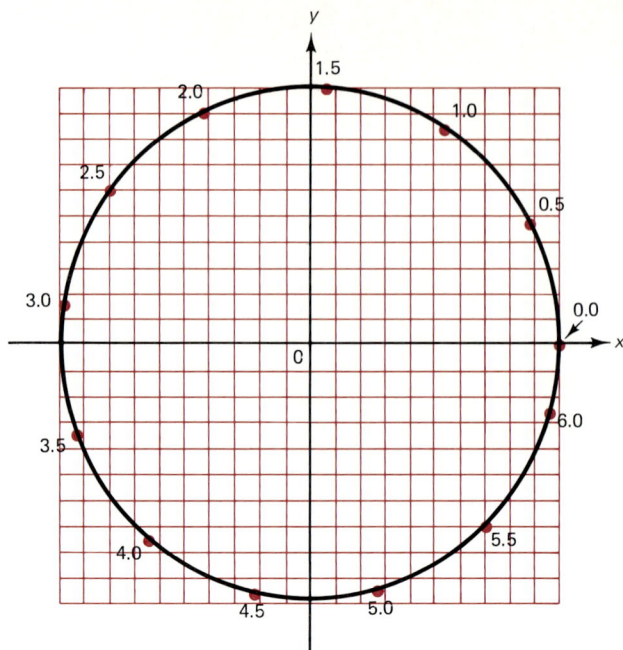

FIGURE 10

Use the unit circle and line segments below to show the following.

57. $\sin \theta = \overline{AC}$

58. $\cos \theta = \overline{OA}$

59. $\tan \theta = \overline{DB}$

60. $\cot \theta = \overline{EF}$

61. $\sec \theta = \overline{OD}$

62. $\csc \theta = \overline{OF}$

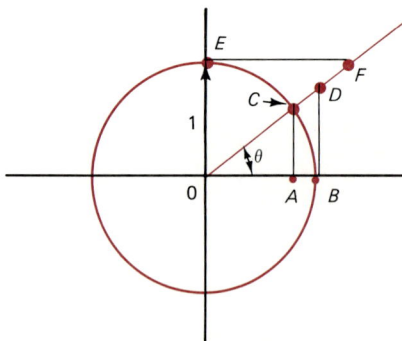

9.3
GRAPHS OF THE SINE AND COSINE FUNCTIONS

The **graphs** of the sine and cosine functions are important in mathematics and science. They can reflect the "up-and-down" nature of sound, weather, business, voltage, and so on. In addition, the graphs give us insight into the functions themselves (as all graphs do).

The Basic Sine and Cosine Curves

We begin with the sine function, $y = \sin x$.

EXAMPLE 9   Graph $y = \sin x$.

Solution   As with any function, we prepare a table of x-y pairs that we plot in making our graph. Recall that the sine repeats itself in cycles of 360° or 2π radians, so we look only at x-values from 0 to 2π.

We can use a unit circle, tables, or a calculator to help us plot the sine curve. (See Figure 11.) Notice that we pick x in multiples of $\pi/6$. (Recall that this is 0°, 30°, 60°, and so on.)

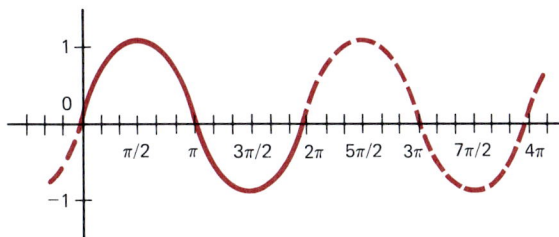

x	0	$\dfrac{\pi}{6}$	$\dfrac{\pi}{3}$	$\dfrac{\pi}{2}$	$\dfrac{2\pi}{3}$	$\dfrac{5\pi}{6}$	π	$\dfrac{7\pi}{6}$	$\dfrac{4\pi}{3}$	$\dfrac{3\pi}{2}$	$\dfrac{5\pi}{3}$	$\dfrac{11\pi}{6}$	2π
$y = \sin x$	0	0.5	0.87	1	0.87	0.5	0	-0.5	-0.87	-1	-0.87	-0.5	0

FIGURE 11

EXAMPLE 10   Graph $y = \cos x$.

Solution   We start with an x-y table similar to that in Figure 11. We look at x-values in intervals of $\pi/6$ (30°). Figure 12 shows the table and the graph.

FIGURE 12

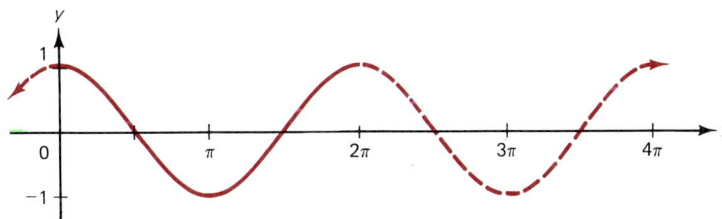

x	0	$\dfrac{\pi}{6}$	$\dfrac{\pi}{3}$	$\dfrac{\pi}{2}$	$\dfrac{2\pi}{3}$	$\dfrac{5\pi}{6}$	π	$\dfrac{7\pi}{6}$	$\dfrac{4\pi}{3}$	$\dfrac{3\pi}{2}$	$\dfrac{5\pi}{3}$	$\dfrac{11\pi}{6}$	2π
$\cos x$	1	0.87	0.5	0	-0.5	-0.87	-1	-0.87	-0.5	0	0.87	0.5	1

Let us make a few observations about the graphs of $y = \sin x$ and $y = \cos x$. (Note the similarities and differences.)

1. In both graphs, the y-coordinate oscillates between -1 and 1.
2. In both graphs, the curve repeats itself every 2π radians.
3. The sine curve is maximum ($\sin x = 1$) at $x = \pi/2, 5\pi/2, 9\pi/2$, and so on; the cosine curve is maximum at $x = 0, 2\pi, 4\pi$, and so on.
4. The sine curve is minimum ($\sin x = -1$) at $x = 3\pi/2, 7\pi/2$, and so on; the cosine curve is minimum at $x = \pi, 3\pi, 5\pi$, and so on.
5. The sine curve crosses the x-axis ($\sin x = 0$) at $x = 0, \pi, 2\pi, 3\pi$, and so on; the cosine curve crosses the x-axis at $x = \pi/2, 3\pi/2, 5\pi/2$, and so on.

A function is called periodic if $f(x + p) = f(x)$ for all x in the domain (that is, the function repeats itself every p units). The smallest such p is called the **period**. For $y = \sin x$ and $y = \cos x$ the period is 2π. The cycle of the curve for x between 0 and 2π is called the **basic sine (cosine) curve**. The rest of the graph are repetitions of the basic sine curve in periods of 2π.

In Examples 9 and 10 and Figures 11 and 12, we used 13 (x, y) pairs to obtain an accurate picture of one cycle of the basic curves. This is an awful lot of points to plot every time. Thus, we restrict our attention to the five **quarter values** of one period:

$$x = 0, \frac{\pi}{2}, \pi, \frac{3\pi}{2}, 2\pi$$

These five values of x cut one period into four equal pieces (hence the name quarter values.) These x-values produce the five **key points** needed to sketch one cycle of the basic curves. Figures 13 and 14 show the development of the basic sine and cosine curves.

FIGURE 13 Sine curve.

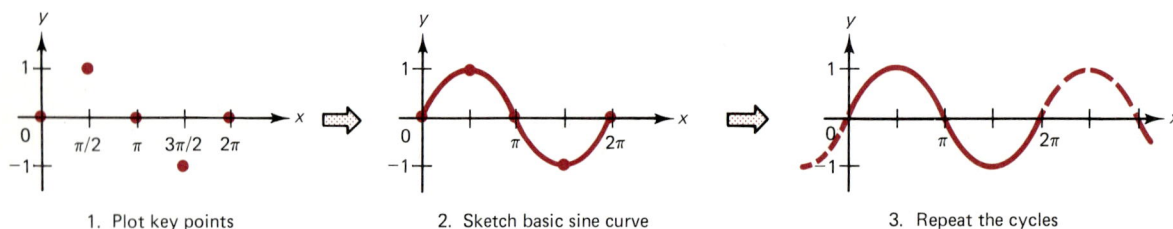

1. Plot key points   2. Sketch basic sine curve   3. Repeat the cycles

FIGURE 14 Cosine curve.

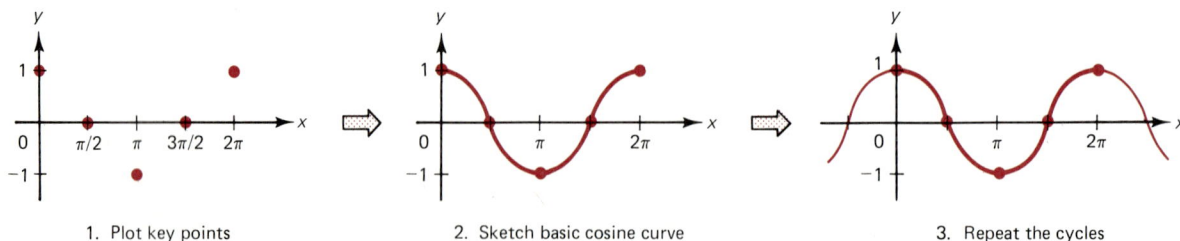

1. Plot key points   2. Sketch basic cosine curve   3. Repeat the cycles

Amplitude

EXAMPLE 11 Graph $y = 2 \sin x$.

Solution This function is similar to that in Example 9. If we consider the quarter values of $x = 0, \pi/2, \pi, 3\pi/2$, and 2π, we obtain the following key points:

$$(0, 0), \quad \left(\frac{\pi}{2}, 2\right), \quad (\pi, 0), \quad \left(\frac{3\pi}{2}, -2\right), \quad (2\pi, 0)$$

Find and plot
key points

Draw sine curve
through key
points

Extend graph

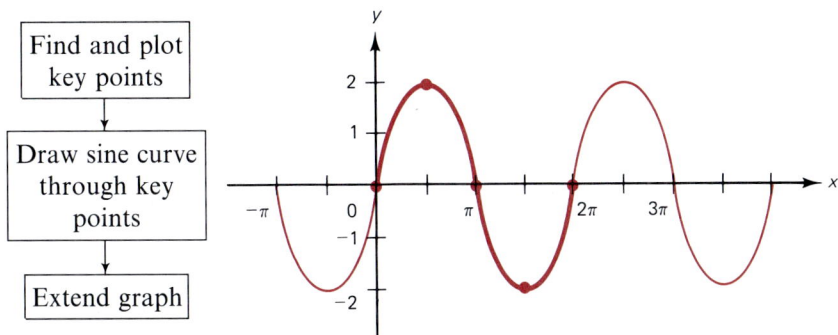

Note that $y = 2 \sin x$ has the exact same shape as $y = \sin x$, except that $y = 2 \sin x$ is *twice as high and twice as deep*. Figure 15 shows one cycle of several graphs all of the form $y = A \sin x$ and $y = A \cos x$, where $|A|$ is a real number called the **amplitude**. (Viewed on an oscilloscope, a pure musical tone appears as a sine curve; the amplitude $|A|$ corresponds to the loudness or volume. Figure 15 might represent different volume levels of the same tone.)

FIGURE 15

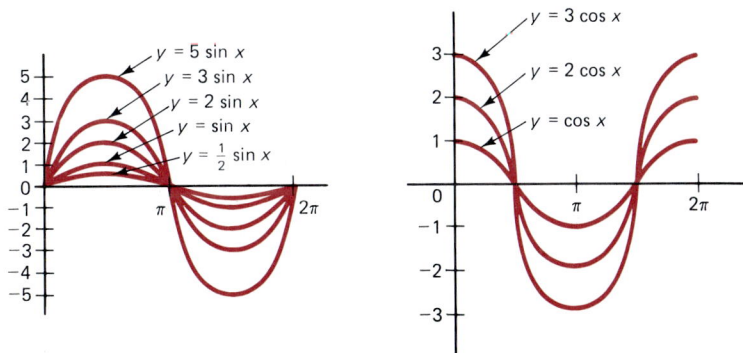

Frequency

EXAMPLE 12 Graph $y = \sin 2x$.

Solution Do *not* confuse this function with $y = 2 \sin x$. They are very different as their graphs show. This function, $y = \sin 2x$, first doubles x and then finds $\sin 2x$.

We must pay special attention to the $2x$ term. When the values $x = 0, \pi/2, \pi, 3\pi/2$, and 2π are doubled, we obtain $2x = 0, \pi, 2\pi, 3\pi$, and 4π, which produces *two* periods. Thus, to get just one period, we look at the interval 0 to π. Now the quarter values are 0, $\pi/4$, $\pi/2$, $3\pi/4$, and π, and the key points are

$$(0, 0), \quad \left(\frac{\pi}{4}, 1\right), \quad \left(\frac{\pi}{2}, 0\right), \quad \left(\frac{3\pi}{4}, -1\right), \quad (\pi, 0)$$

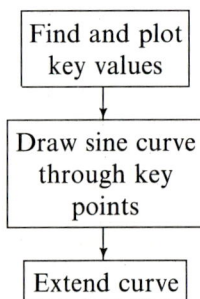

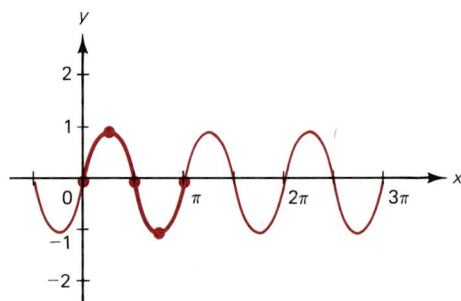

Figure 16 shows several graphs, all of the form $y = \sin Bx$ and $y = \cos Bx$, where $B > 0$. The number B is called the **frequency** and tells us how many cycles of the curve occur between 0 and 2π. (In the case $B = 1/2$, notice that the curve is only halfway through its cycle.)

Referring to our example of the sine curve representing a pure musical tone, the curves in Figure 16 would represent different pitches all at the same loudness.

What is the relation between the frequency B and the period P (the length of one cycle)? Since there are B cycles between 0 and 2π, each cycle is $2\pi/B$ units long. That is,

$$P = \frac{2\pi}{B}$$

For instance, if $B = 2$, then $P = 2\pi/2 = \pi$; if $B = \frac{1}{2}$, then $P = 2\pi/\frac{1}{2} = 4\pi$. See Figure 16.

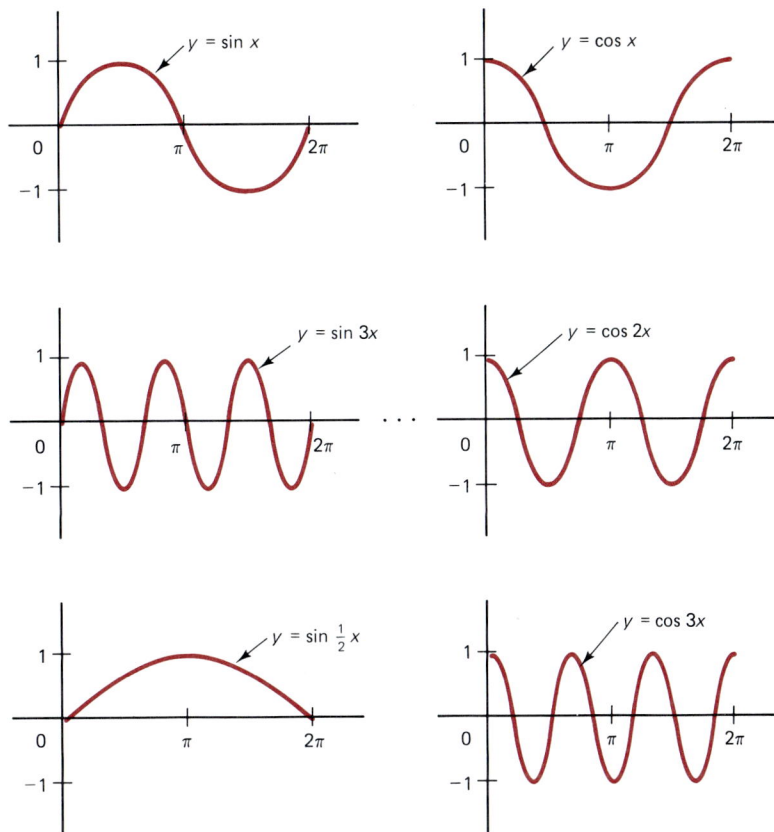

FIGURE 16

Frequency and Amplitude Together

We can now put all this information together to graph functions of the form $y = A \sin Bx$ or $y = A \cos Bx$, where we have

1. Amplitude $|A|$ (oscillates between $-A$ and A)
2. Frequency B (B cycles between 0 and 2π)
3. Period $P = 2\pi/B$ (length of one cycle)

Our graph will have the same shape as the basic sine or cosine curve, except taller or shorter (depending on A) and with longer or shorter cycle (depending on B).

We begin by finding the period, $P = 2\pi/B$, and then the quarter values: $x = 0$, $P/4$, $P/2$, $3P/4$, and P. We use these to find the key points (see the following box). Finally, we sketch the basic curve through these points and extend it in both directions.

KEY POINTS	$y = A \sin Bx$	$y = A \cos Bx$
x-intercepts occur at	$x = 0,\ P/2,\ P$	$x = P/4,\ 3P/4$
Maxima occur at	$x = \begin{cases} P/4 & \text{if } A > 0 \\ 3P/4 & \text{if } A < 0 \end{cases}$	$x = \begin{cases} 0,\ P & \text{if } A > 0 \\ P/2 & \text{if } A < 0 \end{cases}$
Minima occur at	$x = \begin{cases} 3P/4 & \text{if } A > 0 \\ P/4 & \text{if } A < 0 \end{cases}$	$x = \begin{cases} P/2 & \text{if } A > 0 \\ 0,\ P & \text{if } A < 0 \end{cases}$

EXAMPLE 13   Graph $y = 3 \cos 2x$.

Solution   Here the frequency is 2 and the amplitude 3. We find the period and the key points.

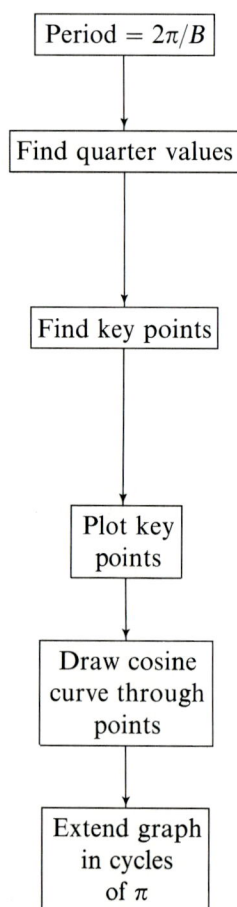

Period $= 2\pi/B$	$\text{Period} = \dfrac{2\pi}{2} = \pi$

Find quarter values	$x = 0,\ \dfrac{\pi}{4},\ \dfrac{\pi}{2},\ \dfrac{3\pi}{4},\ \pi$

Find key points	x-intercepts: $\left(\dfrac{\pi}{4},\ 0\right), \left(\dfrac{3\pi}{4},\ 0\right)$ maxima: $(0,\ 3),\ (\pi,\ 3)$ minimum: $\left(\dfrac{\pi}{2},\ -3\right)$

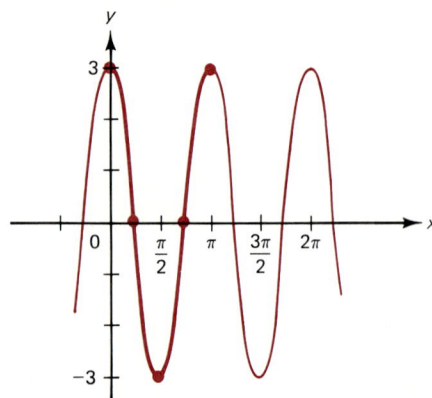

Plot key points

Draw cosine curve through points

Extend graph in cycles of π

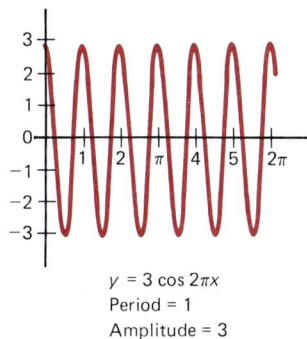

$y = 3 \cos 2\pi x$
Period = 1
Amplitude = 3

$y = 2 \cos \frac{1}{2} x$
Period = 4π
Amplitude = 2

FIGURE 17

Figure 17 shows two more such graphs.

YES	NO
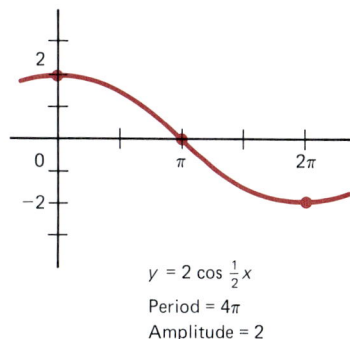	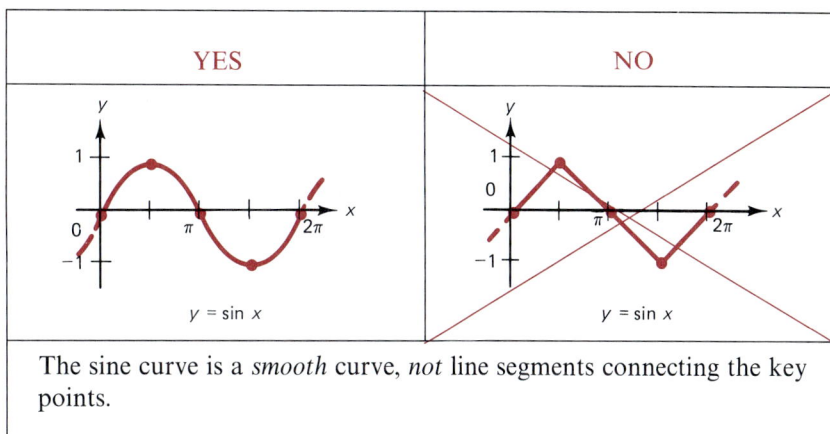
$y = \sin x$	$y = \sin x$

The sine curve is a *smooth* curve, *not* line segments connecting the key points.

Graph the following functions on the interval $-3 \le x \le 3$.

Warm-up Exercises

1. $y = 2x + 5$ **2.** $y = 5 - x$

3. $y = x^2 - 3$ **4.** $y = 5 - 2x^2$

Complete the following tables, using a calculator or trigonometric tables.

Warm-up Exercises

5.

x	$0°$	$15°$	$30°$	$45°$	$60°$	$75°$	$90°$
$\sin x$							

6.

x	$0°$	$15°$	$30°$	$45°$	$60°$	$75°$	$90°$
$\cos x$							

7. Graph the $(x, \sin x)$ pairs in Problem 5.

8. Graph the $(x, \cos x)$ pairs in Problem 6

Graph the following functions.

9. $y = \dfrac{1}{3} \sin x$ **10.** $y = 3 \sin x$

11. $y = \sin 2x$

12. $y = \sin \dfrac{1}{2} x$

13. $y = 4 \sin 3x$

14. $y = -3 \sin 2x$

15. $y = 3 \cos x$

16. $y = 2 \cos x$

17. $y = \cos 2x$

18. $y = \cos 3x$

19. $y = -\cos x$

20. $y = 2 \cos 3x$

21. $y = \dfrac{1}{3} \cos 3x$

22. $y = \dfrac{1}{2} \cos 2x$

23. $y = -3 \cos x$

24. $y = \dfrac{1}{4} \cos \dfrac{1}{4} x$

25. $y = \dfrac{1}{4} \sin x$

26. $y = \sin 4x$

27. $y = -\sin x$

28. $y = \sin \dfrac{1}{3} x$

29. $y = 2 \sin 3x$

30. $y = 3 \sin 2x$

Electrical Application

31. The voltage V in an alternating-current (ac) circuit is a function of time t, as given by

$$V = 110 \sin 120\pi t$$

(a) What is the period?

(b) How many cycles occur in 1 second?

(c) What is the amplitude?

(d) Graph this function for one cycle. (Be sure to scale the axes appropriately.)

Sound Applications

32. A musical tone may be distorted by a sound system that cannot reproduce its full amplitude. What happens is that the sound gets "clipped." Graph the following function that shows a 10-watt sound trying to come out of a 5-watt system. (*Hint:* Graph the entire sine first.)

$$W = \begin{cases} 10 \sin t & \text{if } 0 \le t \le \pi/6 \\ 5 & \text{if } \pi/6 \le t < 5\pi/6 \\ 10 \sin t & \text{if } 5\pi/6 \le t < 7\pi/6 \\ -5 & \text{if } 7\pi/6 \le t < 11\pi/6 \\ 10 \sin t & \text{if } 11\pi/6 \le t < 2\pi \end{cases}$$

33. At one particular instant, the displacement y of a guitar string of length L depends on the distance x from the neck:

$$y = A_n \sin \frac{n\pi x}{L}$$

Graph the first three harmonics ($n = 1, 2, 3$) for $L = 70$ centimeters and $A_1 = 0,4$, $A_2 = 0.4$, and $A_3 = 0.2$ centimeters.

9.4

GRAPHS OF SINE AND COSINE, II

Recall from page 190 that $y = f(x - c)$ has the same graph as $y = f(x)$, except shifted c units to the right (if c is positive). Graphs of trigonometric functions can also be shifted to the right or left.

The graph of $y = A \sin B(x - c)$ [or $y = A \cos B(x - c)$]:

1. Has the same graph as $y = A \sin Bx$ [or $y = A \cos Bx$], except:
2a. Shifted $|c|$ units to the *left* if c is *negative*.
2b. Shifted $|c|$ units to the *right* if c is *positive*.

The number $|c|$ is called the **phase shift**. It measures how much the sine curve is advanced or delayed.

EXAMPLE 14 Graph $y = 3 \sin 2\left(x - \dfrac{\pi}{4}\right)$.

Solution This graph has the same shape as the graph of $y = 3 \sin 2x$, except shifted $\pi/4$ units to the *right*.

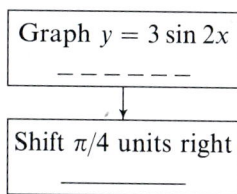

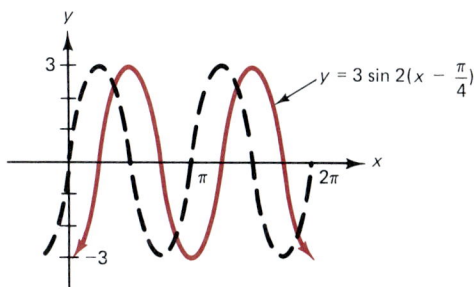

Graph $y = 3 \sin 2x$

- - - - -

↓

Shift $\pi/4$ units right

_____

EXAMPLE 15 Graph $y = 4 \cos\left(\pi x + \dfrac{\pi}{2}\right)$.

Solution We first put this into a standard form by rewriting it

$$y = 4 \cos\left(\pi x + \frac{\pi}{2}\right) = 4 \cos \pi\left(x + \frac{1}{2}\right)$$

We now graph $y = 4 \cos \pi x$, which has amplitude 4 and period $2\pi/\pi = 2$. Finally, we shift $1/2$ unit to the *left*.

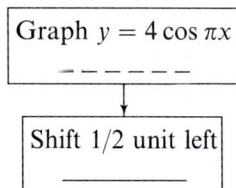

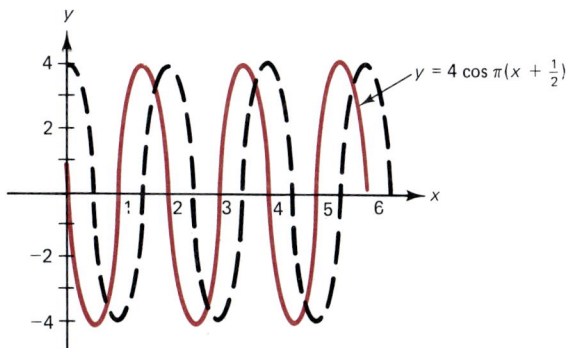

Graph $y = 4 \cos \pi x$

- - - - - -

↓

Shift $1/2$ unit left

_____

YES	NO
$y = 4 \cos\left(\pi x + \dfrac{\pi}{2}\right)$ $= 4 \cos \pi\left(x + \dfrac{1}{2}\right)$ has a shift of 1/2 left.	$y = 4 \cos\left(\pi x + \dfrac{\pi}{2}\right)$ has a shift of $\dfrac{\pi}{2}$ left
The coefficient of x within the parentheses must be 1. If necessary, we factor.	

If we have two functions f and g, we can show their sum graphically by adding the y-coordinates $f(x) + g(x)$ along various vertical lines (see page 175). This is important to us since many complicated-appearing trigonometric functions are in fact sums of simpler sine and cosine functions.

EXAMPLE 16　Graph $y = \dfrac{x}{2} + \cos x$.

Solution　We first graph $y = x/2$ and $y = \cos x$ separately; we then add the y-coordinates along several vertical lines (representing different x-values).

Graph $y = \dfrac{x}{2}$

– – – – – –

Graph $y = \cos x$

· · · · · · · · · · · ·

Add y-values on several vertical lines

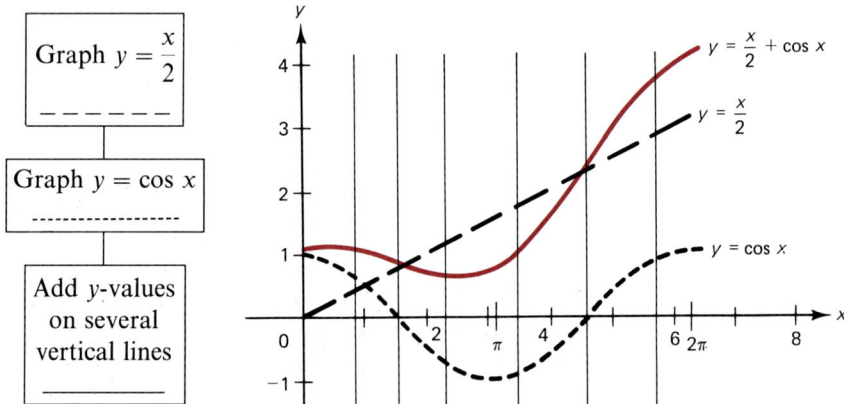

$y = \dfrac{x}{2} + \cos x$

$y = \dfrac{x}{2}$

$y = \cos x$

We see that the sum $y = \dfrac{x}{2} + \cos x$ has properties of both curves: It goes up as $y = x/2$ does, but it also "wiggles" up and down as $y = \cos x$ does.

EXAMPLE 17 Graph $y = \sin^2 x$.

Solution This is a **composite function**; that is, it is one function (sine) followed by another (squaring). To plot the points, we first find the sin x values; then we square the values of sin x. Figure 18 shows the graph and the x-y table. Note that the values of $\sin^2 x$ are all positive (or zero); hence the curve is always above (or on) the x-axis. As a contrast, Figure 19 shows the graph of $y = \sin x^2$ (here, we square first and then find the sine). (Note their differences.)

FIGURE 18 $y = \sin^2 x$.

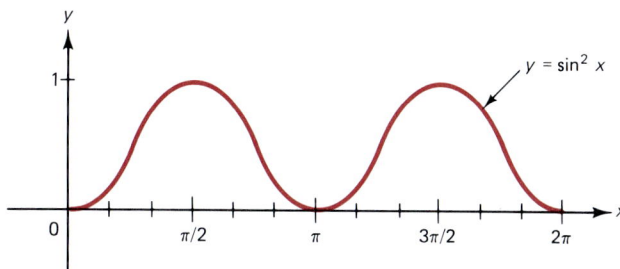

x	0	$\frac{\pi}{6}$	$\frac{\pi}{3}$	$\frac{\pi}{2}$	$\frac{2\pi}{3}$	$\frac{5\pi}{6}$	π	$\frac{7\pi}{6}$	$\frac{4\pi}{3}$	$\frac{3\pi}{2}$	$\frac{5\pi}{3}$	$\frac{11\pi}{6}$	2π
$\sin x$	0	0.5	0.87	1	0.87	0.5	0	-0.5	-0.87	-1	-0.87	-0.5	0
$\sin^2 x$	0	0.25	0.75	1	0.75	0.25	0	0.25	0.75	1	0.75	0.25	0

FIGURE 19 $y = \sin x^2$.

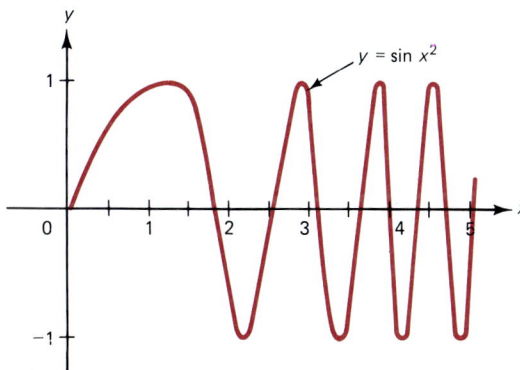

Complete the following table.

		Quadrant for θ	$\sin \theta$	$\cos \theta$	$\tan \theta$	$\cot \theta$	$\sec \theta$	$\csc \theta$
Warm-up	**1.**	I	0.8	?	?	?	?	?
Exercises	**2.**	I	?	0.2	?	?	?	?
	3.	II	0.9	?	?	?	?	?
	4.	III	-0.3	?	?	?	?	?
	5.	III	?	-0.5	?	?	?	?
	6.	IV	?	0.6	?	?	?	?

Graph the following functions.

Warm-up
Exercises

7. $y = (x - 1)^2$

8. $y = (1 + x)^3$

9. $y = 3 \sin 2x$

10. $y = 2 \cos 3x$

Graph each of the following functions for $0 \le x \le 2\pi$.

11. $y = 2 \sin \left(x + \dfrac{\pi}{2} \right)$

12. $y = \cos 2\left(x + \dfrac{\pi}{2} \right)$

13. $y = \cos \left(x - \dfrac{\pi}{2} \right)$

14. $y = \sin \left(x - \dfrac{\pi}{2} \right)$

15. $y = \sin \left(x + \dfrac{\pi}{4} \right)$

16. $y = \cos \left(x + \dfrac{\pi}{4} \right)$

17. $y = 2 \sin 3\left(x - \dfrac{\pi}{2} \right)$

18. $y = 3 \cos 2\left(x - \dfrac{\pi}{4} \right)$

19. $y = 3 \cos (2x + \pi)$

20. $y = 2 \sin (3x + \pi)$

21. $y = \dfrac{x}{2} + 2 \cos x$

22. $y = \dfrac{x}{3} + \sin 3x$

23. $y = \sin x + \cos x$

24. $y = \cos x - \sin x$

25. $y = \cos^3 x$

26. $y = \cos^4 x$

27. $y = \dfrac{1}{\sin x}$

28. $y = \dfrac{1}{\cos x}$

29. $y = x \sin x$

30. $y = x^2 \cos x$

31. $y = \sin^3 x$

32. $y = \sin^4 x$

Electrical
Applications

33. The current i through an alternating-current (ac) circuit is given by

$$i = 10 \sin \left(120\pi t + \frac{\pi}{3} \right)$$

for time t. Graph this function for one cycle.

34. The *sawtooth wave* (shown in Figure 20 and often seen in a cathode ray oscilloscope) is actually made up of sine waves of various frequencies and amplitudes. Graph the following approximation to the sawtooth:

$$y = \sin x + \frac{1}{3} \sin 3x + \frac{1}{5} \sin 5x$$

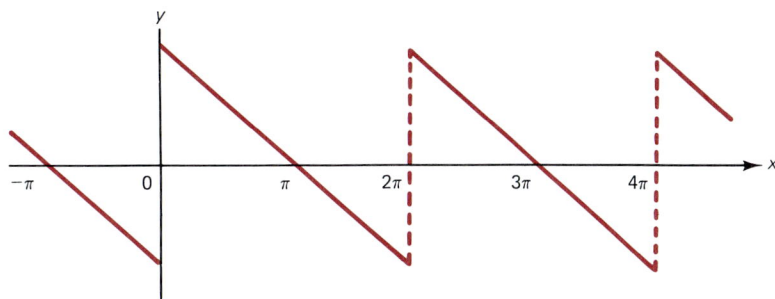

FIGURE 20

(You can surely see the pattern for the rest of the terms; the more terms, the better the approximation.)

Musical Application

35. An interesting musical tone is generally a sum of various related pure tones at different amplitudes (called *harmonics*). Graph the following very simplified versions of two such musical instruments.

$$Violin: \quad y = \sin x + 0.98 \sin 2x + 0.45 \sin 3x$$

$$Piano: \quad y = \sin x + 0.22 \sin 2x + 0.28 \sin 3x$$

Meteorological Application

36. In any city, the seasonal temperature changes tend to have a sinusoidal character to them. Consider the data below for the average temperatures (°F) in various cities throughout the year.
(a) Graph the data for each city (temperature on the *y*-axis, month on the *x*-axis). Make a two-year plot: This will better display the sine nature.
(b) For each city, find the amplitude as follows:

$$A = \frac{\text{high} - \text{low}}{2}$$

(c) Use the amplitude and period (12 months) to express the temperature as a function of the month (January = 1, February = 2, and so on).

Month	Chicago, Illinois	Washington, D.C.	Edinburgh, Scotland	Fairbanks, Alaska	Los Angeles, California
1	26	37	39	−11	54
2	28	38	39	−3	55
3	36	45	41	9	57
4	49	56	45	29	59
5	60	66	49	47	62
6	70	74	55	58	65
7	76	78	58	60	69
8	74	76	58	54	69
9	66	70	54	44	68
10	55	59	49	26	65
11	40	48	43	4	61
12	29	38	40	−8	57

In this section we look at the graphs of the other trigonometric functions: tangent, cotangent, secant, and cosecant. To graph these, we rewrite each function in terms of sine and cosine:

$$\tan x = \frac{\sin x}{\cos x} \qquad \cot x = \frac{\cos x}{\sin x}$$

$$\sec x = \frac{1}{\cos x} \qquad \csc x = \frac{1}{\sin x}$$

Since we are familiar with the sine and cosine functions, we use them to graph the others. However, we must be careful: Each of these four other functions has a denominator ($\sin x$ or $\cos x$) that might be zero, and we *cannot* have a zero denominator. (See the domain restrictions in Table 2, page 316.)

Recall (from studying rational functions) that an x-value that produces a zero denominator generally gives us a **vertical asymptote**, which is a vertical line that the curve approaches but never crosses.

EXAMPLE 18 Graph $y = \tan x$.

Solution We rewrite this as $y = \tan x = \dfrac{\sin x}{\cos x}$. The denominator is zero when $\cos x = 0$. This occurs at $x = \pm \pi/2,\ \pm 3\pi/2,\ \pm 5\pi/2$, and so on. Thus, the vertical asymptotes are at $x = \pm \pi/2,\ \pm 3\pi/2,\ \pm 5\pi/2$, and so on, and these values are excluded from the domain. Therefore, we investigate the function between these asymptotes, say $-\pi/2$ to $\pi/2$. In making our x-y table, we use the tangent tables or the $\boxed{\tan}$ key.

Figure 21 shows this graph. We first notice how the curve shoots up to approach $x = \pi/2$ and down to approach $x = -\pi/2$. If we plot more points, we see that this pattern repeats itself exactly for $\pi/2 < x < 3\pi/2$, for $3\pi/2 < x < 5\pi/2$, and so on.

Unlike the sine or cosine, note that the period of the tangent is π. Thus,

$$\tan x = \tan (x + \pi)$$

YES	NO
The period of the tangent is π.	~~The period of the tangent is 2π.~~

x	$\tan x$
$-\dfrac{\pi}{2}$	Undefined
$-\dfrac{\pi}{3}$	-1.73
$-\dfrac{\pi}{4}$	-1.00
$-\dfrac{\pi}{6}$	-0.57
0	0
$\dfrac{\pi}{6}$	0.57
$\dfrac{\pi}{4}$	1.00
$\dfrac{\pi}{3}$	1.73
$\dfrac{\pi}{2}$	Undefined

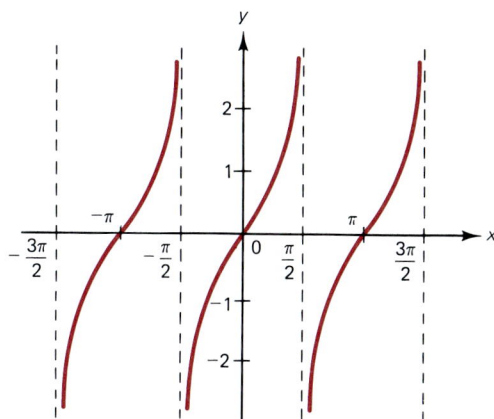

FIGURE 21

EXAMPLE 19 Graph $y = \cot x$.

Solution We rewrite this as $y = \cot x = \dfrac{\cos x}{\sin x}$. Since the denominator is $\sin x$, the asymptotes occur when $\sin x = 0$, or $x = 0,\ \pm\pi,\ \pm 2\pi,\ \pm 3\pi$, and so on. Thus, we investigate the graph $y = \cot x$ between these asymptotes, say 0 to π. In making our x-y table, we can use the trigonometric tables, or the $\boxed{\tan}$ and $\boxed{1/x}$ keys on a calculator.

Figure 22 shows the graph and the table. Note the similarities (and differences) with the graph of $y = \tan x$. The period is π for both,

FIGURE 22

x	$\cot x$
0	Undefined
$\dfrac{\pi}{6}$	1.73
$\dfrac{\pi}{4}$	1.00
$\dfrac{\pi}{3}$	0.58
$\dfrac{\pi}{2}$	0
$\dfrac{2\pi}{3}$	-0.58
$\dfrac{3\pi}{4}$	-1.00
$\dfrac{5\pi}{6}$	-1.73
π	Undefined

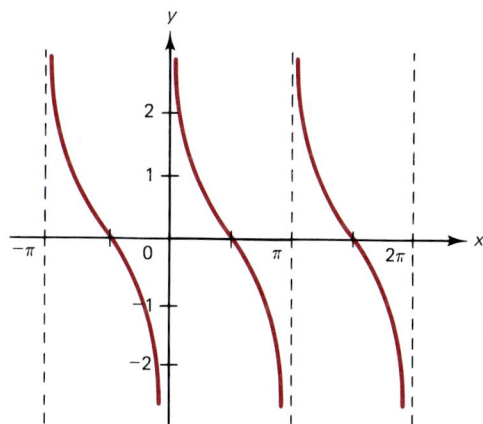

but the asymptotes are at different *x*-values. The graph of $y = \cot x$ is always decreasing, while the graph of $y = \tan x$ is always increasing (from left to right).

As with sines and cosines, the graphs of $y = A \tan Bx$ and $y = A \cot Bx$ are similar to $y = \tan x$ and $y = \cot x$. The following table summarizes this connection.

	The Graph of:	
	$y = A \tan Bx$	$y = A \cot Bx$
Has the same shape as	$y = \tan x$	$y = \cot x$
With period	$\dfrac{\pi}{B}$	$\dfrac{\pi}{B}$
Vertical asymptotes	$x = \pm \dfrac{\pi}{2B}, \pm \dfrac{3\pi}{2B}, \dots$	$x = 0, \pm \dfrac{\pi}{B}, \pm \dfrac{2\pi}{B}, \dots$
Zeros	$x = 0, \pm \dfrac{\pi}{B}, \pm \dfrac{2\pi}{B}, \dots$	$x = \pm \dfrac{\pi}{2B}, \pm \dfrac{3\pi}{2B}, \dots$

EXAMPLE 20 Graph $y = \sec x$.

Solution We rewrite this as $y = \sec x = \dfrac{1}{\cos x}$. Therefore, it has vertical asymptotes where $\cos x = 0$: $x = \pm \pi/2, \pm 3\pi/2, \pm 5\pi/2$, and so on. Rather than make an *x-y* table, we first sketch the graph of $y = \cos x$ with a dotted curve and then locate a few reciprocal values. For instance, if $\cos x = 1/2$, then $\sec x = 2$.

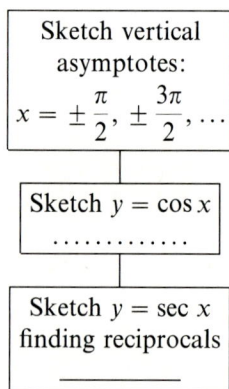

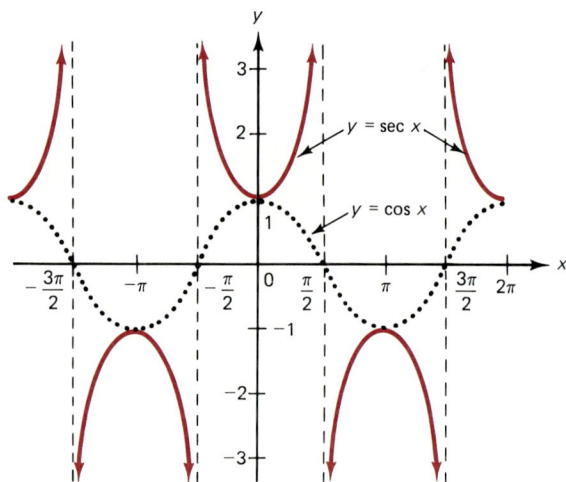

Sketch vertical asymptotes:

$$x = \pm \frac{\pi}{2}, \pm \frac{3\pi}{2}, \dots$$

Sketch $y = \cos x$

. . . .

Sketch $y = \sec x$ finding reciprocals

The graph of $y = \csc x = \dfrac{1}{\sin x}$ is very much like that of $y = \sec x$. We have $|\csc x| \geq 1$ and the period is 2π.

	The Graph of:	
	$y = A \sec Bx$	$y = A \csc Bx$
Has the same shape as	$y = \sec x$	$y = \csc x$
With period	$\dfrac{2\pi}{B}$	$\dfrac{2\pi}{B}$
Asymptotes at	$\pm \dfrac{\pi}{2B}, \pm \dfrac{3\pi}{2B}, \cdots$	$0, \pm \dfrac{\pi}{B}, \pm \dfrac{2\pi}{B}, \cdots$
With $y = A$ at	$0, \pm \dfrac{2\pi}{B}, \pm \dfrac{4\pi}{B}, \cdots$	$\dfrac{\pi}{2B}, -\dfrac{3\pi}{2B}, \dfrac{5\pi}{2B}, \cdots$
With $y = -A$ at	$\pm \dfrac{\pi}{B}, \pm \dfrac{3\pi}{B}, \cdots$	$-\dfrac{\pi}{2B}, \dfrac{3\pi}{2B}, \cdots$

PROBLEM SET
9.5

Each of the following real numbers t is wound around a unit circle, as shown at the left. Find the coordinates of the terminal point (x, y).

1. $t = 2$ **2.** $t = 11.56$ **3.** $t = \pi$

4. $t = \dfrac{3\pi}{2}$ **5.** $t = -4$ **6.** $t = -2.17$

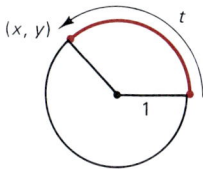

Graph each of the following functions for $0 \le x \le 2\pi$.

Warm-up Exercises

7. $y = 3 \cos 2x$ **8.** $y = 2 \sin 3x$

9. $y = 4 \sin\left(x - \dfrac{\pi}{4}\right)$ **10.** $y = -2 \cos 3\left(x + \dfrac{\pi}{4}\right)$

Graph each of the following functions for $0 \le x \le 2\pi$.

11. $y = \cot 2x$ **12.** $y = \tan 2x$

13. $y = \sec 2x$ **14.** $y = \csc 2x$

15. $y = \tan x$ **16.** $y = \cot x$

17. $y = \tan \dfrac{1}{2} x$ **18.** $y = \cot \dfrac{1}{2} x$

19. $y = 2 \cot 3x$ **20.** $y = 2 \tan 3x$

21. $y = \csc x$ **22.** $y = \sec x$

23. $y = 3 \csc \dfrac{1}{2} x$ **24.** $y = \sec \dfrac{1}{2} x$

25. $y = 2 \sec 3x$ **26.** $y = 2 \csc 3x$

27. $y = -\tan x$ **28.** $y = -\sec x$

29. The relation between the focal length f of a 35-mm camera lens and its angle of view θ is given by

$$f = 21 \cot \frac{\theta}{2}$$

Graph this function for $0 < \theta < \pi$.

Sports
Application

30. Using the angle θ of the maximum lean that a player can have, the *gripping power* of basketball shoes is given by

$$g = \tan \theta$$

Graph this function for $0 \leq \theta < \pi/2$.

Engineering
Application

31. The tension T in a cable holding a 10-kilogram object depends on the angle θ (made with the horizon):

$$T = 10 \csc \theta$$

Graph this function for $0 < \theta < \pi/2$.

Mapmaking
Application

32. In the Mercator projection of the earth (the basic schoolroom world map), the distortion (or relative exaggeration) S depends on the latitude ϕ, as follows:

$$S = \sec^2 \phi$$

Graph this function for $-90° < \phi < 90°$. [This explains why the very northern-most locations (Greenland, upper Canada, and so on) appear so unrealistically large.]

CHAPTER 9
SUMMARY

Important Properties and Definitions

For an angle θ measured in radians,

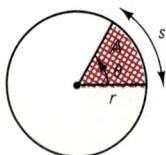

$$\theta = \frac{s}{r} \qquad s = \theta r$$

$$180° = \pi \text{ radians} \quad \text{or} \quad \frac{D \text{ degrees}}{180°} = \frac{R \text{ radians}}{\pi \text{ radians}}$$

Any real number t can be viewed as an arc length starting at $(1, 0)$ and wound about a unit circle, terminating at (x, y). Then we have

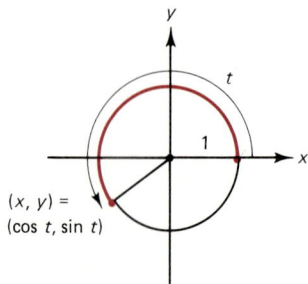

$$\sin t = y \qquad \csc t = \frac{1}{y}$$

$$\cos t = x \qquad \sec t = \frac{1}{x}$$

$$\tan t = \frac{y}{x} \qquad \cot t = \frac{x}{y}$$

$(x, y) =$
$(\cos t, \sin t)$

One cycle of the graphs of each of the trigonometric functions is shown below. (Dashed lines represent vertical asymptotes whose x-values are excluded from the function's domain.)

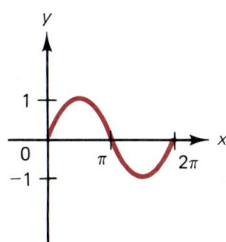

y = sin x

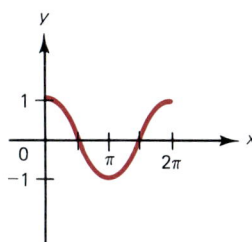

y = cos x

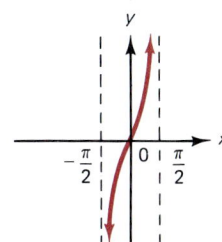

y = tan x

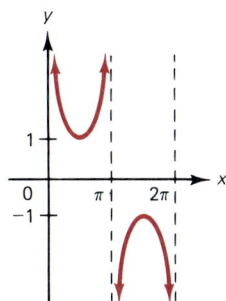

y = csc x

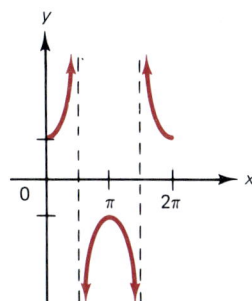

y = sec x

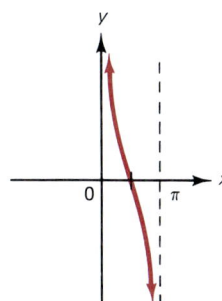

y = cot x

The graphs of $\begin{cases} y = A \sin B(x - c) \\ y = A \cos B(x - c) \\ y = A \sec B(x - c) \\ y = A \csc B(x - c) \end{cases}$ have period $\dfrac{2\pi}{B}$

$\begin{cases} y = A \tan B(x - c) \\ y = A \cot B(x - c) \end{cases}$ have period $\dfrac{\pi}{B}$

All graphs are shifted $|c|$ units $\begin{cases} \text{left if } c < 0 \\ \text{right if } c > 0 \end{cases}$

Review Exercises

Complete the following table for a circle radius r and a central angle θ.

	θ (radians)	θ (degrees)	r	Arc length	sin θ	cos θ	tan θ
1.	$\pi/3$	?	10	?	?	?	?
2.	$3\pi/4$	?	5	?	?	?	?
3.	?	45°	100	?	?	?	?
4.	?	180°	20	?	?	?	?
5.	?	?	10	5π	?	?	?
6.	?	?	50	?	1	0	?

Complete the following table for an angle θ in radians.

	(radians)	Equivalent between 0 and 2π	Quadrant	Reference Angle	$\sin \theta$	$\cos \theta$	$\tan \theta$
7.	0.15	?	?	?	?	?	?
8.	11.30	?	?	?	?	?	?
9.	−0.72	?	?	?	?	?	?
10.	$\dfrac{3\pi}{4}$	?	?	?	?	?	?
11.	?	?	I	?	0.56	?	?
12.	?	?	II	?	?	−0.7	?

Graph the following functions for $0 \le x \le 2\pi$.

13. $y = 2 \sin 3x$

14. $y = -\dfrac{1}{2} \cos 2x$

15. $y = \cos\left(x + \dfrac{\pi}{4}\right)$

16. $y = 3 \sin 2\left(x - \dfrac{\pi}{6}\right)$

17. $y = 2 \sin^2 x$

18. $y = \dfrac{x}{3} + \cos x$

19. $y = \cot x$

20. $y = 3 \tan 2x$

21. $y = \sec x$

22. $y = \csc 2x$

Trigonometric Identities, Equations, and Inverses

Consider the following equations:

(1) $$x + 2 = 10$$

(2) $$(x - 2)(x + 2) = x^2 - 4$$

Equation (1) is true only if $x = 8$. This is called a **conditional** (or **open**) **equation**, since its truth depends on x. Equation (2) is *always* true. This is called an **identity**. In this chapter we study **trigonometric identities**, which are equations and relations that involve the trigonometric functions and are true for *all* angles or real numbers in the domain of the function. (Review Section 8.4—page 279.) We also study **trigonometric equations**, which are true for only some x.

**10.1
VERIFYING
TRIGONOMETRIC
IDENTITIES**

A **trigonometric expression** is an algebraic expression that involves trigonometric functions; for instance,

$$\frac{1}{1 - \sin x} + \frac{1}{1 + \sin x} \qquad \text{and} \qquad \tan x + \cot x$$

are trigonometric expressions. We frequently encounter complicated trigonometric expressions, which we simplify using the standard methods of algebra: factoring, finding an LCD, adding, subtracting, multiplying, dividing, rationalizing fractions, and so on.

EXAMPLE 1 Simplify the expression $T = \dfrac{1}{1 - \sin x} + \dfrac{1}{1 + \sin x}$.

Solution   We treat this expression as we would treat the sum of the algebraic fractions

$$\frac{1}{1 - z} + \frac{1}{1 + z}$$

We find the lowest common denominator (LCD), $(1 - \sin x)(1 + \sin x) = 1 - \sin^2 x$. We then use the other identities to simplify the expression further.

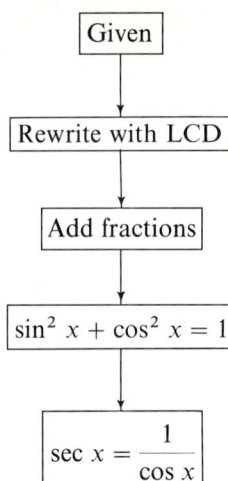

Given	$T = \dfrac{1}{1 - \sin x} + \dfrac{1}{1 + \sin x}$
Rewrite with LCD	$= \dfrac{1 + \sin x}{(1 - \sin x)(1 + \sin x)} + \dfrac{1 - \sin x}{(1 - \sin x)(1 + \sin x)}$
Add fractions	$= \dfrac{2}{1 - \sin^2 x}$
$\sin^2 x + \cos^2 x = 1$	$= \dfrac{2}{\cos^2 x}$
$\sec x = \dfrac{1}{\cos x}$	$= 2 \sec^2 x$

Recall that an **identity** is an equation that is always true (that is, for *all* x in the domain of the function). For example, **negative-angle identities**

$$\cos(-x) = \cos x \qquad \text{and} \qquad \sin(-x) = -\sin x$$

are identities (since they are true for all x-values, not just some x-values). Identities are important to us since they allow us to replace one expression by another expression that might be more convenient.

Our goal in this section is to **verify** (establish the truth of) trigonometric identities. Although no two identities are verified in exactly the same way, we can give a set of general procedures to follow when verifying an identity.

When verifying a trigonometric identity:

1. Know (thoroughly) the basic trigonometric identities (reciprocal, quotient, Pythagorean, and negative-angle). (See page 279.)
2. Choose the easier side of the equation as a target.
3. Try transforming the more complicated side into the simpler side.
4. Try rewriting all functions in terms of sines and cosines.
5. Use the standard methods of algebra for simplifying expressions.
6. Try transforming *both* sides of the identity into an equal third expression.

7. If the identity involves equal fractions, try cross multiplying $\left(\text{since } \dfrac{A}{B} = \dfrac{C}{D} \text{ if and only if } AD = BC \right).$

EXAMPLE 2   Verify that $\csc x = \dfrac{\cot x}{\cos x}$.

Solution The left side is the simpler, so it becomes our target. Thus, we work on the right side of the equation and try to transform it into the left side.

$$\boxed{\cot x = \dfrac{\cos x}{\sin x}}$$

$$\dfrac{\cot x}{\cos x} = \dfrac{\dfrac{\cos x}{\sin x}}{\cos x}$$

$$\boxed{\text{Multiply by } \dfrac{\sin x}{\sin x}}$$

$$= \dfrac{\dfrac{\cos x}{\sin x} \cdot \sin x}{\cos x \sin x}$$

$$\boxed{\text{Simplify}}$$

$$= \dfrac{\cos x}{\cos x \sin x}$$

$$\boxed{\text{Simplify}}$$

$$= \dfrac{1}{\sin x}$$

$$\boxed{\csc x = \dfrac{1}{\sin x}}$$

$$= \csc x$$

We have now transformed the right side into the left side; thus, the identity is verified.

EXAMPLE 3 Verify that $\csc^2 x + \sec^2 x = \dfrac{\sec^2 x}{\sin^2 x}$.

Solution We choose the right side as our target or goal. We now take the left side and try to transform it into the right side.

$$\boxed{\text{Reciprocal relations}}$$

$$\csc^2 x + \sec^2 x = \dfrac{1}{\sin^2 x} + \dfrac{1}{\cos^2 x}$$

$$\boxed{\text{Write with LCD}}$$

$$= \dfrac{\cos^2 x}{\sin^2 x \cos^2 x} + \dfrac{\sin^2 x}{\sin^2 x \cos^2 x}$$

$$\boxed{\text{Add fractions}}$$

$$= \dfrac{\cos^2 x + \sin^2 x}{\sin^2 x \cos^2 x}$$

$$\boxed{\sin^2 x + \cos^2 x = 1}$$

$$= \dfrac{1}{\sin^2 x \cos^2 x}$$

$$\boxed{\sec^2 x = \dfrac{1}{\cos^2 x}}$$

$$= \dfrac{\sec^2 x}{\sin^2 x}$$

Thus, the identity is verified, since the left side has been transformed into the right side. Note that we use the basic identities over and over.

EXAMPLE 4 Verify $\dfrac{\sec x + 1}{\tan x} = \dfrac{\tan x}{\sec x - 1}$.

Solution Here both sides appear equally complicated. Therefore, we work on both sides and try to reduce them to the same expression. We start with the left side and rewrite it in terms of sines and cosines.

$$\boxed{\begin{array}{c}\text{Rewrite with}\\\text{sines and cosines}\end{array}} \qquad \frac{\sec x + 1}{\tan x} = \frac{\dfrac{1}{\cos x} + 1}{\dfrac{\sin x}{\cos x}}$$

$$\boxed{\text{Multiply by } \frac{\cos x}{\cos x}} \qquad = \frac{\left(\dfrac{1}{\cos x} + 1\right)\cos x}{\dfrac{\sin x}{\cos x}\,\cos x}$$

$$\boxed{\text{Simplify}} \qquad = \frac{1 + \cos x}{\sin x}$$

Now we work on the right side. Watch the fourth step very carefully: It is a tricky step, in which we *rationalize* the denominator.

$$\boxed{\begin{array}{c}\text{Rewrite with}\\\text{sines and cosines}\end{array}} \qquad \frac{\tan x}{\sec x - 1} = \frac{\dfrac{\sin x}{\cos x}}{\dfrac{1}{\cos x} - 1}$$

$$\boxed{\text{Multiply by } \frac{\cos x}{\cos x}} \qquad = \frac{\dfrac{\sin x}{\cos x}\,\cos x}{\left(\dfrac{1}{\cos x} - 1\right)\cos x}$$

$$\boxed{\text{Simplify}} \qquad = \frac{\sin x}{1 - \cos x}$$

$$\boxed{\text{Multiply by } \frac{1 + \cos x}{1 + \cos x}} \qquad = \frac{\sin x}{1 - \cos x} \cdot \frac{1 + \cos x}{1 + \cos x}$$

$$\boxed{\text{Multiply}} \qquad = \frac{\sin x(1 + \cos x)}{1 - \cos^2 x}$$

$$\boxed{\sin^2 x = 1 - \cos^2 x} \qquad = \frac{\sin x(1 + \cos x)}{\sin^2 x}$$

$$\boxed{\text{Simplify}} \qquad = \frac{1 + \cos x}{\sin x}$$

Since the two sides of the original equation are equal to the same expression, they are equal to each other. Thus, the identity is verified.

Obvious question: Why did we rationalize in step 4 above? Since we had $1 - \cos x$ in the denominator, multiplying by $1 + \cos x$ produces $1 - \cos^2 x$. This, of course, equals $\sin^2 x$ and helps us simplify the fraction.

EXAMPLE 5 Verify $\dfrac{\sec x + 1}{\tan x} = \dfrac{\tan x}{\sec x - 1}$.

Solution This is a rerun of Example 4. This time, however, we cross multiply in an attempt to show the fractions equal. $\left(\text{Recall that } \dfrac{A}{B} = \dfrac{C}{D} \text{ if and only if } AD = BC.\right)$

Desired identity	$\dfrac{\sec x + 1}{\tan x} = \dfrac{\tan x}{\sec x - 1}$
True if and only if ... (cross multiply)	$(\sec x + 1)(\sec x - 1) = (\tan x)(\tan x)$
True if and only if ...	$\sec^2 x - 1 = \tan^2 x$

which is true (this is one of the Pythagorean identities—page 287).

Note that this verification is much faster than in Example 4. Often cross multiplying two fractions provides a very rapid proof of an identity.

PROBLEM SET
10.1

Warm-up Exercises

Simplify the following expressions as much as possible.

1. $(x + 3)(x - 3)$

2. $(x - \sqrt{5})(x + \sqrt{5})$

3. $\dfrac{2}{x - \sqrt{3}} \cdot \dfrac{x + \sqrt{3}}{x + \sqrt{3}}$

4. $\dfrac{2}{x - 1} + \dfrac{3}{x + 1}$

Write the following basic trigonometric identities.

Warm-up Exercises

5. Write the quotient identities.

6. Write the reciprocal identities.

Simplify the following as much as possible.

7. $\dfrac{4}{\sin x} + \dfrac{7}{\cos x}$

8. $(\sin x + \cos x)^2$

9. $\dfrac{2}{1 + \sin x} \cdot \dfrac{1 - \sin x}{1 - \sin x}$

10. $\dfrac{\dfrac{1}{\sin x}}{\dfrac{\cos x}{\sin x}}$

11. $\dfrac{\csc x}{\tan x + \cot x}$

12. $\dfrac{\tan x + \cot x}{\sec x}$

Verify the following identities.

13. $\sin x = \dfrac{\tan x}{\sec x}$

14. $\cos x = \dfrac{\cot x}{\csc x}$

15. $\tan x = \dfrac{\sec x}{\csc x}$

16. $\cot x = \dfrac{\csc x}{\sec x}$

17. $\sec x = \dfrac{\csc x}{\cot x}$

18. $\csc x = \dfrac{\sec x}{\tan x}$

19. $\cot x = \cos x \csc x$

20. $\sec x = \tan x \csc x$

21. $\sin x + \cos x \cot x = \csc x$

22. $\cos x + \sin x \tan x = \sec x$

23. $\csc^2 x(1 - \sin^2 x) = \cot^2 x$

24. $\sec^2 x(1 - \cos^2 x) = \tan^2 x$

25. $\dfrac{1 + \sin x}{\sin x} = 1 + \csc x$

26. $\sec x - 1 = \dfrac{1 - \cos x}{\cos x}$

27. $\dfrac{1}{\sec^2 x} + \dfrac{1}{\csc^2 x} = 1$

28. $\dfrac{\csc^2 x}{\cot^2 x} - \dfrac{\sec^2 x}{\csc^2 x} = 1$

29. $\dfrac{\csc x}{\sin x} - \cos^2 x \csc^2 x = 1$

30. $\tan x \cot x - \dfrac{\sin x}{\csc x} = \dfrac{\cos x}{\sec x}$

31. $\cos^2 x - \sin^2 x = 1 - 2 \sin^2 x$

32. $\sin^2 x - \cos^2 x = 1 - 2 \cos^2 x$

33. $\tan x + \cot x = \sec x \csc x$

34. $\sec x \csc x = \dfrac{\sec x + \csc x}{\sin x + \cos x}$

35. $\sec x - \cos x = \sin x \tan x$

36. $\csc x - \sin x = \cos x \cot x$

37. $\dfrac{1 + \tan x}{1 - \tan x} = \dfrac{\cot x + 1}{\cot x - 1}$

38. $\dfrac{1 + \sin x}{1 - \sin x} = \dfrac{\csc x + 1}{\csc x - 1}$

39. $\dfrac{\sin x}{1 - \cos x} = \dfrac{1 + \cos x}{\sin x}$

40. $\dfrac{\cos x}{1 + \sin x} = \dfrac{1 - \sin x}{\cos x}$

41. $\dfrac{\sin^2 x}{1 - \sin x} + \dfrac{\sin^2 x}{1 + \sin x} = 2 \tan^2 x$

42. $\dfrac{\sin x}{1 - \cos x} + \dfrac{\sin x}{1 + \cos x} = 2 \csc x$

43. $\dfrac{1 + 2 \sin x \cos x}{\sin x + \cos x} = \sin x + \cos x$

44. $\sec x + \tan x = \dfrac{\cos x}{1 - \sin x}$

45. $\dfrac{1 - \cos x}{\sec x - 1} = \dfrac{1 + \cos x}{1 + \sec x}$

46. $\dfrac{\csc x + 1}{\sin x + 1} = \dfrac{1 - \csc x}{\sin x - 1}$

47. $(\sin x + \cos x)^2 + (\sin x - \cos x)^2 = 2$

48. $(\sec x + \tan x)^2 - (\sec x - \tan x)^2 = 4 \sec x \tan x$

49. $(\csc x + \cot x)^2 = \dfrac{1 + \cos x}{1 - \cos x}$

50. $\sec^2 x - \csc^2 x = \dfrac{\tan^2 x - 1}{\sin^2 x}$

51. $\tan^3 x + \tan^2 x + \tan x + 1 = \sec^2 x(1 + \tan x)$

52. $\cot^3 x + \cot^2 x + \cot x + 1 = \csc^2 x(1 + \cot x)$

53. $\dfrac{2 - \sec^2 x}{2 - \csc^2 x} = -\tan^2 x$

54. $\dfrac{\cos^2 x - \sin^2 x}{1 - 2 \sin x \cos x} = \dfrac{\csc x + \sec x}{\csc x - \sec x}$

55. $\dfrac{\tan x + \cot x}{\tan x - \cot x} = \dfrac{\sec^2 x}{\tan^2 x - 1}$

56. $\dfrac{\cos x}{1 - \tan x} + \dfrac{\sin x}{1 - \cot x} = \sin x + \cos x$

57. $\cot^2 u - \cos^2 u = \cot^2 u \cos^2 u$

58. $1 - \tan^4 x = 2 \sec^2 u - \sec^4 x$

59. $\dfrac{\sin^3 x + \cos^3 x}{\sin x + \cos x} = 1 - \sin x \cos x$

60. $\dfrac{\sin^3 x - \cos^3 x}{\sin x - \cos x} = 1 + \sin x \cos x$

61. $\dfrac{1}{\sec x(1 + \sin x)} = \sec x(1 - \sin x)$

62. $\dfrac{1}{\csc x - \cot x} = \csc x + \cot x$

The following may or may not be identities. Substitute a few x-values (say x = 0, 1, 2, … radians) to test the relations. If the relation is false for any one x-value, the relation is not an identity. If all the test values make the relation true, try to prove that it is an identity. (Remember: One x-value can prove that a relation is not an identity, but 1000 x-values cannot prove that it is!)

63. $\sin x + \cos x = 1$

64. $\dfrac{1}{\sin x - \cos x} = \sin x + \cos x$

65. $\dfrac{1}{\sec x - \tan x} = \sec x + \tan x$

66. $\sin x + \tan x = \sin x(1 + \sec x)$

67. $\dfrac{\cos x + \sin x}{\cos x - \sin x} = \dfrac{\cos x - \sin x}{\cos x + \sin x}$

68. $\dfrac{\sec x + \tan x}{\csc x + \cot x} = \dfrac{\cot x - \csc x}{\tan x - \sec x}$

69. $\dfrac{\cos x + \sin x}{\cos x - \sin x} = \dfrac{\csc x + 2 \cos x}{\csc x - 2 \sin x}$

70. $\dfrac{\tan x - \cot x}{\tan x + \cot x} = 1 - 2 \cos^2 x$

71. $\dfrac{\csc x - \sec x}{\csc x + \sec x} = \dfrac{\cos x + \sin x}{\cos x - \sin x}$

72. $\dfrac{1}{1 - \cos x} + \dfrac{1}{1 + \cos x} = 2 \csc^2 x$

73. $\dfrac{1}{1 - \sin x} + \dfrac{1}{1 + \sin x} = 2 \sec^2 x$

74. $\dfrac{1}{1 - \tan x} + \dfrac{1}{1 + \tan x} = 2 \cos^2 x$

75. $\sec x + \tan x = \sqrt{\dfrac{1 + \sin x}{1 - \sin x}} \quad$ for $0 < x < \dfrac{\pi}{2}$

76. $\csc x + \cot x = \sqrt{\dfrac{1 + \cos x}{1 - \cos x}} \quad$ for $0 < x < \pi$

77. $\sin x + \cos x = \sqrt{\dfrac{\sec x + 2 \sin x}{\sec x}} \quad$ for $0 < x < \dfrac{\pi}{2}$

78. $\sin x + \tan x = \sqrt{\dfrac{\cos x + 2 \cot x}{\cos x}} \quad$ for $0 < x < \dfrac{\pi}{2}$

10.2 SUM-AND-DIFFERENCE-OF-ANGLES IDENTITIES

By now, algebra and trigonometry students should realize that there is a big difference between

$$(2 + 3)^2 \qquad \text{and} \qquad 2^2 + 3^2$$

The number on the left is $5^2 = 25$, while the number on the right is $4 + 9 = 13$. It makes a big difference whether we square first or whether we add first. In general,

$$f(a + b) \neq f(a) + f(b)$$

for a function f and real numbers a and b. (There are a few exceptions.) The trigonometric functions, however, are no exception. For instance,

$$\sin (10° + 45°) = \sin 55° = 0.8192$$

$$\text{while} \quad \sin 10° + \sin 45° = 0.1736 + 0.7071 = 0.8807$$

Note that $\sin (10° + 45°) \neq \sin 10° + \sin 45°$.

It is our goal in this section to develop special formulas for $\cos (B - A)$, $\cos (B + A)$, $\sin (B - A)$, and so on.

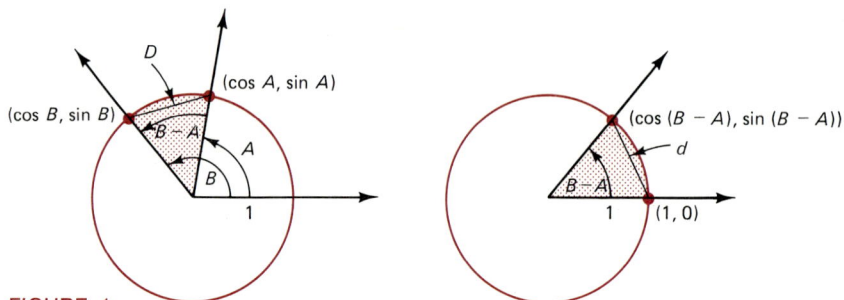

FIGURE 1

Figure 1 shows two unit circles: on the left, angles A and B; and on the right, the angle $B - A$ moved to standard position. On a unit circle, recall the coordinates for the point at angle θ in standard position are $(\cos \theta, \sin \theta)$.

We look at the chord lengths D (on the left) and d (on the right). Since the angles are both $B - A$ and the radii are both 1, we have $D = d$. We use the distance formula to compute D^2 and then d^2. Finally, we set $D^2 = d^2$.

| Compute D^2 using distance formula and $\sin^2 \theta + \cos^2 \theta = 1$ | $\begin{aligned} D^2 &= (\cos A - \cos B)^2 + (\sin A - \sin B)^2 \\ &= \cos^2 A - 2 \cos A \cos B + \cos^2 B \\ &\quad + \sin^2 A - 2 \sin A \sin B + \sin^2 B \\ &= 2 - 2 \cos A \cos B - 2 \sin A \sin B \end{aligned}$ |

$$\downarrow$$

| Distance formula and $\sin^2 \theta + \cos^2 \theta = 1$ | $\begin{aligned} d^2 &= [\cos (B - A) - 1]^2 + [\sin (B - A) - 0]^2 \\ &= \cos^2 (B - A) - 2 \cos (B - A) + 1 + \sin^2 (B - A) \\ &= 2 - 2 \cos (B - A) \end{aligned}$ |

$$\downarrow$$

| Set $d^2 = D^2$ | $d^2 = D^2$ |

$$\downarrow$$

| Subtract 2; divide by -2 | $\begin{aligned} 2 - 2 \cos (B - A) &= 2 - 2 \cos A \cos B - 2 \sin A \sin B \\ -2 \cos (B - A) &= -2 \cos A \cos B - 2 \sin A \sin B \\ \cos (B - A) &= \cos A \cos B + \sin A \sin B \end{aligned}$ |

This is our *cosine formula* for $\cos (B - A)$. We derive a similar formula for $\cos (B + A)$ by first writing $B + A = B - (-A)$ and then using the negative-angle identities.

Cosine formula	$\cos (B + A) = \cos [B - (-A)]$

$$\boxed{\begin{aligned}\cos (-A) &= \cos A \\ \sin (-A) &= -\sin A\end{aligned}}$$

$$= \cos (-A) \cos B + \sin (-A) \sin B$$

$$= \cos A \cos B - \sin A \sin B$$

Let us summarize the important identities that we have derived. For any angles (or real numbers) A and B, we have the following **cosine-of-differences (sums) identity**.

$$\boxed{\begin{aligned}\cos (B - A) &= \cos B \cos A + \sin B \sin A \\ \cos (B + A) &= \cos B \cos A - \sin B \sin A\end{aligned}}$$

Note carefully the sign reversal: For $\cos (B - A)$ we add the terms, and for $\cos (B + A)$ we subtract the terms.

YES	NO
$\cos (B + A) = \cos B \cos A - \sin B \sin A$	$\cos (B + A) = \cos B + \cos A$
$\cos (A + B)$ is a functional notation; we *cannot* use the distributive law.	

EXAMPLE 6   Find $\cos 15°$ and $\cos 75°$ *without* tables or calculator.

Solution   We use the known sine and cosine functions for $30°$ and $45°$ and the cosine identities.

(a) $\cos 15° = \cos (45° - 30°)$

$$= \cos 45° \cos 30° + \sin 45° \sin 30°$$

$$= \left(\frac{\sqrt{2}}{2}\right)\left(\frac{\sqrt{3}}{2}\right) + \left(\frac{\sqrt{2}}{2}\right)\left(\frac{1}{2}\right)$$

$$= \frac{\sqrt{6}}{4} + \frac{\sqrt{2}}{4} = \frac{\sqrt{6} + \sqrt{2}}{4}$$

(b) $\cos 75° = \cos (45° + 30°)$

$$= \cos 45° \cos 30° - \sin 45° \sin 30°$$

$$= \left(\frac{\sqrt{2}}{2}\right)\left(\frac{\sqrt{3}}{2}\right) - \left(\frac{\sqrt{2}}{2}\right)\left(\frac{1}{2}\right)$$

$$= \frac{\sqrt{6}}{4} - \frac{\sqrt{2}}{4} = \frac{\sqrt{6} - \sqrt{2}}{4}$$

We can produce another special identity if we let $B = \pi/2$.

$$\cos\left(\frac{\pi}{2} - A\right) = \cos\frac{\pi}{2}\cos A + \sin\frac{\pi}{2}\sin A$$

$$= (0)\cos A + (1)\sin A$$

$$= \sin A$$

Using this identity, we can get $\sin\left(\dfrac{\pi}{2} - A\right) = \cos A$. Let us summarize these two identities. For any angle (or real number) A, we have the **cofunction identities**:

$$\cos A = \sin\left(\frac{\pi}{2} - A\right)$$

$$\sin A = \cos\left(\frac{\pi}{2} - A\right)$$

We can put the cosine identities and the cofunction identities together to produce the following **sine-of-differences (sums) identities** for any angles or real numbers A and B.

$$\sin(B - A) = \sin B \cos A - \cos B \sin A$$

$$\sin(B + A) = \sin B \cos A + \cos B \sin A$$

Let us prove the sine-of-sums identity (the other identity is similarly proved).

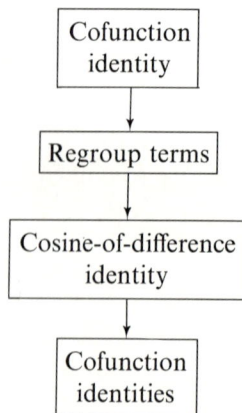

Cofunction identity	$\sin(B + A) = \cos\left(\dfrac{\pi}{2} - B - A\right)$
Regroup terms	$= \cos\left[\left(\dfrac{\pi}{2} - B\right) - A\right]$
Cosine-of-difference identity	$= \cos\left(\dfrac{\pi}{2} - B\right)\cos A + \sin\left(\dfrac{\pi}{2} - B\right)\sin A$
Cofunction identities	$= \sin B \cos A + \cos B \sin A$

EXAMPLE 7   Find $\sin \pi/12$ and $\sin 5\pi/12$ without tables or calculator.

Solution   We are using radian measure here (recall that $\pi/12 = 15°$). We rewrite $\dfrac{\pi}{12} = \dfrac{\pi}{4} - \dfrac{\pi}{6}$ and $\dfrac{5\pi}{12} = \dfrac{\pi}{4} + \dfrac{\pi}{6}$.

(a) $\sin \dfrac{\pi}{12} = \sin\left(\dfrac{\pi}{4} - \dfrac{\pi}{6}\right)$

$$= \sin\frac{\pi}{4}\cos\frac{\pi}{6} - \cos\frac{\pi}{4}\sin\frac{\pi}{6}$$

$$= \left(\frac{\sqrt{2}}{2}\right)\left(\frac{\sqrt{3}}{2}\right) - \left(\frac{\sqrt{2}}{2}\right)\left(\frac{1}{2}\right)$$

$$= \frac{\sqrt{6}}{4} - \frac{\sqrt{2}}{4} = \frac{\sqrt{6} - \sqrt{2}}{4}$$

(b) $\sin \dfrac{5\pi}{12} = \sin\left(\dfrac{\pi}{4} + \dfrac{\pi}{6}\right)$

$$= \sin\frac{\pi}{4}\cos\frac{\pi}{6} + \cos\frac{\pi}{4}\sin\frac{\pi}{6}$$

$$= \left(\frac{\sqrt{2}}{2}\right)\left(\frac{\sqrt{3}}{2}\right) + \left(\frac{\sqrt{2}}{2}\right)\left(\frac{1}{2}\right)$$

$$= \frac{\sqrt{6}}{4} + \frac{\sqrt{2}}{4} = \frac{\sqrt{6} + \sqrt{2}}{4}$$

EXAMPLE 8 Use the cosine and sine identities to verify the identity

$$\tan (A + B) = \frac{\tan A + \tan B}{1 - \tan A \tan B}$$

Solution We establish this important identity using $\tan x = \sin x/\cos x$.

$$\boxed{\tan x = \frac{\sin x}{\cos x}} \qquad \tan (A + B) = \frac{\sin (A + B)}{\cos (A + B)}$$

$$\downarrow$$

$$\boxed{\text{Addition identities}} \qquad\qquad = \frac{\sin A \cos B + \cos A \sin B}{\cos A \cos B - \sin A \sin B}$$

$$\downarrow$$

$$\boxed{\begin{array}{c}\text{Divide top and}\\ \text{bottom by}\\ \cos A \cos B\end{array}} \qquad = \dfrac{\dfrac{\sin A \cos B + \cos A \sin B}{\cos A \cos B}}{\dfrac{\cos A \cos B - \sin A \sin B}{\cos A \cos B}}$$

$$\downarrow$$

$$\boxed{\begin{array}{c}\text{Rewrite fractions}\\ \text{and simplify}\end{array}} \qquad = \dfrac{\dfrac{\sin A \,\cancel{\cos B}}{\cos A \,\cancel{\cos B}} + \dfrac{\cancel{\cos A}\, \sin B}{\cancel{\cos A}\, \cos B}}{\dfrac{\cancel{\cos A\, \cos B}}{\cancel{\cos A\, \cos B}} - \dfrac{\sin A \sin B}{\cos A \cos B}}$$

$$\downarrow$$

$$\boxed{\tan x = \frac{\sin x}{\cos x}} \qquad\qquad = \frac{\tan A + \tan B}{1 - \tan A \tan B}$$

Summarizing, for any angles or real numbers A and B, the following identities hold (where the functions are defined):

$$\tan (A + B) = \frac{\tan A + \tan B}{1 - \tan A \tan B}$$

$$\tan (A - B) = \frac{\tan A - \tan B}{1 + \tan A \tan B}$$

The proof of the subtraction identity is similar to Example 8 and is left as an exercise.

PROBLEM SET 10.2

Use the tables or a calculator to evaluate the following.

Warm-up Exercises

1. $\sin 50° + \sin 10°$ and $\sin 60°$
2. $\sin 5° + \sin 20°$ and $\sin 25°$
3. $\sin 100° + \sin 100°$ and $\sin 200°$
4. $\cos 20° + \cos 50°$ and $\cos 70°$
5. $\cos 100° + \cos 200°$ and $\cos 300°$
6. $\cos 50° + \cos 50°$ and $\cos 100°$

Complete the following without tables or calculator.

		θ (radians)	θ (degrees)	$\sin \theta$	$\cos \theta$	$\tan \theta$
Warm-up	**7.**	$\pi/6$	?	?	?	?
Exercises	**8.**	$\pi/4$	?	?	?	?
	9.	$\pi/3$	?	?	?	?
	10.	$\pi/2$	?	?	?	?
	11.	$2\pi/3$	?	?	?	?
	12.	$3\pi/4$	?	?	?	?
	13.	$5\pi/6$	?	?	?	?
	14.	π	?	?	?	?

		Quadrant for θ	$\sin \theta$	$\cos \theta$	$\tan \theta$
Warm-up	**15.**	I	4/5	?	?
Exercises	**16.**	II	5/13	?	?
	17.	III	?	-0.4	?
	18.	IV	?	0.8	?

Use the known trigonometric values (see Problems 7 to 14) to evaluate the following without tables or a calculator.

19. $\cos 105°$ **20.** $\cos 195°$ **21.** $\cos \dfrac{11\pi}{12}$

22. $\cos \dfrac{17\pi}{12}$ **23.** $\sin \dfrac{7\pi}{12}$ **24.** $\sin \dfrac{13\pi}{12}$

25. $\sin 165°$ **26.** $\sin 255°$ **27.** $\tan 195°$

28. $\tan \dfrac{17\pi}{12}$ **29.** $\tan \dfrac{11\pi}{12}$ **30.** $\tan 105°$

Write each of the following expressions as a trigonometric function of a single angle.
(Do not evaluate.)

31. $\cos 70° \cos 10° - \sin 70° \sin 10°$

32. $\sin 40° \cos 30° - \cos 40° \sin 30°$

33. $\sin \dfrac{\pi}{4} \cos \dfrac{\pi}{7} + \cos \dfrac{\pi}{4} \sin \dfrac{\pi}{7}$

34. $\cos \dfrac{\pi}{5} \cos \dfrac{\pi}{6} + \sin \dfrac{\pi}{5} \sin \dfrac{\pi}{6}$

35. $\dfrac{\tan 20° + \tan 15°}{1 - \tan 20° \tan 15°}$

36. $\dfrac{\tan 50° - \tan 35°}{1 + \tan 50° \tan 35°}$

37. $\cos 40° \cos 40° - \sin 40° \sin 40°$

38. $\cos^2 50° - \sin^2 50°$

Complete the following tables for angles A and B.

	Quadrant for A	$\sin A$	$\cos A$	Quadrant for B	$\sin B$	$\cos B$	$\sin (A+B)$	$\cos (A+B)$	Quadrant for $A+B$
39.	I	3/5	?	I	12/13	?	?	?	?
40.	I	5/13	?	II	?	$-4/5$	?	?	?
41.	II	?	$-1/2$	III	$-\sqrt{2}/2$	?	?	?	?
42.	III	?	-0.6	IV	?	0.8	?	?	?
43.	IV	?	0.2	III	-0.7	?	?	?	?
44.	IV	-0.4	?	II	?	-0.1	?	?	?

Verify the following identities.

45. $\cos (x + 180°) = -\cos x$ **46.** $\cos (x + 90°) = -\sin x$

47. $\cos \left(x - \dfrac{\pi}{2}\right) = \sin x$ **48.** $\cos (\pi - x) = -\cos x$

49. $\sin (\pi + x) - \quad \sin x$ **50.** $\sin \left(\dfrac{\pi}{2} + x\right) = \cos x$

51. $\tan (x + 180°) = \tan x$ **52.** $\cot (A + B) = \dfrac{\cos A \cot B - 1}{\cot A + \cot B}$

53. $\sin (A + B) - \sin (A - B) = 2 \cos A \sin B$

54. $\cos (A + B) + \cos (A - B) = 2 \cos A \cos B$

55. $\cos 2A = \cos (A + A) = \cos^2 A - \sin^2 A$

56. $\sin 2A = \sin (A + A) = 2 \sin A \cos A$

57. $\tan 2A = \tan (A + A) = \dfrac{2 \tan A}{1 - \tan^2 A}$

58. $\cot 2A = \cot (A + A) = \dfrac{\cot^2 A - 1}{2 \cot A}$

59. Derive a formula for $\cos (A + B + C)$.

60. Derive a formula for $\sin (A + B + C)$.

Mathematics Application

61. For the lines shown at the left, recall the relation between the slope and the angle of inclination:

$$m_1 = \tan A \qquad \text{and} \qquad m_2 = \tan B$$

We can use the tangent identity to find θ, the angle between the lines. Since $\theta = A - B$ (why?), we have

$$\tan \theta = \frac{\tan A - \tan B}{1 + \tan A \tan B} = \frac{m_1 - m_2}{1 + m_1 m_2}$$

Use this identity to find the angle θ between each of the following lines.

(a) $y = 2x + 5$
$ y = 3x - 7$

(b) $y = -5x + 7$
$ y = x + 2$

(c) $2x - 3y = 10$
$ x + 2y = 4$

(d) $2x + 7y = 1$
$ y = 5$

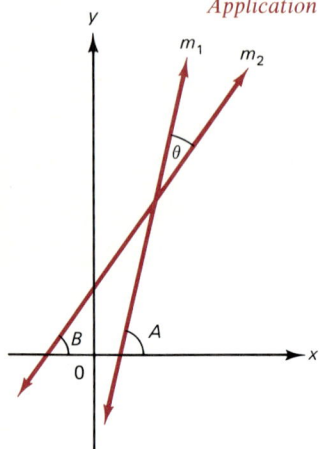

10.3
DOUBLE- AND HALF-ANGLE IDENTITIES

In the preceding section we discussed the following identities for the sums of angles:

$$\sin (A + B) = \sin A \cos B + \cos A \sin B$$

$$\cos (A + B) = \cos A \cos B - \sin A \sin B$$

$$\tan (A + B) = \frac{\tan A + \tan B}{1 - \tan A \tan B}$$

Recall that the formulas for $A - B$ can be found by interchanging "$+$" and "$-$" in the identities above.

In this section we look at the special case of the angle $A + A$, or $2A$. These are called **double-angle identities**, since we are finding the trigonometric functions of twice an angle.

$$\sin (A + A) = \sin A \cos A + \cos A \sin A$$
$$= 2 \sin A \cos A$$

$$\cos (A + A) = \cos A \cos A - \sin A \sin A$$
$$= \cos^2 A - \sin^2 A$$

$$\tan (A + A) = \frac{\tan A + \tan A}{1 - \tan A \tan A}$$
$$= \frac{2 \tan A}{1 - \tan^2 A}$$

Summarizing, we have the following double-angle identities for any angle or real number A (if A and $2A$ are in the domain of the function):

$$\sin 2A = 2 \sin A \cos A$$

$$\cos 2A = \cos^2 A - \sin^2 A$$

$$\tan 2A = \frac{2 \tan A}{1 - \tan^2 A}$$

YES	NO
$\sin 2A = 2 \sin A \cos A$	~~$\sin 2A = 2 \sin A$~~

EXAMPLE 9 Suppose that we know that $\sin 10° = 0.1736$ and $\cos 10° = 0.9848$. We can compute the functions of $20°$:

(a) $\sin 20° = 2 \sin 10° \cos 10° = 2(0.1736)(0.9848) = 0.3419$

(b) $\cos 20° = \cos^2 10° - \sin^2 10°$

$$= (0.9848)^2 - (0.1736)^2 = 0.9397$$

EXAMPLE 10 We can use $\sin^2 A + \cos^2 A = 1$ to rewrite the identity for $\cos 2A$.

(a) $\cos 2A = \cos^2 A - \sin^2 A$

$$= (1 - \sin^2 A) - \sin^2 A \quad (\text{since } \cos^2 A = 1 - \sin^2 A)$$

$$= 1 - 2 \sin^2 A$$

(b) $\cos 2A = \cos^2 A - \sin^2 A$

$$= \cos^2 A - (1 - \cos^2 A) \quad (\text{since } \sin^2 A = 1 - \cos^2 A)$$

$$= 2 \cos^2 A - 1$$

The identities that we found in Example 10 are very important and are usually rewritten in the following forms:

$$\sin^2 A = \frac{1 - \cos 2A}{2}$$

$$\cos^2 A = \frac{1 + \cos 2A}{2}$$

EXAMPLE 11 We can use the double-angle identities to express $\sin 3A$ in terms of $\sin A$.

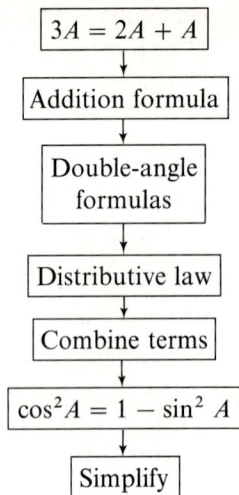

$$\sin 3A = \sin (2A + A)$$

$$= \sin 2A \cos A + \cos 2A \sin A$$

$$= (2 \sin A \cos A) \cos A + (\cos^2 A - \sin^2 A) \sin A$$

$$= 2 \sin A \cos^2 A + \cos^2 A \sin A - \sin^3 A$$

$$= 3 \sin A \cos^2 A - \sin^3 A$$

$$= 3 \sin A (1 - \sin^2 A) - \sin^3 A$$

$$= 3 \sin A - 4 \sin^3 A$$

EXAMPLE 12 Suppose that $\cos \theta = -3/5$ and θ is in quadrant II. Find $\sin 2\theta$, $\cos 2\theta$, and $\tan 2\theta$.

Solution To use the double-angle identities, we must first know both $\sin \theta$ and $\cos \theta$. We construct a reference right triangle with θ in quadrant II and $\cos \theta = -3/5$ in Figure 2. We see that

$$\sin \theta = 4/5 \qquad \cos \theta = -3/5 \qquad \tan \theta = -4/3$$

Using these in the double-angle identities, we get

$$\sin 2\theta = 2 \sin \theta \cos \theta = 2\left(\frac{4}{5}\right)\left(\frac{-3}{5}\right) = \frac{-24}{25}$$

$$\cos 2\theta = \cos^2 \theta - \sin^2 \theta = \left(\frac{-3}{5}\right)^2 - \left(\frac{4}{5}\right)^2 = \frac{9}{25} - \frac{16}{25} = \frac{-7}{25}$$

$$\tan 2\theta = \frac{\sin 2\theta}{\cos 2\theta} = \frac{-24/25}{-7/25} = \frac{24}{7}$$

We could also perform the last calculation using the $\tan 2\theta$ identity. Finally, since its sine and cosine are both negative, 2θ must lie in quadrant III.

FIGURE 2

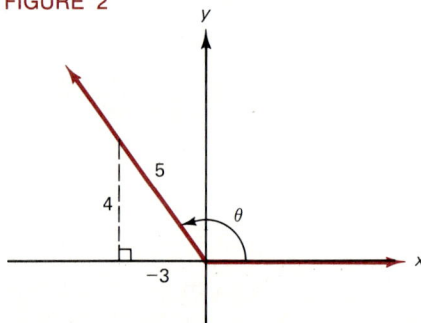

We now wish to develop identities that give the trigonometric functions for $\theta/2$ in terms of functions of θ. These are called **half-angle identities**.

We start with two identities involving $\cos 2A$ (see page 353—Example 10). We then solve these for $\cos A$ and $\sin A$.

$$\cos^2 A = \frac{1 + \cos 2A}{2} \qquad \sin^2 A = \frac{1 - \cos 2A}{2}$$

$$\cos A = \pm \sqrt{\frac{1 + \cos 2A}{2}} \qquad \sin A = \pm \sqrt{\frac{1 - \cos 2A}{2}}$$

If we let $A = \theta/2$ (and then $2A = \theta$), we get the following **half-angle identities**:

$$\cos \frac{\theta}{2} = \pm \sqrt{\frac{1 + \cos \theta}{2}}$$

$$\sin \frac{\theta}{2} = \pm \sqrt{\frac{1 - \cos \theta}{2}}$$

The correct sign ($+$ or $-$) is determined by the quadrant in which $\theta/2$ lies.

EXAMPLE 13   Use the half-angle identities to find the values for $\sin 165°$ and $\cos 165°$.

Solution   The angle $165° = 330°/2$. We need $\cos 330°$. Since $330°$ is equivalent or coterminal to $-30°$, we have

$$\cos 330° = \cos(-30°) = \cos 30° = \frac{\sqrt{3}}{2} \approx 0.866$$

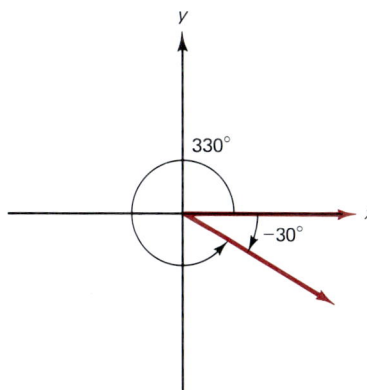

Since 165° is in quadrant II, the sine is positive and the cosine is negative.

$$\boxed{\sin \frac{\theta}{2} = \pm \sqrt{\frac{1 - \cos \theta}{2}}}$$

$$\downarrow$$

$$\boxed{\text{Sine is } +}$$

$$\sin 165° = \sin \frac{330°}{2} = +\sqrt{\frac{1 - \cos 330°}{2}}$$

$$= +\sqrt{\frac{1 - 0.866}{2}}$$

$$= +0.259$$

$$\boxed{\cos \frac{\theta}{2} = \pm \sqrt{\frac{1 + \cos \theta}{2}}}$$

$$\downarrow$$

$$\boxed{\text{Cosine is } -}$$

$$\cos 165° = \cos \frac{330°}{2} = -\sqrt{\frac{1 + \cos 330°}{2}}$$

$$= -\sqrt{\frac{1 + 0.866}{2}}$$

$$= -0.966$$

EXAMPLE 14 Suppose that $\sin A = -3/5$ and $\pi < A < 3\pi/2$ (quadrant III). Find $\sin \frac{A}{2}$, $\cos \frac{A}{2}$, and $\tan \frac{A}{2}$.

Solution We draw a reference right triangle for A in quadrant III (page 357); we can see that $\cos A = -4/5$. Also, we divide by 2 to find a domain for $A/2$:

$$\pi < A < \frac{3\pi}{2} \qquad \text{implies} \qquad \frac{\pi}{2} < \frac{A}{2} < \frac{3\pi}{4}$$

We see that $A/2$ is in quadrant II; thus, the sine is positive and the cosine is negative. We substitute into the half-angle identities. (Be careful when simplifying the fractions under the radical.)

$$\sin \frac{A}{2} = \sqrt{\frac{1 - \left(\frac{-4}{5}\right)}{2}} = \sqrt{\frac{9}{10}} \quad \text{or} \quad \frac{3\sqrt{10}}{10} \quad \text{or} \quad 0.949$$

$$\cos \frac{A}{2} = -\sqrt{\frac{1 + \left(\frac{-4}{5}\right)}{2}} = -\sqrt{\frac{1}{10}} \quad \text{or} \quad \frac{-\sqrt{10}}{10} \quad \text{or} \quad -0.316$$

$$\tan \frac{A}{2} = \frac{\sin \frac{A}{2}}{\cos \frac{A}{2}} = \frac{3\sqrt{10}/10}{-\sqrt{10}/10} = -3$$

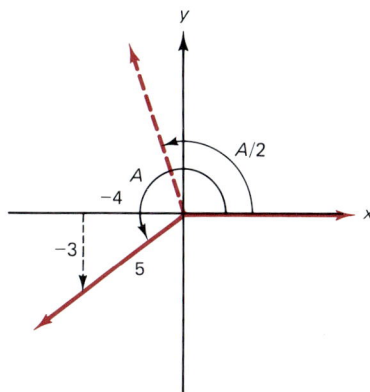

In Example 14, we computed $\tan \dfrac{A}{2}$ as $\sin \dfrac{A}{2}$ divided by $\cos \dfrac{A}{2}$. The following are also identities for $\tan \dfrac{\theta}{2}$.

$$\tan \frac{\theta}{2} = \pm \sqrt{\frac{1 - \cos \theta}{1 + \cos \theta}}$$

$$\tan \frac{\theta}{2} = \frac{1 - \cos \theta}{\sin \theta}$$

PROBLEM SET
10.3

Simplify the following in terms of sin x, sin y, and so on.

Warm-up
Exercises

1. $\sin (x + y)$ **2.** $\cos (x - y)$

3. $\cos (x + y)$ **4.** $\sin (x - y)$

5. $\cos (x + x)$ **6.** $\sin (x + x)$

7. $\tan (x + y)$ **8.** $\tan (x + x)$

Complete the following table.

	Quadrant for θ	$\sin \theta$	$\cos \theta$	$\tan \theta$
9.	I	5/13	?	?
10.	II	?	$-15/17$	?
11.	III	?	$-4/5$	?
12.	IV	$-12/13$	?	?

Warm-up
Exercises

Use only the facts that $\sin 5° = 0.087$ and $\cos 5° = 0.996$ (no tables or trigonometric calculators) to evaluate the following.

13. $\sin 10°$ **14.** $\cos 10°$ **15.** $\cos 20°$

16. $\sin 20°$ **17.** $\sin 15°$ **18.** $\cos 15°$

Simplify each of the following as much as possible (in terms of sin θ or cos θ).

19. $\sin 2\theta$ **20.** $\cos 2\theta$

21. $\sin 4\theta$ **22.** $\cos 5\theta$

23. $\cos 6\theta$ **24.** $\sin 6\theta$

25. $\tan 4\theta$ **26.** $\tan 6\theta$

Complete the following table.

	Quadrant for θ	$\sin \theta$	$\cos \theta$	$\sin 2\theta$	$\cos 2\theta$	Quadrant for 2θ
27.	I	4/5	?	?	?	?
28.	I	?	5/13	?	?	?
29.	II	?	−4/5	?	?	?
30.	II	12/13	?	?	?	?
31.	III	−0.7	?	?	?	?
32.	III	?	−0.3	?	?	?
33.	IV	?	0.5	?	?	?
34.	IV	−0.9	?	?	?	?

Use the half-angle identities to evaluate the following (without tables or a trigonometric calculator).

35. $\sin 15° \left[= \sin \dfrac{1}{2}(30°) \right]$ **36.** $\sin 22.5° \left[= \sin \dfrac{1}{2}(45°) \right]$

37. $\cos 22.5°$ **38.** $\cos 15°$

39. $\cos 45°$ **40.** $\sin 45°$

Complete the following table.

	θ range	$\sin \theta$	$\cos \theta$	$\theta/2$ range	$\sin \theta/2$	$\cos \theta/2$	$\tan \theta/2$
41.	$0° < \theta < 90°$	12/13	?	?	?	?	?
42.	$0° < \theta < 90°$	?	4/5	?	?	?	?
43.	$\pi/2 < \theta < \pi$	?	−5/13	?	?	?	?
44.	$\pi/2 < \theta < \pi$	?	−2/3	?	?	?	?
45.	$180° < \theta < 270°$	−1/3	?	?	?	?	?
46.	$270° < \theta < 360°$	?	1/2	?	?	?	?

Verify the following identities.

47. $\csc 2A = \dfrac{1}{2} \sec A \cdot \csc A$ **48.** $\sec 2A = \dfrac{\csc^2 A \cdot \sec^2 A}{\csc^2 A - \sec^2 A}$

49. $\cos x + \sin x = \dfrac{\cos 2x}{\cos x - \sin x}$ **50.** $\cos x - \sin x - \dfrac{\cos 2x}{\sin x + \cos x}$

51. $\dfrac{\sin 2x}{\sin x} = 2 \cos x$ **52.** $\cos 2x = \cos^4 x - \sin^4 x$

53. $\dfrac{\cos 2x - 1}{\cos x - 1} = 2 \cos x + 2$ **54.** $\cot 2A = \dfrac{\cot^2 A - 1}{2 \cot A}$

55. $\dfrac{2 \tan x}{\tan 2x} = (1 + \tan x)(1 - \tan x)$

56. $2 \cot A \cot 2A = (\cot A + 1)(\cot A - 1)$

57. $\sin^2 \dfrac{x}{2} = \dfrac{\tan x - \sin x}{2 \tan x}$

58. $\cos^2 \dfrac{x}{2} = \dfrac{\tan x + \sin x}{2 \tan x}$

59. $\sec^2 \dfrac{x}{2} = \dfrac{2}{1 + \cos x}$

60. $\csc^2 \dfrac{x}{2} = \dfrac{2}{1 - \cos x}$

61. $\sin^2 \dfrac{A}{2} \cos^2 \dfrac{A}{2} = \dfrac{1}{4} \sin^2 A$

62. $\sin^2 \dfrac{A}{2} \cos^2 \dfrac{A}{2} = \dfrac{1}{8}(1 - \cos 2A)$

63. $\sec^2 \dfrac{x}{2} + \csc^2 \dfrac{x}{2} = 4 \csc^2 x$

64. $\tan \dfrac{A}{2} = \dfrac{1 - \cos 2A}{2 \sin A + \sin 2A}$

65. $\tan \dfrac{A}{2} = \dfrac{1 - \cos A}{\sin A}$

66. $\tan \dfrac{A}{2} = \dfrac{\sin A}{1 + \cos A}$

67. $\cos \dfrac{x}{2} = \dfrac{1}{2} \sqrt{2 + 2 \cos x} \quad$ for $0 \le x \le \pi$

68. $\sin \dfrac{x}{2} = \dfrac{1}{2} \sqrt{2 - 2 \cos x} \quad$ for $0 \le x \le \pi$

69. $\cos \dfrac{x}{4} = \dfrac{1}{2} \sqrt{2 + \sqrt{2 + 2 \cos x}} \quad$ for $0 \le x \le \pi$

70. $\sin \dfrac{x}{4} = \dfrac{1}{2} \sqrt{2 - \sqrt{2 + 2 \cos x}} \quad$ for $0 \le x \le \pi$

Mathematics Application

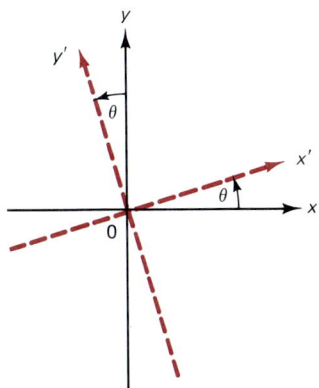

71. Given the equation

$$Ax^2 + Bxy + Cy^2 = F$$

we can rotate the x-y axes to eliminate the xy term. To do this, we let

$$x = x' \cos \theta - y' \sin \theta$$
$$y = x' \sin \theta + y' \cos \theta$$

where θ is the angle through which we rotate the axes, and x' and y' are the new coordinates. Substitute these expressions for x and y into the original equation and simplify to the form

$$x'^2[\quad] + x'y'[\quad] + y'^2[\quad] = F$$

Since our goal is to eliminate the $x'y'$ term, set the contents of the middle bracket to zero and show that

$$\cot 2\theta = \dfrac{A - C}{B}$$

Also, show that

$$x^2 + y^2 = (x')^2 + (y')^2$$

72. For the equations below use Problem 71 to:

(i) Find $\cot 2\theta$.

(ii) Find $\cos 2\theta$ (using $\cot 2\theta$ and a reference triangle).

(iii) Find $\sin \theta$ and $\cos \theta$, using the half-angle identities.

(iv) Using these $\sin \theta$ and $\cos \theta$ values, substitute

$$x = x' \cos \theta - y' \sin \theta$$

$$y = x' \sin \theta + y' \cos \theta$$

into the original equation and simplify to a new equation. (If you do it correctly, there should be no $x'y'$ term.)

(a) $9x^2 + 24xy + 2y^2 = 100$ (b) $3x^2 + \sqrt{3}xy + 2y^2 = 9$

10.4 EVEN MORE IDENTITIES

There are still more trigonometric identities. These involve the sums and products of trigonometric functions and often prove important in more advanced mathematics and science courses. We briefly sketch the proofs.

We start with the identities for the sines and cosines of $A + B$ and $A - B$; we then add and subtract the identities to obtain new identities.

Sine identities

$$\sin (A + B) = \sin A \cos B + \cos A \sin B$$
$$\sin (A - B) = \sin A \cos B - \cos A \sin B$$

Add; subtract

$$\sin (A + B) + \sin (A - B) = 2 \sin A \cos B$$
$$\sin (A + B) - \sin (A - B) = \qquad 2 \cos A \sin B$$

Cosine identities

$$\cos (A - B) = \cos A \cos B + \sin A \sin B$$
$$\cos (A + B) = \cos A \cos B - \sin A \sin B$$

Add; subtract

$$\cos (A - B) + \cos (A + B) = 2 \cos A \cos B$$
$$\cos (A - B) - \cos (A + B) = \qquad 2 \sin A \sin B$$

Dividing each of these relations by 2, we get the four following **product identities**.

$$\sin A \cos B = \frac{1}{2} [\sin (A + B) + \sin (A - B)]$$

$$\cos A \sin B = \frac{1}{2} [\sin (A + B) - \sin (A - B)]$$

$$\cos A \cos B = \frac{1}{2} [\cos (A - B) + \cos (A + B)]$$

$$\sin A \sin B = \frac{1}{2} [\cos (A - B) - \cos (A + B)]$$

EXAMPLE 15 The following examples use these product identities.

(a) $\sin 75° \sin 15° = \dfrac{1}{2}\left[\cos(75° - 15°) - \cos(75° + 15°)\right]$

$$= \dfrac{1}{2}\left[\cos 60° - \cos 90°\right]$$

$$= \dfrac{1}{2}\left[\dfrac{1}{2} - 0\right] = \dfrac{1}{4}$$

(b) $\cos 5x \sin 3x = \dfrac{1}{2}\left[\sin(5x + 3x) - \sin(5x - 3x)\right]$

$$= \dfrac{1}{2}\left[\sin 8x - \sin 2x\right]$$

We can use these identities to derive identities for $\sin u + \sin v$, and so on. We let $u = A + B$ and $v = A - B$. Solving for A and B, we get

$$A = \dfrac{u + v}{2} \qquad \text{and} \qquad B = \dfrac{u - v}{2}$$

Substituting these values into the product identities, we get the following **sum (difference) identities**.

$$\sin u + \sin v = 2\sin\left(\dfrac{u + v}{2}\right)\cos\left(\dfrac{u - v}{2}\right)$$

$$\sin u - \sin v = 2\cos\left(\dfrac{u + v}{2}\right)\sin\left(\dfrac{u - v}{2}\right)$$

$$\cos u + \cos v = 2\cos\left(\dfrac{u + v}{2}\right)\cos\left(\dfrac{u - v}{2}\right)$$

$$\cos u - \cos v = -2\sin\left(\dfrac{u + v}{2}\right)\sin\left(\dfrac{u - v}{2}\right)$$

EXAMPLE 16 The following examples use the sum and difference identities above.

(a) $\cos 105° + \cos 75° = 2\cos\left(\dfrac{105° + 75°}{2}\right)\cos\left(\dfrac{105° - 75°}{2}\right)$

$$= 2\cos 90° \cos 15° = 2(0)\cos 15° = 0$$

(b) $\sin(x + h) - \sin x = 2\cos\left(\dfrac{2x + h}{2}\right)\sin\dfrac{h}{2}$

We often wish to simplify $a \sin x + b \cos x$. We now rewrite it in the form

$$T = a \sin x + b \cos x = c \sin (x + \theta)$$

where we want to find c and θ. Consider the right triangle at the left where $c = \sqrt{a^2 + b^2}$.

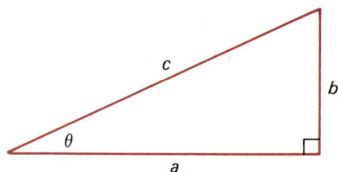

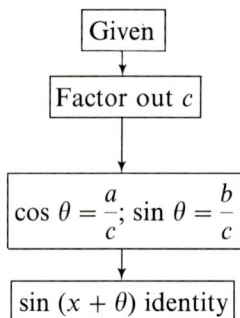

Given	$T = a \sin x + b \cos x$
Factor out c	$= c\left(\dfrac{a}{c} \sin x + \dfrac{b}{c} \cos x\right)$
$\cos \theta = \dfrac{a}{c}; \sin \theta = \dfrac{b}{c}$	$= c(\cos \theta \sin x + \sin \theta \cos x)$
$\sin (x + \theta)$ identity	$= c \sin (x + \theta)$

Thus, we have the **reduction identity**:

$$\boxed{a \sin x + b \cos x = c \sin (x + \theta)}$$

where $c = \sqrt{a^2 + b^2}$, $\cos \theta = a/c$, and $\sin \theta = b/c$. Here θ is a phase shift (see page 327).

EXAMPLE 17 Express $4 \sin x + 3 \cos x$ as $c \sin (x + \theta)$.

Solution Here $c = \sqrt{4^2 + 3^2} = \sqrt{25} = 5$. Also, $\cos \theta = 4/5$ and $\sin \theta = 3/5$; this gives $\theta \approx 36.87°$. Thus,

$$4 \sin x + 3 \cos x = 5 \sin (x + \theta)$$
$$\approx 5 \sin (x + 36.87°)$$

EXAMPLE 18 Express $5 \sin x - 12 \cos x$ as $c \sin (x + \theta)$.

Solution Here $c = \sqrt{5^2 + (-12)^2} = \sqrt{169} = 13$. Also, $\cos \theta = 5/13$ and $\sin \theta = -12/13$. Since the cosine is positive and the sine is negative, θ is in quadrant IV. Thus, $\theta \approx -1.18$ radians and we have

$$5 \sin x - 12 \cos x = 13 \sin (x + \theta)$$
$$\approx 13 \sin (x - 1.18)$$

PROBLEM SET
10.4

Fill in the right side of the following identities with the simplest possible expression.

Warm-up Exercises

1. $\sin^2 t + \cos^2 t =$ _____

2. $1 + \tan^2 x =$ _____

3. $\cos (A - B) =$ _____

4. $\sin (A + B) =$ _____

5. $\sin 2x = \underline{\hspace{1.5cm}}$ **6.** $\cos 2x = \underline{\hspace{1.5cm}}$

7. $\cos \dfrac{t}{2} = \underline{\hspace{1.5cm}}$ **8.** $\sin \dfrac{m}{2} = \underline{\hspace{1.5cm}}$

9. $\sin p \cos q - \cos p \sin q = \underline{\hspace{1.5cm}}$

10. $\cos u \cos v - \sin u \sin v = \underline{\hspace{1.5cm}}$

11. $\sin(-x) = \underline{\hspace{1.5cm}}$ **12.** $\cos\left(\dfrac{\pi}{2} - \theta\right) = \underline{\hspace{1.5cm}}$

For each of the following equations (or pair of equations), find the angle θ.

*Warm-up
Exercises*

13. $\sin \theta = 0.7$ (quadrant I)

14. $\cos \theta = 0.6$ (quadrant IV)

15. $\tan \theta = -1.7$ (quadrant II)

16. $\sin \theta = -0.1$ (quadrant III)

17. $\sin \theta = 0.4000$ and $\cos \theta = -0.9165$

18. $\sin \theta = -0.900$ and $\cos \theta = 0.436$

Use the product identities to rewrite the following as sums.

19. $\sin 45° \sin 45°$ **20.** $\cos 45° \cos 45°$

21. $\cos \dfrac{\pi}{2} \cos \dfrac{\pi}{4}$ **22.** $\sin \dfrac{\pi}{4} \sin \dfrac{\pi}{2}$

23. $\cos \dfrac{\pi}{3} \sin \dfrac{\pi}{4}$ **24.** $\cos \dfrac{\pi}{4} \sin \dfrac{\pi}{3}$

25. $\sin 2x \cos 4x$ **26.** $\cos 8x \cos 10x$

Use the sum identities to rewrite the following as products.

27. $\sin 90° + \sin 90°$ **28.** $\cos 45° + \cos 45°$

29. $\sin \dfrac{\pi}{6} + \sin \dfrac{\pi}{4}$ **30.** $\cos \dfrac{2\pi}{3} - \cos \dfrac{\pi}{4}$

31. $\cos(x + h) - \cos x$ **32.** $\sin(x + h) - \sin x$

33. $\cos 100A + \cos 200A$ **34.** $\sin 50x + \sin 150x$

Use the reduction identity to rewrite the following in the form $c \sin(x + \theta)$, giving θ to the nearest tenth of a degree.

35. $3 \sin x + 4 \cos x$ **36.** $12 \sin x + 5 \cos x$

37. $5 \sin x + 12 \cos x$ **38.** $8 \sin x + 15 \cos x$

39. $\cos x - \sin x$ **40.** $2 \cos x - 3 \sin x$

41. $4 \cos x + \sin x$ **42.** $\cos x - 5 \sin x$

Fill in the missing terms on the right sides of the following equations.

43. $\sin 2x + \sin 8x = (\ \ ?\ \ )(\ \ ?\ \ )$

44. $\cos 4x \sin 6x = (\ \ ?\ \ ) + (\ \ ?\ \ )$

45. $\sin 10x \sin 10y = (\ \ ?\ \ ) + (\ \ ?\ \ )$

46. $\cos x + \sin x = (?) \sin(x + ?)$

47. $\sin x - \cos x = (?) \sin(x + ?)$

48. $\cos(A - B) - \cos(A - C) = (\ \ ?\ \ )(\ \ ?\ \ )$

49. $\cos(x - y) \cos(x - z) = (\ \ ?\ \ ) + (\ \ ?\ \ )$

50. $2 \cos 3x - 3 \sin 3x = (?) \sin(x + ?)$

Verify the following identities.

51. $\tan A \tan B = \dfrac{\cos(A - B) - \cos(A + B)}{\cos(A - B) + \cos(A + B)}$

52. $\cot A \cot B = \dfrac{\cos(A - B) + \cos(A + B)}{\cos(A - B) - \cos(A + B)}$

53. $\tan B = \dfrac{\sin(A + B) - \sin(A - B)}{\cos(A - B) + \cos(A + B)}$

54. $\cot A = \dfrac{\sin(A + B) - \sin(A - B)}{\cos(A - B) - \cos(A + B)}$

55. $\tan\left(\dfrac{u - v}{2}\right) = \dfrac{\sin u - \sin v}{\cos v + \cos u}$

56. $\cot\left(\dfrac{u + v}{2}\right) = \dfrac{\sin u - \sin v}{\cos v - \cos u}$

57. $\sin^2 u - \sin^2 v = \sin(u + v)\sin(u - v)$

58. $\cos^2 u - \cos^2 v = -\sin(u + v)\sin(u - v)$

Musical Applications

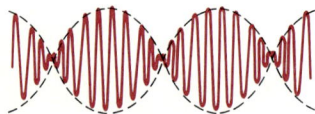

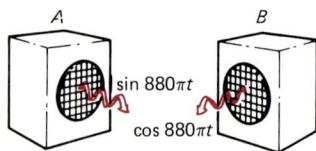

59. When two tones of equal amplitude but with slightly different frequencies are sounded together, there is a *beating effect* that a skilled tuner can hear. The effect appears like a fast wave (the tone) enveloped by a very slow wave (the beating). (A stringed-instrument tuner knows that when the beating disappears, the string is in tune with a standard pitch.) Use the sum identities to rewrite the following. Try to identify the meaning of the resulting terms.
(a) $\sin 880\pi t + \sin 882\pi t$
(b) $\cos 400\pi t + \cos 410\pi t$
(c) $\sin 2\pi f_1 t + \sin 2\pi f_2 t$

60. Two equal tones are coming out of the speakers of a stereo system, somewhat out of phase. That is, speakers A and B have signals

$$A = \sin 880\pi t \qquad \text{and} \qquad B = \cos 880\pi t$$

Rewrite the sum $A + B$ as a single signal, and identify the amplitude and phase angle.

Physical Application

61. A spring is released and then given an extra little push. Its resulting position is now given by

$$y = 12 \sin 4t + 9 \cos 4t$$

Rewrite this as a single sine term, and identify the amplitude and the phase angle.

10.5 TRIGONOMETRIC EQUATIONS

Let us start this section with **simple trigonometric equations**. Such an equation involves only a trigonometric function of x (or some other variable) being equal to a real number. For example, $\sin x = 0.35$ is such an equation. (Here we are given the value of the function, and we want the angle or angles that gives that value.)

Recall from page 287 that we can use the tables or a calculator to find the acute angle that gives a certain trigonometric value.

EXAMPLE 19 Consider the equation $\sin x = 0.35$. We can use the tables to find that $x = 20.5°$. Using a calculator, we have

PRESS	DISPLAY	MEANING
$\boxed{\cdot}\,\boxed{3}\,\boxed{5}\,\boxed{\text{INV}}\,\boxed{\sin}$	$\boxed{\begin{array}{l}20.487315 \\ {\scriptstyle\text{DEG}}\end{array}}$	$\sin 20.487315° = 0.35$

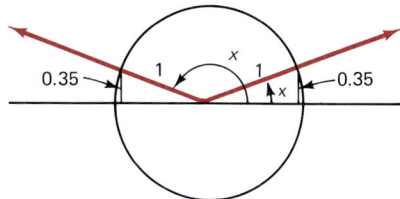

However, $20.5°$ is not the only angle that satisfies $\sin x = 0.35$. The unit circle above shows two such x. In quadrant I we have $x = 20.5°$. In quadrant II we have $x = 180° - 20.5° = 159.5°$. But that's not all! We can add positive and negative multiples of $360°$, and these new angles also satisfy $\sin x = 0.35$. For instance, $380.5°$ and $519.5°$ also satisfy $\sin x = 0.35$, since these are equivalent to $20.5°$ and $159.5°$. Thus, we can write the solutions to $\sin x = 0.35$ as

$$20.5° \pm n \cdot 360°$$

$$159.5° \pm n \cdot 360°$$

Figure 3 shows another view of this problem. We see the graph of $y = \sin x$ and some of the many angles that satisfy $\sin x = 0.35$.

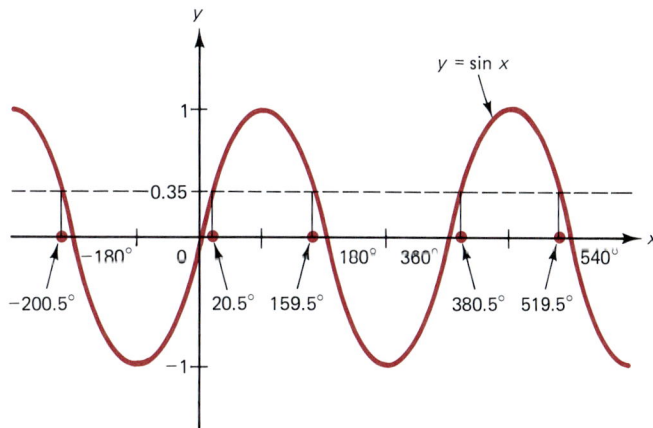

FIGURE 3

Often we want only the solutions between $0°$ and $360°$. We call these the **primary solutions**.

We now consider more complicated trigonometric equations, such as

$$2 \sin^2 x + \sin x = 1 \qquad \text{or} \qquad \sin 2x = 0.38$$

As with proving identities, there is no single method or procedure that works for all trigonometric equations: Each equation must be handled uniquely. (Practice helps.) With an equation such as $2 \sin^2 x + \sin x = 1$, we might first solve for $\sin x$, and then solve for x. With $\sin 2x = 0.38$, we might first solve for $2x$, and then solve for x. [Ultimately, of course, we always solve for x (or whatever letter is used).]

Some suggestions for solving trigonometric equations:

1. If only *one* trigonometric function appears, try solving for it; then solve for x.
2. Try rewriting the equation with a zero on the right; then factor the left side and solve.
3. If two trigonometric functions appear, try using an identity to rewrite one in terms of the other (or both in terms of sine or cosine).
4. Use identities to simplify expressions.
5. If multiple or half-angles appear with other functions, try rewriting them in terms of a single angle.
6. Check (verify) the results in the original equation.

EXAMPLE 20   Solve $2 \sin^2 x + \sin x = 1$, for $0° \leq x < 360°$.

Solution   Since only the sine function appears, we first solve for $\sin x$; then we solve for x.

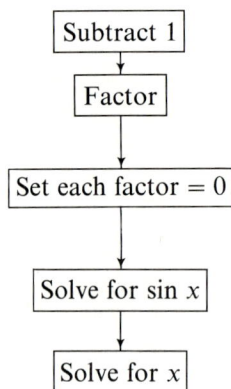

Subtract 1	$2 \sin^2 x + \sin x - 1 = 0$
Factor	$(2 \sin x - 1)(\sin x + 1) = 0$
Set each factor $= 0$	$2 \sin x - 1 = 0 \quad \text{or} \quad \sin x + 1 = 0$
Solve for $\sin x$	$\sin x = \dfrac{1}{2} \quad \text{or} \quad \sin x = -1$
Solve for x	$x = 30°, 150° \quad \text{or} \quad x = 270°$

Thus, there are three primary solutions: $30°$, $150°$, and $270°$.

EXAMPLE 21   Solve $\csc x = 2 \sin x + 1$, for $0 \leq x < 2\pi$.

Solution We rewrite $\csc x = \dfrac{1}{\sin x}$ so that only the sine is present. (Also, we immediately *exclude* $x = 0$ or $x = \pi$ as possible solutions. Why?)

$$\boxed{\csc x = \frac{1}{\sin x}}$$

$$\frac{1}{\sin x} = 2 \sin x + 1$$

$$\boxed{\begin{array}{c}\text{Multiply both sides}\\ \text{by } \sin x\end{array}}$$

$$1 = 2 \sin^2 x + \sin x$$

At this point, the equation is exactly the same as in Example 20, so we know that the solutions are $30°$, $150°$, and $270°$ (or $\pi/6$, $5\pi/6$, and $3\pi/2$ in radians). (This example also teaches another lesson: If possible, a mathematician never solves the same equation twice.)

EXAMPLE 22 Solve $\tan x \sin x = \tan x$, for $0 \le x < 2\pi$.

Solution First thought is to divide both sides by $\tan x$, right? Wrong, since $\tan x$ might be zero. Instead, we move the $\tan x$ to the left side and factor.

$$\boxed{\text{Subtract } \tan x}$$

$$\tan x \sin x - \tan x = 0$$

$$\boxed{\text{Factor}}$$

$$\tan x(\sin x - 1) = 0$$

$$\boxed{\begin{array}{c}\text{Set each factor}\\ = 0; \text{ solve}\end{array}}$$

$$\tan x = 0 \quad \text{or} \quad \sin x - 1 = 0$$

$$\sin x = 1$$

$$\boxed{\text{Solve for } x}$$

$$x = 0, \pi \quad \text{or} \quad x = \frac{\pi}{2}$$

Thus, the primary solutions are 0, $\pi/2$, and π.

YES	NO
$\tan x \sin x = \tan x$	~~$\tan x \sin x = \tan x$~~
$\tan x \sin x - \tan x = 0$	~~$\sin x = 1$~~
If we divide by $\tan x$, we *lose* the solutions 0 and π.	

EXAMPLE 23   Solve $\sin 2x = 0.38$, for $0° \le x < 360°$.

Solution   Since there is only one trigonometric function, the sine, we solve the equation directly for $2x$. Watch carefully: Since we want $0° \le x < 360°$, we must solve for $2x$ in the domain $0° \le 2x < 720°$. When we divide by 2, then x will be in the right domain $0° \le x < 360°$.

Given	$\sin 2x = 0.38$
Solve for $2x$	$2x = 22.3°$ or $157.7°$
Add 360° to angles	$2x = 22.3°$ or $157.7°$ or $382.3°$ or $517.7°$
Divide by 2	$x = 11.2°$ or $78.8°$ or $191.2°$ or $258.8°$

Notice how we found the two extra solutions *before* we divided.

EXAMPLE 24   Solve $\cos 2x - \sin x = 1$, for $0 \le x < 2\pi$.

Solution   We first use a double-angle identity to rewrite $\cos 2x = 1 - 2\sin^2 x$. (We choose this identity since it involves $\sin x$.) We then solve for $\sin x$ and finally x.

$\cos 2x = 1 - 2\sin^2 x$	$1 - 2\sin^2 x - \sin x = 1$
Subtract 1	$-2\sin^2 x - \sin x = 0$
Factor	$-\sin x(2\sin x + 1) = 0$
Set each factor $= 0$; solve	$\sin x = 0$ or $2\sin x + 1 = 0$ $\sin x = \dfrac{-1}{2}$
Solve for x	$x = 0, \pi$ or $x = \dfrac{7\pi}{6}, \dfrac{11\pi}{6}$

The reader should check these four solutions in the original equation. For example, consider $x = \pi$:

$$\cos 2\pi - \sin \pi = 1 - 0 = 1 \qquad \textit{Checks.}$$

EXAMPLE 25   Solve $\sin x + \cos x = 1$, for $0 \le x < 2\pi$.

Solution   Here we use the reduction identity (see page 362) to rewrite

$$\sin x + \cos x = c \sin (x + \theta)$$

where $c = \sqrt{1^2 + 1^2} = \sqrt{2}$, $\sin \theta = 1/\sqrt{2}$, and $\cos \theta = 1/\sqrt{2}$. Thus, $\theta = \pi/4$ or $45°$.

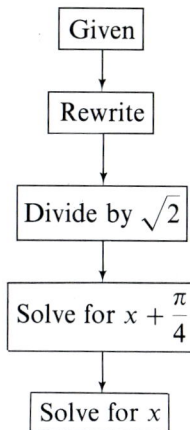

Given	$\sin x + \cos x = 1$
↓	
Rewrite	$\sqrt{2}\,\sin\left(x + \dfrac{\pi}{4}\right) = 1$
↓	
Divide by $\sqrt{2}$	$\sin\left(x + \dfrac{\pi}{4}\right) = \dfrac{1}{\sqrt{2}}$
↓	
Solve for $x + \dfrac{\pi}{4}$	$\left(x + \dfrac{\pi}{4}\right) = \dfrac{\pi}{4}$ or $\dfrac{3\pi}{4}$
↓	
Solve for x	$x = 0$ or $\dfrac{\pi}{2}$

Thus, x is 0 or $\pi/2$.

EXAMPLE 26   Solve $\sin^2 \dfrac{x}{2} - 3\sin \dfrac{x}{2} + 2 = 0$, for $0° \le x < 360°$.

Solution   We factor and solve for $\sin \dfrac{x}{2}$.

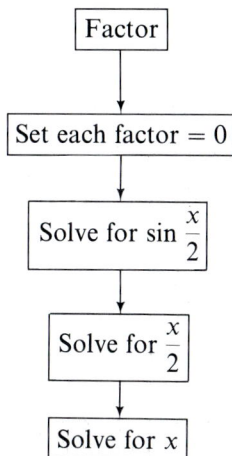

Factor	$\left(\sin \dfrac{x}{2} - 2\right)\left(\sin \dfrac{x}{2} - 1\right) = 0$
↓	
Set each factor $= 0$	$\sin \dfrac{x}{2} - 2 = 0$ or $\sin \dfrac{x}{2} - 1 = 0$
↓	
Solve for $\sin \dfrac{x}{2}$	$\sin \dfrac{x}{2} = 2$ or $\sin \dfrac{x}{2} = 1$
↓	
Solve for $\dfrac{x}{2}$	(no solution) or $\dfrac{x}{2} = 90°$
↓	
Solve for x	$x = 180°$

There is one solution: $180°$ (or π). $\left(\text{Note that we get no solutions from } \sin \dfrac{x}{2} = 2, \text{ since } -1 \le \sin \dfrac{x}{2} \le 1.\right)$

PROBLEM SET 10.5

Solve the following equations. For the trigonometric equations give only the primary solutions ($0 \le x < 2\pi$) to the nearest hundredth of a radian or in terms of π.

Warm-up Exercises

1. $2x + 5 = 13$

2. $3(x - 4) + x = 16$

3. $x^2 - 7x + 10 = 0$

4. $2x^2 + 9x = 5$

5. $x + \dfrac{1}{x} = 2$

6. $\dfrac{1}{x} + \dfrac{1}{2} = \dfrac{x + 2}{6}$

7. $\sin x = \dfrac{1}{2}$

8. $\cos x = \dfrac{1}{\sqrt{2}}$

9. $\cos x = \dfrac{-3}{\sqrt{2}}$

10. $\sin x = \dfrac{-1}{\sqrt{2}}$

11. $\sin x = 0$

12. $\cos x = 0$

13. $\sin x = 0.47$

14. $\cos x = -0.13$

15. $\cos x = -0.95$

16. $\tan x = 1.5$

Solve the following equations. Give only the primary solutions $(0° \le x < 360°)$ *to the nearest tenth of a degree.*

17. $\sin^2 x = \dfrac{3}{4}$

18. $\cos^2 x = \dfrac{1}{2}$

19. $\tan^2 x - \tan x = 0$

20. $\cos^2 x + \cos x = 0$

21. $2 \sin^2 x = \sin x + 1$

22. $\cos^3 x = \cos x$

23. $\sin x + \csc x = \dfrac{5}{2}$

24. $\cos x - \sec x = \dfrac{3}{2}$

25. $2 + \cos x = 2 \sin^2 x$

26. $2 - \sin^2 x = 2 \cos^2 x$

27. $\tan x = \sin x$

28. $\cot x = \cos x$

29. $\tan 2x = 1.98$

30. $\cot 3x = 0.93$

31. $\cos \dfrac{x}{2} = -0.31$

32. $\sin \dfrac{x}{3} = 0.59$

33. $\cos 2x = \cos x$

34. $\cos 2x = \sin x + 1$

35. $\sin 2x = \cos x$

36. $\cos 4x = \sin 2x$

37. $\dfrac{1}{2} \sin x + \dfrac{\sqrt{3}}{2} \cos x = 1$

38. $\dfrac{\sqrt{3}}{2} \sin x - \dfrac{1}{2} \cos x = \dfrac{1}{2}$

39. $2 \cos^2 \dfrac{x}{2} = -\cos \dfrac{x}{2}$

40. $\sin^2 \dfrac{x}{2} + 2 = 3 \sin \dfrac{x}{2}$

41. $\sin^2 x = \sin x$

42. $\cos 2x = \cot x$

43. $\sec x = \cot x$

44. $\cos^3 x = \cos x$

45. $\sin 2x = \tan x$

46. $3 \sin x = \cos x$

47. $2 \sin x = 3 \cos x$

48. $\csc x = \tan x$

49. $\cos \dfrac{x}{3} = 0.35$

50. $\sec^2 x = 2 \tan x$

51. $\tan x = \cos x$

52. $\cot x \csc x = \csc x$

53. $\cos 3x = -0.75$

54. $\tan \dfrac{x}{4} = 2.5$

Optical Application

55. The fraction f of light passing through a polarized lens depends on the acute angle θ between the light and the polarization of the lens:

$$f = \cos^2 \theta$$

Find θ necessary to produce the following percents of light:
  (a) $f = 100\%$ (b) $f = 50\%$ (c) $f = 25\%$ (d) $f = 0\%$

Geological Application

56. The radius r of the earth at latitude ϕ is given by

$$r = p + (a - p) \cos^2 \phi$$

where p is the polar radius (6357 kilometers) and a is the equatorial radius (6378 kilometers). At what latitudes ϕ does the earth have the following radii?

(a) 6360 kilometers   (b) 6365 kilometers

(c) 6370 kilometers   (d) 6375 kilometers

Physical Application

57. An object is thrown (or shot) into the air with speed v at an acute angle θ to the ground. The object falls back to the earth x units from where it was thrown. The value x satisfies

$$x = \frac{v^2}{g} \sin 2\theta$$

For $v = 30$ and $g = 9.8$, what angle(s) θ will produce the following distances?

(a) $x = 20$ (b) $x = 40$ (c) $x = 80$ (d) $x = 100$

(e) x is the maximum possible distance.

10.6 INVERSES OF TRIGONOMETRIC FUNCTIONS

Recall that a **function** can be viewed as a set of ordered pairs in which the first number is assigned to the second.

Given a function f, we can find its **inverse, f^{-1},** by interchanging its x- and y-coordinates (see page 181). For example, if $f(x) = x^2$, then f contains $(3, 9)$, and f^{-1} contains $(9, 3)$. Similarly, if $f(x) = \sin x$, then f contains

$$(0, 0), \left(\frac{\pi}{6}, \frac{1}{2}\right), \left(\frac{\pi}{4}, \frac{\sqrt{2}}{2}\right), \left(\frac{\pi}{3}, \frac{\sqrt{3}}{2}\right), \left(\frac{\pi}{2}, 1\right), \text{ and so on}$$

and f^{-1} contains

$$(0, 0), \left(\frac{1}{2}, \frac{\pi}{6}\right), \left(\frac{\sqrt{2}}{2}, \frac{\pi}{4}\right), \left(\frac{\sqrt{3}}{2}, \frac{\pi}{3}\right), \left(1, \frac{\pi}{2}\right), \text{ and so on}$$

Figure 4 shows the functions, $y = x^2$ and $y = \sin x$, and their inverses, $x = y^2$ and $x = \sin y$. (Note the symmetry of f and f^{-1} about the line $y = x$.)

The graphs of $x = y^2$ and $x = \sin y$ displayed in Figure 4 are *not* functions, since some x-values produce two or more y-values. That is, the graphs fail the vertical-line test (page 148).

FIGURE 4

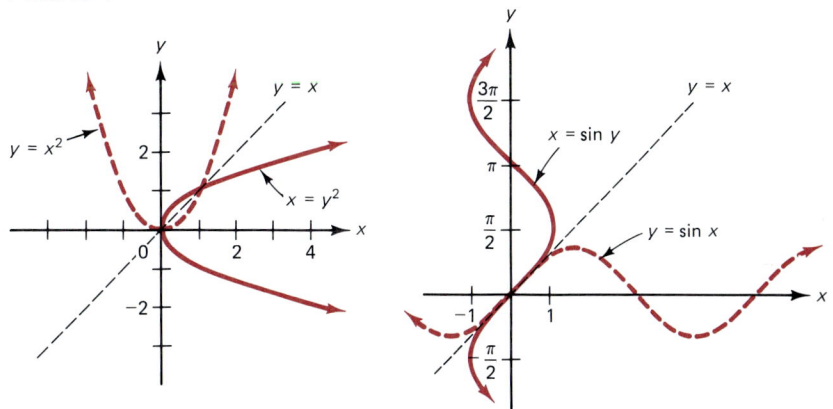

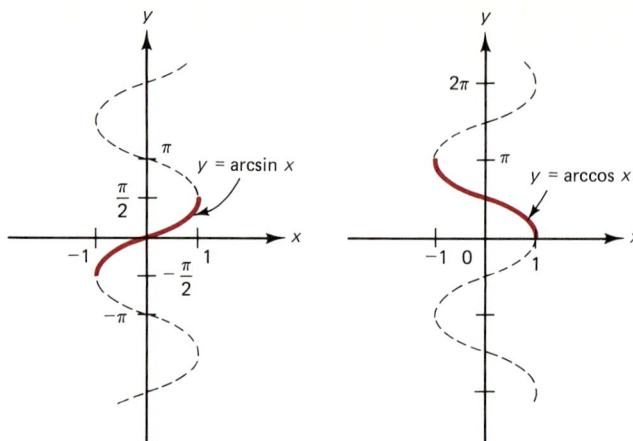

FIGURE 5

We make these inverses into functions by limiting the range so that there is only *one* y-value for each x-value. [That is, we make the functions one-to-one (see page 179). Recall that $y^2 = x$—*not a function*—becomes the function $y = \sqrt{x}$ when we restrict the range to $y \geq 0$.]

Figure 5 shows how we make the inverse-sine and inverse-cosine graphs represent functions by eliminating from the graph those y-values that are repeated and keeping only *one* complete and continuous sweep through all the possible x-values, −1 to 1.

By restricting their ranges, we can now define the **inverse-sine** and **inverse-cosine functions** as follows.

$y = \sin^{-1} x$ means $\sin y = x$ for $\dfrac{-\pi}{2} \leq y \leq \dfrac{\pi}{2}$ and $-1 \leq x \leq 1$

$y = \cos^{-1} x$ means $\cos y = x$ for $0 \leq y \leq \pi$ and $-1 \leq x \leq 1$

YES	NO
$\sin^{-1} x$ is an inverse of sine	$\sin^{-1} x = \dfrac{1}{\sin x}$

Here the notation −1 indicates inverse, *not* an exponent.

The inverse $\sin^{-1} x$ is also written arcsin x; similarly $\cos^{-1} x$ is also written arccos x. Note that there are two common notations for the inverses[1] (although your instructor may have his or her preference). Although there are

---

[1] Some texts also use the notations Sin^{-1} x and Arcsin x, and so on. Most calculus and science texts, however, use the notation that we give.

many solutions to $\sin y = x$, there is only *one* value $\sin^{-1} x$. This is called the **principal value**. (Recall that there are two solutions to $x^2 = 16$, but only *one* $\sqrt{16} = +4$, the principal root.)

EXAMPLE 27 Find $\sin^{-1}(1/\sqrt{2})$.

Solution What angle has a sine of $1/\sqrt{2}$? That is, solve $\sin y = 1/\sqrt{2}$. Actually, there are many such angles ($\pi/4$, $3\pi/4$, $9\pi/4$, $11\pi/4$, and so on), but we want only the principal value (in the range $-\pi/2 \le y \le \pi/2$). Thus, we write

$$\sin^{-1} \frac{1}{\sqrt{2}} = \frac{\pi}{4}$$

EXAMPLE 28 Find $\cos^{-1}(-0.4540)$.

Solution What angle has a cosine of -0.4540? Although there are many such angles ($117°$, $243°$, $477°$, $603°$, and so on), we want only the principal value (in the range $0 \le y \le 180°$). Thus, we have

$$\cos^{-1}(-0.4540) = 117°$$

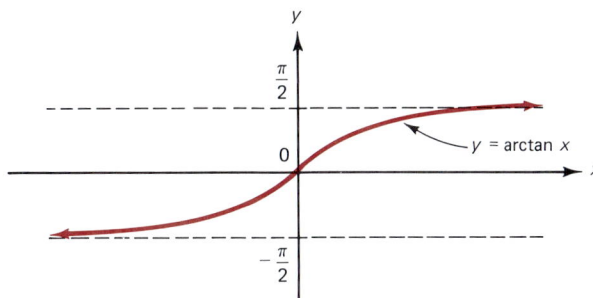

FIGURE 6

Figure 6 shows the graph of $y = \arctan x$, the inverse function of the tangent. Notice that $y = \pi/2$ and $y = -\pi/2$ are horizontal asymptotes. (This is fitting, since $x = \pi/2$ and $x = -\pi/2$ are vertical asymptotes for $y = \tan x$.) Like the inverse sine and cosine functions, the inverse tangent function has a restricted range: $-\pi/2 < y < \pi/2$. The domain is the set of all real numbers.

EXAMPLE 29 Find $\arctan 1$.

Solution What real number satisfies $\tan y = 1$ and $-\pi/2 < y < \pi/2$? The number $\pi/4$. Thus, $\arctan 1 = \pi/4$.

EXAMPLE 30 The inverse functions can easily be evaluated on a hand calculator using the $\boxed{\text{INV}}$ (or second-function) key. *The principal values have been wired right into the calculator*, and this is the only value that is returned. Consider the following.

(a) arcsin 0.73 in degrees.
(b) arccos (−0.21) in degrees.
(c) $\tan^{-1}$ 6.9 in radians.
(d) $\sin^{-1}$ 1.52 in radians.

PRESS	DISPLAY	MEANING
· 7 3 INV sin	46.886393 DEG	arcsin $(0.73) \approx 47°$
· 2 1 +/− INV cos	102.12235 DEG	arccos $(-0.21) \approx 102°$
C DRG	0. RAD	Put into radian mode
6 · 9 INV tan	1.4268709 RAD	$\tan^{-1}$ 6.9 $\approx$ 1.43
1 · 5 2 INV sin	E 0. RAD	Error: $\sin^{-1}$ 1.52 is not defined

With the last entry, we tried to trick the calculator. Since $-1 \le \sin\theta \le 1$, we cannot find $\sin^{-1}$ 1.52. Thus, the calculator answered, "Error."

Although there are inverse functions for cotangent, secant, and cosecant, we do not discuss them. Let us summarize the inverse functions.

Function	Domain	Range	Meaning
$y = \sin^{-1} x$	$-1 \le x \le 1$	$\dfrac{-\pi}{2} \le y \le \dfrac{\pi}{2}$	$\sin y = x$
$y = \cos^{-1} x$	$-1 \le x \le 1$	$0 \le y \le \pi$	$\cos y = x$
$y = \tan^{-1} x$	All real x	$\dfrac{-\pi}{2} < y < \dfrac{\pi}{2}$	$\tan y = x$

EXAMPLE 31   Sketch $\sin^{-1}$ 2/3.

Solution   Let $\theta = \sin^{-1}$ 2/3. Thus, $\sin\theta = 2/3$, and θ is an angle in the first quadrant. (Why?) We sketch the reference right triangle in Figure 7.

FIGURE 7

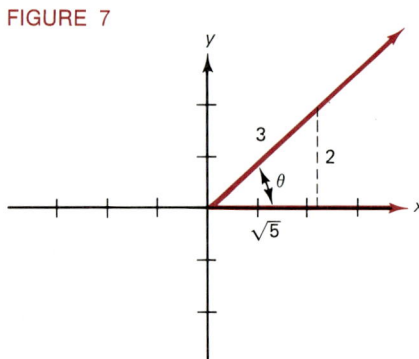

EXAMPLE 32   Find $\cos[\sin^{-1}(-4/7)]$.

Solution   This looks worse than it is. As with most mathematical problems, it is best to break this problem down into smaller pieces. Rewrite this expression as

$$\cos\underbrace{\left[\sin^{-1}\left(\frac{-4}{7}\right)\right]}_{\theta} = \cos\theta \qquad \text{where } \theta = \sin^{-1}\left(\frac{-4}{7}\right)$$

That is, we want $\cos\theta$, where θ is an angle satisfying $\sin\theta = -4/7$. Figure 8 shows the reference right triangle with $\theta = \sin^{-1}(-4/7)$. (Why is θ in quadrant IV?) We now use this triangle to compute the cosine: $\cos\theta = \sqrt{33}/7 \approx 0.821$.

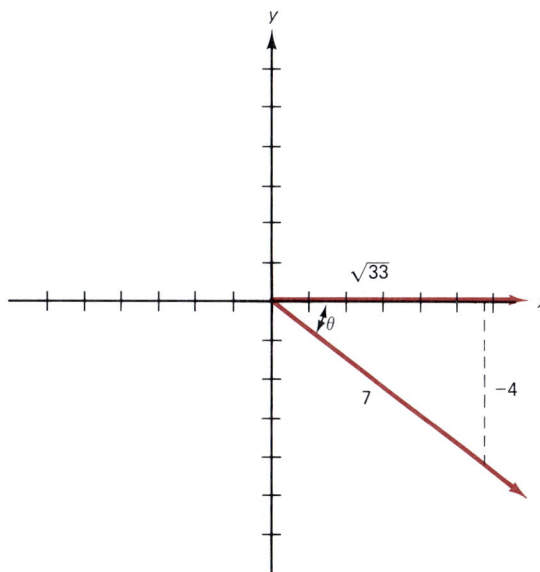

FIGURE 8

We can check this answer on a calculator:

PRESS	DISPLAY	MEANING
4 ÷ 7 +/− =	−0.5714286 DEG	$\dfrac{-4}{7}$
INV sin	−34.849904 DEG	$\sin^{-1}\left(\dfrac{-4}{7}\right) \approx -34.8°$
cos	0.8206518 DEG	$\cos\left[\sin^{-1}\left(\dfrac{-4}{7}\right)\right] \approx 0.821$

EXAMPLE 33   Write $\cot\left(\arccos\dfrac{x}{2}\right)$ as a simpler function of x.

Solution   As with Example 32, we rewrite this

$$\cot\left(\underbrace{\arccos\dfrac{x}{2}}_{\theta}\right) = \cot\theta \qquad \text{where } \theta = \arccos\dfrac{x}{2}$$

That is, we want $\cot\theta$, where θ is the angle that satisfies $\cos\theta = x/2$.

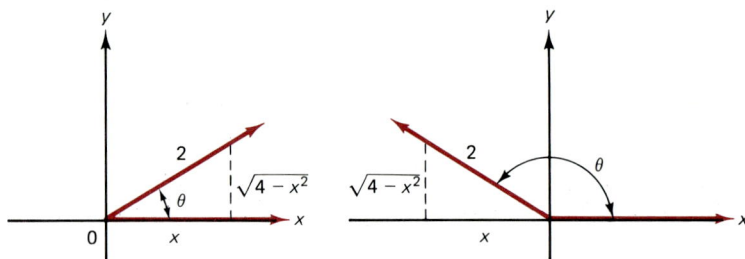

FIGURE 9

Figure 9 shows two possibilities for θ: in quadrants I or II; however, both cases yield the same result:

$$\cot\theta = \dfrac{x}{\sqrt{4-x^2}}$$

EXAMPLE 34   Find $\sin\left(\cos^{-1}\dfrac{4}{5} + \tan^{-1}\dfrac{5}{12}\right)$.

Solution   Again, we break this into smaller parts.

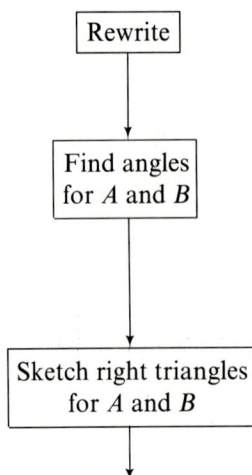

| Rewrite |

$$\sin\left(\cos^{-1}\dfrac{4}{5} + \tan^{-1}\dfrac{5}{12}\right) = \sin(A+B)$$

$$\text{where } A = \cos^{-1}\dfrac{4}{5} \quad \text{and} \quad B = \tan^{-1}\dfrac{5}{12}$$

| Find angles for A and B |

$$\text{or} \quad \cos A = \dfrac{4}{5} \quad \text{and} \quad \tan B = \dfrac{5}{12}$$

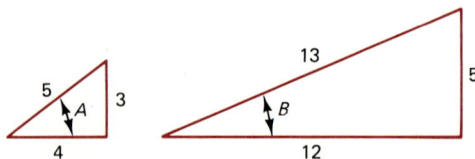

| Sketch right triangles for A and B |

```
Sum-of-angles
   identity
```

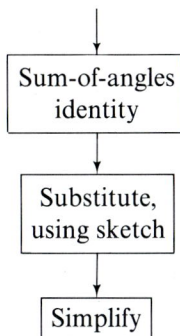

$$\sin(A + B) = \sin A \cos B + \cos A \sin B$$

```
Substitute,
using sketch
```

$$= \left(\frac{3}{5}\right)\left(\frac{12}{13}\right) + \left(\frac{4}{5}\right)\left(\frac{5}{13}\right)$$

```
Simplify
```

$$= \frac{36}{65} + \frac{20}{65} = \frac{56}{65}$$

Since the sine and arcsine functions are inverses of each other (as are cosine and arccosine), we have the following:

$$\sin(\sin^{-1} y) = y \qquad \text{for } -1 \le y \le 1$$

$$\sin^{-1}(\sin x) = x \qquad \text{for } -\frac{\pi}{2} \le x \le \frac{\pi}{2}$$

$$\cos(\cos^{-1} y) = y \qquad \text{for } -1 \le y \le 1$$

$$\cos^{-1}(\cos x) = x \qquad \text{for } 0 \le x \le \pi$$

YES	NO
$\sin^{-1}(\sin 30°) = 30°$	~~$\sin^{-1}(\sin 150°) = 150°$~~

The relation $\sin^{-1}(\sin x) = x$ is valid only for $-90° \le x \le 90°$. Try this on your calculator:

PRESS	DISPLAY	MEANING
1 5 0 sin	DEG 0.5	$\sin 150°$
INV sin	DEG 30.	$\sin^{-1}(\sin 150°) = 30°$

We can now solve certain equations by taking the sine or arcsine (cosine or arccosine) of both sides of an equation.

EXAMPLE 35   Solve $y = 2 \sin(x^3 + 1)$, for x.

Solution   We work to isolate x. First, we divide by 2; then we take $\sin^{-1}$ of both sides.

Divide by 2	$\dfrac{y}{2} = \sin(x^3 + 1)$
↓	
$\sin^{-1}$ of both sides	$\sin^{-1}\left(\dfrac{y}{2}\right) = x^3 + 1$
↓	
Subtract 1	$\sin^{-1}\left(\dfrac{y}{2}\right) - 1 = x^3$
↓	
Cube root	$\sqrt[3]{\sin^{-1}\left(\dfrac{y}{2}\right) - 1} = x$

The domain of $\sin^{-1}$ is between -1 and 1; thus,

$$-1 \le \frac{y}{2} \le 1$$

$$\text{or} \quad -2 \le y \le 2$$

Therefore, this solution for x is valid only for $-2 \le y \le 2$.

PROBLEM SET 10.6

Graph the following functions for $0 \le x \le 2\pi$.

Warm-up Exercises

1. $y = \sin x$ **2.** $y = \cos x$

3. $y = \tan x$ **4.** $y = 3 \sin 2x$

Solve the following equations, giving only the primary solutions.

Warm-up Exercises

5. $\sin y = \dfrac{\sqrt{3}}{2}$ **6.** $\cos y = \dfrac{1}{2}$ **7.** $\tan y = 1$

8. $\sin y = 0.75$ **9.** $\cos y = -0.30$ **10.** $\tan y = 4.31$

For each of the following functions, find the inverse function.

Warm-up Exercises

11. $f(x) = x - 2$ **12.** $f(x) = 3x$

13. $f(x) = 2x + 1$ **14.** $f(x) = x^3$

Find the following indicated values, with or without tables or calculator.

15. $\arcsin 0$ **16.** $\arccos 1$

17. $\sin^{-1}(-1)$ **18.** $\cos^{-1} 0$

19. $\arcsin \dfrac{\sqrt{3}}{2}$ **20.** $\arcsin\left(\dfrac{-1}{\sqrt{2}}\right)$

21. $\cos^{-1}(-0.86)$ **22.** $\sin^{-1}(-0.15)$

23. $\tan^{-1} 0.821$ **24.** $\cos^{-1}(-0.35)$

25. $\arcsin(-0.59)$ **26.** $\arctan \dfrac{1}{\sqrt{2}}$

27. $\sin^{-1} 1$ **28.** $\arccos(-1)$

29. $\arccos\left(\dfrac{-1}{2}\right)$ **30.** $\cos^{-1} 0.592$

31. $\sin^{-1} 0.57 + \cos^{-1} 0.57$

32. $\sin^{-1} 0.123 + \cos^{-1} 0.123$

34. $\tan^{-1}\left(\dfrac{2}{5}\right) + \tan^{-1}\left(\dfrac{5}{2}\right)$

34. $\tan^{-1}\left(\dfrac{-6}{7}\right) + \tan^{-1}\left(\dfrac{-7}{6}\right)$

Using x-axes and y-axes with the same scaling, graph the following pairs of functions on the same plane.

35. $y = \tan x \left(\dfrac{-\pi}{2} < x < \dfrac{\pi}{2}\right)$ and $y = \tan^{-1} x$

36. $y = 2 \sin x \left(\dfrac{-\pi}{2} \le x \le \dfrac{\pi}{2}\right)$ and $y = \sin^{-1} \dfrac{x}{2}$

37. $y = \sin 2x \left(\dfrac{-\pi}{4} \le x \le \dfrac{\pi}{4}\right)$ and $y = \dfrac{1}{2} \arcsin x$

38. $y = \cos 2x \left(0 \le x \le \dfrac{\pi}{2}\right)$ and $y = \dfrac{1}{2} \cos^{-1} x$

Give the domain and range for the following functions.

39. $y = \sin^{-1} x$

40. $y = \cos^{-1} x$

41. $y = \cos^{-1} 2x$

42. $y = \sin^{-1} 3x$

43. $y = \tan^{-1} 2x$

44. $y = \tan^{-1} \dfrac{x}{2}$

45. $y = 2 \sin^{-1} (x - 2)$

46. $y = 4 + \cos^{-1} (x + 1)$

47. $y = \sqrt{\sin^{-1} x}$

48. $y = \sqrt{\cos^{-1} x}$

Sketch the reference right triangle for each of the following angles.

49. $\sin^{-1} \dfrac{3}{5}$

50. $\arcsin \dfrac{5}{13}$

51. $\cos^{-1} \dfrac{8}{17}$

52. $\tan^{-1} \dfrac{4}{3}$

53. $\sin^{-1}\left(\dfrac{-5}{7}\right)$

54. $\arccos \dfrac{9}{10}$

55. $\sin^{-1} x$

56. $\tan^{-1} \dfrac{x}{2}$

Evaluate the following without tables or calculator.

57. $\cos\left(\tan^{-1} \dfrac{5}{12}\right)$

58. $\sin\left(\arctan \dfrac{4}{3}\right)$

59. $\tan (\arccos 0)$

60. $\cos (\sin^{-1} 0)$

61. $\sin\left[\cos^{-1}\left(\dfrac{-3}{5}\right)\right]$

62. $\cot\left[\arcsin\left(\dfrac{-8}{17}\right)\right]$

63. $\tan (\arccos x)$

64. $\sin (\tan^{-1} x)$

65. $\cos (\sin^{-1} x)$

66. $\cot (\arctan x)$

67. $\tan\left[\cos^{-1}\left(\dfrac{-4}{5}\right) - \tan^{-1} \dfrac{5}{12}\right]$

68. $\cos\left[\sin^{-1}\left(\dfrac{-4}{7}\right) + \cos^{-1} \dfrac{1}{3}\right]$

69. $\sin\left(\dfrac{\pi}{2} + \cos^{-1} \dfrac{2}{9}\right)$

70. $\cos\left(\dfrac{\pi}{2} - \tan^{-1} \dfrac{3}{4}\right)$

71. $\sin (2 \sin^{-1} x)$

72. $\cos (2 \cos^{-1} x)$

Solve the following equations for x.

73. $y = \sin x^2$

74. $y = \sin (x + 3)$

75. $y = \cos (x^2 - 1)$

76. $y + 2 = 3 \cos \dfrac{x}{2}$

77. $y = 4 \cos^{-1} (x + 1)$

78. $y = \dfrac{1}{2} \sin^{-1} (x - 1)$

79. $y = 3 \arcsin (x^3 - 1)$

80. $y + 7 = \dfrac{2}{7} \arccos \sqrt{x - 1}$

81. $y = 3 + \dfrac{1}{2} \arccos (1 - x)$

82. $y + 3 = 1 - 2 \arcsin (2x + 1)$

83. $y = 3 \sin (2x^3 + 1)$

84. $y = 5 \sin^{-1} (x + 1)$

85. $y = 2 \cos^{-1} (x - 7)$

86. $y = 4 \left(\cos^{-1} \dfrac{x}{2} \right)^{1/2}$

Electrical Application

87. The phase angle ϕ in an alternating-current (ac) circuit is given by

$$\phi = \tan^{-1} \frac{X}{R}$$

where R is the resistance and X the reactance. Find ϕ in the following cases:

(a) $X = 1000; R = 2000$ (b) $X = 20{,}000; R = 4000$
(c) $X = -3000; R = 5000$ (d) $X = 0; R = 6000$

Mathematics Application

88. The angle of inclination θ of a straight line in the plane is given by

$$\theta = \tan^{-1} m$$

where m is the slope of the line. Find θ in the following cases.

(a) $m = 2$

(b) $m = -\dfrac{1}{4}$

(c) $y = 2 - 3x$

(d) $2x - 5y = 8$

Geological Application

89. The basic equation of paleomagnetism is

$$\tan I = 2 \cot \theta$$

where I is the inclination and θ the colatitude of a site.
(a) Solve for I in terms of θ.
(b) Solve for θ in terms of I.

Physical Application

90. The basic form of a standing wave is

$$y = A \sin (kx - \omega t)$$

Solve this for x.

CHAPTER 10 SUMMARY

Important Properties and Definitions

The following are the major identities discussed in this chapter. The different variables represent all possible angles or real numbers, except where the functions are not defined.

$$\csc x = \frac{1}{\sin x} \qquad \sec x = \frac{1}{\cos x} \qquad \cot x = \frac{1}{\tan x}$$

$$\tan x = \frac{\sin x}{\cos x} \qquad \cot x = \frac{\cos x}{\sin x}$$

$$\sin^2 x + \cos^2 x = 1 \qquad \cot^2 x + 1 = \csc^2 x \qquad \tan^2 x + 1 = \sec^2 x$$

$$\sin(-x) = -\sin x \qquad \cos(-x) = \cos x$$

$$\cos(A + B) = \cos A \cos B - \sin A \sin B$$

$$\cos(A - B) = \cos A \cos B + \sin A \sin B$$

$$\sin(A + B) = \sin A \cos B + \cos A \sin B$$

$$\sin(A - B) = \sin A \cos B - \cos A \sin B$$

$$\tan(A \pm B) = \frac{\tan A \pm \tan B}{1 \mp \tan A \tan B}$$

$$\sin 2A = 2 \sin A \cos A \qquad \cos 2A = \cos^2 A - \sin^2 A$$

$$\sin^2 A = \frac{1 - \cos 2A}{2} \qquad \cos^2 A = \frac{1 + \cos 2A}{2}$$

$$\tan 2A = \frac{2 \tan A}{1 - \tan^2 A}$$

$$\sin \frac{\theta}{2} = \pm \sqrt{\frac{1 - \cos \theta}{2}} \qquad \cos \frac{\theta}{2} = \pm \sqrt{\frac{1 + \cos \theta}{2}}$$

$$\tan \frac{\theta}{2} = \frac{1 - \cos \theta}{\sin \theta}$$

$$\sin A \cos B = \frac{1}{2} [\sin(A + B) + \sin(A - B)]$$

$$\cos A \sin B = \frac{1}{2} [\sin(A + B) - \sin(A - B)]$$

$$\cos A \cos B = \frac{1}{2} [\cos(A - B) + \cos(A + B)]$$

$$\sin A \sin B = \frac{1}{2} [\cos(A - B) - \cos(A + B)]$$

$$\sin u + \sin v = 2 \sin \left(\frac{u + v}{2}\right) \cos \left(\frac{u - v}{2}\right)$$

$$\sin u - \sin v = 2 \cos \left(\frac{u + v}{2}\right) \sin \left(\frac{u - v}{2}\right)$$

$$\cos u + \cos v = 2 \cos \left(\frac{u + v}{2}\right) \cos \left(\frac{u - v}{2}\right)$$

$$\cos u - \cos v = -2 \sin \left(\frac{u + v}{2}\right) \sin \left(\frac{u - v}{2}\right)$$

$$a \sin x + b \cos x = c \sin(x + \theta) \text{ where}$$

$$c = \sqrt{a^2 + b^2}, \quad \sin \theta = b/c, \quad \text{and} \cos \theta = a/c.$$

The inverse trigonometric functions are defined by

$$y = \sin^{-1} x \quad \text{means} \quad \sin y = x \quad \text{for } \frac{-\pi}{2} \le y \le \frac{\pi}{2} \quad \text{and} \quad -1 \le x \le 1$$

$$y = \cos^{-1} x \quad \text{means} \quad \cos y = x \quad \text{for } 0 \le y \le \pi \quad \text{and} \quad -1 \le x \le 1$$

$$y = \tan^{-1} x \quad \text{means} \quad \tan y = x \quad \text{for } \frac{-\pi}{2} < y < \frac{\pi}{2} \quad \text{and} \quad \text{all real } x$$

We also have the relations

$$\sin\,(\sin^{-1} y) = y \qquad \text{for } -1 \le y \le 1$$

$$\sin^{-1}\,(\sin x) = x \qquad \text{for } \frac{-\pi}{2} \le x \le \frac{\pi}{2}$$

$$\cos\,(\cos^{-1} y) = y \qquad \text{for } -1 \le y \le 1$$

$$\cos^{-1}\,(\cos x) = x \qquad \text{for } 0 \le x \le \pi$$

Review Exercises

Complete the following table.

	Quadrant for x	$\sin x$	$\cos x$	$\tan x$	$\cot x$	$\sec x$	$\csc x$
1.	I	?	8/17	?	?	?	?
2.	III	?	?	4/3	?	?	?

Simplify each of the following expressions as much as possible (in terms of sines and cosines).

3. $\dfrac{\dfrac{\cos x}{\sin x} - \dfrac{1}{\sin x}}{\dfrac{\cos x}{\sin x} + \dfrac{1}{\sin x}}$

4. $\dfrac{\csc x}{\tan x + \cot x}$

Verify the following identities.

5. $\sec x = \tan x \csc x$

6. $\sec x - 1 = \dfrac{1 - \cos x}{\cos x}$

7. $\dfrac{\cos x}{1 + \sin x} = \dfrac{1 - \sin x}{\cos x}$

8. $\sec^2 x - \csc^2 x = \dfrac{\tan^2 x - 1}{\sin^2 x}$

9. $\dfrac{\sin^3 x - \cos^3 x}{\sin x - \cos x} = 1 - \sin x \cos x$

10. $\csc x + \cot x = \dfrac{1}{\csc x - \cot x}$

11. $\sin\,(\pi + x) = -\sin x$

12. $\cos\,(x + 45°) = \dfrac{\sqrt{2}}{2}\,(\cos x - \sin x)$

13. $\sec 2A = \dfrac{\csc^2 A \sec^2 A}{\csc^2 A - \sec^2 A}$

14. $\cos 2x = \cos^4 x - \sin^4 x$

15. $\cos \dfrac{x}{2} = \dfrac{\tan x + \sin x}{2 \tan x}$

16. $\tan \dfrac{A}{2} = \dfrac{1 - \cos 2A}{2 \sin A + \sin 2A}$

17. $\cot A \cot B = \dfrac{\cos (A - B) + \cos (A + B)}{\cos (A - B) - \cos (A + B)}$

18. $\cos^2 u - \cos^2 v = -\sin (u + v) \sin (u - v)$

Suppose that $\sin A = 0.2 \left(and\ 0 \le A \le \dfrac{\pi}{2} \right)$ and $\sin B = 0.6 \left(and\ \dfrac{\pi}{2} \le B \le \pi \right)$.

19. Find $\cos A$.

20. Find $\cos B$.

21. Find $\tan A$.

22. Find $\tan B$.

23. Find $\csc A$.

24. Find $\csc B$.

25. Find $\sec A$.

26. Find $\sec B$.

27. Find $\sin (A + B)$.

28. Find $\sin (A - B)$.

29. Find $\cos (A + B)$.

30. Find $\cos (A - B)$.

31. Find $\tan (A + B)$.

32. Find $\tan (A - B)$.

33. Find $\sin 2A$.

34. Find $\sin 2B$.

35. Find $\cos 2B$.

36. Find $\cos 2A$.

37. Find $\sin \dfrac{A}{2}$.

38. Find $\sin \dfrac{B}{2}$.

39. Find $\cos \dfrac{B}{2}$.

40. Find $\cos \dfrac{A}{2}$.

41. Find $\cos 4A$.

42. Find $\sin 3A$.

Write the following in the form $c \sin (x + \theta)$.

43. $5 \sin x - 12 \cos x$

44. $-\cos x - \sin x$

Simplify the following.

45. $\sin (x + h) - \sin x$

46. $\cos (x - h) - \cos x$

Solve the following trigonometric equations. (For Problems 47 to 53 give only the solutions $0° \le x \le 360°$; for Problems 54 to 60, $0 \le x < 2\pi$.)

47. $\cos x = \dfrac{1}{2}$

48. $\sin x = 0$

49. $\tan x = -1$

50. $\cos x = 0.415$

51. $\sin x = -0.77$

52. $\tan x = -3.56$

53. $3 \cos x + 1 = 4$

54. $\tan^2 x + \tan x = 0$

55. $2 - \sin^2 x = 2 \cos^2 x$

56. $\cot x = \cos x$

57. $\cos 2x = 0.53$

58. $\cos 2x = \sin x + 1$

59. $\dfrac{\sqrt{3}}{2} \sin x - \dfrac{1}{2} \cos x = \dfrac{1}{2}$

60. $\cos^2 2x = \dfrac{1}{4}$

Find the following indicated values.

61. $\sin^{-1} \dfrac{1}{2}$

62. $\cos^{-1} 0$

63. $\arctan(-1)$

64. $\arcsin \dfrac{\sqrt{3}}{2}$

65. $\arccos 0.81$

66. $\tan^{-1}(-0.22)$

Graph the following functions.

67. $y = \sin^{-1} x$

68. $y = \cos^{-1} x$

69. $y = \tan^{-1} x$

Evalute the following.

70. $\tan\left(\sin^{-1} \dfrac{4}{5}\right)$

71. $\sin\left(\arccos \dfrac{\sqrt{3}}{2}\right)$

72. $\cos\left(\sin^{-1} \dfrac{x}{2}\right)$

73. $\cos\left(\tan^{-1} \dfrac{1}{5}\right)$

74. $\sin\left(\tan^{-1} \dfrac{3}{4} + \tan^{-1} \dfrac{4}{3}\right)$

75. $\sin(2 \tan^{-1} x)$

Solve the following equations for x.

76. $y = 5 \sin^{-1}(2x + 3)$

77. $y = 3 \sin^2(1 - 3x)$

Triangles and Vectors

In this chapter we return to the roots of trigonometry: solving triangles. We **solve a triangle** when we take partial information about the sides and angles and find any or all of the other sides or angles.

11.1
THE LAW OF COSINES

With this section we now look beyond the right triangle to the general triangle. In addition to the right triangle, we now consider the **acute triangle** (all angles less than 90°) and the **obtuse triangle** (one angle greater than 90°). Figure 1 (page 386) shows these three types of triangles.

To solve any triangle, we always need to know *three* pieces of information about the triangle:

1. Three sides (**SSS**)
2. Two sides and the angle between them (**SAS**)
3. One side and two angles (**ASA** or **AAS**)
4. Two sides and an angle opposite one of them (**SSA**)

We consider cases 1 and 2 in this section, case 3 on page 390, and case 4 on page 394.

To deal with the SSS and SAS cases, we have the following very remarkable theorem: **the law of cosines**.

Theorem (Law of Cosines)

For any triangle (as labeled in Figure 1) we have the following relations:

$$
\begin{aligned}
c^2 &= a^2 + b^2 - 2ab \cos C \\
b^2 &= a^2 + c^2 - 2ac \cos B \\
a^2 &= b^2 + c^2 - 2bc \cos A
\end{aligned}
$$

These may appear to be three different relations. Not really. In words they all say the same thing: *In any triangle, the square of a side is equal*

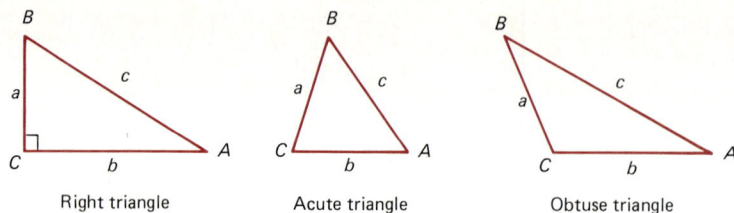

Right triangle   Acute triangle   Obtuse triangle

FIGURE 1

to the sum of the squares of the other two sides, minus twice the product of the other two sides and the cosine of the angle between them.

We now prove the first relation (the others are very similar). We place the triangle so that angle C is in standard position, as seen in either triangle in Figure 2. Notice that the coordinates of the point off the x-axis are $(a \cos C, a \sin C)$. Why? We now compute c^2 using the distance formula with the points $(b, 0)$ and $(a \cos C, a \sin C)$.

| Distance formula | $c^2 = (b - a \cos C)^2 + (0 - a \sin C)^2$ |

| Square | $= b^2 - 2ab \cos C + a^2 \cos^2 C + a^2 \sin^2 C$ |

| Factor | $= b^2 - 2ab \cos C + a^2(\cos^2 C + \sin^2 C)$ |

| $\cos^2 C + \sin^2 C = 1$ | $= a^2 + b^2 - 2ab \cos C$ |

This is the desired relation.

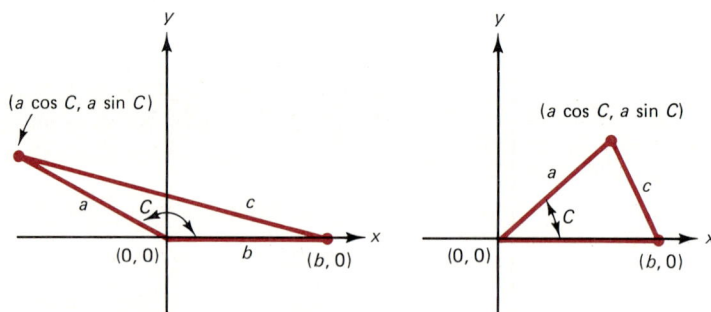

FIGURE 2

EXAMPLE 1   Find the missing side and angles in the triangle in Figure 3.

Solution   This is our first triangle that is *not* a right triangle. We use the law of cosines. This is an SAS (two sides and included angle). First, let us find c.

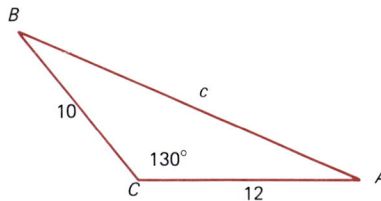

FIGURE 3

$$\boxed{c^2 = a^2 + b^2 - 2ab \cos C} \qquad\qquad c^2 = 10^2 + 12^2 - 2(10)(12) \cos 130°$$

$$\downarrow$$

$$\boxed{\text{Evaluate } \cos 130°} \qquad\qquad = 10^2 + 12^2 - 2(10)(12)(-0.6428)$$

$$\downarrow$$

$$\boxed{\text{Simplify}} \qquad\qquad = 100 + 144 + 154.272 = 398.27$$

$$\downarrow$$

$$\boxed{\text{Square root}} \qquad\qquad c = \sqrt{398.27} = 19.96$$

Thus, $c \approx 20$. Let us show how we can perform this calculation on a calculator.

PRESS	DISPLAY	MEANING
$\boxed{10}\ \boxed{x^2}\ \boxed{+}\ \boxed{12}\ \boxed{x^2}\ \boxed{-}$	$\boxed{\begin{matrix}244.\\ \text{\tiny DEG}\end{matrix}}$	$10^2 + 12^2$ minus
$\boxed{2}\ \boxed{\times}\ \boxed{10}\ \boxed{\times}\ \boxed{12}\ \boxed{\times}\ \boxed{130}\ \boxed{\cos}\ \boxed{=}$	$\boxed{\begin{matrix}398.26903\\ \text{\tiny DEG}\end{matrix}}$	$2(10)(12) \cos 130°$
$\boxed{\sqrt{x}}$	$\boxed{\begin{matrix}19.956679\\ \text{\tiny DEG}\end{matrix}}$	$c \approx 20$

We now have all three sides (SSS) and can again use the law of cosines to find the missing angles. We find angle A using the version of the law of cosines that involves $\cos A$. Here, we substitute the three sides and solve for $\cos A$ and then A.

$$\boxed{a^2 = b^2 + c^2 - 2bc \cos A} \qquad\qquad 10^2 = 12^2 + 20^2 - 2(12)(20) \cos A$$

$$\downarrow$$

$$\boxed{\text{Simplify}} \qquad\qquad 100 = 544 - 480 \cos A$$

$$\downarrow$$

$$\boxed{\text{Solve for } \cos A} \qquad\qquad 0.925 = \cos A$$

$$\downarrow$$

$$\boxed{\text{Solve for } A} \qquad\qquad 22.3° = A$$

Thus, $A = 22.3°$. We find B by subtracting:

$$B = 180° - C - A = 180° - 130° - 22.3° = 27.7°$$

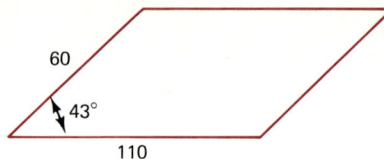

FIGURE 4

EXAMPLE 2 Find the diagonals of the parallelogram in Figure 4.

Solution We first recall some facts about parallelograms: Opposite sides are equal, and the sum of neighboring angles is 180°. Thus, the sides are 60, 110, 60, and 110, and the angles 43°, 137°, 43°, and 137°. Figure 5 shows the parallelogram cut in two ways, each cut giving a different diagonal.

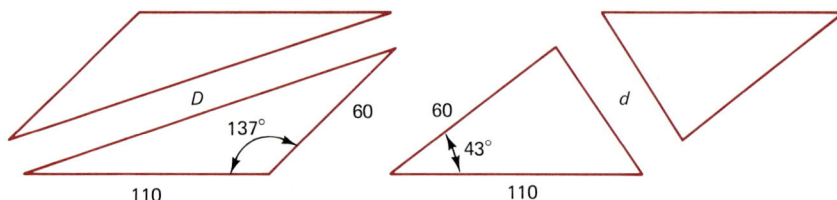

FIGURE 5

To find the diagonals, we use the law of cosines on the triangular halves. We first find the longer diagonal, D.

$$D^2 = 60^2 + 110^2 - 2(60)(110) \cos 137°$$

$\boxed{\text{Law of cosines}}$
$$= 25{,}353.869$$

$$D = \sqrt{25{,}353.869} = 159.2$$

We use the other triangle (on the right) to find d.

$$d^2 = 60^2 + 110^2 - 2(60)(110) \cos 43°$$

$\boxed{\text{Law of cosines}}$
$$= 6046.1311$$

$$d = \sqrt{6046.1311} = 77.8$$

YES	NO
$c^2 = 5^2 + 8^2 - 2(5)(8) \cos 50°$	$c^2 = 5^2 + 8^2 - 2(5)(8) \cos 50°$
To use the law of cosines, the given angle must be *between* the given sides.	

Evaluate the following.

1. $\cos 47°$

2. $\cos 0.1°$

3. $\cos 151°$

4. $\cos 141.7°$

5. $\cos^{-1} 0.7113$

6. $\cos^{-1}(-0.6123)$

Simplify the following.

7. $\sqrt{5^2 + 8^2 - 2(8)(5)(0.71)}$

8. $\sqrt{10^2 + 9^2 - 2(10)(9)(-0.43)}$

Solve the following for x.

9. $10 = 28 - 2x$

10. $76 = 48 - 2x$

11. $8^2 = 7^2 + 6^2 - 2(7)(6)x$

12. $10^2 = 5^2 + 6^2 - 2(5)(6)x$

Complete the following table for the triangle as labeled in Figure 6.

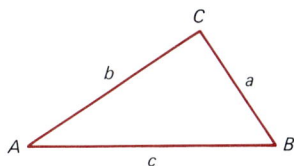

FIGURE 6

	a	b	c	A	B	C
13.	11	?	7	?	89°	?
14.	100	?	65	?	115°	?
15.	?	20	15	90°	?	?
16.	?	12	17	103°	?	?
17.	100	80	70	?	?	?
18.	10	16	18	?	?	?
19.	20	20	20	?	?	?
20.	8	15	17	?	?	?

Complete the following table for the parallelogram labeled in Figure 7.

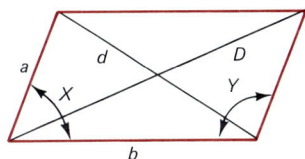

FIGURE 7

	a	b	X	Y	D	d
21.	30	30	130°	?	?	?
22.	40	10	170°	?	?	?
23.	9	12	?	19°	?	?
24.	42	37	?	115°	?	?
25.	70	60	?	?	90	?
26.	65	50	?	?	20	?
27.	120	30	?	?	?	100
28.	150	100	?	?	?	200

29. Show that the Pythagorean theorem is a special case ($C = 90°$) of the law of cosines.

30. Derive a formula for angle C in terms of adjacent sides a and b and opposite side c.

31. An airport is monitoring the paths of two airplanes flying at the same altitude. One plane is 50 miles away from the airport at a bearing of N 15° E; the other is 40 miles away at a bearing of N 65° E. How far apart are the two planes?

32. Rachel is flying an airplane from Chicago to New York (850 miles). Somehow, she manages to veer 15° off course. She discovers the error after traveling 250 miles. At this point, how far is she from New York?

33. On a baseball diamond, the distance from home plate to first base is 90 feet and to the pitcher's rubber is 60.5 feet. How far is first base from the pitcher's rubber?

34. In the following baseball fields, let L be the distance from home plate to the left-field pole and C the distance to center (as measured on a line over second base). Find the distance from the left-field pole to center.

(a) Wrigley Field: $L = 355$; $C = 400$

(b) Astrodome: $L = 340$; $C = 406$

(c) Yankee Stadium: $L = 312$; $C = 417$

Carpentry Application

35. A carpenter is measuring the angle θ that two intersecting walls make. He makes marks of length a on one wall and b on the other wall and measures c (the length between the marks). Compute θ in the following cases:

(a) $a = 3$, $b = 4$, $c = 4.9$

(b) $a = 3$, $b = 4$, $c = 5.1$

(c) $a = 2$, $b = 5$, $c = 5.5$

Surveying Applications

36. Points A and B are across a lake from each other. The distances from A and B to a third point C are 1500 meters and 1200 meters, and angle C is $51.5°$. How far apart are points A and B?

37. Two roads meet at a $63°$ angle. Car A is 1000 meters from the intersection, and car B is 850 meters from the intersection. How far apart are cars A and B?

11.2
THE LAW OF SINES

In the preceding section we used the law of cosines to solve a triangle in the SSS and SAS cases. In this section we consider the ASA and AAS cases by using the **law of sines**.

Theorem (Law of Sines)
In any triangle (as labeled in Figure 8), the sides are proportional to the sines of the angles opposite those sides:

$$\frac{a}{\sin A} = \frac{b}{\sin B} = \frac{c}{\sin C}$$

Our proof uses either the acute or the obtuse triangle shown in Figure 8. In both cases, notice that we drop a perpendicular from point C to $\overline{AB}$ (or its extension).

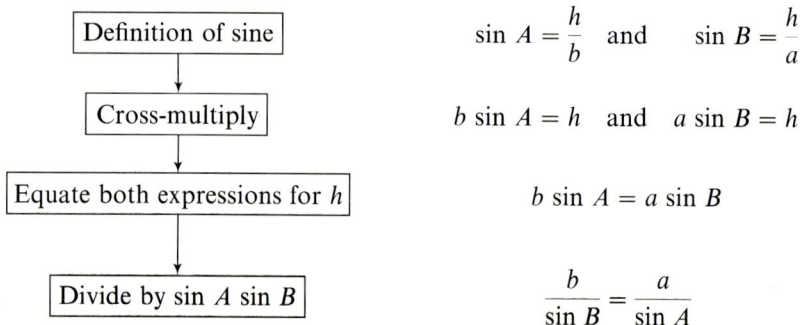

Definition of sine	$\sin A = \dfrac{h}{b}$ and $\sin B = \dfrac{h}{a}$
Cross-multiply	$b \sin A = h$ and $a \sin B = h$
Equate both expressions for h	$b \sin A = a \sin B$
Divide by $\sin A \sin B$	$\dfrac{b}{\sin B} = \dfrac{a}{\sin A}$

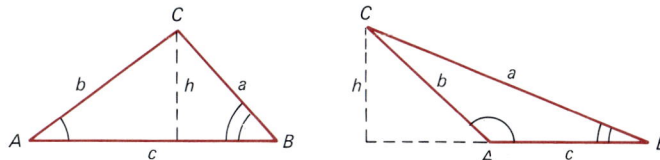

FIGURE 8

In a similar way, we can show that each of these ratios is equal to $c/\sin C$. Thus, the law of sines is proved.

When using the law of sines, notice that there are *four* quantities: two angles and the two sides opposite those angles. We will be given the two angles and one of the sides, and we will want to find the other side.

EXAMPLE 3 Find the missing side x in Figure 9.

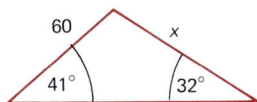

FIGURE 9

Solution The given information is of the type AAS. Including the missing side x, we have two angles and their opposite sides. Thus, we can use the law of sines.

$$\boxed{\frac{a}{\sin A} = \frac{b}{\sin B}}$$

$$\frac{x}{\sin 41°} = \frac{60}{\sin 32°}$$

$\downarrow$

$$\boxed{\text{Evaluate sines}}$$

$$\frac{x}{0.6561} = \frac{60}{0.5299}$$

$\downarrow$

$$\boxed{\text{Multiply by 0.6561}}$$

$$x = 74.29$$

Thus, $x \approx 74$. On a calculator, we might have the following sequence:

PRESS	DISPLAY	MEANING
60 × 41 sin ÷ 32 sin =	74.28215 DEG	$\dfrac{60 \sin 41°}{\sin 32°}$

EXAMPLE 4 Find the missing side x in Figure 10.

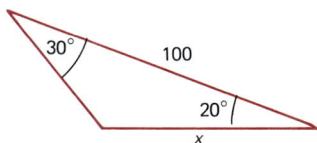

FIGURE 10

Solution Here the given information is of the form ASA. However, we have a slight problem: the given side, 100, is *not* opposite a given angle. Fortunately, this problem is easily cured, since we can find the third angle: $180° - 20° - 30° = 130°$. Figure 11 shows the triangle redrawn with the information that we need to use the law of sines.

FIGURE 11

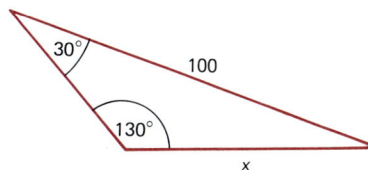

$$\boxed{\frac{a}{\sin A} = \frac{b}{\sin B}}$$

$$\frac{x}{\sin 30°} = \frac{100}{\sin 130°}$$

$\downarrow$

$$\boxed{\text{Evaluate sines}}$$

$$\frac{x}{0.5} = \frac{100}{0.7660}$$

$\downarrow$

$$\boxed{\text{Multiply by 0.5}}$$

$$x = 65.3$$

EXAMPLE 5   A ship is sailing due east. At one point, the bearing from the boat to a lighthouse is N 41° E. After 5 kilometers, the bearing of the boat to the lighthouse is N 24° E. How far is the ship from the lighthouse at the second point?

Solution   Figure 12 shows the situation. On the left are the given bearings. We need to convert these to the angles of the triangle on the right:

$$A = 90° + 24° = 114°$$

$$B = 90° - 41° = 49°$$

$$C = 180° - A - B = 17°$$

Now we can use the law of sines to find the distance b.

$$\frac{b}{\sin 49°} = \frac{5}{\sin 17°}$$

$$\boxed{\text{Law of sines}}$$

$$\frac{b}{0.7547} = \frac{5}{0.2924}$$

$$b = 12.9$$

Thus, the distance is about 13 kilometers.

FIGURE 12

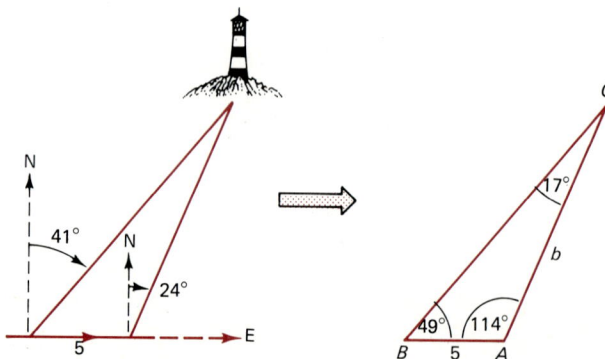

Evaluate the following.

1. $\sin 20°$ **2.** $\sin 41°$ **3.** $\sin 0.1°$

4. $\sin 101.3°$ **5.** $\sin 179.9°$ **6.** $\sin 150.3°$

Solve the following equations for x.

7. $\dfrac{x}{0.60} = \dfrac{60}{0.80}$ **8.** $\dfrac{x}{0.75} = \dfrac{200}{0.40}$

9. $\dfrac{x}{\sin A} = \dfrac{y}{\sin B}$ **10.** $\dfrac{x}{\sin A} = \dfrac{x-1}{\sin B}$

Complete the following table for the triangle in Figure 8.

	a	b	c	A	B	C
11.	10	?	?	40°	60°	?
12.	20	?	?	15°	65°	?
13.	?	150	?	70°	100°	?
14.	?	80	?	25°	110°	?
15.	?	?	25	120°	40°	?
16.	?	?	40	85°	55°	?
17.	?	15	?	?	21.7°	47.4°
18.	?	?	8	?	50.7°	69.9°
19.	10	?	?	θ	2θ	3θ
20.	?	20	?	60	3θ	θ

Complete the following table for the triangle in Figure 8 using the law of sines and/or the law of cosines.

	a	b	c	A	B	C
21.	?	20	?	51°	66°	?
22.	12	?	?	61°	79°	?
23.	30	?	?	?	81°	11.3°
24.	?	?	100	21.7°	80.9°	?
25.	120	90	?	?	?	52°
26.	?	50	55	60°	?	?
27.	10	11	12	?	?	?
28.	14	20	25	?	?	?
29.	20	?	?	39°	?	49°
30.	21	43	?	?	?	56°

31. Recall that the area of a triangle is given by

$$\text{Area} = \frac{1}{2}\,bh$$

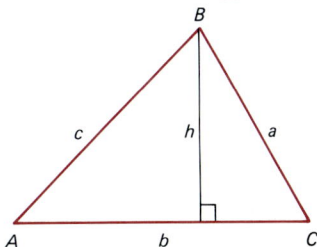

(a) Show $h = a \sin C$.

(b) Show $a = b\,\dfrac{\sin A}{\sin B}$.

(c) Use parts (a) and (b) to show that

$$\text{Area} = \frac{1}{2}b^2\,\frac{\sin A \sin C}{\sin B}$$

*Surveying
Application*

32. A real estate broker measures the frontage of a triangular lot as b, and the front angles as A and C. Use the area formula derived in Problem 31(c) to compute the area in each of the following cases.
(a) $b = 100$, $A = 42°$, $C = 46.5°$.
(b) $b = 150$, $A = 55°$, $C = 23.7°$.
(c) $b = 175$, $A = 37.3°$, $C = 34.9°$.

*Technical
Application*

33. Two guy wires are attached to the top of a pole in a plane with the pole. The wires form acute angles $49.3°$ and $37.2°$ with the ground, and the wires are 200 meters apart at the ground. What is the total amount of wire needed for the wires?

*Navigation
Applications*

34. A ship is sailing due east. At one point, the bearing from the ship to a lighthouse is N $67.8°$ E. After 7 kilometers, this bearing is N $39.1°$ E. How far is the ship from the lighthouse at the second point?

35. Another ship is sailing due west. At one point, the bearing of the ship to a lighthouse is N $31.8°$ E. After 12 miles this bearing is N $79.3°$ E. How far away is the lighthouse at this point?

36. Two observers on the ground, 10,000 meters apart, both see an airplane in a line with them at angles of elevation $51.7°$ and $29.8°$. What is the altitude of the airplane?

11.3
THE AMBIGUOUS
CASE: SSA

Our goal in this chapter is solving triangles. We use the law of cosines in the cases SSS and SAS. We use the law of sines in the cases ASA and AAS. In these cases the triangle is uniquely determined from the information; that is, there is one and only one such triangle that satisfies the given facts.

In this section we consider the last case: SSA. Unlike the other cases, the SSA does *not* always produce a unique triangle. Instead, given a side, a side, and an angle, we might get:

1. One triangle
2. Two triangles
3. No triangle at all

Let us look at *acute angle A* and sides a and b. Figure 13 shows the possible cases:

(a) a is too short: no triangle. $(a < b \sin A)$
(b) a is "just right": one triangle. $(a = b \sin A)$

FIGURE 13 *A is acute.*

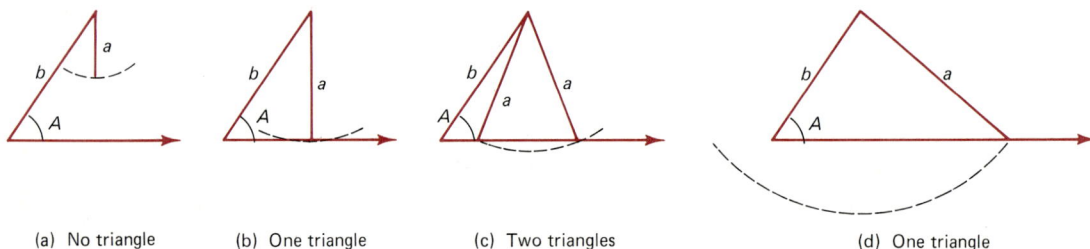

(a) No triangle   (b) One triangle   (c) Two triangles   (d) One triangle

(c) a is a bit longer: two triangles. ($b > a > b \sin A$)

(d) a is much longer: one triangle. ($a \geq b$)

(a) No triangle   (b) One triangle

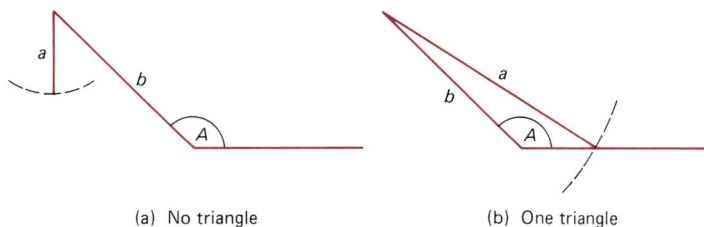

FIGURE 14 A is obtuse.

If angle A is *obtuse*, we have the two cases shown in Figure 14:

(a) a is too short: no triangle. ($a \leq b$)

(b) a is long enough: one triangle. ($a > b$)

Fortunately, there is no need to memorize all these cases. Rather, we can use the law of sines to tells us what is happening.

Rule

If the sides a and b and angle A of a triangle are given (SSA), use the law of sines to find $\sin B$, which yields the following possibilities:

(a) $0 < \sin B < 1$	Two triangles (if A is acute and $a < b$)
	One triangle (if A is obtuse or $a > b$)
(b) $\sin B = 1$	One triangle (in fact, a right triangle)
(c) $\sin B > 1$	No triangle

EXAMPLE 6 A triangle has sides $a = 8$ and $b = 10$ and angle $A = 35°$. Solve the triangle.

Solution This is an SSA case, so we must prepare ourselves for 0, 1, or 2 triangles. We substitute the given information into the law of sines and see what $\sin B$ tells us.

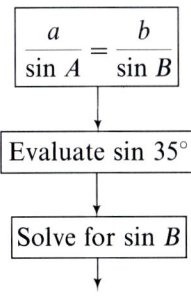

$$\boxed{\frac{a}{\sin A} = \frac{b}{\sin B}}$$

$$\frac{8}{\sin 35°} = \frac{10}{\sin B}$$

$\downarrow$

$$\boxed{\text{Evaluate } \sin 35°}$$

$$\frac{8}{0.5736} = \frac{10}{\sin B}$$

$\downarrow$

$$\boxed{\text{Solve for } \sin B}$$

$$\sin B = \frac{10(0.5736)}{8} = 0.717$$

$\downarrow$

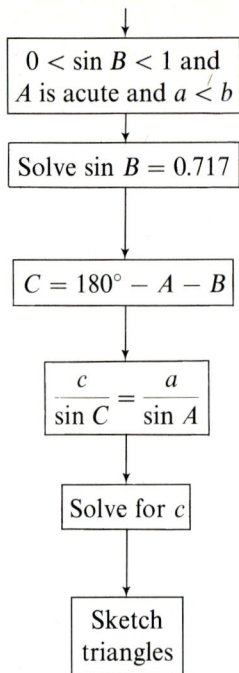

$$0 < \sin B < 1 \text{ and } A \text{ is acute and } a < b$$

Solve $\sin B = 0.717$

$C = 180° - A - B$

$$\dfrac{c}{\sin C} = \dfrac{a}{\sin A}$$

Solve for c

Sketch triangles

There are *two* triangles.

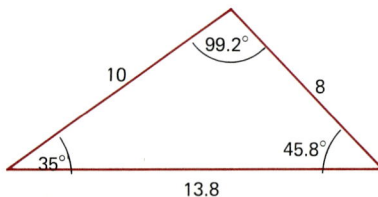

$B = 45.8°$

$C = 180° - 35° - 45.8°$
$= 99.2°$

$$\dfrac{c}{\sin 99.2°} = \dfrac{8}{\sin 35°}$$

$c = 13.8$

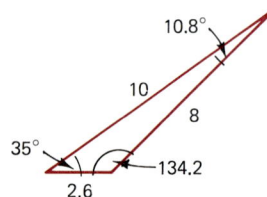

$B = 180° - 45.8°$
$= 134.2°$

$C = 180° - 35° - 134.2°$
$= 10.8°$

$$\dfrac{c}{\sin 10.8°} = \dfrac{8}{\sin 35°}$$

$c = 2.6$

Here, $\sin B = 0.717$ had two valid solutions: 45.8° and 134.2°. Each B yields a different triangle.

EXAMPLE 7   A triangle has sides $a = 10$, $b = 16$, and angle $A = 100°$. Solve the triangle.

Solution   This is again the case SSA. We substitute the given information into the law of sines and see what $\sin B$ tells us.

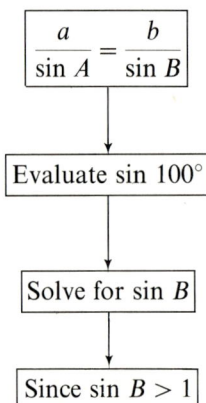

$$\dfrac{a}{\sin A} = \dfrac{b}{\sin B}$$

$$\dfrac{10}{\sin 100°} = \dfrac{16}{\sin B}$$

Evaluate $\sin 100°$

$$\dfrac{10}{0.9848} = \dfrac{16}{\sin B}$$

Solve for $\sin B$

$$\sin B = \dfrac{16(0.9848)}{10} = 1.58$$

Since $\sin B > 1$

There is *no* triangle.

Unlike Example 6 (which produced two triangles), this information produces no triangle. That is, there is no triangle with those sides and angle. (Try to draw one!)

EXAMPLE 8   A triangle has sides $a = 13$ and $b = 10$ and angle $A = 108°$. Solve the triangle.

Solution   We start this SSA case with the law of sines.

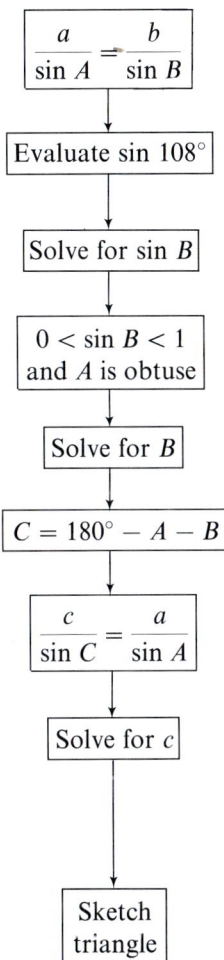

$\dfrac{a}{\sin A} = \dfrac{b}{\sin B}$	$\dfrac{13}{\sin 108°} = \dfrac{10}{\sin B}$
Evaluate $\sin 108°$	$\dfrac{13}{0.9511} = \dfrac{10}{\sin B}$
Solve for $\sin B$	$\sin B = \dfrac{10(0.9511)}{13} = 0.7316$
$0 < \sin B < 1$ and A is obtuse	There is *one* triangle.
Solve for B	$B = 47°$
$C = 180° - A - B$	$C = 180° - 108° - 47° = 25°$
$\dfrac{c}{\sin C} = \dfrac{a}{\sin A}$	$\dfrac{c}{0.4226} = \dfrac{13}{0.9511}$
Solve for c	$c = 5.8$
Sketch triangle	

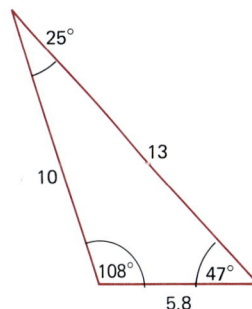

Note that we get only one triangle. Although $\sin B = 0.7316$ has another solution ($133°$), we cannot have a triangle with both $133°$ and $108°$ angles. Why not?

Table 1 summarizes the methods for solving triangles.

TABLE 1 ARBITRARY TRIANGLE

Case	Meaning (Given Information)	Method for Solving
SSS	Three sides	Law of cosines
SAS	Two sides, included angle	Law of cosines
ASA	Two angles, included side	Law of sines
AAS	Two angles, a side	Law of sines
SSA	Two sides, an angle	Law of sines (*Caution:* There may be 0, 1, or 2 triangles.)
AAA	Three angles	*Cannot* be solved without a side.

PROBLEM SET 11.3

Evaluate the following.

Warm-up Exercises

1. sin 105°　　　　　　　　**2.** sin 14.7°

3. cos 66.6°　　　　　　　**4.** cos 109.4°

5. sin^{-1} 0.713　　　　　　**6.** sin^{-1} 1.0

7. sin^{-1} 1.775　　　　　　**8.** sin^{-1} 0.078

9. cos^{-1} 0.817　　　　　**10.** cos^{-1} (−0.587)

State the following theorems.

Warm-up Exercises

11. The law of sines.　　　　**12.** The law of cosines.

Solve the following triangles with the standard labeling—see Figure 8. If there are two triangles, give them both; if there is no triangle, state this.

	a	b	c	A	B	C
13.	10	6	?	42°	?	?
14.	8	11	?	31°	?	?
15.	5	11	?	?	79°	?
16.	12	3	?	?	86°	?
17.	?	10	4	?	?	97°
18.	?	7	2.7	?	107°	?
19.	150	?	300	30°	?	?
20.	80	?	50	?	?	10°
21.	?	12	20	?	80°	?
22.	5	?	1.7	77.1°	?	?

Solve the following triangles, using the law of sines and/or the law of cosines (if possible). Give all possible solutions.

	a	b	c	A	B	C
23.	10	20	?	36°	?	?
24.	20	?	25	?	75°	?
25.	?	18	?	52°	17°	?
26.	?	?	30	18°	81°	?

27. 15	20	22	?	?	?
28. 10	42	?	?	?	32°
29. 40	?	60	?	111°	?
30. ?	?	100	50°	70°	?
31. ?	?	?	40°	60°	80°
32. 100	120	?	70°	?	?
33. ?	24	18	?	100°	?
34. ?	50	?	50°	?	50°

*Technical
Application*

35. A wrecking ball is at the end of an 80-foot steel cord. The cord is attached to the top of a 100-foot crane that forms a 41° angle to the ground. Will the ball hit the ground at all? If so, where? What is the smallest angle to the ground the crane must exceed so that the ball will not touch the ground?

11.4
VECTORS

Many physical quantities, such as time, mass, or energy, can be described by a single number (for example, 10 seconds, 8 kilograms, or 110 joules). These quantities are called **scalars** and are represented by simple variables (t, m, or E).

On the other hand, some physical quantities, such as force, velocity, or magnetic field, require both a *magnitude* and a *direction*. Such quantities are called **vectors** and are represented by letters in boldface or with arrows above them (**F**, **v**, **B**, or $\vec{F}$, $\vec{v}$, $\vec{B}$). Figure 15 illustrates such vectors.

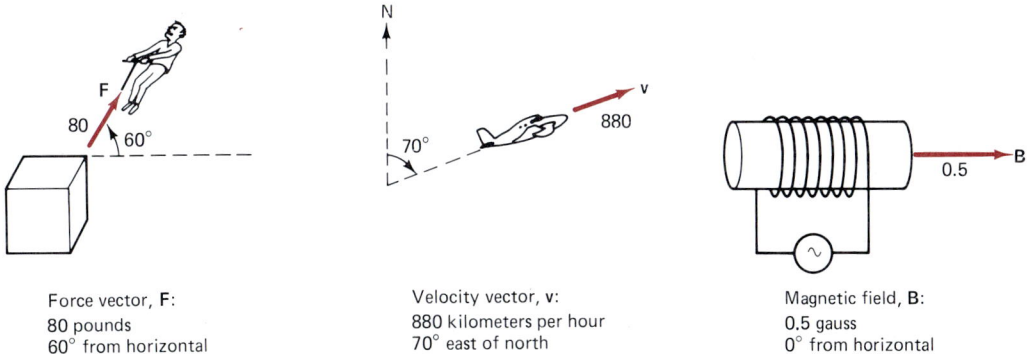

Force vector, **F**:
80 pounds
60° from horizontal

Velocity vector, **v**:
880 kilometers per hour
70° east of north

Magnetic field, **B**:
0.5 gauss
0° from horizontal

FIGURE 15 Vectors.

We can consider a vector **A** in the plane as a directed line segment with two parts: **magnitude** (length), written $|\mathbf{A}|$, and **direction** (angle from positive x-axis), written θ_A. Sometimes, we call the initial point of the vector the **tail** and the terminal point the **tip**.

Two vectors are **equal** if they have the same direction (they are parallel) and magnitude. Also, $\mathbf{B} = -\mathbf{A}$ if **B** and **A** have the same magnitude but opposite direction. Figure 16 (page 400) illustrates these ideas. (This definition of equality allows us to move a vector, as long as we do not change its direction or magnitude—only its starting point.)

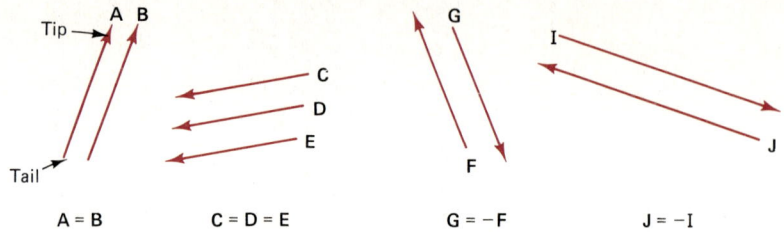

FIGURE 16

Frequently, we have two vectors (such as two forces or two velocities) and we wish to combine them and find a **resultant** or **vector sum**. Figure 17 shows two equivalent ways to add the vectors **A** and **B**:

(a) We move the tail of **B** to the tip of **A** and complete the triangle. (Recall that we can move a vector.) The sum **A** + **B** is the third side.
(b) We put the tails together and form a parallelogram. The sum **A** + **B** is the diagonal from the common tail.

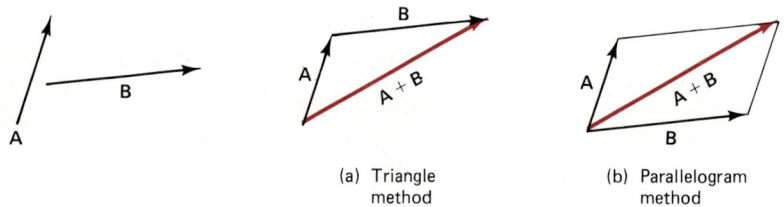

(a) Triangle method

(b) Parallelogram method

FIGURE 17

Notice that the sum **A** + **B** is the same using either method (same direction and same magnitude).

Recall that we define the difference of real numbers $a - b = a + (-b)$. We now define the **difference** of vectors **A** and **B** similarly:

$$\mathbf{A} - \mathbf{B} = \mathbf{A} + (-\mathbf{B})$$

Figure 18 illustrates the difference of two vectors. Notice that **A** − **B** is also the vector going from the tip of **B** to the tip of **A** [that is, **B** + (**A** − **B**) = **A**].

FIGURE 18

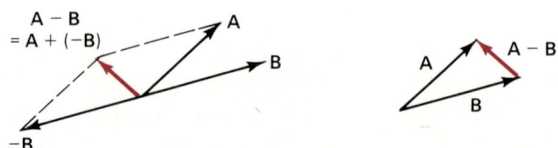

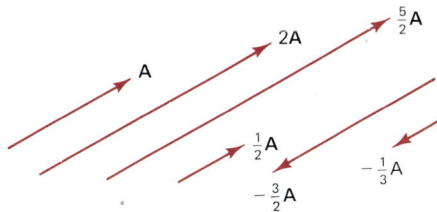

FIGURE 19

Also, if c is a scalar (real number) and **A** is a vector, we define **scalar multiplication** c**A** as the vector with magnitude $|c||\mathbf{A}|$ in the same direction as **A** if $c > 0$, and opposite direction if $c < 0$. Figure 19 shows how scalar multiplication acts like a stretching or a shrinking of **A**.

EXAMPLE 9   A plane is flying due east at an airspeed of 200 miles per hour. There is a 50-mile-per-hour wind to the north. What is the resulting speed and direction of the plane?

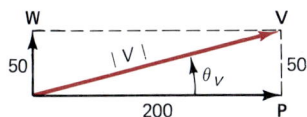

FIGURE 20

Solution   The velocities of the plane (**P**) and the wind (**W**) are vectors. Their sum, $\mathbf{V} = \mathbf{P} + \mathbf{W}$, is the resultant velocity vector of the plane. Since the wind and the plane are at right angles, we can use the Pythagorean theorem. Figure 20 shows this situation.

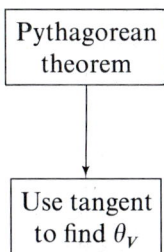

$$|V| = \sqrt{50^2 + 200^2}$$
$$= \sqrt{42{,}500} \approx 206$$

$$\tan \theta_V = \frac{50}{200}$$
$$\theta_V \approx 14°$$

Thus, the plane's velocity is 206 miles per hour (with respect to the ground) in a direction of 14° north of east. (As a bearing from the north, the direction is given as N 76° E.)

EXAMPLE 10   Becky and Lisa are two stubborn mules, each pulling a load in a direction of their own choosing. Becky is pulling with 110 pounds, 60° west of north. Lisa is pulling with 80 pounds, 70° east of north. What is the resultant force (direction and magnitude) of their combined efforts?

Solution   Force is a vector quantity, and the resultant of two forces is their sum **F**. In Figure 21, we first see a sketch of the mules' pulls; next to that, we have filled in the parallelogram (notice that the angle between

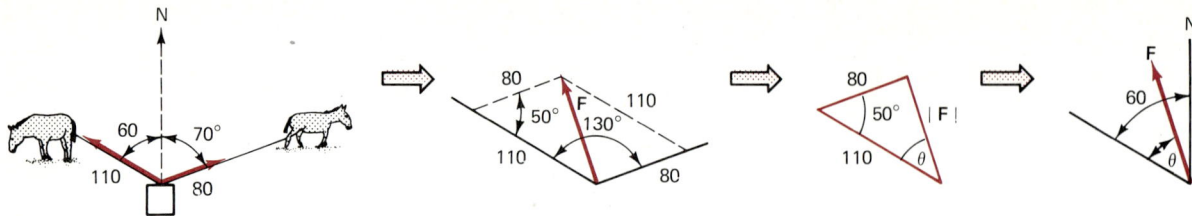

FIGURE 21

the pulls is $60° + 70° = 130°$); then we see the triangle that we use to find $|\mathbf{F}|$ and θ; finally, the last sketch shows that the direction of the resultant pull is $60° - \theta$ from the north.

Use law of cosines to find $|\mathbf{F}|$

$$|\mathbf{F}|^2 = 110^2 + 80^2 - 2(110)(80) \cos 50°$$
$$= 12{,}100 + 6400 - 17{,}600(0.6428)$$
$$= 7186.9$$
$$|\mathbf{F}| = 84.8$$

Thus, together they are exerting 84.8 pounds. (This number is small compared to $110 + 80$ pounds, since they are pulling in quite different directions.) We now find θ.

Use law of sines to find θ

$$\frac{80}{\sin \theta} = \frac{84.8}{\sin 50°}$$

$$\sin \theta = \frac{(0.7660)(80)}{84.8} = 0.7227$$

$$\theta = 46.3°$$

Thus, the resultant is $60° - 46.3° = 13.7°$ west of north. Therefore, the resulting pull is 84.8 pounds, 13.7° west of north.

EXAMPLE 11 A plane is flying at 880 kilometers per hour, 50° east of north. What is the plane's velocity in the north direction and in the east direction?

Solution In this problem, we want to **resolve** the vector into its x- and y-components. To find the north and east components, we use simple right-triangle trigonometry. See Figure 22.

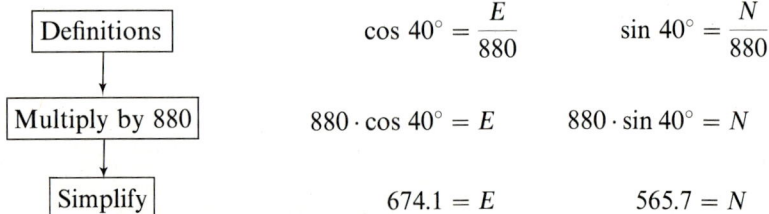

Definitions

$$\cos 40° = \frac{E}{880} \qquad\qquad \sin 40° = \frac{N}{880}$$

Multiply by 880

$$880 \cdot \cos 40° = E \qquad\qquad 880 \cdot \sin 40° = N$$

Simplify

$$674.1 = E \qquad\qquad 565.7 = N$$

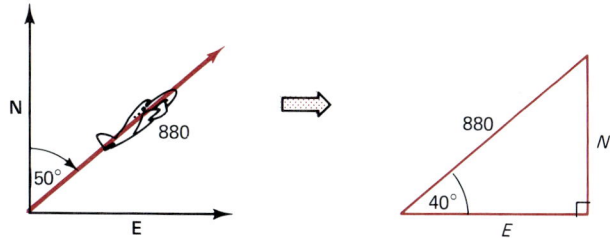

FIGURE 22

Thus, the plane is traveling 674 kilometers per hour east and 566 kilometers per hour north.

Example 11 demonstrates an important concept: resolving a vector **A** into its x- and y-components, which we call A_x and A_y. As seen in Figure 23, a vector **A** with magnitude $|\mathbf{A}|$ and direction θ_A to the x-axis has the following x- and y-components:

$$A_x = |\mathbf{A}| \cos \theta_A$$
$$A_y = |\mathbf{A}| \sin \theta_A$$

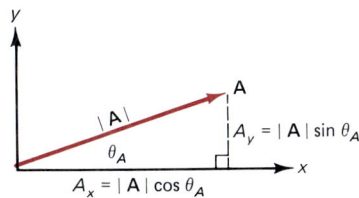

FIGURE 23

Figure 24 (page 404) shows why resolving a vector into its components is so important. We take the sum **S** of two vectors **A** and **B** and resolve all three into their x- and y-components. We can see the following:

If $\mathbf{S} = \mathbf{A} + \mathbf{B}$, then $S_x = A_x + B_x$ and $S_y = A_y + B_y$.

In words: *To add the vectors **A** and **B**, we can add their x- and y-components to obtain the x- and y-components of their sum **S**.*

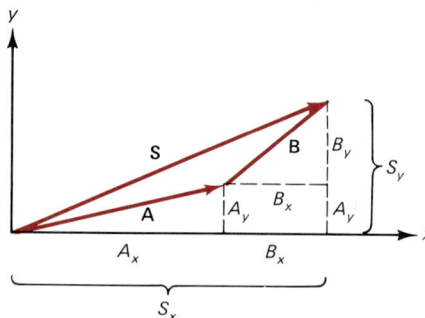

FIGURE 24

EXAMPLE 12   Let us redo Example 10 (the mules' pulls) by resolving each pull into its x- and y-components.

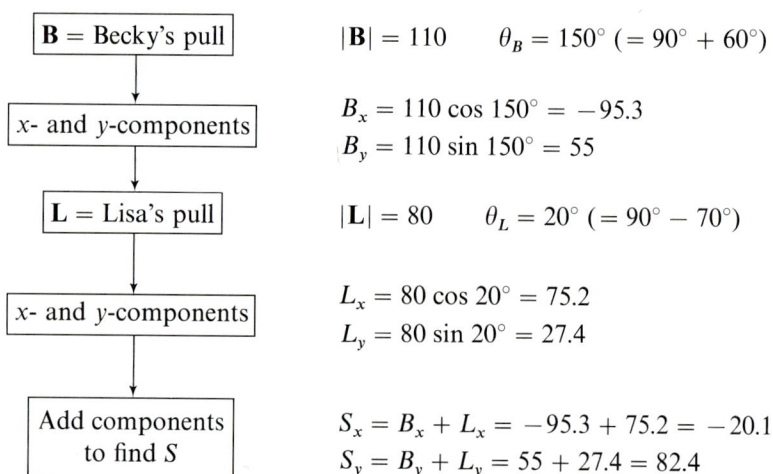

$\boxed{\textbf{B} = \text{Becky's pull}}$	$|\textbf{B}| = 110 \qquad \theta_B = 150° \, (= 90° + 60°)$

$\downarrow$

$\boxed{x\text{- and }y\text{-components}}$	$\begin{aligned} B_x &= 110 \cos 150° = -95.3 \\ B_y &= 110 \sin 150° = 55 \end{aligned}$

$\downarrow$

$\boxed{\textbf{L} = \text{Lisa's pull}}$	$|\textbf{L}| = 80 \qquad \theta_L = 20° \, (= 90° - 70°)$

$\downarrow$

$\boxed{x\text{- and }y\text{-components}}$	$\begin{aligned} L_x &= 80 \cos 20° = 75.2 \\ L_y &= 80 \sin 20° = 27.4 \end{aligned}$

$\downarrow$

$\boxed{\begin{array}{c}\text{Add components}\\\text{to find } S\end{array}}$	$\begin{aligned} S_x &= B_x + L_x = -95.3 + 75.2 = -20.1 \\ S_y &= B_y + L_y = 55 + 27.4 = 82.4 \end{aligned}$

Thus, **S** is -20.1 pounds in the x-direction and 82.4 pounds in the y-direction. (This is the resolved form for **S**.) The reader should check that **S** is 84.8 pounds, 103.7° from the positive x-axis. From the north, this direction is 13.7° west of north. This agrees with Example 10.

PROBLEM SET
11.4

Use the law of sines and/or the law of cosines to solve the following triangles with the standard labeling—see Figure 8.

	a	b	c	A	B	C
1.	100	120	?	?	?	70°
2.	180	100	200	?	?	?
3.	400	?	?	?	50°	60°
4.	50	?	?	40°	70°	?
5.	160	120	?	100°	?	?
6.	?	70	20	?	25°	?

Use the vectors shown in Figure 25 to sketch the following vectors.

7. $\mathbf{A} + \mathbf{B}$

8. $\mathbf{A} + \mathbf{C}$

9. $\mathbf{C} + \mathbf{D}$

10. $\mathbf{D} + \mathbf{E}$

11. $\mathbf{B} + \mathbf{C} + \mathbf{E}$

12. $\mathbf{A} + \mathbf{D} + \mathbf{E}$

13. $-\mathbf{A}$

14. $-\mathbf{B}$

15. $\mathbf{D} + (-\mathbf{E})$

16. $\mathbf{D} + (-\mathbf{A})$

17. $\mathbf{A} - \mathbf{B}$

18. $\mathbf{C} - \mathbf{D}$

19. $\mathbf{B} - \mathbf{A}$

20. $2\mathbf{A}$

21. $-2\mathbf{D}$

22. $-\dfrac{1}{4}\mathbf{C}$

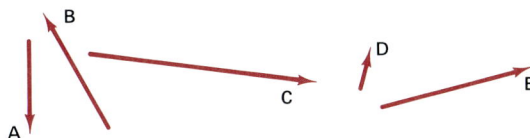

FIGURE 25

For each of the following vectors, $\mathbf{A}$ and $\mathbf{B}$, find their sum $\mathbf{S} = \mathbf{A} + \mathbf{B}$. (Hint: Sketch first.)

	$|\mathbf{A}|$	$\theta_{\mathbf{A}}$	$|\mathbf{B}|$	$\theta_{\mathbf{B}}$	$|\mathbf{S}|$	$\theta_{\mathbf{S}}$
23.	10	0°	20	90°	?	?
24.	100	0°	120	90°	?	?
25.	50	90°	100	180°	?	?
26.	200	180°	50	90°	?	?
27.	250	270°	100	180°	?	?
28.	200	180°	400	270°	?	?
29.	400	37°	500	82°	?	?
30.	40	15.1°	10	60.7	?	?

For each of the following vectors $\mathbf{A}$, its x- and y-components are A_x and A_y. Complete the following table.

	$|\mathbf{A}|$	$\theta_{\mathbf{A}}$	A_x	A_y
31.	100	70°	?	?
32.	500	31.3°	?	?
33.	150	$-33°$	?	?
34.	20	$-71.7°$	?	?
35.	?	?	10	15
36.	?	?	400	150
37.	?	?	50	-20
38.	?	?	-10	-8

For each of the following vectors, **A** and **B**, find **D** = **B** − **A**.

39. $|\mathbf{A}| = 10, \theta_\mathbf{A} = 0°; |\mathbf{B}| = 8, \theta_\mathbf{B} = 90°$

40. $|\mathbf{A}| = 8, \theta_\mathbf{A} = 90°; |\mathbf{B}| = 6, \theta_\mathbf{B} = 180°$

41. $|\mathbf{A}| = 5, \theta_\mathbf{A} = 180°; |\mathbf{B}| = 10, \theta_\mathbf{B} = 90°$

42. $|\mathbf{A}| = 4, \theta_\mathbf{A} = 270°; |\mathbf{B}| = 7, \theta_\mathbf{B} = 0°$

43. $|\mathbf{A}| = 7, \theta_\mathbf{A} = 15°; |\mathbf{B}| = 10, \theta_\mathbf{B} = 25°$

44. $|\mathbf{A}| = 5, \theta_\mathbf{A} = 80°; |\mathbf{B}| = 8, \theta_\mathbf{B} = 42°$

Complete the following table for **S** = **A** + **B**.

	A_x	A_y	B_x	B_y	S_x	S_y	$|\mathbf{S}|$	θ_s
45.	100	−50	80	90	?	?	?	?
46.	40	80	60	70	?	?	?	?
47.	−10	8	−2	−6	?	?	?	?
48.	14	9	6	−5	?	?	?	?
49.	150	−310	−70	110	?	?	?	?
50.	−600	500	200	−300	?	?	?	?

Mathematics Applications

51. Suppose that **A** and **B** are vectors with x- and y-components (A_x, A_y) and (B_x, B_y). We define the *dot product* as

$$\mathbf{A} \cdot \mathbf{B} = A_x B_x + A_y B_y$$

For the following vector, find the x- and y-components and $\mathbf{A} \cdot \mathbf{B}$. $|\mathbf{A}| = 100$, $\theta_A = 41°; |\mathbf{B}| = 150, \theta_B = 94°$.

52. Consider the dot product, as defined in Problem 51. Let d be the distance between the tips of the vectors **A** and **B**.
(a) Find d^2 using the distance formula. (*Hint*: Put tails at origin.)
(b) Find d^2 using the law of cosines. (*Hint*: Put tails at origin.)
(c) Equate these two expressions for d^2 to show

$$\mathbf{A} \cdot \mathbf{B} = |A| \cdot |B| \cos \theta$$

Hint: Use the definitions of $\mathbf{A} \cdot \mathbf{B}, |\mathbf{A}|^2$, and $|\mathbf{B}|^2$.

Physical Applications

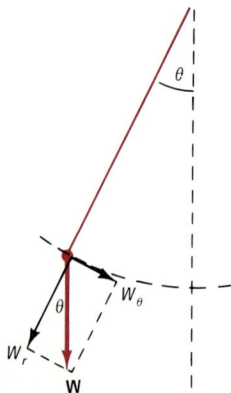

53. The displacement of an object is a vector **r**. The *work W* (energy) used in moving the object against a force **F** is

$$W = -\mathbf{F} \cdot \mathbf{r}$$

(See Problem 52.) Find the work done in the following cases.
(a) $|F| = 100, \theta_F = 50°; |r| = 150, \theta_r = 15°$
(b) $|F| = 300, \theta_F = 110°; |r| = 20, \theta_r = 75°$
(c) $|F| = 2000, \theta_F = 15°; |r| = 600, \theta_r = 105°$

54. The weight **W** of an object on a pendulum is a vector in the downward direction. This vector can be resolved into two components: W_r in the direction of the rope (which is equal to the tension in the rope), and W_θ in the direction perpendicular to the rope (which is the restoring force). Find W_r and W_θ in the following cases.
(a) $|W| = 10, \theta = 5°$
(b) $|W| = 5, \theta = 2°$
(c) $|W| = 10, \theta = 0.01$ radian
(d) $|W| = 100, \theta = 0.05$ radian

55. An airplane's velocity is a vector **v**. The wind's velocity is also a vector **w**. The resulting velocity of the airplane is the sum **v** + **w**. Find the resultant **v** + **w** in the following cases.
(a) $|\mathbf{v}| = 400$, $\theta_v = 50°$; $|\mathbf{w}| = 50$, $\theta_w = 0°$
(b) $|\mathbf{v}| = 300$, $\theta_v = 120°$; $|\mathbf{w}| = 40$, $\theta_w = 35°$

CHAPTER 11 SUMMARY

Important Properties and Definitions

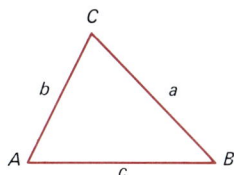

Law of cosines:

$c^2 = a^2 + b^2 - 2ab \cos C$
$a^2 = b^2 + c^2 - 2bc \cos A$
$b^2 = a^2 + c^2 - 2ac \cos B$

Law of sines:

$$\frac{a}{\sin A} = \frac{b}{\sin B} = \frac{c}{\sin C}$$

Review Exercises

Refer to Figure 26 to work Problems 1 to 4.

1. $B = 41°$, $c = 100$; find a and b.
2. $A = 15.3°$, $q = 50$; find d and a.
3. $a = 200$, $A = 15°$, $B = 22°$; find b and p.
4. $B = 31.7°$, $c = 70$, $d = 60$; find a and A.

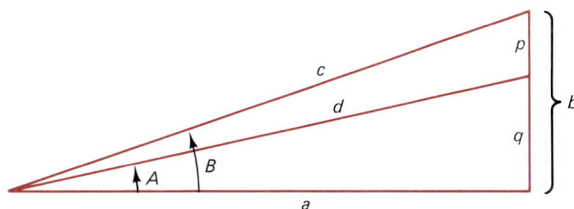

FIGURE 26

Solve the following triangles, using the standard labeling—see Figure 8.

	a	b	c	A	B	C
5.	20	25	?	40°	?	?
6.	100	?	80	?	62°	?
7.	20	30	40	?	?	?
8.	10	?	?	?	51°	62°
9.	?	?	500	?	10°	30.5°
10.	100	40	?	70°	?	?

For each of the following vectors **A** *and* **B**, *find their sums* **S** = **A** + **B**. (*Hint: Sketch first.*)

	$|\mathbf{A}|$	$\theta_\mathbf{A}$	$|\mathbf{B}|$	$\theta_\mathbf{B}$	$|\mathbf{S}|$	$\theta_\mathbf{S}$
11.	200	0°	150	90°	?	?
12.	140	180°	200	90°	?	?
13.	80	50°	110	60°	?	?
14.	150	20°	200	120°	?	?

Complete the following table for vector **A** *and its components* A_x *and* A_y.

	$|\mathbf{A}|$	$\theta_\mathbf{A}$	A_x	A_y
15.	1000	55.5°	?	?
16.	?	?	500	400

Complex Numbers and Polar Coordinates

12.1

COMPLEX NUMBERS

We observe the fact that many expressions such as $\sqrt{-36}$ are not real numbers. [Why? Since $(-6)^2 = 36$ and $(6)^2 = 36$, there is no real number whose square is -36.] Even your hand calculator is wired to reflect this fact:

PRESS	DISPLAY	MEANING
3 6 +/−	−36.	Enter −36
$\sqrt{x}$	E 0.	Error: $\sqrt{-36}$ is not real

Also, we are frustrated trying to solve certain quadratic equations, such as

$$x^2 = -36 \quad \text{or} \quad x^2 - 3x + 11 = 0$$

since their solutions involved the square roots of negative numbers. For instance, the solutions to the equations above are

$$x = \pm\sqrt{-36} \quad \text{and} \quad x = \frac{3 \pm \sqrt{-35}}{2}$$

To overcome this shortcoming of the real numbers, mathematicians invent new symbols and numbers.

Definition

$$i = \sqrt{-1} \quad \text{and} \quad i^2 = -1$$

The symbol i is not a real number; rather, it is called an **imaginary number**. Using this notation, we can compute the square roots of negative numbers.

Definition

For any real number $a \geq 0$,

$$\sqrt{-a} = i\sqrt{a}$$

EXAMPLE 1 The following are examples of imaginary numbers.

(a) $\sqrt{-1} = i$

(b) $\sqrt{-5} = i\sqrt{5}$

(c) $\sqrt{-9} = 3i$

(d) $\sqrt{-24} = i\sqrt{24} = 2i\sqrt{6}$

The **complex numbers C** are defined as the numbers of the form $a + bi$ or $a + ib$ (for real numbers a and b). For example, $2 + 5i$ and $7 - i\sqrt{2}$ are complex numbers. For a complex number $a + bi$, we call a the **real part** and b the **imaginary part**.

EXAMPLE 2 The following table shows the real and imaginary parts of complex numbers. Notice that a real number (like 10) is a complex number whose imaginary part is 0.

	Complex Number	Real Part	Imaginary Part
(a)	$5 + 7i$	5	7
(b)	$\dfrac{-3 - i\sqrt{77}}{4}$	$\dfrac{-3}{4}$	$\dfrac{-\sqrt{77}}{4}$
(c)	10	10	0
(d)	$i\sqrt{5}$	0	$\sqrt{5}$

YES	NO
For $4 + 5i$, the imaginary part is 5.	~~For $4 + 5i$, the imaginary part is 5i.~~
The imaginary part of $a + bi$ is the real number b.	

Two complex numbers are **equal** if their real parts are equal *and* their imaginary parts are equal. For instance,

$$x + yi = 2 + 5i \quad \text{if and only if} \quad x = 2 \text{ and } y = 5$$

Similarly, $3 + 5i \neq 3 - 6i$.

As we mentioned earlier, the set of real numbers is contained within the set of complex numbers (since real number x can be written $x + 0i$). Thus, the complete picture of the number systems can be viewed in Figure 1.

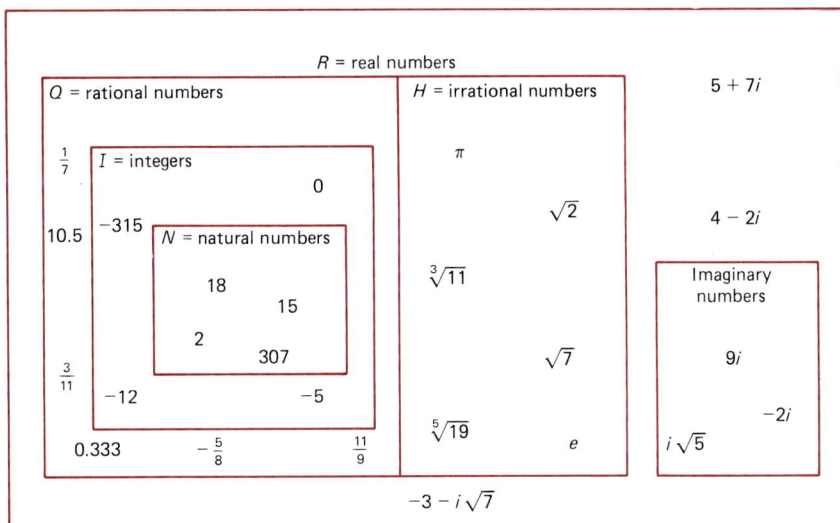

C = complex numbers

R = real numbers

Q = rational numbers

H = irrational numbers

$5 + 7i$

$\frac{1}{7}$

I = integers

0

π

$\sqrt{2}$

$4 - 2i$

10.5 -315

N = natural numbers

$\sqrt[3]{11}$

Imaginary numbers

18

15

2

307

$\sqrt{7}$

$9i$

$\frac{3}{11}$

-12

-5

$-2i$

0.333 $-\frac{5}{8}$ $\frac{11}{9}$

$\sqrt[5]{19}$

e

$i\sqrt{5}$

$-3 - i\sqrt{7}$

FIGURE 1

The complex numbers have their own algebra; that is, we can add, subtract, multiply, and divide them. (It is often convenient to think of complex numbers as if they were first-degree polynomials. For instance, treat $5 - 2i$ like $5 - 2x$. They are *not* the same, but their algebra is very similar.)

Definition

For complex numbers $a + bi$ and $c + di$, we define the **sum** as

$$(a + bi) + (c + di) = (a + c) + (b + d)i$$

EXAMPLE 3 In words, *to add complex numbers, we add the real parts and add the imaginary parts.*

(a) $(5 + 7i) + (2 - 3i) = (5 + 2) + (7 - 3)i = 7 + 4i$

(b) $(4 - i) + (7 + 2i) + (3 - 5i) = (4 + 7 + 3) + (-1 + 2 - 5)i$
$$= 14 - 4i$$

EXAMPLE 4 *To subtract complex numbers, we add the negative of the number to be subtracted.* For example, the negative of $3 - 6i$ is $-3 + 6i$.

(a) $(5 + 7i) - (4 - 2i) = (5 + 7i) + (-4 + 2i) = 1 + 9i$

(b) $(8 - i) - (-9 + 4i) = (8 - i) + (9 - 4i) = 17 - 5i$

Note how we treated the last subtraction almost exactly like the polynomial subtraction $(8 - x) - (-9 + 4x) = 17 - 5x$.

Definition

For complex numbers $a + bi$ and $c + di$, we define the **product** as

$$(a + bi)(c + di) = (ac - bd) + (ad + bc)i$$

Unfortunately, this is not an easy definition to remember. An easier approach is the following.

> To multiply complex numbers:
>
> 1. Multiply them as if they were two binomials (using the distributive law).
> 2. Use $i^2 = -1$ to simplify.

EXAMPLE 5   Multiply $(2 + 3i)(5 - 4i)$.

Solution   We multiply these numbers as though this were the polynomial multiplication $(2 + 3x)(5 - 4x)$ and then use $i^2 = -1$.

| Distributive law |

$(2 + 3i)(5 - 4i) = 10 - 8i + 15i - 12i^2$

| $i^2 = -1$ |

$= 10 - 8i + 15i + 12$

| Combine terms |

$= 22 + 7i$

Now let us look at the powers of i:

$$i^1 = i$$
$$i^2 = -1 \quad \text{(definition of } i\text{)}$$
$$i^3 = i^2 \cdot i = (-1)i = -i$$
$$i^4 = i^2 \cdot i^2 = (-1)(-1) = 1$$
$$i^5 = i^4 \cdot i = 1 \cdot i = i$$
$$i^6 = i^4 \cdot i^2 = 1 \cdot i^2 = -1$$
$$\vdots$$

If we continue this, we note that the sequence repeats itself every four exponents (or, in multiples of four).

Theorem 1

The powers of i are given by the following (k is an integer):

$$i = i^1 = i^5 = i^9 = \cdots = i^{4k+1}$$

$$-1 = i^2 = i^6 = i^{10} = \cdots = i^{4k+2}$$

$$-i = i^3 = i^7 = i^{11} = \cdots = i^{4k+3}$$

$$1 = i^4 = i^8 = i^{12} = \cdots = i^{4k}$$

EXAMPLE 6   Since $i^{4k} = 1$, we always factor out an i^{4k} term.

(a) $i^{15} = i^{12} \cdot i^3 = i^3 = -i$ (b) $i^{62} = i^{60} \cdot i^2 = i^2 = -1$

(c) $i^{25} = i^{24} \cdot i = i$ (d) $i^{32} = 1$

**PROBLEM SET
12.1**

*Warm-up
Exercises*

Simplify the following as much as possible.

1. $\sqrt{16}$ 2. $\sqrt{36}$
3. $\sqrt{12}$ 4. $\sqrt{200}$
5. $(6 + 2x) + (5 - 3x)$ 6. $(4 - x) + (5 + 4x) + (7 - 9x)$
7. $(9 - 4x) - (5 - 2x)$ 8. $(5 + x) - (4 - 2x) - (8 - 3x)$
9. $(2 + x)(3 - 5x)$ 10. $(4 + 3x)^2$
11. $(7 - 5x)(7 + 5x)$ 12. $(8 + x)(8 - x)$

Identify the real and imaginary parts of the following complex numbers.

13. $4 - 5i$ 14. $10 + 7i$ 15. $\dfrac{5 + i\sqrt{3}}{2}$

16. $\dfrac{-4 - i\sqrt{7}}{5}$ 17. -7 18. $4i$

Simplify the following to the form $a + bi$.

19. $\sqrt{-4}$ 20. $\sqrt{-25}$ 21. $\sqrt{-3}$
22. $\sqrt{-11}$ 23. $\sqrt{-20}$ 24. $\sqrt{-180}$
25. $(7 - 2i) + (5 + i)$ 26. $(8 - 7i) + (-9 - 4i)$
27. $2i + 4i + 8i - 10$ 28. $5 + 5i - 7i - 9i$
29. $(7 + 5i) + (6 - 3i) + (-2 + i)$ 30. $(5 - 3i) + (4 + 2i) + (-3 - 8i)$
31. $(5 - 6i) - (7 - 8i)$ 32. $(4 - 9i) - (-10 - i)$
33. $(4 + i) - (3 - i) - (-1 + 2i)$ 34. $(7 - i) - (10 - 4i) - (-2 + 5i)$
35. $(5 + 4i)(1 - 2i)$ 36. $(6 - 3i)(2 + 3i)$
37. $(10 - i)(5 - 3i)$ 38. $(3 + 5i)(2 - 4i)$
39. $(1 + 7i)^2$ 40. $(2 - 5i)^2$
41. $(5 - i)(5 + i)$ 42. $(8 + 5i)(8 - 5i)$
43. i^{11} 44. i^{18}
45. i^{27} 46. i^{32}
47. $(7 + i) + (8 - 3i)$ 48. $(8 + 5i)(10 - 2i)$

49. $(7 - 2i)(6 + 5i)$

50. $(4 - 3i) - (-3 - 6i)$

51. $(8 + 5i)^2$

52. $(8 - 4i) + (-3 + 7i)$

53. $4(5 + 3i)^2 + 2(5 + 3i) + 2i$

54. $2(3 + 4i)^2 - 5(3 + 4i) + 7$

Electrical Applications

55. In electricity, the impedance Z in an alternating-current circuit is given by the complex number $R + jX$. (Here, j is used instead of i, since i usually denotes the current.) The reactance X is given by $L\omega - 1/\omega C$, where L is the inductance, C is the capacitance, R is the resistance, and ω is the angular frequency. Find $Z = R + jX$ in the following cases.
(a) $R = 1000$; $L = 0.5$, $C = 0.0000005$; $\omega = 4000$
(b) $R = 2500$; $L = 0.1$; $C = 0.000005$; $\omega = 1000$
(c) $R = 500$; $L = 0.2$; $C = 0.000005$; $\omega = 1000$

56. Ohm's law for the voltage V across an impedance Z gives

$$V = I \cdot Z$$

where I is the current (all three of these quantities are complex numbers). Find V in the following cases.
(a) $I = 0.00002 + j0.00001$; $Z = 2400 - j3300$
(b) $I = 0.00005 - j0.00002$; $Z = 3500 + j1500$

12.2 COMPLEX CONJUGATES

We often consider complex numbers in pairs. If $z = x + yi$ is a complex number, then we say that

$$\bar{z} = x - yi$$

is its **conjugate**.

EXAMPLE 7 Consider the following complex numbers and their conjugates. Notice that we find the negative of the imaginary part of the number.

Number	Conjugate
(a) $4 + 5i$	$4 - 5i$
(b) $9 - i$	$9 + i$
(c) $\dfrac{4}{3} + \dfrac{i\sqrt{7}}{3}$	$\dfrac{4}{3} - \dfrac{i\sqrt{7}}{3}$
(d) 6	6
(e) $-3i$	$3i$

We have the following useful facts about conjugate pairs.

Theorem 2

If z and $\bar{z}$ are conjugates of each other, then

C1 $z + \bar{z}$ is a real number.

C2 $z \cdot \bar{z}$ is a nonnegative real number.

We demonstrate these with the following examples.

EXAMPLE 8 We add the following conjugate pairs and always get a real sum.

(a) $(4 + 5i) + (4 - 5i) = 8$ (b) $(a + bi) + (a - bi) = 2a$

EXAMPLE 9 We multiply the following conjugate pairs and always get a real product. [We use the special product $(x + y)(x - y) = x^2 - y^2$ and the fact that $i^2 = -1$.]

(a) $(8 + 5i)(8 - 5i) = 64 - 25i^2 = 64 + 25 = 89$

(b) $(3 - 11i)(3 + 11i) = 9 - 121i^2 = 9 + 121 = 130$

(c) $(a + bi)(a - bi) = a^2 - b^2i^2 = a^2 + b^2$

(d) $(7i)(-7i) = -49i^2 = 49$

Note that the product is *always a nonnegative real number.*

Theorem 3

If $z_1 = a + bi$ and $z_2 = c + di$ are two complex numbers, then

---

C3 $\overline{z_1 + z_2} = \bar{z_1} + \bar{z_2}$

C4 $\overline{z_1 \cdot z_2} = \bar{z_1} \cdot \bar{z_2}$

C5 $\overline{z_1^k} = \bar{z_1}^k$

In words, this theorem says that if we first add (multiply or raise to a power) and then take the conjugate, we get the same result as if we first take the conjugates and then add (multiply or raise to a power).

We can use the conjugate to help us divide complex numbers. Note the similarity to simplifying radical expressions by rationalizing.

To divide complex numbers:

1. Find the conjugate of the denominator.

2. Multiply the fraction by $1 = \dfrac{\text{conjugate}}{\text{conjugate}}$.

3. Simplify.

EXAMPLE 10 Divide $\dfrac{7 + 3i}{2 - i}$.

Solution The conjugate of the denominator is $2 + i$.

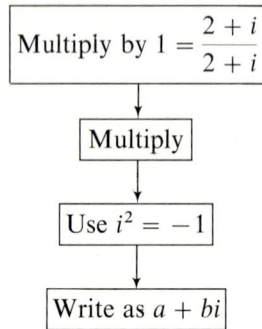

$$\boxed{\text{Multiply by } 1 = \frac{2 + i}{2 + i}}$$

$$\downarrow$$

$$\boxed{\text{Multiply}}$$

$$\downarrow$$

$$\boxed{\text{Use } i^2 = -1}$$

$$\downarrow$$

$$\boxed{\text{Write as } a + bi}$$

$$\frac{7 + 3i}{2 - i} = \frac{7 + 3i}{2 - i} \cdot \frac{2 + i}{2 + i}$$

$$= \frac{14 + 7i + 6i + 3i^2}{4 - i^2}$$

$$= \frac{11 + 13i}{5}$$

$$= \frac{11}{5} + \frac{13}{5}i$$

Note that when we multiply the denominator by its conjugate, the denominator becomes the real number 5.

PROBLEM SET
12.2

Simplify the following to the form $a + bi$.

Warm-up
Exercises

1. $\sqrt{-100}$ **2.** $-\sqrt{-20}$
3. i^7 **4.** i^5
5. $(5 + 3i) + (7 - 2i)$ **6.** $(7 - i) - (-9 - 2i)$
7. $(7 + i)(6 - 3i)$ **8.** $(2 - 5i)^2$
9. $(5 - 4i)(5 + 4i)$ **10.** $(6 + i)(6 - i)$

Simplify the following by rationalizing the denominators.

Warm-up
Exercises

11. $\dfrac{1}{\sqrt{2}}$ **12.** $\dfrac{7}{\sqrt{3}}$

13. $\dfrac{6}{5 - \sqrt{7}}$ **14.** $\dfrac{3}{4 + \sqrt{2}}$

For each of the following complex numbers, give its conjugate.

15. $5 + 6i$ **16.** $9 - 2i$ **17.** $\dfrac{5 - 3i}{4}$

18. $\dfrac{7 + 8i}{6}$ **19.** $7i$ **20.** $-i$

21. -2 **22.** $\dfrac{8}{3}$

For each of the following complex numbers z, find $z + \bar{z}$ and $z \cdot \bar{z}$. That is, find the sum and the product of the number and its conjugate.

23. $3 + 4i$ **24.** $5 - i$ **25.** $8 - 2i$
26. $2 + 7i$ **27.** $4 + i\sqrt{3}$ **28.** $6 - i\sqrt{2}$
29. $10i$ **30.** $-3i$

For Problems 31 to 36, let $z_1 = 4 + i$, $z_2 = 5 - 2i$, $z_3 = 2 - 3i$, and $z_4 = -3 + i\sqrt{7}$.

31. (a) Find $\overline{z_1 + z_2}$. (b) Find $\overline{z_1} + \overline{z_2}$. (c) Are they equal?

32. (a) Find $\overline{z_1 + z_3}$. (b) Find $\overline{z_1} + \overline{z_3}$. (c) Are they equal?

33. (a) Find $\overline{z_1 \cdot z_2}$. (b) Find $\overline{z_1} \cdot \overline{z_2}$. (c) Are they equal?

34. (a) Find $\overline{z_2 \cdot z_3}$. (b) Find $\overline{z_2} \cdot \overline{z_3}$. (c) Are they equal?

35. (a) Find $\overline{z_1^2}$. (b) Find $\overline{z_1}^2$. (c) Are they equal?

36. (a) Find $\overline{z_4^2}$. (b) Find $\overline{z_4}^2$. (c) Are they equal?

Simplify the following to the form $x + yi$.

37. $\dfrac{2 + 3i}{4 - 2i}$ **38.** $\dfrac{5 - 2i}{3 + 4i}$ **39.** $\dfrac{5 - 7i}{6 - 2i}$

40. $\dfrac{7}{6 + i}$ **41.** $\dfrac{-3i}{4 - i}$ **42.** $\dfrac{10 + i}{5 + i}$

43. $\dfrac{4 + i}{i}$ **44.** $\dfrac{5 + 2i}{-7i}$ **45.** $\dfrac{1}{i}$

46. $\dfrac{1}{-i}$ **47.** $\dfrac{5}{i^3}$ **48.** $\dfrac{-1}{i^{15}}$

The following exercises yield the proofs in Theorem 3. Let $z_1 = a + bi$ and $z_2 = c + di$.

49. Find $\overline{z_1} + \overline{z_2}$ and $\overline{z_1 + z_2}$. Are they equal?

50. Find $\overline{z_1}\,\overline{z_2}$ and $\overline{z_1 z_2}$. Are they equal?

51. Find $\overline{z_1^2}$ and $\overline{z_1}^2$. Are they equal?

52. Find $\overline{z_2^2}$ and $\overline{z_2}^2$. Are they equal?

Solve the following equations for z (in the form $a + bi$).

53. $z + 2 + 3i = 5 - 4i$ **54.** $z - (9 - 2i) = 10 + 5i$

55. $(2 - i)z = 5 + i$ **56.** $(4 + 3i)z = 6i$

Electrical Applications

57. As seen in Problem 55, page 414, the impedance in an alternating-current circuit is $Z = R + jX$, where $X = L\omega - 1/\omega C$. For $R = 1000$ ohms, $L = 0.5$ henry, $C = 0.0000002$ farad, and $\omega = 2000$ radians per second, find:

(a) Z (b) $\overline{Z}$ (c) $Z \cdot \overline{Z}$ (d) $\sqrt{Z \cdot \overline{Z}}$

58. As seen in Problem 56, page 414, Ohm's law is

$$V = I \cdot Z$$

Find the current I (in the form $a + jb$) for the following voltages V and impedances Z.

(a) $V = 110$; $Z = 2500 + j1500$

(b) $V = 150 + j150$; $Z = 3500 - j4500$

12.3 TRIGONOMETRIC (OR POLAR) FORM OF A COMPLEX NUMBER

Complex numbers can be represented graphically. Since complex numbers have two parts (real and imaginary), we need a two-dimensional plane. The **complex plane** is pictured very much like the x-y plane that we have seen throughout this text.

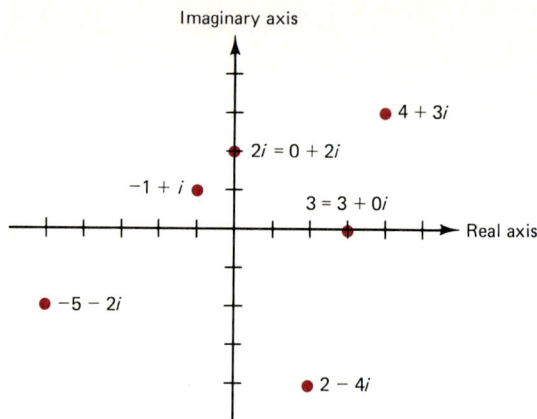

Imaginary axis

● 4 + 3i

● 2i = 0 + 2i

−1 + i ●

3 = 3 + 0i

Real axis

● −5 − 2i

● 2 − 4i

FIGURE 2

1. We put the real part on the *x*-axis (called the **real axis**).
2. We put the imaginary part on the *y*-axis (called the **imaginary axis**).

EXAMPLE 12   In Figure 2 we show the complex numbers $4 + 3i$, $2 - 4i$, $-1 + i$, and $-5 - 2i$. [Note how we graph these exactly as we would the ordered pairs $(4, 3)$, $(2, -4)$, $(-1, 1)$, and $(-5, -2)$.] The real number $3 = 3 + 0i$ lies on the real axis. The imaginary number $2i = 0 + 2i$ lies on the imaginary axis.

In Figure 3 we see the complex number $z = x + iy$ pictured on the plane. We have a trigonometric form for describing a complex number in the plane. This method involves r (the **absolute value** of z) and θ (the **argument** of z). As we will see in the next section, it is usually easier to multiply and find the roots of complex numbers in this form; this is a major reason that we introduce this form.

Recall that $\cos \theta = x/r$ and $\sin \theta = y/r$. (These were the original definitions!) We can solve for x and y:

$$x = r \cos \theta \qquad \text{and} \qquad y = r \sin \theta$$

FIGURE 3

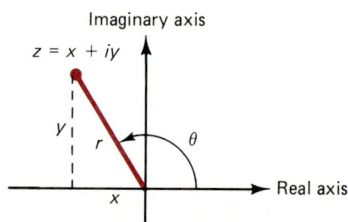

Imaginary axis

$z = x + iy$

y

r

θ

Real axis

x

Let us now rewrite $z = x + iy$, using these values.

| Substitute | $x + iy = r \cos \theta + ir \sin \theta$ |

| Factor r | $x + iy = r(\cos \theta + i \sin \theta)$ |

We call this the **trigonometric** or **polar form**[1] of a complex number:

$$x + iy = r(\cos \theta + i \sin \theta)$$

where $x = r \cos \theta$ and $r = \sqrt{x^2 + y^2}$

$y = r \sin \theta$ $\tan \theta = y/x$

The signs of x and y determine the quadrant of θ.

EXAMPLE 13   Write $z = 12 + 5i$ in polar form.

Solution   We show $12 + 5i$ in Figure 4. We sketch in the right triangle to help us find r and θ.

| Pythagorean theorem | $r = \sqrt{12^2 + 5^2} = \sqrt{169} = 13$ |

| θ is in quadrant I | $\tan \theta = \dfrac{5}{12}$ means $\theta \approx 22.6°$ |

| $z = r(\cos \theta + i \sin \theta)$ | Thus, $12 + 5i = 13(\cos 22.6° + i \sin 22.6°)$ |

FIGURE 4

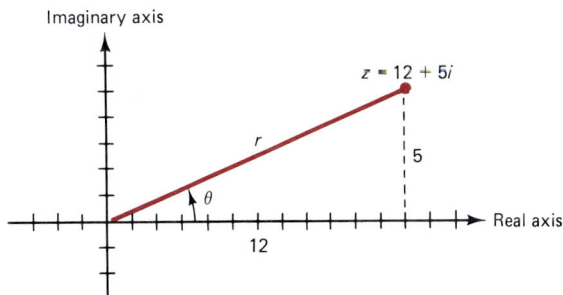

---

[1] Some textbooks use the notation cis $\theta = \cos \theta + i \sin \theta$.

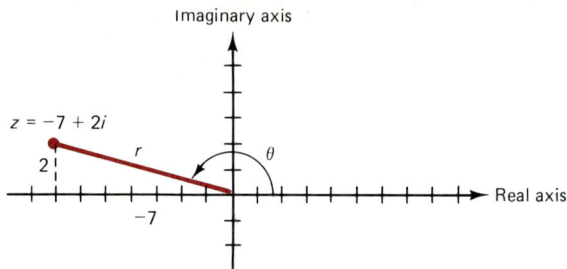

FIGURE 5

EXAMPLE 14 Write $z = -7 + 2i$ in trigonometric form.

Solution As Figure 5 shows, $-7 + 2i$ lies in quadrant II. We use this fact to find θ.

| Pythagorean theorem | $r = \sqrt{(-7)^2 + 2^2} = \sqrt{53} \approx 7.3$ |

| θ is in quadrant II | $\tan \theta = \dfrac{2}{-7}$ means $\theta \approx 164.1°$ |

| $z = r(\cos \theta + i \sin \theta)$ | Thus, $-7 + 2i = 7.3(\cos 164.1° + i \sin 164.1°)$ |

In finding θ, the calculator gives $-15.9°$ when we punch $\boxed{\text{INV}}$ $\boxed{\tan}$. We have to add $180°$ to put θ in quadrant II.

EXAMPLE 15 Write -2 in trigonometric form.

Solution As a complex number, $-2 = -2 + 0i$. This point lies on the negative x-axis (real axis), so $\theta = \pi$. Also, $r = |-2| = 2$. Thus,

$$-2 = 2(\cos \pi + i \sin \pi)$$

EXAMPLE 16 Sketch $z = 7(\cos 200° + i \sin 200°)$ and write it in the form $x + iy$.

Solution Figure 6 shows how we sketch z: we rotate $200°$ and then mark off 7 units from the origin. To write this in the form $x + iy$, we expand this expression and then evaluate the trigonometric functions.

| Distributive law | $7(\cos 200° + i \sin 200°) = 7 \cos 200° + 7i \sin 200°$ |

| Evaluate functions | $= 7(-0.9397) + 7i(-0.3420)$ |

| Simplify | $= -6.58 - 2.39i$ |

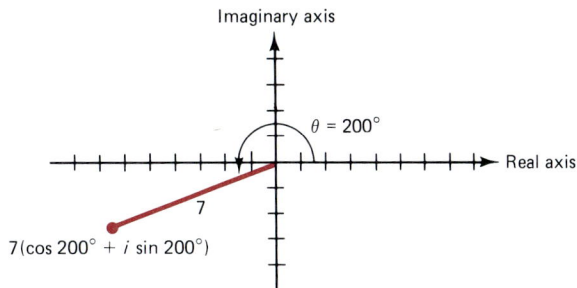

$\theta = 200°$

7

$7(\cos 200° + i \sin 200°)$

FIGURE 6

PROBLEM SET
12.3

Let $A = (4, 1)$, $B = (0, 7)$, $C = (-1, 5)$, and $D = (2, -3)$ on the Cartesian coordinate plane.

Warm-up
Exercises

1. Graph A. **2.** Graph B.

3. Graph C. **4.** Graph D.

Identify the complex numbers shown in Figure 7.

5. Point A **6.** Point B **7.** Point C

8. Point D **9.** Point E **10.** Point F

11. Point G **12.** Point H **13.** Point I

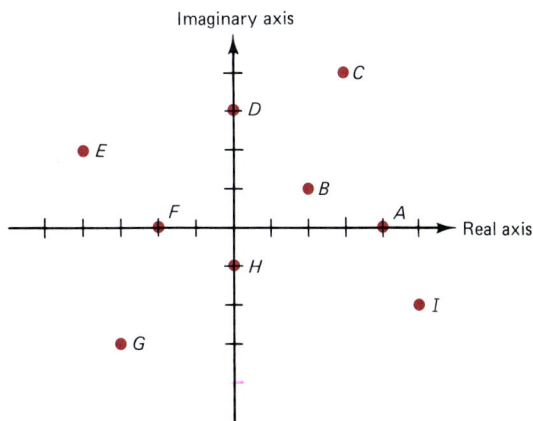

FIGURE 7

Locate the following complex numbers on the complex plane.

14. $3 + 2$ **15.** $2 - 2i$ **16.** $-2i$

17. 1 **18.** $-3 + 4i$ **19.** 0

20. $-1 - 2i$ **21.** i **22.** -3

For each of the following pairs of complex numbers, z_1 and z_2:
(a) *Find the sum $z_1 + z_2$.*
(b) *Graph the points 0, z_1, z_2, and $z_1 + z_2$.*
(c) *Connect the points in the order 0, z_1, $z_1 + z_2$, z_2, and 0.*
(d) *What figure is formed by connecting the points?*

23. $z_1 = 2 + i$ and $z_2 = 1 + 3i$

24. $z_1 = 3 + 4i$ and $z_2 = 5 + i$

25. $z_1 = 3 + 2i$ and $z_2 = 2 - 4i$

26. $z_1 = 2 + 3i$ and $z_2 = -3 + i$

For the following complex numbers z, as shown in Figure 8, find:
(a) r (b) θ

27. $z = 3 + 4i$ **28.** $z = 7 + i$

29. $z = -6 + 2i$ **30.** $z = -5 + 2i$

31. $z = -2 - 7i$ **32.** $z = -10 - i$

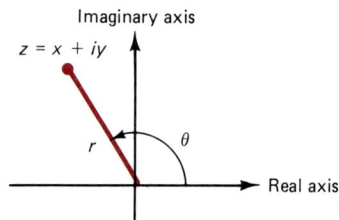

FIGURE 8

Write each of the following complex numbers in polar (or trigonometric) form.

33. $3 + 4i$ **34.** $5 + 12i$

35. $1 + i\sqrt{3}$ **36.** $-2 + 3i$

37. $-4 - 5i$ **38.** $-10 - 5i$

39. $2 - 9i$ **40.** $-\sqrt{5} + i\sqrt{7}$

41. $6i$ **42.** $-4i$

43. 5 **44.** -6

Express the following complex numbers in the form $x + iy$.

45. $2(\cos 60° + i \sin 60°)$ **46.** $3(\cos 120° + i \sin 120°)$

47. $10(\cos 45° + i \sin 45°)$ **48.** $5(\cos 225° + i \sin 225°)$

49. $4\left(\cos \dfrac{\pi}{4} + i \sin \dfrac{\pi}{4} \right)$ **50.** $6\left(\cos \dfrac{4\pi}{3} + i \sin \dfrac{4\pi}{3} \right)$

51. $8(\cos 110° + i \sin 110°)$ **52.** $2(\cos 305° + i \sin 305°)$

53. $3\left(\cos \dfrac{\pi}{2} + i \sin \dfrac{\pi}{2} \right)$ **54.** $4(\cos \pi + i \sin \pi)$

55. $5[\cos (-10°) + i \sin (-10°)]$ **56.** $10[\cos (-25°) + i \sin (-25°)]$

Complete the following table.

	Rectangular form $x + iy$	Polar form $r(\cos \theta + i \sin \theta)$
57.	$6 - 4i$	?
58.	$-7 - i$	?
59.	?	$2(\cos 11.1° + i \sin 11.1°)$
60.	?	$10(\cos 101° + i \sin 101°)$
61.	$-10i$	?
62.	?	$4\left(\cos \dfrac{\pi}{4} + i \sin \dfrac{\pi}{4}\right)$
63.	?	$5\left(\cos \dfrac{3\pi}{4} + i \sin \dfrac{3\pi}{4}\right)$
64.	-2	?

Electrical Application

65. We have seen that in an alternating-current (ac) circuit, the impedance (Z), current (I), and voltage (V) can be represented by complex numbers $a + jb$. Actually, it is more common to see these quantities expressed in polar form: $r(\cos \theta + i \sin \theta)$ or $r\underline{/\theta}$. (For example, $1000 + j1000 = 1414\underline{/45°}$.) Complete the following table.

	$a + jb$	$r\underline{/\theta}$
(a)	$100 + j150$	?
(b)	?	$0.003\underline{/40°}$
(c)	$7000 + j4200$	?
(d)	?	$110\underline{/75°}$
(e)	$0.0002 - j0.0001$	?
(f)	?	$1800\underline{/-52°}$

12.4 DE MOIVRE'S THEOREM

In the preceding section we introduced the polar (or trigonometric) form of a complex number, $z = r(\cos \theta + i \sin \theta)$. The rectangular form $x + iy$ is more convenient when we add (since we just add the real parts and add the imaginary parts); however, for multiplying, the polar form is more convenient.

Consider two complex numbers in polar form: $r(\cos A + i \sin A)$ and $s(\cos B + i \sin B)$. Let us compute their product P.

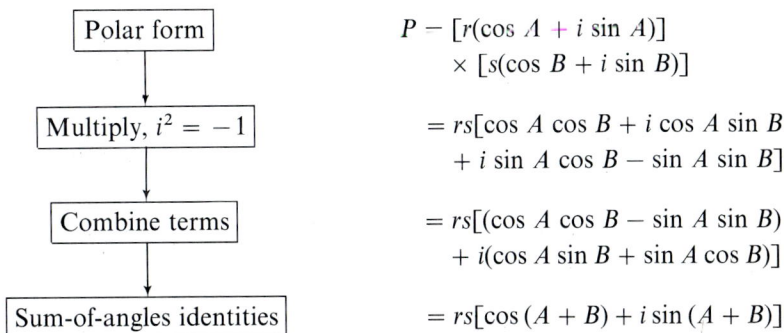

Polar form	$P = [r(\cos A + i \sin A)]$ $\times [s(\cos B + i \sin B)]$
Multiply, $i^2 = -1$	$= rs[\cos A \cos B + i \cos A \sin B$ $+ i \sin A \cos B - \sin A \sin B]$
Combine terms	$= rs[(\cos A \cos B - \sin A \sin B)$ $+ i(\cos A \sin B + \sin A \cos B)]$
Sum-of-angles identities	$= rs[\cos (A + B) + i \sin (A + B)]$

Stating this in words, we have:

> *To multiply complex numbers in polar form:*
>
> 1. We multiply their absolute values.
> 2. Add the arguments (angles).

EXAMPLE 17 We multiply the following numbers in polar form. (Note that we multiply the absolute values and add the angles.)

(a) $[10(\cos 40° + i \sin 40°)][5(\cos 30° + i \sin 30°)]$

$\qquad = 50(\cos 70° + i \sin 70°)$

(b) $[3(\cos 25° + i \sin 25°)]^2$

$\qquad = [3(\cos 25° + i \sin 25°)][3(\cos 25° + i \sin 25°)]$

$\qquad = 9(\cos 50° + i \sin 50°)$

EXAMPLE 18 In this example we divide two complex numbers in polar form. Note that we divide the absolute values and subtract the arguments (angles).

$$\frac{10(\cos 50° + i \sin 50°)}{2(\cos 30° + i \sin 30°)} = 5(\cos 20° + i \sin 20°)$$

Note what happened in Example 17(b): When squaring a number in polar form, we square the absolute value and double the angle. In fact, we have **De Moivre's theorem** for any power of a complex number in polar form.

De Moivre's Theorem
For any complex number $r(\cos \theta + i \sin \theta)$ and real number n,

$$[r(\cos \theta + i \sin \theta)]^n = r^n(\cos n\theta + i \sin n\theta)$$

EXAMPLE 19 The following demonstrate De Moivre's theorem.

(a) $[2(\cos 10° + i \sin 10°)]^5 = 2^5(\cos 5 \cdot 10° + i \sin 5 \cdot 10°)$

$\qquad\qquad\qquad\qquad\qquad = 32(\cos 50° + i \sin 50°)$

(b) $[3(\cos 150° + i \sin 150°)]^4 = 81(\cos 600° + i \sin 600°)$

$\qquad\qquad\qquad\qquad\qquad\quad = 81(\cos 240° + i \sin 240°)$

Note that we replaced $4(150°) = 600°$ by $240°$, its equivalent between $0°$ and $360°$.

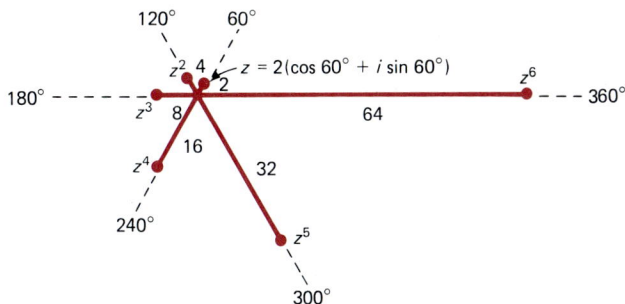

FIGURE 9 $z = 2 (\cos 60° + i \sin 60°)$.

(c) Figure 9 shows graphically the first six powers of $z = 2(\cos 60° + i \sin 60°)$.

EXAMPLE 20   Find $(1 + i\sqrt{3})^9$.

Solution   This would be a monster to do in its present form. Instead, we rewrite the base in polar form; then we use De Moivre's theorem to find the ninth power; finally, we convert back to the $x + iy$ form.

$$\boxed{\begin{array}{l} r = \sqrt{1 + 3} = 2 \\ \theta = \tan^{-1}\sqrt{3} = \dfrac{\pi}{3} \end{array}}$$

$$(1 + i\sqrt{3})^9 = \left[2\left(\cos \frac{\pi}{3} + i \sin \frac{\pi}{3} \right) \right]^9$$

$$\boxed{\text{DeMoivre's theorem}}$$

$$= 2^9(\cos 3\pi + i \sin 3\pi)$$

$$\boxed{\begin{array}{l} \cos 3\pi = -1 \\ \sin 3\pi = 0 \end{array}}$$

$$= -512$$

De Moivre's theorem can also be used to find roots. Recall that $x^{1/k} = \sqrt[k]{x}$. Substituting $n = 1/k$ in De Moivre's theorem, we get

$$[r(\cos \theta + i \sin \theta)]^{1/k} = r^{1/k}\left(\cos \frac{\theta}{k} + i \sin \frac{\theta}{k} \right)$$

Here, we must be careful, since there are generally k distinct roots. Thus, we consider the angles θ such that

$$0° \le \frac{\theta}{k} < 360°$$

$$\text{or} \quad 0° \le \theta < k \cdot 360°$$

We consider all the angles equivalent to θ between $0°$ and $k \cdot 360°$.

$$
\text{The } k\text{th roots of } r(\cos \theta + i \sin \theta) \text{ are } r^{1/k}\left(\cos \frac{A}{k} + i \sin \frac{A}{k}\right)
$$
$$
\text{where } A = \theta, \theta + 360°, \theta + 720°, \dots, \theta + (k-1)360°.
$$

EXAMPLE 21 Find the fourth roots of $16(\cos 40° + i \sin 40°)$.

Solution We expect *four* roots. Thus, we consider four equivalents of $40°$.

Four equivalents of $40°$ → A: $40°$, $400°$, $760°$, $1120°$

Find each $\dfrac{A}{4}$ → $\dfrac{A}{4}$: $10°$, $100°$, $190°$, $280°$

$16^{1/4} = 2$

Compute

$$r^{1/4}\left(\cos \frac{A}{4} + i \sin \frac{A}{4}\right)$$

The roots are:
$$r_1 = 2(\cos 10° + i \sin 10°)$$
$$r_2 = 2(\cos 100° + i \sin 100°)$$
$$r_3 = 2(\cos 190° + i \sin 190°)$$
$$r_4 = 2(\cos 280° + i \sin 280°)$$

Figure 10 shows how these roots are positioned on a circle of radius 2 in the complex plane. (Why are they all on circle of radius 2, and why are they all $90°$ apart?)

FIGURE 10

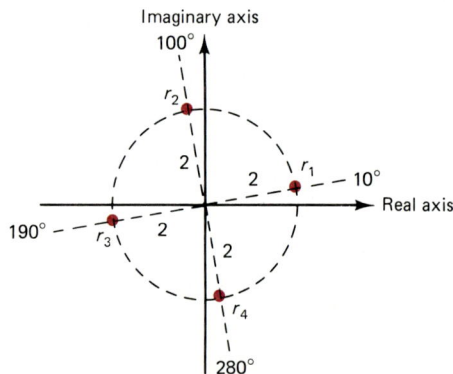

EXAMPLE 22   Find the cube roots of 64.

Solution   Right off, we know that one root is 4. However, there should be two more somewhere. Let us rewrite 64 as a complex number in polar form. Since $64 = 64 + 0i$ is on the x-axis ($0°$), we have $64 = 64(\cos 0° + i \sin 0°)$.

Three equivalents of $0°$	A: $0°$, $360°$, $720°$
Find each $\dfrac{A}{3}$	$\dfrac{A}{3}$: $0°$, $120°$, $240°$

The roots are:

$$r_1 = 4(\cos 0° + i \sin 0°)$$

Compute

$$r^{1/3}\left(\cos \frac{A}{3} + i \sin \frac{A}{3}\right)$$

$$r_2 = 4(\cos 120° + i \sin 120°)$$

$$r_3 = 4(\cos 240° + i \sin 240°)$$

We can also write these roots as 4, $-2 + 2\sqrt{3}i$, $-2 - 2\sqrt{3}i$. (Why?) We can picture these roots on a circle of radius 4. See Figure 11.

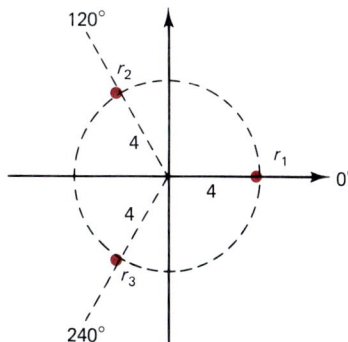

FIGURE 11

PROBLEM SET 12.4

Multiply the following complex numbers.

Warm-up Exercises
1. $(3 + 4i)(5 - 2i)$
2. $(1 + i)(7 - 3i)$
3. $(5 - i)^2$
4. $(5 - 2i)(5 + 2i)$

Write the following complex numbers in polar (trigonometric) form.

Warm-up Exercises
5. $7 + 3i$
6. $2 - 5i$
7. 8
8. $16i$

Write the following complex numbers in the form x + iy.

Warm-up
Exercises

9. $4(\cos 30° + i \sin 30°)$

10. $2(\cos 15° + i \sin 15°)$

11. $3(\cos 180° + i \sin 180°)$

12. $10\left(\cos \dfrac{\pi}{2} + i \sin \dfrac{\pi}{2}\right)$

Evaluate the following.

Warm-up
Exercises

13. $27^{1/3}$

14. $32^{1/5}$

15. $10,000^{1/4}$

16. $0.000064^{1/6}$

Simplify the following as much as possible.

17. $[2(\cos 10° + i \sin 10°)][3(\cos 20° + i \sin 20°)]$

18. $[5(\cos 15° + i \sin 15°)][8(\cos 25° + i \sin 25°)]$

19. $\dfrac{12(\cos 100° + i \sin 100°)}{4(\cos 25° + i \sin 25°)}$

20. $\dfrac{6(\cos \pi/2 + i \sin \pi/2)}{3(\cos \pi/3 + i \sin \pi/3)}$

21. $[3(\cos 100° + i \sin 100°)][4(\cos 165° + i \sin 165°)]$

22. $[7(\cos 175° + i \sin 175°)][6(\cos 230° + i \sin 230°)]$

23. $\dfrac{24(\cos 20° + i \sin 20°)}{8(\cos 5° + i \sin 5°)}$

24. $\dfrac{56(\cos \pi + i \sin \pi)}{14(\cos \pi/3 + i \sin \pi/3)}$

25. $\left[\dfrac{1}{2}(\cos(-10°) + i \sin(-10°))\right]^5$

26. $\left[\dfrac{1}{5}(\cos 260° + i \sin 260°)\right]^2$

27. $(2 - i)^7$

28. $(1 - 2i)^5$

29. $[3(\cos 40° + i \sin 40°)][5(\cos 95° + i \sin 95°)]$

30. $[4(\cos 14° + i \sin 14°)]^3$

31. $[2(\cos 25° + i \sin 25°)]^3 [5(\cos 10° + i \sin 10°)]$

32. $[3(\cos 75° + i \sin 75°)]^2 [2(\cos 120° + i \sin 120°)]^4$

33. $[2(\cos 27° + i \sin 27°)][3(\cos 10° + i \sin 10°)][5(\cos 8° + i \sin 8°)]$

34. $[12(\cos 50° + i \sin 50°)]\left[\dfrac{1}{3}(\cos(-20°) + i \sin(-20°))\right]\left[\dfrac{1}{2}(\cos 5° + i \sin 5°)\right]$

35. $(1 + \sqrt{3}i)[4(\cos 12° + i \sin 12°)] \div [2(\cos 7° + i \sin 7°)]$

36. $[10(\cos 210° + i \sin 210°)](1 - i) \div [2(\cos 9° + i \sin 9°)]$

37. $(5 - 2i)^2 [2(\cos 17° + i \sin 17°)]$

38. $(1 + 3i)^3 [3(\cos 41° + i \sin 41°)]$

Find all the indicated roots of the following complex numbers.

39. Fifth roots of $(\cos 40° + i \sin 40°)$

40. Square roots of $25(\cos 170° + i \sin 170°)$

41. Fourth roots of $81(\cos 200° + i \sin 200°)$

42. Third roots of $(\cos 300° + i \sin 300°)$

43. Square roots of $4i$

44. Cube roots of -8

45. Cube roots of $-27i$

46. Square roots of $-9i$

47. Square roots $1 + i$

48. Square roots of $2 - i$

Solve the following equations.

49. $x^2 = -4$

50. $x^2 = 36i$

51. $x^{1/3} = 2(\cos 10° + i \sin 10°)$

52. $x^{1/4} = 3(\cos 20° + i \sin 20°)$

53. $x^3 = 8(\cos 60° + i \sin 60°)$

54. $x^4 = 81(\cos 100° + i \sin 100°)$

Electrical Application

55. Recall that Ohm's law is

$$V = I \cdot Z$$

These quantities (voltage V, current I, and impedance Z) are best expressed as complex numbers in polar form $r\underline{/\theta}$. Complete the following table.

	V	I	Z
(a)	$110\underline{/72°}$	$0.0002\underline{/39°}$	?
(b)	?	$0.0005\underline{/51°}$	$4000\underline{/11°}$
(c)	$70\underline{/31°}$	?	$3500\underline{/42°}$
(d)	$150\underline{/-23°}$	$0.006\underline{/79°}$	?

12.5 POLAR COORDINATES

We generally label points in the plane with a rectangular coordinate system framed by two perpendicular axes. However, in some situations, it is often more convenient to use another system: the polar coordinate system. (*Note:* We are back on the real plane.)

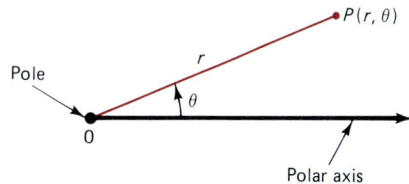

FIGURE 12

In the **polar coordinate system**, we have a point 0 called the **pole** (like the origin) and a ray from the pole called the **polar axis** (like the positive x-axis). Figure 12 shows the general point P, labeled (r, θ). The point P lies on the terminal side of θ and is r units from 0. (Note the similarity to a complex number in polar form.)

In the Cartesian coordinate system, each point has one and only one pair of coordinates. However, in the polar-coordinate system, a point may have many (infinitely many) possible coordinates. Figure 13 shows the *same* point with four different polar coordinates.

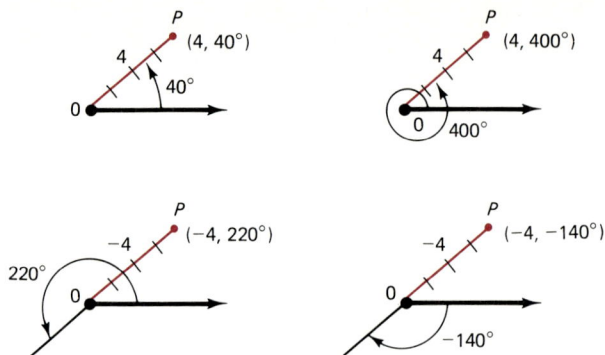

FIGURE 13

In Figure 13, notice that two of the polar-coordinate pairs have negative r: $(-4, 220°)$ and $(-4, -140°)$. Here we find the angle $220°$ or $-140°$ and then move 4 units from the pole in the *opposite* direction.

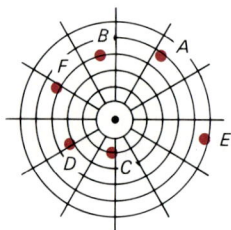

FIGURE 14

EXAMPLE 23 Plot the following points on a polar coordinate system:

(a) $A(5, 50°)$ (b) $B(4, 100°)$ (c) $C(2, -90°)$

(d) $D(-3, 30°)$ (e) $E(-6, 170°)$ (f) $F(-4, -30°)$

Solution Figure 14 shows these points plotted on a typical bit of polar-coordinate graph paper (available in many stationery stores). Notice how this paper differs from rectangular graph paper: instead of having up-and-down, right-and-left grid lines, we now have concentric circles and "spokes" at various angles. With points D, E, and F that have negative r, we first locate θ and then measure $|r|$ units in the opposite direction.

It is often necessary to convert from polar coordinates to rectangular coordinates, and vice versa. The relations between the coordinate systems are exactly the same as those between complex numbers in $x + iy$ form and in $r(\cos \theta + i \sin \theta)$ form. See Figure 15 and Table 1.

FIGURE 15

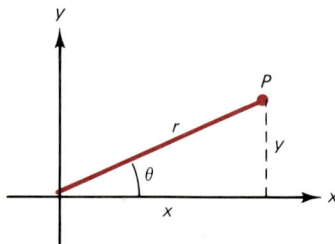

TABLE 1

Polar to Rectangular	Rectangular to Polar
$x = r \cos \theta$	$r = \sqrt{x^2 + y^2}$
$y = r \sin \theta$	$\tan \theta = \dfrac{y}{x}$
	(Signs of x and y determine θ.)

EXAMPLE 24 Write the following polar-coordinate pairs in rectangular coordinates.

(a) $A(5, 50°)$ (b) $B(4, 100°)$ (c) $C\left(-3, \dfrac{\pi}{6}\right)$

Solution   We use the relations $x = r \cos \theta$ and $y = r \sin \theta$.

(a) $x = 5 \cos 50° \approx 3.21$ and $y = 5 \sin 50° \approx 3.83$
Thus, the rectangular coordinates of A are approximately $(3.21, 3.83)$.

(b) $x = 4 \cos 100° \approx -0.69$ and $y = 4 \sin 100° \approx 3.94$
Thus, the rectangular coordinates of B are approximately $(-0.69, 3.94)$.

(c) $x = (-3) \cos \dfrac{\pi}{6} \approx -2.60$ and $y = (-3) \sin \dfrac{\pi}{6} = -1.50$
Thus, the rectangular coordinates of C are approximately $(-2.6, -1.5)$.

EXAMPLE 25 Write the following rectangular-coordinate pairs in polar coordinates.

(a) $D(2, 7)$ (b) $E(-3, 5)$ (c) $F(-4, -9)$

Solution   We use the relations $r = \sqrt{x^2 + y^2}$ and $\tan \theta = y/x$. (The quadrant of the point will help determine θ.)

(a) $r = \sqrt{2^2 + 7^2} \approx 7.3$ and $\tan \theta = \dfrac{7}{2}; \theta \approx 74.1°$ (D is in quadrant I.)
Thus, the polar coordinates of D are approximately $(7.3, 74.1°)$.

(b) $r = \sqrt{(-3)^2 + 5^2} \approx 5.8$ and $\tan \theta = \dfrac{5}{-3}; \theta \approx 121.0°$ (E is in quadrant II.)
Thus, the polar coordinates of E are approximately $(5.8, 121.0°)$.

(c) $r = \sqrt{(-4)^2 + (-9)^2} \approx 9.8$ and $\tan \theta = \dfrac{-9}{-4}; \theta \approx 246.0°$ (F is in quadrant III.)
Thus, the polar coordinates of F are approximately $(9.8, 246.0°)$.

Sometimes, we need to convert an equation in rectangular coordinates to one in polar coordinates, and vice versa. (That is, we want to describe the same points in the plane but in the other coordinate system.) We use the relations in Table 1 to help us.

EXAMPLE 26   Write $x^2 - 4x + y^2 = 5$ in polar coordinates.

Solution   We want an equation in polar coordinates that describes the same points in the plane as this equation does. We use the relations $x = r \cos \theta$ and $y = r \sin \theta$. Also, we use $r^2 = x^2 + y^2$.

| Given | $x^2 - 4x + y^2 = 5$ |

$$\begin{array}{c} x^2 + y^2 = r^2 \\ x = r \cos \theta \end{array}$$

$$r^2 - 4r \cos \theta = 5$$

This is the relation in polar coordinates.

EXAMPLE 27   Write $r = 2 \cos \theta$ in rectangular coordinates.

Solution   Here no obvious substitution jumps out at us; however, we can produce a substitution if we multiply both sides of the equation by r.

| Given | $r = 2 \cos \theta$ |

| Multiply by r | $r^2 = 2r \cos \theta$ |

$$\begin{array}{c} r^2 = x^2 + y^2 \\ x = r \cos \theta \end{array}$$

$$x^2 + y^2 = 2x$$

This is the relation in rectangular coordinates.

**PROBLEM SET
12.5**

Plot the following points on a rectangular coordinate system.

*Warm-up
Exercises*

1. (5, 1) **2.** (−7, 2) **3.** (−4, −3)
4. (5, −1) **5.** (0, 3) **6.** (−2, 0)

Write the following complex numbers in polar (trigonometric) form.

*Warm-up
Exercises*

7. $4 + 7i$ **8.** $5 - 2i$
9. $-3 + 5i$ **10.** $-6 - i$

Write the following complex numbers in the form $x + iy$.

*Warm-up
Exercises*

11. $3(\cos 10° + i \sin 10°)$ **12.** $2(\cos 55° + i \sin 55°)$

13. $-4\left(\cos \dfrac{\pi}{3} + i \sin \dfrac{\pi}{3}\right)$ **14.** $-5(\cos 180° + i \sin 180°)$

Plot the following points on a polar coordinate system.

15. $A(3, 30°)$ **16.** $B(5, 45°)$ **17.** $C\left(2, \dfrac{\pi}{2}\right)$

18. $D\left(4, \dfrac{2\pi}{3}\right)$ **19.** $E(1, \pi)$ **20.** $F(3, 270°)$

21. $G(2, 400°)$ **22.** $H(3, 5\pi)$ **23.** $I\left(-2, \dfrac{\pi}{4}\right)$

24. $J(-3, 120°)$ **25.** $K(-4, -30°)$ **26.** $L\left(-5, \dfrac{-3\pi}{4}\right)$

Express the following polar coordinates in rectangular coordinates.

27. $(3, 30°)$ **28.** $\left(5, \dfrac{\pi}{4}\right)$ **29.** $\left(2, \dfrac{\pi}{2}\right)$

30. $(4, 300°)$ **31.** $\left(-2, \dfrac{\pi}{4}\right)$ **32.** $(-2, 120°)$

33. $(-4, -30°)$ **34.** $\left(-5, \dfrac{-3\pi}{4}\right)$

Express the following rectangular coordinates in polar coordinates.

35. $(2, 3)$ **36.** $(5, 1)$ **37.** $(-2, 4)$
38. $(-1, 3)$ **39.** $(-4, -3)$ **40.** $(-1, -4)$
41. $(4, -7)$ **42.** $(8, -1)$

Rewrite the following equations with polar coordinates.

43. $y = 7$ **44.** $x = -2$
45. $x + y = 5$ **46.** $2x + 3y = 1$
47. $x^2 + y^2 = 4$ **48.** $x^2 + 2x + y^2 = 8$

49. $y = \dfrac{1}{x}$ **50.** $y = 2x$

51. $4x^2 + 9y^2 = 36$ **52.** $x^2 + xy + y^2 = 4$

Rewrite the following equations with rectangular coordinates.

53. $r \cos \theta = 5$ **54.** $r \sin \theta = -1$
55. $r = 2 \cos \theta$ **56.** $r = 3 \sin \theta$

57. $r = \theta$ **58.** $r = \dfrac{1}{\theta}$

59. $r = 4 + 4 \cos \theta$ **60.** $r = 5 + 5 \sin \theta$

61. $r = \dfrac{4}{1 - \cos \theta}$ **62.** $r = \dfrac{2}{2 - \cos \theta}$

63. $r = \dfrac{3}{1 - 2 \cos \theta}$ **64.** $r = \dfrac{1}{1 - 3 \cos \theta}$

The following points are expressed in polar coordinates. Rewrite each of them: (a) With a negative r. (b) With a negative θ. (c) With a θ between 2π and 4π.

65. $(2, 30°)$ **66.** $(4, 70°)$ **67.** $\left(20, \dfrac{\pi}{6}\right)$

68. $\left(40, \dfrac{2\pi}{3}\right)$ **69.** $(5, 150°)$ **70.** $(10, 220°)$

For each of the following equations in polar coordinates, complete the following table. Then plot the points.

θ	0	$\dfrac{\pi}{4}$	$\dfrac{\pi}{2}$	$\dfrac{3\pi}{4}$	π	$\dfrac{5\pi}{4}$	$\dfrac{3\pi}{2}$	$\dfrac{7\pi}{4}$	2π
r									

71. $r = \cos\theta$ **72.** $r = \sin\theta$

73. $r = \sin 2\theta$ **74.** $r = 1 + \cos\theta$

Physical Application

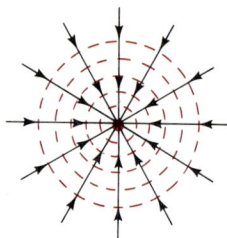

75. The sketch at the left shows the lines of gravitational (or electrical) force toward a point mass (or negative charge). The dashed lines are surfaces of equipotential (that is, all the points on each circle have the same energy). For this situation, which coordinate system would seem more logical: polar or rectangular?

Radar Application

76. The screen of a radar scanner is gridded in polar coordinates, with the pole representing the radar receiver. If each circle represents 50 miles, give the distance and bearing of objects A and B in the sketch.

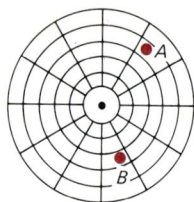

12.6 GRAPHING POLAR EQUATIONS

In the preceding section we saw how points in the plane are described by polar coordinates (r, θ). Just as we can graph equations written in rectangular coordinates, so can we graph equations involving polar coordinates: We plot all the (r, θ) pairs that make the equation true; that is, we plot the solutions.

    Generally, the equations that we see in polar coordinates involve r given as a function of θ, $r = f(\theta)$. *Note the switch*: The independent variable is now the *second* variable θ. In rectangular coordinates, the independent variable is, generally, the first variable x.

EXAMPLE 28   Graph $r = 4\cos\theta$.

Solution   We look for the (r, θ) pairs that satisfy the equation. We make a table (with θ-values in 30° intervals) and evaluate $\cos\theta$. Finally, we compute $r = 4\cos\theta$. Figure 16 shows the table and the graph on polar graph paper. We see that the curve is a circle of radius 2. Note that for $\theta > 180°$, the curve repeats itself. [As an exercise, the student should

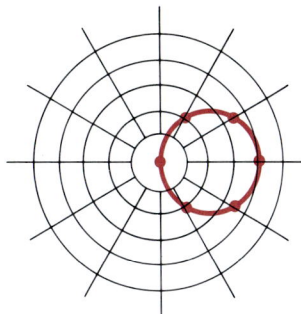

θ	0°	30°	60°	90°	120°	150°	180°	210°	240°	270°	300°	330°	360°
$\cos \theta$	1	0.87	0.5	0	−0.5	−0.87	−1	−0.87	−0.5	0	−0.5	−0.87	1
$r = 4 \cos \theta$	4	3.5	2	0	−2	−3.5	−4	−3.5	−2	0	−2	−3.5	4

FIGURE 16

show that this equation is equivalent to $(x - 2)^2 + y^2 = 2^2$: a circle radius 2 centered at $(2, 0)$.]

EXAMPLE 29 Graph $r = 2(1 + \sin \theta)$.

Solution We choose θ in $\pi/6$ intervals. We then evaluate $\sin \theta$ and finally $r = 2(1 + \sin \theta)$. Figure 17 shows this heart-shaped figure, called a **cardioid**. (Like Example 28, this equation can be rewritten in rectangular coordinates; however, it is a very awkward relation.)

FIGURE 17

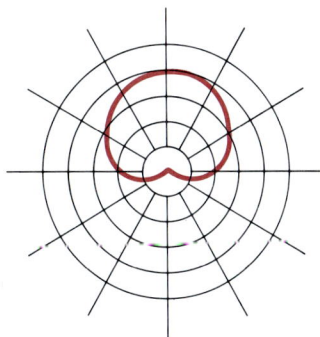

θ	0	$\frac{\pi}{6}$	$\frac{\pi}{3}$	$\frac{\pi}{2}$	$\frac{2\pi}{3}$	$\frac{5\pi}{6}$	π	$\frac{7\pi}{6}$	$\frac{4\pi}{3}$	$\frac{3\pi}{2}$	$\frac{5\pi}{3}$	$\frac{11\pi}{6}$	2π
$\sin \theta$	0	0.5	0.87	1	0.87	0.5	0	−0.5	−0.87	−1	−0.87	−0.5	0
$2(1 + \sin \theta)$	2	3	3.7	4	3.7	3	2	1	0.27	0	0.27	1	2

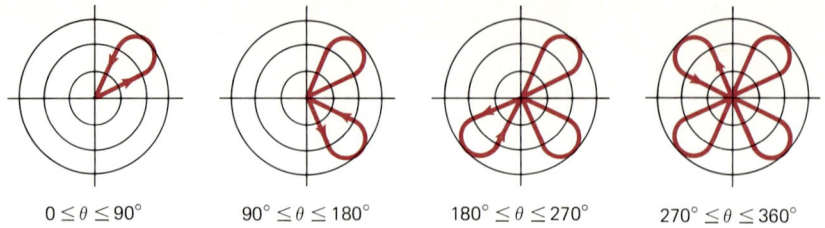

θ	0°	15°	30°	45°	60°	75°	90°	105°	120°	135°	150°	165°	180°
2θ	0°	30°	60°	90°	120°	150°	180°	210°	240°	270°	300°	330°	360°
$3\sin 2\theta$	0	1.5	2.6	3	2.6	1.5	0	−1.5	−2.6	−3	−2.6	−1.5	0

FIGURE 18

EXAMPLE 30   Graph $r = 3 \sin 2\theta$.

Solution   Here, we choose θ in intervals of 15° (since we double θ). The curve is called a **rose** (here, a four-leafed rose). Figure 18 shows part of the table and how the leaves of the rose grow (each 90° interval for θ produces another leaf).

EXAMPLE 31   Graph $r = \dfrac{4}{2 - \cos \theta}$.

Solution   Again, we choose θ in 30° intervals. Figure 19 shows the graph, which is an ellipse.

FIGURE 19

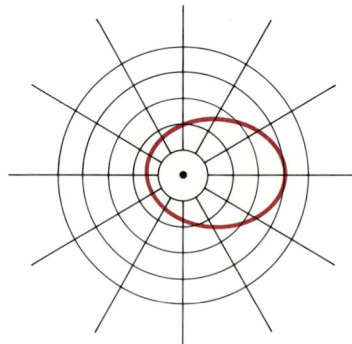

θ	0°	30°	60°	90°	120°	150°	180°	210°	240°	270°	300°	330°	360°
$\cos \theta$	1	0.87	0.5	0	−0.5	−0.87	−1	−0.87	−0.5	0	0.5	0.87	1
$\dfrac{4}{2 - \cos \theta}$	4	3.5	2.7	2	1.6	1.4	1.3	1.4	1.6	2	2.7	3.5	4

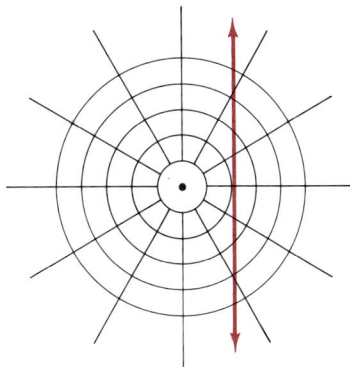

θ	0	$\dfrac{\pi}{6}$	$\dfrac{\pi}{3}$	$\dfrac{\pi}{2}$	$\dfrac{2\pi}{3}$	$\dfrac{5\pi}{6}$	π	$\dfrac{7\pi}{6}$	$\dfrac{4\pi}{3}$	$\dfrac{3\pi}{2}$	$\dfrac{5\pi}{3}$	$\dfrac{11\pi}{6}$	2π
$\cos \theta$	1	0.87	0.5	0	−0.5	−0.87	−1	−0.87	−0.5	0	0.5	0.87	1
$r = \dfrac{2}{\cos \theta}$	2	2.3	4	—	−4	−2.3	−2	−2.3	−4	—	4	2.3	2

FIGURE 20

EXAMPLE 32   Graph $r \cos \theta = 2$.

Solution   Let us write this equation in the form

$$r = \frac{2}{\cos \theta}$$

where $\theta \neq \pi/2,\ 3\pi/2$, and so on, since these θ-values produce $\cos \theta = 0$. Figure 20 shows the graph and the (r, θ) table. Notice that we obtain a vertical line. Looking back at the original equation and recalling that $x = r \cos \theta$, we realize that we have the vertical line $x = 2$.

**PROBLEM SET
12.6**

Plot the following polar coordinates and also write them in rectangular coordinates.

*Warm-up
Exercises*

1. $(5, 30°)$ **2.** $\left(6, \dfrac{2\pi}{3} \right)$ **3.** $\left(-2, \dfrac{\pi}{4} \right)$

4. $(-3, 60°)$ **5.** $\left(5, \dfrac{-\pi}{6} \right)$ **6.** $(-4, -150°)$

Express the following rectangular coordinates in polar coordinates.

*Warm-up
Exercises*

7. $(2, 5)$ **8.** $(-3, 7)$

9. $(-2, -1)$ **10.** $(5, -1)$

Graph the following equations in rectangular coordinates.

*Warm-up
Exercises*

11. $y = 2x + 1$ **12.** $y = 1 - 3x$

13. $y = \dfrac{1}{x}$ **14.** $y = \sin 2x$

Graph the following equations in polar coordinates.

15. $r = 2 \sin \theta$

16. $r = 3 \cos \theta$

17. $r = 3(1 - \cos \theta)$

18. $r = 4(1 + \cos \theta)$

19. $r = \cos 2\theta$

20. $r = 2 \sin 3\theta$

21. $r = \dfrac{2}{1 - \cos \theta}$

22. $r = \dfrac{3}{1 - 2 \cos \theta}$

23. $r \cos \theta = -3$

24. $r \sin \theta = 4$

25. $r = -2 \cos \theta$

26. $r = -4 \sin \theta$

27. $r = -3(1 + \sin \theta)$

28. $r = -2(1 - \sin \theta)$

29. $r = \dfrac{4}{3 - \cos \theta}$

30. $r = \dfrac{5}{1 - 3 \cos \theta}$

31. $r = 5 \cos \theta$

32. $r = 3 \sin 3\theta$

33. $r = 2 \cos 2\theta$

34. $r = 2(2 + \cos \theta)$

35. $r = \theta$

36. $r = \dfrac{3}{2 - \cos \theta}$

37. $r = 3 + 2 \sin \theta$

38. $r = 2 + 3 \cos \theta$

39. $r^2 = \theta$

40. $r = \sin^2 \theta$

41. $r = 4$

42. $r = -2$

43. $\theta = 30°$

44. $\theta = \dfrac{\pi}{4}$

Acoustical Applications

45. A *cardioid microphone* has a "live front" and a "dead back." The volume depends on the angle θ ($\theta = 0$ is the front) and has the following *acceptance pattern*:

$$r = 1 + \cos \theta$$

Graph this equation in polar coordinates and see how the microphone got its name.

46. A *bidirectional microphone* has two "live" sides and two "dead" sides. There are four orders of acceptance patterns for such microphones. Graph them in polar coordinates:

$$
\begin{array}{ll}
\text{First order:} & r = |\cos \theta| \\
\text{Second order:} & r = |\cos^2 \theta| \\
\text{Third order:} & r = |\cos^3 \theta| \\
\text{Fourth order:} & r = |\cos^4 \theta|
\end{array}
$$

47. A bidirectional microphone that has a higher acceptance in one direction than the other is called *unidirectional* and has the following *limaçon* acceptance pattern:

$$r = |1 + 2 \cos \theta|$$

Graph this pattern.

Physical Application

48. The path of an object passing by a point force (gravitational or electrical) at the origin (pole) is given by

$$r = \dfrac{1}{-K + \sqrt{K^2 + 2E} \cos \theta}$$

where E is the particle's energy and K is a force constant (other constants have been set to 1 for simplicity). There are several cases:

$$K = 0 \quad \text{No force} \qquad\qquad E < 0 \quad \text{Trapped particle}$$
$$K < 0 \quad \text{Attractive force} \qquad E > 0 \quad \text{Free particle}$$
$$K > 0 \quad \text{Repulsive force}$$

The paths are *conic sections* (depending on E and K). Graph the following (where E and K have been set to simple numbers for ease of graphing).
(a) $K = 0$, $E = 2$: line (a free particle)
(b) $K = -2$, $E = -1.5$: ellipse (like a planet about the sun)
(c) $K = -1$, $E = 0$: parabola (like a comet)
(d) $K = -1$, $E = 1.5$: hyperbola (take only branch $-120° < \theta < 120°$) (like a scattered particle)
(e) $K = 1$, $E = 1.5$: hyperbola (take only branch $-60° < \theta < 60°$) (like a scattered particle)

CHAPTER 12 SUMMARY

Important Definitions and Theorems

$$(a + bi) + (c + di) = (a + c) + (b + d)i$$
$$(a + bi)(c + di) = (ac - bd) + (ad + bc)i$$
$$i = i^1 = i^5 = i^9 = \cdots$$
$$-1 = i^2 = i^6 = i^{10} = \cdots$$
$$-i = i^3 = i^7 = i^{11} = \cdots$$
$$1 = i^4 = i^8 = i^{12} = \cdots$$

If $z = a + bi$, then $\bar{z} = a - bi$ is its conjugate.

$z + \bar{z}$ is a real number.

$z \cdot \bar{z}$ is a positive real number.

$$\overline{z_1 + z_2} = \overline{z_1} + \overline{z_2} \qquad \overline{z_1 \cdot z_2} = \overline{z_1} \cdot \overline{z_2} \qquad \overline{z_1^k} = \overline{z_1}^k$$

The relations between the complex forms $x + iy$ and $r(\cos \theta + i \sin \theta)$ are

$$\begin{aligned} x &= r \cos \theta \\ y &= r \sin \theta \end{aligned} \quad \text{and} \quad \begin{aligned} r &= \sqrt{x^2 + y^2} \\ \tan \theta &= \frac{y}{x}; \text{ signs of } x \text{ and } y \\ &\text{determine quadrant for } \theta \end{aligned}$$

Rectangular and polar coordinates satisfy the same relations (above).

$$[r(\cos \theta + i \sin \theta)]^n = r^n(\cos n\theta + i \sin n\theta)$$

The kth roots of $r(\cos \theta + i \sin \theta)$ are $r^{1/k}\left(\cos \dfrac{A}{k} + i \sin \dfrac{A}{k}\right)$, where $A = \theta, \theta + 360°$, $\theta + 720°, \ldots, \theta + (n - 1)360°$.

Review Exercises

Identify the real and imaginary parts of the following.

1. $7 - 3i$

2. $\dfrac{3 + i\sqrt{11}}{4}$

Simplify the following to the form a + bi.

3. $\sqrt{-4}$

4. $(7 + i) + (8 - 2i)$

5. $(3 - i) - (4 + 5i) - (i + 2)$

6. $(2 + 7i)(1 - 5i)$

7. $(2 - 5i)^2$

8. i^{13}

9. $(3 - 7i)(3 + 7i)$

10. $(3 - 7i) + (3 + 7i)$

11. $\dfrac{4 - i}{3 + i}$

Use $z_1 = 3 - i$ and $z_2 = 4 + 2i$ in Problems 12 to 14.

12. Show $\overline{z_1 + z_2} = \overline{z_1} + \overline{z_2}$.

13. Show $\overline{z_1 \cdot z_2} = \overline{z_1} \cdot \overline{z_2}$.

14. Show $\overline{z_1^2} = \overline{z_1}^2$.

Locate the following points on the complex plane.

15. $2 + 3i$

16. $-3 + i$

17. $-5i$

18. -2

Complete the following table for complex numbers.

	Rectangular form $x + iy$	Polar form $r(\cos\theta + i\sin\theta)$
19.	$-2 + 5i$	?
20.	$4i$	?
21.	?	$3(\cos 70° + i \sin 70°)$
22.	?	$-2[\cos(-15°) + i \sin(-15°)]$

Simplify the following as much as possible.

23. $[4(\cos 15° + i \sin 15°)][2(\cos 80° + i \sin 80°)]$

24. $[2(\cos 20° + i \sin 20°)]^5$

25. $(1 - i)^8$

26. Third roots of $8(\cos 75° + i \sin 75°)$.

27. Fifth roots of $(1 - i)$.

Complete the following table.

	Rectangular coordinates	Polar coordinates
28.	$(3, -7)$	?
29.	$(-4, 0)$	?
30.	?	$(6, 75°)$
31.	?	$\left(-5, \dfrac{\pi}{2}\right)$
32.	$x + y = 7$	?
33.	?	$r = 6 \cos\theta$

Graph the following equations in polar coordinates.

34. $r = 2 \sin\theta$

35. $r = 3$

36. $r = 3 \cos 2\theta$

37. $r = 2 - \sin\theta$

38. $r = \dfrac{4}{2 + \cos\theta}$

39. $r \sin\theta = 3$

Systems of Equations and Inequalities

Consider the equations

$$A: \quad x + y = 7$$

$$B: \quad x - y = 3$$

The ordered pair (6, 1) satisfies (is a solution of) equation A, but not B. Similarly, (9, 6) satisfies equation B, but not A. However, the ordered pair (5, 2) satisfies *both* equations A and B.

A set of two or more equations for which we seek a common (or simultaneous) solution is called a **system of equations**. In the example above, equations A and B form a system, and (5, 2) is its solution.

13.1
SYSTEMS OF LINEAR EQUATIONS IN TWO VARIABLES

The simplest system to study is made up of linear equations. Such a system is called a **linear system**. In this section we consider linear systems with just two equations in two variables. Recall (page 125) that a linear equation, such as $2x + 3y = 7$, has as its graph a straight line. Thus, a system of two such equations corresponds to two straight lines, l_1 and l_2, in the plane. As shown in Figure 1, there are three possible cases.

Case I: The two lines intersect in a single point (a, b). This ordered pair (a, b) is the solution to the system. We call this system **consistent**. It has one (and only one) solution.

FIGURE 1

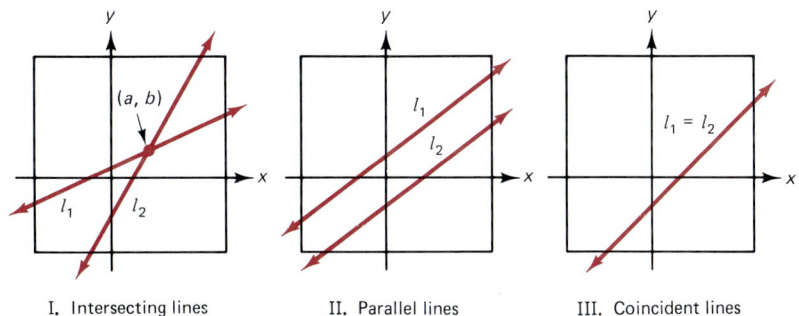

I. Intersecting lines   II. Parallel lines   III. Coincident lines

Case II: The two lines are parallel and, of course, do not intersect. There is *no* solution. We call this system **inconsistent**.

Case III: The two lines are really the same line (or **coincident**). The solution is either line (or equation). We call this system **dependent**. There are infinitely many solutions.

One possible method of solving these systems of linear equations is by graphing the equations and locating the intersection. This method, however, can be very time consuming and inaccurate. Therefore, we concentrate on algebraic methods.

To solve a system of equations, we change the system into another (simpler to solve) system with the same solutions. To do this we use the following operations, which are based on the properties of equality discussed in Chapter 1.

Property

For any system of equations, we have the following operations:

S1 We can multiply both sides of any equation by a nonzero constant.

S2 We can add the corresponding sides of two equations.

S3 We can substitute equals for equals in any equation.

One of the methods that uses these operations is the **elimination method**. This method gets its name because we *eliminate* one of the variables (x or y) to obtain one equation with only one variable; we then solve that equation for its variable (the remaining variable).

EXAMPLE 1 Solve the system

$$A: \quad x + y = 10$$
$$B: \quad x - y = \ 4$$

Solution This is a simple system. Using S2, we can add equations A and B together (which we abbreviate A + B), and this will eliminate y.

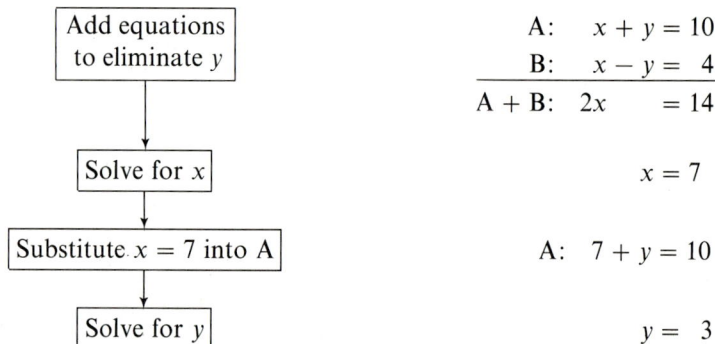

Add equations to eliminate y	$A:$	$x + y = 10$
	$B:$	$x - y = \ 4$
	$A + B:$	$2x \quad = 14$

Solve for x 　　　　　　　　　　　　　　　 $x = 7$

Substitute $x = 7$ into A 　　　　　　　 $A: \quad 7 + y = 10$

Solve for y 　　　　　　　　　　　　　　 $y = \ 3$

Therefore, the solution is the ordered pair (7, 3). This is shown graphically in Figure 2(a). Here we substituted $x = 7$ into one of the equations to find y. It is good practice to check the (x, y) pair in the other equation.

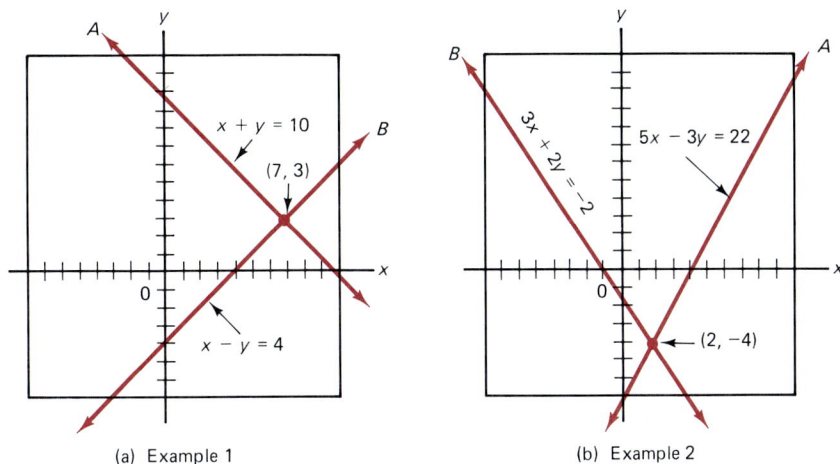

(a) Example 1

(b) Example 2

(c) Example 3

FIGURE 2

EXAMPLE 2 Solve the system

$$\text{A:} \quad 5x - 3y = 22$$

$$\text{B:} \quad 3x + 2y = -2$$

Solution Unlike Example 1, the variable to be eliminated does not jump out at us; rather, we have to do a little work first. Using S1, we can multiply both sides of equation A by 2 (producing a $-6y$) and then multiply both sides of equation B by 3 (producing a $6y$). The $-6y$ and $6y$ will cancel when we add the new equations. (We use the abbreviation 2A + 3B to denote this new equation.)

Multiply both sides of A by 2, both sides of B by 3	2A: $10x - 6y = 44$
	3B: $\quad 9x + 6y = -6$
Add to eliminate y	2A + 3B: $19x \quad\quad = 38$
Solve for x	$x = 2$
Substitute $x = 2$ into A	A: $5(2) - 3y = 22$
	$-3y = 12$
Solve for y	$y = -4$

The solution is $(2, -4)$. This is shown graphically in Figure 2(b). (*Note*: If we substitute $x = 2$ into B, we also get $y = -4$.)

Eliminating a variable may produce two unusual results:

1. A *false* result, such as $0 = 4$, means that the lines are parallel, and there is *no* solution.
2. A result that is always *true*, such as $0 = 0$, means that the lines are the same. Thus, either equation is the solution.

EXAMPLE 3 Solve the system

$$\text{A:} \qquad \frac{x}{6} - \frac{y}{2} = 1$$

$$\text{B:} \qquad \frac{-2x}{7} + \frac{6y}{7} = 1$$

Solution We must first clear the fractions by multiplying each equation by its LCD. Then we eliminate a variable.

Multiply by LCDs to clear fractions	6A: $\quad x - 3y = 6$
	7B: $\quad -2x + 6y = 7$
Multiply 6A by 2	2(6A): $\quad 2x - 6y = 12$
	7B: $\quad -2x + 6y = 7$
Add to eliminate y	$0 = 19$
Result is false	The lines are parallel.

This false result $0 = 19$ indicates that there is *no solution* (the lines are parallel). This is shown graphically in Figure 2(c).

Another method for solving systems is the **substitution method**. Here one of the variables is easily replaced in one of the equations.

EXAMPLE 4 Solve the system

$$\text{A:} \quad 2x - 5y = 22$$

$$\text{B:} \quad y = 2x - 14$$

Solution Here equation B gives y in terms of x. Using S3, we can substitute this expression for y into equation A.

Substitute $y = 2x - 14$ into A	A: $2x - 5y = 22$
	$2x - 5(2x - 14) = 22$
Distributive law	$2x - 10x + 70 = 22$
	$-8x = -48$
Solve for x	$x = 6$
Substitute $x = 6$ into B	$y = 2(6) - 14 = -2$

Thus, the solution is $(6, -2)$.

EXAMPLE 5 How many milliliters of a 20% HCl solution and a 40% HCl solution must be mixed to produce 600 milliliters of a 25% HCl solution?

Solution With verbal applications like this, it is often helpful to make a table with our information. (Review page 72.) We let x and y be the unknown HCl solution quantities.

	Solution	Strength	Pure Acid
20% solution	x	20%	$0.20x$
40% solution	y	40%	$0.40y$
Mixture (25%)	600	25%	150 ($=25\%$ of 600)

From this table we get two equations: one for the total amount of solution, and one for the amount of pure acid.

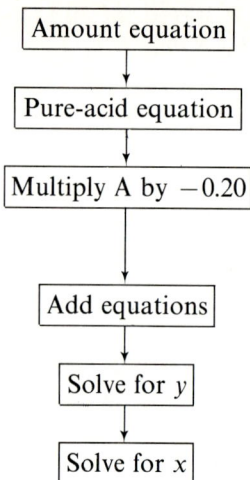

Amount equation	A: $x + y = 600$
↓	
Pure-acid equation	B: $0.20x + 0.40y = 150$
↓	
Multiply A by -0.20	-0.20A: $-0.20x - 0.20y = -120$
	B: $0.20x + 0.40y = 150$
↓	
Add equations	$0.20y = 30$
↓	
Solve for y	$y = 150$
↓	
Solve for x	$x = 450$

Therefore, we need 450 milliliters of the 20% solution and 150 milliliters of the 40% solution.

PROBLEM SET
13.1

Solve the following equations.

Warm-up
Exercises

1. $x + 7 = 5$ **2.** $x - 6 = -2$

3. $3x = 150$ **4.** $-4x = 100$

5. $2x - 3 = -15$ **6.** $-5x + 3 = -2$

7. $4(x + 3) - 8 = 6x + 8$ **8.** $5x - 2(4 - 3x) = 3(2 - x)$

9. $\dfrac{2x}{3} + \dfrac{x - 1}{5} = \dfrac{17}{15}$ **10.** $\dfrac{10}{x} - \dfrac{24}{x^2} = \dfrac{-2}{x}$

11. $ax + b = c$ (for x)   **12.** $u(x - v) - w = t$ (for x)

Solve the following linear systems.

13. $x - y = 10$
      $-x - y = 20$

14. $-x + y = -3$
      $x + y = -11$

15. $x + y = 7$
      $2x - y = -1$

16. $3x + 3y = 10$
      $3x + 3y = 30$

17. $2x - 3y = 14$
      $5x + 2y = 35$

18. $7x - 5y = 16$
      $2x + 3y = 9$

19. $12x + 10y = 3$
      $-6x - 5y = 5$

20. $11x - 2y = -28$
      $7x + 5y = 1$

21. $4x - 7y = 24$
      $5x - 4y = 11$

22. $3x - 8y = 8$
      $2x - 7y = 7$

23. $0.02x + 0.1y = 3$
      $0.05x - 0.2y = 3$

24. $0.08x - 0.5y = 21$
      $0.01x + 0.2y = 0$

25. $\dfrac{2x}{5} - \dfrac{y}{7} = \dfrac{-23}{35}$
    $\dfrac{x}{6} - \dfrac{y}{2} = \dfrac{1}{6}$

26. $\dfrac{3x}{4} - \dfrac{y}{5} = \dfrac{9}{2}$
    $\dfrac{x}{7} + \dfrac{3y}{2} = \dfrac{6}{7}$

27. $4x - y = 9$
$x = y - 3$

28. $6x + 3y = 69$
$x = 2y - 1$

29. $y = 3x + 6$
$y = 4x + 1$

30. $y = 8x - 3$
$y = 4x + 5$

31. $\dfrac{x}{2} + \dfrac{y}{4} = 5$

$\dfrac{x}{3} + \dfrac{y}{6} = 7$

32. $4x - 3y = 19$
$5x - 7y = 27$

33. $3x + 4y = -5$
$2x - 5y = 12$

34. $2x - 3y = 1$
$x = 3y - 1$

35. $y = 5x - 3$
$y = 3(x + 1) + 2$

36. $3x - 4y = 5$

$x - \dfrac{y}{2} = 0$

37. $\dfrac{x}{3} + \dfrac{y}{5} = \dfrac{14}{15}$

$2x - 5y = -13$

38. $\dfrac{x}{9} - \dfrac{y}{2} = \dfrac{1}{3}$

$\dfrac{x}{7} + \dfrac{y}{3} = \dfrac{3}{7}$

39. $2x - 9y = 11$
$9x - 2y = 11$

40. $x + y = 10$
$y = x + 2$

41. $\dfrac{x}{3} + \dfrac{y}{4} = \dfrac{29}{12}$

$y = 3x + 1$

42. $\dfrac{4x}{5} + \dfrac{y}{3} = \dfrac{1}{5}$

$x = 2y - 7$

43. $\dfrac{1}{x} + \dfrac{1}{y} = \dfrac{5}{6}$

$\dfrac{2}{x} - \dfrac{1}{y} = \dfrac{2}{3}$

44. $\dfrac{1}{x} - \dfrac{1}{y} = \dfrac{-3}{10}$

$\dfrac{3}{x} + \dfrac{2}{y} = \dfrac{8}{5}$

(*Hint*: For Problems 43 and 44 let $u = 1/x$ and $v = 1/y$; then solve for u and v; finally, solve for x and y.)

45. $ax + by = e$
$cx + dy = f$
(for x and y)

46. $px - qy = u$
$rx - sy = v$
(for x and y)

47. The sum of two numbers is 91. Their difference is 23. Find the numbers.

48. The sum of two numbers is 110. Their difference is 100. Find the numbers.

49. One number is 2 more than 6 times another. Their sum is 23. Find the numbers.

50. Mr. Gelt invests $100,000: some at 10% interest and the rest at 12%. The total interest for 1 year is $11,100. How much is invested at each rate?

51. Carlin has 29 coins (nickels and dimes). Their value is $1.90 (190 cents). How many of each coin does he have?

52. A theater group sells 700 tickets: some at $3 and the rest at $5. If the total revenue is $3100, how many of each ticket did they sell?

Consumer Application

53. The annual cost to operate a certain used car is given by $C = 900 + 0.10m$, where m is miles driven. The annual cost of a certain new car is given by $C = 1500 + 0.07m$. The Clarks are considering both cars. Find the intersection of the two equations. This is the number of miles where the cost is the same.

Population Application

54. The population of Westfield is 10,000 and growing by 400 people per year. Its neighbor, Eastfield, is 19,000 but losing 500 people per year. If x is years and y is population, solve the following system to find when the populations will be the same.

$$y = 10,000 + 400x$$
$$y = 19,000 - 500x$$

Business Applications

55. The cost to produce x number of an item is $y = 250,000 + 4x$. The total revenue from their sales is given by $y = 9x$. Solve these two equations to find the *break-even* point, where the cost equals the revenue.

56. In a certain state, the federal and state corporate income taxes are interdependent; that is, the state tax can be deducted from the federal tax, and vice versa. A certain company earned \$1,000,000 and is in the 40% federal tax bracket and the 5% state tax bracket. If x is the state tax and y the federal tax, solve the system

$$x = 0.05(1,000,000 - y)$$
$$y = 0.40(1,000,000 - x)$$

Engineering Application

57. The stresses on a certain column satisfy the following system:

$$x + y = 49,285$$
$$x - y = 49,615$$

Solve this for the stresses x and y.

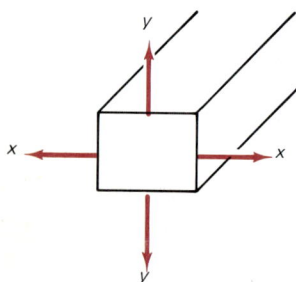

13.2
SYSTEMS OF LINEAR EQUATIONS IN THREE VARIABLES

Consider the system

$$\text{A:} \quad x + y + z = 2$$
$$\text{B:} \quad x + 4y - z = -10$$
$$\text{C:} \quad x - 2y + z = 8$$

Here we have three linear equations (A, B, and C) and three variables (x, y, and z). A solution is an **ordered triple** that satisfies all three equations at the same time.

Systems with three or more variables and equations are admittedly harder to solve than the two-equation, two-variable systems; however, the general procedure (the **elimination method**) is the same (only longer).

---

1. Using two of the equations, eliminate one of the variables.
2. Using another pair of equations, eliminate the *same* variable.
3. Solve these two resulting equations.
4. Substitute for the other variables.
5. Verify that the ordered triple satisfies each of the original equations.

Although we do examples with only three equations and three variables, the same elimination technique can be used for higher-order systems.

EXAMPLE 6 Solve the system

$$A: \quad x + y + z = 2$$
$$B: \quad x + 4y - z = -10$$
$$C: \quad x - 2y + z = 8$$

Solution We choose a variable to eliminate. Here we eliminate z *twice*, and then solve the resulting system.

Add A + B to eliminate z; call new equation D		$A:$	$x + y + z =$	2
		$B:$	$x + 4y - z = -10$	
		$D:$	$2x + 5y \quad = -8$	

Add B + C to eliminate z; call new equation E		$B:$	$x + 4y - z = -10$
		$C:$	$x - 2y + z = 8$
		$E:$	$2x + 2y \quad = -2$

Add D + (−E) to eliminate x

$$D: \quad 2x + 5y \quad = -8$$
$$-E: \quad -2x - 2y \quad = 2$$
$$\overline{\qquad\qquad 3y \quad = -6}$$
$$\qquad\qquad\quad y \quad = -2$$

Solve for y

Substitute $y = -2$ into E

$$2x - 4 = -2$$

Solve for x

$$x = 1$$

Substitute $x = 1$ and $y = -2$ into A

$$1 - 2 + z = 2$$

Solve for z

$$z = 3$$

Therefore, the solution is (1, −2, 3). (The reader should check that this triple does indeed satisfy all three of the original equations.)

EXAMPLE 7 Solve the system

$$A: \quad 2x + 3y = 6$$
$$B: \quad 3x - 4z = 29$$
$$C: \quad 5y - 2z = 10$$

Solution We must first rewrite these equations slightly since each equation is missing a variable. (This system is actually easier to solve—some of the variables are already eliminated.) Equation B does not have a y-term, and we use equations A and C to eliminate y.

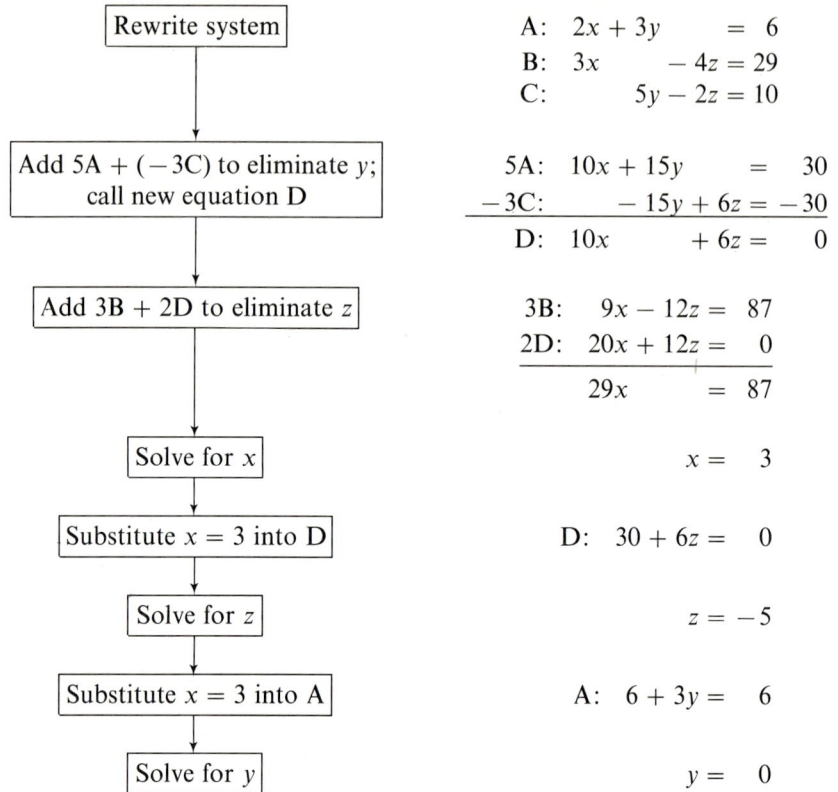

Rewrite system	A: $2x + 3y \qquad = 6$
	B: $3x \qquad - 4z = 29$
	C: $\qquad 5y - 2z = 10$

$$
\begin{array}{rl}
\text{5A:} & 10x + 15y \qquad = 30 \\
-3C: & \qquad -15y + 6z = -30 \\
\hline
\text{D:} & 10x \qquad + 6z = 0
\end{array}
$$

Add $5A + (-3C)$ to eliminate y; call new equation D

Add $3B + 2D$ to eliminate z

$$
\begin{array}{rl}
\text{3B:} & 9x - 12z = 87 \\
\text{2D:} & 20x + 12z = 0 \\
\hline
& 29x \qquad = 87
\end{array}
$$

Solve for x

$$x = 3$$

Substitute $x = 3$ into D

D: $30 + 6z = 0$

Solve for z

$$z = -5$$

Substitute $x = 3$ into A

A: $6 + 3y = 6$

Solve for y

$$y = 0$$

Thus, the solution is $(3, 0, -5)$. The reader should verify that this triple satisfies the original equations.

As with the two-variable systems, there may be no solution or an infinite number of solutions. (See page 444.)

PROBLEM SET
13.2   *Solve the following systems.*

Warm-up
Exercises

1. $x + y = 11$
     $x - y = 3$

2. $x - y = 20$
     $-x - y = 14$

3. $2x + y = 7$
     $3x + 2y = 11$

4. $8x - y = 33$
     $5x + 3y = 17$

5. $y = 3x + 11$
     $y = 5x + 15$

6. $2x + 5y = 29$
     $y = 3x - 1$

7. $\dfrac{x}{2} + y = 5$

     $\dfrac{x}{5} + \dfrac{y}{15} = 1$

8. $\dfrac{x}{5} - \dfrac{y}{2} = \dfrac{27}{10}$

     $\dfrac{x}{7} + \dfrac{y}{4} = \dfrac{-31}{28}$

Solve the following systems.

9.
$$\begin{aligned} x + y - z &= -3 \\ x + 2y + z &= 7 \\ -x + 4y + z &= 3 \end{aligned}$$

10.
$$\begin{aligned} x - y - z &= -1 \\ 2x + y - z &= -2 \\ x + 2y + 3z &= 8 \end{aligned}$$

11.
$$\begin{aligned} x + 2y + 3z &= -6 \\ 2x + y - z &= 3 \\ x - y - z &= 0 \end{aligned}$$

12.
$$\begin{aligned} 3x - y + 2z &= 19 \\ 2x + y - z &= 0 \\ x + 3y + 5z &= 17 \end{aligned}$$

13.
$$\begin{aligned} x + 2y &= -4 \\ y - 3z &= -5 \\ 2x + 5z &= 12 \end{aligned}$$

14.
$$\begin{aligned} 2x - 5y &= -13 \\ 3x + 4z &= -16 \\ 4y - 3z &= 7 \end{aligned}$$

15.
$$\begin{aligned} x + y + z &= 7 \\ x - y - z &= -5 \\ x + y - z &= 0 \end{aligned}$$

16.
$$\begin{aligned} x + y &= 9 \\ x - z &= 5 \\ y - z &= 10 \end{aligned}$$

17.
$$\begin{aligned} 2x - 5y &= -9 \\ 5y + 7z &= 26 \\ 9x - 4z &= -30 \end{aligned}$$

18.
$$\begin{aligned} 2x + y - z &= 9 \\ 3x + 2y + z &= 16 \\ x + 3y + 2z &= 7 \end{aligned}$$

19.
$$\begin{aligned} \frac{x}{6} - \frac{y}{4} &= \frac{2}{3} \\ \frac{y}{5} + \frac{z}{3} &= \frac{17}{15} \\ \frac{x}{8} - \frac{z}{4} &= 1 \end{aligned}$$

20.
$$\begin{aligned} \frac{x}{10} - \frac{y}{15} &= \frac{7}{10} \\ \frac{y}{6} + \frac{z}{3} &= \frac{1}{6} \\ \frac{x}{8} - \frac{z}{6} &= \frac{1}{6} \end{aligned}$$

21.
$$\begin{aligned} x + y + z + t &= 10 \\ x - y + z + t &= 6 \\ x + y + z + 2t &= 14 \\ x - y - z - t &= -8 \end{aligned}$$

22.
$$\begin{aligned} x + y - z + t &= 4 \\ 2x + y - z + t &= 5 \\ 3x + 2y + z + t &= 4 \\ 2x + y + 2z - t &= -2 \end{aligned}$$

23.
$$\begin{aligned} x - y &= -1 \\ y + z &= 2 \\ x + t &= 5 \\ t - w &= -3 \\ z + w &= 5 \end{aligned}$$

24.
$$\begin{aligned} x + y + z &= 2 \\ x - y + t &= 1 \\ y + t + w &= 0 \\ z + t + w &= 3 \\ x + t - w &= 2 \end{aligned}$$

25. The sum of three numbers is 66. The sum of the first two less the third is 8. Three times the first less the second is 3. Find the numbers.

26. In a triangle, angle x is 13 times angle z, and angle x is 90° more than angle y. Find the angles. (Recall that the sum of the angles in a triangle is 180°.)

27. Jamie has 21 coins in nickels, dimes, and quarters. Their value is $2.20 (220 cents). The sum of the number of dimes and quarters less the number of nickels is 1. How many of each coin does she have?

28. A theater sells 800 tickets: some at $2, some at $4, and the rest at $5. Their total revenue from the tickets is $2500. The number of $2 tickets sold equals the sum of $4 and $5 tickets sold. How many of each ticket was sold?

Electrical Application

29. The circuit shown has currents x, y, and z through the three resistors shown on page 452. Kirchhoff's and Ohm's laws lead to the following system:

$$\begin{aligned} x + y + z &= 0 \\ 10x - 30y &= 110 \\ -30y + 50z &= 140 \end{aligned}$$

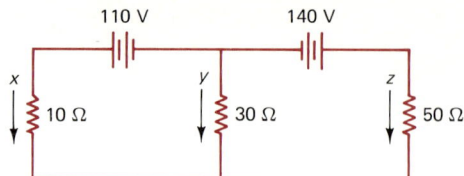

Solve this system for the currents. (A negative result means that the current is opposite to the arrow in the sketch.)

Business Application

30. Three companies (*X*, *Y*, and *Z*) are strongly interconnected:

Company *X* is worth \$20,000,000, plus 1/3 of *Y*, plus 1/5 of *Z*.
Company *Y* is worth \$50,000,000, plus 1/4 of *X*, plus 1/2 of *Z*.
Company *Z* is worth \$200,000,000, plus 1/2 of *X*, plus 1/3 of *Y*.

(a) Let *x*, *y*, and *z* be the value of companies *X*, *Y*, and *Z*. Translate the information above into three equations.

(b) Solve the resulting system.

13.3
NONLINEAR SYSTEMS OF EQUATIONS

A **nonlinear system of equations** is a system of equations that has at least one term that is not of first degree (or linear). For example,

$$\text{A:} \quad y = -x^2 - 2x + 5 \qquad\qquad \text{A:} \quad x^2 + y^2 = 25$$
$$\text{B:} \quad y = x^2 - 6x - 1 \qquad\qquad \text{B:} \quad x - 3y = -5$$

are nonlinear systems of equations. In addition to squares, these systems may also involve cubes, square roots, logarithms, and so on. Graphically, we now have the intersection of two curves; thus, we often get two or more intersection points.

If one of the equations can easily be solved in terms of one of the variables, we use the **substitution method** (as seen on page 445).

To solve a nonlinear system using the substitution method:

1. Solve one equation for one of the variables.
2. Substitute this expression into the other equation and solve.
3. Substitute to find the first variable.

Note: There will probably be at least two points, so you may have to do step 3 more than once.

EXAMPLE 8 Solve the system

$$\text{A:} \quad y = -x^2 - 2x + 5$$
$$\text{B:} \quad y = x^2 - 6x - 1$$

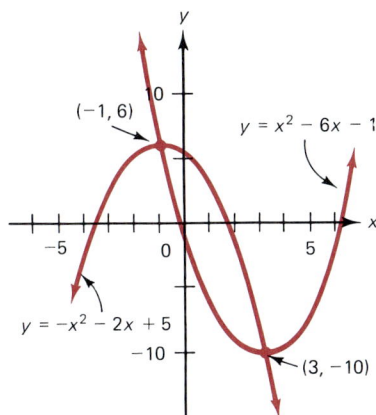

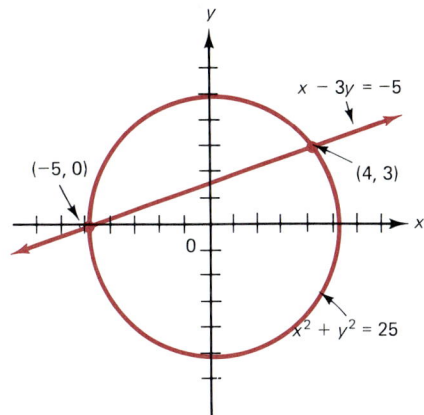

(a) Example 8   (b) Example 9

FIGURE 3

Solution Both equations are given in terms of y. Here we substitute one expression for y into the other equation. (Or we set the y-expressions equal to each other.)

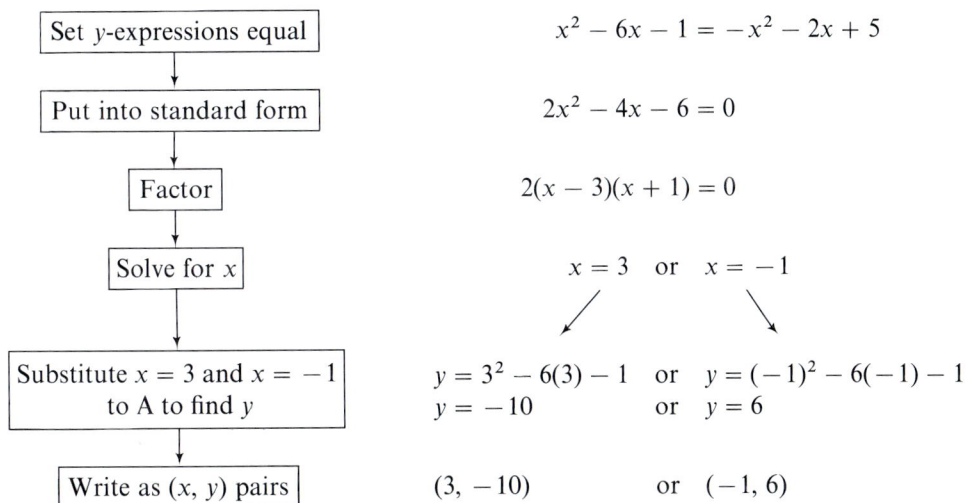

There are *two* solutions: $\{(3, -10), (-1, 6)\}$. Figure 3(a) shows these points as the intersection of two parabolas.

EXAMPLE 9 Solve the system

$$\text{A: } x^2 + y^2 = 25$$
$$\text{B: } x - 3y = -5$$

Solution We can easily solve equation B for x. We then substitute this expression into equation A and solve for y.

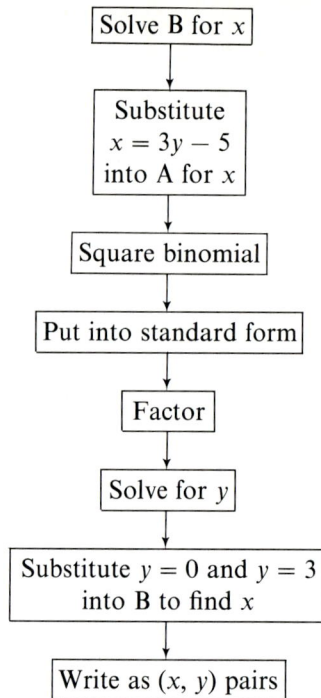

Solve B for x	$x = 3y - 5$
↓	
Substitute $x = 3y - 5$ into A for x	$x^2 + y^2 = 25$ $(3y - 5)^2 + y^2 = 25$
↓	
Square binomial	$9y^2 - 30y + 25 + y^2 = 25$
↓	
Put into standard form	$10y^2 - 30y = 0$
↓	
Factor	$10y(y - 3) = 0$
↓	
Solve for y	$y = 0$ or $y = 3$
↓	
Substitute $y = 0$ and $y = 3$ into B to find x	$x - 0 = -5$ or $x - 9 = -5$ $x = -5$ or $x = 4$
↓	
Write as (x, y) pairs	$(-5, 0)$ or $(4, 3)$

The solution set is $\{(-5, 0), (4, 3)\}$. Figure 3(b) shows these points as the intersection of a circle and a straight line.

EXAMPLE 10 Solve the system

$$A: \quad x^2 + 4y^2 = 16$$
$$B: \quad 4x^2 - 4y^2 = 4$$

Solution A system such as this with only second-degree terms can be solved nicely by the elimination method. Just as with linear systems, we add equations A and B to eliminate y; then we solve for x.

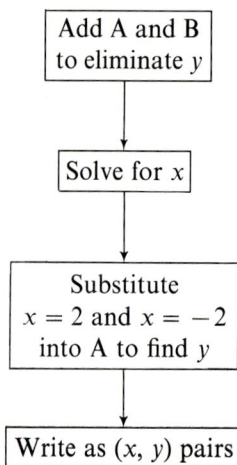

Add A and B to eliminate y	$\begin{aligned} A: \quad & x^2 + 4y^2 = 16 \\ B: \quad & 4x^2 - 4y^2 = 4 \\ \hline & 5x^2 = 20 \end{aligned}$
↓	
Solve for x	$x^2 = 4$ $x = 2$ or $x = -2$
↓	
Substitute $x = 2$ and $x = -2$ into A to find y	$\begin{aligned} 4 + 4y^2 &= 16 \\ 4y^2 &= 12 \\ y^2 &= 3 \\ y &= \pm\sqrt{3} \end{aligned}$ or $\begin{aligned} 4 + 4y^2 &= 16 \\ 4y^2 &= 12 \\ y^2 &= 3 \\ y &= \pm\sqrt{3} \end{aligned}$
↓	
Write as (x, y) pairs	$\{(2, \sqrt{3}), (2, -\sqrt{3}), (-2, \sqrt{3}), (-2, -\sqrt{3})\}$

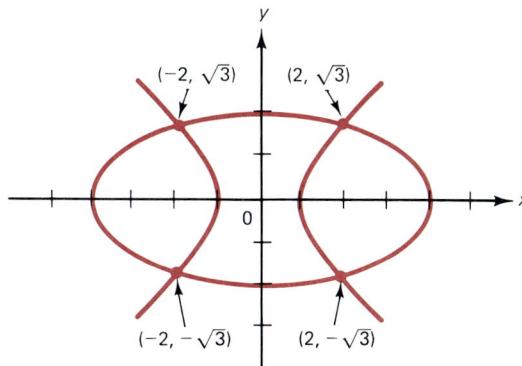

FIGURE 4

There are *four* solutions. Figure 4 shows these points as the intersection of an ellipse and a hyperbola.

Solve the following equations.

1. $2x + 5 = 19$

2. $-3x - 1 = 17$

3. $x^2 + 30 = 11x$

4. $(x - 3)^2 = 2$

5. $(x + 1)(x - 2) = 18$

6. $x^3 = x$

Solve the following linear systems.

7. $x + y = 17$
    $x - y = 5$

8. $2x - y = 9$
    $3x + 2y = -11$

9. $2x + 3y = 11$
    $y = 6x - 3$

10. $7x - 2y = -25$
    $x = -5y + 2$

11. $y = 1 - 4x$
    $y = 5x - 17$

12. $x = 10y + 7$
    $x = -3y + 33$

Solve the following nonlinear systems.

13. $y = x^2 + 1$
    $y = -x^2 + 3$

14. $y = -x^2 - x + 5$
    $y = x^2 + x + 1$

15. $y = x^3 - x^2 + 2x + 5$
    $y = x^3 + x^2 + 2x + 3$

16. $y = x^4 - x^2 - 2$
    $y = x^4 + x^2 - 8$

17. $x = y^2 + 2y + 7$
    $x = 2y^2 - 5y + 13$

18. $x = 5y^2 - y - 5$
    $x = 2y^2 + 7y - 9$

19. $2x + y = 5$
    $x^2 + 2y^2 = 19$

20. $x - 3y = 2$
    $x^2 - y^2 = 24$

21. $4x - y = 2$
    $y = x^2 + 1$

22. $3x - y = 2$
    $y = -x^2 + 12$

23. $3x^2 + y^2 = 31$
  $x^2 - 2y^2 = 1$

24. $x^2 + 4y^2 = 9$
  $2x^2 - y^2 = 18$

25. $y = x^2 + 7x - 13$
  $y = 4x - 3$

26. $x^2 - y^2 = 35$
  $x + y = 7$

27. $x^2 + y^2 = 29$
  $x - y = 3$

28. $x^2 + y^2 = 50$
  $x^2 - y^2 = 48$

29. $x = y^2 - 7y - 7$
  $x = 2y^2 - 20y + 33$

30. $2x^2 + 5y^2 = 37$
  $2x + y = 9$

31. $2x^2 + 3y^2 = 5$
  $3x^2 + 2y^2 = 5$

32. $x = 2y^2 - 3y - 6$
  $x = 8y^2 - 2y - 8$

33. $x + y = 5$
  $xy = 6$

34. $x - 4y = -4$
  $xy = 8$

35. $y = x^2$
  $x = y^2$

36. $y = 3/x^2$
  $y = x^2 - 2$

37. The sum of two numbers is 11. Their product is 18. Find the numbers.

38. The difference of two numbers is 5. Their product is 36. Find the numbers.

39. The sum of two numbers is 10. The sum of their squares is 58. Find the numbers.

40. The sum of the squares of two positive numbers is 89. The difference of their squares is 39. Find the numbers.

Business Applications

41. The total cost to produce x items of a certain product is $y = x^2 + 100x + 40,000$. The total revenue from the sales of x items is $y = 600x$. Find the intersection points (break-even points).

42. The relation between the demand x and the price y of an item is given by $y = 10,000/x$. The relation between the supply x and the price y is $-7x + 1000y = 3000$. Solve this nonlinear system to find the *equilibrium point* where supply = demand.

13.4
SYSTEMS OF INEQUALITIES

On page 210 we graphed inequalities, such as $x + y \geq 4$ and $x^2 + y^2 \leq 25$. [Recall that we first graph the corresponding equation and then see if a test point satisfies the inequality—see Figure 5(a) and (b).]

FIGURE 5

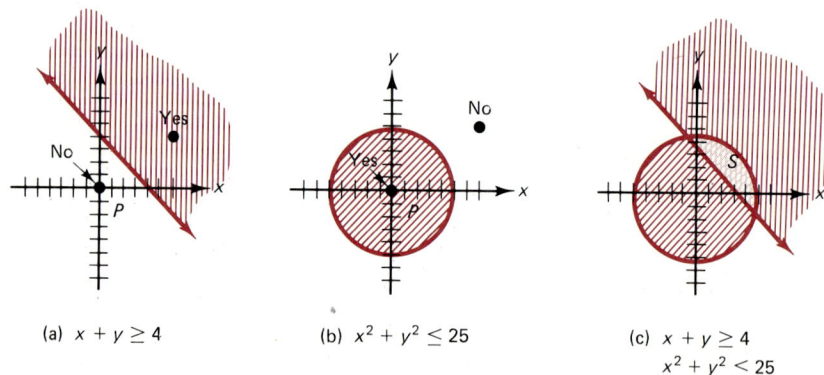

(a) $x + y \geq 4$

(b) $x^2 + y^2 \leq 25$

(c) $x + y \geq 4$
    $x^2 + y^2 \leq 25$

We now wish to consider two or more inequalities, such as

$$A: \quad x + y \geq 4$$
$$B: \quad x^2 + y^2 \leq 25$$

taken together. We call this a **system of inequalities**; its solution set is the intersection of the solution sets of each inequality [see Figure 5(c)]. We have denoted the solution set S.

To graph a system of inequalities:

1. Graph each inequality, using a different direction of shading for each region. (Review pages 210 to 215, if necessary.)
2. Find the intersection (or overlap) of the regions.

EXAMPLE 11 Graph the system

$$A: \quad 2x + 3y \leq 12$$
$$B: \quad 2x + \ y \geq \ 8$$

Solution We first graph the corresponding equations (two lines). We determine each inequality region, and then we find their intersection.

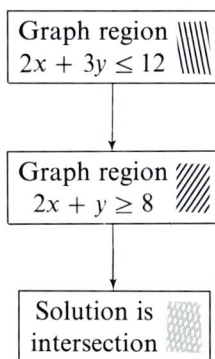

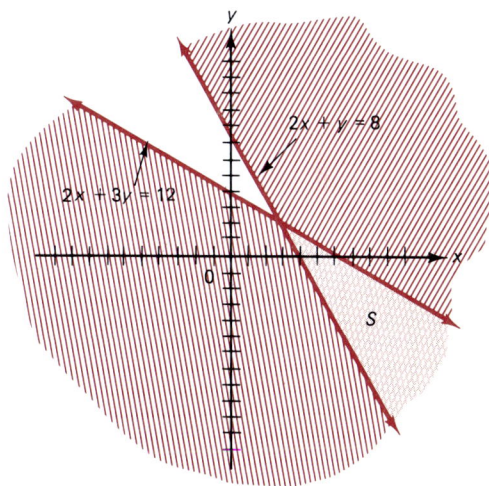

EXAMPLE 12 Graph the system

$$A: \quad x^2 + y^2 \leq 4$$
$$B: \quad \quad\quad y \geq x^2 - 1$$
$$C: \quad \quad\quad x \geq 0$$

Solution We begin by graphing the corresponding equations (a circle, a parabola, and a line).

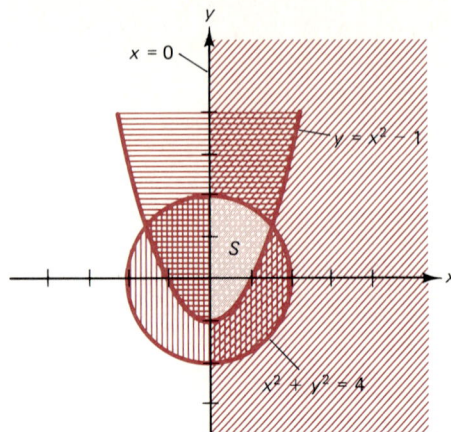

Graph the three regions:

$$x^2 + y^2 \leq 4$$
$$y \geq x^2 - 1$$
$$x \geq 0$$

Find the intersection

EXAMPLE 13 Graph the system

$$\text{A:} \quad x^2 + 9y^2 \leq 9$$
$$\text{B:} \quad x^2 - y^2 \leq 4$$
$$\text{C:} \quad x \geq 0$$
$$\text{D:} \quad y \geq 0$$

Solution We first graph each equation (an ellipse, a hyperbola, and two lines). Inequalities C and D tell us that the final region lies in the first quadrant.

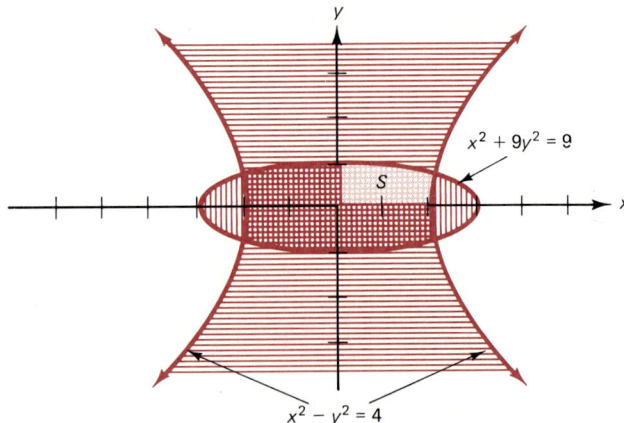

Graph regions:

$$x^2 + 9y^2 \leq 9$$
$$x^2 - y^2 \leq 4$$

Inequalities C and D indicate first quadrant

Find intersection

PROBLEM SET
13.4

Graph the following equations.

Warm-up Exercises

1. $x + y = 5$ **2.** $2x - 3y = 12$
3. $y = x^2 + 1$ **4.** $4x^2 + 9y^2 = 36$

Graph the following inequalities.

Warm-up Exercises

5. $x + y \leq 5$ **6.** $2x - 3y < 12$
7. $y > x^2 + 1$ **8.** $4x^2 + 9y^2 \leq 36$

Graph the following systems of inequalities.

9. $x + y \leq 4$
  $x - y \geq 3$

10. $x - y \geq 6$
  $x + y \leq 2$

11. $2x + y \leq 8$
  $x + 2y \leq 8$

12. $3x - 4y > 12$
  $x + 3y < 6$

13. $x + 5y \leq 5$
  $6x + y \geq 6$
  $x \quad \geq 0$
  $y \geq 0$

14. $2x + 5y < 20$
  $5x + 3y < 15$
  $x \quad > 0$
  $y > 0$

15. $y \geq x^2 + 1$
  $4x^2 + 9y^2 \leq 36$

16. $x^2 + y^2 \leq 25$
  $x^2 - y^2 \geq 4$
  $y \leq 1$
  $y \geq -1$

17. $x^2 + y^2 \geq 1$
  $y \leq 6$
  $y \geq -3$

18. $4x^2 + y^2 \leq 16$
  $y \leq 1$
  $y \geq -1$

19. $x + y > 7$
  $x - y < 2$

20. $y > x^2$
  $y < 4 - x^2$

21. $x^2 + 4y^2 \geq 4$
  $9x^2 + 4y^2 \leq 36$
  $y \geq 0$

22. $4x^2 - y^2 > 4$
  $x^2 + 9y^2 < 9$
  $x \quad > 0$

23. $y > x^2 + 1$
  $y < -x^2 - 1$

24. $y < x^2 + 5x + 6$
  $y > x + 3$

25. $x + y \geq 3$
  $x - y \leq 2$
  $x^2 + y^2 \leq 16$

26. $x \geq -3$
  $x \leq 3$
  $y \leq 4$
  $y \geq -4$

27. $x^2 + 4y^2 \leq 16$
  $y \leq x$
  $y \geq -x$
  $x^2 + y^2 \geq 4$

28. $x^2 - y^2 \leq 4$
  $-x^2 + y^2 \leq 4$
  $x \quad \geq 0$
  $y \geq 0$

29. $x^2 < 4$
  $y^2 > 9$

30. $|x| \geq 3$
  $|y| \leq 2$

Health Application

31. A patient must have at least 42 grams of protein, but less than 21 grams of fat. An ounce of hamburger provides 7 grams of protein and 3 grams of fat; a slice of wheat bread provides 3 grams of protein and 1 gram of fat. A dietitian using ounces of hamburger x and slices of bread y arrives at the following system:

$$7x + 3y \geq 42$$
$$3x + y \leq 21$$
$$x \quad \geq 0$$
$$y \geq 0$$

Graph this system of inequalities.

Environmental Application

32. A company manufactures two products. Product A (x) produces 15 grams of SO_2 and 40 grams of particulate matter per item. Product B (y) produces 30 grams of SO_2 and 20 grams of particulate matter. The EPA allows them at most 60,000 grams of SO_2 and 100,000 grams of particulate matter. Graph this system of inequalities that determines the allowable amounts of products A and B.

$$15x + 30y \leq 60{,}000$$
$$40x + 20y \leq 100{,}000$$
$$x \geq 0$$
$$y \geq 0$$

Business Application

33. A company has two radio stations: One has a transmitter at $(0, 0)$, and the other has a transmitter at $(20, 10)$. (These distances are in miles.) Both transmitters can broadcast 15 miles.

(a) Write an inequality for the first transmitter: everything within 15 miles of $(0, 0)$.

(b) Write an inequality for the second transmitter: everything within 15 miles of $(20, 10)$.

(c) Graph the system of inequalities and show the intersection that would receive transmission from both stations.

13.5
LINEAR PROGRAMMING

One of the new and important applications of systems of inequalities is **linear programming**, a field of mathematics developed in the 1940s during the war.

The word "programming" does not refer to a computer, but rather to a systematic method of maximizing a quantity (such as profit) or minimizing a quantity (such as cost). The quantity to be optimized (such as profit or cost) is called the **objective function** and has as typical forms

$$P = 100x + 70y \qquad \text{or} \qquad C = 500x + 800y$$

where x and y are the variables of the problem.

EXAMPLE 14   A certain hospital makes a \$2000 profit on a major surgical procedure and a \$1000 profit on a minor surgical procedure. The average major procedure takes 3 hours and the minor procedure, 2 hours; the hospital has 24 hours available for surgery. At most, 9 total surgical procedures can be performed per day, and at most 6 minor procedures. How many of each type of surgery should be scheduled to maximize the hospital's profit?

Solution   We must first sort all this information into equations and inequalities (this is often the hardest part). A good place to begin is to identify the variables; then we translate the rest of the information.

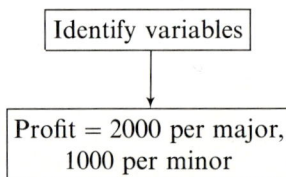

| Identify variables | Let $x =$ number of major surgeries |
| | $y =$ number of minor surgeries |

| Profit = 2000 per major, 1000 per minor | Maximize $P = 2000x + 1000y$ |

Can we choose just any x and y? No, there are certain limitations: time, total number, and so on. We now translate these **constraints** (or limitations) into equations or inequalities.

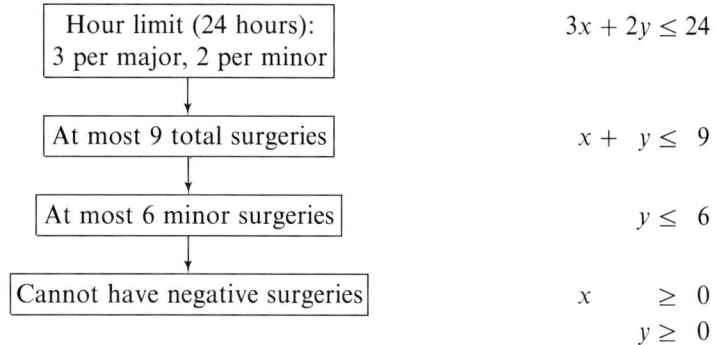

Hour limit (24 hours): 3 per major, 2 per minor	$3x + 2y \leq 24$
↓	
At most 9 total surgeries	$x + y \leq 9$
↓	
At most 6 minor surgeries	$y \leq 6$
↓	
Cannot have negative surgeries	$x \geq 0$ $y \geq 0$

All these inequalities must be satisfied; thus, they form a system of inequalities, as seen in the preceding section. The solution to this system is called the **feasible region**: Its (x, y) pairs represent the only suitable or allowable pairs of major and minor surgeries.

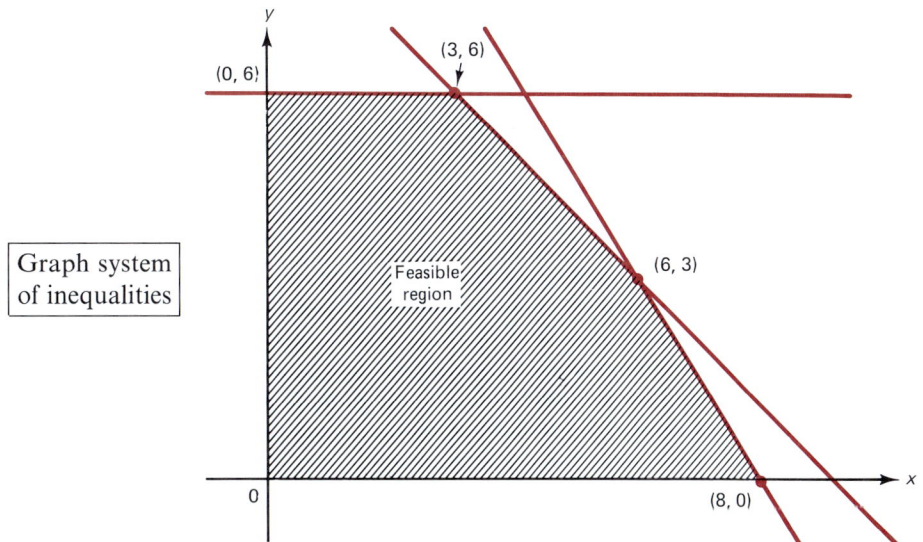

Graph system of inequalities

Even though we have limited the number of possible (x, y) values to this feasible region, there are still too many points to check for the one producing the maximum profit. Fortunately, we need not check every point of the feasible region. One of the major theorems of linear programming states that *an optimal value for the objective function occurs at a corner*. Thus, we find the corners of the feasible region (at the intersection of the lines) and evaluate the objective function for these (x, y) pairs.

Evaluate corners in objective function $P = 2000x + 1000y$

⬇

Choose largest P

Corner	$P = 2000x + 1000y$
$(0, 0)$	$0 + 0 = 0$
$(8, 0)$ ✓	$16{,}000 + 0 = 16{,}000$ ✓
$(6, 3)$	$12{,}000 + 3000 = 15{,}000$
$(3, 6)$	$6000 + 6000 = 12{,}000$
$(0, 6)$	$0 + 6000 = 6000$

The largest profit 16,000 is at (8, 0), or eight major surgeries and zero minor surgeries.

Let us summarize what Example 14 has shown about linear programming.

To solve a linear programming application graphically:

1. Identify the objective function to be maximized or minimized.
2. Translate the constraints on the variables into equations or inequalities.
3. Graph the system of inequalities. (This is the feasible region.)
4. Find the corners of the feasible region.
5. Evaluate the coordinates of each corner in the objective function.
6. Choose the corner point that produces the largest (smallest) value for the objective function.

EXAMPLE 15 Solve the following linear programming problem graphically:

$$\text{Minimize} \quad C = 3x + 5y$$
$$\text{subject to} \quad 2x + y \geq 40$$
$$x + y \geq 30$$
$$x + 3y \geq 50$$
$$x \quad\quad \geq 0$$
$$y \geq 0$$

Solution This problem is already formulated; however, unlike Example 14, this one is a minimization problem. We begin by graphing the set of constraints; we then find the corners of the feasible regions; finally, we evaluate each corner in the objective function and choose the corner producing the lowest C.

Graph system of inequalities

Find corners by finding intersections of lines

Evaluate corners in $C = 3x + 5y$

Choose minimum

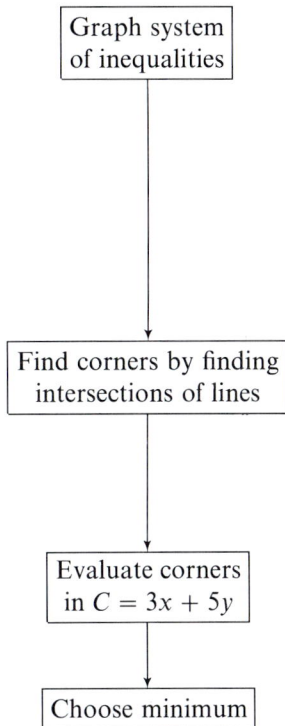

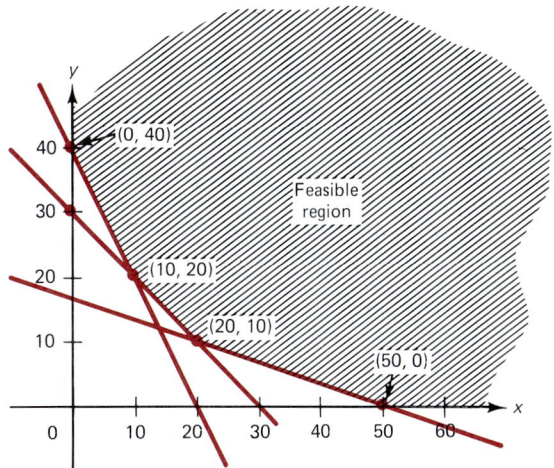

Corner	$C = 3x + 5y$
(0, 40)	$0 + 200 = 200$
(10, 20)	$30 + 100 = 130$
(20, 10) $\checkmark$	$60 + 50 = 110 \checkmark$ (lowest)
(50, 0)	$150 + 0 = 150$

The solution is (20, 10).

Thus, if we choose $x = 20$ and $y = 10$, we get the minimal C, which is 110.

PROBLEM SET 13.5

Solve the following systems of equations.

Warm-up Exercises

1. $x + y = 15$
$x - y = 7$

2. $2x + y = 16$
$x \quad\quad = 6$

3. $2x + 5y = 9$
$x + y = 3$

4. $4x + 3y = 26$
$3x + 4y = 23$

Graph the following systems of inequalities.

Warm-up Exercises

5. $x + y \leq 8$
$x + 2y \leq 10$
$x \quad\quad \geq 0$
$y \geq 0$

6. $x + 2y \leq 6$
$2x + y \leq 6$
$x \quad\quad \geq 0$
$y \geq 0$

Solve the following linear programming problems.

7. Maximize $P = 60x + 90y$
subject to $3x + y \leq 9$
$y \leq 6$
$x \quad\quad \geq 0$
$y \geq 0$

8. Maximize $P = 40x + 50y$
subject to $2x + y \leq 8$
$x \quad\quad \leq 3$
$x \quad\quad \geq 0$
$y \geq 0$

9. Maximize $P = 60x + 50y$
subject to $\quad 5x + y \leq 30$
$$x + y \leq 10$$
$$x \quad\quad \geq 0$$
$$y \geq 0$$

10. Maximize $P = 500x + 400y$
subject to $\quad 3x + 2y \leq 24$
$$x + y \leq 9$$
$$x \quad\quad \geq 0$$
$$y \geq 0$$

11. Minimize $C = 150x + 200y$
subject to $\quad 3x + 5y \geq 30$
$$x + y \geq 8$$
$$x \quad\quad \geq 2$$
$$y \geq 0$$

12. Minimize $C = 40x + 70y$
subject to $\quad x + 2y \geq 10$
$$x + y \geq 6$$
$$x \quad\quad \geq 0$$
$$y \geq 1$$

13. Minimize $C = 80x + 110y$
subject to $\quad 3x + 5y \geq 300$
$$5x + 3y \geq 300$$
$$x + y \geq 80$$
$$x \quad\quad \geq 0$$
$$y \geq 0$$

14. Maximize $P = 5000x + 8000y$
subject to $\quad 7x + 3y \leq 210$
$$x + y \leq 100$$
$$3x + 7y \leq 210$$
$$x \quad\quad \geq 0$$
$$y \geq 0$$

Business Applications

15. An automobile manufacturing company has two new cars: Raccoons (x) and Beavers (y). The profit from a Raccoon is $1000 and from a Beaver $1500. Each Raccoon takes 150 hours of labor to build, and each Beaver 250 hours. The company has at most 75,000 hours available. Each Raccoon uses 1200 pounds of steel and each Beaver uses 1200 pounds. The company has at most 480,000 pounds. Find how many of each car to make to maximize their profit. (Remember, x and y must be zero or positive.)

16. A farmer has some land on which to plant corn (x) and soybeans (y). The revenue from an acre of corn is $500 and from an acre of soybeans $400. An acre of corn costs $3 in seed and an acre of soybeans costs $5. The farmer has at most $7500 to spend on seed. An acre of corn costs $30 in labor and an acre of soybeans costs $15. The farmer has at most $30,000 to spend on labor. Find how much of each crop should be planted to maximize profit. (Remember, x and y must be zero or positive.)

17. A company is reducing its pollutants using taller stacks and filters. The unit cost of a taller stack is 100, and of a filter 50. The SO_2 pollution is reduced by 30 for each unit of taller stack and by 10 for each unit of filter. The company must reduce SO_2 by 60,000. The particulate pollution is reduced by 150 for each unit of taller stack and by 300 for each unit of filter. The company must reduce particulate pollution by 600,000. What combination of taller stack and filter will meet the requirements and minimize costs?

18. A company makes chairs and tables. The profit on a chair is $20 and on a table $25. Sawing and gluing a chair requires 30 minutes, and a table 20 minutes. Finishing a chair requires 20 minutes, and a table 30 minutes. The company can make no more than 1500 chairs. They have available 60,000 minutes for sawing and gluing and 60,000 minutes for finishing. How many chairs and tables should they produce to maximize profit?

Health Application

19. A dietitian is planning a diet for a hospital patient based on two foods: slices of wheat bread (x) and cups of yogurt (y). A slice of bread provides 65 calories, and the yogurt provides 130 calories per cup. The bread provides 3 grams of protein per slice, and the yogurt provides 10 grams of protein per cup. The patient must have at least 30 grams of protein. The bread provides 15 grams of carbohydrates per slice, and the yogurt provides 10 grams of carbohydrates per cup. The patient can have no more than 60 grams of carbohydrates. How much should the patient eat to minimize the calorie intake. (Remember, x and y must be zero or positive.)

CHAPTER 13 SUMMARY

Important Properties

For any system of equations:

S1. We can multiply both sides of an equation by a nonzero constant.
S2. We can add the corresponding sides of any equations.
S3. We can substitute equals for equals in any equation.

Review Exercises

Solve the following systems or problems involving systems.

1. $x + y = 9$
$x - y = 15$

2. $2x - 7y = 37$
$5x + 4y = -15$

3. $\dfrac{x}{2} + \dfrac{y}{3} = \dfrac{29}{6}$

$\dfrac{x}{4} - \dfrac{y}{5} = \dfrac{-3}{20}$

4. $y = 5x + 4$
$y = 2x + 25$

5. Denise has 17 coins (dimes and quarters). Their worth is 290 cents. How many of each coin does she have?

6. $x + y + z = 6$
$x - y + z = 4$
$x - y - z = 8$

7. $3x + y - z = 8$
$x + 2y + z = -4$
$x - y + 4z = -19$

8. $\dfrac{x}{2} + \dfrac{y}{3} = \dfrac{-1}{6}$

$\dfrac{x}{5} - \dfrac{z}{4} = \dfrac{-21}{20}$

$\dfrac{y}{3} + \dfrac{z}{4} = \dfrac{7}{12}$

9. A family invests $10,000: some at 10%, some at 12%, and the rest at 14%. The amount invested at 10% equals the sum of the money invested at 12% and 14%. The total interest for 1 year is $1140. How much was invested at each rate?

10. $y = x^2 - 3x + 5$
$y = x^2 + 3x - 3$

11. $x + 2y = 1$
$3x^2 + y^2 = 79$

12. $x^2 - 4y^2 = 5$
$3x^2 + 2y^2 = 29$

13. $y = 2^{x+4}$
$y = 4^{2x-1}$

14. $x + 2y \le 6$
$x + y \le 4$
$x \ge 1$
$y \ge 0$

15. $y > x^2 + 1$
$y < -x^2 + 5$

16. $x^2 + y^2 \ge 9$
$4x^2 + y^2 \le 16$
$y \ge 0$

17. $x^2 \ge 16$
$y^2 \le 25$

18. Maximize $P = 50x + 30y$
subject to $\quad 3x + y \leq 21$
$\quad\quad\quad\quad x + y \leq 13$
$\quad\quad\quad\quad x \quad\quad \geq 0$
$\quad\quad\quad\quad\quad\quad y \geq 0$

19. Minimize $C = 100x + 70y$
subject to $\quad 2x + 3y \geq 18$
$\quad\quad\quad\quad x + y \geq 7$
$\quad\quad\quad\quad x \quad\quad \geq 0$
$\quad\quad\quad\quad\quad\quad y \geq 0$

20. A company manufactures two items: widgets and flanges. Their profit on a widget is $10 and on a flange, $15. They have 75,000 minutes of labor available; a widget requires 15 minutes, and a flange 25 minutes. They have 48,000 pounds of steel; a widget requires 12 pounds of steel, and a flange 12 pounds. How many widgets and flanges should they produce to maximize their profit?

Matrices
and Determinants

In this chapter we study a special mathematical idea: the matrix. A **matrix** (plural: **matrices**) is a rectangular array of numbers. For instance,

$$\begin{bmatrix} 2 & 4 & -1 \\ 0 & 6 & 12 \end{bmatrix} \qquad \begin{bmatrix} 1 \\ 0 \\ -3 \\ 8 \end{bmatrix} \qquad \begin{bmatrix} 5 & -2 & 0 \\ 8 & 0 & 3 \\ 1 & 0 & 0 \end{bmatrix}$$

are examples of matrices. One of the major uses of matrices is to store information that comes in blocks.

EXAMPLE 1   A certain baseball team is keeping very close track of its three highest-paid players. It obtains the following data:

	1985				1986			
	Singles	Doubles	Triples	Homers	Singles	Doubles	Triples	Homers
Polk	82	12	3	4	85	10	2	5
Fillmore	75	9	2	15	79	10	3	7
Garfield	101	14	6	2	112	12	7	4

We can rewrite this table as two matrices (which we label with capital letters): A for the 1985 statistics and B for the 1986 statistics.

$$A = \begin{bmatrix} 82 & 12 & 3 & 4 \\ 75 & 9 & 2 & 15 \\ 101 & 14 & 6 & 2 \end{bmatrix} \qquad B = \begin{bmatrix} 85 & 10 & 2 & 5 \\ 79 & 10 & 3 & 7 \\ 112 & 12 & 7 & 4 \end{bmatrix}$$

These matrices are compact forms of the original table. Each row represents a different player and each column represents a different type of hit.

Every matrix has its **size**, which is

$$r \text{ by } c$$

where r is the number of rows and c the number of columns. Each element has an address: row first, column second. For example, in matrix A, $a_{2,4} = 15$ is the entry in the second row and fourth column. (We label the elements of a matrix with lowercase letters.)

EXAMPLE 2   (a) Matrices A and B of Example 1 are both 3 by 4 (3 rows and 4 columns).

(b) $C = \begin{bmatrix} 2 & -3 & 4 & 8 \\ 0 & 6 & -1 & 2 \end{bmatrix}$ has size 2 by 4.

Some of its entries are $c_{2,1} = 0$ and $c_{1,3} = 4$.

(c) $D = \begin{bmatrix} 2 \\ -3 \\ 0 \\ 8 \\ -7 \end{bmatrix}$ has size 5 by 1.

Some of its entries are $d_{2,1} = -3$ and $d_{4,1} = 8$.

Matrices have their own algebra; that is, in some cases, we can add, subtract, and multiply them.

We say that two matrices are **equal** if they have the same size and all their corresponding elements are equal. For example,

$$[x \quad y \quad z] = [3 \quad 6 \quad -8]$$

if and only if $x = 3$, $y = 6$, and $z = -8$. Also,

$$\begin{bmatrix} 2 & 3 & -4 \\ 5 & -6 & 7 \end{bmatrix} \neq \begin{bmatrix} 2 & 3 \\ 5 & -6 \end{bmatrix}$$

are unequal matrices.

If matrices A and B have the *same size*, then we can find their **sum** $A + B$ by simply adding the corresponding elements of A and B.

EXAMPLE 3   Some matrices can be added; some cannot. (They must have the *same size*.)

(a) $\begin{bmatrix} 2 & 3 & 5 \\ -4 & 0 & 9 \end{bmatrix} + \begin{bmatrix} 3 & 10 & -1 \\ 0 & 8 & 2 \end{bmatrix} = \begin{bmatrix} 5 & 13 & 4 \\ -4 & 8 & 11 \end{bmatrix}$

(b) $\begin{bmatrix} 2 & 3 & 5 \\ -4 & 0 & 9 \end{bmatrix}$ and $\begin{bmatrix} 1 \\ -3 \\ 8 \end{bmatrix}$ cannot be added together.

EXAMPLE 4   Find the 2-year totals for the players in Example 1.

Solution   Since A and B are the same size, we can add them to find a matrix for the 2-year totals. We let $S = A + B$.

$$S = A + B = \begin{bmatrix} 82 & 12 & 3 & 4 \\ 75 & 9 & 2 & 15 \\ 101 & 14 & 6 & 2 \end{bmatrix} + \begin{bmatrix} 85 & 10 & 2 & 5 \\ 79 & 10 & 3 & 7 \\ 112 & 12 & 7 & 4 \end{bmatrix}$$

$$= \begin{bmatrix} 167 & 22 & 5 & 9 \\ 154 & 19 & 5 & 22 \\ 213 & 26 & 13 & 6 \end{bmatrix}$$

From this sum, we can see for example that $s_{1,3} = 5$, which means that player 1 (Polk) got a total of 5 triples in the 2 years.

We can multiply every entry of a matrix by a constant (or scalar); this is called **scalar multiplication**.

EXAMPLE 5   The following shows two scalar multiplications. Notice that the **negative** of a matrix is given by $-A = (-1)A$ (as is the case with real numbers).

(a) $2\begin{bmatrix} 1 & 3 & -5 \\ -8 & 0 & 7 \end{bmatrix} = \begin{bmatrix} 2 & 6 & -10 \\ -16 & 0 & 14 \end{bmatrix}$

(b) $-\begin{bmatrix} 1 & -9 \\ 4 & 2 \end{bmatrix} = (-1)\begin{bmatrix} 1 & -9 \\ 4 & 2 \end{bmatrix} = \begin{bmatrix} -1 & 9 \\ -4 & -2 \end{bmatrix}$

We **subtract** matrices as we subtract real numbers: We add the negative of the matrix to be subtracted. If A and B are the same size,

$$\boxed{A - B = A + (-B)}$$

EXAMPLE 6   When subtracting, we change the signs in *all* the entries of the second matrix. As with addition, the matrices must be the *same size*.

(a) $\begin{bmatrix} 9 & 4 \\ 0 & -3 \end{bmatrix} - \begin{bmatrix} 5 & -1 \\ -6 & 7 \end{bmatrix} = \begin{bmatrix} 9 & 4 \\ 0 & -3 \end{bmatrix} + \begin{bmatrix} -5 & 1 \\ 6 & -7 \end{bmatrix} = \begin{bmatrix} 4 & 5 \\ 6 & -10 \end{bmatrix}$

(b) $\begin{bmatrix} 1 \\ 3 \\ -9 \end{bmatrix}$ and $\begin{bmatrix} -2 & 7 \\ 0 & -5 \end{bmatrix}$ cannot be subtracted from each other.

EXAMPLE 7 Find the difference between the players' performances in 1986 and 1985 (in Example 1).

Solution Here we subtract the matrices; we let $D = B - A$.

$$D = B - A = \begin{bmatrix} 85 & 10 & 2 & 5 \\ 79 & 10 & 3 & 7 \\ 112 & 12 & 7 & 4 \end{bmatrix} - \begin{bmatrix} 82 & 12 & 3 & 4 \\ 75 & 9 & 2 & 15 \\ 101 & 14 & 6 & 2 \end{bmatrix}$$

$$= \begin{bmatrix} 3 & -2 & -1 & 1 \\ 4 & 1 & 1 & -8 \\ 11 & -2 & 1 & 2 \end{bmatrix}$$

The positive entries are increases, and the negative entries are decreases. Thus, $d_{2,1} = 4$ means that player 2 got 4 *more* singles; $d_{3,2} = -2$ means that player 3 got 2 *fewer* doubles.

There is also **matrix multiplication**, but it is somewhat more difficult than addition. First, matrices A and B must be **compatible**; this means that the number of columns of A must equal the number of rows of B.

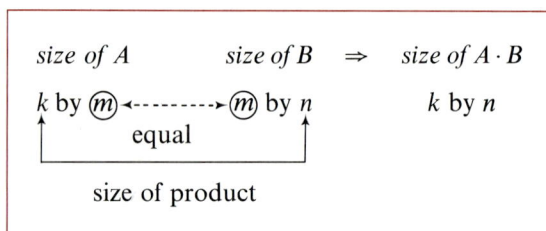

> *size of A*   *size of B* $\Rightarrow$ *size of A · B*
>
> k by $\textcircled{m}$ ⟵ - - - - - - - - - ⟶ $\textcircled{m}$ by n k by n
>
>   equal
>
>   size of product

The following table shows examples of matrices that are compatible (and not compatible) and the size of their product (if it exists).

Size of A	Size of B	Compatible?	Size of $A · B$
2 by $\textcircled{3}$ ⟵ - - - - - ⟶ $\textcircled{3}$ by 4		Yes	2 by 4
1 by $\textcircled{2}$ ⟵ - - - - - ⟶ $\textcircled{2}$ by 5		Yes	1 by 5
2 by 4	3 by 4	No	—
5 by $\textcircled{2}$ ⟵ - - - - - ⟶ $\textcircled{2}$ by 3		Yes	5 by 3
4 by 1	2 by 4	No	—

To find entry i, j of the product $A · B$:

1. Multiply the elements of row i of A by the corresponding elements of column j of B.
2. Add these products.

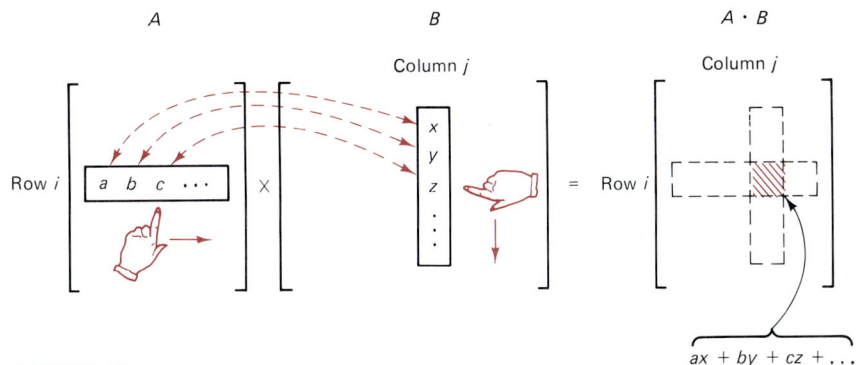

FIGURE 1

Figure 1 gives a sketch of how this matrix multiplication works. (In practice, students often use their fingers to help them: The left index finger goes across row i of A, while the right index finger goes down column j of B.)

EXAMPLE 8 Multiply the matrices

$$\begin{bmatrix} 2 & -3 \\ 1 & 5 \end{bmatrix} \times \begin{bmatrix} -1 & 4 \\ 0 & -6 \end{bmatrix}$$

Solution Since A has two columns and B has two rows, A and B are compatible for multiplication. We show this multiplication in "slow motion."

$$\begin{bmatrix} 2 & -3 \\ 1 & 5 \end{bmatrix} \times \begin{bmatrix} -1 & 4 \\ 0 & -6 \end{bmatrix} \rightarrow (2)(-1) + (-3)(0) = -2 \rightarrow \begin{bmatrix} -2 & \\ & \end{bmatrix}$$

$$\begin{bmatrix} 2 & -3 \\ 1 & 5 \end{bmatrix} \times \begin{bmatrix} -1 & 4 \\ 0 & -6 \end{bmatrix} \rightarrow (2)(4) + (-3)(-6) = 26 \rightarrow \begin{bmatrix} -2 & 26 \\ & \end{bmatrix}$$

$$\begin{bmatrix} 2 & -3 \\ 1 & 5 \end{bmatrix} \times \begin{bmatrix} -1 & 4 \\ 0 & -6 \end{bmatrix} \rightarrow (1)(-1) + (5)(0) = -1 \rightarrow \begin{bmatrix} -2 & 26 \\ -1 & \end{bmatrix}$$

$$\begin{bmatrix} 2 & -3 \\ 1 & 5 \end{bmatrix} \times \begin{bmatrix} -1 & 4 \\ 0 & -6 \end{bmatrix} \rightarrow (1)(4) + (5)(-6) = -26 \rightarrow \begin{bmatrix} -2 & 26 \\ -1 & -26 \end{bmatrix}$$

Thus, the product $A \cdot B = \begin{bmatrix} -2 & 26 \\ -1 & -26 \end{bmatrix}$.

EXAMPLE 9 For the players in Example 1, find their total bases (1 for each single, 2 for each double, and so on) for 1985.

Solution Let us use a *value matrix*

$$V = \begin{bmatrix} 1 \\ 2 \\ 3 \\ 4 \end{bmatrix} \begin{matrix} \text{single} \\ \text{double} \\ \text{triple} \\ \text{homer} \end{matrix}$$

which gives the base value of each type of hit. Matrix A is 3 by 4 and V is 4 by 1, so they are compatible. We compute $T = A \cdot V$.

$$T = A \cdot V = \begin{bmatrix} 82 & 12 & 3 & 4 \\ 75 & 9 & 2 & 15 \\ 101 & 14 & 6 & 2 \end{bmatrix} \cdot \begin{bmatrix} 1 \\ 2 \\ 3 \\ 4 \end{bmatrix}$$

$$= \begin{bmatrix} 82(1) + 12(2) + 3(3) + 4(4) \\ 75(1) + 9(2) + 2(3) + 15(4) \\ 101(1) + 14(2) + 6(3) + 2(4) \end{bmatrix} = \begin{bmatrix} 131 \\ 159 \\ 155 \end{bmatrix}$$

Thus, $t_{1,1} = 131$ means that Polk had 131 total bases, $t_{2,1} = 159$ means that Fillmore had 159 total bases, and $t_{3,1} = 155$ means that Garfield had 155 total bases.

EXAMPLE 10 The matrices

$$E = \begin{bmatrix} 0 & 2 & 3 & -1 \\ -4 & -7 & 8 & 5 \end{bmatrix} \quad \text{and} \quad F = \begin{bmatrix} 1 \\ 4 \\ -7 \end{bmatrix}$$

cannot be multiplied together, since they are not compatible.

**PROBLEM SET
14.1**

Simplify the following.

*Warm-up
Exercises*

1. $4 + (-7)$

2. $-11 + (-10)$

3. $-2 - 10$

4. $-4 - (-3)$

5. $(-4)(-2)$

6. $(-5)(2)$

7. $2 \cdot 7 + 3 \cdot 8$

8. $3 \cdot 2 - 4 \cdot 5$

9. $(-2)(5) + (-3)(-6)$

10. $6(-2) + (-3)(-5)$

Rewrite each of the following tables as two matrices.

11.

Salesperson	1985 Sales New Cars	1985 Sales Used Cars	1986 Sales New Cars	1986 Sales Used Cars
H. Hackenbush	45	52	49	63
O. Driftwood	39	78	25	80
R. Firefly	41	29	49	50
Q. Wagstaff	53	59	50	60
J. Loophole	27	46	35	35

12.

Dieter	Diet 1 Weight loss Month 1	Month 2	Month 3	Diet 2 Weight loss Month 1	Month 2	Month 3
Paunchworth	7.5	5.2	1.3	4.5	4.7	3.9
MacSchmaltz	10.3	6.3	0.2	6.4	5.9	5.1

For each of the following matrices, give its size and the indicated entries.

13. $A = \begin{bmatrix} 2 & -3 & 7 \\ 8 & 9 & 0 \end{bmatrix}$ Find $a_{1,3}$ and $a_{2,2}$.

14. $B = \begin{bmatrix} 4 & -3 \\ -5 & 7 \\ 9 & 0 \\ 6 & -8 \end{bmatrix}$ Find $b_{2,1}$ and $b_{1,2}$.

15. $C = \begin{bmatrix} 1 & -2 & 3 & -4 & 5 \end{bmatrix}$ Find $c_{1,3}$ and $c_{1,4}$.

16. $D = \begin{bmatrix} 1 & 2 & -3 \\ 4 & -5 & 6 \\ -7 & 8 & 9 \end{bmatrix}$ Find $d_{3,1}$ and $d_{1,2}$.

Complete the following table.

	Size of A	Size of B	Compatible?	Size of $A \cdot B$
17.	3 by 5	5 by 6	?	?
18.	2 by 7	7 by 2	?	?
19.	3 by 2	2 by 4	?	?
20.	4 by 3	4 by 3	?	?
21.	5 by 1	2 by 5	?	?
22.	6 by 1	1 by 3	?	?

Perform the following matrix operations (where possible).

23. $\begin{bmatrix} 4 & -7 & 8 \\ 0 & 1 & 3 \end{bmatrix} + \begin{bmatrix} 9 & 0 & -1 \\ 5 & -1 & 5 \end{bmatrix}$

24. $\begin{bmatrix} 1 & -2 \\ -3 & 4 \\ 5 & -6 \end{bmatrix} + \begin{bmatrix} 0 & 0 \\ 0 & 0 \\ 0 & 0 \end{bmatrix}$

25. $\begin{bmatrix} 1 & -4 \\ 8 & 0 \end{bmatrix} + \begin{bmatrix} 7 & 8 & -9 \end{bmatrix}$

26. $\begin{bmatrix} 1 & 2 & -3 & 4 \end{bmatrix} + \begin{bmatrix} 10 & -11 & 12 & 13 \end{bmatrix}$

27. $\begin{bmatrix} 9 & 8 & -7 \\ 6 & -5 & 4 \\ -3 & 2 & 1 \end{bmatrix} + \begin{bmatrix} -9 & -8 & 7 \\ -6 & 5 & -4 \\ 3 & -2 & -1 \end{bmatrix}$

28. $\begin{bmatrix} 4 & 0 \\ -5 & 1 \\ 2 & 7 \end{bmatrix} + \begin{bmatrix} 1 & 2 & -3 \\ 4 & -5 & 6 \end{bmatrix}$

29. $7 \begin{bmatrix} 2 & 8 \\ 3 & 0 \end{bmatrix}$

30. $5 \begin{bmatrix} 6 \\ -4 \\ -1 \end{bmatrix}$

31. $-2 \begin{bmatrix} 2 & 3 & -7 \\ 3 & -5 & 0 \end{bmatrix}$

32. $0 \begin{bmatrix} 9 \\ -4 \end{bmatrix}$

33. $\begin{bmatrix} 1 & 2 \\ -4 & 5 \end{bmatrix} - \begin{bmatrix} 5 \\ 9 \end{bmatrix}$

34. $\begin{bmatrix} 1 & -2 \\ 3 & 8 \\ -5 & 10 \end{bmatrix} - \begin{bmatrix} 6 & -3 \\ 4 & 10 \\ -2 & 7 \end{bmatrix}$

35. $\begin{bmatrix} 4 & -5 & 6 \\ 2 & 1 & 9 \\ 8 & 10 & -3 \end{bmatrix} - \begin{bmatrix} 0 & 0 & 0 \\ 0 & 0 & 0 \\ 0 & 0 & 0 \end{bmatrix}$

36. $\begin{bmatrix} 7 & & 7 \\ 8 & - & 8 \\ -9 & & -9 \end{bmatrix}$

37. $\begin{bmatrix} 3 \\ 1 \end{bmatrix} \cdot [-5 \quad 2]$

38. $[-2 \quad 10] \cdot \begin{bmatrix} 1 \\ 3 \end{bmatrix}$

39. $\begin{bmatrix} -2 & 0 & 3 \\ 1 & 0 & 4 \end{bmatrix} \begin{bmatrix} 1 & 0 \\ 3 & -2 \end{bmatrix}$

40. $\begin{bmatrix} 2 & -5 \\ 0 & 1 \end{bmatrix} \cdot \begin{bmatrix} 2 & 1 & -7 \\ 3 & 0 & -2 \end{bmatrix}$

41. $\begin{bmatrix} 2 & -5 \\ -6 & 7 \end{bmatrix} \cdot \begin{bmatrix} 1 & 0 \\ 0 & 1 \end{bmatrix}$

42. $\begin{bmatrix} 1 & -2 & 3 \\ -4 & 5 & -6 \\ 7 & -8 & 9 \end{bmatrix} \cdot \begin{bmatrix} 1 & 0 & 0 \\ 0 & 1 & 0 \\ 0 & 0 & 1 \end{bmatrix}$

43. $\begin{bmatrix} 2 & -3 & 4 \\ 8 & 1 & -2 \end{bmatrix} + \begin{bmatrix} 4 & 0 & 3 \\ -5 & 1 & 2 \end{bmatrix} - \begin{bmatrix} 1 & 2 & -3 \\ 4 & -5 & 6 \end{bmatrix}$

44. $\begin{bmatrix} 1 & 0 & 2 \\ 0 & 1 & 3 \\ 0 & -1 & 4 \end{bmatrix} - \begin{bmatrix} 0 & 0 & 1 \\ 0 & 5 & 1 \\ 0 & 6 & 0 \end{bmatrix} - \begin{bmatrix} 0 & -7 & 0 \\ -1 & 0 & 0 \\ 0 & 0 & 5 \end{bmatrix}$

45. $\begin{bmatrix} 2 & 1 \\ 0 & 3 \end{bmatrix} \cdot \begin{bmatrix} -3 & 0 \\ 1 & 1 \end{bmatrix} \cdot \begin{bmatrix} 1 & 0 \\ 0 & 2 \end{bmatrix}$

46. $[1 \quad -2] \cdot \begin{bmatrix} 2 & 3 & 0 \\ -1 & 0 & 4 \end{bmatrix} \cdot \begin{bmatrix} -1 \\ 0 \\ 2 \end{bmatrix}$

47. $\begin{bmatrix} 1 & -3 \\ 2 & 4 \end{bmatrix} \cdot \begin{bmatrix} 4 & 0 \\ -1 & 3 \end{bmatrix} + \begin{bmatrix} 7 & -3 \\ -5 & 9 \end{bmatrix}$

48. $\begin{bmatrix} 4 & -3 \\ 0 & 10 \end{bmatrix} + \begin{bmatrix} 1 & 0 \\ 3 & -2 \end{bmatrix} \cdot \begin{bmatrix} 0 & -4 \\ 5 & 1 \end{bmatrix}$

49. Add the matrices from Problem 11. (What does this matrix mean?)

50. Subtract the matrices from Problem 11. (What does this matrix mean?)

Solve for x and y in the following matrix equations.

51. $\begin{bmatrix} 2 & x \\ 5 & 7 \end{bmatrix} = \begin{bmatrix} 2 & 8 \\ y & 7 \end{bmatrix}$

52. $[3 \quad x \quad -10] = [y \quad 7 \quad -10]$

53. $\begin{bmatrix} 2x + 5 \\ 4 \end{bmatrix} = \begin{bmatrix} 11 \\ y - 3 \end{bmatrix}$

54. $\begin{bmatrix} 2x - 1 \\ 2y + 1 \end{bmatrix} = \begin{bmatrix} 5x + 8 \\ y - 5 \end{bmatrix}$

55. $\begin{bmatrix} x \\ y \end{bmatrix} = \begin{bmatrix} y + 2 \\ 4 - x \end{bmatrix}$

56. $\begin{bmatrix} y \\ 2x \end{bmatrix} = \begin{bmatrix} 14 - x \\ 16 - y \end{bmatrix}$

57. $\begin{bmatrix} 2 & 1 \\ 3 & 2 \end{bmatrix} \cdot \begin{bmatrix} x \\ y \end{bmatrix} = \begin{bmatrix} 5 \\ 7 \end{bmatrix}$

58. $\begin{bmatrix} 4 & -1 \\ 3 & 1 \end{bmatrix} \cdot \begin{bmatrix} x \\ y \end{bmatrix} = \begin{bmatrix} 7 \\ 7 \end{bmatrix}$

(Hint: In Problems 57 and 58, first multiply the matrices on the left side of each equation.)

59. A dietitian is monitoring the fat, protein, and carbohydrate intakes of three patients. The results are given in the following table. (All units are in grams.)

Patient	Monday			Tuesday		
	Fat	Protein	Carbohydrate	Fat	Protein	Carbohydrate
O'Connor	55	60	110	50	65	100
Manocotti	70	115	225	75	120	195
Chang	45	40	120	45	50	130

(a) Write each table as a matrix.
(b) What is the size of each matrix?
(c) Find the sum of the matrices. (What does this matrix mean?)
(d) Find the difference of the matrices. (What does this matrix mean?)
(e) Multiply each matrix by a calorie matrix

$$C = \begin{bmatrix} 9 \\ 4 \\ 4 \end{bmatrix}$$

(There are 9 calories in a gram of fat, 4 in a gram of protein, and 4 in a gram of carbohydrate.) What do these products mean?

60. In most high-level computer languages, blocklike data can be stored in *arrays*, which are matrices. For example, in **BASIC**

```
10 DIM A(2, 3)
20 A(1, 1) =   5
30 A(1, 2) = −7
40 A(1, 3) = −4
50 A(2, 1) =   0
60 A(2, 2) =   8
70 A(2, 3) =   3
```

declares an array (matrix) A of size 2 by 3. The assignment A(1, 1) = 5 means $a_{1,1} = 5$, and so on. Write the matrix A given by the statements 20 to 70.

14.2
MATRIX SOLUTIONS
TO LINEAR SYSTEMS

In the preceding section we studied the matrix, its algebra, and its use for storing blocklike information. In this section we look at another very important application of matrices: the solution of linear systems of equations.

After solving many systems of equations in Chapter 13, perhaps you noticed that most of the calculations were performed only on the coefficients. The variables (x, y, and z) act almost as "placeholders." We have a special matrix that holds the coefficients and constants of a linear system of equations. We call this the **augmented matrix.**

EXAMPLE 11 The following examples are systems and their augmented matrices. Notice that they are called *augmented* (meaning *increased*) since they contain the coefficients and the right-hand-side terms.

System		Augmented Matrix
(a) $\begin{aligned} 2x + 5y &= 7 \\ 8x - 3y &= -4 \end{aligned}$	$\Leftrightarrow$	$\begin{bmatrix} 2 & 5 & \vdots & 7 \\ 8 & -3 & \vdots & -4 \end{bmatrix}$
(b) $\begin{aligned} 3x - y + 5z &= 8 \\ 8x + 9y - 2z &= -12 \\ -x + 4y - z &= 1 \end{aligned}$	$\Leftrightarrow$	$\begin{bmatrix} 3 & -1 & 5 & \vdots & 8 \\ 8 & 9 & -2 & \vdots & -12 \\ -1 & 4 & -1 & \vdots & 1 \end{bmatrix}$
(c) $\begin{aligned} x + y &= 5 \\ x - z &= 7 \\ y + 5z &= -1 \end{aligned}$	$\Leftrightarrow$	$\begin{bmatrix} 1 & 1 & 0 & \vdots & 5 \\ 1 & 0 & -1 & \vdots & 7 \\ 0 & 1 & 5 & \vdots & -1 \end{bmatrix}$

The last example above, especially, shows how each column represents a variable (x is first, y is second, z is third, and the right-hand side is fourth). If a variable is missing, we must insert a zero to hold its place.

Recall that the solution to a linear system of equations might be given as $x = k$, $y = m$, and $z = n$; that is, (k, m, n) is the solution. Let us convert this to an augmented matrix:

$$\begin{aligned} x &= k \\ y &= m \\ z &= n \end{aligned} \Leftrightarrow \begin{bmatrix} 1 & 0 & 0 & \vdots & k \\ 0 & 1 & 0 & \vdots & m \\ 0 & 0 & 1 & \vdots & n \end{bmatrix}$$

We call the matrix on the right a **final matrix**, since we can easily read the solution from it.

EXAMPLE 12 The following final matrices tell us the solutions immediately.

(a) $\begin{bmatrix} 1 & 0 & \vdots & 7 \\ 0 & 1 & \vdots & -3 \end{bmatrix}$ means $x = 7$ and $y = -3$.

(b) $\begin{bmatrix} 1 & 0 & 0 & \vdots & 8 \\ 0 & 1 & 0 & \vdots & -4 \\ 0 & 0 & 1 & \vdots & 1/2 \end{bmatrix}$ means $x = 8$, $y = -4$, and $z = 1/2$.

Our goal is to transform a starting augmented matrix (as seen in Example 11) into a final matrix (as seen in Example 12). To help us do this, we have the following operation rules that allow us to change an augmented matrix into another augmented matrix and still have the same solutions. These rules mirror the rules for systems of equations (see page 442), since augmented matrices are really just systems of equations in shorthand.

Rule	Example
AM1 We can multiply (or divide) any row by a nonzero constant.	$R_1 \begin{bmatrix} 1 & 2 & \vdots & 3 \\ 4 & 5 & \vdots & 6 \end{bmatrix} \Rightarrow \begin{bmatrix} 10 & 20 & \vdots & 30 \\ 4 & 5 & \vdots & 6 \end{bmatrix}$ (Multiply R_1 by 10.)
AM2 We can add a constant times one row to another row.	$R_1 \begin{bmatrix} 1 & 2 & \vdots & 3 \\ 4 & 5 & \vdots & 6 \end{bmatrix} \Rightarrow \begin{bmatrix} 1 & 2 & \vdots & 3 \\ 0 & -3 & \vdots & -6 \end{bmatrix}$ (Add $-4R_1$ to R_2.)
AM3 We can interchange any two rows.	$R_1 \begin{bmatrix} 1 & 2 & \vdots & 3 \\ 4 & 5 & \vdots & 6 \end{bmatrix} \Rightarrow \begin{bmatrix} 4 & 5 & \vdots & 6 \\ 1 & 2 & \vdots & 3 \end{bmatrix}$ (Switch R_1 and R_2.)

YES	NO
AM2: $\begin{bmatrix} 1 & 2 & \vdots & 3 \\ 4 & 5 & \vdots & 6 \end{bmatrix} \Rightarrow \begin{bmatrix} 1 & 2 & \vdots & 3 \\ 0 & -3 & \vdots & -6 \end{bmatrix}$	AM2: $\begin{bmatrix} 1 & 2 & \vdots & 3 \\ 4 & 5 & \vdots & 6 \end{bmatrix} \Rightarrow \begin{bmatrix} -4 & -8 & \vdots & -12 \\ 0 & -3 & \vdots & -6 \end{bmatrix}$

When using Property AM2, we add a constant times row i to row j ($i \neq j$), but *we leave row i unchanged*.

Of these three rules, AM2 is probably the trickiest to work with; on the other hand, it is probably the most useful. Following is a general procedure or plan for transforming an augmented matrix into a final matrix.

Strategy for transforming an augmented matrix to a final matrix:

1. Try to get a 1 in the first-row, first-column position $a_{1,1}$. We use division (AM1) or interchange rows (AM3).
2. "Zero-out" the rest of the first column, using AM2. (For example, if the first element of R_2 is a 3, then we add $-3R_1$ to R_2 to produce a zero.)
3. Try to get a 1 in the second-row, second-column position $a_{2,2}$ (as in step 1), and zero-out the rest of the column (as in step 2).
4. Try to get a 1 in the third-row, third-column position $a_{3,3}$ (as in step 1), and zero-out the rest of the column (as in step 2).
5. If possible, continue until you have a final matrix.

Recall from the preceding chapter that not all systems have nice, unique solutions (the systems may be dependent or inconsistent). In these cases we cannot produce a final matrix.

EXAMPLE 13 Use augmented matrices to solve the system

$$x + 3y = 1$$
$$2x - 5y = 13$$

Solution We first rewrite this system in augmented-matrix form. We then use the rules for row operations to reduce this to a final matrix. We already have a 1 in the first position.

Convert to matrix form;
$a_{1,1} = 1$

$$x + 3y = 1 \Rightarrow \begin{bmatrix} 1 & 3 & \vdots & 1 \\ 2 & -5 & \vdots & 13 \end{bmatrix} \begin{matrix} R_1 \\ R_2 \end{matrix}$$

"Zero" rest of column 1:
Add $-2R_1$ to R_2 (AM2)

$$\Rightarrow \begin{bmatrix} 1 & 3 & \vdots & 1 \\ 0 & -11 & \vdots & 11 \end{bmatrix} \begin{matrix} R_1 \\ R_2 \end{matrix}$$

Get 1 in $a_{2,2}$ position:
Divide R_2 by -11 (AM1)

$$\Rightarrow \begin{bmatrix} 1 & 3 & \vdots & 1 \\ 0 & 1 & \vdots & -1 \end{bmatrix} \begin{matrix} R_1 \\ R_2 \end{matrix}$$

"Zero" rest of column 2:
Add $-3R_2$ to R_1 (AM2)

$$\Rightarrow \begin{bmatrix} 1 & 0 & \vdots & 4 \\ 0 & 1 & \vdots & -1 \end{bmatrix} \begin{matrix} R_1 \\ R_2 \end{matrix}$$

Read solutions

$$x = 4 \quad \text{and} \quad y = -1$$

EXAMPLE 14 Use augmented matrices to solve the system

$$x - 3y = 16$$
$$2x + z = 4$$
$$y - z = -7$$

Solution We first rewrite this system as an augmented matrix, carefully using zeros for the missing variables. We then work to reduce this to a final matrix.

Convert to matrix form

$$\begin{matrix} x - 3y = 16 \\ 2x + z = 4 \\ y - z = -7 \end{matrix} \Rightarrow \begin{bmatrix} 1 & -3 & 0 & \vdots & 16 \\ 2 & 0 & 1 & \vdots & 4 \\ 0 & 1 & -1 & \vdots & -7 \end{bmatrix} \begin{matrix} R_1 \\ R_2 \\ R_3 \end{matrix}$$

$a_{1,1}$ is already 1;
"zero" rest of column 1:
Add $-2R_1$ to R_2 (AM2)

$$\Rightarrow \begin{bmatrix} 1 & -3 & 0 & \vdots & 16 \\ 0 & 6 & 1 & \vdots & -28 \\ 0 & 1 & -1 & \vdots & -7 \end{bmatrix} \begin{matrix} R_1 \\ R_2 \\ R_3 \end{matrix}$$

<table>
<tr><td>Make $a_{2,2}=1$:
Switch R_2 and R_3 (AM3)</td><td>$\Rightarrow \begin{bmatrix} 1 & -3 & 0 & \vdots & 16 \\ 0 & 1 & -1 & \vdots & -7 \\ 0 & 6 & 1 & \vdots & -28 \end{bmatrix} \begin{matrix} R_1 \\ R_2 \\ R_3 \end{matrix}$</td></tr>
<tr><td>"Zero" rest of column 2:
Add $3R_2$ to R_1 (AM2);
add $-6R_2$ to R_3 (AM2)</td><td>$\Rightarrow \begin{bmatrix} 1 & 0 & -3 & \vdots & -5 \\ 0 & 1 & -1 & \vdots & -7 \\ 0 & 0 & 7 & \vdots & 14 \end{bmatrix} \begin{matrix} R_1 \\ R_2 \\ R_3 \end{matrix}$</td></tr>
<tr><td>Make $a_{3,3}=1$:
Divide R_3 by 7 (AM1)</td><td>$\Rightarrow \begin{bmatrix} 1 & 0 & -3 & \vdots & -5 \\ 0 & 1 & -1 & \vdots & -7 \\ 0 & 0 & 1 & \vdots & 2 \end{bmatrix} \begin{matrix} R_1 \\ R_2 \\ R_3 \end{matrix}$</td></tr>
<tr><td>"Zero" the rest of column 3:
Add R_3 to R_2 (AM2);
add $3R_3$ to R_1 (AM2)</td><td>$\Rightarrow \begin{bmatrix} 1 & 0 & 0 & \vdots & 1 \\ 0 & 1 & 0 & \vdots & -5 \\ 0 & 0 & 1 & \vdots & 2 \end{bmatrix}$</td></tr>
<tr><td>Read solutions</td><td>$x = 1, \quad y = -5, \quad \text{and} \quad z = 2$</td></tr>
</table>

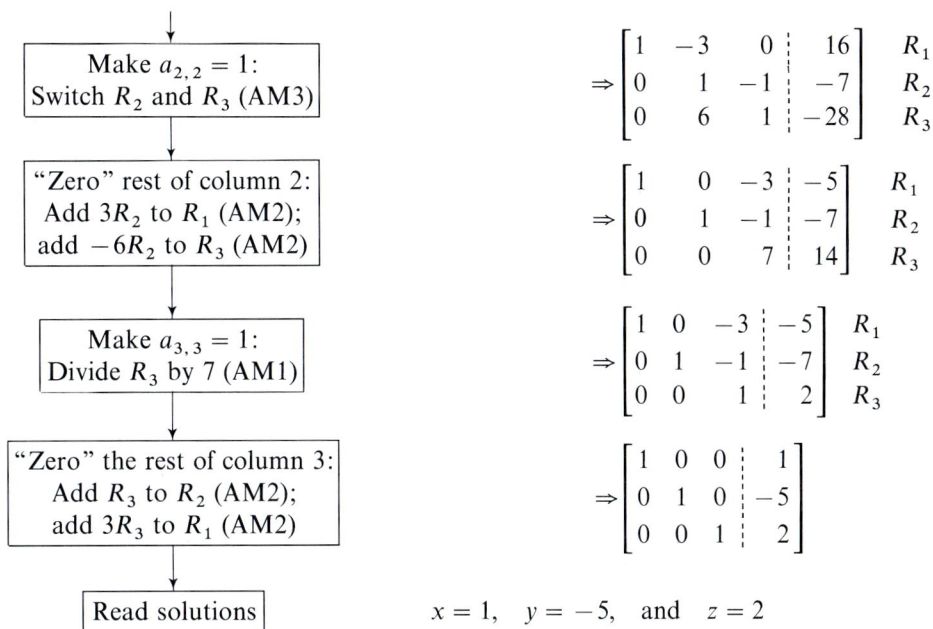

EXAMPLE 15 Use augmented matrices to solve the system

$$2x - 5y = 8$$
$$-4x + 10y = 4$$

Solution We first convert to matrix form and then put a 1 in the first position.

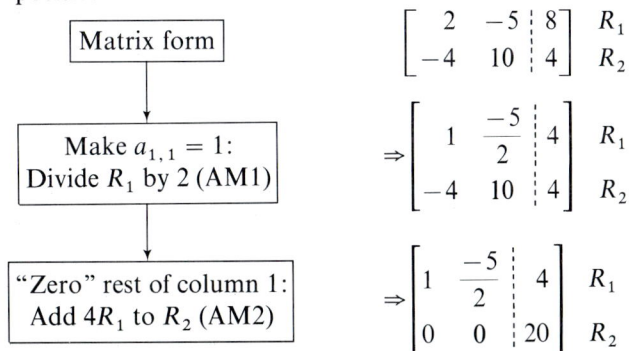

Matrix form
$$\begin{bmatrix} 2 & -5 & \vdots & 8 \\ -4 & 10 & \vdots & 4 \end{bmatrix} \begin{matrix} R_1 \\ R_2 \end{matrix}$$

Make $a_{1,1}=1$: Divide R_1 by 2 (AM1)
$$\Rightarrow \begin{bmatrix} 1 & \dfrac{-5}{2} & \vdots & 4 \\ -4 & 10 & \vdots & 4 \end{bmatrix} \begin{matrix} R_1 \\ R_2 \end{matrix}$$

"Zero" rest of column 1: Add $4R_1$ to R_2 (AM2)
$$\Rightarrow \begin{bmatrix} 1 & \dfrac{-5}{2} & \vdots & 4 \\ 0 & 0 & \vdots & 20 \end{bmatrix} \begin{matrix} R_1 \\ R_2 \end{matrix}$$

The last row $[0 \quad 0 \mid 20]$ translates as $0 = 20$. This false equation means that the system is inconsistent, or that there is *no* solution.

Let us summarize the possible final states of a matrix row.

Case	Meaning
$[0 \quad 1 \quad 0 \mid 3]$	Consistent: $y = 3$
$[0 \quad 0 \quad 0 \mid 0]$	Dependent system
$[0 \quad 0 \quad 0 \mid 5]$	Inconsistent: no solution

Let $A = \begin{bmatrix} 1 & -3 \\ 5 & 7 \end{bmatrix}$, $B = \begin{bmatrix} 0 & 4 \\ -2 & 9 \end{bmatrix}$, and $C = \begin{bmatrix} -1 & 0 & 5 \\ 2 & -4 & 0 \end{bmatrix}$.

Warm-up
Exercises

1. What is the size of A?

2. What is the size of C?

3. Find $a_{1,2}$.

4. Find $b_{2,1}$.

5. Find $A + B$.

6. Find $A - B$.

7. Find $A \cdot B$.

8. Find $B \cdot A$.

9. Find $A \cdot C$.

10. Find $C - A \cdot C$.

Solve the following linear systems.

Warm-up
Exercises

11. $x + y = 10$
$x - y = \ 2$

12. $2x + 3y = \ 5$
$3x - 4y = 16$

13. $x + y + z = 2$
$x + y - z = 6$
$x - y - z = 6$

14. $x + y = 6$
$x + z = 7$
$y + z = 3$

Read the solutions from the following final matrices.

15. $\begin{bmatrix} 1 & 0 & | & -2 \\ 0 & 1 & | & 5 \end{bmatrix}$

16. $\begin{bmatrix} 1 & 0 & | & 0 \\ 0 & 1 & | & -9 \end{bmatrix}$

17. $\begin{bmatrix} 1 & 0 & 0 & | & 1 \\ 0 & 1 & 0 & | & 2 \\ 0 & 0 & 1 & | & -3 \end{bmatrix}$

18. $\begin{bmatrix} 1 & 0 & 0 & | & 1 \\ 0 & 1 & 0 & | & 0 \\ 0 & 0 & 1 & | & -5 \end{bmatrix}$

Write the following systems as augmented matrices (but do not solve).

19. $2x - 3y + 4z = \ \ 8$
$x + \ y + 3z = \ \ 1$
$5x - \ y + 2z = -2$

20. $x - \ y = \ \ 7$
$-x + 2z = -3$
$-y - 4z = -1$

21. $x + \ y + 2z + 4t = \ \ 3$
$x - \ y + 7z - \ t = -2$
$2x + \ y - \ z + 2t = \ \ 0$
$5x + 4y + \ z - \ t = -8$

22. $x - \ y = \ \ 7$
$x + \ z = \ \ 3$
$y + 2t = -2$
$z - 8t = -1$

Write the following augmented matrices as linear systems (but do not solve).

23. $\begin{bmatrix} 2 & -1 & | & 3 \\ 8 & 3 & | & -2 \end{bmatrix}$

24. $\begin{bmatrix} 7 & 0 & | & -8 \\ 1 & 2 & | & 4 \end{bmatrix}$

25. $\begin{bmatrix} 1 & 0 & 1 & | & 4 \\ 0 & -1 & 1 & | & -3 \\ 1 & -2 & 0 & | & 2 \end{bmatrix}$

26. $\begin{bmatrix} 0 & 4 & 1 & | & -8 \\ 1 & -2 & 7 & | & 0 \\ 1 & 0 & 0 & | & 3 \end{bmatrix}$

Solve the following systems using augmented matrices.

27. $2x + 3y = -13$
$x - 2y = \ \ 4$

28. $3x - 4y = \ \ 8$
$x + 7y = -14$

29. $x - y = -6$
$x + z = -4$
$y - z = \ 12$

30. $x - 2y = 8$
$x + 3z = 2$
$y - \ z = 2$

31. $2x - 3y = \ \ 14$
$3x + 5y = -17$

32. $4x - 3y = -27$
$5x + 7y = \ \ 20$

33. $\quad x + 2y + 3z = -7$
$\quad\quad 2x + 3y + 4z = -8$
$\quad\quad 3x - 4y + 5z = -9$

34. $\quad 2x - 5y + 7z = 34$
$\quad\quad\quad x - 4y - 5z = -7$
$\quad\quad 3x - 4y + 8z = 40$

35. $\quad x - y = 3$
$\quad\quad y + z = 1$
$\quad\quad x + t = -3$
$\quad\quad z + t = 7$

36. $\quad x - z = 8$
$\quad\quad y + z = 1$
$\quad\quad x - t = 3$
$\quad\quad y - t = 6$

37. $\quad x^2 - y^2 = 7$
$\quad\quad x^2 + y^2 = 25$

38. $\quad x^2 + 2y^2 = 27$
$\quad\quad 3x^2 - y^2 = 74$

(For Problems 37 and 38, use matrices to solve for x^2 and y^2; then find x and y, taking all positive and negative combinations.)

Electrical Application

39. A circuit has currents x, y, and z through its three resistors. Kirchhoff's and Ohm's laws lead to the following system:

$$\begin{array}{rcl} x + y + z &=& 0 \\ 10x - 5y &=& 85 \\ -5y + 20z &=& 75 \end{array}$$

Solve this system using an augmented matrix.

Business Application

40. A store is mixing together two varieties of coffee: $2 per pound and $3 per pound. How much of each should be mixed to produce 100 pounds worth $2.70 per pound?

14.3
SQUARE MATRICES AND MULTIPLICATIVE INVERSES

On page 470 we saw that we can multiply matrices A and B if the number of columns of A equals the number of rows of B. In this section we look at a special type of matrix: the **square matrix**, which has the same number of rows as columns: for example,

$$\begin{bmatrix} 1 & 2 \\ -3 & 3 \end{bmatrix} \quad \begin{bmatrix} 1 & 0 & 7 \\ -6 & 2 & 4 \\ 8 & -\dfrac{1}{2} & 1 \end{bmatrix} \quad \begin{bmatrix} 1 & 0 & 0 & 1 \\ 0 & 2 & 3 & 0 \\ 0 & 0 & 1 & 0 \\ 0 & 1 & 5 & 0 \end{bmatrix}$$

are square matrices. Multiplication of square matrices has a nice property.

> The product of two n by n matrices is an n by n matrix.

EXAMPLE 16 The product of two 2 by 2 matrices is another 2 by 2 matrix, and so on.

(a) $\begin{bmatrix} 1 & 2 \\ -3 & 0 \end{bmatrix} \cdot \begin{bmatrix} -2 & 0 \\ 1 & 5 \end{bmatrix} = \begin{bmatrix} 0 & 10 \\ 6 & 0 \end{bmatrix}$

(b) $\begin{bmatrix} 1 & 1 & 0 \\ 0 & 2 & 0 \\ 0 & 0 & -1 \end{bmatrix} \cdot \begin{bmatrix} 1 & 2 & 0 \\ 5 & 0 & -1 \\ 0 & -3 & 0 \end{bmatrix} = \begin{bmatrix} 6 & 2 & -1 \\ 10 & 0 & -2 \\ 0 & 3 & 0 \end{bmatrix}$

For every size, n by n, there is a special square matrix called the **identity matrix** (written I_n), which has 1's down the main diagonal (that is, $a_{ii} = 1$, for all i) and 0's everywhere else: for example,

$$I_2 = \begin{bmatrix} 1 & 0 \\ 0 & 1 \end{bmatrix} \qquad I_3 = \begin{bmatrix} 1 & 0 & 0 \\ 0 & 1 & 0 \\ 0 & 0 & 1 \end{bmatrix} \qquad I_4 = \begin{bmatrix} 1 & 0 & 0 & 0 \\ 0 & 1 & 0 & 0 \\ 0 & 0 & 1 & 0 \\ 0 & 0 & 0 & 1 \end{bmatrix}$$

Recall that the final matrix in the solution of linear system of equations has the form

$$\begin{bmatrix} 1 & 0 & \vdots & m \\ 0 & 1 & \vdots & n \end{bmatrix}$$

where (m, n) is the solution of the system. Notice that the left side of the augmented matrix is simply an identity matrix.

EXAMPLE 17   Note what happens when we multiply a square matrix by the identity matrix: The other matrix is unchanged.

(a) $\begin{bmatrix} -2 & 1 \\ 7 & -9 \end{bmatrix} \cdot \begin{bmatrix} 1 & 0 \\ 0 & 1 \end{bmatrix} = \begin{bmatrix} -2 & 1 \\ 7 & -9 \end{bmatrix}$

(b) $\begin{bmatrix} 1 & 0 & 0 \\ 0 & 1 & 0 \\ 0 & 0 & 1 \end{bmatrix} \cdot \begin{bmatrix} 1 & -2 & 3 \\ 4 & 5 & -6 \\ -7 & 8 & 9 \end{bmatrix} = \begin{bmatrix} 1 & -2 & 3 \\ 4 & 5 & -6 \\ -7 & 8 & 9 \end{bmatrix}$

In general, if A is an n by n matrix,

$$\boxed{A \cdot I_n = A \qquad \text{and} \qquad I_n \cdot A = A}$$

That is, the matrix I_n acts as the number 1 does for real numbers. This is why it is called an identity.

Recall that every real number a (except 0) has a multiplicative inverse (or reciprocal) $1/a$ such that $a(1/a) = 1$; for example,

$$2\left(\frac{1}{2}\right) = 1 \qquad \frac{2}{5}\left(\frac{5}{2}\right) = 1 \qquad \frac{-7}{9}\left(\frac{-9}{7}\right) = 1$$

Similarly, most square matrices A have a **multiplicative inverse**, written A^{-1}, which satisfies

$$\boxed{A \cdot A^{-1} = I_n \qquad \text{and} \qquad A^{-1} \cdot A = I_n}$$

(Some square matrices do *not* have such an inverse.)

EXAMPLE 18   Let us look at the matrices

$$A = \begin{bmatrix} 3 & 7 \\ 2 & 5 \end{bmatrix} \qquad \text{and} \qquad A^{-1} = \begin{bmatrix} 5 & -7 \\ -2 & 3 \end{bmatrix}$$

We claim that A^{-1} is the inverse of A:

$$A \cdot A^{-1} = \begin{bmatrix} 3 & 7 \\ 2 & 5 \end{bmatrix} \cdot \begin{bmatrix} 5 & -7 \\ -2 & 3 \end{bmatrix} = \begin{bmatrix} 1 & 0 \\ 0 & 1 \end{bmatrix} = I_n$$

$$A^{-1} \cdot A = \begin{bmatrix} 5 & -7 \\ -2 & 3 \end{bmatrix} \cdot \begin{bmatrix} 3 & 7 \\ 2 & 5 \end{bmatrix} = \begin{bmatrix} 1 & 0 \\ 0 & 1 \end{bmatrix} = I_n$$

We can use augmented matrices and the row operations to help find the inverse of a matrix A.

$$\underset{\substack{\uparrow \quad\; \uparrow \\ A \qquad I_2}}{\begin{bmatrix} 3 & 7 & \vdots & 1 & 0 \\ 2 & 5 & \vdots & 0 & 1 \end{bmatrix}} \xrightarrow{\text{row operations}} \underset{\substack{\uparrow \qquad \uparrow \\ I_2 \qquad A^{-1}}}{\begin{bmatrix} 1 & 0 & \vdots & 5 & -7 \\ 0 & 1 & \vdots & -2 & 3 \end{bmatrix}}$$

To find the inverse of a square matrix A:

1. Augment A with a copy of I_n to get $[A \vdots I_n]$.
2. Use row operations to reduce the left-side matrix to I_n.
3. The right-side matrix is then A^{-1}. (The final matrix is $[I_n \vdots A^{-1}]$.)

Note: If a row of all zeros appears on the left, then there is no inverse.

EXAMPLE 19   Find the inverse of

$$A = \begin{bmatrix} 1 & 5 \\ 2 & 11 \end{bmatrix}$$

Solution   We augment A with a copy of I_2 to get $[A \vdots I_2]$. We then try to reduce this augmented matrix to the final form $[I_2 \vdots A^{-1}]$.

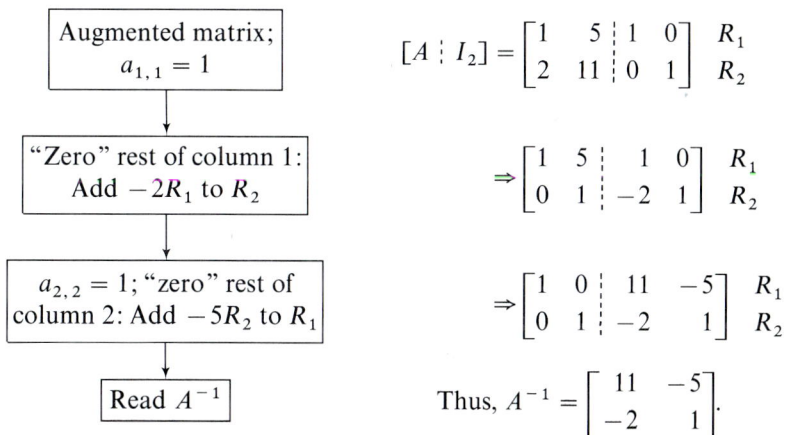

| Augmented matrix; $a_{1,1} = 1$ | $[A \vdots I_2] = \begin{bmatrix} 1 & 5 & \vdots & 1 & 0 \\ 2 & 11 & \vdots & 0 & 1 \end{bmatrix} \begin{matrix} R_1 \\ R_2 \end{matrix}$ |

"Zero" rest of column 1: Add $-2R_1$ to R_2

$$\Rightarrow \begin{bmatrix} 1 & 5 & \vdots & 1 & 0 \\ 0 & 1 & \vdots & -2 & 1 \end{bmatrix} \begin{matrix} R_1 \\ R_2 \end{matrix}$$

$a_{2,2} = 1$; "zero" rest of column 2: Add $-5R_2$ to R_1

$$\Rightarrow \begin{bmatrix} 1 & 0 & \vdots & 11 & -5 \\ 0 & 1 & \vdots & -2 & 1 \end{bmatrix} \begin{matrix} R_1 \\ R_2 \end{matrix}$$

Read A^{-1}

Thus, $A^{-1} = \begin{bmatrix} 11 & -5 \\ -2 & 1 \end{bmatrix}$.

The reader should check that A and A^{-1} are, in fact, inverses.

EXAMPLE 20 Find the inverse of

$$A = \begin{bmatrix} 4 & 5 \\ 6 & 7 \end{bmatrix}$$

Solution Again, we begin with the augmented matrix $[A \mid I_2]$ and try to reduce it to $[I_2 \mid A^{-1}]$.

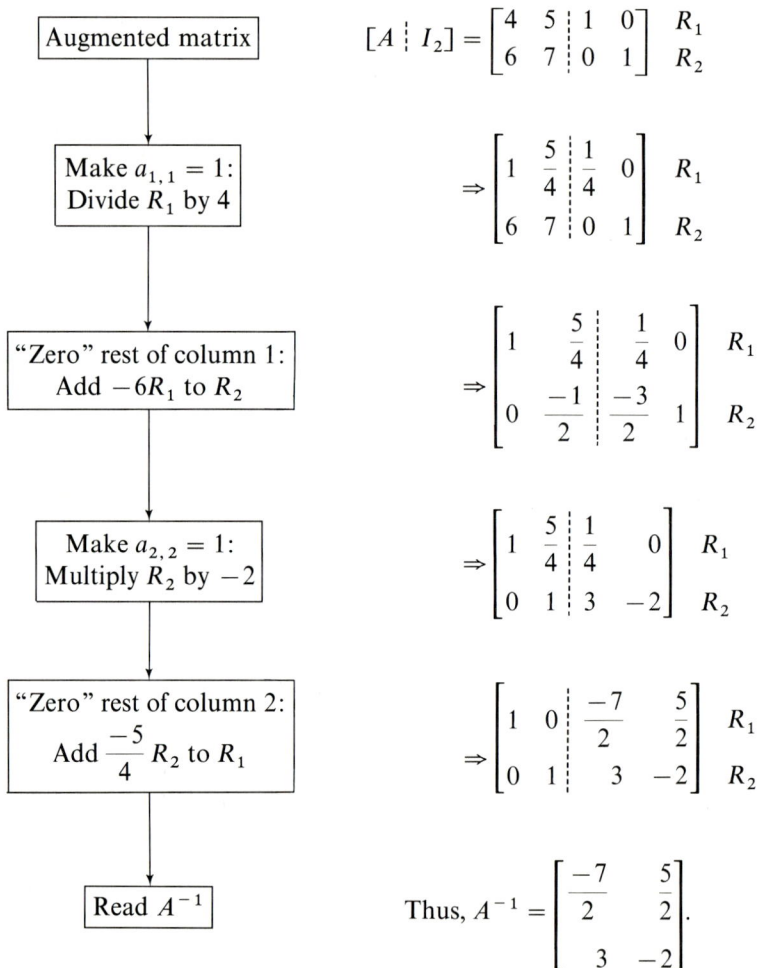

| Augmented matrix | $[A \mid I_2] = \begin{bmatrix} 4 & 5 & 1 & 0 \\ 6 & 7 & 0 & 1 \end{bmatrix} \begin{matrix} R_1 \\ R_2 \end{matrix}$ |

| Make $a_{1,1} = 1$: Divide R_1 by 4 | $\Rightarrow \begin{bmatrix} 1 & \dfrac{5}{4} & \dfrac{1}{4} & 0 \\ 6 & 7 & 0 & 1 \end{bmatrix} \begin{matrix} R_1 \\ R_2 \end{matrix}$ |

| "Zero" rest of column 1: Add $-6R_1$ to R_2 | $\Rightarrow \begin{bmatrix} 1 & \dfrac{5}{4} & \dfrac{1}{4} & 0 \\ 0 & \dfrac{-1}{2} & \dfrac{-3}{2} & 1 \end{bmatrix} \begin{matrix} R_1 \\ R_2 \end{matrix}$ |

| Make $a_{2,2} = 1$: Multiply R_2 by -2 | $\Rightarrow \begin{bmatrix} 1 & \dfrac{5}{4} & \dfrac{1}{4} & 0 \\ 0 & 1 & 3 & -2 \end{bmatrix} \begin{matrix} R_1 \\ R_2 \end{matrix}$ |

| "Zero" rest of column 2: Add $\dfrac{-5}{4} R_2$ to R_1 | $\Rightarrow \begin{bmatrix} 1 & 0 & \dfrac{-7}{2} & \dfrac{5}{2} \\ 0 & 1 & 3 & -2 \end{bmatrix} \begin{matrix} R_1 \\ R_2 \end{matrix}$ |

| Read A^{-1} | Thus, $A^{-1} = \begin{bmatrix} \dfrac{-7}{2} & \dfrac{5}{2} \\ 3 & -2 \end{bmatrix}$. |

The reader should check that $A \cdot A^{-1} = I_2$.

EXAMPLE 21 Find the inverse of the matrix

$$A = \begin{bmatrix} 1 & -3 & 6 \\ 0 & -1 & 4 \\ -4 & 5 & 8 \end{bmatrix}$$

Solution   We start with the augmented matrix.

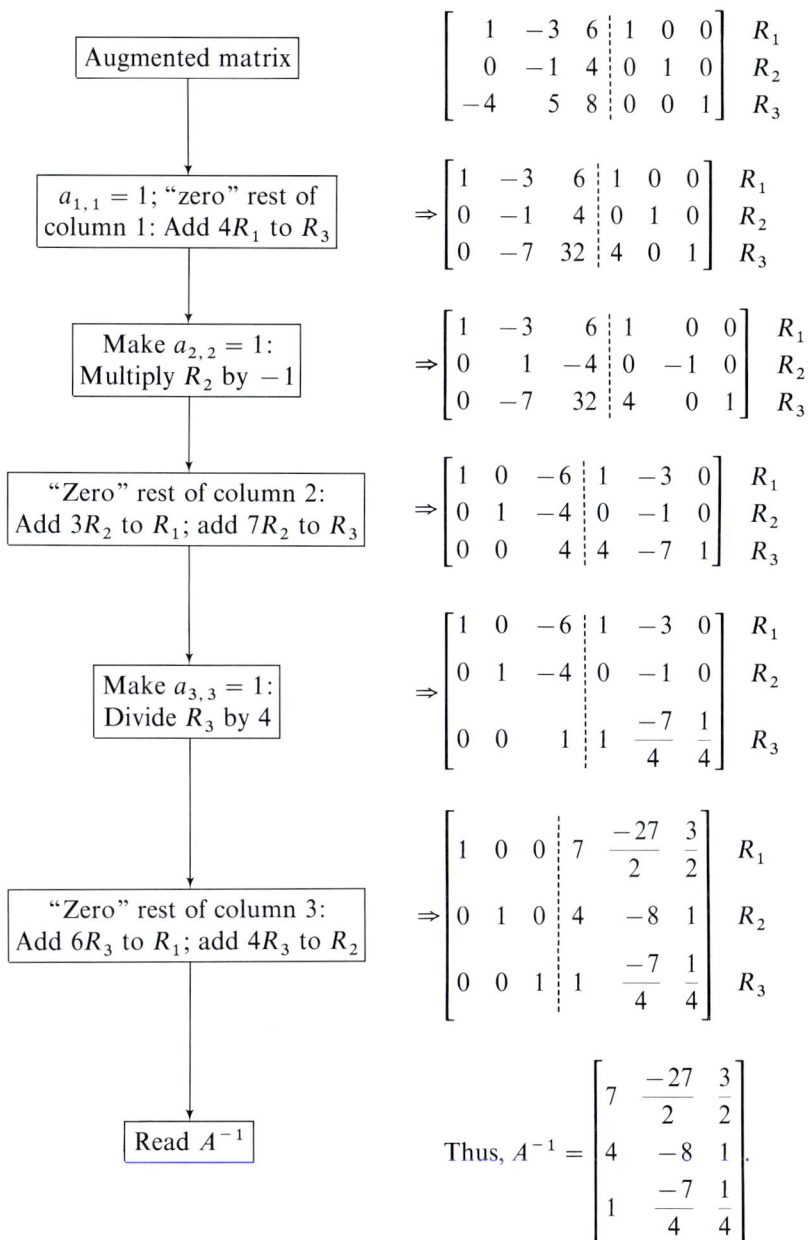

Augmented matrix

$$\begin{bmatrix} 1 & -3 & 6 & \vdots & 1 & 0 & 0 \\ 0 & -1 & 4 & \vdots & 0 & 1 & 0 \\ -4 & 5 & 8 & \vdots & 0 & 0 & 1 \end{bmatrix} \begin{matrix} R_1 \\ R_2 \\ R_3 \end{matrix}$$

$a_{1,1} = 1$; "zero" rest of column 1: Add $4R_1$ to R_3

$$\Rightarrow \begin{bmatrix} 1 & -3 & 6 & \vdots & 1 & 0 & 0 \\ 0 & -1 & 4 & \vdots & 0 & 1 & 0 \\ 0 & -7 & 32 & \vdots & 4 & 0 & 1 \end{bmatrix} \begin{matrix} R_1 \\ R_2 \\ R_3 \end{matrix}$$

Make $a_{2,2} = 1$: Multiply R_2 by -1

$$\Rightarrow \begin{bmatrix} 1 & -3 & 6 & \vdots & 1 & 0 & 0 \\ 0 & 1 & -4 & \vdots & 0 & -1 & 0 \\ 0 & -7 & 32 & \vdots & 4 & 0 & 1 \end{bmatrix} \begin{matrix} R_1 \\ R_2 \\ R_3 \end{matrix}$$

"Zero" rest of column 2: Add $3R_2$ to R_1; add $7R_2$ to R_3

$$\Rightarrow \begin{bmatrix} 1 & 0 & -6 & \vdots & 1 & -3 & 0 \\ 0 & 1 & -4 & \vdots & 0 & -1 & 0 \\ 0 & 0 & 4 & \vdots & 4 & -7 & 1 \end{bmatrix} \begin{matrix} R_1 \\ R_2 \\ R_3 \end{matrix}$$

Make $a_{3,3} = 1$: Divide R_3 by 4

$$\Rightarrow \begin{bmatrix} 1 & 0 & -6 & \vdots & 1 & -3 & 0 \\ 0 & 1 & -4 & \vdots & 0 & -1 & 0 \\ 0 & 0 & 1 & \vdots & 1 & \dfrac{-7}{4} & \dfrac{1}{4} \end{bmatrix} \begin{matrix} R_1 \\ R_2 \\ R_3 \end{matrix}$$

"Zero" rest of column 3: Add $6R_3$ to R_1; add $4R_3$ to R_2

$$\Rightarrow \begin{bmatrix} 1 & 0 & 0 & \vdots & 7 & \dfrac{-27}{2} & \dfrac{3}{2} \\ 0 & 1 & 0 & \vdots & 4 & -8 & 1 \\ 0 & 0 & 1 & \vdots & 1 & \dfrac{-7}{4} & \dfrac{1}{4} \end{bmatrix} \begin{matrix} R_1 \\ R_2 \\ R_3 \end{matrix}$$

Read A^{-1}

Thus, $A^{-1} = \begin{bmatrix} 7 & \dfrac{-27}{2} & \dfrac{3}{2} \\ 4 & -8 & 1 \\ 1 & \dfrac{-7}{4} & \dfrac{1}{4} \end{bmatrix}$.

The reader should check that $A \cdot A^{-1} = I_3$.

EXAMPLE 22   Let us find the inverse of

$$A = \begin{bmatrix} 1 & -3 \\ -2 & 6 \end{bmatrix}$$

with an augmented matrix.

Augmented matrix

$$\begin{bmatrix} 1 & -3 & \vdots & 1 & 0 \\ -2 & 6 & \vdots & 0 & 1 \end{bmatrix} \begin{matrix} R_1 \\ R_2 \end{matrix}$$

"Zero" first column:
Add $2R_1$ to R_2

$$\Rightarrow \begin{bmatrix} 1 & -3 & \vdots & 1 & 0 \\ 0 & 0 & \vdots & 2 & 1 \end{bmatrix} \begin{matrix} R_1 \\ R_2 \end{matrix}$$

We cannot continue: The last row $[0 \quad 0 \ \vdots \ 2 \quad 1]$ indicates that there is *no inverse* for the matrix A. This example points out the fact that not all square matrices have a multiplicative inverse.

PROBLEM SET
14.3

Perform the following matrix operations.

Warm-up Exercises

1. $\begin{bmatrix} 1 & 2 & -3 \\ 4 & -5 & 6 \end{bmatrix} + \begin{bmatrix} 7 & -8 & 9 \\ 3 & 4 & -5 \end{bmatrix}$

2. $\begin{bmatrix} 1 \\ 6 \\ -9 \end{bmatrix} - \begin{bmatrix} 2 \\ -4 \\ -6 \end{bmatrix}$

3. $\begin{bmatrix} 1 \\ -2 \end{bmatrix} \cdot [3 \quad -4 \quad 5]$

4. $[6 \quad 0 \quad -1] \cdot \begin{bmatrix} 1 & 2 \\ -3 & 4 \\ 5 & -6 \end{bmatrix}$

5. $\begin{bmatrix} 1 & 0 \\ 0 & 1 \end{bmatrix} \cdot \begin{bmatrix} 2 \\ -5 \end{bmatrix}$

6. $[1 \quad 4 \quad -9] \cdot \begin{bmatrix} 1 & 0 & 0 \\ 0 & 1 & 0 \\ 0 & 0 & 1 \end{bmatrix}$

Write the following systems as augmented matrices and solve by matrix row operations.

Warm-up Exercises

7. $x + 2y = 4$
$3x - 5y = 23$

8. $2x + 3y = -1$
$5x - 2y = 7$

9. $x + y + z = 6$
$x + y - z = 2$
$x - y - z = 0$

10. $x + y = 5$
$y + z = 2$
$x + z = -1$

Determine whether each of the following pairs are multiplicative inverses of each other.

11. $A = \begin{bmatrix} 1 & 2 \\ 3 & 4 \end{bmatrix}$; $B = \begin{bmatrix} -4 & 2 \\ 3 & -1 \end{bmatrix}$

12. $A = \begin{bmatrix} 1 & 4 & 3 \\ 0 & 5 & 2 \\ 0 & 7 & 3 \end{bmatrix}$; $B = \begin{bmatrix} 1 & 9 & -7 \\ 0 & 3 & -2 \\ 0 & -7 & 5 \end{bmatrix}$

13. $A = \begin{bmatrix} 2 & 0 & 3 \\ 1 & 0 & 0 \\ -5 & 3 & 0 \end{bmatrix}$; $B = \begin{bmatrix} 2 & 0 & 3 \\ 0 & 3 & 0 \\ -1 & 0 & -2 \end{bmatrix}$

14. $A = \begin{bmatrix} 3 & 0 & 1 \\ 0 & \frac{1}{2} & 0 \\ -2 & 0 & -1 \end{bmatrix}$; $B = \begin{bmatrix} 1 & 0 & 1 \\ 0 & 2 & 0 \\ -2 & 0 & -3 \end{bmatrix}$

Find the multiplicative inverse of each of the following matrices.

15. $\begin{bmatrix} 4 & 3 \\ 7 & 5 \end{bmatrix}$

16. $\begin{bmatrix} 1 & 4 \\ 3 & 11 \end{bmatrix}$

17. $\begin{bmatrix} 2 & 4 \\ 4 & 9 \end{bmatrix}$

18. $\begin{bmatrix} 6 & 7 \\ 4 & 5 \end{bmatrix}$

19. $\begin{bmatrix} 3 & 5 \\ -1 & 1 \end{bmatrix}$

20. $\begin{bmatrix} 4 & -2 \\ -9 & 5 \end{bmatrix}$

21. $\begin{bmatrix} 1 & 0 & -2 \\ 0 & 1 & 5 \\ -1 & 0 & 3 \end{bmatrix}$

22. $\begin{bmatrix} 1 & 1 & 4 \\ 0 & 2 & 1 \\ 1 & -2 & 3 \end{bmatrix}$

23. $\begin{bmatrix} 1 & 0 & 1 \\ 0 & 4 & 3 \\ -2 & 2 & -1 \end{bmatrix}$

24. $\begin{bmatrix} 1 & 0 & 1 \\ -1 & 1 & 1 \\ 0 & -1 & 1 \end{bmatrix}$

25. $\begin{bmatrix} 2 & 0 \\ 0 & 3 \end{bmatrix}$

26. $\begin{bmatrix} 0 & 4 \\ 5 & 0 \end{bmatrix}$

27. $\begin{bmatrix} a & 0 \\ 0 & b \end{bmatrix}$

28. $\begin{bmatrix} a & b \\ c & d \end{bmatrix}$

29. $\begin{bmatrix} 2 & 0 & 0 \\ 0 & 3 & 0 \\ 0 & 0 & 4 \end{bmatrix}$

30. $\begin{bmatrix} a & 0 & 0 \\ 0 & b & 0 \\ 0 & 0 & c \end{bmatrix}$

31. $\begin{bmatrix} 1 & 0 \\ 0 & 1 \end{bmatrix}$

32. $\begin{bmatrix} -1 & 0 \\ 0 & -1 \end{bmatrix}$

33. $\begin{bmatrix} 2 & 4 \\ 3 & 6 \end{bmatrix}$

34. $\begin{bmatrix} 1 & -5 \\ -2 & 10 \end{bmatrix}$

35. $\begin{bmatrix} 0 & 1 & 0 \\ 2 & 0 & 0 \\ 0 & 0 & 3 \end{bmatrix}$

36. $\begin{bmatrix} 0 & 0 & 4 \\ 3 & 0 & 0 \\ 0 & -2 & 0 \end{bmatrix}$

37. $\begin{bmatrix} 1 & 1 & 0 \\ 1 & 0 & 1 \\ 0 & 1 & 1 \end{bmatrix}$

38. $\begin{bmatrix} 0 & 1 & 1 \\ 1 & 0 & 1 \\ 1 & 1 & 0 \end{bmatrix}$

14.4
SOLVING LINEAR SYSTEMS OF EQUATIONS WITH MATRIX INVERSES

In this section we use matrix inverses to solve linear systems that are written as matrix equations. Recall that two matrices are **equal** if they are the same size and *all* corresponding entries are equal. For example,

$$\begin{bmatrix} x \\ y \end{bmatrix} = \begin{bmatrix} 2 \\ -3 \end{bmatrix}$$

means $x = 2$ and $y = -3$.

EXAMPLE 23   In the following examples, we rewrite systems of linear equations as matrix equations, using a matrix of unknowns.

(a) $\begin{aligned} 2x - 5y &= 7 \\ 4x + 3y &= -8 \end{aligned}$ $\Leftrightarrow$ $\begin{bmatrix} 2 & -5 \\ 4 & 3 \end{bmatrix} \cdot \begin{bmatrix} x \\ y \end{bmatrix} = \begin{bmatrix} 7 \\ -8 \end{bmatrix}$

(b)
$$\begin{aligned} x + y - z &= 12 \\ 2x - y + z &= 11 \\ 3x + y + 4z &= 10 \end{aligned} \Leftrightarrow \begin{bmatrix} 1 & 1 & -1 \\ 2 & -1 & 1 \\ 3 & 1 & 4 \end{bmatrix} \cdot \begin{bmatrix} x \\ y \\ z \end{bmatrix} = \begin{bmatrix} 12 \\ 11 \\ 10 \end{bmatrix}$$

The student should multiply the matrices on the right and see that these products yield the equations on the left.

EXAMPLE 24   Consider the system

$$\begin{aligned} 4x + 5y &= -2 \\ 6x + 7y &= 8 \end{aligned}$$

We first rewrite this as a matrix equation and solve it parallel to the simple linear equation $\frac{2}{5}x = 16$. We do this to show the obvious similarity: We multiply both sides by the matrix inverse (or reciprocal). We use the matrix inverse found in Example 20.

	Linear equation	*Matrix equation*
Given	$\dfrac{2}{5}x = 16$	$\begin{bmatrix} 4 & 5 \\ 6 & 7 \end{bmatrix} \cdot \begin{bmatrix} x \\ y \end{bmatrix} = \begin{bmatrix} -2 \\ 8 \end{bmatrix}$
Multiply by inverse	$\dfrac{5}{2} \cdot \dfrac{2}{5}x = \dfrac{5}{2} \cdot 16$	$\begin{bmatrix} -\frac{7}{2} & \frac{5}{2} \\ 3 & -2 \end{bmatrix} \cdot \begin{bmatrix} 4 & 5 \\ 6 & 7 \end{bmatrix} \cdot \begin{bmatrix} x \\ y \end{bmatrix} = \begin{bmatrix} -\frac{7}{2} & \frac{5}{2} \\ 3 & -2 \end{bmatrix} \cdot \begin{bmatrix} -2 \\ 8 \end{bmatrix}$
Multiply	$1 \cdot x = 40$	$\begin{bmatrix} 1 & 0 \\ 0 & 1 \end{bmatrix} \cdot \begin{bmatrix} x \\ y \end{bmatrix} = \begin{bmatrix} 27 \\ -22 \end{bmatrix}$
Identity property	$x = 40$	$\begin{bmatrix} x \\ y \end{bmatrix} = \begin{bmatrix} 27 \\ -22 \end{bmatrix}$

Thus, $x = 27$ and $y = -22$.

Notice how the inverse acts exactly as the reciprocal does. (We multiply on the *left* by the matrix inverse.) Generally, the elimination and augmented-matrix methods are the most convenient for solving linear systems; however, using the matrix inverse is most useful when solving many systems, all with the same coefficients (but different right-hand-side constants). In that case we find the inverse of the coefficient matrix once and multiply it by all the different right-hand-side matrices.

EXAMPLE 25   Solve the system

$$\begin{aligned} x + 5y &= 6 \\ 2x + 11y &= -2 \end{aligned}$$

Solution   We first rewrite this system as a matrix equation; then we multiply both sides by the inverse of the coefficient matrix (which we found in Example 19).

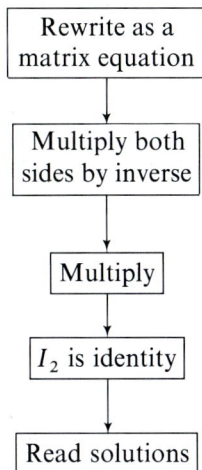

Rewrite as a matrix equation	$\begin{bmatrix} 1 & 5 \\ 2 & 11 \end{bmatrix} \cdot \begin{bmatrix} x \\ y \end{bmatrix} = \begin{bmatrix} 6 \\ -2 \end{bmatrix}$
Multiply both sides by inverse	$\begin{bmatrix} 11 & -5 \\ -2 & 1 \end{bmatrix} \cdot \begin{bmatrix} 1 & 5 \\ 2 & 11 \end{bmatrix} \cdot \begin{bmatrix} x \\ y \end{bmatrix} = \begin{bmatrix} 11 & -5 \\ -2 & 1 \end{bmatrix} \cdot \begin{bmatrix} 6 \\ -2 \end{bmatrix}$
Multiply	$\begin{bmatrix} 1 & 0 \\ 0 & 1 \end{bmatrix} \cdot \begin{bmatrix} x \\ y \end{bmatrix} = \begin{bmatrix} 76 \\ -14 \end{bmatrix}$
I_2 is identity	$\begin{bmatrix} x \\ y \end{bmatrix} = \begin{bmatrix} 76 \\ -14 \end{bmatrix}$
Read solutions	Thus, $x = 76$ and $y = -14$.

EXAMPLE 26   Solve the system

$$\begin{aligned} x - 3y + 6z &= 3 \\ -y + 4z &= 4 \\ -4x + 5y + 8z &= -8 \end{aligned}$$

Solution   We first rewrite this system as a matrix equation; then we multiply both sides by the inverse (which we found in Example 21).

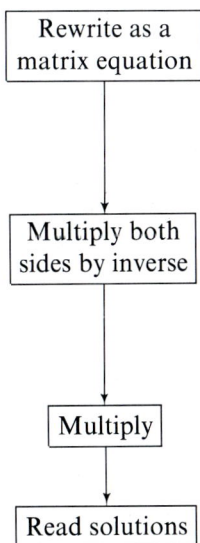

Rewrite as a matrix equation	$\begin{bmatrix} 1 & -3 & 6 \\ 0 & -1 & 4 \\ -4 & 5 & 8 \end{bmatrix} \cdot \begin{bmatrix} x \\ y \\ z \end{bmatrix} = \begin{bmatrix} 3 \\ 4 \\ -8 \end{bmatrix}$
Multiply both sides by inverse	$\begin{bmatrix} 7 & \dfrac{-27}{2} & \dfrac{3}{2} \\ 4 & -8 & 1 \\ 1 & \dfrac{-7}{4} & \dfrac{1}{4} \end{bmatrix} \cdot \begin{bmatrix} 1 & -3 & 6 \\ 0 & -1 & 4 \\ -4 & 5 & 8 \end{bmatrix} \cdot \begin{bmatrix} x \\ y \\ z \end{bmatrix} = \begin{bmatrix} 7 & \dfrac{-27}{2} & \dfrac{3}{2} \\ 4 & -8 & 1 \\ 1 & \dfrac{-7}{4} & \dfrac{1}{4} \end{bmatrix} \cdot \begin{bmatrix} 3 \\ 4 \\ -8 \end{bmatrix}$
Multiply	$\begin{bmatrix} 1 & 0 & 0 \\ 0 & 1 & 0 \\ 0 & 0 & 1 \end{bmatrix} \cdot \begin{bmatrix} x \\ y \\ z \end{bmatrix} = \begin{bmatrix} -45 \\ -28 \\ -6 \end{bmatrix}$
Read solutions	Thus, $x = -45$, $y = -28$, and $z = -6$.

Multiply the following matrices.

1. $\begin{bmatrix} 1 & 2 \\ 3 & 4 \end{bmatrix} \cdot \begin{bmatrix} -2 \\ 3 \end{bmatrix}$

2. $\begin{bmatrix} 1 & -2 \\ 0 & 3 \end{bmatrix} \cdot \begin{bmatrix} -4 \\ 1 \end{bmatrix}$

3. $\begin{bmatrix} 3 & 4 \\ 2 & 3 \end{bmatrix} \cdot \begin{bmatrix} 3 & -4 \\ -2 & 3 \end{bmatrix}$

4. $\begin{bmatrix} 7 & 3 \\ 5 & 2 \end{bmatrix} \cdot \begin{bmatrix} -2 & 3 \\ 5 & -7 \end{bmatrix}$

5. $\begin{bmatrix} 1 & 2 \\ 3 & 4 \end{bmatrix} \cdot \begin{bmatrix} x \\ y \end{bmatrix}$

6. $\begin{bmatrix} 3 & 5 \\ -7 & 9 \end{bmatrix} \cdot \begin{bmatrix} u \\ v \end{bmatrix}$

Solve the following equations.

7. $6x = 42$

8. $-5x = 20$

9. $\dfrac{-2}{3} \cdot x = 14$

10. $\dfrac{3}{10} \cdot x = \dfrac{27}{20}$

Find the multiplicative inverse of each of the following matrices.

11. $\begin{bmatrix} 5 & 3 \\ 7 & 4 \end{bmatrix}$

12. $\begin{bmatrix} 1 & 2 & 5 \\ -3 & 4 & 8 \\ 0 & 3 & 7 \end{bmatrix}$

For Problems 13 to 20, rewrite the systems as matrix equations, and the matrix equations as systems (but do not solve).

13. $2x - 3y + 4z = 5$
$x + y + 3z = 1$
$5x - y + 2z = -2$

14. $9x - 7y + z = 2$
$3x + 4y + 7z = -1$
$4x - 3y + 11z = 15$

15. $\begin{bmatrix} 2 & -1 \\ 8 & 3 \end{bmatrix} \begin{bmatrix} x \\ y \end{bmatrix} = \begin{bmatrix} 3 \\ -2 \end{bmatrix}$

16. $\begin{bmatrix} 7 & 0 \\ 1 & 2 \end{bmatrix} \begin{bmatrix} x \\ y \end{bmatrix} = \begin{bmatrix} -8 \\ 4 \end{bmatrix}$

17. $x - 4z = 0$
$x + 5y = -2$
$y + 7z = 3$

18. $x - y = 7$
$-x + 2z = -3$
$-y - 4z = -1$

19. $\begin{bmatrix} 1 & 0 & 1 \\ 0 & -1 & 1 \\ 1 & -2 & 0 \end{bmatrix} \begin{bmatrix} x \\ y \\ z \end{bmatrix} = \begin{bmatrix} 4 \\ -3 \\ 2 \end{bmatrix}$

20. $\begin{bmatrix} 0 & 4 & 1 \\ 1 & -2 & 7 \\ 1 & 0 & 0 \end{bmatrix} \begin{bmatrix} x \\ y \\ z \end{bmatrix} = \begin{bmatrix} -8 \\ 0 \\ 3 \end{bmatrix}$

Rewrite the following systems as matrix equations and solve using the multiplicative inverse (see Problems 15 to 38 on page 487).

21. $4x + 3y = -2$
$7x + 5y = 1$

22. $x + 4y = 0$
$3x + 11y = 2$

23. $2x + 4y = -6$
$4x + 9y = 5$

24. $6x + 7y = 9$
$4x + 5y = -1$

25. $3x + 5y = 8$
$-x + y = 3$

26. $4x - 2y = 3$
$-9x + 5y = 2$

27. $x - 2z = 1$
$y + 5z = 0$
$-x + 3z = 3$

28. $x + y + 4z = 2$
$2y + z = -3$
$x - 2y + 3z = 5$

29.
$$x \qquad + z = \quad 4$$
$$4y + 3z = -2$$
$$-2x + 2y - \quad z = \quad 7$$

30.
$$x \qquad + z = \quad 1$$
$$-x + y + z = -2$$
$$-y + z = 3$$

31. $2x \qquad = \quad 6$
$$3y = 12$$

32. $\qquad 4y = 20$
$$5x \qquad = 30$$

33. $ax \qquad = c$
$$by = d$$

34. $ax + by = e$
$$cx + dy = f$$

35. $2x \qquad = 10$
$$3y \qquad = 18$$
$$4z = \quad 8$$

36. $ax \qquad = d$
$$by \qquad = e$$
$$cz = f$$

37. $x \qquad = \quad 3$
$$y = -4$$

38. $-x \qquad = \quad 5$
$$-y = -2$$

39. $2x + 4y = 7$
$$3x + 6y = 5$$

40. $x - \quad 5y = 10$
$$-2x + 10y = \quad 3$$

41.
$$y \qquad = \quad 5$$
$$2x \qquad = \quad 8$$
$$3z = -6$$

42.
$$z = \quad 4$$
$$3x \qquad = -12$$
$$-2y \qquad = \quad 8$$

43. $x + y \qquad = 3$
$$x \qquad + z = 4$$
$$y + z = 5$$

44. $\qquad y + z = -2$
$$x + \qquad z = \quad 2$$
$$x + y \qquad = \quad 4$$

Electrical Application

45. Consider the system in Problem 29 on page 451.
  (a) Write this system as a matrix equation.
  (b) Find the multiplicative inverse of the coefficient matrix.
  (c) Solve the system using this inverse.

Business Application

46. A company invests $100,000 in two ventures: one at 10% interest and the other at 12%. Their interest for 1 year was $10,240.
  (a) Translate this information into a system.
  (b) Write this system as a matrix equation.
  (c) Find the multiplicative inverse of the coefficient matrix.
  (d) Use this inverse to solve this system.

14.5
DETERMINANTS

A **determinant** is a real number associated with a square matrix. A determinant can often help us compute the solutions to systems of linear equations.

Definition
The **2 by 2 determinant** is given by

$$\begin{vmatrix} a & b \\ c & d \end{vmatrix} = \begin{vmatrix} a & b \\ c & d \end{vmatrix} = ad - bc$$

EXAMPLE 27 The 2 by 2 determinant takes a 2 by 2 matrix and returns a real number. Note that we *subtract* the cross product.

(a) $\begin{vmatrix} 2 & 5 \\ 3 & 10 \end{vmatrix} = \begin{vmatrix} 2 & 5 \\ 3 & 10 \end{vmatrix} = 2(10) - 3(5) = 20 - 15 = 5$

(b) $\begin{vmatrix} -1 & 3 \\ 4 & 5 \end{vmatrix} = \begin{vmatrix} -1 & 3 \\ 4 & 5 \end{vmatrix} = -1(5) - 4(3) = -5 - 12 = -17$

YES	NO
$\begin{vmatrix} 1 & 2 \\ 3 & 4 \end{vmatrix} = 4 - 6 = -2$	$\begin{vmatrix} 1 & 2 \\ 3 & 4 \end{vmatrix} = 4 + 6 = 10$
We *subtract* cross products.	

The 3 by 3 determinant is a bit harder. There are several ways to evaluate a 3 by 3 determinant; one method is as follows.

Definition
The **3 by 3 determinant** is defined by

$$\begin{vmatrix} a_1 & b_1 & c_1 \\ a_2 & b_2 & c_2 \\ a_3 & b_3 & c_3 \end{vmatrix} = \oplus a_1 \begin{vmatrix} b_2 & c_2 \\ b_3 & c_3 \end{vmatrix} \ominus a_2 \begin{vmatrix} b_1 & c_1 \\ b_3 & c_3 \end{vmatrix} \oplus a_3 \begin{vmatrix} b_1 & c_1 \\ b_2 & c_2 \end{vmatrix}$$

Notice that we multiplied each element of the first column (with alternating signs) by a 2 by 2 determinant called its **minor**.

EXAMPLE 28 Expand the determinant

$$D = \begin{bmatrix} 1 & 2 & -3 \\ 4 & -5 & 6 \\ 2 & -1 & 4 \end{bmatrix}$$

Solution We multiply each element of the first column by its minor. We get each element's minor by "knocking out" its row and column.

Knock out row and column to find minor

$$D = \oplus \begin{vmatrix} 1 & 2 & -3 \\ 4 & -5 & 6 \\ 2 & -1 & 4 \end{vmatrix} \ominus \begin{vmatrix} 1 & 2 & -3 \\ 4 & -5 & 6 \\ 2 & -1 & 4 \end{vmatrix} \oplus \begin{vmatrix} 1 & 2 & -3 \\ 4 & -5 & 6 \\ 2 & -1 & 4 \end{vmatrix}$$

| Element times minor | | $= \oplus 1 \begin{vmatrix} -5 & 6 \\ -1 & 4 \end{vmatrix} \ominus 4 \begin{vmatrix} 2 & -3 \\ -1 & 4 \end{vmatrix} \oplus 2 \begin{vmatrix} 2 & -3 \\ -5 & 6 \end{vmatrix}$ |

$$= 1(-14) - 4(5) + 2(-3)$$

Expand minors

Simplify

$$= -14 - 20 - 6 = -40$$

Note the alternating sign pattern in Example 28. We can, in fact, use any row or column and its minors to expand a 3 by 3 determinant. We use the following sign pattern:

$$\begin{vmatrix} + & - & + \\ - & + & - \\ + & - & + \end{vmatrix}$$ If the sum of the row and column is even, the sign is $+$; if the sum is odd, the sign is $-$.

EXAMPLE 29   Expand the determinant

$$D = \begin{vmatrix} 7 & -2 & 3 \\ 0 & 0 & 4 \\ 5 & 1 & 7 \end{vmatrix}$$

Solution   A row (or column) with several zeros (such as row 2) makes an easier expansion (since $0 \cdot a = 0$). The sign pattern is $- + -$.

Expand using row 2

$$D = \ominus \begin{vmatrix} 7 & -2 & 3 \\ 0 & 0 & 4 \\ 5 & 1 & 7 \end{vmatrix} \oplus \begin{vmatrix} 7 & -2 & 3 \\ 0 & 0 & 4 \\ 5 & 1 & 7 \end{vmatrix} \ominus \begin{vmatrix} 7 & -2 & 3 \\ 0 & 0 & 4 \\ 5 & 1 & 7 \end{vmatrix}$$

$$= -0 + 0 - 4 \begin{vmatrix} 7 & -2 \\ 5 & 1 \end{vmatrix}$$

Use $- + -$ pattern

$$= -4(17) = -68$$

Often we can simplify a determinant somewhat before we actually expand it. Below are some handy properties of determinants that can save us some time and effort. We use the notation $|A|$ to mean the determinant of the square matrix A.

Property

For square matrices A and B (with determinants $|A|$ and $|B|$):

Property	Example
D1 If every entry of a row or column of A is 0, then $$|A| = 0$$	$$\begin{vmatrix} -2 & 7 & 8 \\ 0 & 0 & 0 \\ -9 & 1 & 5 \end{vmatrix} = 0$$ (R_2 has all zeros.)
D2 If B is formed by interchanging two rows (or columns) of A, then $$|B| = -|A|$$	$$\begin{vmatrix} 1 & 2 & 3 \\ 4 & 5 & 6 \\ 7 & 8 & 9 \end{vmatrix} = -\begin{vmatrix} 1 & 2 & 3 \\ 7 & 8 & 9 \\ 4 & 5 & 6 \end{vmatrix}$$ (We interchanged R_2 and R_3.)
D3 If two rows (or columns) of A are identical, then $$|A| = 0$$	$$\begin{vmatrix} 1 & 2 & 3 \\ 4 & 5 & 6 \\ 4 & 5 & 6 \end{vmatrix} = 0$$ (R_2 and R_3 are identical.)
D4 If B is formed by multiplying every element of a row (or column) of A by a constant k, then $$|B| = k|A|$$	$$\begin{vmatrix} 10 & 20 & 30 \\ 4 & 5 & 6 \\ 7 & 8 & 9 \end{vmatrix} = 10\begin{vmatrix} 1 & 2 & 3 \\ 4 & 5 & 6 \\ 7 & 8 & 9 \end{vmatrix}$$ (We multiplied R_1 by 10.)
D5 If B is formed by adding one row (or column) times a constant to another row (or column) of A, then $$|B| = |A|$$	$$\begin{vmatrix} 1 & 2 & 3 \\ 4 & 5 & 6 \\ 7 & 8 & 9 \end{vmatrix} = \begin{vmatrix} 1 & 2 & 3 \\ 2 & 1 & 0 \\ 7 & 8 & 9 \end{vmatrix}$$ (We added $-2R_1$ to R_2.)

Of all these properties, D5 is probably the most useful, since we can use it to put zeros in the determinant. Compare this property to AM2 (page 477).

EXAMPLE 30 Use Property D5 to help expand the determinant

$$\begin{vmatrix} 1 & 2 & 3 \\ 4 & -5 & 6 \\ -7 & 8 & -9 \end{vmatrix}$$

Solution We can put two zeros into column 3 by adding $-2R_1$ to R_2 and $3R_1$ to R_3. This will make the determinant easier to expand.

$$\boxed{\begin{array}{l} \text{Add } -2R_1 \text{ to } R_2; \\ \text{add } 3R_1 \text{ to } R_3 \end{array}} \quad \begin{vmatrix} 1 & 2 & 3 \\ 4 & -5 & 6 \\ -7 & 8 & -9 \end{vmatrix} = \begin{vmatrix} 1 & 2 & 3 \\ 2 & -9 & 0 \\ -4 & 14 & 0 \end{vmatrix}$$

$$\boxed{\begin{array}{c} \text{Expand down } C_3; \text{ only} \\ \text{nonzero entry (3)} \\ \text{contributes} \end{array}} \qquad \begin{aligned} &= \oplus 3 \begin{vmatrix} 2 & -9 \\ -4 & 14 \end{vmatrix} = 3(-8) \\ &= -24 \end{aligned}$$

Using the square matrices $A = \begin{bmatrix} 4 & 7 \\ 3 & 5 \end{bmatrix}$, $B = \begin{bmatrix} 3 & 1 \\ 5 & 2 \end{bmatrix}$, *and* $C = \begin{bmatrix} 5 & 4 \\ 0 & 3 \end{bmatrix}$, *perform the following operations.*

Warm-up
Exercises

1. $A + B$

2. $A - C$

3. $A \cdot B$

4. $B \cdot C$

5. A^2

6. B^2

7. $AB - BA$

8. $AC - CA$

9. A^{-1}

10. B^{-1}

11. Solve $A \begin{bmatrix} x \\ y \end{bmatrix} = \begin{bmatrix} 1 \\ 6 \end{bmatrix}$.

12. Solve $B \begin{bmatrix} x \\ y \end{bmatrix} = \begin{bmatrix} -3 \\ 5 \end{bmatrix}$.

For each of the following statements, state the determinant property (D1 to D5) that it shows.

13. $\begin{vmatrix} 7 & 1 & 3 \\ 0 & 0 & 0 \\ 1 & -5 & 4 \end{vmatrix} = 0$

14. $\begin{vmatrix} 40 & 20 & 100 \\ 4 & -1 & 7 \\ 8 & 4 & -5 \end{vmatrix} = 20 \begin{vmatrix} 2 & 1 & 5 \\ 4 & -1 & 7 \\ 8 & 4 & -5 \end{vmatrix}$

15. $\begin{vmatrix} -1 & 2 & 2 \\ 8 & -4 & -4 \\ 5 & 9 & 9 \end{vmatrix} = 0$

16. $\begin{vmatrix} 1 & 2 & -3 \\ 5 & 2 & 1 \\ 8 & 7 & 6 \end{vmatrix} = - \begin{vmatrix} 5 & 2 & 1 \\ 1 & 2 & -3 \\ 8 & 7 & 6 \end{vmatrix}$

17. $\begin{vmatrix} 4 & 7 & -9 \\ -4 & 8 & 1 \\ 7 & -2 & 10 \end{vmatrix} = \begin{vmatrix} 4 & 7 & -9 \\ 0 & 15 & -8 \\ 7 & -2 & 10 \end{vmatrix}$

18. $\begin{vmatrix} 1 & -4 & 0 \\ 2 & 8 & 0 \\ 9 & 7 & 0 \end{vmatrix} = 0$

19. $\begin{vmatrix} 1 & 2 & 3 \\ 4 & 5 & 6 \\ 7 & 8 & 9 \end{vmatrix} = 3 \begin{vmatrix} 1 & 2 & 1 \\ 4 & 5 & 2 \\ 7 & 8 & 3 \end{vmatrix}$

20. $\begin{vmatrix} 1 & 2 & 3 \\ 4 & 5 & 6 \\ -2 & 5 & 4 \end{vmatrix} = \begin{vmatrix} 1 & 2 & 3 \\ 4 & 5 & 6 \\ 0 & 9 & 10 \end{vmatrix}$

Evaluate the following determinants.

21. $\begin{vmatrix} 4 & -1 \\ 3 & 2 \end{vmatrix}$ **22.** $\begin{vmatrix} -3 & 5 \\ 4 & -7 \end{vmatrix}$ **23.** $\begin{vmatrix} -5 & 2 \\ -3 & 1 \end{vmatrix}$

24. $\begin{vmatrix} 1 & 0 \\ 0 & 1 \end{vmatrix}$ **25.** $\begin{vmatrix} 3 & 0 \\ 0 & -5 \end{vmatrix}$ **26.** $\begin{vmatrix} 3 & -5 \\ 0 & 2 \end{vmatrix}$

27. $\begin{vmatrix} 0 & 1 \\ 1 & 0 \end{vmatrix}$ **28.** $\begin{vmatrix} 4 & 6 \\ -5 & 0 \end{vmatrix}$ **29.** $\begin{vmatrix} 7 & 1 \\ 1 & 0 \end{vmatrix}$

30. $\begin{vmatrix} 1 & -1 & 2 \\ -1 & 2 & 3 \\ -1 & 5 & -4 \end{vmatrix}$ **31.** $\begin{vmatrix} 1 & 6 & -5 \\ 5 & 2 & 2 \\ 3 & -4 & -4 \end{vmatrix}$ **32.** $\begin{vmatrix} 1 & 3 & -5 \\ 4 & 0 & 3 \\ 0 & 0 & 0 \end{vmatrix}$

33. $\begin{vmatrix} 2 & -5 & 7 \\ -2 & 4 & 1 \\ 2 & 3 & -5 \end{vmatrix}$ **34.** $\begin{vmatrix} 10 & 40 & 50 \\ 0 & 3 & 7 \\ -1 & -2 & 6 \end{vmatrix}$ **35.** $\begin{vmatrix} 2 & 3 & 1 \\ 4 & -5 & 7 \\ -6 & 8 & 1 \end{vmatrix}$

36. $\begin{vmatrix} 1 & -8 & 3 \\ -2 & 7 & 5 \\ 3 & -4 & 8 \end{vmatrix}$ **37.** $\begin{vmatrix} 1 & 4 & 7 \\ 2 & 4 & 7 \\ 6 & 4 & 7 \end{vmatrix}$ **38.** $\begin{vmatrix} 0 & 1 & 0 \\ 1 & 0 & 0 \\ 0 & 0 & 1 \end{vmatrix}$

39. $\begin{vmatrix} 1 & 0 & 0 \\ 0 & 1 & 0 \\ 0 & 0 & 1 \end{vmatrix}$ **40.** $\begin{vmatrix} 2 & 0 & 0 \\ 0 & 3 & 0 \\ 0 & 0 & 4 \end{vmatrix}$ **41.** $\begin{vmatrix} a & 0 & 0 \\ 0 & b & 0 \\ 0 & 0 & c \end{vmatrix}$

42. $\begin{vmatrix} 2 & 0 & 0 \\ 0 & 0 & 3 \\ 0 & 4 & 0 \end{vmatrix}$ **43.** $\begin{vmatrix} i & j & k \\ 1 & 2 & 3 \\ 4 & 5 & 6 \end{vmatrix}$ **44.** $\begin{vmatrix} i & j & k \\ 2 & -2 & 5 \\ 1 & 0 & 3 \end{vmatrix}$

45. Consider the matrix

$$M = \begin{bmatrix} a & b \\ c & d \end{bmatrix}$$

which has determinant $D = ad - bc$. If $D \neq 0$, show that the matrix

$$N = \begin{bmatrix} d/D & -b/D \\ -c/D & a/D \end{bmatrix}$$

is the inverse of M. That is, show that $M \cdot N = I_2$.

46. Use the result of Problem 45 to find the inverse of each of the following matrices.

(a) $\begin{bmatrix} 2 & 4 \\ 5 & 7 \end{bmatrix}$ (b) $\begin{bmatrix} -1 & 3 \\ 5 & 2 \end{bmatrix}$ (c) $\begin{bmatrix} 4 & -3 \\ 5 & -1 \end{bmatrix}$

Solve the following for x.

47. $\begin{vmatrix} 4-x & 3 \\ 2 & 5-x \end{vmatrix} = 0$ **48.** $\begin{vmatrix} 5-x & 12 \\ 4 & 7-x \end{vmatrix} = 0$

14.6
CRAMER'S RULE

Up to this point, we have studied various different ways to solve linear systems of equations:

1. Elimination method
2. Substitution method
3. Augmented-matrix method
4. Matrix-inverse method

In this section we consider another method that often works better than the others when the coefficients are awkward. We call this method **Cramer's rule**.

Consider the system

$$A: \quad ax + by = e$$
$$B: \quad cx + dy = f$$

We can solve these literal equations for x and y by elimination.

Multiply A by d; multiply B by $-b$	dA:	$adx + bdy = ed$
	$-bB$:	$-bcx - bdy = -bf$

$$\text{Add to eliminate } y \qquad\qquad (ad - bc)x \qquad = ed - bf$$

$$\text{Solve for } x \qquad\qquad\qquad x = \frac{ed - bf}{ad - bc}$$

$$\text{Similarly, find } y \qquad\qquad\qquad y = \frac{af - ce}{ad - bc}$$

Notice that both denominators are $ad - bc$, which is the determinant D of the coefficient matrix:

$$D = \begin{vmatrix} a & b \\ c & d \end{vmatrix} = ad - bc$$

The numerators are also special determinants, as we see in the following theorem.

Theorem (Cramer's Rule)

The solutions to the system

$$ax + by = e$$
$$cx + dy = f$$

are given by

$$x = \frac{D_x}{D} \qquad \text{and} \qquad y = \frac{D_y}{D}$$

where D, D_x, and D_y are given by

$$D = \begin{vmatrix} a & b \\ c & d \end{vmatrix} \neq 0 \qquad D_x = \begin{vmatrix} e & b \\ f & d \end{vmatrix} \qquad D_y = \begin{vmatrix} a & e \\ c & f \end{vmatrix}$$

In using Cramer's rule:

1. D is the determinant of the coefficient matrix.
2. D_x is the determinant formed by replacing the x-coefficients (first column) of D by the right-hand-side terms (e and f).
3. D_y is the determinant formed by replacing the y-coefficients (second column) of D by the right-hand-side terms (e and f).
4. If $D = 0$, there are no unique solutions.

EXAMPLE 31   Use Cramer's rule to solve the system

$$5x - 11y = 7$$
$$8x + 13y = -2$$

Solution   The elimination method or other matrix methods would be messy here. We first find D, and then use the right-hand-side terms (7 and -2) to find D_x and D_y.

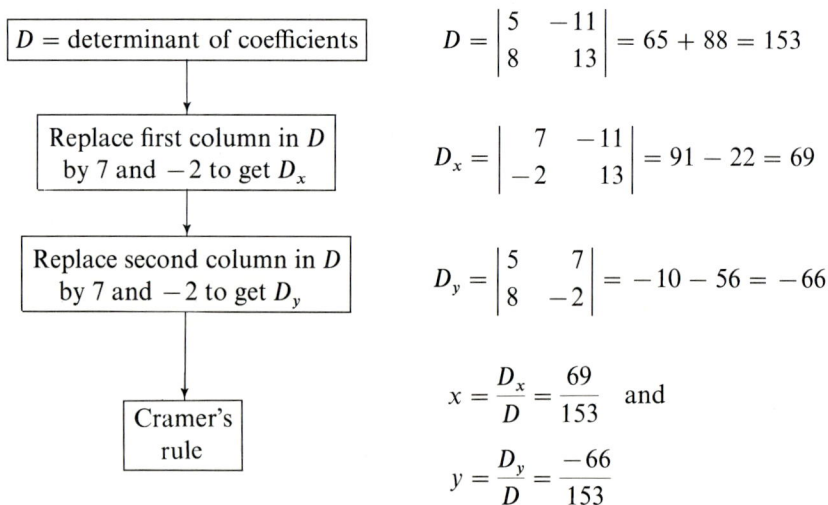

| D = determinant of coefficients | $D = \begin{vmatrix} 5 & -11 \\ 8 & 13 \end{vmatrix} = 65 + 88 = 153$ |

| Replace first column in D by 7 and -2 to get D_x | $D_x = \begin{vmatrix} 7 & -11 \\ -2 & 13 \end{vmatrix} = 91 - 22 = 69$ |

| Replace second column in D by 7 and -2 to get D_y | $D_y = \begin{vmatrix} 5 & 7 \\ 8 & -2 \end{vmatrix} = -10 - 56 = -66$ |

Cramer's rule

$$x = \frac{D_x}{D} = \frac{69}{153} \quad \text{and}$$

$$y = \frac{D_y}{D} = \frac{-66}{153}$$

Using a hand calculator, we can approximate the solutions as

$$x = 0.45098 \quad \text{and} \quad y = -0.43137$$

We also have a version of Cramer's rule for linear systems with three equations.

Theorem (Cramer's Rule)

The solutions to the system

$$a_1 x + b_1 y + c_1 z = d_1$$
$$a_2 x + b_2 y + c_2 z = d_2$$
$$a_3 x + b_3 y + c_3 z = d_3$$

are given by

$$x = \frac{D_x}{D} \qquad y = \frac{D_y}{D} \qquad z = \frac{D_z}{D}$$

where

$$D = \begin{vmatrix} a_1 & b_1 & c_1 \\ a_2 & b_2 & c_2 \\ a_3 & b_3 & c_3 \end{vmatrix} \neq 0 \qquad D_x = \begin{vmatrix} d_1 & b_1 & c_1 \\ d_2 & b_2 & c_2 \\ d_3 & b_3 & c_3 \end{vmatrix}$$

$$D_y = \begin{vmatrix} a_1 & d_1 & c_1 \\ a_2 & d_2 & c_2 \\ a_3 & d_3 & c_3 \end{vmatrix} \qquad D_z = \begin{vmatrix} a_1 & b_1 & d_1 \\ a_2 & b_2 & d_2 \\ a_3 & b_3 & d_3 \end{vmatrix}$$

As with the two-variable systems, D is again the determinant of the coefficient matrix, and D_x, D_y, and D_z are formed by replacing the first, second, or third column of D by the right-hand-side terms.

EXAMPLE 32   Use Cramer's rule to solve the system

$$3x - 5y + 2z = 10$$
$$7x + 4y + 5z = -3$$
$$4x - 2y - 7z = 8$$

Solution   We start by finding D; then we find D_x, D_y, and D_z using the right-hand-side terms (10, -3, and 8).

$$D = \begin{vmatrix} 3 & -5 & 2 \\ 7 & 4 & 5 \\ 4 & -2 & -7 \end{vmatrix}$$

D = determinant of coefficients
Expand with minors

$$= \oplus 3 \begin{vmatrix} 4 & 5 \\ -2 & -7 \end{vmatrix} \ominus 7 \begin{vmatrix} -5 & 2 \\ -2 & -7 \end{vmatrix} \oplus 4 \begin{vmatrix} -5 & 2 \\ 4 & 5 \end{vmatrix}$$

$$= 3(-18) - 7(39) + 4(-33)$$

$$= -459$$

Replace *first* column in D by 10, -3, 8 to get D_x

$$D_x = \begin{vmatrix} 10 & -5 & 2 \\ -3 & 4 & 5 \\ 8 & -2 & -7 \end{vmatrix} = -327$$

$$D_y = \begin{vmatrix} 3 & 10 & 2 \\ 7 & -3 & 5 \\ 4 & 8 & -7 \end{vmatrix} = 769$$

Similarly,
find D_y and D_z

$$D_z = \begin{vmatrix} 3 & -5 & 10 \\ 7 & 4 & -3 \\ 4 & -2 & 8 \end{vmatrix} = 118$$

Now we use Cramer's rule.

Cramer's
rule

$$x = \frac{D_x}{D} = \frac{-327}{-459}$$

$$y = \frac{D_y}{D} = \frac{769}{-459}$$

$$z = \frac{D_z}{D} = \frac{118}{-459}$$

Approximate
as decimals

$$x \approx 0.7124 \qquad y \approx -1.6754 \qquad z \approx -0.2571$$

EXAMPLE 33 Use Cramer's rule to solve the system

$$\begin{aligned} x + 7y - 10z &= 1 \\ 5x + y - 2z &= 5 \\ 4x - 6y + 8z &= -13 \end{aligned}$$

Solution We begin by evaluating D (by first simplifying it using determinant property D5).

D = determinant
of coefficients

$$D = \begin{vmatrix} 1 & 7 & -10 \\ 5 & 1 & -2 \\ 4 & -6 & 8 \end{vmatrix}$$

Add $-5R_1$ to R_2;
add $-4R_1$ to R_3

$$= \begin{vmatrix} 1 & 7 & -10 \\ 0 & -34 & 48 \\ 0 & -34 & 48 \end{vmatrix}$$

Since $R_2 = R_3$

$$= 0$$

Since $D = 0$, Cramer's rule will not produce unique solutions; thus, we stop.

For the matrices

$$A = \begin{bmatrix} 1 & 2 \\ -3 & 4 \end{bmatrix} \quad B = \begin{bmatrix} 10 & -2 \\ 4 & 3 \end{bmatrix} \quad C = \begin{bmatrix} 1 & -2 & 3 \\ 4 & 5 & -6 \\ -7 & 8 & 0 \end{bmatrix} \quad D = \begin{bmatrix} 1 & 7 & -2 \\ 0 & -5 & 1 \\ -1 & 2 & 3 \end{bmatrix}$$

find the following.

Warm-up
Exercises

1. $A + B$ **2.** $B - A$ **3.** AB

4. BA **5.** $C - D$ **6.** $C + D$

7. CD **8.** DC **9.** $|A|$

10. $|B|$ **11.** $|C|$ **12.** $|D|$

Use Cramer's rule to solve the following systems.

13. $5x - 6y = 7$
   $4x + 9y = 11$

14. $10x - 3y = 5$
   $7x + 2y = 6$

15. $20x - 10y = 7$
   $5x + 9y = 8$

16. $15x + 7y = 13$
   $11x + 10y = -7$

17. $8x + 5y = 4$
   $4x - 9y = 1$

18. $9x + 11y = 3$
   $5x + 6y = 4$

19. $3x + 7y = 12$
   $6x + 14y = 7$

20. $8x - 14y = 3$
   $7x + 9y = 10$

21. $2x - 3y - z = 1$
   $x + 7y + z = -3$
   $3x + 4y + 5z = 10$

22. $5x - y + 2z = 2$
   $2x + 3y + 7z = -1$
   $-3x - 5y + 8z = 5$

23. $-2x + 3y + 4z = 7$
   $4x - y - z = 8$
   $3x - 2y + 5z = 2$

24. $4x - 5y + 6z = 11$
   $3x + 7y - 2z = -3$
   $-5x - 3y + z = 0$

25. $x + 5y = 7$
   $2x - 7z = 10$
   $4y + 9z = -3$

26. $4x - 7z = -5$
   $8x + 9y = 10$
   $2y - 7z = 6$

27. $x + 3y = e$
   $x + 4y = f$

28. $ax = e$
   $by = f$

29. $x + y = e$
   $x + z = f$
   $y + z = g$

30. $x + y + z = a$
   $x + y - z = 0$
   $x - y - z = 0$

CHAPTER 14
SUMMARY   **Important Properties and Definitions**

Augmented matrices:

1. Correspond to linear equations as follows:

$$\begin{matrix} ax + by = e \\ cx + dy = f \end{matrix} \quad \Leftrightarrow \quad \begin{bmatrix} a & b & | & e \\ c & d & | & f \end{bmatrix}$$

2. Can, in general, be transformed into a final matrix:

$$\begin{bmatrix} 1 & 0 & \vdots & m \\ 0 & 1 & \vdots & n \end{bmatrix} \quad \text{which means} \quad x = m \text{ and } y = n$$

AM1   We can multiply (or divide) any row by a nonzero constant.
AM2   We can add a constant times one row to another row.
AM3   We can interchange any two rows.

The identity matrix $I_n = \begin{bmatrix} 1 & 0 & 0 & \cdots & 0 \\ 0 & 1 & 0 & \cdots & \\ 0 & 0 & 1 & \cdots & 0 \\ 0 & & & 0 & 1 \end{bmatrix}$ satisfies $I_n \cdot A = A$ for any n by n matrix.

$$\begin{vmatrix} a & b \\ c & d \end{vmatrix} = ad - bc$$

$$\begin{vmatrix} a_1 & b_1 & c_1 \\ a_2 & b_2 & c_2 \\ a_3 & b_3 & c_3 \end{vmatrix} = a_1 \begin{vmatrix} b_2 & c_2 \\ b_3 & c_3 \end{vmatrix} - a_2 \begin{vmatrix} b_1 & c_1 \\ b_3 & c_3 \end{vmatrix} + a_3 \begin{vmatrix} b_1 & c_1 \\ b_2 & c_2 \end{vmatrix}$$

For square matrices A and B (with determinants $|A|$ and $|B|$):

D1   If every entry of a row or column of A is 0, then $|A| = 0$.
D2   If B is formed by interchanging two rows (or columns) of A, then $|B| = -|A|$.
D3   If two rows (or columns) of A are identical, then $|A| = 0$.
D4   If B is formed by multiplying every element of a row (or column) of A by a constant k, then $|B| = k|A|$.
D5   If B is formed by adding one row (or column) times a constant to another row (or column) of A, then $|B| = |A|$.

Cramer's rule:

The solutions to

$$ax + by = e$$
$$cx + dy = f$$

are given by

$$x = \frac{D_x}{D} \quad \text{and} \quad y = \frac{D_y}{D}$$

where

$$D = \begin{vmatrix} a & b \\ c & d \end{vmatrix} \neq 0 \qquad D_x = \begin{vmatrix} e & b \\ f & d \end{vmatrix} \qquad D_y = \begin{vmatrix} a & e \\ c & f \end{vmatrix}$$

Review Exercises

Consider the following table (all units are grams of food value).

	Monday			Tuesday		
	Fat	Protein	Carbohydrate	Fat	Protein	Carbohydrate
Breakfast	10	15	13	12	14	17
Lunch	21	32	45	19	35	50
Dinner	17	25	60	20	22	55

1. Rewrite the table as two matrices.
2. Give the size of each matrix in Problem 1.
3. Add the matrices in Problem 1.
4. Subtract the matrices in Problem 1.
5. Multiply each matrix in Problem 1 by the calorie matrix

$$\begin{array}{r}\text{Fat} \\ \text{Protein} \\ \text{Carbohydrate}\end{array} \begin{bmatrix} 9 \\ 4 \\ 4 \end{bmatrix}$$

Write each of the following augmented matrices as a linear system.

6. $\left[\begin{array}{ccc|c} 2 & -4 & 5 & 3 \\ 7 & 8 & 1 & -8 \\ 9 & 3 & 0 & 10 \end{array}\right]$

7. $\left[\begin{array}{ccc|c} 1 & 0 & 0 & -3 \\ 0 & 1 & 0 & 5 \\ 0 & 0 & 1 & 7 \end{array}\right]$

Write each of the following systems as an augmented matrix, and solve.

8. $\begin{aligned} x - 11y &= -6 \\ 3x + 4y &= 19 \end{aligned}$

9. $\begin{aligned} x - 3y - 4z &= -6 \\ 2x + y &= 7 \\ y - z &= 7 \end{aligned}$

For each of the following pairs of matrices A and B, determine if B is A^{-1}.

10. $A = \begin{bmatrix} 4 & 5 \\ 3 & 4 \end{bmatrix}; \quad B = \begin{bmatrix} 4 & -5 \\ -3 & 4 \end{bmatrix}$

11. $A = \begin{bmatrix} 1 & 0 & 2 \\ 0 & 1 & 0 \\ 0 & 2 & 1 \end{bmatrix}; \quad B = \begin{bmatrix} 3 & 0 & -2 \\ 0 & 1 & 0 \\ -1 & 0 & 1 \end{bmatrix}$

Find the inverse of each of the following matrices.

12. $\begin{bmatrix} 7 & 9 \\ 3 & 4 \end{bmatrix}$

13. $\begin{bmatrix} 1 & 3 & 4 \\ 0 & 4 & 5 \\ 2 & 1 & 2 \end{bmatrix}$

Write each of the following systems as a matrix equation, and solve. (See Problems 12 and 13 for the inverses.)

14. $\begin{aligned} 7x + 9y &= 1 \\ 3x + 4y &= -2 \end{aligned}$

15. $\begin{aligned} x + 3y + 4z &= 8 \\ 4y + 5z &= -3 \\ 2x + y + 2z &= -1 \end{aligned}$

Evaluate the following determinants.

16. $\begin{vmatrix} 3 & 1 \\ -7 & 8 \end{vmatrix}$

17. $\begin{vmatrix} 3 & -4 & 5 \\ 1 & 2 & 4 \\ 2 & -1 & 8 \end{vmatrix}$

Solve the following systems using Cramer's rule.

18. $\begin{aligned} 3x + y &= 13 \\ -7x + 8y &= -3 \end{aligned}$

19. $\begin{aligned} 3x - 4y + 5z &= -9 \\ x + 2y + 4z &= 9 \\ 2x - y + 8z &= 1 \end{aligned}$

CHAPTER 15

Polynomial Functions and Their Zeros

Recall from page 158 that a **polynomial function** has the general form

$$P(x) = a_n x^n + a_{n-1} x^{n-1} + \cdots + a_2 x^2 + a_1 x + a_0$$

where the a-terms are constant coefficients (real or complex). We devote this chapter to finding the **zeros** (or **roots**) of such a polynomial; that is, we seek the numbers c such that $P(c) = 0$ (where is the polynomial equal to zero?).

The zeros of first- and second-degree polynomials are easy to find (review Chapter 3). For instance,

$$P(x) = 7x + 9 = 0 \qquad\qquad Q(x) = x^2 - 7x + 10 = 0$$

$$7x = -9 \qquad\qquad\qquad (x - 5)(x - 2) = 0$$

$$x = \frac{-9}{7} \qquad\qquad\qquad\qquad x = 5 \quad \text{or} \quad 2$$

are the zeros of typical first- and second-degree polynomials. That is, $P(-9/7) = 0$, $Q(5) = 0$, and $Q(2) = 0$. Unfortunately, it is not as easy to find the zeros of polynomials of degree three or more.

15.1 DIVISION OF POLYNOMIALS (AND SYNTHETIC DIVISION)

On page 32 we discussed the **division of polynomials**, which is used in finding the zeros of polynomials.

EXAMPLE 1   The division of polynomials mirrors the division of whole numbers. Consider the following two divisions, side by side.

$$
\begin{array}{r}
12 \\
413)\overline{4987} \\
413 \\
\hline
857 \\
826 \\
\hline
31
\end{array}
\qquad
\begin{array}{r}
x + 2 \\
4x^2 + x + 3)\overline{4x^3 + 9x^2 + 8x + 7} \\
4x^3 + x^2 + 3x \\
\hline
8x^2 + 5x + 7 \\
8x^2 + 2x + 6 \\
\hline
3x + 1
\end{array}
$$

The similarity is very clear: Each power of x has its own column just as each power of 10 does in whole-number division. (Polynomial division, of course, may also involve negative coefficients.) Here $x + 2$ is the **quotient** and $3x + 1$ is the **remainder** (as we have in whole-number division). Recall that we can now write

$$4x^3 + 9x^2 + 8x + 7 = (4x^2 + x + 3)(x + 2) + (3x + 1)$$

Let us now look at a very special case for polynomial division: where the divisor is of the form $x + c$ or $x - c$.

EXAMPLE 2 Divide $(2x^3 - 9x^2 + 4x + 8) \div (x - 3)$.

Solution We first set this up as a long division (as in Example 1). Then we set about to find a shortcut for divisions of this type.

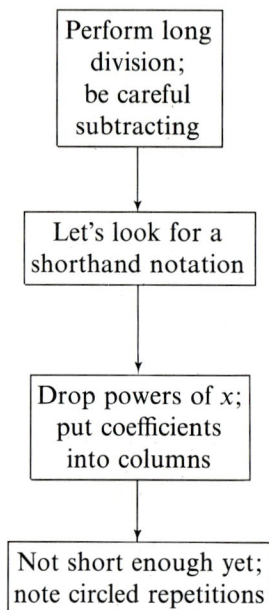

| Perform long division; be careful subtracting |

$$
\begin{array}{r}
2x^2 - 3x\ - 5 \\
x - 3 \overline{)2x^3 - 9x^2 + 4x +\ 8} \\
\underline{2x^3 - 6x^2} \\
-3x^2 + 4x \\
\underline{-3x^2 + 9x} \\
-5x +\ 8 \\
\underline{-5x + 15} \\
-\ 7
\end{array}
$$

| Let's look for a shorthand notation |

| Drop powers of x; put coefficients into columns |

$$
\begin{array}{r}
②\ ⊝\ ⊝ \\
1 - 3 \overline{)②\ \ -9\ \ ④\ \ ⑧} \\
②\ \ -6 \\
⊝\ \ ④ \\
-3\ \ \ 9 \\
⊝\ \ ⑧ \\
-5\ \ 15 \\
-7
\end{array}
$$

| Not short enough yet; note circled repetitions |

As we see, there are a lot of repeated numbers when we divide by first-degree terms such as $x - 3$. We can "squeeze out" the repetition and produce a very compact division technique, called **synthetic division**. This method is easier to show than to express in words; however, let us first make a few notes.

1. If the divisor is $x + c$, we put only $-c$ outside.
2. If the divisor is $x - c$, we put only $+c$ outside.
3. We always bring down the first coefficient.
4. After multiplying, we *add* (since we changed the sign of the c-term).
5. The last term is the remainder.

Note: This method works only if the divisor is of the form $x + c$ or $x - c$.

EXAMPLE 3 Divide: $(3x^2 + 4x^2 - 9x + 6) \div (x + 2)$.

Solution Since the divisor is $x + 2$, we can use synthetic division. When we rewrite the division in compact notation, notice that we put -2 outside and bring down the first coefficient 3.

| Rewrite in compact form |

$$-2 \rfloor \quad 3 \qquad 4 \qquad -9 \qquad 6$$

| Bring down first coefficient |

$$-2 \rfloor \quad 3 \qquad 4 \qquad -9 \qquad 6$$
$$\qquad \quad 3$$

| $-2(3) = -6$; put in next column; add: $4 + (-6) = -2$ |

$$-2 \rfloor \quad 3 \qquad 4 \qquad -9 \qquad 6$$
$$\qquad \qquad \quad -6$$
$$\qquad \quad 3 \qquad -2$$

| $-2(-2) = 4$; put in next column; add: $-9 + 4 = -5$ |

$$-2 \rfloor \quad 3 \qquad 4 \qquad -9 \qquad 6$$
$$\qquad \qquad \quad -6 \qquad 4$$
$$\qquad \quad 3 \qquad -2 \qquad -5$$

| $-2(-5) = 10$; put in next column; add: $6 + 10 = 16$ |

$$-2 \rfloor \quad 3 \qquad 4 \qquad -9 \qquad 6$$
$$\qquad \qquad \quad -6 \qquad 4 \qquad 10$$
$$\qquad \quad 3 \qquad -2 \qquad -5 \qquad \rfloor 16$$

| Write quotient and remainder |

$$Quotient = 3x^2 - 2x - 5$$

$$Remainder = 16$$

Notice that the last term is the remainder and all the others are the quotient (which starts one power less than the original dividend).

EXAMPLE 4 Divide: $(2x^3 - 11x^2 + 28) \div (x - 5)$.

Solution Since the divisor is $x - 5$, we can use synthetic division (but now the outside term is $+5$). Note that we use a zero as a placeholder for the missing x-term.

| Rewrite in compact form |

$$5 \rfloor \quad 2 \qquad -11 \qquad 0 \qquad 28$$

Bring down first coefficient

Multiply each new term by 5; then add

Write quotient and remainder

$$
\begin{array}{r|rrrr}
5 & 2 & -11 & 0 & 28 \\
 & \downarrow & 10 & -5 & 25 \\
\hline
 & 2 & -1 & -5 & \boxed{3}
\end{array}
$$

$Quotient = 2x^2 - x - 5$

$Remainder = 3$

EXAMPLE 5 Divide: $(2x^3 - 3x^2 + 2x - 3) \div (x + i)$.

Solution Here the divisor contains the imaginary number i. But we still use synthetic division, except that we must be more careful with the complex numbers. (Recall that $i^2 = -1$.)

Rewrite in compact form

$$
\begin{array}{r|rrrr}
-i & 2 & -3 & 2 & -3
\end{array}
$$

Bring down first coefficient

$$
\begin{array}{r|rrrr}
-i & 2 & -3 & 2 & -3
\end{array}
$$

Multiply each new term by $-i$; then add

$$
\begin{array}{r|rrrr}
 & & -2i & -2 + 3i & 3 \\
\hline
 & 2 & -3 - 2i & 3i & \boxed{0}
\end{array}
$$

Write quotient and remainder

$Quotient = 2x^2 + (-3 - 2i)x + 3i$

$Remainder = 0$

PROBLEM SET 15.1

Solve the following equations.

Warm-up Exercises

1. $2x + 5 = 0$ **2.** $-3x + 7 = 0$

3. $x^2 + 4x - 5 = 0$ **4.** $2x^2 - 13x - 7 = 0$

Simplify the following polynomials.

Warm-up Exercises

5. $(2x - 5) + (-7x - 3)$ **6.** $(5x - 1) - (3x - 2)$

7. $(x + 7)(x^2 - 5x + 1)$ **8.** $(2x - 6)(3x^2 - x + 9)$

Perform the following long divisions, showing the quotient and remainder of each.

9. $19 \div 3$

10. $82 \div 7$

11. $357 \div 12$

12. $1234 \div 92$

13. $20,035 \div 103$

14. $50,009 \div 608$

Perform the following long divisions, showing quotient and remainder.

15. $(2x^3 + x^2 + 2x - 14) \div (2x - 3)$

16. $(6x^3 - 7x^2 + 11x + 1) \div (3x - 2)$

17. $(3x^3 - 11x^2 - 14x + 8) \div (3x + 4)$

18. $(12x^3 - 29x^2 - 6x + 31) \div (4x - 7)$

19. $(2x^4 - 5x^3 - 11x^2 + 27x - 17) \div (x^2 - 4x + 3)$

20. $(2x^4 - 10x^3 + 11x^2 - 2x - 21) \div (2x^2 - 4x + 7)$

21. $(x^5 - x + 1) \div (x^2 - 1)$

22. $(x^6 - 1) \div (x^3 - 1)$

Use synthetic division to perform the following divisions, showing quotient and remainder.

23. $(4x^3 + 9x^2 - 3x - 10) \div (x + 2)$

24. $(3x^3 - 32x^2 + 19x + 6) \div (x - 10)$

25. $(5x^3 + 28x^2 - 9x + 14) \div (x + 6)$

26. $(4x^3 - 21x^2 - 2x + 25) \div (x - 5)$

27. $(x^4 - 5x^3 + 9x^2 - 13x + 12) \div (x - 3)$

28. $(5x^4 - x^3 - 7x^2 + 5x - 10) \div (x - 1)$

29. $(2x^4 + 3x^3 - 30x^2 + 22x - 17) \div (x + 5)$

30. $(3x^4 + 15x^3 - 47x^2 - 34x + 1) \div (x + 7)$

31. $(x^4 + 16) \div (x + 2)$ **32.** $(x^3 + 125) \div (x + 5)$

33. $x^4 \div (x + 1)$ **34.** $x^5 \div (x - 1)$

35. $(2x^3 - 17x + 15) \div (x - 3)$

36. $(4x^3 + 7x^2 - 6) \div (x + 2)$

37. $(2x^3 - 6ix^2 + 3x - 9i) \div (x - 3i)$

38. $[2x^3 + (1 + 4i)x^2 + (2 + 2i)x + 4i] \div (x + 2i)$

39. $(x^3 - x^2 + 4x - 2) \div (x - 1 - i)$

40. $(x^3 + 2x^2 - 3x - 10) \div (x + 2 - i)$

41. $(x^3 - 6x^2 + 11x - 1) \div (x - 4)$

42. $(x^5 + 1) \div (x - 1)$

43. $(x^2 - x + 1) \div (x + i)$

44. $x^4 \div (x + 2)$

15.2

REMAINDER AND FACTOR THEOREMS

Recall that we check whole-number division by multiplying the quotient by the divisor and adding the remainder to get the original dividend. This same property also holds for polynomial division.

EXAMPLE 6   Consider the following divisions.

(a)
$$
\begin{array}{r}
26 \\
19\overline{)512} \\
38 \\
\overline{132} \\
114 \\
\overline{18}
\end{array}
$$
means $512 = (19)(26) + 18$

quotient   remainder

(b)
$$
\begin{array}{r}
2x + 3 \\
x + 1\overline{)2x^2 + 5x + 7} \\
2x^2 + 2x \\
\overline{3x + 7} \\
3x + 3 \\
\overline{4}
\end{array}
$$
means $2x^2 + 5x + 7 = (x + 1)(2x + 3) + 4$

Let us state this as a theorem that we use throughout the chapter.

Theorem 1 (Division Algorithm)
For any polynomial $P(x)$ and divisor $x - c$, we can find a polynomial
quotient $Q(x)$ and **remainder** R that satisfy

$$P(x) = (x - c)Q(x) + R$$

The quotient $Q(x)$ has degree 1 less that $P(x)$, and the remainder R is
simply a constant.

Let us take Theorem 1 and substitute $x = c$.

Theorem 1	$P(x) = (x - c)Q(x) + R$
Substitute $x = c$	$P(c) = (c - c)Q(c) + R$
Since $c - c = 0$	$P(c) = R$

Theorem 2 (Remainder Theorem)

For a polynomial $P(x) = (x - c)Q(x) + R$, we have
$$P(c) = R$$

This is a very important theorem, since it gives us a choice:

1. We can find R by evaluating $P(c)$.
2. We can find $P(c)$ by dividing by $x - c$ and taking R.

In other words, we find whichever is easier, $P(c)$ or R.

EXAMPLE 7 For $P(x) = 3x^4 - 5x^3 + 2x^2 - 7x + 4$:

    (a) Find $P(3)$.   (b) Find $P(-1)$.

Solution Here is a situation where Theorem 2 comes in handy. Rather than evaluate $P(x)$ directly, we divide synthetically and keep the remainder.

(a) To find $P(3)$, we divide by $x - 3$ and then take the remainder.

Synthetic division:	$3 \rfloor$	3	-5	2	-7	4
$P(x) \div (x - 3)$			9	12	42	105
		3	4	14	35	$\lfloor 109$

Since $R = 109$ Thus, $P(3) = 109$.

(b) To find $P(-1)$, we divide by $x - (-1) = x + 1$.

Synthetic division:	$-1 \rfloor$	3	-5	2	-7	4
$P(x) \div (x + 1)$			-3	8	-10	17
		3	-8	10	-17	$\lfloor 21$

Since $R = 21$ Thus, $P(-1) = 21$.

With whole numbers, we say that *m is a factor of n* if we can write $n = k \cdot m$ (k is a whole number) with a *remainder of zero*. As examples,

    (a) $100 = 2(50)$ Thus, 2 is a factor of 100, since $R = 0$.

    (b) $89 = 5(17) + 4$ Thus, 5 is *not* a factor of 89, since $R \neq 0$.

Similarly with polynomials, we say that

> $x - c$ **factors** $P(x)$ if and only if
>
> $P(x) = (x - c)Q(x)$ (and $R = 0$)

This definition says that $x - c$ factors $P(x)$ if and only if $R = 0$. Theorem 2 tells us that $R = P(c)$. Putting these two statements together, we get the following important theorem.

Theorem 3 (Factor Theorem)

> The polynomial $x - c$ factors $P(x)$ if and only if $P(c) = 0$.
> Equivalently, $x + c$ factors $P(x)$ if and only if $P(-c) = 0$.

> To show $x - c$ factors $P(x)$, we have a choice:
>
> 1. Use synthetic division and show that $R = 0$.
> 2. Show $P(c) = 0$; that is, c is a zero of $P(x)$.

EXAMPLE 8   Is $x - 3$ a factor of $P(x) = -2x^3 + 9x^2 - 13x + 12$?

Solution   We show both methods.

(a) We use synthetic division with divisor $x - 3$.

| Synthetic division |

$$
\begin{array}{r|rrrr}
3 & -2 & 9 & -13 & 12 \\
  &   & -6 & 9 & -12 \\
\hline
  & -2 & 3 & -4 & \;\;0
\end{array}
$$

| Since $R = 0$ |

Thus, $x - 3$ is a factor of $P(x)$.

(b) We evaluate $P(3)$, since the divisor is $x - 3$.

| Evaluate $P(3)$ |

$$
\begin{aligned}
P(3) &= -2(3)^3 + 9(3)^2 - 13(3) + 12 \\
&= -54 + 81 - 39 + 12 \\
&= 0
\end{aligned}
$$

| Since $P(3) = 0$ |

Thus, $x - 3$ is a factor of $P(x)$.

Both methods give the same result (as they should!): $x - 3$ is a factor of $P(x)$. In our problems we choose the method that seems the simpler.

EXAMPLE 9   Is $x + 1$ a factor of $P(x) = x^{19} + 1$?

Solution   Synthetic division would be too cumbersome here, so we evaluate $P(-1)$ instead.

| $(-1)^{19} = -1$
| since 19 is odd |

$$
\begin{aligned}
P(-1) &= (-1)^{19} + 1 \\
&= -1 + 1 \\
&= 0
\end{aligned}
$$

| Since $P(-1) = 0$ |

Thus, $x + 1$ is a factor of $P(x)$.

EXAMPLE 10   Is $x + 4$ a factor of $P(x) = 5x^3 + 16x^2 - 10x + 21$?

Solution   Here we use synthetic division and look for a zero remainder.

$$\boxed{\text{Synthetic division}}$$

$$\underline{-4|} \quad \begin{array}{rrrr} 5 & 16 & -10 & 21 \\ & -20 & 16 & -24 \\ \hline 5 & -4 & 6 & \underline{|-3} \end{array}$$

$$\boxed{\text{Since } R = -3 \neq 0}$$

Thus, $x + 4$ is *not* a factor of $P(x)$.

PROBLEM SET
15.2

Simplify the following.

Warm-up
Exercises

1. $2^2 + 3(2) - 7$

2. $4(-3)^3 - 5(-3)^2 - 6(-3) + 11$

3. $(-1)^{10} + 1$

4. $(-1)^{15} + 1$

5. $(-1)^{21} - 2(-1)^{15} + (-1)^8$

6. $-1 - (-1)^2 - (-1)^3 - (-1)^4$

Use synthetic division to divide the following.

Warm-up
Exercises

7. $(x^2 - 7x + 4) \div (x - 3)$

8. $(x^2 - 5x - 11) \div (x + 2)$

9. $(x^3 - 4x^2 + 9x - 7) \div (x - 2)$

10. $(2x^3 + 9x^2 + 4x - 23) \div (x + 4)$

11. $(x^5 - 7) \div (x + 1)$

12. $(x^9 - 1) \div (x - 1)$

For each of the following polynomials, find $Q(x)$ and R so that $P(x) = (x - 2)Q(x) + R$.

Warm-up
Exercises

13. $P(x) = x^2 - 9x + 14$

14. $P(x) = x^2 - 11x + 3$

15. $P(x) = 2x^2 - 5x - 3$

16. $P(x) = 3x^2 + 5x - 20$

17. $P(x) = 4x^3 - 3x^2 + 7x - 8$

18. $P(x) = 2x^3 + x^2 - 7x + 2$

Use the remainder theorem to evaluate the following polynomials.

19. $P(x) = x^3 - x^2 + 5x - 11$; find $P(3)$ and $P(-3)$.

20. $P(x) = 2x^3 + 4x^2 - 11x - 7$; find $P(6)$ and $P(-2)$.

21. $P(x) = 5x^4 - 12x^3 + 8x^2 + 11x - 7$; find $P(3)$ and $P(-1)$.

22. $P(x) = -2x^4 + 5x^3 - 7x^2 - 8x + 12$; find $P(2)$ and $P(-3)$.

Determine if the second polynomial is a factor of the first.

23. $x^3 + 4x^2 - 6x + 6$; $x + 2$

24. $2x^3 - 5x^2 - 4x + 3$; $x - 3$

25. $3x^3 + 2x^2 - 7x + 2$; $x - 1$

26. $2x^3 - 3x^2 - 23x - 10$; $x - 4$

27. $x^4 + 7x^3 + 13x^2 + 19x + 20$; $x + 5$

28. $x^4 - x^3 - 4x^2 + 5x + 4$; $x + 2$

29. $x^5 - 1$; $x - 1$

30. $x^7 + 1$; $x + 1$

31. $x^8 + 1$; $x + 1$

32. $x^{10} - 1$; $x - 1$

33. $x^{100} - 2x^{41} + 1$; $x - 1$

34. $x^{100} - 2x^{61} + 1$; $x + 1$

35. $x^{100} - 3x^{70} + 3x^{40} + 1$; $x - 1$

36. $x^{100} + 3x^{80} + 3x^{30} + 1$; $x + 1$

37. $x^5 - a^5$; $x - a$

38. $x^7 + b^7$; $x + b$

39. $x^8 + c^8$; $x + c$

40. $x^{10} - d^{10}$; $x - d$

41. $x^4 + ax^3 + a^2x^2 + a^3x + a^4$; $x - a$

42. $x^4 - bx^3 + b^2x^2 - b^3x + b^4$; $x + b$

Evaluate each of the following polynomial functions for the given values. Then use these evaluations to list any possible factors of the polynomial.

43. $f(x) = x^3 + x^2 - 3x + 9$; find $f(-1)$, $f(2)$, and $f(-3)$.

44. $f(x) = x^3 + 2x^2 - x - 2$; find $f(-1)$, $f(1)$, and $f(-2)$.

45. $f(x) = x^7 - x$; find $f(0)$, $f(1)$, and $f(-1)$.

46. $P(x) = x^{10} + 2x^5 + 1$; find $P(0)$, $P(1)$, and $P(-1)$.

47. $f(x) = x^3 + 3ax^2 + 3a^2x + a^3$; find $f(a)$, $f(-a)$, and $f(0)$.

48. $P(x) = x^3 + bx^2 - a^2x - a^2b$; find $P(a)$, $P(-a)$, and $P(-b)$.

15.3 FUNDAMENTAL THEOREM OF ALGEBRA

We have seen that some quadratic equations such as

$$x^2 + 1 = 0 \qquad \text{or} \qquad x^2 - 2x + 7 = 0$$

$$(\textit{Solution: } x = \pm i) \qquad\qquad (\textit{Solution: } x = 1 \pm i\sqrt{6})$$

have no real solutions, but *do* have complex solutions (or zeros). We might ask: Does every polynomial have at least one zero (if not real, then complex)?

Thanks to the brilliant work of Karl Gauss in 1799 (when he was only 20 years old!), the answer is "yes." The result is called the **fundamental theorem of algebra**.

Theorem 4 (Fundamental Theorem of Algebra)

> Every polynomial $P(x)$ of degree $n \geq 1$ has at least one zero (real or complex); that is, we can always find a c so that $P(c) = 0$.

A *constant* polynomial, such as $P(x) = 5$, does *not* have a zero (since it is always 5, it can never be zero). However, *every* other polynomial with a variable always has at least one zero. (Note that this theorem states that a zero exists, but does not state what the zero actually is.)

The fundamental theorem tells of only one zero, c_1. However, since $P(c_1) = 0$, we know that $x - c_1$ factors $P(x)$ [see page 512, the factor theorem]. We then repeat the fundamental theorem with the quotients as follows.

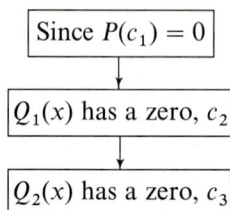

Since $P(c_1) = 0$	$P(x) = (x - c_1)Q_1(x)$
↓	
$Q_1(x)$ has a zero, c_2	$= (x - c_1)\overbrace{(x - c_2)Q_2(x)}$
↓	
$Q_2(x)$ has a zero, c_3	$= (x - c_1)(x - c_2)\overbrace{(x - c_3)Q_3(x)}$

We repeat this process until the last quotient Q_n is a constant (such as 5, which has no zero). Thus, we get the following theorem.

Theorem 5

> Every polynomial $P(x)$ of degree $n \geq 1$ can be factored as
> $$P(x) = a(x - c_1)(x - c_2) \cdots (x - c_n)$$
> where a is a constant and $c_1, c_2, \ldots,$ and c_n are the zeros of $P(x)$. (Notice that P has exactly n zeros.)

EXAMPLE 11 Consider the polynomial $P(x) = 5x^3 - 10x^2 + 35x$. We find the zeros by factoring and then using the quadratic formula.

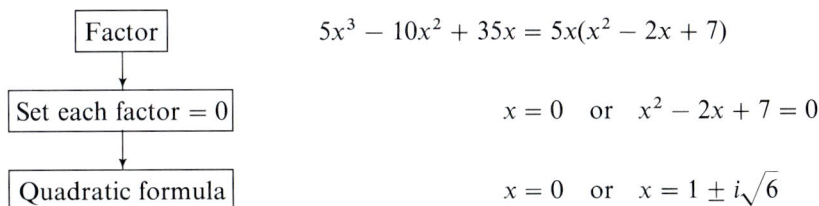

Factor	$5x^3 - 10x^2 + 35x = 5x(x^2 - 2x + 7)$
Set each factor $= 0$	$x = 0$ or $x^2 - 2x + 7 = 0$
Quadratic formula	$x = 0$ or $x = 1 \pm i\sqrt{6}$

Thus, the zeros are 0, $1 + i\sqrt{6}$, and $1 - i\sqrt{6}$, and our polynomial can be written as the product

$$P(x) = 5(x - 0)[x - (1 + i\sqrt{6})][x - (1 - i\sqrt{6})]$$

or

$$P(x) = 5x(x - 1 - i\sqrt{6})(x - 1 + i\sqrt{6})$$

EXAMPLE 12 Find a polynomial $P(x)$ of lowest degree with zeros 1 and -4, with $P(0) = 8$.

Solution By Theorem 5, $P(x)$ has the form

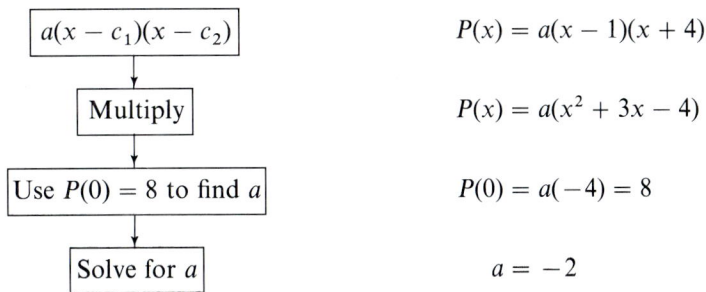

$a(x - c_1)(x - c_2)$	$P(x) = a(x - 1)(x + 4)$
Multiply	$P(x) = a(x^2 + 3x - 4)$
Use $P(0) = 8$ to find a	$P(0) = a(-4) = 8$
Solve for a	$a = -2$

Thus, the desired polynomial is $P(x) = -2x^2 - 6x + 8$. Note that we use the zeros in Theorem 5 and then solve for the constant a.

Note that first-degree polynomials [such as $P(x) = 2x + 5$] have one zero, second-degree polynomials [such as $P(x) = x^2 - 5x + 1$] have two zeros, and third-degree polynomials [such as $P(x) = 5x^3 - 10x^2 + 35$] have three zeros. Does this pattern continue? Yes.

Theorem 6

> A polynomial $P(x)$ of degree $n \geq 1$ has *at most n* distinct zeros.

Theorems 5 and 6 tell us that there are at most n zeros. However, the n zeros may not all be different; some may be **repeated zeros**. The number of times a zero is repeated is called its **multiplicity**.

EXAMPLE 13 Consider the following polynomials. Each zero and its multiplicity are given at the right.

Polynomial	Zeros	Multiplicity
(a) $P(x) = 5x^3 + 5x^2 - 10x = 5x(x - 1)(x + 2)$	0 1 -2	1 1 1
(b) $P(x) = x^6 - 2x^5 + x^4 = x^4(x - 1)^2$	$0, 0, 0, 0$ $1, 1$	4 2
(c) $P(x) = 8(x - 5)(x + 4)^2(x - i)^3$	5 $-4, -4$ i, i, i	1 2 3

Whenever we use the quadratic formula and get complex zeros, we notice that they always come in conjugate pairs; for instance,

$$x^2 - 2x + 3 = 0 \qquad\qquad x^2 + x + 4 = 0$$

$$(\text{Solution: } x = 1 \pm i\sqrt{2}) \qquad \left(\text{Solution: } x = \frac{-1}{2} \pm \frac{i\sqrt{15}}{2}\right)$$

Do complex zeros always come in conjugate pairs like this? The answer is "yes" if all the coefficients of the polynomial are real.

Theorem 7 (Conjugate Pair Theorem)

> Let $P(x)$ be a polynomial with *real* coefficients. If $z = a + bi$ is a zero, then $\bar{z} = a - bi$ is also a zero.

We sketch the proof of this in the problem set.

EXAMPLE 14 Suppose that we know that $-2 + i\sqrt{3}$ is a zero of a polynomial with only real coefficients. Then we know that $-2 - i\sqrt{3}$ is also a zero.

EXAMPLE 15 Find a polynomial with real coefficients of lowest degree which has zeros 2 and $5i$, and $P(0) = 200$.

Solution Since $5i$ is a zero and the polynomial has real coefficients, then the conjugate $-5i$ is also a zero. We work this like Example 12. We write out the general form and then solve for a.

$$\boxed{a(x - c_1)(x - c_2)(x - c_3)}$$

$\downarrow$

$$\boxed{\text{Multiply}}$$

$\downarrow$

$$\boxed{\text{Multiply}}$$

$\downarrow$

$$\boxed{\begin{array}{c}\text{Use } P(0) = 200 \\ \text{to find } a\end{array}}$$

$P(x) = a(x - 2)(x - 5i)(x + 5i)$

$P(x) = a(x - 2)(x^2 + 25)$

$P(x) = a(x^3 - 2x^2 + 25x - 50)$

$P(0) = a(-50) = 200$

$a = -4$

Thus, $P(x) = -4x^3 + 8x^2 - 100x + 200$.

PROBLEM SET 15.3

Solve the following equations.

Warm-up Exercises

1. $2x - 6 = 0$

2. $5x + 1 = 0$

3. $x^2 - 4 = 0$

4. $x^2 + 3x - 10 = 0$

5. $x^2 - x + 5 = 0$

6. $x^2 + 16 = 0$

7. $x^3 - x = 0$

8. $x^4 + x^3 + x^2 = 0$

Multiply the following.

Warm-up Exercises

9. $a(x - 2)(x + 3)$

10. $a(x + 1)(x + 2)(x + 3)$

11. $ax^2(x - 3i)(x + 3i)$

12. $ax^3(x - 3 - i)(x - 3 + i)$

For each of the following polynomials, write $P(x)$ as a product

$$P(x) = (x - c_1)(x - c_2) \cdots (x - c_n) \qquad \text{(Let } a = 1.\text{)}$$

13. $P(x)$ has zeros 2, 3, and -4.

14. $P(x)$ has zeros 0, 4, and -7.

15. $P(x)$ has zeros $2i$ and $-2i$.

16. $P(x)$ has zeros 1, i, and $-i$.

17. $P(x)$ has zeros $1 + 3i$ and $1 - 3i$.

18. $P(x)$ has zeros 0, $2 - i$, and $2 + i$

19. $P(x) = x^3 + 4x^2 - 5x$

20. $P(x) = x^3 + 5x^2 + 6x$

21. $P(x) = x^2 - 2x + 6$

22. $P(x) = x^2 + 2x + 2$

Give the zeros and their multiplicities for the following polynomials.

23. $P(x) = x^2 - 9x + 14$

24. $P(x) = x^2 + 2x - 15$

25. $P(x) = x^5 - x^4$

26. $P(x) = x^7 - x^5$

27. $P(x) = x^3(x-1)^2(x+2)^5$

28. $P(x) = x^4(x-7)^3(x+3)^2(x-8)$

Find the polynomial with real coefficients satisfying each of the following conditions.

29. With zeros 1, 2, and -3, and $P(0) = -18$.

30. With zeros 1, 1, and 4, and $P(0) = 20$.

31. With zeros -1 and $3i$, and $P(0) = 45$.

32. With zeros 2 and $2i$, and $P(0) = -48$.

33. With zero $2 - i$, and $P(2) = 3$.

34. With zero $1 + 2i$, and $P(1) = -8$.

For each of the following polynomials with only real coefficients:
(a) Give all the zeros.
(b) Give the multiplicity of each zero.
(c) Write as a product $a(x - c_1) \cdots (x - c_n)$.

35. $P(x) = 2x + 8$

36. $P(x) = -3x + 10$

37. $P(x) = x^2 - 6x + 8$

38. $P(x) = 2x^2 - 18$

39. $P(x) = 6x^2 - 5x - 4$

40. $P(x) = 15x^2 + x - 2$

41. $P(x)$ has zeros $2i$ and 1, and $P(0) = 4$.

42. $P(x)$ has zeros $1 - i$ and -3, and $P(0) = 5$.

43. $P(x)$ has zeros $3i$, $2 + i$, 0, 0, and 0, and $P(1) = 40$.

44. $P(x)$ has zeros $-2i$, $6 - i$, 0, 0, 1, and 1, and $P(2) = 1$.

45. Supply the reasons in the following proof.

Theorem 7: Let $P(x)$ be a real polynomial. If $z = a + bi$ is a zero, then $\bar{z} = a - bi$ is also zero. (*Hint:* Review complex conjugates—page 414.)

Statement	Reason
(a) $P(z) = a_n z^n + \cdots + a_1 z + a_0 = 0$	(a) _____
(b) $P(\bar{z}) = a_n \bar{z}^n + \cdots + a_1 \bar{z} + a_0$	(b) _____
(c) $\quad = \overline{a_n z^n} + \cdots + \overline{a_1 z} + a_0$	(c) _____
(d) $\quad = \overline{a_n}\,\overline{z^n} + \cdots + \overline{a_1}\,\overline{z} + \overline{a_0}$	(d) _____
(e) $\quad = \overline{a_n z^n} + \cdots + \overline{a_1 z} + \overline{a_0}$	(e) _____
(f) $\quad = \overline{a_n z^n + \cdots + a_1 z + a_0}$	(f) _____
(g) $\quad = \overline{P(z)}$	(g) _____
(h) $\quad = 0$	(h) _____
(i) Thus, $\bar{z}$ is a zero of $P(x)$.	(i) _____

15.4
COUNTING AND BOUNDING REAL ZEROS

The fundamental theorem of algebra (page 514) guarantees us zeros to polynomials. However, it does not tell us how many are real or where to look for them. In this section we consider two very handy theorems for isolating the real zeros of polynomials with real coefficients.

Our first theorem is called **Descartes' rule of signs**, which involves the number of variations in sign. In a polynomial (with zero coefficients deleted), a **variation in sign** occurs when consecutive coefficients are of opposite sign; that is, the coefficients go from $+$ to $-$, or $-$ to $+$. As examples,

$$2x^6 - 3x^4 - 8x^3 + 7x^2 - 1 \qquad \text{has 3 variations in sign}$$

$$x^5 + x^4 + x^3 + x^2 + x - 1 \qquad \text{has 1 variation in sign}$$

Theorem 8 (Descartes' Rule of Signs)

Let $P(x)$ be a polynomial with *real* coefficients.

1. The number of *positive* real zeros of $P(x)$ is equal to the number of variations in sign in $P(x)$, or an even number less.
2. The number of *negative* real zeros of $P(x)$ is equal to the number of variations in sign in $P(-x)$, or an even number less. [*Note:* $P(-x)$ can be found by changing the signs on all terms in $P(x)$ with odd exponents.]

EXAMPLE 16 Consider $P(x) = 2x^4 - 3x^3 - 7x^2 + 10x - 2$.

(a) *Positive zeros*:

$$P(x) = 2x^4 - 3x^3 - 7x^2 + 10x - 2$$

Since there are 3 variations in sign, the number of *positive* real zeros is 3 or 1.

(b) *Negative zeros*:

$$P(-x) = 2x^4 + 3x^3 - 7x^2 - 10x - 2$$

Since there is 1 variation in sign, the number of negative real zeros is 1. Notice that we computed $P(-x)$ by changing the signs in the x^3- and x-terms.

EXAMPLE 17 Consider $P(x) = 5x^7 - 8x^5 - 4x^2 - x + 10$.

(a) *Positive zeros*:

$$P(x) = 5x^7 - 8x^5 - 4x^2 - x + 10$$

Since there are 2 variations in sign, the number of positive real zeros is 2 or 0.

(b) *Negative zeros*:

$$P(-x) = -5x^7 + 8x^5 - 4x^2 + x + 10$$

Since there are 3 variations in sign, the number of negative real zeros is 3 or 1.

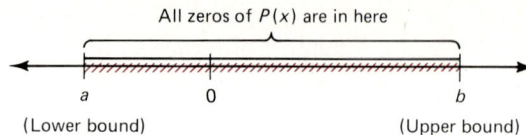

All zeros of $P(x)$ are in here

a 0 b

(Lower bound)   (Upper bound)

FIGURE 1

Our next theorem helps to put bounds on the location of the real zeros. A number a is a **lower bound** of the zeros of $P(x)$ if there is no zero less than a. Similarly, b is an **upper bound** of the zeros of $P(x)$ if there is no zero greater than b. Figure 1 illustrates these bounds. (These bounds are not unique; that is, there are many upper and lower bounds.) The following theorem tells us when we have found one (or both) of the bounds.

Theorem 9 (*Bound Theorem*)

Suppose that we divide a polynomial $P(x)$ by $x - r$, using synthetic division. We look at the bottom row of this division:

1. If $r > 0$, and all numbers in the bottom row are the same sign (or zero), then r is an *upper bound* for the zeros of $P(x)$; that is, no zeros are greater than r.
2. If $r < 0$, and the numbers in the bottom row alternate in sign (zero can be $+$ or $-$), then r is a *lower bound*; that is, no zeros are less than r.

EXAMPLE 18   Consider $P(x) = 2x^3 - 7x^2 + 4x + 1$.

(a) *Positive zeros:* There are 2 variations in sign, so there are 2 or 0 positive zeros. To find an upper bound, we use trial and error ($x = 2, 4, 6$, and so on) until we get a bottom row of *all nonnegative numbers.*

$x = 2$:
$$\underline{2|} \quad \begin{array}{rrrr} 2 & -7 & 4 & 1 \\ & 4 & -6 & -4 \\ \hline 2 & -3 & -2 & \boxed{-3} \end{array} \rightarrow 2 \text{ is } not \text{ an upper bound}$$

$x = 4$:
$$\underline{4|} \quad \begin{array}{rrrr} 2 & -7 & 4 & 1 \\ & 8 & 4 & 32 \\ \hline 2 & 1 & 8 & \boxed{33} \end{array} \rightarrow 4 \text{ } is \text{ an upper bound}$$
$$\text{(all positives)}$$

(b) *Negative zeros:* $P(-x) = -2x^3 - 7x^2 - 4x + 1$ has 1 variation in sign, so there is 1 negative zero. To find a lower bound, we again use trial and error ($x = -2, -4, -6$, and so on) until we get a bottom row of alternating signs (0 can be $+$ or $-$).

$$x = -2: \quad \underline{-2\rfloor} \quad 2 \qquad -7 \qquad 4 \qquad 1$$

$$\phantom{x = -2: \quad \underline{-2\rfloor} \quad 2 \qquad} -4 \qquad 22 \qquad -52$$

$$\phantom{x = -2: \quad \underline{-2\rfloor} \quad} 2 \qquad -11 \qquad 26 \quad \underline{\rfloor -51} \; \rightarrow \; -2 \text{ is a lower bound}$$

$$\text{(signs alternate)}$$

All real zeros

We now know that all real zeros are between -2 and 4. Although -2 and 4 are bounds for the real zeros, they are not the only bounds. For example, -3 and 6 are also bounds for the real zeros.

It is often convenient to perform several synthetic divisions using the same table. We write the top row (the original polynomial), and we write the bottom row (the quotient and remainder). We do the middle row mentally.

EXAMPLE 19   Consider $P(x) = 2x^3 - 9x^2 - 5x + 2$.
Using Descartes' rule of signs, we can see that there are 2 or 0 positive real zeros and 1 negative real zero. We now use a compact table for doing many synthetic divisions at once. (We do the middle steps mentally.)

	2	-9	-5	2	$\rightarrow$ original polynomial
$1\rfloor$	2	-7	-12	-10	$\rightarrow$ not a bound
$2\rfloor$	2	-5	-15	-28	$\rightarrow$ not a bound
$3\rfloor$	2	-3	-14	-40	$\rightarrow$ not a bound
$4\rfloor$	2	-1	-9	-34	$\rightarrow$ not a bound
$5\rfloor$	2	1	0	2	$\rightarrow$ *is an upper bound!*
$-1\rfloor$	2	-11	6	-4	$\rightarrow$ *is a lower bound!*

Thus, the real zeros are all between -1 and 5.

All real zeros

Divide the following polynomials using synthetic division.

Warm-up
Exercises

1. $(x^2 - 7x - 3) \div (x + 2)$ **2.** $(x^2 + 3x + 11) \div (x - 1)$

3. $(2x^3 - x^2 - x + 1) \div (x - 2)$ **4.** $(3x^3 - 2x^2 + x - 1) \div (x + 3)$

5. $(x^4 - 1) \div (x - 1)$ **6.** $(x^5 + 1) \div (x + 1)$

For each of the following polynomials, find $P(-x)$.

7. $P(x) = x^2 - 7x + 2$

8. $P(x) = 2x^3 - 7x^2 - 8x + 3$

9. $P(x) = x^4 + x^3 + x^2 + x + 1$

10. $P(x) = x^5 - x^4 - x^3 + x^2 + x - 1$

For each of the following polynomials:
(a) Determine the number of positive real zeros.
(b) Determine the number of negative real zeros.
(c) Find an upper bound for the real zeros.
(d) Find a lower bound for the real zeros.

11. $P(x) = x^2 - 11x - 3$

12. $P(x) = 2x^2 + 7x + 5$

13. $P(x) = x^2 + 17x - 1$

14. $P(x) = 5x^2 - x + 3$

15. $P(x) = x^3 - x^2 + x - 1$

16. $P(x) = x^3 + x^2 - x - 1$

17. $P(x) = x^3 - x + 1$

18. $P(x) = x^3 - 3x^2 + 7$

19. $P(x) = x^4 - x^2 + 8$

20. $P(x) = x^4 + x^2 - 6$

21. $P(x) = x^3 - x^2 + 7x + 8$

22. $P(x) = 4x^3 - x^2 - 2x - 2$

23. $P(x) = x^3 + 2x^2 + 3x + 4$

24. $P(x) = x^3 - 2x^2 - 3x + 4$

25. $P(x) = x^3 - 1$

26. $P(x) = x^4 - 1$

27. $P(x) = x^5 - x^3 - 1$

28. $P(x) = x^5 - x^3 + 1$

29. $P(x) = 6x^6 - 5x^5 - 4x^4 + 3x^3 + 2x^2 - x - 1$

30. $P(x) = x^7 + 2x^6 - 3x^5 + 4x^4 - 5x^3 - 6x^2 + 7x - 8$

15.5
RATIONAL ZEROS OF POLYNOMIALS

In solving quadratic equations, we observe that some equations have nice rational solutions and others involve irrational numbers (radicals). As examples, we have

$$2x^2 + 9x - 5 = 0 \qquad x^2 - 2x - 4 = 0$$

$$(\textit{Solution: } x = \tfrac{1}{2}, -5) \qquad (\textit{Solution: } x = 1 \pm \sqrt{5})$$

How can we tell what the **rational zeros** of a polynomial are (or might be)? The following theorem gives all the possible candidates for the rational zeros.

> **Theorem 10 (Rational-Zero Theorem)**
> Let $P(x) = a_n x^n + \cdots + a_1 x + a_0$ be a polynomial with integer coefficients.

> If a rational number p/q (reduced to lowest terms) is a zero of $P(x)$, we have:
>
> 1. The numerator p is a factor of a_0.
> 2. The denominator q is a factor of a_n.

This theorem produces all the rational zeros, but it may also produce a lot of numbers that are *not* zeros; therefore, we have to check each candidate, using synthetic division.

Recall that if c is a zero of $P(x)$ [that is, $P(c) = 0$], then dividing by $x - c$ produces a 0 remainder (the factor theorem). For example, since 1 is a zero of $P(x) = 2x^3 - 5x^2 + 2x + 1$, we get

$$
\begin{array}{r|rrrr}
1 & 2 & -5 & 2 & 1 \\
  &   & 2 & -3 & -1 \\
\hline
  & 2 & -3 & -1 & 0
\end{array}
$$

$\longrightarrow$ remainder

quotient
(reduced polynomial)

In searching for more zeros, we can now restrict our attention to the quotient, $2x^2 - 3x - 1$. We call this quotient the **reduced polynomial** for $P(x)$.

EXAMPLE 20   Find all the rational zeros of

$$P(x) = 2x^3 - 3x^2 - 2x + 3$$

Solution   We first use the rational-zero theorem to find all candidates p/q for the rational zeros and Descartes' rule of signs to count the number of zeros to expect.

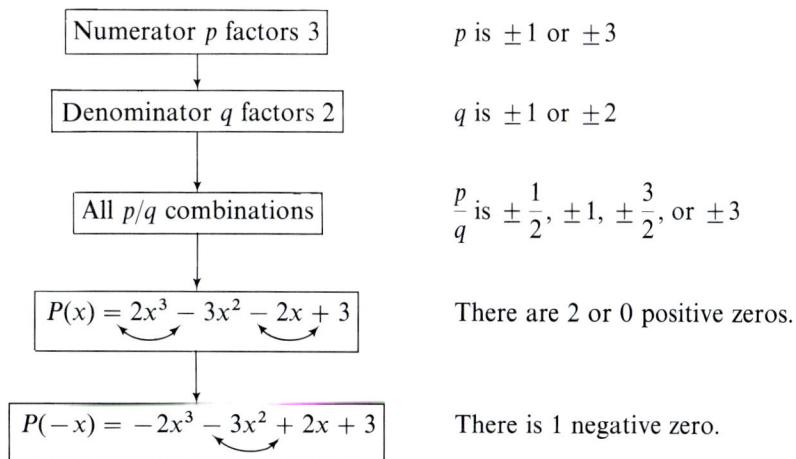

| Numerator p factors 3 | p is ± 1 or ± 3 |

| Denominator q factors 2 | q is ± 1 or ± 2 |

| All p/q combinations | $\dfrac{p}{q}$ is $\pm\dfrac{1}{2}$, ± 1, $\pm\dfrac{3}{2}$, or ± 3 |

| $P(x) = 2x^3 - 3x^2 - 2x + 3$ | There are 2 or 0 positive zeros. |

| $P(-x) = -2x^3 - 3x^2 + 2x + 3$ | There is 1 negative zero. |

Now we check the eight candidates for rational zeros using synthetic division, reducing the polynomial each time we find a zero. We use the compact notation (as in Example 19) for doing more than one synthetic division at one time. We begin with the positive candidates: 1/2, 1, 3/2, and 3; then the negative candidates: $-1/2$, -1, $-3/2$, and -3. We continue until we reach upper and lower bounds.

Positive zeros (expect 2 or 0)

$$\begin{array}{r} 2 \quad -3 \quad -2 \quad 3 \\ \frac{1}{2} \overline{\left) 2 \quad -2 \quad -3 \quad \frac{3}{2}\right.} \\ 1 \overline{\left) 2 \quad -1 \quad -3 \quad \boxed{0} \right.} \end{array}$$

$2 \quad -3 \quad -2 \quad 3 \quad \rightarrow$ original polynomial

$\frac{1}{2} \overline{\left) 2 \quad -2 \quad -3 \quad \frac{3}{2} \right.} \quad \rightarrow$ not a zero

$1 \overline{\left) 2 \quad -1 \quad -3 \quad \boxed{0} \right.} \quad \rightarrow$ 1 *is a zero!*

Look for another positive zero

$2 \quad -1 \quad -3 \quad \rightarrow$ reduced polynomial

$\frac{3}{2} \overline{\left) 2 \quad 2 \quad \boxed{0} \right.} \quad \rightarrow \frac{3}{2}$ *is a zero and upper bound!*

Look for the negative zero

$2 \quad 2 \quad \rightarrow$ further reduced polynomial

$\frac{-1}{2} \overline{\left) 2 \quad 1 \right.} \quad \rightarrow$ not a zero

$-1 \overline{\left) 2 \quad \boxed{0} \right.} \quad \rightarrow -1$ *is the last zero!*

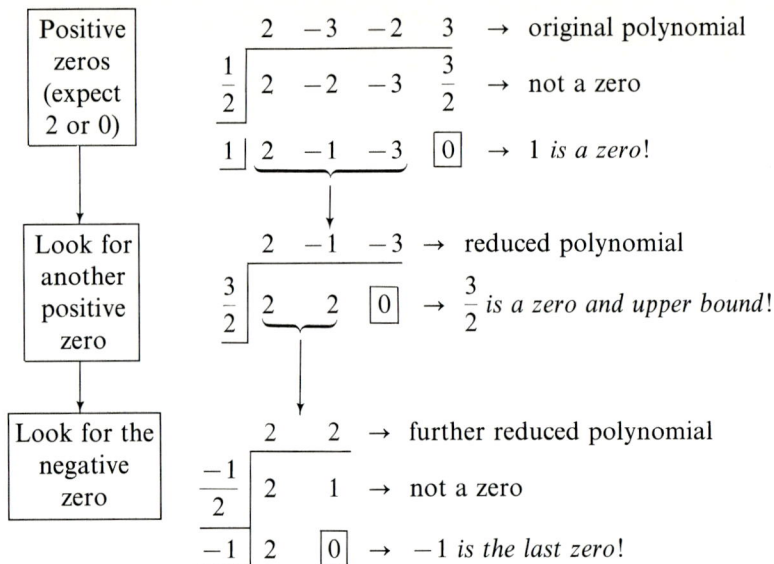

Since we expected two positive zeros and one negative zero, we know that we have them all: 1, 3/2, and −1. Notice how reducing the polynomial simplifies finding the next zero. Also, the original polynomial can now be written

$$P(x) = 2(x - 1)\left(x - \frac{3}{2}\right)(x + 1)$$

(The "2" remaining in the synthetic-division table provides the leading coefficient above.)

EXAMPLE 21   Find the rational zeros of

$$P(x) = 2x^5 - 3x^4 - 7x^3 + 7x^2 - 9x + 10$$

Solution   We first find all the rational-zero candidates p/q.

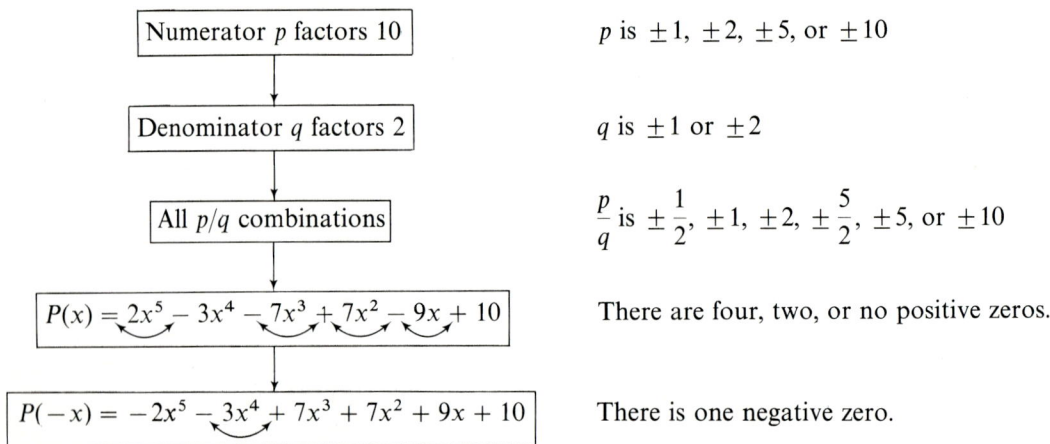

Numerator p factors 10

p is ± 1, ± 2, ± 5, or ± 10

Denominator q factors 2

q is ± 1 or ± 2

All p/q combinations

$\dfrac{p}{q}$ is $\pm\dfrac{1}{2}$, ± 1, ± 2, $\pm\dfrac{5}{2}$, ± 5, or ± 10

$P(x) = 2x^5 - 3x^4 - 7x^3 + 7x^2 - 9x + 10$

There are four, two, or no positive zeros.

$P(-x) = -2x^5 - 3x^4 + 7x^3 + 7x^2 + 9x + 10$

There is one negative zero.

Again, we use a compact synthetic division, looking for a remainder of 0. After finding a zero, we continue with the reduced polynomial (the quotient) until we find the upper and lower bounds.

$$\begin{array}{rrrrrr} 2 & -3 & -7 & 7 & -9 & 10 \end{array} \rightarrow \text{original polynomial}$$

Look for positive zero (expect 4, 2, or 0)	

$\dfrac{1}{2}$: $\begin{array}{rrrrrr} 2 & -2 & -8 & 3 & -\dfrac{15}{2} & \dfrac{25}{4} \end{array}$ → not a zero

1 : $\begin{array}{rrrrrr} 2 & -1 & -8 & -1 & -10 & \boxed{0} \end{array}$ → 1 *is a zero!*

Look for another positive zero	

$\begin{array}{rrrrr} 2 & -1 & -8 & -1 & -10 \end{array} \rightarrow \text{reduced polynomial}$

2 : $\begin{array}{rrrrr} 2 & 3 & -2 & -5 & -20 \end{array}$ → not a zero

$\dfrac{5}{2}$: $\begin{array}{rrrrr} 2 & 4 & 2 & 4 & \boxed{0} \end{array}$ → $\dfrac{5}{2}$ *is a zero and upper bound!* (no more positive zeros)

Look for the negative zero	

$\begin{array}{rrrr} 2 & 4 & 2 & 4 \end{array} \rightarrow \text{further reduced polynomial}$

$-\dfrac{1}{2}$: $\begin{array}{rrrr} 2 & 3 & \dfrac{1}{2} & \dfrac{15}{4} \end{array}$ → not a zero

-1 : $\begin{array}{rrrr} 2 & 2 & 0 & 4 \end{array}$ → not a zero

-2 : $\begin{array}{rrrr} 2 & 0 & 2 & \boxed{0} \end{array}$ → -2 *is the negative zero!*

The last quotient $2x^2 + 2 = 2(x^2 + 1)$ has *no* real zeros (the zeros are $\pm i$). The rational zeros are thus 1, 5/2, and -2. Also, we can use this to factor the original polynomial as follows:

$$P(x) = (x - 1)\left(x - \frac{5}{2}\right)(x + 2)(2x^2 + 2)$$

or

$$P(x) = 2(x - 1)\left(x - \frac{5}{2}\right)(x + 2)(x - i)(x + i)$$

PROBLEM SET
15.5

For each of the following polynomials with only real coefficients, write $P(x)$ as a product

$$a(x - c_1) \cdots (x - c_n)$$

Warm-up Exercises

1. $P(x)$ has zeros 1 and -3, and $P(0) = 2$.

2. $P(x)$ has zeros -1, 2, 2, and 2, and $P(0) = 4$.

3. $P(x)$ has zeros 2 and i, and $P(0) = 3$.

4. $P(x)$ has zeros $2i$ and $1 + i$, and $P(0) = 1$.

For each of the following polynomials:
(a) Determine the number of positive and negative real zeros.
(b) Find a lower and upper bound for the real zeros.

Warm-up
Exercises

5. $P(x) = x^2 - 7x - 1$

6. $P(x) = 2x^2 - 9x + 1$

7. $P(x) = x^3 - x^2 + x - 1$

8. $P(x) = x^4 - x^3 - x^2 + x - 1$

Find all rational zeros of the following polynomials.

9. $P(x) = x^4 + 5x^3 + 5x^2 - 5x - 6$

10. $P(x) = x^4 + 6x^3 + 7x^2 - 6x - 8$

11. $P(x) = 2x^3 - x^2 - 13x - 6$

12. $P(x) = 2x^3 + x^2 - 25x + 12$

13. $P(x) = 6x^3 - 7x^2 - 9x - 2$

14. $P(x) = 6x^3 - 5x^2 - 29x + 10$

15. $P(x) = 6x^3 + 35x^2 + 21x - 20$

16. $P(x) = 10x^3 - 23x^2 + 10x + 3$

17. $P(x) = x^4 - 2x^3 - 3x^2 + 4x + 4$

18. $P(x) = x^4 + 4x^3 - 2x^2 - 12x + 9$

For each of the following polynomials, find all zeros (rational, irrational, and complex):
(a) First find all rational zeros.
(b) Find the zeros of the remaining reduced polynomial.

19. $P(x) = x^3 - x^2 + x - 1$

20. $P(x) = x^3 + x^2 - 2x - 2$

21. $P(x) = x^3 - 2x^2 - 5x + 10$

22. $P(x) = x^3 + 3x^2 + 4x + 12$

23. $P(x) = x^3 - 4x^2 + 8$

24. $P(x) = x^3 - 7x^2 + 10x - 4$

25. $P(x) = x^4 - 1$

26. $P(x) = x^3 - 1$

27. $P(x) = x^4 + 15x + 14$

28. $P(x) = x^4 - 2x^2 + 16x - 15$

15.6
APPROXIMATING REAL ZEROS

When studying algebra, students usually encounter nice equations with nice solutions. Often, however, the real world is not as kind: The equations may have hard-to-find irrational solutions.

In this section we look briefly at finding or approximating the real zeros of polynomials with real coefficients. The following theorem helps us close in on these zeros.

Theorem 11 (*Location Theorem*)

Let $P(x)$ be a polynomial with real coefficients. If $P(a)$ and $P(b)$ have opposite signs (one + and the other −), then there is at least one zero for $P(x)$ between a and b.

Figure 2 illustrates this theorem. Since a polynomial function has a continuous, unbroken graph, it cannot go from + to − (or vice versa) without passing through zero. To use this theorem, we locate (by trial and error) points that produce opposite signs in $P(x)$. We then know a zero lies between these points.

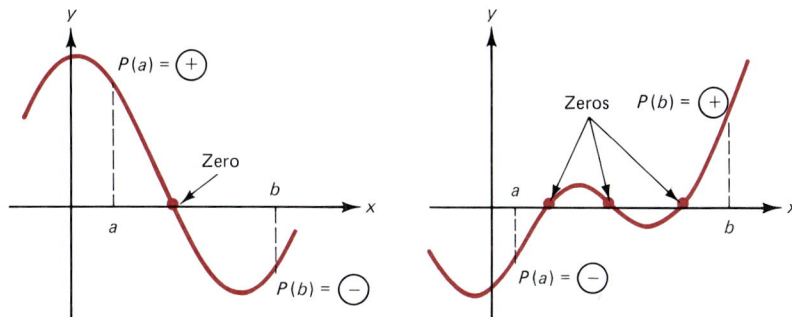

FIGURE 2

Recall from the remainder theorem (see page 510) that we can find $P(c)$ as the remainder of division by $x - c$. Thus, we can evaluate $P(x)$ for a number of x-values using the compact version of synthetic division.

EXAMPLE 22 Locate the zeros of $P(x) = x^3 + 3x^2 - 7x + 1$.

Solution We narrow the location of the zeros to intervals with integer endpoints, such as $[1, 2]$.

(a) *Positive zeros:* $P(x)$ has two variations in sign, so there are two or no positive zeros. We use synthetic division [using the remainder column as $P(c)$] looking for sign changes ($+$ to $-$, or $-$ to $+$) in $P(c)$.

$$
\begin{array}{c}
c \\
\downarrow
\end{array}
\qquad\qquad\qquad
\begin{array}{c}
P(c) \\
\downarrow
\end{array}
$$

	1	3	-7	1	
0	1	3	-7	1	zero between 0 and 1
1	1	4	-3	-2	zero between 1 and 2
2	1	5	3	7	

Positive zeros

Thus, we have located intervals for the two positive zeros: between 0 and 1 and between 1 and 2.

(b) *Negative zeros:* Since $P(-x) = -x^3 + 3x^2 + 7x + 1$ has one variation in sign, there is one negative zero. We now check the negative integers, looking for a sign change in $P(x)$.

$$
\begin{array}{c}
c \\
\downarrow
\end{array}
\qquad\qquad\qquad
\begin{array}{c}
P(c) \\
\downarrow
\end{array}
$$

	1	3	-7	1	
0	1	3	-7	1	
-1	1	2	-9	10	
-2	1	1	-9	19	
-3	1	0	-7	22	
-4	1	-1	-3	13	zero between -5 and -4
-5	1	-2	3	-14	

Negative zero

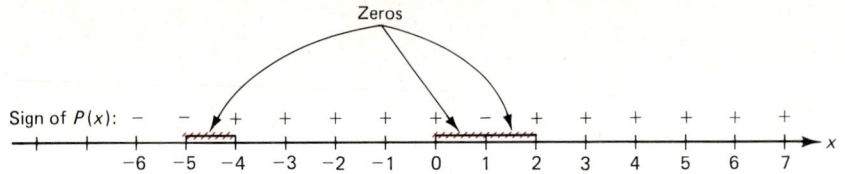

FIGURE 3

The negative zero is therefore between -5 and -4. Figure 3 summarizes the locations of these three zeros [always between $+$ and $-$ values of $P(x)$].

EXAMPLE 23   For $P(x) = x^3 + 3x^2 - 7x + 1$ (Example 22), find the zero that is between 1 and 2, to the nearest tenth.

Solution   We know that there is a zero z somewhere between 1 and 2, so we try a number (such as 1.5) between 1 and 2. Depending on the sign of $P(1.5)$, we make our next choice. We then continue narrowing down until our interval is about 0.05 in length. [We now use a calculator to help find $P(c)$.] Remember that our zero z is always between a $+$ and $-$ value of $P(c)$.

Find $P(c)$	Use sign of $P(c)$	Conclusion

Try 1.5	$P(1.5) = 0.625$		$1.0 < z < 1.5$
Try 1.3	$P(1.3) = -0.833$		$1.3 < z < 1.5$
Try 1.4	$P(1.4) = -0.176$		$1.4 < z < 1.5$
Try 1.45	$P(1.45) = 0.20612$		$1.4 < z < 1.45$

Each time, we pick a number between the ends of the interval. We then know that our zero is in the half of the interval that has $+$ and $-$ as endpoints [its values of $P(x)$]. Here we have narrowed the location of our zero to the interval $(1.4, 1.45)$. Thus, to the nearest tenth, we can say that our zero is 1.4.

For each of the following polynomials:

(a) Find the number of positive and negative real zeros.
(b) Find the lower and upper bounds for the real zeros.
(c) Find all the rational zeros.

1. $P(x) = 2x + 5$

2. $P(x) = 3x - 7$

3. $P(x) = x^2 - 5x + 4$

4. $P(x) = 6x^2 - 11x - 7$

5. $P(x) = x^3 - 2x + 1$

6. $P(x) = x^3 + x^2 - 7x - 10$

7. $P(x) = x^4 - 5x^2 + 4$

8. $P(x) = x^4 - 10x^3 + 35x^2 - 50x + 24$

Locate all the intervals in which the zeros of these polynomials lie.

9. $P(x) = x^3 - 4x^2 - x + 5$

10. $P(x) = x^3 - 3x^2 - 5x + 12$

11. $P(x) = x^3 - 5x^2 + x + 6$

12. $P(x) = x^3 - 3x^2 - x + 2$

13. $P(x) = x^3 - 3x^2 + 4x - 6$

14. $P(x) = x^3 - 2x^2 + 3x - 1$

15. $P(x) = x^4 - 5x^2 + 3$

16. $P(x) = x^5 - 6x^3 + 4$

Find the real zero (to the nearest tenth) in the indicated interval.

17. $P(x) = x^3 - 2x^2 + 5x - 3; [0, 1]$

18. $P(x) = x^3 - 4x^2 - 2x + 4; [4, 5]$

19. $P(x) = x^3 - 3x^2 - 5x + 12; [1, 2]$

20. $P(x) = x^3 - 2x^2 - 3x + 3; [3, 4]$

21. $P(x) = x^3 - 2x^2 - 3x + 3; [-2, -1]$

22. $P(x) = x^3 - 4x^2 + 5x + 1; [-1, 0]$

For each of the following polynomials, find all real zeros to the nearest tenth.

23. $P(x) = x^3 - 7$

24. $P(x) = x^4 - 11$

25. $P(x) = x^3 + x + 1$

26. $P(x) = x^3 - x^2 - 1$

27. $P(x) = x^3 - 2x^2 + 1$

28. $P(x) = x^4 - 2x^3 + 3x^2 - 4x + 5$

For each of the following polynomials, find all zeros (rational, irrational, or complex) and give their multiplicity (if greater than one). (Approximate the real zeros to the nearest tenth)

29. $P(x) = x^3 - 1$

30. $P(x) = x^3 + 8$

31. $P(x) = x^4 - 2x^3 + 5x^2 - 16x + 12$

32. $P(x) = x^4 - x^3 - 5x^2 + 4x + 4$

33. $P(x) = x^3 + x^2 + x + 1$

34. $P(x) = x^3 - x^2 + x - 1$

35. $P(x) = x^3 - 6x^2 + 11x - 5$

36. $P(x) = x^3 + 2x^2 - 5x - 5$

Important Theorems

Theorem 1 (Division Algorithm). For a polynomial $P(x)$ and divisor $x - c$, there is a quotient $Q(x)$ and remainder R such that

$$P(x) = (x - c)Q(x) + R$$

Theorem 2 (Remainder Theorem). For a polynomial $P(x) = (x - c)Q(x) + R$, we have $P(c) = R$.

Theorem 3 (Factor Theorem). The polynomial $x - c$ factors $P(x)$ if and only if $P(c) = 0$.

Theorem 4 (Fundamental Theorem of Algebra). Every polynomial of degree $n \geq 1$ has at least one zero (real or complex).

Theorem 5. Every polynomial of degree $n \geq 1$ can be factored as

$$P(x) = a(x - c_1)(x - c_2) \cdots (x - c_n)$$

where $c_1, c_2, \ldots, c_n$ are the zeros of $P(x)$.

Theorem 6. Every polynomial of degree $n \geq 1$ has at most n distinct zeros.

Theorem 7 (Conjugate Pair Theorem). Let $P(x)$ be a polynomial with real coefficients. If $z = a + bi$ is a zero of $P(x)$, then $\bar{z} = a - bi$ is also a zero of $P(x)$.

Theorem 8 (Descartes' Rule of Signs). Let $P(x)$ be a polynomial with real coefficients. Then:
  (a) The number of positive real zeros is equal to the number of variations in sign in $P(x)$, or an even number less.
  (b) The number of negative real zeros is equal to the number of variations in sign in $P(-x)$, or an even number less.

Theorem 9 (Bound Theorem). If we divide $P(x)$ by $x - r$ synthetically and consider the bottom row:
  (a) If $r \geq 0$ and all the numbers are the same sign (or 0), then r is an *upper bound*; that is, there are no zeros greater than r.
  (b) If $r \leq 0$ and the numbers alternate in sign (0 can be $+$ or $-$, as needed), then r is a *lower bound* for the zeros; that is, there are no zeros lower than r.

Theorem 10 (Rational-Zero Theorem). Let $P(x) = a_n x^n + \cdots + a_1 x + a_0$ be a polynomial with *integer* coefficients. If a reduced rational number p/q is a zero of $P(x)$:
  (a) The numerator p is a factor of a_0.
  (b) The denominator q is a factor of a_n.

Theorem 11 (Location Theorem). Let $P(x)$ be a polynomial with real coefficients. If $P(a)$ and $P(b)$ have opposite signs, then there is at least one zero for $P(x)$ between a and b.

Review Exercises

Perform the following divisions, showing quotient and remainder. Use synthetic division in Problems 4 to 6.

1. $1507 \div 14$

2. $(2x^3 + 9x^2 + 4x - 13) \div (2x + 5)$

3. $(x^8 - 1) \div (x^4 - 1)$

4. $(2x^3 - 3x^2 - 13x + 17) \div (x - 3)$

5. $(4x^4 + 7x^3 + x^2 - x + 1) \div (x + 1)$

6. $(x^4 + 81) \div (x + 3)$

Use the remainder theorem to evaluate $P(x) = x^3 - 4x^2 + 6x - 1$ *for the given values.*

7. Find $P(2)$.

8. Find $P(-1)$.

For each of the following pairs of polynomials, determine if the second is a factor of the first.

9. $x^4 - 7x^2 + 10x - 8$; $x - 2$

10. $x^{10} + 1$; $x + 1$

Using the given information, write the following polynomials in the form $P(x) = (x - c_1)(x - c_2) \cdots (x - c_n)$.

11. $P(x)$ has zeros 4, i, $-i$.

12. $P(x) = x^2 + x + 3$

For each of the following polynomials, give its zeros and their multiplicities.

13. $P(x) = x^6 - x^4$

14. $P(x) = (x - 1)(x + 2)^3(x - 3)^5$

Using the given information, find the polynomial with real coefficients of lowest degree.

15. With zeros 2 and -5, and $P(0) = 30$.

16. With zero $1 + i$, and $P(1) = 3$.

For each of the following polynomials:
(a) Find the number of positive and negative real zeros.
(b) Find upper and lower bounds for the real zeros.

17. $P(x) = 5x^2 - 2x + 3$

18. $P(x) = 2x^3 - 3x^2 - 5x - 9$

Find all rational zeros of the following polynomials. Then find all other zeros of the reduced polynomial.

19. $P(x) = 2x^4 - 7x^3 - 18x^2 + 13x + 10$

20. $P(x) = x^4 + 3x^3 + 4x^2 + 7x - 15$

For each of the following polynomials, find the largest real zero to the nearest tenth.

21. $P(x) = x^3 - 5x^2 + x + 6$

22. $P(x) = x^3 - 11$

CHAPTER 16

Sequences and Series

16.1
SEQUENCES

A **sequence** is a function, such as $s(n) = 1/n$, in which the domain for n is the natural numbers, 1, 2, 3, 4, and so on. For example, the sequence $s(n) = 1/n$ appears as

$$\frac{1}{1}, \frac{1}{2}, \frac{1}{3}, \frac{1}{4}, \ldots$$

A sequence is like a string of numbers, usually defined by a rule or formula. The numbers in the sequence are called the **terms**. We usually write the terms in a sequence as

$$a_1, a_2, a_3, a_4, \ldots, a_n, \ldots$$

which we read "a sub 1," "a sub 2," and so on. The term a_1 is called the first term, a_2 is called the second term, and so on. The term a_n is called the nth term, or the **general term**.

EXAMPLE 1   Write the first five terms of the sequence $a_n = n^2 + 1$.

Solution   We substitute into the formula $a_n = n^2 + 1$ with $n = 1, 2, 3, 4,$ and 5.

Substitute $n = 1, 2, 3, 4, 5$ into formula

$$a_1 = 1^2 + 1 = 2$$
$$a_2 = 2^2 + 1 = 5$$
$$a_3 = 3^2 + 1 = 10$$
$$a_4 = 4^2 + 1 = 17$$
$$a_5 = 5^2 + 1 = 26$$

Thus, the sequence starts 2, 5, 10, 17, 26, ...

EXAMPLE 2   Write the first five terms of the sequence

$$a_n = \frac{(-1)^n}{2^n}$$

Solution   Again we substitute $n = 1, 2, 3, 4$, and 5 into the formula. We must be careful with the term $(-1)^n$. Note how this alternates between -1 (when n is odd) and $+1$ (when n is even).

$$a_1 = \frac{(-1)^1}{2^1} = \frac{-1}{2} \qquad a_2 = \frac{(-1)^2}{2^2} = \frac{1}{4}$$

$$a_3 = \frac{(-1)^3}{2^3} = \frac{-1}{8} \qquad a_4 = \frac{(-1)^4}{2^4} = \frac{1}{16}$$

$$a_5 = \frac{(-1)^5}{2^5} = \frac{-1}{32}$$

Thus, the sequence starts $\dfrac{-1}{2}, \dfrac{1}{4}, \dfrac{-1}{8}, \dfrac{1}{16}, \dfrac{-1}{32}, \ldots$

EXAMPLE 3   Find the general term for the sequence

$$\frac{1}{3}, \frac{1}{6}, \frac{1}{9}, \frac{1}{12}, \frac{1}{15}, \ldots$$

Solution   This is the reverse of Examples 1 and 2. Here we have the sequence, and we want the formula for a_n. We must look for a pattern in the sequence.

We see that each numerator is 1, and each denominator is a multiple of 3: $3 \cdot 1, \ 3 \cdot 2, \ 3 \cdot 3, \ 3 \cdot 4, \ldots, \ 3 \cdot n$. Thus, the general term is given by

$$a_n = \frac{1}{3n}$$

EXAMPLE 4   Find the general term for the sequence

$$\frac{3}{1}, \frac{9}{4}, \frac{27}{9}, \frac{81}{16}, \frac{243}{25}, \ldots$$

Solution   As in Example 3, we look for the pattern. We note that the numerators are all powers of 3: $3^1, 3^2, 3^3, 3^4, \ldots, 3^n$. The denominators are all perfect squares: $1^2, 2^2, 3^2, 4^2, \ldots, n^2$. Thus, the general term is

$$a_n = \frac{3^n}{n^2}$$

PROBLEM SET 16.1

Evaluate the following.

Warm-up Exercises

1. $(-3)^2$

2. $10 - 3(2)$

3. $\dfrac{14 + 6}{14 - 4}$

4. $\dfrac{1}{2^5}$

5. $(-1)^2$

6. $(-1)^3$

Write the first five terms of the following sequences.

7. $a_n = 3n - 1$
8. $a_n = 10n + 3$
9. $a_n = 101 - n$

10. $a_n = (-1)^n$
11. $a_n = (-1)^n n^3$
12. $a_n = 3^n$

13. $a_n = \dfrac{1}{2^n}$
14. $a_n = \dfrac{1}{n^2}$
15. $a_n = \dfrac{n}{n+1}$

16. $a_n = \dfrac{n+1}{n^2}$
17. $a_n = \dfrac{n+1}{n+5}$
18. $a_n = \dfrac{(-1)^n}{n^3}$

19. $a_n = \dfrac{(-1)^n}{n+1}$
20. $a_n = x^n$
21. $a_n = \dfrac{1}{x^n}$

22. $a_n = \left(1 + \dfrac{1}{n}\right)^n$

Find the general term a_n for the following sequences.

23. $1, 8, 27, 64, 125, \ldots$
24. $1, 4, 9, 16, 25, \ldots$

25. $2, 4, 8, 16, 32, \ldots$
26. $5, 25, 125, 625, \ldots$

27. $-3, 6, -9, 12, -15, \ldots$
28. $-10, 100, -1000, 10000, \ldots$

29. $\dfrac{1}{1}, \dfrac{1}{2}, \dfrac{1}{3}, \dfrac{1}{4}, \dfrac{1}{5}, \ldots$
30. $\dfrac{1}{100}, \dfrac{1}{200}, \dfrac{1}{300}, \dfrac{1}{400}, \ldots$

31. $\dfrac{1}{2}, \dfrac{2}{3}, \dfrac{3}{4}, \dfrac{4}{5}, \ldots$
32. $\dfrac{1}{1}, \dfrac{2}{8}, \dfrac{3}{27}, \dfrac{4}{81}, \dfrac{5}{243}, \ldots$

33. $\dfrac{-1}{1}, \dfrac{1}{4}, \dfrac{-1}{9}, \dfrac{1}{16}, \dfrac{-1}{25}, \ldots$
34. $\dfrac{-1}{3}, \dfrac{1}{9}, \dfrac{-1}{27}, \dfrac{1}{81}, \dfrac{-1}{243}, \ldots$

For each of the following sequences:
(a) If the first four terms are given, find the next term and a_n.
(b) If a_n is given, find the first four terms.

35. $a_n = \sqrt{n+1} - \sqrt{n}$
36. $a_n = 2^{n+1} - 2^n$

37. $2, 6, 18, 54, \ldots$
38. $100, 10, 1, \dfrac{1}{10}, \ldots$

39. $a_n = \log n$
40. $a_n = \log_n 10 \quad (n \geq 2)$

41. $10, 6, 2, -2, \ldots$
42. $20, 200, 2000, 20000, \ldots$

43. $a_n = i^n \; (i = \sqrt{-1})$
44. $a_n = 2^n i^n \; (i = \sqrt{-1})$

45. $\dfrac{-2}{10}, \dfrac{4}{100}, \dfrac{-6}{1000}, \dfrac{8}{10,000}, \ldots$
46. $\dfrac{-1}{100}, \dfrac{4}{1000}, \dfrac{-9}{10,000}, \dfrac{16}{100,000}, \ldots$

Population Application

47. One version of *Pareto's law* is that the population a_n of the nth largest city is given by

$$a_n = \frac{P}{n}$$

where $P = 7,000,000$, the population of New York (the largest city). Assuming this is true, find the populations of the first four largest cities in the United States.

48. In calculus, the following sequence is very important:

$$a_n = \left(1 + \frac{1}{n}\right)^n$$

(a) Find $a_1, a_2, a_5, a_{10}, a_{100}, a_{1000}$, and $a_{10,000}$.
(b) What number does this sequence seem to approach?

49. A very famous sequence is the *Fibonacci sequence*:

$$a_1 = 1 \qquad a_2 = 1 \qquad a_n = a_{n-1} + a_{n-2}$$

For example, $a_3 = a_2 + a_1 = 1 + 1 = 2$; $a_4 = a_3 + a_2 = 2 + 1 = 3$; and so on. Find a_5, a_6, a_7, a_8, and a_9.

50. If a machine depreciates at a constant rate d each year, the amount D_n of depreciation in year n is given by

$$D_n = dV(1 - d)^{n-1}$$

where V is the original value. Find the first 4 years (terms) for the following situations:
(a) $V = \$100,000, d = 10\%$ (b) $\$500,000, d = 20\%$

16.2
SERIES

Consider the sequence $1, 4, 9, 16, 25, 36, \ldots$. From this sequence we can form a new sequence based on the sums of the terms:

$$S_1 = 1$$
$$S_2 = 1 + 4 \, (= 5)$$
$$S_3 = 1 + 4 + 9 \, (= 14)$$
$$S_4 = 1 + 4 + 9 + 16 \, (= 30)$$
$$S_5 = 1 + 4 + 9 + 16 + 25 \, (= 55)$$

and so on.

Such a sequence of sums is called a **series**.

We have a special notation for dealing with sums and series. It is called **sigma notation** and uses the Greek letter $\sum$ (**sigma**). An example of this notation is

$$\overset{\text{finish}}{\underset{\text{start}}{\sum}} \quad \overset{\text{general term}}{}$$
$$\sum_{n=1}^{6} a_n = a_1 + a_2 + a_3 + a_4 + a_5 + a_6$$

The $\sum$ stands for sum; the $n = 1$ below $\sum$ means start the sum at $n = 1$; the 6 above $\sum$ means add the terms of the sequence until $n = 6$; finally, the letter n is called the **index of summation** (although any letter will do).

EXAMPLE 5   The following table shows how $\sum$ notation works.

$\sum$ Notation	Expanded Notation
$\sum\limits_{n=1}^{5} n^2$	$1^2 + 2^2 + 3^2 + 4^2 + 5^2 \quad (=55)$
$\sum\limits_{n=1}^{4} \dfrac{1}{2^n}$	$\dfrac{1}{2} + \dfrac{1}{4} + \dfrac{1}{8} + \dfrac{1}{16} \quad \left(=\dfrac{15}{16}\right)$
$\sum\limits_{n=2}^{5} \dfrac{x^n}{n^3}$	$\dfrac{x^2}{8} + \dfrac{x^3}{27} + \dfrac{x^4}{64} + \dfrac{x^5}{125}$
$\sum\limits_{n=1}^{5} \dfrac{(-1)^n}{n}$	$\dfrac{-1}{1} + \dfrac{1}{2} - \dfrac{1}{3} + \dfrac{1}{4} - \dfrac{1}{5} \quad \left(=\dfrac{-47}{60}\right)$
$\sum\limits_{n=3}^{7} 10n$	$30 + 40 + 50 + 60 + 70 \quad (=250)$

EXAMPLE 6   Write the series

$$\frac{1}{10} + \frac{8}{100} + \frac{27}{1000} + \frac{64}{10,000} + \frac{125}{100,000}$$

in $\sum$ notation.

Solution   We first discover a pattern for the general term. Notice that the numerators are perfect cubes, n^3, and the denominators are powers of ten, 10^n. Let us now rewrite the series

$$\frac{1^3}{10^1} + \frac{2^3}{10^2} + \frac{3^3}{10^3} + \frac{4^3}{10^4} + \frac{5^3}{10^5}$$

The general term is $n^3/10^n$, while n runs from 1 to 5. Thus, the series can be written

$$\sum_{n=1}^{5} \frac{n^3}{10^n}$$

EXAMPLE 7   In statistics, the *mean m* and *variance σ^2* of the data x_1, $x_2, \ldots, x_{50}$ are given by

$$m = \frac{1}{50}[x_1 + x_2 + \cdots + x_{50}]$$

$$\sigma^2 = \frac{1}{50}[(x_1 - m)^2 + (x_2 - m)^2 + \cdots + (x_{50} - m)^2]$$

Write these in $\sum$ notation.

Solution For the mean m, the general term is x_n; for the variance σ^2, the general term is $(x_n - m)^2$. In both cases the index runs from 1 to 50. Thus, we get

$$m = \frac{1}{50} \sum_{n=1}^{50} x_n$$

$$\sigma^2 = \frac{1}{50} \sum_{n=1}^{50} (x_n - m)^2$$

In high-level computer languages, sequences are stored in *arrays* (or matrices), and series are added using *loops*. A variable, such as SUM, is set to zero, and then each element of the array (sequence) is added to the old SUM to form the new SUM. The following compares $\sum$ notation, BASIC, and Pascal.

$\sum$ notation	BASIC	Pascal
$\sum_{n=1}^{100} a_n$	SUM = 0 FOR N = 1 TO 100 SUM = SUM + A(N) NEXT N	SUM := 0; FOR N := 1 TO 100 DO SUM := SUM + A[N];

PROBLEM SET 16.2

Write the first five terms of the following sequences.

Warm-up Exercises

1. $a_n = n^3$　　　　**2.** $a_n = 10n - 4$　　　　**3.** $a_n = (-1)^n$

4. $a_n = \dfrac{1}{3^n}$　　　**5.** $a_n = \dfrac{n-1}{n+1}$　　　**6.** $a_n = \dfrac{(-1)^{n-1}}{n^2}$

Find the general term a_n for the following sequences.

Warm-up Exercises

7. 3, 6, 9, 12, 15, ...　　　　　　**8.** 1, 5, 25, 125, 625, ...

9. $-2, 4, -8, 16, -32, ...$　　　**10.** $\dfrac{-2}{3}, \dfrac{4}{9}, \dfrac{-6}{27}, \dfrac{8}{81}, \dfrac{-10}{243}, ...$

Expand the following series and simplify.

11. $\sum_{n=1}^{4} n^3$　　　　**12.** $\sum_{n=2}^{5} (n^2 + 1)$　　　　**13.** $\sum_{n=3}^{6} (n + 5)$

14. $\sum_{n=1}^{5} \dfrac{1}{n^3}$　　　**15.** $\sum_{n=1}^{4} \dfrac{1}{n}$　　　**16.** $\sum_{n=1}^{3} \dfrac{1}{3^n}$

17. $\sum_{n=1}^{6} (-1)^n n$　　**18.** $\sum_{n=2}^{5} \dfrac{(-1)^n}{n+1}$　　**19.** $\sum_{n=1}^{5} \dfrac{n+1}{n+2}$

20. $\sum_{n=1}^{4} x^n$　　　**21.** $\sum_{n=1}^{5} n^x$　　　**22.** $\sum_{n=3}^{7} \dfrac{x^n}{2^n}$

23. $\sum_{n=1}^{7} x_n^2$　　　**24.** $\sum_{n=1}^{5} \dfrac{(-1)^n}{x^n}$

Write the following series using $\sum$ notation.

25. $9 + 16 + 25 + 36 + 49$　　　　**26.** $8 + 27 + 64 + 125$

27. $\dfrac{1}{4} + \dfrac{1}{8} + \dfrac{1}{16} + \dfrac{1}{32}$　　　**28.** $\dfrac{1}{10} + \dfrac{1}{100} + \dfrac{1}{1000} + \dfrac{1}{10,000}$

29. $-3 + 4 - 5 + 6 - 7$

30. $-3 + 9 - 27 + 81$

31. $\dfrac{1}{2} + \dfrac{2}{3} + \dfrac{3}{4} + \dfrac{4}{5}$

32. $\dfrac{1}{16} + \dfrac{1}{25} + \dfrac{1}{36} + \dfrac{1}{49} + \dfrac{1}{64}$

33. $x^3 + x^4 + x^5 + x^6 + x^7$

34. $\dfrac{x^2}{5} + \dfrac{x^3}{6} + \dfrac{x^4}{7} + \dfrac{x^5}{8}$

35. $a_1^3 + a_2^3 + a_3^3 + a_4^3$

36. $(x_5 - 10)^4 + (x_6 - 10)^4 + (x_7 - 10)^4$

For the following sequences, find the first five terms of the sequence of partial sums:

$$S_1 = \sum_{n=1}^{1} a_n, \ S_2 = \sum_{n=1}^{2} a_n, \ S_3 = \sum_{n=1}^{3} a_n, \text{ and so on}$$

(If possible, give the general term for the series S_n.)

37. $1, 2, 3, 4, 5, \ldots$

38. $2, 5, 8, 11, 14, \ldots$

39. $2, 4, 8, 16, 32, \ldots$

40. $3, 9, 27, 81, 243, \ldots$

41. $-1, 1, -1, 1, -1, \ldots$

42. $1, 0, 1, 0, 1, \ldots$

43. $\dfrac{1}{2}, \dfrac{1}{4}, \dfrac{1}{8}, \dfrac{1}{16}, \dfrac{1}{32}, \ldots$

44. $\dfrac{1}{3}, \dfrac{1}{9}, \dfrac{1}{27}, \dfrac{1}{81}, \dfrac{1}{243}, \ldots$

Calculus Application

45. An approximate series for the natural logarithm is given by

$$\ln(1 + x) \approx x - \frac{x^2}{2} + \frac{x^3}{3} - \frac{x^4}{4} + \cdots - \frac{x^{50}}{50}$$

for $-1 < x < 1$. Write this in $\sum$ notation.

Business Application

46. In studying consumer attitudes toward a product, marketers use the formula

$$A = \sum_{i=1}^{n} b_i a_i$$

where A is the attitude toward the product, b_i the ith belief about the product, and a_i an evaluation of the ith belief. If there are $n = 5$ beliefs, expand this formula.

Information Theory Application

47. The *information I* contained in a message using five symbols that have frequencies $p_1, \ldots, p_5$ is given by

$$I = \sum_{n=1}^{5} p_n \log_2 \frac{1}{p_n}$$

Find I if $p_1 = 1/2$, $p_2 = 1/4$, $p_3 = 1/8$, and $p_4 = p_5 = 1/16$.

Computer Application

48. As seen on page 538, series are added in high-level computer languages using loops. Complete the following table.

$\sum$ Notation	BASIC	Pascal
?	SUM $= 0$ FOR K $= 10$ TO 60 SUM $=$ SUM $+$ B(K) NEXT K	?
$\displaystyle\sum_{j=3}^{15} a_j$	?	?

The sequence 20, 25, 30, 35, 40, 45, 50, ... is an example of a special kind of sequence called an arithmetic sequence. An **arithmetic sequence** (or **progression**) is a sequence for which the *difference* between any term (after the first) and the one before it is a constant. This difference is called the **common difference**, *d*. In the sequence above, the common difference is 5.

EXAMPLE 8   The following are arithmetic sequences. We find the common difference by subtracting any term minus the previous term.

Arithmetic Sequence	Common Difference	
(a) 2, 5, 8, 11, 14, ...	3	(Each term is 3 *more* than the one before.)
(b) 5, 15, 25, 35, 45, ...	10	(Each term is 10 *more* than the one before.)
(c) 10, 8, 6, 4, 2, ...	-2	(Each term is 2 *less* than the one before.)
(d) 4, -3, -10, -17, -24, ...	-7	(Each term is 7 *less* than the one before.)

With only the first term a_1 and the common difference d, we can find the general term a_n.

> Each term is d more than the previous term

$$1 \to a_1$$
$$2 \to a_2 = a_1 + d$$
$$3 \to a_3 = a_2 + d = a_1 + 2d$$
$$4 \to a_4 = a_3 + d = a_1 + 3d$$
$$\vdots \qquad \vdots$$
$$n \to a_n = a_1 + (n-1)d$$

Property 1

The general term a_n of an arithmetic sequence with first term a_1 and common difference d is given by

$$a_n = a_1 + (n-1)d$$

EXAMPLE 9   Find a_{20} and a_n for the following sequence:

$$4, 7, 10, 13, 16, \ldots$$

Solution   The first term a_1 is clearly 4. The common difference d is 3. We now use Property 1.

$$\boxed{a_n = a_1 + (n-1)d}$$

> Simplify

$$a_{20} = 4 + (20 - 1)3$$
$$= 4 + 19(3)$$
$$= 61$$

To find the general term a_n, we substitute $a_1 = 4$ and $d = 3$ and simplify.

| Substitute and simplify |

$$a_n = 4 + (n - 1)3$$
$$= 4 + 3n - 3$$
$$= 1 + 3n$$

EXAMPLE 10 Find a_{25} and a_n for the following sequence:

$$7, -2, -11, -20, -29, \ldots$$

Solution The first term a_1 is 7. The common difference is -9. (*Be careful:* Decreasing sequences always have a negative common difference.)

| $a_n = a_1 + (n - 1)d$ |

| Simplify |

$$a_{25} = 7 + (25 - 1)(-9)$$
$$= 7 + 24(-9)$$
$$= -209$$

| Substitute and simplify |

$$a_n = 7 + (n - 1)(-9)$$
$$= 7 - 9n + 9$$
$$= 16 - 9n$$

Now we wish to find the *sum* of an arithmetic sequence. We call this an **arithmetic series**.

Consider the task of adding the numbers from 1 to 100. Folklore has it that the mathematician Karl Gauss used the following procedure when he was 7 years old. (Some say that he was 10, others say that the numbers were 1 to 1000, but this is not important. The procedure is.)

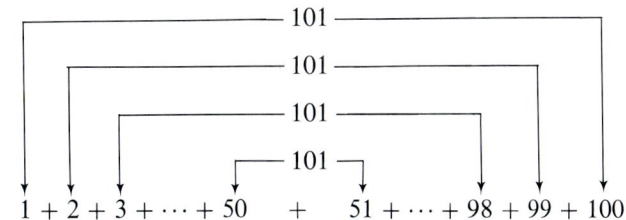

$$
\begin{array}{c}
101 \\
101 \\
101 \\
101 \\
1 + 2 + 3 + \cdots + 50 \quad + \quad 51 + \cdots + 98 + 99 + 100
\end{array}
$$

Instead of adding $1 + 2 + 3 + \ldots$, young Gauss added $1 + 100 = 101$, $2 + 99 = 101$, $3 + 98 = 101, \ldots, 50 + 51 = 101$. He had 50 such pairs each adding to 101. In total, this makes $(50)(101) = 5050$.

This suggests the rule of *adding the first and last terms and then multiplying by half the number of terms.*

Property 2
The sum S_n of an arithmetic series with first term a_1 and last term a_n is given by

$$S_n = \frac{n}{2}(a_1 + a_n)$$

EXAMPLE 11 Find the sum of the arithmetic series

$$2 + 8 + 14 + 20 + 26 + 32 + 38 + 44$$

Solution We have eight terms; also, $a_1 = 2$ and $a_8 = 44$. We use Property 2.

$$\boxed{S_n = \frac{n}{2}(a_1 + a_n)}$$

$$S_8 = \frac{8}{2}(2 + 44)$$

$$\boxed{\text{Simplify}}$$

$$= 4(46) = 184$$

EXAMPLE 12 Find the sum of the first 30 terms of the series

$$10 + 7 + 4 + 1 + (-2) + \cdots$$

Solution Here $n = 30$ and $a_1 = 10$, but we have to do a little work to get a_{30}. We use Property 1 with the common difference $d = -3$.

$$\boxed{a_n = a_1 + (n - 1)d}$$

$$a_{30} = 10 + (30 - 1)(-3)$$
$$= 10 + (29)(-3) = -77$$

We now use Property 2 to find the sum, S_{30}.

$$\boxed{S_n = \frac{n}{2}(a_1 + a_n)}$$

$$S_{30} = \frac{30}{2}[10 + (-77)]$$

$$= 15(-67) = -1005$$

PROBLEM SET 16.3

For the following sequences, find the next term and the general term a_n.

Warm-up Exercises

1. 1, 4, 9, 16, ...

2. 3, 9, 27, 81, ...

3. $-2, 4, -8, 16, \ldots$

4. $\dfrac{1}{1}, \dfrac{-1}{2}, \dfrac{1}{3}, \dfrac{-1}{4}, \ldots$

5. 3, 5, 7, 9, ...

6. 1, 8, 15, 22, ...

7. 10, 8, 6, 4, ...

8. 20, 15, 10, 5, ...

Write the following sums in $\sum$ notation.

Warm-up Exercises

9. $1 + 8 + 27 + 64$

10. $9 + 16 + 25 + 36 + 49$

11. $\dfrac{1}{2} + \dfrac{1}{4} + \dfrac{1}{8} + \dfrac{1}{16}$

12. $\dfrac{1}{6} - \dfrac{1}{7} + \dfrac{1}{8} - \dfrac{1}{9} + \dfrac{1}{10}$

13. $5 + 8 + 11 + 14$

14. $20 + 25 + 30 + 35$

For each of the following arithmetic sequences:
(a) Give the next term.
(b) Give the common difference.

15. 1, 9, 17, 25, 33, ...

16. 3, 14, 25, 36, 47, ...

17. 10, 6, 2, -2, -6, ...

18. 19, 14, 9, 4, -1, ...

19. $75, 25, -25, -75, \ldots$

20. $100, 87, 74, 61, \ldots$

21. $5.1, 5.3, 5.5, 5.7, \ldots$

22. $10, 9\frac{2}{3}, 9\frac{1}{3}, 9, \ldots$

For each of the following arithmetic sequences:
(a) Write the first four terms.
(b) Find the general term a_n.

23. $a_1 = 9; d = -4$

24. $a_1 = 40; d = -11$

25. $a_1 = -6; d = 4$

26. $a_1 = -17; d = 6$

27. $a_1 = 5; d = 0.6$

28. $a_1 = 8; d = -3/4$

Find the indicated terms for the arithmetic sequences in Problems 15 to 22.

29. Find a_{19} in Problem 15.

30. Find a_{30} in Problem 16.

31. Find a_{40} in Problem 17.

32. Find a_{35} in Problem 18.

33. Find a_{50} in Problem 19.

34. Find a_{60} in Problem 20.

35. Find a_{100} in Problem 21.

36. Find a_{200} in Problem 22.

Find the indicated sums for the arithmetic series in Problems 15 to 22. (Use Problems 29 to 36 for the last term.)

37. Find S_{19} in Problem 15.

38. Find S_{30} in Problem 16.

39. Find S_{40} in Problem 17.

40. Find S_{35} in Problem 18.

41. Find S_{50} in Problem 19.

42. Find S_{60} in Problem 20.

43. Find S_{100} in Problem 21.

44. Find S_{200} in Problem 22.

For the indicated arithmetic series:
(a) Write the series in $\sum$ notation.
(b) Find the sum.

45. The sum of the integers from 1 to 1000.

46. The sum of the integers from 1 to 1,000,000.

47. The first ten terms of $2 + 6 + 10 + 14 + \cdots$.

48. The first 20 terms of $5 + 10 + 15 + 20 + \cdots$.

49. The first 50 terms of $10 + 30 + 50 + 70 + \cdots$.

50. The first 100 terms of $1000 + 998 + 996 + 994 + \cdots$.

51. The series $2 + 5 + 8 + \cdots + 47$.

52. The series $5 + 15 + 25 + \cdots + 165$.

53. The series $200 + 197 + 194 + \cdots + 110$.

54. The series $100 + 98 + 96 + \cdots + (-100)$.

Business Applications

55. To guarantee increased insurance coverage, a person pays increasing premiums every year: $100, $105, $110, and so on.
(a) Write a formula for the nth year's premium.
(b) What is the premium for year 20?
(c) What is the sum of the premiums for the first 20 years?

56. A company buys a machine and pays it off as follows: $10,000 the first year, $9500 the second, $9000 the third, and so on.
(a) Write a formula for the payment in the nth year.
(b) What is the payment in year 10?
(c) What is the total payment after 10 years?

Computer Application

57. In using a *bubble sort* to alphabetize a list of n names, the computer makes $n - 1$ comparisons of adjacent names on the first pass through the list. On the second pass, $n - 2$ comparisons are made; on the third pass, $n - 3$ comparisons; and so on down to 1 comparison on the last pass (there are $n - 1$ passes). How many total comparisons does this program require?

Health Application

58. A patient is given medicine in doses that increase in an arithmetic sequence: 20 milligrams, 25 milligrams, 30 milligrams, and so on.
(a) Write a formula for the nth dosage.
(b) What is the twelfth dosage?
(c) After 12 doses, how much total medicine has been taken?

Consumer Applications

59. Sue Sanchez starts a job at $18,000 per year with an annual raise of $800.
(a) Write the annual pay for the first 4 years.
(b) How much will Sue be earning in year 10?
(c) What will be Sue's total earnings for the first 10 years?

60. Mr. and Mrs. Ulanowski buy a $500 TV set and agree to pay for it as follows: $100 the first month, $95 the second month, and $5 less every month for 12 payments.
(a) What is the twelfth payment?
(b) What is the total payment made over the 12 months?

16.4 GEOMETRIC SEQUENCES AND SERIES

The sequence 1, 2, 4, 8, 16, 32, 64,... is an example of a special kind of sequence called a geometric sequence. A **geometric sequence** (or **progression**) is a sequence for which the ratio of any term (after the first) to the term before it is a constant. We call this ratio the **common ratio**, *r*. In the sequence above, the common ratio *r* is 2.

EXAMPLE 13 The following are examples of geometric sequences. We find the common ratio by dividing any term by the term before it.

Geometric Sequence	Common Ratio
(a) 10, 20, 40, 80, 160,...	2 (Each term is 2 times previous term.)
(b) 2, 20, 200, 2000,...	10 (Each term is 10 times previous term.)
(c) 30, 10, $\frac{10}{3}$, $\frac{10}{9}$, $\frac{10}{27}$,...	$\frac{1}{3}$ $\left(\text{Each term is } \frac{1}{3} \text{ times previous term.}\right)$
(d) 1000, -200, 40, -8,...	$\frac{-1}{5}$ $\left(\text{Each term is } \frac{-1}{5} \text{ times previous term.}\right)$

As with arithmetic sequences, we can find the general term for a geometric sequence with only the first term a_1 and the common ratio r.

| Each term is r times the previous term |

$$1 \rightarrow a_1$$
$$2 \rightarrow a_2 = a_1 r$$
$$3 \rightarrow a_3 = (a_1 r)r = a_1 r^2$$
$$4 \rightarrow a_4 = (a_1 r^2)r = a_1 r^3$$
$$\vdots \quad \vdots$$
$$n \rightarrow a_n = a_1 r^{n-1}$$

Property 3

The general term a_n of a geometric sequence with first term a_1 and common ratio r is given by

$$a_n = a_1 r^{n-1}$$

EXAMPLE 14 Find a_{10} and a_n for the following sequence:

$$20, 40, 80, 160, \ldots$$

Solution The first term a_1 is 20. The common ratio is 2. We now use Property 3.

$$a_n = a_1 r^{n-1}$$

$$a_{10} = 20(2)^9$$
$$= 20(512) = 10{,}240$$

The general term a_n is given by

$$a_n = a_1 r^{n-1}$$

$$a_n = 20(2)^{n-1}$$

EXAMPLE 15 Cassandra buys a $7000 car that depreciates 15% per year. (That is, each year it is only worth 85% of last year's value.) How much is the car worth in year 8?

Solution This is a geometric sequence with common ratio 0.85 and first term 7000:

$$\text{Year 1:} \quad a_1 = 7000$$
$$\text{Year 2:} \quad a_2 = 7000(0.85) = 5950$$
$$\text{Year 3:} \quad a_3 = 7000(0.85)^2 = 5057.50$$
$$\vdots \qquad \vdots$$
$$\text{Year 8:} \quad a_8 = 7000(0.85)^7 = 2244.04$$

Thus, the car is worth about $2244 in year 8. (We used a hand calculator to compute the exponent.)

If your calculator has a constant key $\boxed{\text{K}}$, you can produce this sequence as follows.

PRESS	DISPLAY	MEANING
$\boxed{\cdot}\ \boxed{8}\ \boxed{5}\ \boxed{\times}\ \boxed{\text{K}}$	0.85	Enter 0.85 as a constant multiplier (common ratio)
$\boxed{7}\ \boxed{0}\ \boxed{0}\ \boxed{0}$	7000.	First term (Year 1)
$\boxed{=}$	5950.	Year 2
$\boxed{=}$	5057.5	Year 3
$\boxed{=}$	4298.875	Year 4
$\vdots$	$\vdots$	$\vdots$
$\boxed{=}$	2244.0394	Year 8

We use $\boxed{\times}$ and $\boxed{K}$ to enter the common ratio; then we enter the first term; finally, pressing $\boxed{=}$ repeatedly generates the terms of the geometric sequence.

As with arithmetic sequences, we want to find the sum of the first n terms of a geometric sequence. We call this a **geometric series**. Let S_n be the sum of the first n terms of a geometric series.

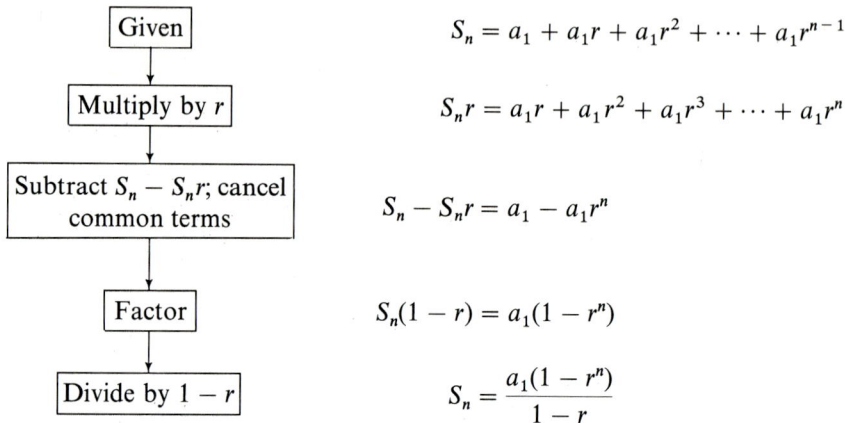

$$\boxed{\text{Given}} \qquad\qquad S_n = a_1 + a_1 r + a_1 r^2 + \cdots + a_1 r^{n-1}$$

$$\boxed{\text{Multiply by } r} \qquad\qquad S_n r = a_1 r + a_1 r^2 + a_1 r^3 + \cdots + a_1 r^n$$

$$\boxed{\begin{array}{c}\text{Subtract } S_n - S_n r; \text{ cancel} \\ \text{common terms}\end{array}} \qquad S_n - S_n r = a_1 - a_1 r^n$$

$$\boxed{\text{Factor}} \qquad\qquad S_n(1 - r) = a_1(1 - r^n)$$

$$\boxed{\text{Divide by } 1 - r} \qquad\qquad S_n = \frac{a_1(1 - r^n)}{1 - r}$$

Property 4

The sum S_n of the first n terms of a geometric series with first term a_1 and common ratio r is given by

$$\boxed{S_n = \frac{a_1(1 - r^n)}{1 - r}}$$

EXAMPLE 16 Find the sum of the first 10 terms of the geometric sequence

$$3 + 6 + 12 + 24 + 48 + \cdots$$

Solution The first term a_1 is 3. The common ratio r is 2. Finally, $n = 10$. We now use Property 4.

$$\boxed{S_n = \frac{a_1(1 - r^n)}{1 - r}} \qquad\qquad S_{10} = \frac{3(1 - 2^{10})}{1 - 2}$$

$$= \frac{3(1 - 1024)}{-1}$$

$$\boxed{\text{Simplify}} \qquad\qquad = \frac{3(-1023)}{-1} = 3069$$

Thus, the sum is 3069.

EXAMPLE 17 Mr. Poulikakos deposits $2000 every year in an IRA (Individual Retirement Account) for 15 years at 12% interest. The total accumulation after 15 years is given by

$$2000 + 2000(1.12) + 2000(1.12)^2 + \cdots + 2000(1.12)^{14}$$

Find the total.

Solution This is a geometric series with $a_1 = 2000$, $r = 1.12$, and $n = 15$. [The common ratio $r = 1.12$ comes from the fact that every year's investment is multiplied by 1 (itself) and 0.12 (the interest).] Using Property 4 and a hand calculator, we get

$$\boxed{S_n = \frac{a_1(1 - r^n)}{1 - r}}$$

$$S_{15} = \frac{2000(1 - 1.12^{15})}{1 - 1.12}$$

$$\downarrow$$

$$\boxed{\text{Simplify}}$$

$$= \frac{2000(1 - 5.474)}{-0.12}$$

$$= \frac{2000(-4.474)}{-0.12} = 74,559.5$$

Thus, saving $2000 per year for 15 years at 12% interest accumulates to about $74,560.

PROBLEM SET 16.4

For the following arithmetic sequences:
(a) Give the next term.
(b) Give the common difference.
(c) Give a_{10}.
(d) Give the sum of the first 10 terms.

Warm-up
Exercises

1. 1, 2, 3, 4, $\cdots$
2. 1, 3, 5, 7, $\ldots$
3. 4, 7, 10, 13, $\ldots$
4. 40, 50, 60, 70, $\ldots$
5. 2, 2.5, 3, 3.5, $\ldots$
6. 8, $8\frac{1}{4}$, $8\frac{1}{2}$, $8\frac{3}{4}$, $\ldots$
7. 10, 6, 2, -2, $\ldots$
8. 100, 85, 70, 55, $\ldots$

For each of the following geometric sequences:
(a) Give the next term.   *(b) Find the common ratio.*

9. 2, 12, 72, 432, $\ldots$
10. 4, 60, 900, 13500, $\ldots$
11. 400, 200, 100, 50, $\ldots$
12. 100, 20, 4, 0.8, $\ldots$
13. 8, 7.2, 6.48, 5.832, $\ldots$
14. 81, 54, 36, 24, $\ldots$
15. $\dfrac{1}{2}, \dfrac{1}{4}, \dfrac{1}{8}, \dfrac{-1}{16}, \ldots$
16. 5, -10, 20, -40, $\ldots$

For each of the following geometric sequences, write the first four terms and find the eighth term a_8.

17. $a_1 = 18; r = \dfrac{1}{3}$
18. $a_1 = 100; r = \dfrac{1}{5}$
19. $a_1 = 7; r = -10$
20. $a_1 = 3; r = -2$
21. $a_1 = \dfrac{1}{5}; r = \dfrac{1}{2}$
22. $a_1 = \dfrac{2}{9}; r = \dfrac{-3}{4}$

Find the indicated terms for the geometric sequences in Problems 9 to 16.

23. Find a_7 in Problem 9. **24.** Find a_6 in Problem 10.

25. Find a_9 in Problem 11. **26.** Find a_8 in Problem 12.

27. Find a_7 in Problem 13. **28.** Find a_8 in Problem 14.

29. Find a_9 in Problem 15. **30.** Find a_{10} in Problem 16.

Find the indicated sums for the geometric series in Problem 9 to 16.

31. Find S_7 in Problem 9. **32.** Find S_6 in Problem 10.

33. Find S_9 in Problem 11. **34.** Find S_8 in Problem 12.

35. Find S_7 in Problem 13. **36.** Find S_8 in Problem 14.

37. Find S_9 in Problem 15. **38.** Find S_{10} in Problem 16.

39. Show that S_n can also be given by

$$S_n = \frac{a_1 - a_{n+1}}{1 - r}$$

40. Consider the geometric series

$$\frac{1}{2} + \frac{1}{4} + \frac{1}{8} + \frac{1}{16} + \cdots$$

(a) Find S_5. (b) Find S_{10}.

(c) Find S_{20}. (d) Find S_{25}.

(e) What do you think happens to S_n as n gets very large?

For the indicated geometric series:
(a) Write the series in $\sum$ notation.
(b) Find the sum.

41. The first six terms of $2, 4, 8, 16, \ldots$.

42. The first seven terms of $3, 30, 300, \ldots$.

43. The first five terms of $\dfrac{1}{3}, \dfrac{1}{9}, \dfrac{1}{27}, \ldots$.

44. The first eight terms of $10{,}000, 5000, 2500, \ldots$.

45. $1 + 3 + 9 + \cdots + 729$

46. $50 + 500 + 5000 + \cdots + 5{,}000{,}000$

47. $\dfrac{1}{2} + \dfrac{1}{4} + \dfrac{1}{8} + \cdots + \dfrac{1}{128}$

48. $\dfrac{1}{10} - \dfrac{1}{100} + \dfrac{1}{1000} - \cdots - \dfrac{1}{1{,}000{,}000}$

Life Science Application
49. The eighteenth-century scientist Malthus claimed that populations grow geometrically (in a geometric sequence). The annual populations of a certain animal are $50{,}000$, $55{,}000$, $60{,}500$, $66{,}550$, and so on.

(a) What is the common ratio?

(b) Find a_8.

Musical Application
50. The frets on the neck of a guitar form a geometric sequence. If the length from the bridge to the neck is L, then the frets are at lengths Lr, Lr^2, Lr^3, and so on.

(a) Find a formula for a_{12}, the length at the twelfth fret.

(b) If the length at fret 12 is half the original length ($a_{12} = L/2$), solve for r.

51. If P dollars are invested every year for n years at i rate of interest, then after n years it will accumulate to

$$P + P(1 + i) + P(1 + i)^2 + \cdots + P(1 + i)^{n-1}$$

(a) What is the common ratio of this sequence?
(b) What is the sum as a formula?
(c) Find the sum if $P = \$1000$, $n = 20$, and $i = 9\%$ ($=0.09$).
(d) Find the sum if $P = \$500$, $n = 30$, and $i = 10\%$ ($=0.10$).

52. Rebecca opens an IRA, depositing $\$1500$ every year for 30 years at 11% interest. What is the total accumulation after the 30 years? (See Problem 51.)

53. Lisa opens a savings account, depositing $\$50$ every month for 20 years at 12% interest. (See Problem 51.)

(a) Compute the total accumulation as if this were a $\$600$ annual deposit.
(b) Compute the total accumulation of $\$50$ per month for 240 months at 1% interest per month. (This is monthly compounding.)

16.5
INFINITE GEOMETRIC SERIES

What happens to a geometric series with an infinite number of terms (that is, no last term)? Can it be summed? That depends on the common ratio, r. Let us look at the different cases.

Case I: If $|r| < 1$, the terms of the sequence approach zero. For example, if $r = 1/5$ and $a_1 = 20$, then

$$20, 4, 0.8, 0.16, 0.032, 0.0064, \ldots \to 0$$

The terms get closer and closer to zero. This series *can* be summed, as we will see.

Case II: If $|r| > 1$, the terms of the sequence get larger and larger without bound. For example, if $r = 10$ and $a_1 = 4$, then we get

$$4, 40, 400, 4000, 40000, \ldots$$

These terms get larger and larger, and the sum grows without bound. This series *cannot* be summed.

Case III: If $r = 1$, the sequence is constant. For example, if $a_1 = 6$, then we get

$$6, 6, 6, 6, 6, \ldots$$

The terms are always 6 and *cannot* be summed, since the sum grows without bound.

Case IV: If $r = -1$, the sequence oscillates. For example, if $a_1 = 5$, then we get

$$5, -5, 5, -5, 5, -5, \ldots$$

The terms jump back and forth between 5 and -5. This series *cannot* be summed, since the sum jumps back and forth between 0 and 5.

Case I is the only one for which we can add all the terms of an infinite geometric series.

EXAMPLE 18 Consider the series

$$1000 + 500 + 250 + 125 + 62.5 + \cdots$$

Here $a_1 = 1000$ and $r = 1/2$ (or 0.5). Let us compute some of the sums and try to discover a pattern. (We use Property 4 and a hand calculator for the larger n-values.)

$$S_1 = 1000$$

$$S_2 = 1500$$

$$S_3 = 1750$$

$$S_4 = 1875$$

$$S_5 = 1937.5$$

$$S_{10} = \frac{1000(1 - 0.5^{10})}{1 - 0.5} = 1998.0468$$

$$S_{20} = \frac{1000(1 - 0.5^{20})}{1 - 0.5} = 1999.9980927$$

$$S_{50} = \frac{1000(1 - 0.5^{50})}{1 - 0.5} = 1999.999999999998224$$

Clearly, the sums are getting very, very close to 2000 (but they never quite reach 2000). This is because the term 0.5^n is getting smaller and smaller.

Recall that the sum of the first n terms of a geometric series is given by

$$S_n = \frac{a_1(1 - r^n)}{1 - r}$$

In Case I, where $|r| < 1$, the term r^n gets smaller and smaller as n gets very large (as happened in Example 18). We call S_∞ the **sum of an infinite geometric series**. In Example 18 we would say that $S_\infty = 2000$, since this the number the sums are approaching.

We find S_∞ as follows, using Property 4.

r^n approaches 0	$S_\infty = \dfrac{a_1(1 - 0)}{1 - r}$
↓	
Simplify	$= \dfrac{a_1}{1 - r}$

Property 5

The sum S_∞ of an infinite geometric series with first term a_1 and common ratio r ($|r| < 1$) is given by

$$\boxed{S_\infty = \frac{a_1}{1 - r}}$$

EXAMPLE 19 Let us reconsider the series in Example 18 and use Property 5. We use $a_1 = 1000$ and $r = 0.5$.

$$S_\infty = \frac{a_1}{1 - r}$$

$$S_\infty = \frac{1000}{1 - 0.5}$$

$$= \frac{1000}{0.5} = 2000$$

The sum is 2000, as we concluded earlier.

EXAMPLE 20 Find the sum of the infinite geometric series

$$9 - 3 + 1 - \frac{1}{3} + \frac{1}{9} - \frac{1}{27} + \cdots$$

Solution The common ratio is $-1/3$; since $|r| < 1$, we can use Property 5 with $a_1 = 9$.

$$S_\infty = \frac{a_1}{1 - r}$$

$$S_\infty = \frac{9}{1 - (-1/3)}$$

$$= \frac{9}{4/3} = \frac{27}{4}$$

EXAMPLE 21 Find a fraction that is equal to the repeating decimal $0.717171\ldots$.

Solution We first rewrite this decimal as an infinite geometric series.

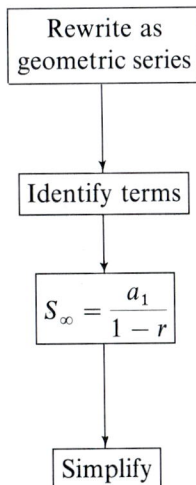

| Rewrite as geometric series | $0.717171\ldots = 0.71 + 0.0071 + 0.000071 + \cdots$ |

$$= 0.71 + 0.71\left(\frac{1}{100}\right) + 0.71\left(\frac{1}{10,000}\right) + \cdots$$

| Identify terms | $a_1 = 0.71 = \frac{71}{100}; \qquad r = \frac{1}{100}$ |

| $S_\infty = \frac{a_1}{1 - r}$ | $S_\infty = \dfrac{\dfrac{71}{100}}{1 - \dfrac{1}{100}}$ |

| Simplify | $= \dfrac{\left(\dfrac{71}{100}\right)100}{\left(1 - \dfrac{1}{100}\right)100} = \dfrac{71}{99}$ |

We can check this answer by converting 71/99 to a decimal in the usual manner: Divide 71 by 99 to get $0.7171\ldots$.

EXAMPLE 22   Find the sum of the infinite series

$$10, 50, 250, 1250, \ldots$$

Solution   Since the common ratio $r = 5$, which is greater than 1, we cannot use Property 5. There is *no* sum.

PROBLEM SET
16.5   *For the following sequences:*
(a) *State whether it is an arithmetic sequence, a geometric sequence, or neither. Give the common difference (or ratio).*
(b) *Give the next term and the general term.*
(c) *Give the sum of the first eight terms.*

Warm-up
Exercises

1. $1, 2, 4, 8, \ldots$

2. $2, 6, 10, 14, \ldots$

3. $21, 17, 13, 9, \ldots$

4. $10, 30, 90, 270, \ldots$

5. $10, -10, 10, -10, \ldots$

6. $1, 4, 9, 16, \ldots$

7. $1, \dfrac{1}{3}, \dfrac{1}{9}, \dfrac{1}{27}, \ldots$

8. $5, 25, 45, 65, \ldots$

9. $1, 8, 27, 64, \ldots$

10. $1600, -400, 100, -25, \ldots$

11. $8, -4, 2, -1, \ldots$

12. $100, 80, 60, 40, \ldots$

For each of the following geometric series:
(a) *Find* S_3.   (b) *Find* S_5.   (c) *Find* S_{10}.   (d) *Find* S_∞.

13. $1000 + 600 + 360 + 216 + \cdots$

14. $1000 + 800 + 640 + 512 + \cdots$

15. $2 + 1 + \dfrac{1}{2} + \dfrac{1}{4} + \cdots$

16. $81 + 54 + 36 + 24 + \cdots$

Find the sums of the following infinite geometric series. (Be careful: Some may not have sums — see Example 22.)

17. $\dfrac{1}{2} + \dfrac{1}{4} + \dfrac{1}{8} + \dfrac{1}{16} + \cdots$

18. $1 + \dfrac{1}{3} + \dfrac{1}{9} + \dfrac{1}{27} + \cdots$

19. $\dfrac{1}{2} - \dfrac{1}{4} + \dfrac{1}{8} - \dfrac{1}{16} + \cdots$

20. $1 - \dfrac{1}{3} + \dfrac{1}{9} - \dfrac{1}{27} + \cdots$

21. $4 + \dfrac{1}{2} + \dfrac{1}{16} + \dfrac{1}{128} + \cdots$

22. $2 + 3 + \dfrac{9}{2} + \dfrac{27}{4} + \cdots$

23. $100 + 40 + 16 + \cdots$

24. $1 + \dfrac{1}{10} + \dfrac{1}{100} + \dfrac{1}{1000} + \cdots$

25. $5 + 10 + 20 + 40 + \cdots$

26. $\dfrac{1}{4} - \dfrac{1}{5} + \dfrac{4}{25} - \dfrac{16}{125} + \cdots$

Write each of the following repeating decimals as a fraction.

27. $0.55555\ldots$

28. $0.2222\ldots$

29. $0.070707\ldots$

30. $0.858585\ldots$

31. $0.235235\ldots$

32. $0.194194\ldots$

For each of the following infinite geometric series:
(a) Write it in $\sum$ notation (using ∞ for the upper limit of the sum).
(b) Find the sum.

33. $\dfrac{1}{4} + \dfrac{1}{16} + \dfrac{1}{64} + \cdots$

34. $\dfrac{1}{6} + \dfrac{1}{36} + \dfrac{1}{216} + \cdots$

35. $1600 + 800 + 400 + \cdots$

36. $8100 + 2700 + 900 + \cdots$

37. $1000 - 500 + 250 - \cdots$

38. $2400 - 600 + 150 - \cdots$

*Business
Application*

39. With a *perpetual annuity* a fixed amount (for instance, $1000) is sent to a person at the end of every year *forever*. If the interest rate is r, the amount needed to fund this is given by

$$\frac{1000}{1+r} + \frac{1000}{(1+r)^2} + \frac{1000}{(1+r)^3} + \cdots$$

(a) What is the first term of this series?
(b) What is the common ratio? (Be careful.)
(c) Find the sum of this infinite geometric series.

*Health
Application*

40. A patient's first dosage of a certain medicine is 100 milligrams. In an effort to taper off her intake, each subsequent dose is one-half the previous dose.
(a) Write the sequence of the first five doses.
(b) How much total medicine will she be given if this procedure is continued forever?

*Physical
Application*

41. A ball is dropped from a height of 4 meters. Each time it rebounds to 0.8 of its previous height (see the sketch).
(a) Write the downward trips as the infinite geometric series $4 + 3.2 + 2.56 + 2.048 + \cdots$ and find the sum for total downward motion.
(b) Write the upward trips as the infinite geometric series $3.2 + 2.56 + 2.048 + \cdots$ and find the sum for total upward motion.
(c) Add the downward and upward total to find the total distance traveled.

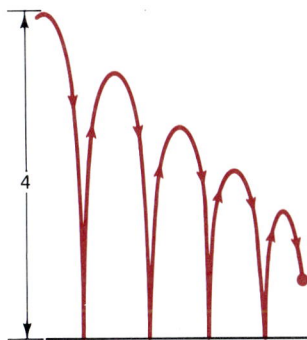

*Economics
Application*

42. The government is considering a $400-per-family tax rebate. The government feels that each family will spend 75% of this amount ($=$300$). The people receiving this $300 will spend 75% of it ($=$225$), and so on. The resulting increase to the economy is

$$400 + 300 + 225 + 168.75 + \cdots$$

This is called the *multiplier effect* and is an infinite geometric series. Find its sum, which is the total effect the original $400 has on the economy.

For an arithmetic sequence or series:
common difference: $d = a_{n+1} - a_n$
general term: $a_n = a_1 + (n-1)d$
sum of first n terms: $S_n = \dfrac{n}{2}(a_1 + a_n)$

For a geometric sequence or series:
common ratio: $r = a_{n+1}/a_n$
general term: $a_n = a_1 r^{n-1}$
sum of first n terms: $S_n = \dfrac{a_1(1 - r^n)}{1 - r}$
sum of infinite series: $S_\infty = \dfrac{a_1}{1 - r}$ (only if $|r| < 1$)

Review Exercises

Write the first five terms of the following sequences.

1. $a_n = 20 - 3n$

2. $a_n = \dfrac{(-1)^n}{n+1}$

Find the general term a_n for each of the following sequences.

3. $2, 6, 18, 54, \ldots$

4. $\dfrac{-x}{10}, \dfrac{x^2}{100}, \dfrac{-x^3}{1000}, \dfrac{x^4}{10{,}000}$

Expand each of the following series, and simplify.

5. $\displaystyle\sum_{n=1}^{6} \dfrac{1}{n}$

6. $\displaystyle\sum_{n=1}^{4} \dfrac{(-1)^n}{n^2}$

Write the following in $\sum$ notation.

7. $1 + 4 + 9 + 16 + 25$

8. $\dfrac{3}{x} + \dfrac{6}{x^2} + \dfrac{9}{x^3} + \dfrac{12}{x^4} + \dfrac{15}{x^5}$

For the arithmetic sequence $50, 46, 42, 38, \ldots$:

9. Find the common difference d.   **10.** Find a_{10}.
11. Find S_{10}.   **12.** Find the general term a_n.

For the geometric sequence $5, 10, 20, 40, \ldots$:

13. Find the common ratio r.   **14.** Find a_{10}.
15. Find S_{10}.   **16.** Find the general term a_n.

Find the sum of each of the following infinite geometric series.

17. $1000 + 300 + 90 + 27 + \cdots$

18. $1 + \dfrac{2}{5} + \dfrac{4}{25} + \dfrac{8}{125} + \cdots$

19. Write $0.292929\ldots$ as a fraction.

CHAPTER 17

Further Topics in Algebra

17.1
PERMUTATIONS

Often we want to know in how many ways we can do a certain task (or make an arrangement) when there are several sets of options. Consider the following example.

EXAMPLE 1   The Spee-Dee Food Restaurant has four sandwiches and three drinks. How many sandwich/drink possibilities are there?

Sandwich	Drink
Ham	Soda
Beef	Coffee
Chicken	Shake
Taco	

Solution   We can see all the possibilities in the following table.

Drink	Sandwich			
	Ham	Beef	Chicken	Taco
Soda	Ham, Soda	Beef, Soda	Chicken, Soda	Taco, Soda
Coffee	Ham, Coffee	Beef, Coffee	Chicken, Coffee	Taco, Coffee
Shake	Ham, Shake	Beef, Shake	Chicken, Shake	Taco, Shake

Altogether, four sandwiches and three drinks produce 12 total possibilities. The result, 12, is of course the product of 3 and 4. These 12 possibilities can also be pictured in a **tree diagram** (as in Figure 1 on page 556).

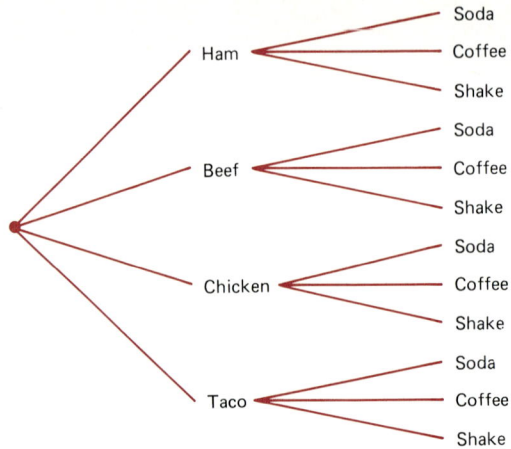

Soda
Ham
Coffee
Shake

Soda
Beef
Coffee
Shake

Soda
Chicken
Coffee
Shake

Soda
Taco
Coffee
Shake

FIGURE 1

Although Example 1 is very simple, it shows how we count possibilities when we have different sets of options: *We multiply the numbers of options together.*

Property 1 (Counting Principle)
Suppose that we have a sequence of tasks to do (or choices to make): Task A can be done in a ways, task B can be done in b ways, task C can be done in c ways, and so on. Then the total number of ways to do the sequence A, B, C, ... is given by

$$a \times b \times c \cdots$$

EXAMPLE 2   Suppose that the Spee-Dee Foods Restaurant expands its offering, so that now its menu looks as follows:

Sandwich	Drink	Side Order	Dessert
Hamburger	Cola	Fries	Ice cream
Hotdog	Tea	Onion rings	Cookies
Ham	Coffee		Pie
Beef	Milk		
Chicken	Shake		
Taco	Lemonade		
Ribs			

How many possible ways are there to order a sandwich, a drink, a side order, and a dessert?

Solution We simply multiply all the numbers of options together.

Counting Principle	$\underset{\text{Sandwich}}{7}$	×	$\underset{\text{Drink}}{6}$	×	$\underset{\text{Side order}}{2}$	×	$\underset{\text{Dessert}}{3}$	=	$\underset{\text{Total}}{252}$

Thus, there are 252 total ways to order one of each type.

EXAMPLE 3 In a certain state there are three types of license plates available using digits (0 to 9) and letters (A to Z):

> Type (a) Six digits
> Type (b) Two letters, four digits
> Type (c) Three letters, three digits

How many different possibilities are available?

Solution We consider each case separately. In each slot we fill in the number of possible characters (26 for letters, 10 for digits). (For the sake of counting, we consider a license such as HZ 731 to be the same as HZ0731: that is, we consider blanks as zeros.)

Type (a) $\underset{\text{Digit}}{10}$ × $\underset{\text{Digit}}{10}$ × $\underset{\text{Digit}}{10}$ × $\underset{\text{Digit}}{10}$ × $\underset{\text{Digit}}{10}$ × $\underset{\text{Digit}}{10}$ = 1,000,000

Type (b) $\underset{\text{Letter}}{26}$ × $\underset{\text{Letter}}{26}$ × $\underset{\text{Digit}}{10}$ × $\underset{\text{Digit}}{10}$ × $\underset{\text{Digit}}{10}$ × $\underset{\text{Digit}}{10}$ = 6,760,000

Type (c) $\underset{\text{Letter}}{26}$ × $\underset{\text{Letter}}{26}$ × $\underset{\text{Letter}}{26}$ × $\underset{\text{Digit}}{10}$ × $\underset{\text{Digit}}{10}$ × $\underset{\text{Digit}}{10}$ = 17,576,000

$$\text{Total} = \overline{25,336,000}$$

EXAMPLE 4 Fifteen children participate in a contest. In how many ways can a first-, second-, and third-place prize be given?

Solution There are 15 children who can win first place, 14 for second place, and 13 for third place.

$$\underset{\text{First}}{15} \times \underset{\text{Second}}{14} \times \underset{\text{Third}}{13} = \underset{\text{Total}}{2730}$$

Thus, there are 2730 ways to distribute the prizes.

Example 4 shows us a special situation in which we choose members from a set so that the *order is important* and we *cannot choose a member twice*. We call such an arrangement a **permutation**.

We use the symbol **$P(n, r)$** to be the **permutation of n things taken r at a time** (that is, r things out of a group of n). Since there are n ways to choose the first item, $n - 1$ ways to choose the second, and so on, we get the following property.

Property 2

The permutation (or arrangement with order) of n things taken r at a time (or, r things out of a set of n) is given by

$$P(n, r) = n(n - 1)(n - 2) \cdots (n - r + 1)$$

We have a special notation that we use with counting problems. It is called factorial notation.

Definition

We define **n factorial** as

$$n! = n(n - 1)(n - 2) \cdots 3 \cdot 2 \cdot 1 \qquad \text{(for } n \geq 1)$$

$$0! = 1$$

EXAMPLE 5 The following use the permutation and factorial notations.
  (a) $P(8, 3) = 8 \cdot 7 \cdot 6 = 336$ (8 things, 3 at a time)
  (b) $P(50, 4) = 50 \cdot 49 \cdot 48 \cdot 47 = 5{,}527{,}200$ (50 things, 4 at a time)
  (c) $P(100, 2) = 100 \cdot 99 = 9900$ (100 things, 2 at a time)
  (d) $0! = 1$ (by special definition)
  (e) $1! = 1$
  (f) $2! = 2 \cdot 1 = 2$
  (g) $3! = 3 \cdot 2 \cdot 1 = 6$
  (h) $4! = 4 \cdot 3 \cdot 2 \cdot 1 = 24$
  (i) $5! = 5 \cdot 4 \cdot 3 \cdot 2 \cdot 1 = 120$

EXAMPLE 6 A committee has 4 members. In how many ways can a chairperson, a treasurer, and a secretary be chosen?

Solution This is $P(4, 3)$, or the permutation of 3 things from a set of 4. We compute this $4 \cdot 3 \cdot 2 = 24$. If we call the members a, b, c, and d, we can write out all the possibilities (chairperson first, then treasurer, and finally secretary).

abc	*abd*	*bcd*	*acd*
acb	*adb*	*bdc*	*adc*
bca	*bad*	*cdb*	*cda*
bac	*bda*	*cbd*	*cad*
cab	*dba*	*dcb*	*dac*
cba	*dab*	*dbc*	*dca*

EXAMPLE 7 How many ways are there to arrange seven speakers at a conference?

Solution   Here we are arranging all seven out of seven. This is

$$P(7, 7) = 7! = 7 \cdot 6 \cdot 5 \cdot 4 \cdot 3 \cdot 2 \cdot 1 = 5040$$

Finally, let us note that the permutation of n things taken r at a time can also be given by

$$P(n, r) = \frac{n!}{(n - r)!}$$

Although this formula is more compact than that in Property 2, the factorials here can get quite large and make this formula very awkward.

Most calculators have an $\boxed{x!}$ (or $\boxed{n!}$) key, which is generally used with the $\boxed{\text{INV}}$ (or $\boxed{\text{2ndF}}$) key.

PRESS	DISPLAY	MEANING
$\boxed{5}\ \boxed{x!}$	$\boxed{120.}$	$5! = 120$
$\boxed{2}\ \boxed{3}\ \boxed{x!}$	$\boxed{2.5852\ \ 22}$	$23! \approx 2.5852 \times 10^{22}$
$\boxed{6}\ \boxed{9}\ \boxed{x!}$	$\boxed{1.7112\ \ 98}$	$69! \approx 1.7112 \times 10^{98}$
$\boxed{7}\ \boxed{0}\ \boxed{x!}$	$\boxed{\text{E} \quad 0.}$	*Overflow*: 70! is too big for calculator
$\boxed{2}\ \boxed{.}\ \boxed{5}\ \boxed{x!}$	$\boxed{\text{E} \quad 0.}$	*Error*: Domain of $x!$ is set of whole numbers

Note that the factorial function grows so quickly that even 70! exceeds the calculator's capacity. Note also that $x!$ is undefined unless x is a whole number.

PROBLEM SET 17.1

Warm-up
Exercises

Evaluate the following.

1. $4 \cdot 5 \cdot 6$
2. $200 \cdot 4000$
3. $10 \cdot 9 \cdot 8 \cdot 7$
4. $7 \cdot 6 \cdot 5 \cdot 4 \cdot 3 \cdot 2 \cdot 1$
5. $3000 \cdot 2000 \cdot 600 \cdot 0$
6. $5 \cdot 5 \cdot 5 \cdot 5 \cdot 5$

Evaluate the following.

7. $P(6, 2)$	**8.** $P(5, 4)$	**9.** $P(10, 3)$
10. $P(20, 5)$	**11.** $P(50, 3)$	**12.** $P(1000, 2)$
13. 2!	**14.** 3!	**15.** 5!
16. 6!	**17.** 8!	**18.** 11!

Work the following problems.

19. A coin has two sides (heads and tails).
   (a) How many possible outcomes are there for 2 flips?
   (b) How many possible outcomes are there for 3 flips?

(c) How many possible outcomes are there for 10 flips?

(d) How many possible outcomes are there for 20 flips?

20. In a certain state, all license plates are made as follows. There are six characters: The first is a letter, the second is a letter or digit, and the last four are digits. How many possible license plates are there?

21. Ron is choosing options for his new car: 6 models, 20 colors, 3 transmissions, 4 interior packages, 3 exterior packages, 5 types of radio/stereo, air conditioning or not, power steering or not, power windows or not. How many different cars can Ron design?

22. A menu planner has four types of soup, three types of salad, seven main courses, six types of vegetable, three types of potato, and eight types of dessert. How many total meals are there?

23. Twenty athletes are competing in an event. In how many ways can they win gold, silver, and bronze medals?

24. If the letters of the alphabet are scrambled, how many different codes are possible?

25. In how many different ways can six people be seated in six chairs in a row?

26. In how many ways can seven cousins be placed in a line for a family photograph?

27. There are 100 members of the U.S. Senate. In how many ways can the chairmanships of three committees (assuming no repeats and ignoring senority rules) be arranged?

28. Sheri has 5 records. In how many ways can she arrange them to be played on a phonograph?

29. A restaurant has 8 sandwiches, 3 side orders, 5 drinks, and 4 desserts.
    (a) How many ways are there to order one of each?
    (b) How many total ways are there to order (if one may or may not order from any category)?
    (c) How many different ways can a man and woman each order a different sandwich?

30. A certain ice-cream parlor boasts 31 flavors. A woman wants to order a 3-scoop cone.
    (a) How many possibilities are there (if repeat flavors are allowed)?
    (b) How many possibilities are there (if no repeat flavors are allowed)?

31. Eight children are playing musical chairs at a party. In how many ways can the children be eliminated until there is a winner?

32. On a certain Sunday, there are 13 professional baseball games being played. How many possible outcomes are there (including rain-outs)?

Computer Application
33. In a computer, a *bit* has two possibilities: 0 or 1. On a certain computer, a *word* is 32 bits. How many possible words are there?

Health Application
34. A patient is categorized by race (white, black, Oriental), by sex (male, female), by blood type (A, B, AB, O), by RH factor ($+$, $-$), by blood pressure (low, average, high), by weight (low, average, high), and by age (under 21, 21 to 50, over 50). How many possible classifications are there?

Educational Application
35. Ten questions on a test are multiple-choice (with 5 choices). Another 20 questions are true/false.
    (a) How many answer possibilities are there for the multiple-choice section?
    (b) How many answer possibilities are there for the true/false section?
    (c) How many total possibilities are there for all 30 questions?

36. A hotel has 100 rooms and 70 reservations. In how many ways can the reservations be filled?

37. A retail chain has fallen on hard times and must close 7 of its 20 stores, one at a time. In how many ways can this be done?

38. Your hand calculator should use scientific notation and have an $\boxed{x!}$ key. Compute the following (if possible):

    (a) 10!   (b) 39!   (c) 64!   (d) 76!

17.2 COMBINATIONS

The preceding section dealt with permutations for which order was very important. In this section we discuss **combinations**, for which *order is not important*.

EXAMPLE 8   In how many ways can a group of 3 be drawn from a group of 4?

Solution   Here we want a group of 3, but we do *not* care about order. Let us first write out the permutations of 4 things taken 3 at a time. There are $P(4, 3) = 4 \cdot 3 \cdot 2 = 24$.

    Let the original group be $\{a, b, c, d\}$ (see Example 6—page 558).

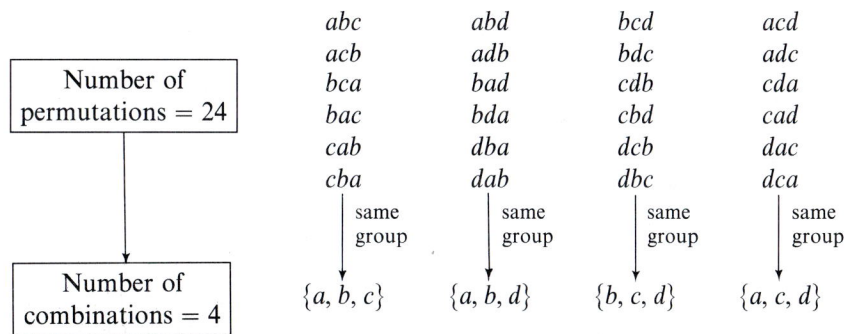

	abc	abd	bcd	acd
Number of permutations = 24	acb	adb	bdc	adc
	bca	bad	cdb	cda
	bac	bda	cbd	cad
	cab	dba	dcb	dac
	cba	dab	dbc	dca
	same group	same group	same group	same group
Number of combinations = 4	$\{a, b, c\}$	$\{a, b, d\}$	$\{b, c, d\}$	$\{a, c, d\}$

Note that the 24 permutations collapse into 4 combinations. This is because there are 6 different ways to rearrange the same group such as $\{a, b, c\}$. Since we do not care about order, we divide the 24 permutations by the 6 different ways to rearrange each group of three; this is how we get 4 groups.

Example 8 shows how combinations are related to permutations. The permutations give *all* the arrangements, while the combinations give only the distinct groupings. As suggested by Example 8, we divide the number of permutations of the whole group by the number of permutations of the subgroup size.

We define the symbol **C(n, r)** to be the **combinations of n things taken r at a time**. Another symbol for the same thing is $\binom{n}{r}$.

Property 3
The combination (or unordered arrangement) of n things taken r at a time is given by

$$\binom{n}{r} = C(n, r) = \frac{P(n, r)}{r!} = \frac{n!}{r!(n-r)!}$$

EXAMPLE 9   The following examples show the combination notation and Property 3.

(a) $C(8, 3) = \dfrac{P(8, 3)}{3!} = \dfrac{8 \cdot 7 \cdot 6}{1 \cdot 2 \cdot 3} = 56$

(b) $\dbinom{20}{4} = \dfrac{P(20, 4)}{4!} = \dfrac{20 \cdot 19 \cdot 18 \cdot 17}{1 \cdot 2 \cdot 3 \cdot 4} = 4845$

(c) $\dbinom{7}{7} = \dfrac{P(7, 7)}{7!} = \dfrac{7 \cdot 6 \cdot 5 \cdot 4 \cdot 3 \cdot 2 \cdot 1}{1 \cdot 2 \cdot 3 \cdot 4 \cdot 5 \cdot 6 \cdot 7} = 1$

(d) $C(100, 3) = \dfrac{P(100, 3)}{3!} = \dfrac{100 \cdot 99 \cdot 98}{1 \cdot 2 \cdot 3} = 161{,}700$

EXAMPLE 10   In how many ways can a 4-person committee be chosen from a 17-person group?

Solution   Since a committee is unordered, this is a combination of 17 things taken 4 at a time; this is

$$\binom{17}{4} = \frac{P(17, 4)}{4!} = \frac{17 \cdot 16 \cdot 15 \cdot 14}{1 \cdot 2 \cdot 3 \cdot 4} = 2380$$

Thus, there are 2380 combinations.

YES	NO
$\dbinom{12}{3} = \dfrac{12 \cdot 11 \cdot 10}{1 \cdot 2 \cdot 3} = 220$	$\dbinom{12}{3} = \dfrac{12}{3} = 4$

The combination symbol $\binom{n}{r}$ should *not* be confused with the fraction $\dfrac{n}{r}$.

EXAMPLE 11   There are 52 cards in a standard deck of playing cards.
  (a) How many 2-card blackjack hands are there?
  (b) How many 5-card poker hands are there?
  (c) How many 13-card bridge hands are there?

Solution   A hand of cards has *no* order, so we consider the combinations:
  (a) *Blackjack*: 52 things taken 2 at a time.

$$\binom{52}{2} = \frac{52 \cdot 51}{1 \cdot 2} = 1326$$

  (b) *Poker*: 52 things taken 5 at a time.

$$\binom{52}{5} = \frac{52 \cdot 51 \cdot 50 \cdot 49 \cdot 48}{1 \cdot 2 \cdot 3 \cdot 4 \cdot 5} = 2,598,960$$

  (c) *Bridge*: 52 things taken 13 at a time.

$$\binom{52}{13} = \frac{52 \cdot 51 \cdot 50 \cdot 49 \cdot 48 \cdot 47 \cdot 46 \cdot 45 \cdot 44 \cdot 43 \cdot 42 \cdot 41 \cdot 40}{1 \cdot 2 \cdot 3 \cdot 4 \cdot 5 \cdot 6 \cdot 7 \cdot 8 \cdot 9 \cdot 10 \cdot 11 \cdot 12 \cdot 13}$$

$$\approx 6.35 \times 10^{11} \quad \text{(with help from a calculator)}$$

The difference between permutation and combination can best be remembered as follows.

If order *is* important, use a *permutation*.
If order is *not* important, use a *combination*.

**PROBLEM SET
17.2**

Evaluate the following.

Warm-up
Exercises
  1. $P(7, 4)$ **2.** $P(15, 3)$ **3.** $P(40, 4)$
  4. $4!$ **5.** $6!$ **6.** $8!$

Work the following problems.

Warm-up
Exercises
  7. A dating service has 106 men and 124 women on file. How many possible man–woman dates can be arranged?

  8. A certain automatic garage-door opener has 9 coding slots, each with 3 positions. How many possible codes are there?

  9. In how many ways can the nine Supreme Court justices be lined up for a photograph? What if the Chief Justice must be in the middle?

  10. A club has 20 members. In how many ways can a president, a treasurer, and a secretary be chosen?

Evaluate the following.

  11. $C(7, 4)$ **12.** $C(10, 2)$ **13.** $C(15, 3)$

SECTION 17.2 Combinations   **563**

14. $\dbinom{40}{4}$ **15.** $\dbinom{100}{2}$ **16.** $\dbinom{8}{8}$

Work the following problems.

17. There are 250 business majors at a certain college.
   (a) In how many ways can a committee of 2 be chosen?
   (b) In how many ways can a committee of 3 be chosen?
   (c) In how many ways can a committee of 5 be chosen?

18. A basketball coach has 12 players on his squad. In how many ways can he pick the starting 5 players?

19. A softball coach has 15 players on her squad. In how many ways can she pick the starting 10 players?

20. A woman has 7 children. In how many ways can she have 3 boys?

21. On a 10-item test, how many ways are there to get 4 right?

22. A man has 10 shirts in his closet. He must choose 3 to take with him on a trip. How many combinations are there?

23. A group contains 12 men and 10 women.
   (a) In how many ways can a subgroup of 3 men be chosen?
   (b) In how many ways can a subgroup of 4 women be chosen?
   (c) In how many ways can a group of 7 (3 men, 4 women) be chosen?

24. A certain Senate committee contains 10 Democrats and 11 Republicans.
   (a) In how many ways can 4 Democrats be chosen?
   (b) In how many ways can 5 Republicans be chosen?
   (c) In how many ways can a subcommittee of 9 (4 Democrats, 5 Republicans) be chosen?

25. How many 10-card gin rummy hands can be drawn from a 52-card deck?

26. How many 7-card gin rummy hands can be drawn from a 52-card deck?

27. How many 5-card poker hands can be drawn without any aces or face cards?

28. How many 13-card bridge hands can be drawn without any aces or face cards?

29. Consider a group with 5 members.
   (a) In how many ways can a subgroup of size 0 be chosen?
   (b) In how many ways can a subgroup of size 1 be chosen?
   (c) In how many ways can a subgroup of size 2 be chosen?
   (d) In how many ways can a subgroup of size 3 be chosen?
   (e) In how many ways can a subgroup of size 4 be chosen?
   (f) In how many ways can a subgroup of size 5 be chosen?
   (g) How many total subgroups are there?

30. Consider a group of size 6. In how many ways can subgroups of size 0, 1, 2, 3, 4, 5, and 6 be chosen? How many total subgroups are there?

Business Application

31. The ABC Bulb Company markets a package with 8 light bulbs. In how many ways can there be 3 defective bulbs?

Health Application

32. A dietitian has 8 meats, 5 vegetables, and 4 starches from which to choose. She is making a menu with 5 meats, 3 vegetables, and 2 starches.
   (a) In how many ways can she choose the 5 meats?
   (b) In how many ways can she choose the 3 vegetables?
   (c) In how many ways can she choose the 2 starches?
   (d) How many possible menus are there?

Sports Application

33. A league with ten teams is planning a round robin (every team plays every other team once). How many games must be scheduled?

Will it rain tomorrow?
Will I win the state lottery?
Will the interest rates drop?

These are uncertain events, so we cannot give "yes" or "no" answers. Instead, we give the event a number between 0 (no) and 1 (yes) to show the likelihood of the event happening. We call this number the **probability** of the event.

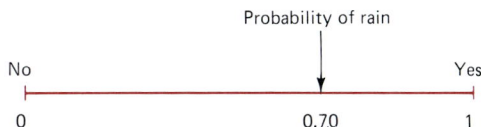

Probability of rain

No   Yes

0   0.70   1

When the weatherman says "The chance of rain is 70%," we can picture this on the number line above as falling somewhat closer to "yes" than to "no."

We label events with capital letters, such as A, and write the probability of A as $P(A)$. The set of all outcomes that we are considering is the **sample space**. An **event** is a set of outcomes. Probabilities satisfy the following.

P1   The probability of an event is between 0 and 1. $0 \leq P(A) \leq 1$

P2   The sum of the probabilities of all outcomes in a $\sum_i P(A_i) = 1$
sample space is 1.

EXAMPLE 12   Pictorially, we can view the sample space as a rectangle, with the events as subsets. Figure 2 shows three sample spaces with the probabilities given below each event. Note that the probabilities within each sample space add up to 1 (or 100%).

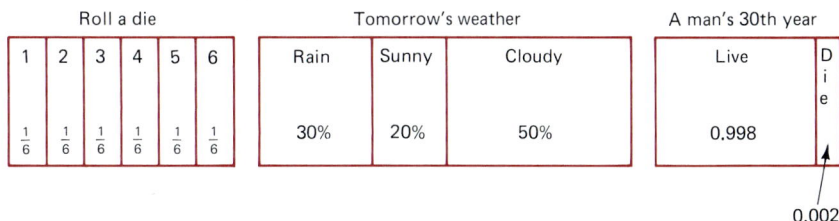

Roll a die						Tomorrow's weather			A man's 30th year	
1	2	3	4	5	6	Rain	Sunny	Cloudy	Live	D i e
$\frac{1}{6}$	$\frac{1}{6}$	$\frac{1}{6}$	$\frac{1}{6}$	$\frac{1}{6}$	$\frac{1}{6}$	30%	20%	50%	0.998	

0.002

FIGURE 2

Since all probabilities add up to 1, we see that the relation between an event happening and not happening is given by

$$P(\text{not } A) = 1 - P(A)$$

EXAMPLE 13 (a) If P(rolling a 1) = 1/6, then P(not rolling a 1) = 5/6.
(b) If P(rain) = 0.30, then P(not rain) = 0.70.
(c) If P(live) = 0.998, then P(die) = 0.002.

How do we compute the probability of an event? We have three basic methods.

(a) **Counting** Where all outcomes are *equally likely* (coins, dice, lotteries, and so on), we use the formula

$$P(A) = \frac{\text{number of outcomes in } A}{\text{number of total cases}}$$

(b) **Scientific** If the outcomes are not equally likely (life/death, rain/sun, and so on), but we have scientific data, we use the formula

$$P(A) = \frac{\text{number of times } A \text{ occurred}}{\text{number of total cases}}$$

(c) **Guesstimate** When we have no data from the past (a race, a Super Bowl, and so on), we make our best guess based on our intuition.

EXAMPLE 14 Karen buys 7 lottery tickets out of 5000 sold. What is her probability of winning? What is her probability of losing?

Solution This is a *counting probability* (assuming that each ticket is equally likely to win). We get

$$P(\text{Karen wins}) = \frac{7}{5000} = 0.0014$$

$$P(\text{Karen loses}) = 1 - 0.0014 = 0.9986$$

EXAMPLE 15 There are about 230,000,000 Americans, and about 50,000 will die this year in automobile crashes. Based on this, what is the probability of a given American dying in an automobile crash next year?

Solution This is an example of *scientific* (or *historical*) *probability*. We use the data to get

$$P(\text{dying in a crash}) = \frac{50,000}{230,000,000} = 0.00022$$

EXAMPLE 16 Setting probabilities for a horse race (or any other one-shot event) is based on an intelligent guess. For example,

$$P(\text{Popcorn wins fourth race}) = \frac{2}{7}$$

is a *guesstimate probability*.

EXAMPLE 17 What is the probability of being dealt a five-card poker hand without any face cards or aces?

Solution This is a counting probability. We use the formula

$$P(\text{no face cards or aces}) = \frac{\text{hands without face cards or aces}}{\text{total number of hands}}$$

We must use combinations (page 561) to find each of these terms. The denominator (total number) was computed in Example 11(b) (page 563) as

$$\boxed{\text{Denominator}} \qquad \binom{52}{5} = \frac{52 \cdot 51 \cdot 50 \cdot 49 \cdot 48}{1 \cdot 2 \cdot 3 \cdot 4 \cdot 5} = 2{,}598{,}960$$

The numerator is computed similarly. There are 36 cards that are not face cards or aces (2, 3, 4, 5, 6, 7, 8, 9, and 10 of the four suits). The number of 5-card hands that can be drawn from these 36 cards is

$$\boxed{\text{Numerator}} \qquad \binom{36}{5} = \frac{36 \cdot 35 \cdot 34 \cdot 33 \cdot 32}{1 \cdot 2 \cdot 3 \cdot 4 \cdot 5} = 376{,}992$$

Putting these together, we get

$$P(\text{no face cards or aces}) = \frac{376{,}992}{2{,}598{,}960} = 0.145$$

Thus, there is about a 14.5% chance of not getting an ace or face card. We can also see that the chance of getting at least one ace or face card is $100\% - 14.5\% = 85.5\%$.

PROBLEM SET 17.3

Solve the following equations.

Warm-up Exercises

1. $x + 0.7 = 1$

2. $x + \dfrac{2}{7} = 1$

3. $x + \dfrac{1}{2} + \dfrac{1}{5} + \dfrac{1}{10} = 1$

4. $x + 71.3\% + 13.2\% = 100\%$

Evaluate the following.

Warm-up Exercises

5. $5!$

6. $7!$

7. $\dbinom{10}{3}$

8. $C(20, 4)$

9. $P(11, 5)$

10. $P(12, 3)$

Work the following problems.

Warm-up Exercises

11. In how many ways can a 3-person committee be chosen from a group of 12?

12. In how many ways can first-, second-, and third-place prizes be chosen from a group of 20 people?

For the each of the following sample spaces, fill in the missing probability.

13.

Snow	Rain	Sun
0.25	0.30	?

14.

A	B	C	D
?	0.3	0.2	0.1

15.

W	X	Y	Z
$\frac{1}{8}$	$\frac{1}{4}$	$\frac{1}{4}$	?

16.

1st	2nd	3rd	4th	5th
?	15%	20%	25%	35%

Complete the following table

	$P(A)$	$P(\text{not } A)$		$P(A)$	$P(\text{not } A)$
17.	1/3	?	**18.**	96%	?
19.	2/9	?	**20.**	?	0.18
21.	0.83	?	**22.**	?	1/500

Compute each of the following probabilities and state whether it is counting, scientific, or guesstimate.

23. A card is pulled from a shuffled 52-card deck. What is the probability that the card is:
   (a) A king?   (b) The king of spades?
   (c) A black card?   (d) A red face card?

24. A church sells 20,000 lottery tickets. What is Andy's probability of winning if he buys:
   (a) 1 ticket?   (b) 10 tickets?   (c) 100 tickets?

25. A single die is rolled. What is the probability of rolling a(n):
   (a) 2?   (b) 1 or 2?   (c) Even number?

26. In a recent year, 2,290,000 students graduated from high school. Four years later, 666,710 students graduated from college. Use these figures to approximate the probability of a given high school senior finishing college.

27. In one year, there were 8,762,306 50-year-olds. By the next year, there were only 8,689,404 51-year-olds. Using these data, what is the probability of a given 50-year-old surviving to be 51?

28. Drawing 5 red cards in a 5-card hand. (There are 26 red cards.)

29. There are 13 people in a group: 7 women and 6 men. A subcommittee of 4 is chosen. What is the probability that
(a) The subcommittee is all women?
(b) The subcommittee is all men?

30. Five cousins are asked to line up for a photograph. What is the probability that they will line up by increasing heights, left to right?

31. What is the probability that there will be a nuclear disarmament next year?

32. A committee consists of 15 conservatives, 19 moderates, and 12 liberals. If a chairperson is chosen by lot, what is the probability the chairperson is:
(a) A conservative?   (b) A moderate?
(c) A liberal?   (d) Not a liberal?

33. A jar contains 7 red balls, 6 white balls, 5 blue balls, and 4 green balls. What is the probability of pulling:
(a) A red ball?   (b) A blue ball?
(c) Two red balls?   (d) Two white balls?
[In parts (c) and (d), two balls are chosen at once.]

34. What is the probability that the next president will be a woman?

Business Applications **35.** In a sample of 175 radios, a manufacturer finds that 6 are defective. Based on this, what is the probability of finding a defective radio?

36. A dealer has 100 cars, 10 of which are substandard. If three cars are sold, what is the probability that all three are substandard?

Health Application **37.** In a certain town, there were 112 absentees from school out of 2107 students. Based on this, what is the probability that a student will attend school on a given day?

Life Science Application **38.** In a field, a botanist finds 51 red sweet peas, 98 pink sweet peas, and 48 white sweet peas. Based on this, what is the probability of finding:
(a) A red sweet pea?   (b) A white sweet pea?
(c) A pink sweet pea?

Computer Application **39.** In 13,586 transmissions of information, 27 had errors in the transmission. What is the probability that a transmission will be error-free?

Sports Application **40.** Dave Swinginmiss has been at bat 450 times and has 135 hits. What is the probability that he will get a hit the next time at bat (to three decimal places)?

17.4
BINOMIAL THEOREM

Recall from Chapter 2 that a **binomial** is a sum (or difference) of two terms, such as $a + b$ or $x - 5$. We are now interested in powers of binomials such as $x + y$; that is, we consider $(x + y)^n$. Notice the pattern that develops for the first few values of n.

$$(x + y)^0 = 1$$

$$(x + y)^1 = x + y$$

$$(x + y)^2 = x^2 + 2xy + y^2$$

$$(x + y)^3 = x^3 + 3x^2y + 3xy^2 + y^3$$

$$(x + y)^4 = x^4 + 4x^3y + 6x^2y^2 + 4xy^3 + y^4$$

$$(x + y)^5 = x^5 + 5x^4y + 10x^3y^2 + 10x^2y^3 + 5xy^4 + y^5$$

We can see several patterns as this table continues. One pattern involves the coefficients, as follows:

$$
\begin{array}{c}
\text{Row 0} \longrightarrow 1 \\
\text{Row 1} \longrightarrow 1 \quad 1 \\
\text{Row 2} \longrightarrow 1 \quad 2 \quad 1 \\
\text{Row 3} \longrightarrow 1 \quad 3 \quad 3 \quad 1 \\
\text{Row 4} \longrightarrow 1 \quad 4 \quad 6 \quad 4 \quad 1 \\
\text{Row 5} \longrightarrow 1 \quad 5 \quad 10 \quad 10 \quad 5 \quad 1 \\
\text{Row 6} \longrightarrow 1 \quad 6 \quad 15 \quad 20 \quad 15 \quad 6 \quad 1 \\
\text{Row 7} \longrightarrow 1 \quad 7 \quad 21 \quad 35 \quad 35 \quad 21 \quad 7 \quad 1
\end{array}
$$

This triangle is called **Pascal's triangle**. Each entry is the sum of the two entries above it. For example, the 10 in row 5 is the sum of the 4 and 6 above it.

Another pattern in the expansion of $(x + y)^n$ is that the sum of the exponents in each term is n. For example, for $n = 7$, we have x^7, x^6y, x^5y^2, x^4y^3, and so on. Row n of Pascal's triangle gives the coefficients in the terms of $(x + y)^n$.

EXAMPLE 18 Use Pascal's triangle to expand $(x + y)^7$.

Solution We use row 7 of Pascal's triangle and arrange the terms so that the sums of the exponents are always 7.

$$(x + y)^7 = x^7 + 7x^6y + 21x^5y^2 + 35x^4y^3 + 35x^3y^4 + 21x^2y^5 + 7xy^6 + y^7$$

EXAMPLE 19 Expand $(m - 2n)^5$.

Solution Here we use row 5 of Pascal's triangle to find the coefficients:

$$1 \quad 5 \quad 10 \quad 10 \quad 5 \quad 1$$

We let $x = m$ and $y = -2n$.

$$
\begin{aligned}
(m - 2n)^5 &= m^5 + 5m^4(-2n) + 10m^3(-2n)^2 + 10m^2(-2n)^3 + 5m(-2n)^4 + (-2n)^5 \\
&= m^5 + 5m^4(-2n) + 10m^3(4n^2) + 10m^2(-8n^3) + 5m(16n^4) + (-32n^5) \\
&= m^5 - 10m^4n + 40m^3n^2 - 80m^2n^3 + 80mn^4 - 32n^5
\end{aligned}
$$

Recall from page 562 that the combination symbol is given by

$$\binom{n}{r} = \frac{n(n - 1)(n - 2) \cdots (n - r + 1)}{1 \cdot 2 \cdot 3 \cdot 4 \cdots r} = \frac{n!}{r!(n - r)!}$$

We can use this symbol to help write a general theorem for expanding $(x + y)^n$. It is known as the **binomial theorem**.

Theorem (Binomial Theorem)
For any real numbers x and y and positive integer n,

$$(x + y)^n = x^n + \binom{n}{1}x^{n-1}y + \binom{n}{2}x^{n-2}y^2 + \binom{n}{3}x^{n-3}y^3 + \cdots + y^n$$

Using $\sum$ notation, this formula can also be written

$$(x + y)^n = \sum_{k=0}^{n} \binom{n}{k} x^{n-k} y^k$$

EXAMPLE 20   Use the binomial theorem to expand $(a + b)^5$.

Solution   Here $n = 5$. Thus, the sums of the exponents are 5. Also, all the coefficients are all of the form $\binom{5}{k}$, where k is the exponent of b.

$$(a + b)^5 = a^5 + \binom{5}{1}a^4b + \binom{5}{2}a^3b^2 + \binom{5}{3}a^2b^3 + \binom{5}{4}ab^4 + b^5$$

$$= a^5 + \frac{5}{1}a^4b + \frac{5 \cdot 4}{1 \cdot 2}a^3b^2 + \frac{5 \cdot 4 \cdot 3}{1 \cdot 2 \cdot 3}a^2b^3 + \frac{5 \cdot 4 \cdot 3 \cdot 2}{1 \cdot 2 \cdot 3 \cdot 4}ab^4 + b^5$$

$$= a^5 + 5a^4b + 10a^3b^2 + 10a^2b^3 + 5ab^4 + b^5$$

EXAMPLE 21   Write the first four terms of $(p - 3q)^{15}$.

Solution   We use the binomial theorem with $x = p$, $y = -3q$, and $n = 15$. (We use special care with the negative terms.)

$$(p - 3q)^{15} = p^{15} + \binom{15}{1}p^{14}(-3q) + \binom{15}{2}p^{13}(-3q)^2 + \binom{15}{3}p^{12}(-3q)^3 + \cdots$$

$$= p^{15} + \frac{15}{1}p^{14}(-3q) + \frac{15 \cdot 14}{1 \cdot 2}p^{13}(9q^2) + \frac{15 \cdot 14 \cdot 13}{1 \cdot 2 \cdot 3}p^{12}(-27q^3) + \cdots$$

$$= p^{15} - 45p^{14}q + 945p^{13}q^2 - 12{,}285p^{12}q^3 + \cdots$$

EXAMPLE 22   Write the first three terms of $(a + b)^{100}$.

Solution   We use the binomial theorem again, here with $n = 100$.

$$(a + b)^{100} = a^{100} + \binom{100}{1}a^{99}b + \binom{100}{2}a^{98}b^2 + \cdots$$

$$= a^{100} + \frac{100}{1}a^{99}b + \frac{100 \cdot 99}{1 \cdot 2}a^{99}b^2 + \cdots$$

$$= a^{100} + 100a^{99}b + 4950a^{98}b^2 + \cdots$$

**PROBLEM SET
17.4**   *Simplify the following.*

Warm-up   **1.** $(a - b)^2$ **2.** $(2x + 3y)^2$ **3.** $(-x)^7$
Exercises   **4.** $(-a)^{10}$ **5.** $(-2t)^8$ **6.** $(-3k)^5$

Evaluate the following.

Warm-up Exercises

7. $\binom{10}{4}$ **8.** $\binom{6}{2}$ **9.** $\binom{7}{6}$

10. $\binom{20}{3}$ **11.** $\binom{50}{2}$ **12.** $\binom{200}{3}$

Use Pascal's triangle to expand the following expressions completely.

13. $(x + 1)^4$ **14.** $(a + b)^6$ **15.** $(y - 2)^3$

16. $(t + 3)^4$ **17.** $(p - q)^5$ **18.** $(x + 2y)^4$

19. $(m - 5)^3$ **20.** $(2x + 3y)^4$

Use the binomial theorem to write the first four terms of the following expressions.

21. $(x + y)^{14}$ **22.** $(a + 2)^{10}$ **23.** $(x - 3)^{20}$

24. $(u - v)^{12}$ **25.** $(2m + 5n)^9$ **26.** $(t + 10)^7$

27. $(x - 7)^8$ **28.** $(k - 4)^{10}$ **29.** $(2u + v)^{13}$

30. $\left(m + \dfrac{1}{2}\right)^{15}$ **31.** $(u^3 - v^5)^{20}$ **32.** $(a^7 + b^9)^5$

Life Science Application

33. In a certain family, there is a 1/4 chance of a child having blue eyes. If the family has three children, the probabilities for blue-eyed children are given by

$$\left(\frac{1}{4} + \frac{3}{4}\right)^3 = \left(\frac{1}{4}\right)^3 + 3\left(\frac{1}{4}\right)^2\left(\frac{3}{4}\right) + 3\left(\frac{1}{4}\right)\left(\frac{3}{4}\right)^2 + \left(\frac{3}{4}\right)^3$$

$$= \frac{1}{64} + \frac{9}{64} + \frac{27}{64} + \frac{27}{64}$$

    ↑   ↑   ↑   ↑

3 blue-eyed children   2 blue-eyed children   1 blue-eyed child   0 blue-eyed children

Follow the same procedure (using Pascal's triangle or the binomial theorem) to find the probabilities of blue-eyed children in the following cases.

(a) Four children: $\left(\dfrac{1}{4} + \dfrac{3}{4}\right)^4$

(b) Five children: $\left(\dfrac{1}{4} + \dfrac{3}{4}\right)^5$

(c) Six children: $\left(\dfrac{1}{4} + \dfrac{3}{4}\right)^6$

Business Application

34. There is a 99/100 chance that a stereo made by Boombox will work properly (and a 1/100 chance it will be defective). The company is shipping out 50 of these stereos and wants the probabilities for the number of good stereos out of the 50. Write the first four terms of

$$\left(\frac{99}{100} + \frac{1}{100}\right)^{50}$$

(see Problem 33) to find the probabilities of there being 50, 49, 48, or 47 good radios out of the 50 shipped out.

Suppose that we started adding up the odd integers:

> First odd integer: 1
> First 2 odd integers: $1 + 3 = 4$
> First 3 odd integers: $1 + 3 + 5 = 9$
> First 4 odd integers: $1 + 3 + 5 + 7 = 16$
> First 5 odd integers: $1 + 3 + 5 + 7 + 9 = 25$
> First 6 odd integers: $1 + 3 + 5 + 7 + 9 + 11 = 36$

At some point, we notice a pattern: The sums are all perfect squares. We then ask: Is the sum of the first n odd integers always equal to n^2? That is, is the following true?

> First n odd integers: $1 + 3 + 5 + 7 + \cdots + (2n - 1) = n^2$

Mathematicians have a technique for proving such statements that involve all the natural numbers: 1, 2, 3, 4, and so on. We call this technique **mathematical induction**.

Property (Mathematical Induction)
Suppose that we want to prove a sequence of statements S_1, S_2, S_3, and so on. Every S_n is true if we can do the following:

1. We prove that S_1 is true (that is, prove the first statement).
2. Assuming that S_k is true, we prove that S_{k+1} is true (that is, prove that S_k implies S_{k+1}).

How does this show that all S_n are true?

1. We first show that S_1 is true.
2. Since S_1 is true, then S_2 is true.
3. Since S_2 is true, then S_3 is true.
4. And so on through all S_n.

EXAMPLE 23 We now use mathematical induction to prove the conjecture with which we opened the section:

$$S_n: 1 + 3 + 5 + 7 + 9 + \cdots + (2n - 1) = n^2$$

Solution In proving this, we need to look at three key statements:

$$S_1: \quad 1 = 1^2 \quad \text{(Replace } n \text{ by 1.)}$$

$$S_k: \quad 1 + 3 + 5 + \cdots + (2k - 1) = k^2 \quad \text{(Replace } n \text{ by } k.\text{)}$$

$$S_{k+1}: \quad 1 + 3 + 5 + \cdots + (2k - 1) + (2k + 1) = (k + 1)^2$$

$$\text{(Replace } n \text{ by } k + 1)$$

Now we use the procedure for mathematical induction. We have to take S_k and show that it implies S_{k+1}.

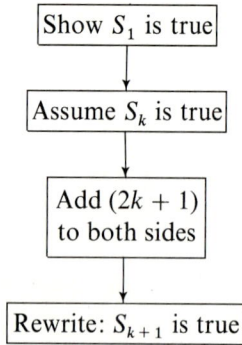

S_1: $1 = 1^2$ This is true.

S_k: $1 + 3 + 5 + \cdots + (2k - 1) = k^2$

$1 + 3 + 5 + \cdots + (2k - 1) + (2k + 1) = k^2 + 2k + 1$

$1 + 3 + 5 + \cdots + (2k - 1) + (2k + 1) = (k + 1)^2$ (This is S_{k+1}.)

This concludes our proof. S_1 is true (this is usually a fairly easy step). We then assume that S_k is true. By adding the next odd number $(2k + 1)$ to both sides, we get S_{k+1}. These are the steps needed to prove the conjecture.

Mathematical induction is similar to a sequence of falling dominos (see Figure 3). To knock over all the dominos:

1. We must push over the first domino.
2. We must know that if domino k falls that domino $k + 1$ will also fall.

Comparing this to mathematical induction, proving S_1 is like pushing domino 1. Proving that S_k implies S_{k+1} is like domino k pushing over domino $k + 1$. (Some mathematicians like to compare mathematical induction to climbing a ladder: We climb rung 1; then if we climb rung k, we climb rung $k + 1$. In this way we climb all the rungs of the ladder.)

FIGURE 3

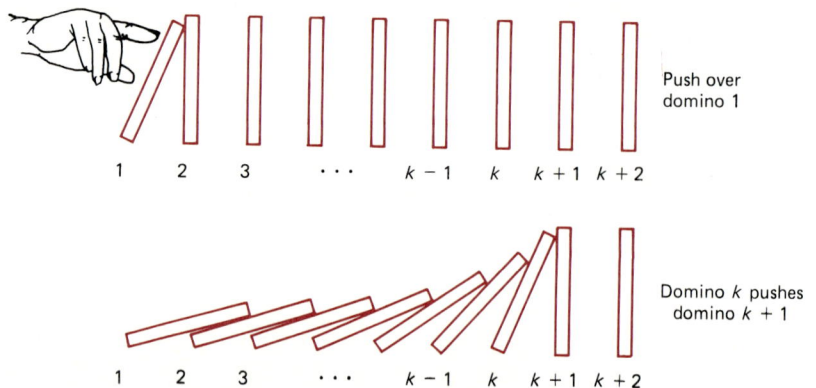

EXAMPLE 24   Prove $S_n: \dfrac{1}{2} + \dfrac{1}{4} + \dfrac{1}{8} + \cdots + \dfrac{1}{2^n} = \dfrac{2^n - 1}{2^n}$ for all n.

Solution   We use mathematical induction. The key statements that we look at are:

$$S_1: \quad \frac{1}{2} = \frac{2^1 - 1}{2^1} \qquad \text{(Replace } n \text{ by 1.)}$$

$$S_k: \quad \frac{1}{2} + \frac{1}{4} + \cdots + \frac{1}{2^k} = \frac{2^k - 1}{2^k} \qquad \text{(Replace } n \text{ by } k.)$$

$$S_{k+1}: \quad \frac{1}{2} + \frac{1}{4} + \cdots + \frac{1}{2^k} + \frac{1}{2^{k+1}} = \frac{2^{k+1} - 1}{2^{k+1}} \qquad \text{(Replace } n \text{ by } k + 1.)$$

We first show that S_1 is true. Then we show that S_k implies S_{k+1}.

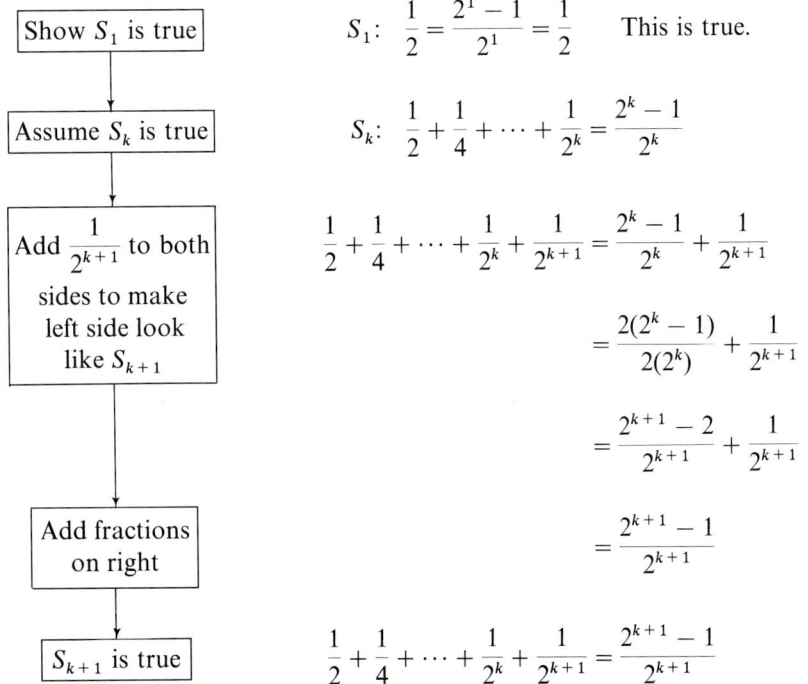

Show S_1 is true	$S_1: \quad \dfrac{1}{2} = \dfrac{2^1 - 1}{2^1} = \dfrac{1}{2}$ This is true.
Assume S_k is true	$S_k: \quad \dfrac{1}{2} + \dfrac{1}{4} + \cdots + \dfrac{1}{2^k} = \dfrac{2^k - 1}{2^k}$
Add $\dfrac{1}{2^{k+1}}$ to both sides to make left side look like S_{k+1}	$\dfrac{1}{2} + \dfrac{1}{4} + \cdots + \dfrac{1}{2^k} + \dfrac{1}{2^{k+1}} = \dfrac{2^k - 1}{2^k} + \dfrac{1}{2^{k+1}}$ $= \dfrac{2(2^k - 1)}{2(2^k)} + \dfrac{1}{2^{k+1}}$ $= \dfrac{2^{k+1} - 2}{2^{k+1}} + \dfrac{1}{2^{k+1}}$
Add fractions on right	$= \dfrac{2^{k+1} - 1}{2^{k+1}}$
S_{k+1} is true	$\dfrac{1}{2} + \dfrac{1}{4} + \cdots + \dfrac{1}{2^k} + \dfrac{1}{2^{k+1}} = \dfrac{2^{k+1} - 1}{2^{k+1}}$

Since S_1 is true and S_k implies S_{k+1}, S_n is true for all n.

EXAMPLE 25   Prove S_n: If $1 < a$, then $a^{n-1} < a^n$ for all n.

Solution   We again use mathematical induction. The key statements that we need are:

$$S_1: \quad a^0 < a^1 \quad \text{(Replace } n \text{ by 1.)}$$

$$S_k: \quad a^{k-1} < a^k \quad \text{(Replace } n \text{ by } k.)$$

$$S_{k+1}: \quad a^k < a^{k+1} \quad \text{(Replace } n \text{ by } k + 1.)$$

We first show that S_1 is true; then we show that S_k implies S_{k+1}.

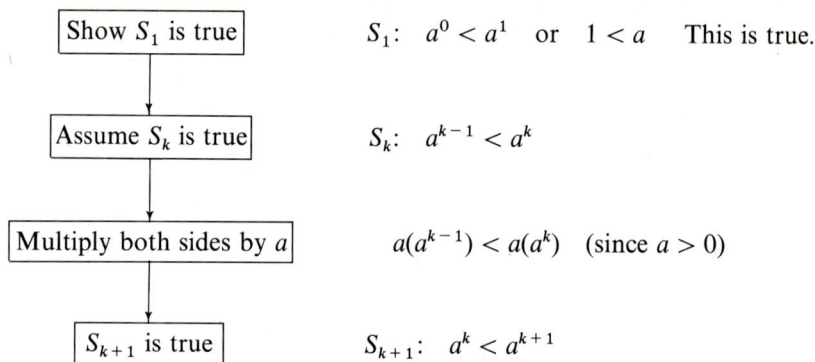

Show S_1 is true	S_1: $a^0 < a^1$ or $1 < a$ This is true.
Assume S_k is true	S_k: $a^{k-1} < a^k$
Multiply both sides by a	$a(a^{k-1}) < a(a^k)$ (since $a > 0$)
S_{k+1} is true	S_{k+1}: $a^k < a^{k+1}$

S_1 is true, and S_k implies S_{k+1}. Thus, S_n is true for all n.

PROBLEM SET
17.5

Simplify the following as much as possible (factor, if necessary).

Warm-up Exercises

1. $n(n + 1) + 2(n + 1)$

2. $\dfrac{n^2 + n}{2} + n + 1$

3. $2^{n+1} + 2^{n+1} - 2$

4. $3^{n+1} + 3^{n+1} - 3$

5. $\dfrac{n}{n+1} + \dfrac{1}{(n+1)(n+2)}$

6. $\dfrac{n^2 + 3n^2 + 2n}{3} + n^2 + 3n + 2$

Use mathematical induction to prove the following.

7. S_n: $1 + 2 + 3 + \cdots + n = \dfrac{n(n+1)}{2}$

8. S_n: $2 + 4 + 8 + \cdots + 2^n = 2^{n+1} - 2$

9. S_n: $3 + 9 + 27 + \cdots + 3^n = \dfrac{3^{n+1} - 3}{2}$

10. S_n: $\dfrac{1}{1 \cdot 2} + \dfrac{1}{2 \cdot 3} + \dfrac{1}{3 \cdot 4} + \cdots + \dfrac{1}{n(n+1)} = \dfrac{n}{n+1}$

11. S_n: $1 \cdot 2 + 2 \cdot 3 + 3 \cdot 4 + \cdots + n(n+1) = \dfrac{n(n+1)(n+2)}{3}$

12. S_n: For $0 < a < 1$, $a^n < 1$

13. S_n: $2^n > n$

14. S_n: $1 + 2n \leq 3^n$

15. S_n: $n^2 + n$ is an even integer.

16. S_n: $n^3 + 2n$ is a multiple of 3.

17. S_n: $n^2 \leq 2^n$ (for $n \geq 4$)

18. S_n: For $a > 0$, $(1 + a)^n \geq 1 + na$

Counting Principle: If we have a sequence of tasks $A, B, C, \ldots$ that can be done in $a, b,$ $c, \ldots$ ways, the total number ways to do the sequence $A, B, C, \ldots$ is

$$a \cdot b \cdot c \cdot \ldots$$

$$n! = 1 \cdot 2 \cdot 3 \cdot \ldots \cdot (n-1) \cdot n \qquad 0! = 1$$

$$P(n, r) = n(n-1)(n-2) \cdot \ldots \cdot (n-r+1) = \frac{n!}{(n-r)!}$$

$$C(n, r) = \binom{n}{r} = \frac{P(n, r)}{r!} = \frac{n!}{r!(n-r)!}$$

For the outcomes A_i in a sample space

$0 \le P(A_i) \le 1$ (The probability is between 0 and 1.)

$\sum_i P(A_i) = 1$ (The sum of the probabilities is 1.)

$P(\text{not } A) = 1 - P(A)$

For equally likely outcomes,

$$P(A) = \frac{\text{number of outcomes in } A}{\text{total number of outcomes}}$$

Binomial Theorem:

$$(x + y)^n = x^n + \binom{n}{1}x^{n-1}y + \binom{n}{2}x^{n-2}y^2 + \cdots + y^n$$

To prove a sequence of statements S_1, S_2, S_3, and so on, using mathematical induction:

1. Prove that S_1 is true.
2. Assuming that S_k is true, prove that S_{k+1} is true.

Review Exercises

Evaluate the following.

1. $6!$ **2.** $0!$ **3.** $P(10, 2)$

4. $P(4, 4)$ **5.** $C(8, 3)$ **6.** $\binom{7}{2}$

7. A restaurant offers three soups, two salads, ten main courses, three side dishes, eight desserts, and seven drinks. How many dinner combinations are there?

8. A true/false test has 10 questions. How many possible sets of answers are there?

9. How many ways are there for 5 children to seat themselves in 4 chairs?

10. A group has 10 people. In how many ways can a 4-person committee be chosen?

11. How many possible 6-card cribbage hands are there (from a 52-card deck)?

12. A baseball coach has 20 players on the squad. In how many ways can he choose his 9 starters?

13. Fill in the missing probability:

A	B	C
$\frac{1}{6}$	$\frac{1}{4}$	?

14. If $P(A) = 0.87$, what is $P(\text{not } A)$?

15. If a card is pulled from a 52-card deck, what is the probability of its being a black card?

16. In an experiment with 150 mice, 20 of them survive a certain shock. Based on these figures, what is the probability that a mouse will survive this shock?

17. What is the probability of drawing two face cards in a 2-card blackjack hand?

Expand the following expressions completely.

18. $(x + 2)^5$

19. $(a - 1)^6$

Write the first four terms of the following.

20. $(x - 2)^{15}$

21. $\left(z + \dfrac{1}{2}\right)^{10}$

Prove the following using mathematical induction.

22. $S_n: 3 + 6 + 9 + \cdots + 3n = \dfrac{3n(n + 1)}{2}$

23. $S_n: \dfrac{2}{3} + \dfrac{2}{9} + \dfrac{2}{27} + \cdots + \dfrac{2}{3^n} = \dfrac{3^n - 1}{3^n}$

24. $S_n:$ If $1 < a < b$, then $a^n < b^n$.

Appendix

	.00	.01	.02	.03	.04	.05	.06	.07	.08	.09
1.0	0.0000	0.0100	0.0198	0.0296	0.0392	0.0488	0.0583	0.0677	0.0770	0.0862
1.1	0.0953	0.1044	0.1133	0.1222	0.1310	0.1398	0.1484	0.1570	0.1655	0.1740
1.2	0.1823	0.1906	0.1989	0.2070	0.2151	0.2231	0.2311	0.2390	0.2469	0.2546
1.3	0.2624	0.2700	0.2776	0.2852	0.2927	0.3001	0.3075	0.3148	0.3221	0.3293
1.4	0.3365	0.3436	0.3507	0.3577	0.3646	0.3716	0.3784	0.3853	0.3920	0.3988
1.5	0.4055	0.4121	0.4187	0.4253	0.4318	0.4383	0.4447	0.4511	0.4574	0.4637
1.6	0.4700	0.4762	0.4824	0.4886	0.4947	0.5008	0.5068	0.5128	0.5188	0.5247
1.7	0.5306	0.5365	0.5423	0.5481	0.5539	0.5596	0.5653	0.5710	0.5766	0.5822
1.8	0.5878	0.5933	0.5988	0.6043	0.6098	0.6152	0.6206	0.6259	0.6313	0.6366
1.9	0.6419	0.6471	0.6523	0.6575	0.6627	0.6678	0.6729	0.6780	0.6831	0.6881
2.0	0.6931	0.6981	0.7031	0.7080	0.7130	0.7178	0.7227	0.7275	0.7324	0.7372
2.1	0.7419	0.7467	0.7514	0.7561	0.7608	0.7655	0.7701	0.7747	0.7793	0.7839
2.2	0.7885	0.7930	0.7975	0.8020	0.8065	0.8109	0.8154	0.8198	0.8242	0.8286
2.3	0.8329	0.8372	0.8416	0.8459	0.8502	0.8544	0.8587	0.8629	0.8671	0.8713
2.4	0.8755	0.8796	0.8838	0.8879	0.8920	0.8961	0.9002	0.9042	0.9083	0.9123
2.5	0.9163	0.9203	0.9243	0.9282	0.9322	0.9361	0.9400	0.9439	0.9478	0.9517
2.6	0.9555	0.9594	0.9632	0.9670	0.9708	0.9746	0.9783	0.9821	0.9858	0.9895
2.7	0.9933	0.9969	1.0006	1.0043	1.0080	1.0116	1.0152	1.0188	1.0225	1.0260
2.8	1.0296	1.0332	1.0367	1.0403	1.0438	1.0473	1.0508	1.0543	1.0578	1.0613
2.9	1.0647	1.0682	1.0716	1.0750	1.0784	1.0818	1.0852	1.0886	1.0919	1.0953
3.0	1.0986	1.1019	1.1053	1.1086	1.1119	1.1151	1.1184	1.1217	1.1249	1.1282
3.1	1.1314	1.1346	1.1378	1.1410	1.1442	1.1474	1.1506	1.1537	1.1569	1.1600
3.2	1.1632	1.1663	1.1694	1.1725	1.1756	1.1787	1.1817	1.1848	1.1878	1.1909
3.3	1.1939	1.1970	1.2000	1.2030	1.2060	1.2090	1.2119	1.2149	1.2179	1.2208
3.4	1.2238	1.2267	1.2296	1.2326	1.2355	1.2384	1.2413	1.2442	1.2470	1.2499
3.5	1.2528	1.2556	1.2585	1.2613	1.2641	1.2669	1.2698	1.2726	1.2754	1.2782
3.6	1.2809	1.2837	1.2865	1.2892	1.2920	1.2947	1.2975	1.3002	1.3029	1.3056
3.7	1.3083	1.3110	1.3137	1.3164	1.3191	1.3218	1.3244	1.3271	1.3297	1.3324
3.8	1.3350	1.3376	1.3403	1.3429	1.3455	1.3481	1.3507	1.3533	1.3558	1.3584
3.9	1.3610	1.3635	1.3661	1.3686	1.3712	1.3737	1.3762	1.3788	1.3813	1.3838
4.0	1.3863	1.3888	1.3913	1.3938	1.3962	1.3987	1.4012	1.4036	1.4061	1.4085
4.1	1.4110	1.4134	1.4159	1.4183	1.4207	1.4231	1.4255	1.4279	1.4303	1.4327
4.2	1.4351	1.4375	1.4398	1.4422	1.4446	1.4469	1.4493	1.4516	1.4540	1.4563
4.3	1.4586	1.4609	1.4633	1.4656	1.4679	1.4702	1.4725	1.4748	1.4770	1.4793
4.4	1.4816	1.4839	1.4861	1.4884	1.4907	1.4929	1.4952	1.4974	1.4996	1.5019
4.5	1.5041	1.5063	1.5085	1.5107	1.5129	1.5151	1.5173	1.5195	1.5217	1.5239
4.6	1.5261	1.5282	1.5304	1.5326	1.5347	1.5369	1.5390	1.5412	1.5433	1.5454
4.7	1.5476	1.5497	1.5518	1.5539	1.5560	1.5581	1.5602	1.5623	1.5644	1.5665
4.8	1.5686	1.5707	1.5728	1.5748	1.5769	1.5790	1.5810	1.5831	1.5851	1.5872
4.9	1.5892	1.5913	1.5933	1.5953	1.5974	1.5994	1.6014	1.6034	1.6054	1.6074
5.0	1.6094	1.6114	1.6134	1.6154	1.6174	1.6194	1.6214	1.6233	1.6253	1.6273
5.1	1.6292	1.6312	1.6332	1.6351	1.6371	1.6390	1.6409	1.6429	1.6448	1.6467
5.2	1.6487	1.6506	1.6525	1.6544	1.6563	1.6582	1.6601	1.6620	1.6639	1.6658
5.3	1.6677	1.6696	1.6715	1.6734	1.6753	1.6771	1.6790	1.6808	1.6827	1.6845
5.4	1.6864	1.6882	1.6901	1.6919	1.6938	1.6956	1.6974	1.6993	1.7011	1.7029

$$\ln(N \cdot 10^m) = \ln N + m \ln 10, \quad \ln 10 = 2.3026$$

TABLE A   NATURAL LOGARITHMS

	.00	.01	.02	.03	.04	.05	.06	.07	.08	.09
5.5	1.7047	1.7066	1.7084	1.7102	1.7120	1.7138	1.7156	1.7174	1.7192	1.7210
5.6	1.7228	1.7246	1.7263	1.7281	1.7299	1.7317	1.7334	1.7352	1.7370	1.7387
5.7	1.7405	1.7422	1.7440	1.7457	1.7475	1.7492	1.7509	1.7527	1.7544	1.7561
5.8	1.7579	1.7596	1.7613	1.7630	1.7647	1.7664	1.7682	1.7699	1.7716	1.7733
5.9	1.7750	1.7766	1.7783	1.7800	1.7817	1.7834	1.7851	1.7867	1.7884	1.7901
6.0	1.7918	1.7934	1.7951	1.7967	1.7984	1.8001	1.8017	1.8034	1.8050	1.8066
6.1	1.8083	1.8099	1.8116	1.8132	1.8148	1.8165	1.8181	1.8197	1.8213	1.8229
6.2	1.8245	1.8262	1.8278	1.8294	1.8310	1.8326	1.8342	1.8358	1.8374	1.8390
6.3	1.8406	1.8421	1.8437	1.8453	1.8469	1.8485	1.8500	1.8516	1.8532	1.8547
6.4	1.8563	1.8579	1.8594	1.8610	1.8625	1.8641	1.8656	1.8672	1.8687	1.8703
6.5	1.8718	1.8733	1.8749	1.8764	1.8779	1.8795	1.8810	1.8825	1.8840	1.8856
6.6	1.8871	1.8886	1.8901	1.8916	1.8931	1.8946	1.8961	1.8976	1.8991	1.9006
6.7	1.9021	1.9036	1.9051	1.9066	1.9081	1.9095	1.9110	1.9125	1.9140	1.9155
6.8	1.9169	1.9184	1.9199	1.9213	1.9228	1.9242	1.9257	1.9272	1.9286	1.9301
6.9	1.9315	1.9330	1.9344	1.9359	1.9373	1.9387	1.9402	1.9416	1.9430	1.9445
7.0	1.9459	1.9473	1.9488	1.9502	1.9516	1.9530	1.9544	1.9559	1.9573	1.9587
7.1	1.9601	1.9615	1.9629	1.9643	1.9657	1.9671	1.9685	1.9699	1.9713	1.9727
7.2	1.9741	1.9755	1.9769	1.9782	1.9796	1.9810	1.9824	1.9838	1.9851	1.9865
7.3	1.9879	1.9892	1.9906	1.9920	1.9933	1.9947	1.9961	1.9974	1.9988	2.0001
7.4	2.0015	2.0028	2.0042	2.0055	2.0069	2.0082	2.0096	2.0109	2.0122	2.0136
7.5	2.0149	2.0162	2.0176	2.0189	2.0202	2.0215	2.0229	2.0242	2.0255	2.0268
7.6	2.0282	2.0295	2.0308	2.0321	2.0334	2.0347	2.0360	2.0373	2.0386	2.0399
7.7	2.0412	2.0425	2.0438	2.0451	2.0464	2.0477	2.0490	2.0503	2.0516	2.0528
7.8	2.0541	2.0554	2.0567	2.0580	2.0592	2.0605	2.0618	2.0631	2.0643	2.0656
7.9	2.0669	2.0681	2.0694	2.0707	2.0719	2.0732	2.0744	2.0757	2.0769	2.0782
8.0	2.0794	2.0807	2.0819	2.0832	2.0844	2.0857	2.0869	2.0882	2.0894	2.0906
8.1	2.0919	2.0931	2.0943	2.0956	2.0968	2.0980	2.0992	2.1005	2.1017	2.1029
8.2	2.1041	2.1054	2.1066	2.1078	2.1090	2.1102	2.1114	2.1126	2.1138	2.1150
8.3	2.1163	2.1175	2.1187	2.1199	2.1211	2.1223	2.1235	2.1247	2.1258	2.1270
8.4	2.1282	2.1294	2.1306	2.1318	2.1330	2.1342	2.1353	2.1365	2.1377	2.1389
8.5	2.1401	2.1412	2.1424	2.1436	2.1448	2.1459	2.1471	2.1483	2.1494	2.1506
8.6	2.1518	2.1529	2.1541	2.1552	2.1564	2.1576	2.1587	2.1599	2.1610	2.1622
8.7	2.1633	2.1645	2.1656	2.1668	2.1679	2.1691	2.1702	2.1713	2.1725	2.1736
8.8	2.1748	2.1759	2.1770	2.1782	2.1793	2.1804	2.1815	2.1827	2.1838	2.1849
8.9	2.1861	2.1872	2.1883	2.1894	2.1905	2.1917	2.1928	2.1939	2.1950	2.1961
9.0	2.1972	2.1983	2.1994	2.2006	2.2017	2.2028	2.2039	2.2050	2.2061	2.2072
9.1	2.2083	2.2094	2.2105	2.2116	2.2127	2.2138	2.2148	2.2159	2.2170	2.2181
9.2	2.2192	2.2203	2.2214	2.2225	2.2235	2.2246	2.2257	2.2268	2.2279	2.2289
9.3	2.2300	2.2311	2.2322	2.2332	2.2343	2.2354	2.2364	2.2375	2.2386	2.2396
9.4	2.2407	2.2418	2.2428	2.2439	2.2450	2.2460	2.2471	2.2481	2.2492	2.2502
9.5	2.2513	2.2523	2.2534	2.2544	2.2555	2.2565	2.2576	2.2586	2.2597	2.2607
9.6	2.2618	2.2628	2.2638	2.2649	2.2659	2.2670	2.2680	2.2690	2.2701	2.2711
9.7	2.2721	2.2732	2.2742	2.2752	2.2762	2.2773	2.2783	2.2793	2.2803	2.2814
9.8	2.2824	2.2834	2.2844	2.2854	2.2865	2.2875	2.2885	2.2895	2.2905	2.2915
9.9	2.2925	2.2935	2.2946	2.2956	2.2966	2.2976	2.2986	2.2996	2.3006	2.3016

TABLE B COMMON LOGARITHMS

n	0	1	2	3	4	5	6	7	8	9
1.0	.0000	.0043	.0086	.0128	.0170	.0212	.0253	.0294	.0334	.0374
1.1	.0414	.0453	.0492	.0531	.0569	.0607	.0645	.0682	.0719	.0755
1.2	.0792	.0828	.0864	.0899	.0934	.0969	.1004	.1038	.1072	.1106
1.3	.1139	.1173	.1206	.1239	.1271	.1303	.1335	.1367	.1399	.1430
1.4	.1461	.1492	.1523	.1553	.1584	.1614	.1644	.1673	.1703	.1732
1.5	.1761	.1790	.1818	.1847	.1875	.1903	.1931	.1959	.1987	.2014
1.6	.2041	.2068	.2095	.2122	.2148	.2175	.2201	.2227	.2253	.2279
1.7	.2304	.2330	.2355	.2380	.2405	.2430	.2455	.2480	.2504	.2529
1.8	.2553	.2577	.2601	.2625	.2648	.2672	.2695	.2718	.2742	.2765
1.9	.2788	.2810	.2833	.2856	.2878	.2900	.2923	.2945	.2967	.2989
2.0	.3010	.3032	.3054	.3075	.3096	.3118	.3139	.3160	.3181	.3201
2.1	.3222	.3243	.3263	.3284	.3304	.3324	.3345	.3365	.3385	.3404
2.2	.3424	.3444	.3464	.3483	.3502	.3522	.3541	.3560	.3579	.3598
2.3	.3617	.3636	.3655	.3674	.3692	.3711	.3729	.3747	.3766	.3784
2.4	.3802	.3820	.3838	.3856	.3874	.3892	.3909	.3927	.3945	.3962
2.5	.3979	.3997	.4014	.4031	.4048	.4065	.4082	.4099	.4116	.4133
2.6	.4150	.4166	.4183	.4200	.4216	.4232	.4249	.4265	.4281	.4298
2.7	.4314	.4330	.4346	.4362	.4378	.4393	.4409	.4425	.4440	.4456
2.8	.4472	.4487	.4502	.4518	.4533	.4548	.4564	.4579	.4594	.4609
2.9	.4624	.4639	.4654	.4669	.4683	.4698	.4713	.4728	.4742	.4757
3.0	.4771	.4786	.4800	.4814	.4829	.4843	.4857	.4871	.4886	.4900
3.1	.4914	.4928	.4942	.4955	.4969	.4983	.4997	.5011	.5024	.5038
3.2	.5051	.5065	.5079	.5092	.5105	.5119	.5132	.5145	.5159	.5172
3.3	.5185	.5198	.5211	.5224	.5237	.5250	.5263	.5276	.5289	.5302
3.4	.5315	.5328	.5340	.5353	.5366	.5378	.5391	.5403	.5416	.5428
3.5	.5441	.5453	.5465	.5478	.5490	.5502	.5514	.5527	.5539	.5551
3.6	.5563	.5575	.5587	.5599	.5611	.5623	.5635	.5647	.5658	.5670
3.7	.5682	.5694	.5705	.5717	.5729	.5740	.5752	.5763	.5775	.5786
3.8	.5798	.5809	.5821	.5832	.5843	.5855	.5866	.5877	.5888	.5899
3.9	.5911	.5922	.5933	.5944	.5955	.5966	.5977	.5988	.5999	.6010
4.0	.6021	.6031	.6042	.6053	.6064	.6075	.6085	.6096	.6107	.6117
4.1	.6128	.6138	.6149	.6160	.6170	.6180	.6191	.6201	.6212	.6222
4.2	.6232	.6243	.6253	.6263	.6274	.6284	.6294	.6304	.6314	.6325
4.3	.6335	.6345	.6355	.6365	.6375	.6385	.6395	.6405	.6415	.6425
4.4	.6435	.6444	.6454	.6464	.6474	.6484	.6493	.6503	.6513	.6522
4.5	.6532	.6542	.6551	.6561	.6571	.6580	.6590	.6599	.6609	.6618
4.6	.6628	.6637	.6646	.6656	.6665	.6675	.6684	.6693	.6702	.6712
4.7	.6721	.6730	.6739	.6749	.6758	.6767	.6776	.6785	.6794	.6803
4.8	.6812	.6821	.6830	.6839	.6848	.6857	.6866	.6875	.6884	.6893
4.9	.6902	.6911	.6920	.6928	.6937	.6946	.6955	.6964	.6972	.6981
5.0	.6990	.6998	.7007	.7016	.7024	.7033	.7042	.7050	.7059	.7067
5.1	.7076	.7084	.7093	.7101	.7110	.7118	.7126	.7135	.7143	.7152
5.2	.7160	.7168	.7177	.7185	.7193	.7202	.7210	.7218	.7226	.7235
5.3	.7243	.7251	.7259	.7267	.7275	.7284	.7292	.7300	.7308	.7316
5.4	.7324	.7332	.7340	.7348	.7356	.7364	.7372	.7380	.7388	.7396

TABLE B COMMON LOGARITHMS

n	0	1	2	3	4	5	6	7	8	9
5.5	.7404	.7412	.7419	.7427	.7435	.7443	.7451	.7459	.7466	.7474
5.6	.7482	.7490	.7497	.7505	.7513	.7520	.7528	.7536	.7543	.7551
5.7	.7559	.7566	.7574	.7582	.7589	.7597	.7604	.7612	.7619	.7627
5.8	.7634	.7642	.7649	.7657	.7664	.7672	.7679	.7686	.7694	.7701
5.9	.7709	.7716	.7723	.7731	.7738	.7745	.7752	.7760	.7767	.7774
6.0	.7782	.7789	.7796	.7803	.7810	.7818	.7825	.7832	.7839	.7846
6.1	.7853	.7860	.7868	.7875	.7882	.7889	.7896	.7903	.7910	.7917
6.2	.7924	.7931	.7938	.7945	.7952	.7959	.7966	.7973	.7980	.7987
6.3	.7993	.8000	.8007	.8014	.8021	.8028	.8035	.8041	.8048	.8055
6.4	.8062	.8069	.8075	.8082	.8089	.8096	.8102	.8109	.8116	.8122
6.5	.8129	.8136	.8142	.8149	.8156	.8162	.8169	.8176	.8182	.8189
6.6	.8195	.8202	.8209	.8215	.8222	.8228	.8235	.8241	.8248	.8254
6.7	.8261	.8267	.8274	.8280	.8287	.8293	.8299	.8306	.8312	.8319
6.8	.8325	.8331	.8338	.8344	.8351	.8357	.8363	.8370	.8376	.8382
6.9	.8388	.8395	.8401	.8407	.8414	.8420	.8426	.8432	.8439	.8445
7.0	.8451	.8457	.8463	.8470	.8476	.8482	.8488	.8494	.8500	.8506
7.1	.8513	.8519	.8525	.8531	.8537	.8543	.8549	.8555	.8561	.8567
7.2	.8573	.8579	.8585	.8591	.8597	.8603	.8609	.8615	.8621	.8627
7.3	.8633	.8639	.8645	.8651	.8657	.8663	.8669	.8675	.8681	.8686
7.4	.8692	.8698	.8704	.8710	.8716	.8722	.8727	.8733	.8739	.8745
7.5	.8751	.8756	.8762	.8768	.8774	.8779	.8785	.8791	.8797	.8802
7.6	.8808	.8814	.8820	.8825	.8831	.8837	.8842	.8848	.8854	.8859
7.7	.8865	.8871	.8876	.8882	.8887	.8893	.8899	.8904	.8910	.8915
7.8	.8921	.8927	.8932	.8938	.8943	.8949	.8954	.8960	.8965	.8971
7.9	.8976	.8982	.8987	.8993	.8998	.9004	.9009	.9015	.9020	.9025
8.0	.9031	.9036	.9042	.9047	.9053	.9058	.9063	.9069	.9074	.9079
8.1	.9085	.9090	.9096	.9101	.9106	.9112	.9117	.9122	.9128	.9133
8.2	.9138	.9143	.9149	.9154	.9159	.9165	.9170	.9175	.9180	.9186
8.3	.9191	.9196	.9201	.9206	.9212	.9217	.9222	.9227	.9232	.9238
8.4	.9243	.9248	.9253	.9258	.9263	.9269	.9274	.9279	.9284	.9289
8.5	.9294	.9299	.9304	.9309	.9315	.9320	.9325	.9330	.9335	.9340
8.6	.9345	.9350	.9355	.9360	.9365	.9370	.9375	.9380	.9385	.9390
8.7	.9395	.9400	.9405	.9410	.9415	.9420	.9425	.9430	.9435	.9440
8.8	.9445	.9450	.9455	.9460	.9465	.9469	.9474	.9479	.9484	.9489
8.9	.9494	.9499	.9504	.9509	.9513	.9518	.9523	.9528	.9533	.9538
9.0	.9542	.9547	.9552	.9557	.9562	.9566	.9571	.9576	.9581	.9586
9.1	.9590	.9595	.9600	.9605	.9609	.9614	.9619	.9624	.9628	.9633
9.2	.9638	.9643	.9647	.9652	.9657	.9661	.9666	.9671	.9675	.9680
9.3	.9685	.9689	.9694	.9699	.9703	.9708	.9713	.9717	.9722	.9727
9.4	.9731	.9736	.9741	.9745	.9750	.9754	.9759	.9763	.9768	.9773
9.5	.9777	.9782	.9786	.9791	.9795	.9800	.9805	.9809	.9814	.9818
9.6	.9823	.9827	.9832	.9836	.9841	.9845	.9850	.9854	.9859	.9863
9.7	.9868	.9872	.9877	.9881	.9886	.9890	.9894	.9899	.9903	.9908
9.8	.9912	.9917	.9921	.9926	.9930	.9934	.9939	.9943	.9948	.9952
9.9	.9956	.9961	.9965	.9969	.9974	.9978	.9983	.9987	.9991	.9996

TABLE C TRIGONOMETRIC FUNCTIONS (DEGREES)

Deg.	Sin	Tan	Cot	Cos		Deg.	Sin	Tan	Cot	Cos	
0.0	0.00000	0.00000	∞	1.0000	**90.0**	**6.0**	0.10453	0.10510	9.514	0.9945	**84.0**
.1	.00175	.00175	573.0	1.0000	89.9	.1	.10626	.10687	9.357	.9943	83.9
.2	.00349	.00349	286.5	1.0000	.8	.2	.10800	.10863	9.205	.9942	.8
.3	.00524	.00524	191.0	1.0000	.7	.3	.10973	.11040	9.058	.9940	.7
.4	.00698	.00698	143.24	1.0000	.6	.4	.11147	.11217	8.915	.9938	.6
.5	.00873	.00873	114.59	1.0000	.5	.5	.11320	.11394	8.777	.9936	.5
.6	.01047	.01047	95.49	0.9999	.4	.6	.11494	.11570	8.643	.9934	.4
.7	.01222	.01222	81.85	.9999	.3	.7	.11667	.11747	8.513	.9932	.3
.8	.01396	.01396	71.62	.9999	.2	.8	.11840	.11924	8.386	.9930	.8
.9	.01571	.01571	63.66	.9999	89.1	.9	.12014	.12101	8.264	.9928	83.1
1.0	0.01745	0.01746	57.29	0.9998	**89.0**	**7.0**	0.12187	0.12278	8.144	0.9925	**83.0**
.1	.01920	.01920	52.08	.9998	88.9	.1	.12360	.12456	8.028	.9923	82.9
.2	.02094	.02095	47.74	.9998	.8	.2	.12533	.12633	7.916	.9921	.8
.3	.02269	.02269	44.07	.9997	.7	.3	.12706	.12810	7.806	.9919	.7
.4	.02443	.02444	40.92	.9997	.6	.4	.12880	.12988	7.700	.9917	.6
.5	.02618	.02619	38.19	.9997	.5	.5	.13053	.13165	7.596	.9914	.5
.6	.02792	.02793	35.80	.9996	.4	.6	.13226	.13343	7.495	.9912	.4
.7	.02967	.02968	33.69	.9996	.3	.7	.13399	.13521	7.396	.9910	.3
.8	.03141	.03143	31.82	.9995	.2	.8	.13572	.13698	7.300	.9907	.2
.9	.03316	.03317	30.14	.9995	88.1	.9	.13744	.13876	7.207	.9905	82.1
2.0	0.03490	0.03492	28.64	0.9994	**88.0**	**8.0**	0.13917	0.14054	7.115	0.9903	**82.0**
.1	.03664	.03667	27.27	.9993	87.9	.1	.14090	.14232	7.026	.9900	81.9
.2	.03839	.03842	26.03	.9993	.8	.2	.14263	.14410	6.940	.9898	.8
.3	.04013	.04016	24.90	.9992	.7	.3	.14436	.14588	6.855	.9895	.7
.4	.04188	.04191	23.86	.9991	.6	.4	.14608	.14767	6.772	.9893	.6
.5	.04362	.04366	22.90	.9990	.5	.5	.14781	.14945	6.691	.9890	.5
.6	.04536	.04541	22.02	.9990	.4	.6	.14954	.15124	6.612	.9888	.4
.7	.04711	.04716	21.20	.9989	.3	.7	.15126	.15302	6.535	.9885	.3
.8	.04885	.04891	20.45	.9988	.2	.8	.15299	.15481	6.460	.9882	.2
.9	.05059	.05066	19.74	.9987	87.1	.9	.15471	.15660	6.386	.9880	81.1
3.0	0.05234	0.05241	19.081	0.9986	**87.0**	**9.0**	0.15643	0.15838	6.314	0.9877	**81.0**
.1	.05408	.05416	18.464	.9985	86.9	.1	.15816	.16017	6.243	.9874	80.9
.2	.05582	.05591	17.886	.9984	.8	.2	.15988	.16196	6.174	.9871	.8
.3	.05756	.05766	17.343	.9983	.7	.3	.16160	.16376	6.107	.9869	.7
.4	.05931	.05941	16.832	.9982	.6	.4	.16333	.16555	6.041	.9866	.6
.5	.06105	.06116	16.350	.9981	.5	.5	.16505	.16734	5.976	.9863	.5
.6	.06279	.06291	15.895	.9980	.4	.6	.16677	.16914	5.912	.9860	.4
.7	.06453	.06467	15.464	.9979	.3	.7	.16849	.17093	5.850	.9857	.3
.8	.06627	.06642	15.056	.9978	.2	.8	.17021	.17273	5.789	.9854	.2
.9	.06802	.06817	14.669	.9977	86.1	.9	.17193	.17453	5.730	.9851	80.1
4.0	0.06976	0.06993	14.301	0.9976	**86.0**	**10.0**	0.1736	0.1763	5.671	0.9848	**80.0**
.1	.07150	.07168	13.951	.9974	85.9	.1	.1754	.1781	5.614	.9845	79.9
.2	.07324	.07344	13.617	.9973	.8	.2	.1771	.1799	5.558	.9842	.8
.3	.07498	.07519	13.300	.9972	.7	.3	.1788	.1817	5.503	.9839	.7
.4	.07672	.07695	12.996	.9971	.6	.4	.1805	.1835	5.449	.9836	.6
.5	.07846	.07870	12.706	.9969	.5	.5	.1822	.1853	5.396	.9833	.5
.6	.08020	.08046	12.429	.9968	.4	.6	.1840	.1871	5.343	.9829	.4
.7	.08194	.08221	12.163	.9966	.3	.7	.1857	.1890	5.292	.9826	.3
.8	.08368	.08397	11.909	.9965	.2	.8	.1874	.1908	5.242	.9823	.2
.9	.08542	.08573	11.664	.9963	85.1	.9	.1891	.1926	5.193	.9820	79.1
5.0	0.08716	0.08749	11.430	0.9962	**85.0**	**11.0**	0.1908	0.1944	5.145	0.9816	**79.0**
.1	.08889	.08925	11.205	.9960	84.9	.1	.1925	.1962	5.079	.9813	78.9
.2	.09063	.09101	10.988	.9959	.8	.2	.1942	.1980	5.050	.9810	.8
.3	.09237	.09277	10.780	.9957	.7	.3	.1959	.1998	5.005	.9806	.7
.4	.09411	.09453	10.579	.9956	.6	.4	.1977	.2016	4.959	.9803	.6
.5	.09585	.09629	10.385	.9954	.5	.5	.1994	.2035	4.915	.9799	.5
.6	.09758	.09805	10.199	.9952	.4	.6	.2011	.2053	4.872	.9796	.4
.7	.09932	.09981	10.019	.9951	.3	.7	.2028	.2071	4.829	.9792	.3
.8	.10106	.10158	9.845	.9949	.2	.8	.2045	.2089	4.787	.9789	.2
.9	.10279	.10334	9.677	.9947	84.1	.9	.2062	.2107	4.745	.9785	78.1
6.0	0.10453	0.10510	9.514	0.9945	**84.0**	**12.0**	0.2079	0.2126	4.705	0.9781	**78.0**
	Cos	Cot	Tan	Sin	Deg.		Cos	Cot	Tan	Sin	Deg.

TABLE C TRIGONOMETRIC FUNCTIONS (DEGREES)

Deg.	Sin	Tan	Cot	Cos		Deg.	Sin	Tan	Cot	Cos	
12.0	0.2079	0.2126	4.705	0.9781	**78.0**	**18.0**	0.3090	0.3249	3.078	0.9511	**72.0**
.1	.2096	.2144	4.665	.9778	77.9	.1	.3107	.3269	3.060	.9505	71.9
.2	.2113	.2162	4.625	.9774	.8	.2	.3123	.3288	3.042	.9500	.8
.3	.2130	.2180	4.586	.9770	.7	.3	.3140	.3307	3.024	.9494	.7
.4	.2147	.2199	4.548	.9767	.6	.4	.3156	.3327	3.006	.9489	.6
.5	.2164	.2217	4.511	.9763	.5	.5	.3173	.3346	2.989	.9483	.5
.6	.2181	.2235	4.474	.9759	.4	.6	.3190	.3365	2.971	.9478	.4
.7	.2198	.2254	4.437	.9755	.3	.7	.3206	.3385	2.954	.9472	.3
.8	.2215	.2272	4.402	.9751	.2	.8	.3223	.3404	2.937	.9466	.2
.9	.2233	.2290	4.366	.9748	77.1	.9	.3239	.3424	2.921	.9461	71.1
13.0	0.2250	0.2309	4.331	0.9744	**77.0**	**19.0**	0.3256	0.3443	2.904	0.9455	**71.0**
.1	.2267	.2327	4.297	.9740	76.9	.1	.3272	.3463	2.888	.9449	70.9
.2	.2284	.2345	4.264	.9736	.8	.2	.3289	.3482	2.872	.9444	.8
.3	.2300	.2364	4.230	.9732	.7	.3	.3305	.3502	2.856	.9438	.7
.4	.2317	.2382	4.198	.9728	.6	.4	.3322	.3522	2.840	.9432	.6
.5	.2334	.2401	4.165	.9724	.5	.5	.3338	.3541	2.824	.9426	.5
.6	.2351	.2419	4.134	.9720	.4	.6	.3355	.3561	2.808	.9421	.4
.7	.2368	.2438	4.102	.9715	.3	.7	.3371	.3581	2.793	.9415	.3
.8	.2385	.2456	4.071	.9711	.2	.8	.3387	.3600	2.778	.9409	.2
.9	.2402	.2475	4.041	.9707	76.1	.9	.3404	.3620	2.762	.9403	70.1
14.0	0.2419	0.2493	4.011	0.9703	**76.0**	**20.0**	0.3420	0.3640	2.747	0.9397	**70.0**
.1	.2436	.2512	3.981	.9699	75.9	.1	.3437	.3659	2.733	.9391	69.9
.2	.2453	.2530	3.952	.9694	.8	.2	.3453	.3679	2.718	.9385	.8
.3	.2470	.2549	3.923	.9690	.7	.3	.3469	.3699	2.703	.9379	.7
.4	2487	.2568	3.895	.9686	.6	.4	.3486	.3719	2.689	.9373	.6
.5	.2504	.2586	3.867	.9681	.5	.5	.3502	.3739	2.675	.9367	.5
.6	.2521	.2605	3.839	.9677	.4	.6	.3518	.3759	2.660	.9361	.4
.7	.2538	.2623	3.812	.9673	.3	.7	.3535	.3779	2.646	.9354	.3
.8	.2554	.2642	3.785	.9668	.2	.8	.3551	.3799	2.633	.9348	.2
.9	.2571	.2661	3.758	.9664	75.1	.9	.3567	.3819	2.619	.9342	69.1
15.0	0.2588	0.2679	3.732	0.9659	**75.0**	**21.0**	0.3584	0.3839	2.605	0.9336	**69.0**
.1	.2605	.2698	3.706	.9655	74.9	.1	.3600	.3859	2.592	.9330	68.9
.2	.2622	.2717	3.681	.9650	.8	.2	.3616	.3879	2.578	.9323	.8
.3	.2639	.2736	3.655	.9646	.7	.3	.3633	.3899	2.565	.9317	.7
.4	.2656	.2754	3.630	.9641	.6	.4	.3649	.3919	2.552	.9311	.6
.5	.2672	.2773	3.606	.9636	.5	.5	.3665	.3939	2.539	.9304	.5
.6	.2689	.2792	3.582	.9632	.4	.6	.3681	.3959	2.526	.9298	.4
.7	.2706	.2811	3.558	.9627	.3	.7	.3697	.3979	2.513	.9291	.3
.8	.2723	.2830	3.534	.9622	.2	.8	.3714	.4000	2.500	.9285	.2
.9	.2740	.2849	3.511	.9617	74.1	.9	.3730	.4020	2.488	.9278	68.1
16.0	0.2756	0.2867	3.487	0.9613	**74.0**	**22.0**	0.3746	0.4040	2.475	0.9272	**68.0**
.1	.2773	.2886	3.465	.9608	73.9	.1	.3762	.4061	2.463	.9265	67.9
.2	.2790	.2905	3.442	.9603	.8	.2	.3778	.4081	2.450	.9259	.8
.3	.2807	.2924	3.420	.9598	.7	.3	.3795	.4101	2.438	.9252	.7
.4	.2823	.2943	3.398	.9593	.6	.4	.3811	.4122	2.426	.9245	.6
.5	.2840	.2962	3.376	.9588	.5	.5	.3827	.4142	2.414	.9239	.5
.6	.2857	.2981	3.354	.9583	.4	.6	.3843	.4163	2.402	.9232	.4
.7	.2874	.3000	3.333	.9578	.3	.7	.3859	.4183	2.391	.9225	.3
.8	.2890	.3019	3.312	.9573	.2	.8	.3875	.4204	2.379	.9219	.2
.9	.2907	.3038	3.291	.9568	73.1	.9	.3891	.4224	2.367	.9212	67.1
17.0	0.2924	0.3057	3.271	0.9563	**73.0**	**23.0**	0.3907	0.4245	2.356	0.9205	**67.0**
.1	.2940	.3076	3.251	.9558	72.9	.1	.3923	.4265	2.344	.9198	66.9
.2	.2957	.3096	3.230	.9553	.8	.2	.3939	.4286	2.333	.9191	.8
.3	.2974	.3115	3.211	.9548	.7	.3	.3955	.4307	2.322	.9184	.7
.4	.2990	.3134	3.191	.9542	.6	.4	.3971	.4327	2.311	.9178	.6
.5	.3007	.3153	3.172	.9537	.5	.5	.3987	.4348	2.300	.9171	.5
.6	.3024	.3172	3.152	.9532	.4	.6	.4003	.4369	2.289	.9164	.4
.7	.3040	.3191	3.133	.9527	.3	.7	.4019	.4390	2.278	.9157	.3
.8	.3057	.3211	3.115	.9521	.2	.8	.4035	.4411	2.267	.9150	.2
.9	.3074	.3230	3.096	.9516	72.1	.9	.4051	.4431	2.257	.9143	66.1
18.0	0.3090	0.3249	3.078	0.9511	**72.0**	**24.0**	0.4067	0.4452	2.246	0.9135	**66.0**
	Cos	Cot	Tan	Sin	Deg.		Cos	Cot	Tan	Sin	Deg.

TABLE C TRIGONOMETRIC FUNCTIONS (DEGREES)

Deg.	Sin.	Tan	Cot	Cos		Deg.	Sin	Tan	Cot	Cos	
24.0	0.4067	0.4452	2.246	0.9135	**66.0**	**30.0**	0.5000	0.5774	1.7321	0.8660	**60.0**
.1	.4083	.4473	2.236	.9128	65.9	.1	.5015	.5797	1.7251	.8652	59.9
.2	.4099	.4494	2.225	.9121	.8	.2	.5030	.5820	1.7182	.8643	.8
.3	.4115	.4515	2.215	.9114	.7	.3	.5045	.5844	1.7113	.8634	.7
.4	.4131	.4536	2.204	.9107	.6	.4	.5060	.5867	1.7045	.8625	.6
.5	.4147	.4557	2.194	.9100	.5	.5	.5075	.5890	1.6977	.8616	.5
.6	.4163	.4578	2.184	.9092	.4	.6	.5090	.5914	1.6909	.8607	.4
.7	.4179	.4599	2.174	.9085	.3	.7	.5105	.5938	1.6842	.8599	.3
.8	.4195	.4621	2.164	.9078	.2	.8	.5120	.5961	1.6775	.8590	.2
.9	.4210	.4642	2.154	.9070	65.1	.9	.5135	.5985	1.6709	.8581	59.1
25.0	0.4226	0.4663	2.145	0.9063	**65.0**	**31.0**	0.5150	0.6009	1.6643	0.8572	**59.0**
.1	.4242	.4684	2.135	.9056	64.9	.1	.5165	.6032	1.6577	.8563	58.9
.2	.4258	.4706	2.125	.9048	.8	.2	.5180	.6056	1.6512	.8554	.8
.3	.4274	.4727	2.116	.9041	.7	.3	.5195	.6080	1.6447	.8545	.7
.4	.4289	.4748	2.106	.9033	.6	.4	.5210	.6104	1.6383	.8536	.6
.5	.4305	.4770	2.097	.9026	.5	.5	.5225	.6128	1.6319	.8526	.5
.6	.4321	.4791	2.087	.9018	.4	.6	.5240	.6152	1.6255	.8517	.4
.7	.4337	.4813	2.078	.9011	.3	.7	.5255	.6176	1.6191	.8508	.3
.8	.4352	.4834	2.069	.9003	.2	.8	.5270	.6200	1.6128	.8499	.2
.9	.4368	.4856	2.059	.8996	64.1	.9	.5284	.6224	1.6066	.8490	58.1
26.0	0.4384	0.4887	2.050	0.8988	**64.0**	**32.0**	0.5299	0.6249	1.6003	0.8480	**58.0**
.1	.4399	.4899	2.041	.8980	63.9	.1	.5314	.6273	1.5941	.8471	57.9
.2	.4415	.4921	2.032	.8973	.8	.2	.5329	.6297	1.5880	.8462	.8
.3	.4431	.4942	2.023	.8965	.7	.3	.5344	.6322	1.5818	.8453	.7
.4	.4446	.4964	2.014	.8957	.6	.4	.5358	.6346	1.5757	.8443	.6
.5	.4462	.4986	2.006	.8949	.5	.5	.5373	.6371	1.5697	.8434	.5
.6	.4478	.5008	1.997	.8942	.4	.6	.5388	.6395	1.5637	.8425	.4
.7	.4493	.5029	1.988	.8934	.3	.7	.5402	.6420	1.5577	.8415	.3
.8	.4509	.5051	1.980	.8926	.2	.8	.5417	.6445	1.5517	.8406	.2
.9	.4524	.5073	1.971	.8918	63.1	.9	.5432	.6469	1.5458	.8396	57.1
27.0	0.4540	0.5095	1.963	0.8910	**63.0**	**33.0**	0.5446	0.6494	1.5399	0.8387	**57.0**
.1	.4555	.5117	1.954	.8902	62.9	.1	.5461	.6519	1.5340	.8377	56.9
.2	.4571	.5139	1.946	.8894	.8	.2	.5476	.6544	1.5282	.8368	.8
.3	.4586	.5161	1.937	.8886	.7	.3	.5490	.6569	1.5224	.8358	.7
.4	.4602	.5184	1.929	.8878	.6	.4	.5505	.6594	1.5166	.8348	.6
.5	.4617	.5206	1.921	.8870	.5	.5	.5519	.6619	1.5108	.8339	.5
.6	.4633	.5228	1.913	.8862	.4	.6	.5534	.6644	1.5051	.8329	.4
.7	.4648	.5250	1.905	.8854	.3	.7	.5548	.6669	1.4994	.8320	.3
.8	.4664	.5272	1.897	.8846	.2	.8	.5563	.6694	1.4938	.8310	.2
.9	.4679	.5295	1.889	.8838	62.1	.9	.5577	.6720	1.4882	.8300	56.1
28.0	0.4695	0.5317	1.881	0.8829	**62.0**	**34.0**	0.5592	0.6745	1.4826	0.8290	**56.0**
.1	.4710	.5340	1.873	.8821	61.9	.1	.5606	.6771	1.4770	.8281	55.9
.2	.4726	.5362	1.865	.8813	.8	.2	.5621	.6796	1.4715	.8271	.8
.3	.4741	.5384	1.857	.8805	.7	.3	.5635	.6822	1.4659	.8261	.7
.4	.4756	.5407	1.849	.8796	.6	.4	.5650	.6847	1.4605	.8251	.6
.5	.4772	.5430	1.842	.8788	.5	.5	.5664	.6873	1.4550	.8241	.5
.6	.4787	.5452	1.834	.8780	.4	.6	.5678	.6899	1.4496	.8231	.4
.7	.4802	.5475	1.827	.8771	.3	.7	.5693	.6924	1.4442	.8221	.3
.8	.4818	.5498	1.819	.8763	.2	.8	.5707	.6950	1.4388	.8211	.2
.9	.4833	.5520	1.811	.8755	61.1	.9	.5721	.6976	1.4335	.8202	55.1
29.0	0.4848	0.5543	1.804	0.8746	**61.0**	**35.0**	0.5736	0.7002	1.4281	0.8192	**55.0**
.1	.4863	.5566	1.797	.8738	60.9	.1	.5750	.7028	1.4229	.8181	54.9
.2	.4879	.5589	1.789	.8729	.8	.2	.5764	.7054	1.4176	.8171	.8
.3	.4894	.5612	1.782	.8721	.7	.3	.5779	.7080	1.4124	.8161	.7
.4	.4909	.5635	1.775	.8712	.6	.4	.5793	.7107	1.4071	.8151	.6
.5	.4924	.5658	1.767	.8704	.5	.5	.5807	.7133	1.4019	.8141	.5
.6	.4939	.5681	1.760	.8695	.4	.6	.5821	.7159	1.3968	.8131	.4
.7	.4955	.5704	1.753	.8686	.3	.7	.5835	.7186	1.3916	.8121	.3
.8	.4970	.5727	1.746	.8678	.2	.8	.5850	.7212	1.3865	.8111	.2
.9	.4985	.5750	1.739	.8669	60.1	.9	.5864	.7239	1.3814	.8100	54.1
30.0	0.5000	0.5774	1.732	0.8660	**60.0**	**36.0**	0.5878	0.7265	1.3764	0.8090	**54.0**
	Cos	Cot	Tan	Sin	Deg.		Cos	Cot	Tan	Sin	Deg.

TABLE C TRIGONOMETRIC FUNCTIONS (DEGREES)

Deg.	Sin	Tan	Cot	Cos		Deg.	Sin	Tan	Cot	Cos	
36.0	0.5878	0.7265	1.3764	0.8090	**54.0**	**40.5**	0.6494	0.8541	1.1708	0.7604	**49.5**
.1	.5892	.7292	1.3713	.8080	53.9	.6	.6508	.8571	1.1667	.7593	.4
.2	.5906	.7319	1.3663	.8070	.8	.7	.6521	.8601	1.1626	.7581	.3
.3	.5920	.7346	1.3613	.8059	.7	.8	.6534	.8632	1.1585	.7570	.2
.4	.5934	.7373	1.3564	.8049	.6	.9	.6547	.8662	1.1544	.7559	49.1
.5	.5948	.7400	1.3514	.8039	.5	**41.0**	0.6561	0.8693	1.1504	0.7547	**49.0**
.6	.5962	.7427	1.3465	.8028	.4	.1	.6574	.8724	1.1463	.7536	48.9
.7	.5976	.7454	1.3416	.8018	.3	.2	.6587	.8754	1.1423	.7524	.8
.8	.5990	.7481	1.3367	.8007	.2	.3	.6600	.8785	1.1383	.7513	.7
.9	.6004	.7508	1.3319	.7997	53.1	.4	.6613	.8816	1.1343	.7501	.6
37.0	0.6018	0.7536	1.3270	0.7986	**53.0**	.5	.6626	.8847	1.1303	.7490	.5
.1	.6032	.7563	1.3222	.7976	52.9	.6	.6639	.8878	1.1263	.7478	.4
.2	.6046	.7590	1.3175	.7965	.8	.7	.6652	.8910	1.1224	.7466	.3
.3	.6060	.7618	1.3127	.7955	.7	.8	.6665	.8941	1.1184	.7455	.2
.4	.6074	.7646	1.3079	.7944	.6	.9	.6678	.8972	1.1145	.7443	48.1
.5	.6088	.7673	1.3032	.7934	.5	**42.0**	0.6691	0.9004	1.1106	0.7431	**48.0**
.6	.6101	.7701	1.2985	.7923	.4	.1	.6704	.9036	1.1067	.7420	47.9
.7	.6115	.7729	1.2938	.7912	.3	.2	.6717	.9067	1.1028	.7408	.8
.8	.6129	.7757	1.2892	.7902	.2	.3	.6730	.9099	1.0990	.7396	.7
.9	.6143	.7785	1.2846	.7891	52.1	.4	.6743	.9131	1.0951	.7385	.6
38.0	0.6157	0.7813	1.2799	0.7880	**52.0**	.5	.6756	.9163	1.0913	.7373	.5
.1	.6170	.7841	1.2753	.7869	51.9	.6	.6769	.9195	1.0875	.7361	.4
.2	.6184	.7869	1.2708	7859	.8	.7	.6782	.9228	1.0837	.7349	.3
.3	.6198	.7898	1.2662	.7848	.7	.8	.6794	.9260	1.0799	.7337	.2
.4	.6211	.7926	1.2617	.7837	.6	.9	.6807	.9293	1.0761	.7325	47.1
.5	.6225	.7954	1.2572	.7826	.5	**43.0**	0.6820	0.9325	1.0724	0.7314	**47.0**
.6	.6239	.7983	1.2527	.7815	.4	.1	.6833	.9358	1.0686	.7302	46.9
.7	.6252	.8012	1.2482	.7804	.3	.2	.6845	.9391	1.0649	.7290	.8
.8	.6266	.8040	1.2437	.7793	.2	.3	.6858	.9424	1.0612	.7278	.7
.9	.6280	.8069	1.2393	.7782	51.1	.4	.6871	.9457	1.0575	.7266	.6
39.0	0.6293	0.8098	1.2349	0.7771	**51.0**	.5	.6884	.9490	1.0538	.7254	.5
.1	.6307	.8127	1.2305	.7760	50.9	.6	.6896	.9523	1.0501	.7242	.4
.2	.6320	.8156	1.2261	.7749	.8	.7	.6909	.9556	1.0464	.7230	.3
.3	.6334	.8185	1.2218	.7738	.7	.8	.6921	.9590	1.0428	.7218	.2
.4	.6347	.8214	1.2174	.7727	.6	.9	.6934	.9623	1.0392	.7206	46.1
.5	.6361	.8243	1.2131	.7716	.5	**44.0**	0.6947	0.9657	1.0355	0.7193	**46.0**
.6	.6374	.8273	1.2088	.7705	.4	.1	.6959	.9691	1.0319	.7181	45.9
.7	.6388	.8302	1.2045	.7694	.3	.2	.6972	.9725	1.0283	.7169	.8
.8	.6401	.8332	1.2002	.7683	.2	.3	.6984	.9759	1.0247	.7157	.7
.9	.6414	.8361	1.1960	.7672	50.1	.4	.6997	.9793	1.0212	.7145	.6
40.0	0.6428	0.8391	1.1918	0.7660	**50.0**	.5	.7009	.9827	1.0176	.7133	.5
.1	.6441	.8421	1.1875	.7649	49.9	.6	.7022	.9861	1.0141	.7120	.4
.2	.6455	.8451	1.1833	.7638	.8	.7	.7034	.9896	1.0105	.7108	.3
.3	.6468	.8481	1.1792	.7627	.7	.8	.7046	.9930	1.0070	.7096	.2
.4	.6481	.8511	1.1750	.7615	.6	.9	.7059	.9965	1.0035	.7083	45.1
40.5	0.6494	0.8541	1.1708	0.7604	**49.5**	**45.0**	0.7071	1.0000	1.0000	0.7071	**45.0**
	Cos	Cot	Tan	Sin	Deg.		Cos	Cot	Tan	Sin	Deg.

TABLE D TRIGONOMETRIC FUNCTIONS (RADIANS)

Rad.	Sin	Tan	Cot	Cos	Rad.	Sin	Tan	Cot	Cos
.00	.00000	.00000	∞	1.00000	.50	.47943	.54630	1.8305	.87758
.01	.01000	.01000	99.997	0.99995	.51	.48818	.55936	1.7878	.87274
.02	.02000	.02000	49.993	.99980	.52	.49688	.57256	1.7465	.86782
.03	.03000	.03001	33.323	.99955	.53	.50553	.58592	1.7067	.86281
.04	.03999	.04002	24.987	.99920	.54	.51414	.59943	1.6683	.85771
.05	.04998	.05004	19.983	.99875	.55	.52269	.61311	1.6310	.85252
.06	.05996	.06007	16.647	.99820	.56	.53119	.62695	1.5950	.84726
.07	.06994	.07011	14.262	.99755	.57	.53963	.64097	1.5601	.84190
.08	.07991	.08017	12.473	.99680	.58	.54802	.65517	1.5263	.83646
.09	.08988	.09024	11.081	.99595	.59	.55636	.66956	1.4935	.83094
.10	.09983	.10033	9.9666	.99500	.60	.56464	.68414	1.4617	.82534
.11	.10978	.11045	9.0542	.99396	.61	.57287	.69892	1.4308	.81965
.12	.11971	.12058	8.2933	.99281	.62	.58104	.71391	1.4007	.81388
.13	.12963	.13074	7.6489	.99156	.63	.58914	.72911	1.3715	.80803
.14	.13954	.14092	7.0961	.99022	.64	.59720	.74454	1.3431	.80210
.15	.14944	.15114	6.6166	.98877	.62	.60519	.76020	1.3154	.79608
.16	.15932	.16138	6.1966	.98723	.66	.61312	.77610	1.2885	.78999
.17	.16918	.17166	5.8256	.98558	.67	.62099	.79225	1.2622	.78382
.18	.17903	.18197	5.4954	.98384	.68	.62879	.80866	1.2366	.77757
.19	.18886	.19232	5.1997	.98200	.69	.63654	.82534	1.2116	.77125
.20	.19867	.20271	4.9332	.98007	.70	.64422	.84229	1.1872	.76484
.21	.20846	.21314	4.6917	.97803	.71	.65183	.85953	1.1634	.75836
.22	.21823	.22362	4.4719	.97590	.72	.65938	.87707	1.1402	.75181
.23	.22798	.23414	4.2709	.97367	.73	.66687	.89492	1.1174	.74517
.24	.23770	.24472	4.0864	.97134	.74	.67429	.91309	1.0952	.73847
.25	.24740	.25534	3.9163	.96891	.75	.68164	.93160	1.0734	.73169
.26	.25708	.26602	3.7591	.96639	.76	.68892	.95045	1.0521	.72484
.27	.26673	.27676	3.6133	.96377	.77	.69614	.96967	1.0313	.71791
.28	.27636	.28755	3.4776	.96106	.78	.70328	.98926	1.0109	.71091
.29	.28595	.29841	3.3511	.95824	.79	.71035	1.0092	.99084	.70385
.30	.29552	.30934	3.2327	.95534	.80	.71736	1.0296	.97121	.69671
.31	.30506	.32033	3.1218	.95233	.81	.72429	1.0505	.95197	.68950
.32	.31457	.33139	3.0176	.94924	.82	.73115	1.0717	.93309	.68222
.33	.32404	.34252	2.9195	.94604	.83	.73793	1.0934	.91455	.67488
.34	.33349	.35374	2.8270	.94275	.84	.74464	1.1156	.89635	.66746
.35	.34290	.36503	2.7395	.93937	.85	.75128	1.1383	.87848	.65998
.36	.35227	.37640	2.6567	.93590	.86	.75784	1.1616	.86091	.65244
.37	.36162	.38786	2.5782	.93233	.87	.76433	1.1853	.84365	.64483
.38	.37092	.39941	2.5037	.92866	.88	.77074	1.2097	.82668	.63715
.39	.38019	.41105	2.4328	.92491	.89	.77707	1.2346	.80998	.62941
.40	.38942	.42279	2.3652	.92106	.90	.78333	1.2602	.79355	.62161
.41	.39861	.43463	2.3008	.91712	.91	.78950	1.2864	.77738	.61375
.42	.40776	.44657	2.2393	.91309	.92	.79560	1.3133	.76146	.60582
.43	.41687	.45862	2.1804	.90897	.93	.80162	1.3409	.74578	.59783
.44	.42594	.47078	2.1241	.90475	.94	.80756	1.3692	.73034	.58979
.45	.43497	.48306	2.0702	.90045	.95	.81342	1.3984	.71511	.58168
.46	.44395	.49545	2.0184	.89605	.96	.81919	1.4284	.70010	.57352
.47	.45289	.50797	1.9686	.89157	.97	.82489	1.4592	.68531	.56530
.48	.46178	.52061	1.9208	.88699	.98	.83050	1.4910	.67071	.55702
.49	.47063	.53339	1.8748	.88233	.99	.83603	1.5237	.65631	.54869
.50	.47943	.54630	1.8305	.87758	1.00	.84147	1.5574	.64209	.54030
Rad.	Sin	Tan	Cot	Cos	Rad.	Sin	Tan	Cot	Cos

TABLE D TRIGONOMETRIC FUNCTIONS (RADIANS)

Rad.	Sin.	Tan	Cot	Cos	Rad.	Sin	Tan	Cot	Cos
1.00	.84147	1.5574	.64209	.54030	**1.50**	.99749	14.101	.07091	.07074
1.01	.84683	1.5922	.62806	.53186	1.51	.99815	16.428	.06087	.06076
1.02	.85211	1.6281	.61420	.52337	1.52	.99871	19.670	.05084	.05077
1.03	.85730	1.6652	.60051	.51482	1.53	.99917	24.498	.04082	.04079
1.04	.86240	1.7036	.58699	.50622	1.54	.99953	32.461	.03081	.03079
1.05	.86742	1.7433	.57362	.49757	1.55	.99978	48.078	.02080	.02079
1.06	.87236	1.7844	.56040	.48887	1.56	.99994	92.621	.01080	.01080
1.07	.87720	1.8270	.54734	.48012	1.57	1.00000	1255.8	.00080	.00080
1.08	.88196	1.8712	.53441	.47133	1.58	.99996	− 108.65	− .00920	− .00920
1.09	.88663	1.9171	.52162	.46249	1.59	.99982	− 52.067	− .01921	− .01920
1.10	.89121	1.9648	.50897	.45360	**1.60**	.99957	− 34.233	− .02921	− .02920
1.11	.89570	2.0143	.49644	.44466	1.61	.99923	− 25.495	− .03922	− .03919
1.12	.90010	2.0660	.48404	.43568	1.62	.99879	− 20.307	− .04924	− .04918
1.13	.90441	2.1198	.47175	.42666	1.63	.99825	− 16.871	− .05927	− .05917
1.14	.90863	2.1759	.45959	.41759	1.64	.99761	− 14.427	− .06931	− .06915
1.15	.91276	2.2345	.44753	.40849	1.65	.99687	− 12.599	− .07937	− .07912
1.16	.91680	2.2958	.43558	.39934	1.66	.99602	− 11.181	− .08944	− .08909
1.17	.92075	2.3600	.42373	.39015	1.67	.99508	− 10.047	− .09953	− .09904
1.18	.92461	2.4273	.41199	.38092	1.68	.99404	− 9.1208	− .10964	− .10899
1.19	.92837	2.4979	.40034	.37166	1.69	.99290	− 8.3492	− .11977	− .11892
1.20	.93204	2.5722	.38878	.36236	**1.70**	.99166	− 7.6966	− .12993	− .12884
1.21	.93562	2.6503	.37731	.35302	1.71	.99033	− 7.1373	− .14011	− .13875
1.22	.93910	2.7328	.36593	.34365	1.72	.98889	− 6.6524	− .15032	− .14865
1.23	.94249	2.8198	.35463	.33424	1.73	.98735	− 6.2281	− .16056	− .15853
1.24	.94578	2.9119	.34341	.32480	1.74	.98572	− 5.8535	− .17084	− .16840
1.25	.94898	3.0096	.33227	.31532	1.75	.98399	− 5.5204	− .18115	− .17825
1.26	.95209	3.1133	.32121	.30582	1.76	.98215	− 5.2221	− .19149	− .18808
1.27	.95510	3.2236	.31021	.29628	1.77	.98022	− 4.9534	− .20188	− .19789
1.28	.95802	3.3413	.29928	.28672	1.78	.97820	− 4.7101	− .21231	− .20768
1.29	.96084	3.4672	.28842	.27712	1.79	.97607	− 4.4887	− .22278	− .21745
1.30	.96356	3.6021	.27762	.26750	**1.80**	.97385	− 4.2863	− .23330	− .22720
1.31	.96618	3.7471	.26687	.25785	1.81	.97153	− 4.1005	− .24387	− .23693
1.32	.96872	3.9033	.25619	.24818	1.82	.96911	− 3.9294	− .25449	− .24663
1.33	.97115	4.0723	.24556	.23848	1.83	.96659	− 3.7712	− .26517	− .25631
1.34	.97348	4.2556	.23498	.22875	1.84	.96398	− 3.6245	− .27590	− .26596
1.35	.97572	4.4552	.22446	.21901	1.85	.96128	− 3.4881	− .28669	− .27559
1.36	.97786	4.6734	.21398	.20924	1.86	.95847	− 3.3608	− .29755	− .28519
1.37	.97991	4.9131	.20354	.19945	1.87	.95557	− 2.2419	− .30846	− .29476
1.38	.98185	5.1774	.19315	.18964	1.88	.95258	− 3.1304	− .31945	− .30430
1.39	.98370	5.4707	.18279	.17981	1.89	.94949	− 3.0257	− 33.051	− .31381
1.40	.98545	5.7979	.17248	.16997	**1.90**	.94630	− 2.9271	− .34164	− .32329
1.41	.98710	6.1654	.16220	.16010	1.91	.94302	− 2.8341	− .35284	− .33274
1.42	.98865	6.5811	.15195	.15023	1.92	.93965	− 2.7463	− .36413	.34215
1.43	.99010	7.0555	.14173	.14033	1.93	.93618	− 2.6632	− .37549	− .35153
1.44	.99146	7.6018	.13155	.13042	1.94	.93262	− 2.5843	− .38695	− .36087
1.45	.99271	8.2381	.12139	.12050	1.95	.92896	− 2.5095	− .39849	− .37018
1.46	.99387	8.9886	.11125	.11057	1.96	.92521	− 2.4383	− .41012	− .37945
1.47	.99492	9.8874	.10114	.10063	1.97	.92137	− 2.3705	− .42185	− .38868
1.48	.99588	10.983	.09105	.09067	1.98	.91744	− 2.3058	− .43368	− .39788
1.49	.99674	12.350	.08097	.08071	1.99	.91341	− 2.2441	− .44562	− .40703
1.50	.99749	14.101	.07091	.07074	**2.00**	.90930	− 2.1850	− .45766	− .41615
Rad.	Sin	Tan	Cot	Cos	Rad.	Sin	Tan	Cot	Cos

Answers to Selected Exercises

PROBLEM SET 1.1

1. 122   **3.** $\dfrac{7}{12}$ **5.** $\dfrac{6}{35}$ **7.** 2.71   **9.** 0.0062   **11.** {March, May}   **13.** {2, 3, 4, 5, 6, 7, 8, 9, 10}

15. {$x|x$ is a natural number between $\frac{1}{2}$ and $6\frac{1}{2}$}   **17.** {$x|x$ is an even integer between 3 and 17}   **19.** *I, Q, R*

21. *Q, R*   **23.** *I, Q, R*   **25.** *H, R*   **27.** *Q, R*   **29.** *N, I, Q, R*   **31.** True   **33.** False

35. True (0.5 = 0.5000 ...)   **37.** True   **39.** False   **41.** 1   **43.** 0   **45.** -1.2 **47.** $5 < 8$ **49.** $-2 > x$

51. $10 \geq r$ **53.** $-2 < y < 3$ **55.** ◄―――――――――――► (0, 5)   **57.** ◄―――――――――――► (-4, 0)

59. ◄―――――――――――► (-4, 0, 3)   **61.** ◄―――――――――――► (-3, 0, 5)

63. 1.27 and 1.23 ◄―――――――――――► (1.23, 1.27)   **65.** Yes, they can.

PROBLEM SET 1.2

1. *I, Q, R*   **3.** *H, R*   **5.** *Q, R*   **7.** Associative   **9.** Identity   **11.** Distributive   **13.** Closure

15. Commutative   **17.** Inverse   **19.** Substitution   **21.** Multiplication property of equality

23. Zero-multiplier   **25.** Zero-product   **27.** Addition property of equality   **29.** Subtraction definition

31. Multiplication property   **33.** Double negative   **35.** Addition property   **37.** Addition of quotients

39. Division of quotients   **41.** Equality of quotients   **43.** Division property   **45.** 11   **47.** 0   **49.** 3

51. Real number   **53.** 1   **55.** 4^2 **57.** $\dfrac{6}{7}$ **59.** Identity; inverse; associative; inverse; identity

PROBLEM SET 1.3

1. 12   **3.** 6   **5.** $\dfrac{1}{8}$ **7.** $\dfrac{1}{12}$ **9.** 2   **11.** -11.4 **13.** -7 **15.** 2   **17.** 80   **19.** 60   **21.** -50

23. 5   **25.** -1 **27.** 16   **29.** $\dfrac{-1}{8}$ **31.** 0.343   **33.** -17 **35.** -68 **37.** 50   **39.** -42 **41.** 81

43. 20   **45.** 502   **47.** 5   **49.** 86   **51.** 289   **53.** 4   **55.** 3   **57.** 0   **59.** 5   **61.** 0.04; 0.02

63. $6105.10; $35,097.50

CHAPTER 1   REVIEW EXERCISES

1. {5, 10, 15, 20, 25, 30}   **2.** {$x|x$ is an even integer between 7 and 21}   **3.** *I, Q, R*   **4.** Commutative

5. Distributive   **6.** Zero-product   **7.** Addition property of equality   **8.** Double negative

9. Division of quotients   **10.** $A = -3, B = 1$ **11.** $x < 5$ **12.**

13.

   14. 3   **15.** -1 **16.** 24   **17.** -16 **18.** 7   **19.** 41   **20.** 54

21. 7   **22.** 3   **23.** 4   **24.** 5   **25.** 14

PROBLEM SET 2.1

1. 8   **3.** 40   **5.** 16   **7.** -64 **9.** 1000   **11.** $\dfrac{1}{32}$ **13.** 5.2×10^5 **15.** 1.3×10^6 **17.** 5×10^{-3}

19. 1.23×10^{-6} **21.** 70,000   **23.** 63,100,000   **25.** 0.082   **27.** 0.65   **29.** z^{26} **31.** $(a + b)^6$

33. $216x^{12}y^9$ **35.** a^4 **37.** $\dfrac{x^6y^9}{125}$ **39.** $\dfrac{1}{25}$ **41.** $\dfrac{1}{k^2}$ **43.** $-4a^5b^6c^3$ **45.** $-12x^{7n}y^{4k}$ **47.** $\dfrac{m^{16}}{2n^{12}p^{26}}$

49. x^{15} **51.** $16c^{12}$ **53.** $\dfrac{32r^5}{243}$ **55.** 1   **57.** $\dfrac{1}{9c^4}$ **59.** $32a^5b^{10}c^{15}d^{20}$ **61.** $\dfrac{a^5}{bc^9}$

63. 3.2×10^3 **65.** t^{13} **67.** 2×10^{-9} **69.** $8xy^{12}$ **71.** r^{15} **73.** 5×10^4 **75.** $(x - y)^9$ **77.** $\dfrac{1}{x^{20}}$

79. a^{30} **81.** 1   **83.** 2.2444×10^6 **85.** 2.0731×10^{24} **87.** 2^{46} **89.** 10^{-3} seconds

91. 5×10^5; 2.27×10^{-9} **93.** 5×10^{-7}

PROBLEM SET 2.2

1. -5 **3.** 12   **5.** $6x + 42$ **7.** z^5 **9.** $4x^4y^4$ **11.** 4; 7; binomial   **13.** 5; 2; trinomial

15. 10; 4; polynomial   **17.** -1; 5; monomial   **19.** $-3pq^2$ **21.** $14t^3 + 6t^2 + t + 11$

23. $2x^4 + 4x^3 + 2x^2 + 10x + 1$ **25.** $5v^3 - 9v^2 + 3v - 9$ **27.** $x^2 + 2x - 10$ **29.** $-y^3 + 4y^2 - 8y + 2$

31. $-6a^3b^5 + 9a^4b^4$ **33.** $4t^3 - 3t^2 + 7t + 2$ **35.** $k^4 - 6k^3 + 10k^2 - 3k - 2$ **37.** $2y^4 + 3y^3 - 19y^2 - 22y - 6$

39. $18t^2 + 71t - 45$ **41.** $35y^2 + 36y - 20$ **43.** $g^2 - 36$ **45.** $4t^2 - 49$ **47.** $16m^4 - 25n^2$ **49.** $x^2 + 4x + 4$

51. $4a^2 - 4a + 1$ **53.** $16u^4 - 40u^2v + 25v^2$ **55.** $-t$ **57.** $y^4 - 6y^3 - 3y^2 + 2y - 9$ **59.** $15x^2 - 17x - 4$

61. $-a^3b^5 + 3a^7b^2 + a^2b^8$ **63.** $a^3 + a^2 - 18a + 18$ **65.** $18x^2 + 3x - 5$ **67.** $2x^4 - 4x^3 - 6x^2 + 12x$

69. $1 + 2t^5s^3 - 3t^{10}s^6$ **71.** $12r^3 - 4r^2t - 3rt - 2t^2 + 14r^4 + 7r^3t + r^2$ **73.** $5x^3 + x^2 + 14x$ **75.** $14x^2 + 42x - 7$

77. $4a^2 + b^2 + c^2 + 4ab - 4ac - 2bc$ **79.** $24x^3 + 26x^2 - 13x - 10$ **81.** $x^3 + 15x^2 + 75x + 125$

83. $-0.01Q^2 + 6800Q - 150{,}000$; $-0.006Q^2 + 2700Q - 74{,}000$; $0.03Q^2 + 43Q + 2400$; $0.007Q^2 + 12Q + 41{,}000$

85. $2t^4 + 40t^3 + 252t^2 + 1510t + 100$ **87.** πr^2; $\pi r^2 + 2\pi rh + \pi h^2$; $2\pi rh + \pi h^2$

PROBLEM SET 2.3

1. 13 R 3   **3.** 274 R 3   **5.** $-21x^6y^{10}$ **7.** $-2a + b$ **9.** $2x^2 - 5x - 22$ **11.** $4x^3 - 3x^2 + 6x$

13. $2m^2n - m^3 + 3m^7n^5$ **15.** $a^2 - 5a - 2$ **17.** $5b^2 + 8b + 21$ R 65   **19.** $x^2 + x + 1$

21. $u^3 + 2$ R 4   **23.** $h^2 - 2h - 3$ **25.** $6p^4 - 5pq + 3q^7$ **27.** $2z^3 + 3z^2 + 14z + 41$ R 122   **29.** $t + 7$

31. $2y^2 + 5y - 3$ **33.** $b^5 - b^4 + b^3 - b^2 + b - 1$ R 2

PROBLEM SET 2.4

1. $8x^2 - 14x$ **3.** $9x^2 - 4$ **5.** $9r^2 - 30rs + 25s^2$ **7.** $a^2 + a - 20$ **9.** $20r^2 + 13rs - 21s^2$

11. $x^2(2x^2 - 3x - 13)$ **13.** $7x^2yz(3z^3 - 2xy)$ **15.** $(7 - y)(ax - 1)$ **17.** $(4m + 5n)(4m - 5n)$

19. $\left(\dfrac{1}{3}uv - ab^2\right)\left(\dfrac{1}{3}uv + ab^2\right)$ **21.** $(m + n + a + b)(m + n - a - b)$ **23.** $(a - 5b)(a^2 + 5b + 25b^2)$

25. $\left(r - \dfrac{1}{4}\right)\left(r^2 + \dfrac{1}{4}r + \dfrac{1}{16}\right)$ **27.** $(a + 8)(a - 4)$ **29.** $(r - 15s)(r + 4s)$ **31.** $(u - 9v)(u + 6v)$

33. $(x^2 + 6)(x^2 + 1)$ **35.** $(5p - q)(2p + 5q)$ **37.** $(4t - 3s)(2t + 7s)$ **39.** $(4r^2 + 7)(2r^2 + 1)$ **41.** $(a + 2b)(c - 5)$

43. $(6x - 1)(4y + 9)$ **45.** Prime   **47.** $p^2(p + 1)(p^2 - p + 1)$ **49.** $3ab^2(a + b)(a^2 - ab + b^2)(a - b)(a^2 + ab + b^2)$

51. $(m + n)(m^2 - mn + n^2)(m - n)(m^2 + mn + n^2)$ **53.** $3a^2(a + 5)(a - 3)$ **55.** $u(u - v)(w + t)$ **57.** $(t - 4u)(t + 3u)$
59. $(5a - 2b)(3a - 4b)$ **61.** $8(u - 5v)(u^2 + 5uv + 25v^2)$ **63.** $4r^3(r - 5)(r + 4)$ **65.** $2k(1 - 5k^6)$
67. $(x - 3)(x - 5)$ **69.** $(1 - 5t)(1 + 5t + 25t^2)$ **71.** $(a - b)(a - 3)$ **73.** $x(1 - x^2 + x^6 - x^{22})$
75. $3x(x^2 + 4x - 7)$ **77.** $(m + n)(m^2 - mn + n^2)(m - n)(m^2 + mn + n^2)(m^2 + n^2)(m^4 - m^2n^2 + n^4)$ **79.** Prime

81. $(7y - 2z)(5y - z)$ **83.** $(10a + 3b)(100a^2 - 30ab + 9b^2)$ **85.** $\dfrac{-wx}{24EI}(x^3 - 2Lx^2 + L^3)$ **87.** Year n: $P(1 + r)^n$

PROBLEM SET 2.5

1. $\dfrac{35}{24}$ **3.** $\dfrac{7}{18}$ **5.** $\dfrac{14}{15}$ **7.** $(x + 3)(x - 3)$ **9.** $(3t + 2)(2t - 1)$ **11.** $\dfrac{5b^4c^3}{6a}$ **13.** $\dfrac{8}{b}$ **15.** $\dfrac{x - y}{y}$

17. $\dfrac{4x^7y^7u^5z}{25t^5}$ **19.** $a^2(a - 7)$ **21.** $\dfrac{a^3(a - 5)}{(a - 6)(a + 6)}$ **23.** $\dfrac{-3z^{11}}{2x^5y^2u^2v^3w^8}$ **25.** $\dfrac{a - 6}{(a + 6)^2(a + 1)}$ **27.** $\dfrac{x - 7}{x - 8}$

29. $\dfrac{x^8 + x^3 + 1}{x^{10}}$ **31.** $\dfrac{2ab - b^3 - a^3}{a^2b^2}$ **33.** $\dfrac{2a^2 + 3a - 40}{(a - 5)(2a - 5)}$ **35.** $\dfrac{3m + 75}{(m - 7)(m - 5)(m + 5)}$ **37.** $\dfrac{6x^2 - 4x + 5}{2x^2}$

39. $\dfrac{15z}{14x}$ **41.** $\dfrac{9t^2 - 1}{6t^2 - t}$ **43.** $\dfrac{7ab - 4bc + 2ac}{abc^2 - 5ab^2c + 3a^2bc}$ **45.** $\dfrac{1}{8}$ **47.** $\dfrac{x + y}{xy}$ **49.** $\dfrac{t^2 - t + 1}{t}$ **51.** $\dfrac{y^2 + x^2}{x^4y^3}$

53. $\dfrac{-1}{x(x + h)}$ **55.** $\dfrac{-2x - h}{x^2(x + h)^2}$ **57.** $\dfrac{4 - x}{1 + y}$ **59.** $\dfrac{x^2 - 3x}{x + 5}$ **61.** $\dfrac{9x - 37}{(x - 5)(x - 6)(x - 4)}$ **63.** $\dfrac{7(x + 7)}{x^3(x - 2)}$

65. $\dfrac{a^2 + 5a - 45}{(a - 9)(a + 9)}$ **67.** $16t + 32$ **69.** $-x^2 - 200x + 3000$ **71.** $\dfrac{18{,}000^2k^2}{18{,}000k^2 + L^2}$

PROBLEM SET 2.6

1. $x^2 - 9$ **3.** $9a^2b^4c^6$ **5.** 3 **7.** 7 **9.** $\dfrac{1}{2}$ **11.** $\dfrac{3}{5}$ **13.** 0.1 **15.** m^2n^3 **17.** $\dfrac{1}{x^3}$ **19.** $\dfrac{4m}{k^2n^4}$

21. $\dfrac{3\sqrt[3]{3}}{5}$ **23.** 3 **25.** $\sqrt[24]{2}$ **27.** $2ab^2c^3\sqrt[5]{10a^2b^2c^2}$ **29.** $\dfrac{6\sqrt[3]{5x^2}}{5x}$ **31.** $\dfrac{\sqrt[4]{15xy^2}}{3y}$ **33.** $\dfrac{\sqrt{91} + \sqrt{7t}}{13 - t}$

35. $7\sqrt{5} - 14$ **37.** $\dfrac{2\sqrt{11} + \sqrt{22}}{4}$ **39.** $\dfrac{14}{10 + 2\sqrt{11}}$ **41.** $\dfrac{1}{\sqrt{x} + \sqrt{y}}$ **43.** $\dfrac{1}{\sqrt{2x + t + 1} + \sqrt{2x + 1}}$ **45.** $3\sqrt{3}$

47. $\dfrac{3\sqrt[4]{2}}{2}$ **49.** $6\sqrt{2}$ **51.** $(12 + 12x)\sqrt[3]{3x}$ **53.** $4xy^5\sqrt{x}$ **55.** $\dfrac{2x}{yz^2}\sqrt[4]{4y^2z}$ **57.** 4.4911 **59.** 12.9703

61. 0.947473 **63.** 1.25893 **65.** 1.00693 **67.** $\dfrac{C - \sqrt[n]{SC^{n-1}}}{C}$ **69.** $\dfrac{\sqrt{2V_{max}}}{2}$ **71.** $\sqrt{\dfrac{Tm}{2Lm}}$

PROBLEM SET 2.7

1. 10 **3.** $\dfrac{2}{3}$ **5.** 2.15443 **7.** $\dfrac{1}{9}$ **9.** x^9 **11.** $\dfrac{1}{u^2v^3}$ **13.** 5 **15.** $\dfrac{1}{4}$ **17.** 4 **19.** 625 **21.** $\dfrac{1}{3}$

23. $\dfrac{25}{4}$ **25.** $\sqrt[9]{x}$ **27.** $\sqrt{k^3}$ **29.** $\sqrt[6]{\dfrac{1}{x + y}}$ **31.** $(m + n)^{1/2}$ **33.** $u^{5/3}$ **35.** $z^{-4/3}$ **37.** $a^{7/6}$

39. $k^{1/4}$ **41.** $x^{5/6} + x^{17/15}$ **43.** $t^{1/6} + \dfrac{1}{t^{5/6}}$ **45.** $a^{1/8}b^{1/6}$ **47.** $\dfrac{u^{1/8}w^{1/16}}{v^{1/12}}$ **49.** $a^{1/3}b^{17/72}$ **51.** 3

53. $\dfrac{1}{t^{1/3}}$ **55.** $\dfrac{1}{3}$ **57.** $t^{1/6} - \dfrac{1}{t^{8/15}}$ **59.** $\dfrac{b^{1/12}}{a^{1/8}}$ **61.** $\dfrac{125}{8}$ **63.** $x^{2/15}$ **65.** $k^{11/4}$ **67.** $\dfrac{1}{(x + h)^{1/2} + x^{1/2}}$

69. 200 **71.** $a^{7/30}$ **73.** $\dfrac{b^{1/8}}{a^{1/10}c^{1/6}}$ **75.** $\dfrac{m^{19/30}}{n^{5/24}}$ **77.** 46.4159 **79.** 1.01366 **81.** 0.1140515

83. 3.7283×10^3 **85.** $R = \sqrt[3]{2^g/1024} = \dfrac{1}{16}\sqrt[3]{2^{g+2}}$ **87.** $6{,}000{,}000$; $280{,}000{,}000$

CHAPTER 2   REVIEW EXERCISES

1. $\dfrac{1}{1000}$ **2.** x^{21} **3.** 2^{3n} or 8^n **4.** $49a^4b^6c^{2k}$ **5.** $\dfrac{1}{a^4}$ **6.** $\dfrac{a^4b^{12}}{16}$ **7.** $40p^{11}q^8$ **8.** $\dfrac{1}{27}$ **9.** a^2b^{10}

10. $\dfrac{y}{18x^2z^7}$ **11.** 8.76×10^3 **12.** 7.82×10^{-4} **13.** $-11ab^2$ **14.** $17x^2 - 17x - 1$ **15.** $-k^2 - 2k - 6$

16. $30a^3b^5c - 35a^4b^4c^3$ **17.** $6k^2 + 11k - 35$ **18.** $6p^4 - p^3 + 4p^2 + 5p - 6$ **19.** $2xy^4 + 6x^4y^5 - 1$

20. $2x^2 + x - 5$ $R(-3)$ **21.** $t^4 - t^3 + t^2 - t + 1$ **22.** $6x^3y^2(x^6 - 2x^2y^3 + 3y^2)$ **23.** $\left(4x - \dfrac{6}{7}y\right)\left(4x + \dfrac{6}{7}y\right)$

24. $(a - 5b)(a^2 + 5b + 25b^2)$ **25.** $(z + 7)(z - 2)$ **26.** $(3t - 8)(2t - 5)$ **27.** $(2x + 1)(3x - 5y)$

28. $3x^3(x + 5)(x - 3)$ **29.** $\dfrac{a + 2}{a^2 + 2a + 4}$ **30.** $\dfrac{3x^2yz^4u^3}{2t^5v^2}$ **31.** $\dfrac{t + 1}{t + 5}$ **32.** $\dfrac{a - 3}{a + 5}$ **33.** $\dfrac{6a^2 + 8a - 5}{a^3}$

34. $\dfrac{2y + 35}{(y - 5)(y + 1)(y + 4)}$ **35.** $\dfrac{3k - 18}{7k^2 + 27k}$ **36.** $\dfrac{k^3 + k^2}{k - 1}$ **37.** $\dfrac{a - 1}{a^4}$ **38.** $\dfrac{4m^2}{n^4k^6}$ **39.** $\dfrac{4\sqrt{3}}{7}$ **40.** $10bcd^2\sqrt{3ac}$

41. $\dfrac{\sqrt[4]{104x}}{2x}$ **42.** $\dfrac{\sqrt{x} + \sqrt{3}}{x - 3}$ **43.** $\dfrac{9a^2b\sqrt{b}}{4c}$ **44.** $\dfrac{5}{2}$ **45.** $z^{5/6} - z^{3/4}$ **46.** $\dfrac{b^{1/6}c^{1/8}}{a^{1/4}}$ **47.** $\dfrac{y^{1/6}}{x^{1/4}}$

PROBLEM SET 3.1

1. 3   **3.** 14   **5.** 5   **7.** -6 **9.** -2 **11.** 1   **13.** $\dfrac{-7}{3}$ **15.** 5   **17.** 6   **19.** $\dfrac{41}{7}$ **21.** $\dfrac{-23}{5}$

23. 37   **25.** 3   **27.** 3   **29.** 0   **31.** $\dfrac{c}{a - b}$ **33.** $\dfrac{3b + a - 5y}{8}$ **35.** $\dfrac{ab}{a - b}$ **37.** $\dfrac{a - tc}{td + b}$ **39.** 1

41. $\dfrac{b - a}{2}$ **43.** 3   **45.** 9   **47.** 3   **49.** $\dfrac{3yt}{3y - 3t - yt}$ **51.** 2   **53.** 10   **55.** 4   **57.** $\dfrac{df - ac}{ab - de}$

59. $\dfrac{80}{7}$ **61.** $\dfrac{-25}{13}$ **63.** $\dfrac{de - c - ab}{a - d}$ **65.** 5   **67.** 4.764   **69.** 2.004   **71.** 30

73. (a) 109.77; (b) 62.62; (c) $\oplus : \dfrac{331(f_s - f_o)}{f_o}$; $\ominus : \dfrac{331(f_o - f_s)}{f_o}$ **75.** \$865.63; $\dfrac{20{,}000M_{33}}{N_{33} - R_{33}}$ **77.** 900; $\dfrac{12D_c}{D_a - D_c}$

PROBLEM SET 3.2

1. $5 + 2n$ **3.** $x = 3y - 6$ **5.** $n + \dfrac{1}{n}$ **7.** 3   **9.** 200   **11.** $\dfrac{23}{3}$ **13.** 5000   **15.** 28, 29, 30   **17.** 13

19. 7; 8   **21.** $11\% : 12{,}000$; $12\% : 18{,}000$ **23.** 10%; 11% **25.** $20\% : 20$; $40\% : 60$ **27.** 1

29. Car : 200; Train : 1800   **31.** 40 mph : 10; 50 mph : 12   **33.** $\dfrac{60}{37}$ **35.** $\dfrac{15}{2}$ **37.** 21, 22, 23

39. $W = 100$; $L = 220$ **41.** Nickels $= 12$; Dimes $= 7$ **43.** 4   **45.** 9%, 10%, 11% **47.** 60   **49.** 2000

51. $\dfrac{42}{5}$ **63.** Nigel: \$250,000; Clarence: \$200,000; Kent: \$150,000

PROBLEM SET 3.3

1. -2 **3.** 1   **5.** 3   **7.** $(x + 2)(x - 2)$ **9.** $(x - 4)(x - 4)$ **11.** $(x - 2)(x - 5)$ **13.** 7, -10

15. 7, 1   **17.** 10, -10 **19.** $\dfrac{3 \pm \sqrt{5}}{10}$ **21.** 3, -7 **23.** $\dfrac{-5 \pm \sqrt{17}}{2}$ **25.** $5 \pm \sqrt{22}$ **27.** 3, 10

29. 6, -6 **31.** 1, 6   **33.** 5, -2 **35.** -9, 3   **37.** $\dfrac{-1}{2}$, 6   **39.** $\pm\sqrt{5}$ **41.** 4, -6 **43.** $7 \pm \sqrt{15}$

45. $1 \pm \sqrt{5}$ **47.** $\dfrac{-1 \pm \sqrt{21}}{2}$ **49.** $\dfrac{-2 \pm \sqrt{22}}{2}$ **51.** $\dfrac{-3 \pm \sqrt{417}}{12}$ **53.** $\dfrac{5 \pm \sqrt{105}}{4}$ **55.** $-1 \pm \sqrt{14}$

57. $1, \dfrac{-5}{3}$ 59. $\dfrac{-15 \pm \sqrt{2145}}{40}$ 61. $\dfrac{3 \pm \sqrt{53}}{2}$ 63. $y(-1 \pm \sqrt{2})$ 65. Two distinct real solutions: $1.449, -3.449$

67. Two distinct real solutions: $7.275, -0.275$ 69. One real solution: 0.250 71. $\dfrac{4}{5}$ or $\dfrac{5}{4}$ 73. 4 by 12

75. 1   77. 5   79. $\dfrac{9 + \sqrt{77}}{2}$ 81. $\dfrac{1}{2\pi\sqrt{LC}}$

PROBLEM SET 3.4

1. $x^2 + 8x + 16$ 3. $x + 5$ 5. $x + 11 + 6\sqrt{x + 2}$ 7. -3 9. ± 7 11. 3, 5   13. 22

15. No real solution   17. 4   19. 1   21. 4   23. $143 - 2\sqrt{4655}$ 25. No real solution   27. ± 3

29. 3   31. 6   33. $\sqrt[3]{6}, \dfrac{\sqrt[3]{4}}{2}$ 35. $\pm\sqrt[4]{7}, \pm\sqrt[4]{9}$ 37. $\pm\sqrt{3}$ 39. $\dfrac{3}{2}, \dfrac{-1}{5}$ 41. 16, 9   43. 80   45. 64

47. 18   49. $\pm\sqrt{5}, \pm\sqrt{2}$ 51. ± 6 53. $\dfrac{1}{5}, \dfrac{-1}{4}$ 55. 3, 11   57. $\sqrt{6}$ 59. $\pm\sqrt{3}$ 61. $4f^2L^2m$

63. $(476.19N)^{1.1696}$

PROBLEM SET 3.5

1. 4   3. 7   5.   7.   9. $(2, 8)$

11. $(6, \infty)$ 13. $(-\infty, 4)$ 15. $(-2, -1]$ 17. $4 < x < 7$ 19. $2 < x \le 8$ 21. $x < 3$ 23. $x \ge -2$

25. $(-\infty, -2]$ 27. $[5, \infty)$ 29. $(1, \infty)$ 31. $(-\infty, 3]$ 33. $\left[-\infty, \dfrac{-5}{8}\right]$ 35. $(-\infty, 4]$ 37. $(1, 3]$

39. $(-4, 6)$ 41. $\left[\dfrac{13}{2}, 14\right)$ 43. $[5, 11]$ 45. $(6, 10)$ 47. $[1, 7)$ 49. $(-\infty, 1] \cup [5, \infty)$

51. $(-\infty, 2) \cup (4, \infty)$ 53. $(-\infty, -5)$ 55. $(2, 5)$ 57. $[1, 4]$ 59. $[-4, -1]$ 61. $(-\infty, -8)$

63. All reals   65. $[3, 15]$ 67. $\left(\dfrac{-50}{14}, \infty\right)$ 69. $[30, \infty)$ 71. $(-\infty, 44{,}000]$ 73. $[2, 4]$

PROBLEM SET 3.6

1. $(x + 2)(x - 2)$ 3. $(k - 2)(k + 7)$ 5. $\dfrac{y + x}{yx}$ 7. $\dfrac{-u - 1}{u + 3}$ 9. $(-1, \infty)$ 11. $(-2, \infty)$ 13. $\left(\dfrac{1}{2}, 6\right)$

15. $(-\infty, -4] \cup [3, \infty)$ 17. $(-9, 3)$ 19. $(0, 3)$ 21. $\left(-\dfrac{5}{3}, \dfrac{3}{2}\right]$ 23. $\left(0, \dfrac{1}{3}\right]$ 25. $(-2, 7)$

27. $(-\infty, -3) \cup [-1, \infty)$ 29. $[-4, 5]$ 31. $(-\infty, -5) \cup (-4, \infty)$ 33. $\left(0, \dfrac{10}{3}\right)$ 35. $(-\infty, -4] \cup [2, \infty)$

37. $\left[\dfrac{-11}{2}, 2\right)$ 39. $\left(-\infty, \dfrac{-5}{2}\right) \cup \left(\dfrac{7}{3}, \infty\right)$ 41. $\left(-\infty, \dfrac{25}{3}\right)$ 43. $(-\infty, -2] \cup [-1, 0]$ 45. $(-2, -1)$

47. $[-3, -2) \cup (-1, \infty)$ 49. $[10, 50]$ 51. $[3, \infty)$

PROBLEM SET 3.7

1. 3   3. 0   5. 4   7. 10   9. -7 11. $(-\infty, -1)$ 13. $(-\infty, 2)$ 15. $\left(\dfrac{a - 1}{5}, \infty\right)$

17. $-8, -2$ 19. $-2, 3$ 21. $\dfrac{5}{3}$ 23. $[-1, 7]$ 25. $\left(-\dfrac{1}{3}, 2\right)$ 27. $\dfrac{-3}{2}$ 29. $(-\infty, -6] \cup [-2, \infty)$

31. $\left(-\infty, \dfrac{-2}{7}\right) \cup \left(\dfrac{4}{7}, \infty\right)$ 33. All reals   35. $(-a - 1, a - 1)$ 37. $(-\infty, -b - 5] \cup [b - 5, \infty)$ 39. $(-8, -2)$

41. $2, -5$ **43.** $(-\infty, -4] \cup [12, \infty)$ **45.** $\left(-\infty, \dfrac{-a + 1}{2}\right) \cup \left(\dfrac{a + 1}{2}, \infty\right)$ **47.** No solution

49. $(-\infty, 0) \cup (1, \infty)$ **51.** $\left[\dfrac{-5}{2}, \infty\right)$ **53.** $\dfrac{-b}{a}, \dfrac{b}{a}$ **55.** $[-t - s, t - s]$ **57.** $[1.753, 1.757]$

CHAPTER 3   REVIEW EXERCISES

1. -12 **2.** -10 **3.** -1 **4.** -8 **5.** $\dfrac{yz}{y - z}$ **6.** $22, 24, 26$ **7.** 3 **8.** 10%: 1000; 9%: 4000

9. 50%: 200; 20%: 100   **10.** 40 mph: 400; 50 mph: 500   **11.** $\dfrac{12}{7}$ **12.** $2, -5$ **13.** $5, 7$ **14.** $-9, 5$

15. $1 \pm 2\sqrt{2}$ **16.** $3 \pm \sqrt{3}$ **17.** $\dfrac{3 \pm \sqrt{105}}{4}$ **18.** $\dfrac{2}{5}$ or $\dfrac{5}{2}$ **19.** 4 **20.** 9 **21.** 3 **22.** ± 1

23. $625; 1296$ **24.** $\sqrt{2}, 1$ **25.** $(-\infty, -2]$ **26.** $(-6, \infty)$ **27.** $(-19, 9)$ **28.** $(-2, 8)$

29. $(-\infty, -6] \cup \left[\dfrac{1}{2}, \infty\right)$ **30.** $\left(\dfrac{1}{2}, 3\right)$ **31.** $(-\infty, -3) \cup (1, \infty)$ **32.** $(-\infty, 6) \cup (6, \infty)$ **33.** $-8, -2$

34. $[-2, 3]$ **35.** $(-\infty, -2) \cup \left(\dfrac{10}{3}, \infty\right)$ **36.** $(-t - 2, t - 2)$

PROBLEM SET 4.1

1. 3 **3.** -1 **5.** 0 **7.** -5 **9.** 170 **11–21.**   **23.** $(3, 1)$

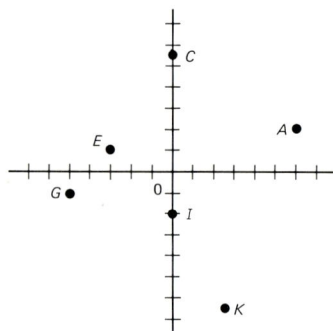

25. $(0, 5)$ **27.** $(-1\frac{1}{2}, 1)$ **29.** $(-2\frac{1}{2}, -2)$ **31.** $(0, -3\frac{1}{2})$ **33.** $(4\frac{1}{2}, -1)$ **35.** $10; (5, 5)$ **37.** $\sqrt{29}; (2\frac{1}{2}, 5)$
39. $\sqrt{52}; (-1, 4)$ **41.** $\sqrt{136}; (1, -1)$ **43.** (a) $A(10, 1000); B(20, 2500)$; (b) 1500.03; (c) $(15, 1750)$

45. (a) $D(35, 1500); E(50, 3500)$; (b) 2000.06; (c) $(42.5, 2500)$ **47.** (a) $B(20, 2500); C(25, -1000)$; (b) 3500.004; (c) $(22.5, 750)$

49. $(3, 2); (0, 5); (5, 0); (4, 1)$ **51.** $(0, 0); (1, 1); (2, 4); (3, 9)$ **53.**

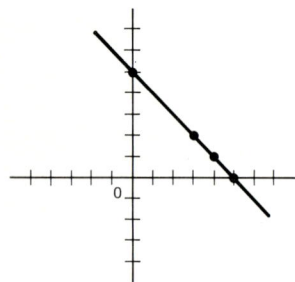

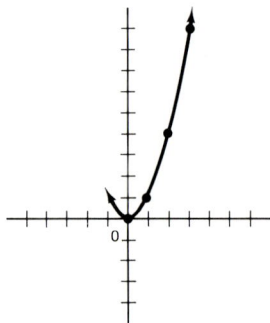

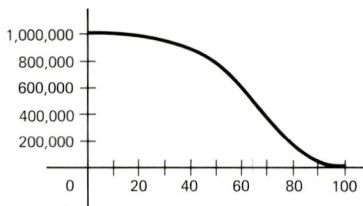

55. $A(1, 40,000); B(3, 30,000); C(5, 35,000); D(6, 5000); E(8, 0)$

PROBLEM SET 4.2

1–5.

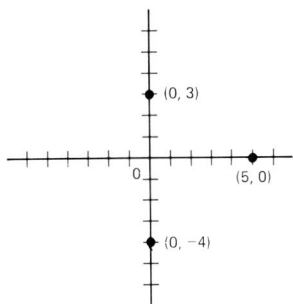
(0, 3)
(5, 0)
(0, −4)

7. 4   **9.** 3   **11.** 12; 6; 4; 5   **13.**

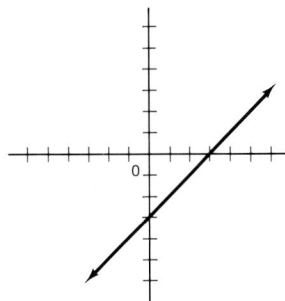

15.

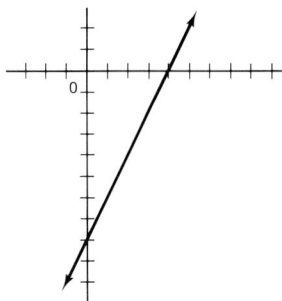

17.

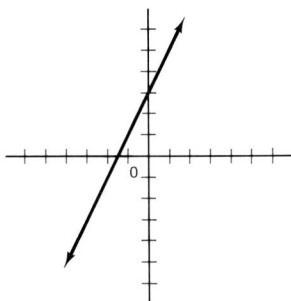

19.

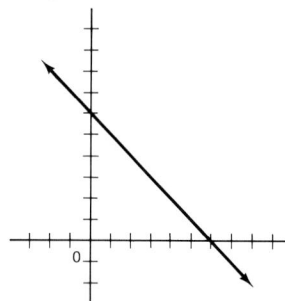

21.

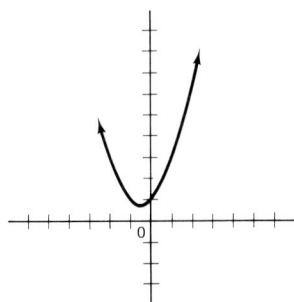

23.

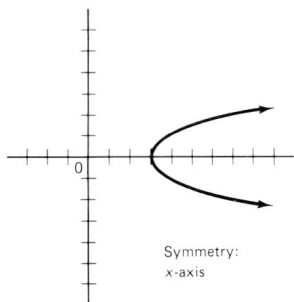

Symmetry:
x-axis

25.

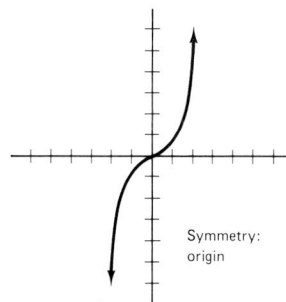

Symmetry:
origin

27.

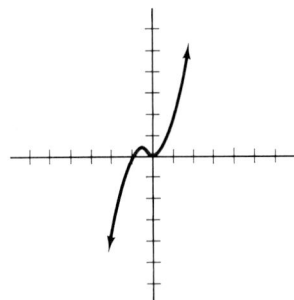

29.

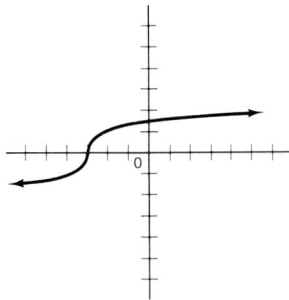

31.

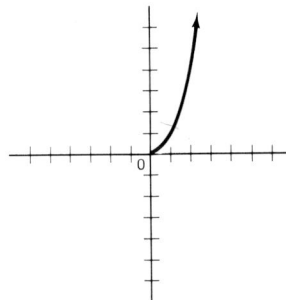

33.

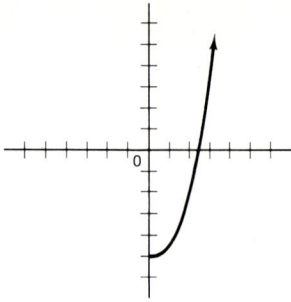

35.

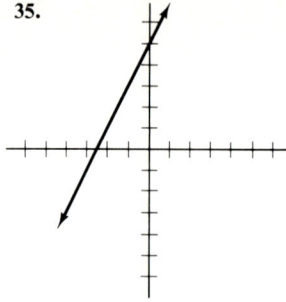

37.

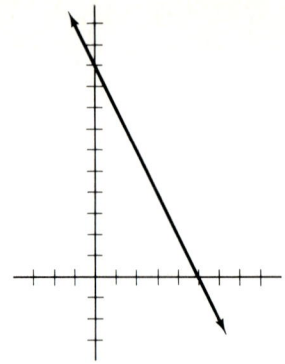

39.

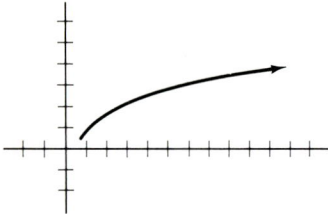

41.

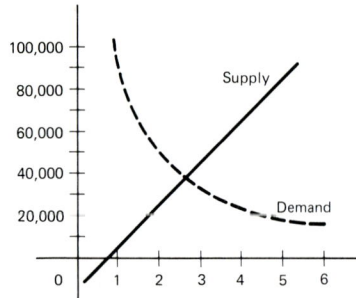

PROBLEM SET 4.3

1. 4   **3.** −6   **5.** −2   **7.** 6/7   **9.**

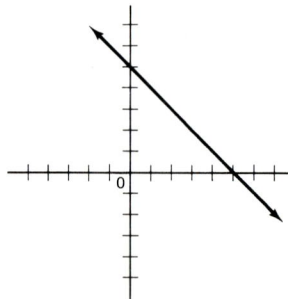

11.

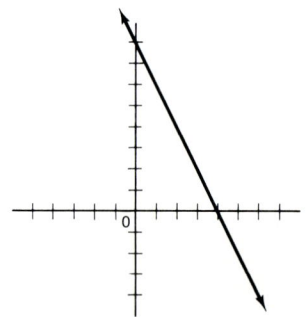

13.

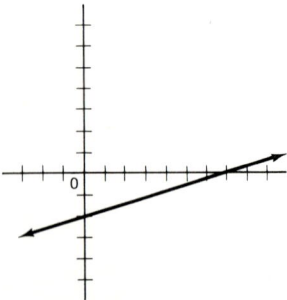

15.

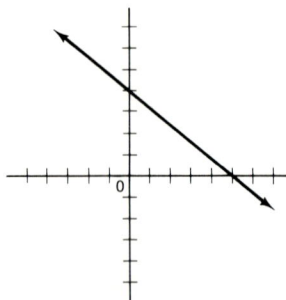

17.

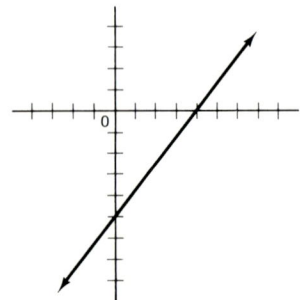

19.

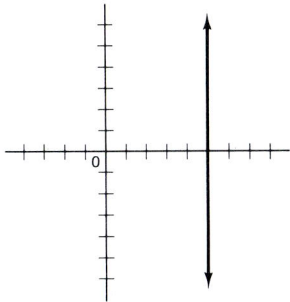

21.

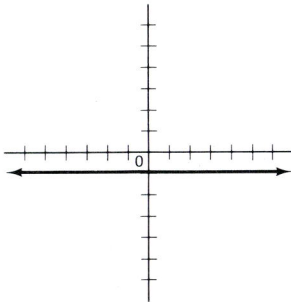

23.

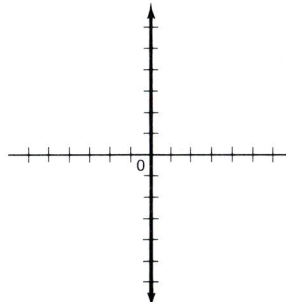

25. 3   **27.** $\dfrac{-10}{3}$ **29.** 0   **31.** No slope   **33.** -3.865 **35.** 250   **37.** Parallel: 2; perpendicular: $\dfrac{-1}{2}$

39. Parallel: $\dfrac{1}{3}$; perpendicular: -3 **41.** Parallel: $\dfrac{-9}{4}$; perpendicular: $\dfrac{4}{9}$ **43.** $\overline{AB}$ perpendicular to $\overline{CD}$ **45.** None

47. $\dfrac{-1}{10}$ **49.** $\dfrac{-1}{5}$ **51.** -10 **53.** 2   **55.** 0.00008   **57.**   **59.** 0.08

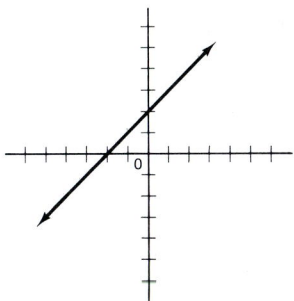

PROBLEM SET 4.4

1.

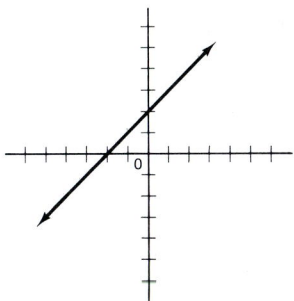

3.

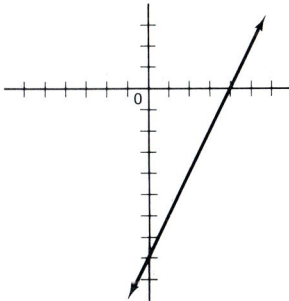

5. $3; \dfrac{-1}{3}$ **7.** $-3; \dfrac{1}{3}$

9. $m = 5$; y-int $= -3$ **11.** $m = \dfrac{2}{3}$; y-int $= -3$ **13.** $m = -2$; y-int $= 5$ **15.** $m = \dfrac{-3}{5}$; y-int $= \dfrac{-1}{5}$ **17.** $y = 2x$

19. $y = \dfrac{-1}{3}x - \dfrac{10}{3}$ **21.** $y = \dfrac{15}{4}x - \dfrac{31}{4}$ **23.** $y = \dfrac{-1}{3}x - \dfrac{1}{3}$ **25.** $3x - 5y = 18$ **27.** $y = 4x - 7$

29. $y = 2x + 6$ **31.** $y = 5x - 23$ **33.** $y = \dfrac{1}{3}x - \dfrac{17}{3}$ **35.** $y = 2x + 4$ **37.** $4x + 3y = 12$ **39.** $y = \dfrac{-1}{3}x + \dfrac{5}{3}$

41. $x = 2$ **43.** $y = -8500x + 80,000$; $8500 loss in value each year   **45.** $y = 200x + 200,000$

47. $y = 0.21x + 0.44$; 21 cents per print; 44 cents to develop roll   **49.** $y = 0.107x + 2.336$; 4.6151 (exact: 4.6151923)

CHAPTER 4   REVIEW EXERCISES

1.

2. $\sqrt{68}$ **3.** $\sqrt{109}$ **4.** $\left(\dfrac{1}{2}, 6\right)$ **5.** $(2, 1)$

6.

7.

8.

9.

Symmetry:
y-axis

10.

11.

12. 4   **13.** $\dfrac{-2}{5}$ **14.** No   **15.** $y = -3x$

16. $y = \dfrac{-1}{4}x + \dfrac{13}{2}$ **17.** $2x - 3y = -9$ **18.** $y = \dfrac{-2}{5}x + \dfrac{31}{5}$ **19.** $m = \dfrac{-4}{5}$; *y*-int $= \dfrac{7}{5}$

PROBLEM SET 5.1

1. 7   **3.** $2x + 3$ **5.** $\dfrac{5}{3}$ **7.** $\left[\dfrac{-2}{5}, \infty\right)$ **9, 11.**

13. $27, 64, -125, \dfrac{1}{8}$

15. 3, 4, 5, 0   **17.** 3; 2; 5; 23; $t + 3$; $h + k + 3$ **19.** 1; 2; 5; 401; $t^2 + 1$; $h^2 + 2hk + k^2 + 1$

21. $-1; \dfrac{-1}{2}; 1; \dfrac{1}{19}; \dfrac{1}{t-1}; \dfrac{1}{h+k-1}$ **23.** 1; 0; $\sqrt{3}$; $\sqrt{21}$; $\sqrt{t+1}$; $\sqrt{h+k+1}$ **25.** 0; 1; 2; 20; $|t|$; $|h + k|$

27. $3a + 3h$; $3h$; 3   **29.** $5a + 5h - 3$; $5h$; 5   **31.** $a^2 + 2ah + h^2 + 6$; $2ah + h^2$; $2a + h$

33. $2a^3 + 6a^2h + 6ah^2 + 2h^3$; $6a^2h + 6ah^2 + 2h^3$; $6a^2 + 6ah + 2h^2$ **35.** $\dfrac{3}{a+h}; \dfrac{-3h}{a(a+h)}; \dfrac{-3}{a(a+h)}$

37. $\sqrt{a+h}$; $\sqrt{a+h} - \sqrt{a} = \dfrac{h}{\sqrt{a+h} + \sqrt{a}}; \dfrac{1}{\sqrt{a+h} + \sqrt{a}}$ **39.** {all reals}   **41.** {all reals, $x \neq 0$}

43. {all reals, $x \neq 2$}   **45.** {all reals, $x \geq 0$}   **47.** {all reals, $x \neq \pm 5$}   **49.** {all reals, $|x| \geq 1$}
51. {all reals, $x \geq 0$ and $x \neq 3$}   **53.** 9, 11, 1; $(x, 2x + 7)$, $(1, 9)$, $(2, 11)$, $(-3, 1)$; $f(x) = 2x + 7$, $f(1) = 9$, $f(2) = 11$, $f(-3) = 1$
55. $x \to 5x^2 - 1$, $1 \to 4$, $2 \to 19$, $-3 \to 44$; $(x, 5x^2 - 1)$, $(1, 4)$, $(2, 19)$, $(-3, 44)$; 4, 19, 44

57. 5, 2, 1, 2, 5; all reals

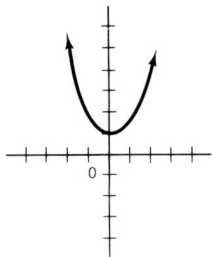

59. $4, 2, 1, \dfrac{1}{2}, \dfrac{1}{5}$; all reals, $x \neq 0$

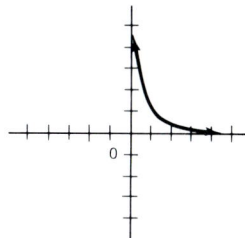

61. (a) 2000; 1000; 100; (b) $\dfrac{100{,}000}{a+h}; \dfrac{-100{,}000h}{a(a+h)}; \dfrac{-100{,}000}{a(a+h)}$ **63.** 5; $\dfrac{40}{3}; \dfrac{125}{3}$

65. (a) 0; 64; 96; (b) 0; 5; (c) $80a + 80h - 16a^2 - 32ah - 16h^2$; $80h - 32ah - 16h^2$; $80 - 32a - 16h$ **67.** 12; 9; 19; 1

PROBLEM SET 5.2

1–5.

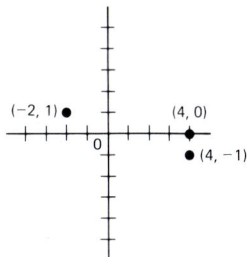

7.

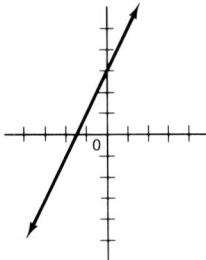

9.

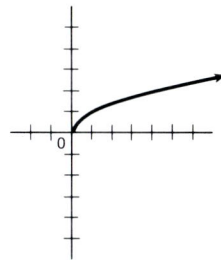

11. 5; 4; 3; 6   **13.** 0; 1; 2; 3   **15.** Function   **17.** Not a function   **19.** Not a function

21.

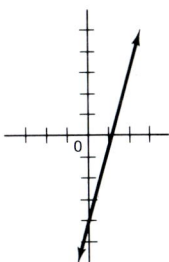

23.

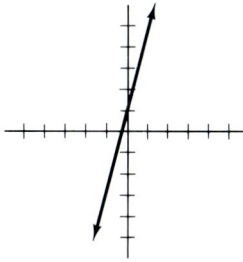

25.

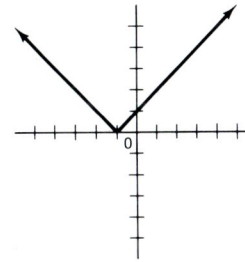

27.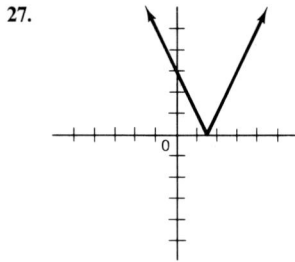

29.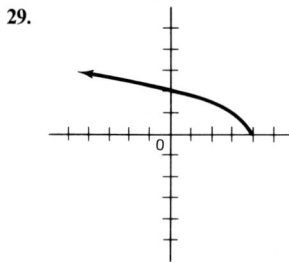

31.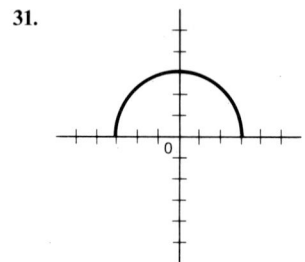

33.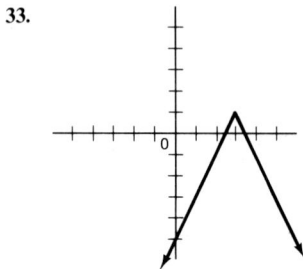

35.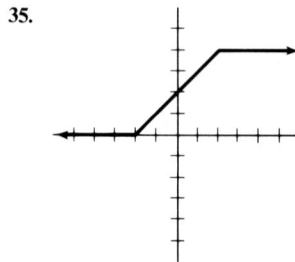

37.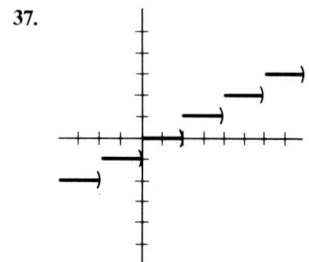

39.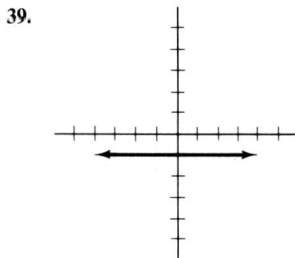

41.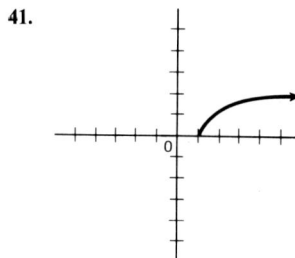

43.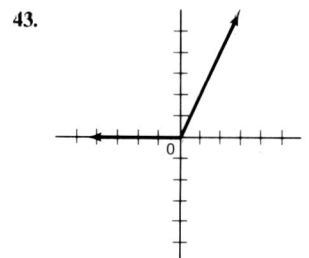

45.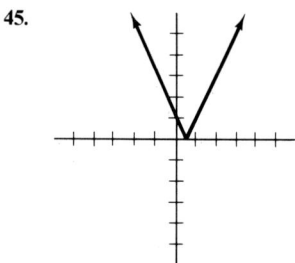

47.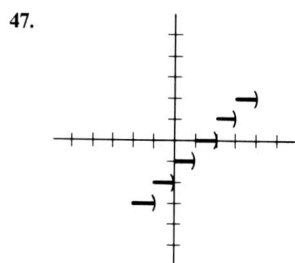

49.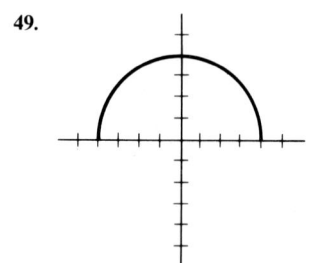

51. $y = \begin{cases} 2 & 0 < x < 1/2 \\ 2.50 & 1/2 \leq x < 1 \\ 3.00 & 1 \leq x < 3/2 \\ \text{and so on} \end{cases}$

53. $y = \begin{cases} 22 & 0 < x < 1 \\ 39 & 1 \leq x < 2 \\ 56 & 2 \leq x < 3 \\ \text{and so on} \end{cases}$

55. 9.5   **57.** -4 **59.** 25   **61.** 0

63. 31   **65.**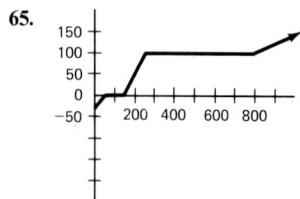

67.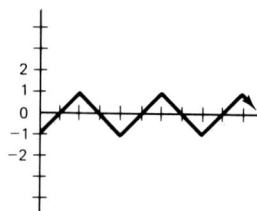

PROBLEM SET 5.3

1. $(x - 4)(x + 4)$ **3.** $x^3(x - 5)(x + 5)$ **5.**

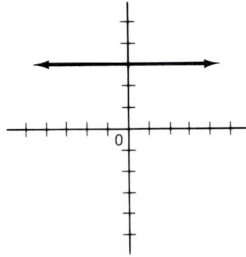

7.

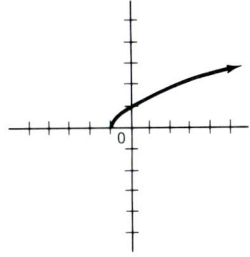

9. Odd   **11.** Neither   **13.** Even   **15.** Odd   **17.**

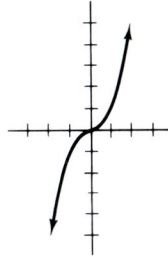

19.

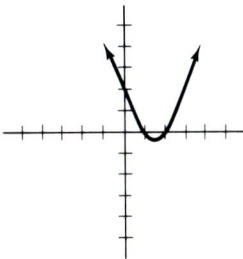

21.

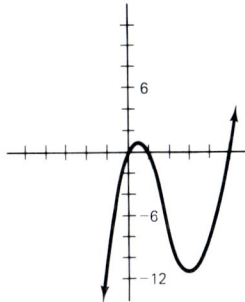

23.

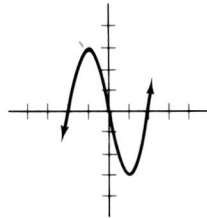

25. Neither

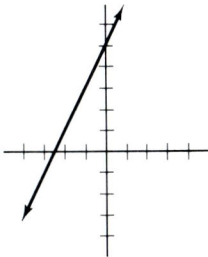

27. Even; $(0, 4)$

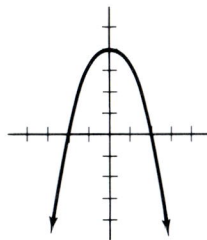

29. Even

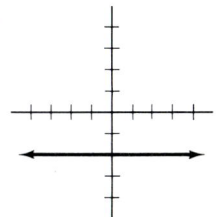

31. Neither; $(3.5, -6.25)$

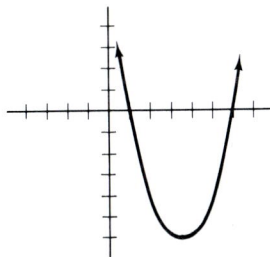

33. Neither; $(0, 0), \left(\dfrac{2}{3}, \dfrac{-4}{27}\right)$

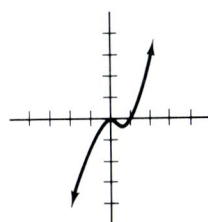

35. Neither

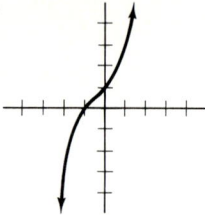

37. Odd; $(0.7, 0.35)$, $(-0.7, -0.35)$

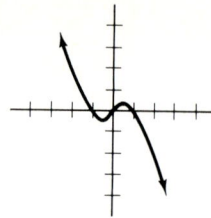

39. Neither

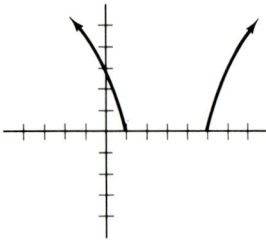

41. Neither; $\left(\dfrac{-1}{2}, 6\dfrac{1}{4}\right)$

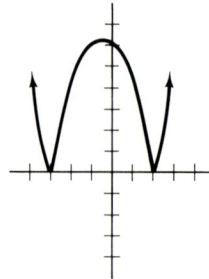

43. $A = x(50 - x)$; $x = 25$, $A = 625$

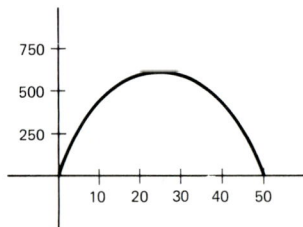

45. (a) $t = 3$; $h = 144$; (b) 0, 6   **47.** $x = 1500$; $P = 7750$

PROBLEM SET 5.4

1. 7   **3.** $-2, 2$ **5.** $(-\infty, -4)$ **7.** {all reals}   **9.**

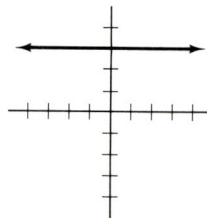

11.

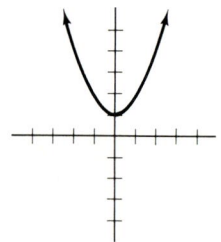

13. $\dfrac{1}{x^3}$ **15.** $\dfrac{x+1}{x^2}$ **17.** Vertical: $x = -2$; horizontal: $y = 1$ **19.** Vertical: $x = 3$, $x = -3$; horizontal: $y = 0$

21. Vertical: $x = 1$, $x = -1$; horizontal: $y = \dfrac{2}{3}$ **23.**

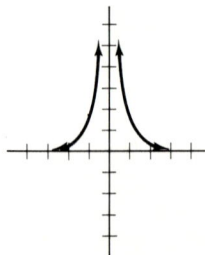

25.

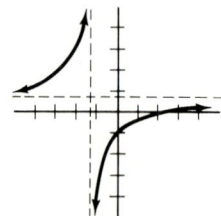

27.

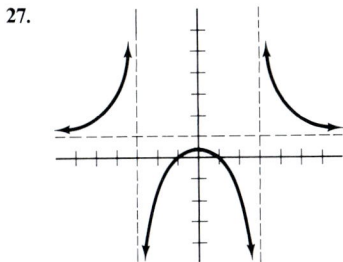

29.

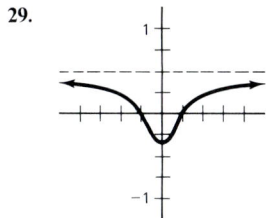

31.

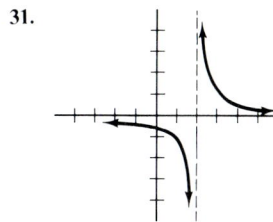

33.

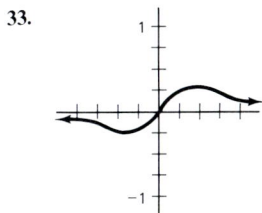

35.

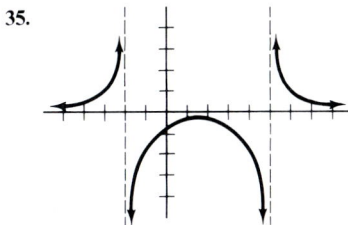

37.

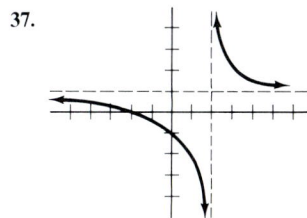

39.

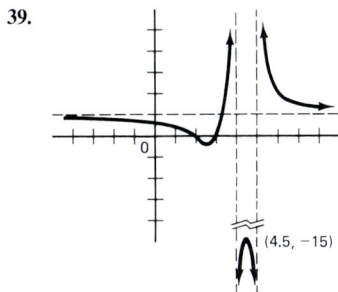

$(4.5, -15)$

41.

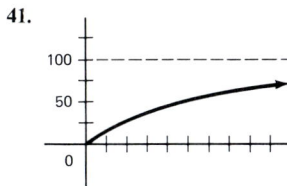

43.

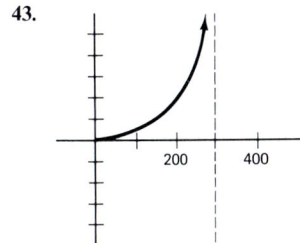

PROBLEM SET 5.5

1. $15; 2a + 2h + 7; 2h; 2$ **3.** $64; a^3 + 3a^2h + 3ah^2 + h^3; 3a^2h + 3ah^2 + h^3; 3a^2 + 3ah + h^2$

5.

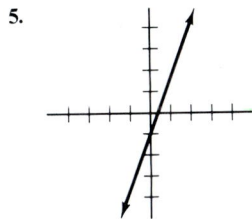

7.

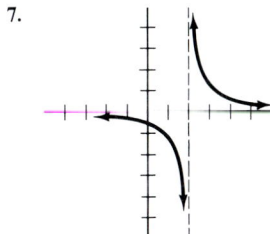

9. $2x^2 + 5x - 7 + \sqrt[3]{x}$; domain $= \{$all reals$\}$

11. $(2x^2 + 5x - 7)(2x + 3)$; domain $= \{$all reals$\}$ **13.** $\dfrac{\sqrt[3]{x}}{2x + 3}$; domain $= \left\{$all reals, $x \neq \dfrac{-3}{2}\right\}$

15. $\sqrt[3]{x}(2x + 3)$; domain $= \{$all reals$\}$ **17.** $\dfrac{(2x^2 + 5x - 7)\sqrt[3]{x}}{2x + 3}$; domain $= \left\{$all reals, $x \neq \dfrac{-3}{2}\right\}$

19.

21.

23.

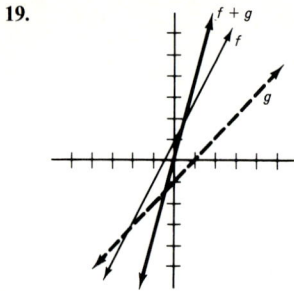

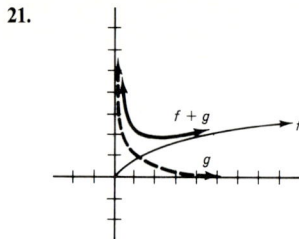

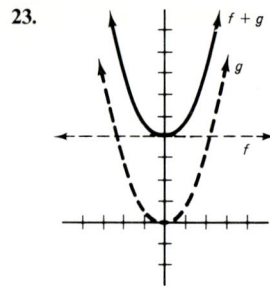

25. $(\sqrt[3]{x})^2 - 1$ **27.** $\dfrac{1}{x^2 - 1}$ **29.** $5\sqrt[3]{x} + 2$ **31.** $\left(\dfrac{1}{x}\right)^2 - 1$ **33.** $\left(\sqrt[3]{\dfrac{1}{x}}\right)^2 - 1$ **41.** $x^5 + x + 1 + \dfrac{1}{x^2}$

43. $\sqrt[6]{x^{25/4}}$ **45.** $\sqrt[6]{x^4}$ **47.** $(x^5 + x + 1)^4$ **49.** $\sqrt[6]{x + x^9 + x^5 + x^4}$ **51.** $\dfrac{1}{[(\sqrt[6]{x})^5 + \sqrt[6]{x} + 1]^2}$ **53.** x^4

55. x^{64} **57.**

PROBLEM SET 5.6

1. $2(x^2 - 7) + 1 = 2x^2 - 13$ **3.** $5\left(\dfrac{x - 1}{5}\right) + 1 = x$ **5.** $y = x - 1$ **7.** $y = \sqrt[3]{\dfrac{x + 3}{2}}$ **9.** $y = 3 + \dfrac{2}{x}$

11. One-to-one   **13.** Not one-to-one; restrict $x > 0$ **15.** Double; halve; $f^{-1}(x) = \dfrac{x}{2}$

17. Subtract 3; add 3; $f^{-1}(x) = x + 3$ **19.** Cube root; cube; $f^{-1}(x) = x^3$

21. $f \circ f^{-1}(x) = (\sqrt[5]{x - 2})^5 + 2 = x - 2 + 2 = x$ **23.** $f \circ f^{-1}(x) = -(-x) = x$ **25.** $f^{-1}(x) = \dfrac{1}{x}$

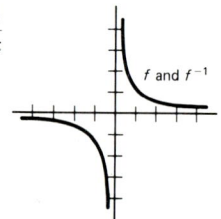

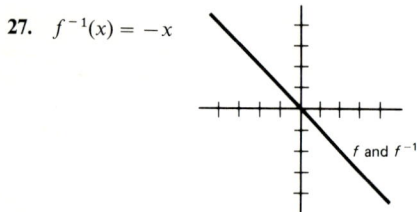

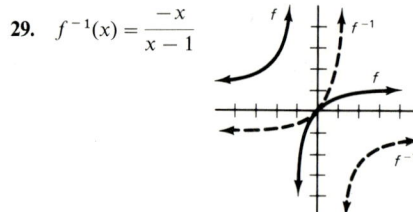

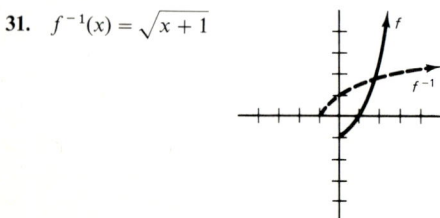

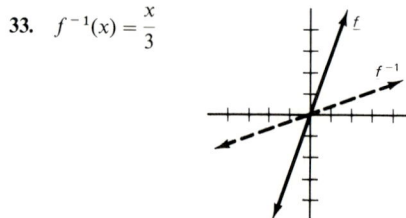

27. $f^{-1}(x) = -x$

29. $f^{-1}(x) = \dfrac{-x}{x - 1}$

31. $f^{-1}(x) = \sqrt{x + 1}$

33. $f^{-1}(x) = \dfrac{x}{3}$

35. $f^{-1}(x) = \sqrt{x-2}$ on $[0, \infty)$

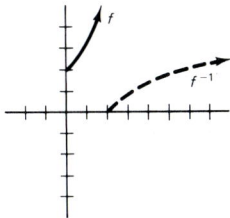

37. $f^{-1}(x) = x^2 + 1$

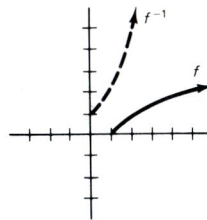

39. $f^{-1}(x) = \dfrac{4}{x}$

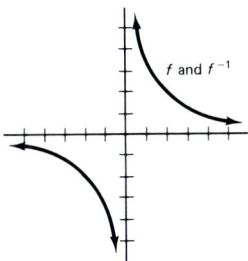

41. $C(F) = \dfrac{5}{9}(F - 32)$

43. $a(S) = \dfrac{S^2}{100{,}000{,}000}$

45. $h(d) = \dfrac{d^2}{1.4884}$

CHAPTER 5   REVIEW EXERCISES

1. $\dfrac{1}{2}$ **2.** $\dfrac{5}{2}$ **3.** $\dfrac{-1}{3}$ **4.** $\dfrac{1}{a+h}$ **5.** $\dfrac{-h}{a(a+h)}$ **6.** $\dfrac{-1}{a(a+h)}$ **7.** $\{\text{all reals, } x \neq 0\}$ **8.** Function

9. Not a function

10.

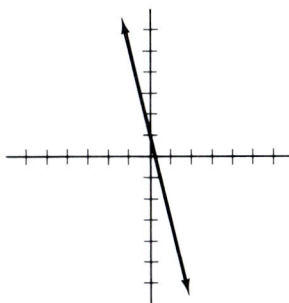

11.

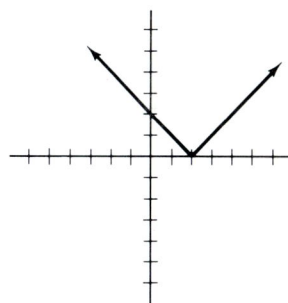

12.

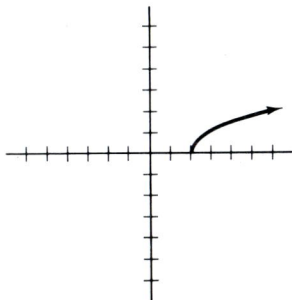

13.

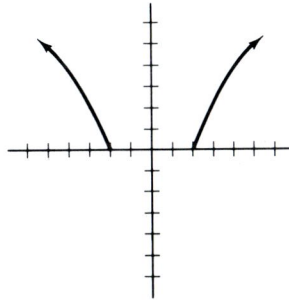

14.

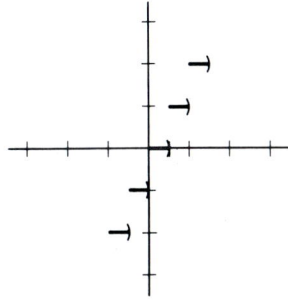

15.

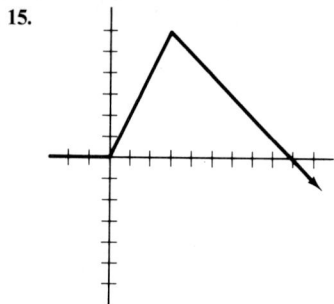

16.

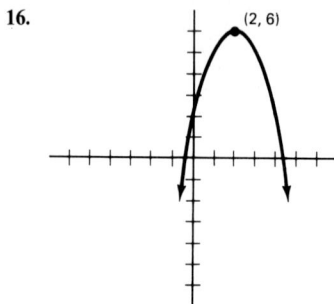

(2, 6)

17.

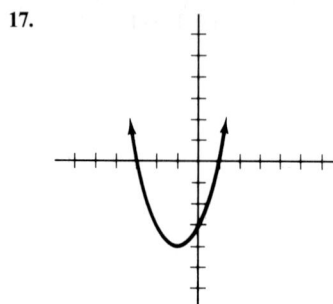

18.

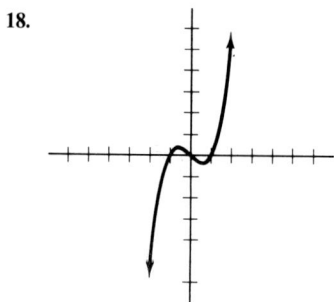

19.

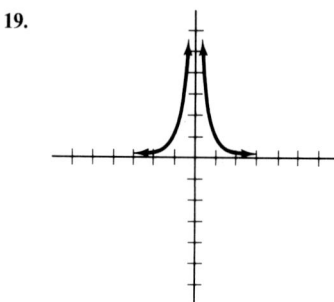

20.

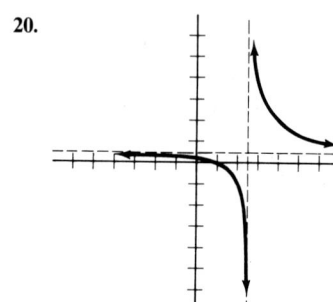

21.

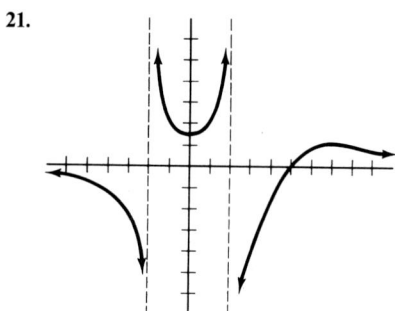

22. $x^2 + 5 + \dfrac{1}{x}$ **23.** $x^2 + 5 - \sqrt[5]{x}$ **24.** $\dfrac{1}{x}\sqrt[5]{x}$ **25.** $\dfrac{1}{x(x^2 + 5)}$

26. $\dfrac{1}{x^2 + 5}$ **27.** $(\sqrt[5]{x})^2 + 5$ **28.** Not one-to-one   **29.** One-to-one

30. $f^{-1}(x) = \dfrac{x + 2}{3}$

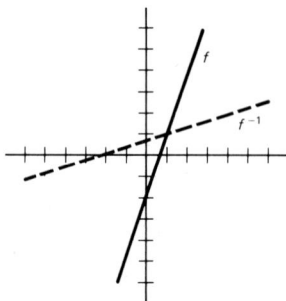

31. $f^{-1}(x) = \dfrac{-2}{x - 1}$

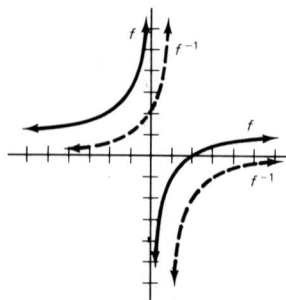

PROBLEM SET 6.1

1.

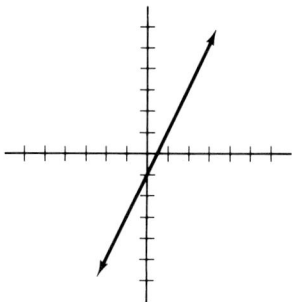

3.

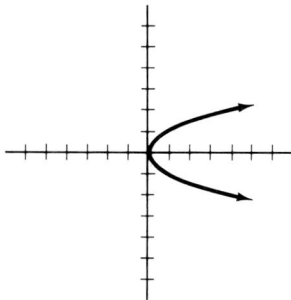

5. 10, −4

7.

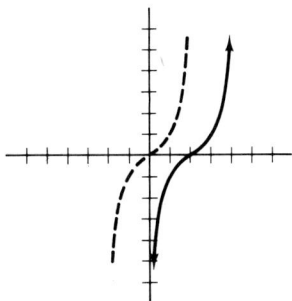

9.

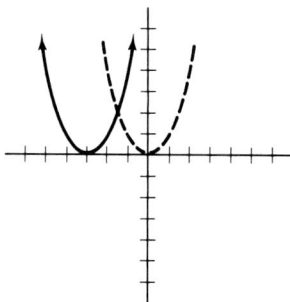

11.

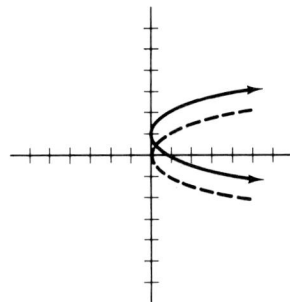

13.

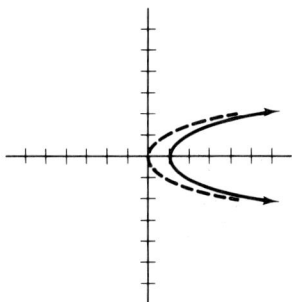

15.

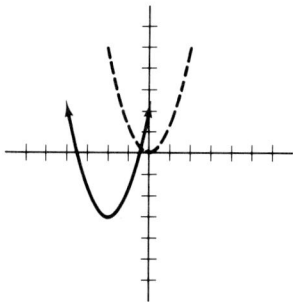

17.

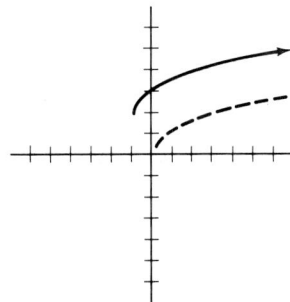

19. $y = (x + 2)^2$ **21.** $y - 5 = \sqrt{x}$ **23.** $y + 2 = (x + 3)^2 + (x + 3)$ **25.** $x + 4 = 4(y - 3)^2$

27. $(x - 4)^2 + (y - 2)^2 = 25$ **29.** $\dfrac{(x - 1)^2}{4} - \dfrac{(y + 8)^2}{25} = 1$ **31.** $y + 1 = x^2$ **33.** $y = (x - 2)^2$

35. $(x + 1)^2 + y^2 = 4$ **37.** $(x + 4)^2 - (y + 4)^2 = 4$

39. (a) $y = f(x) + a$ is equivalent to $y - a = f(x)$, which is a translation a units up.

(b)

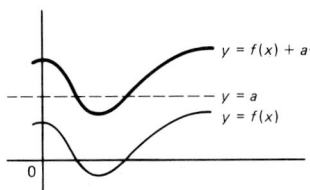

PROBLEM SET 6.2

1. $x^2 - 6x + 9$ **3.** $y^2 + \dfrac{1}{2}y + \dfrac{1}{16}$ **5.** $4, -2$ **7.** $5, -1$ **9.**

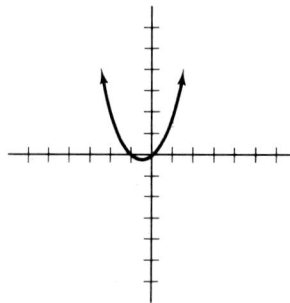

11.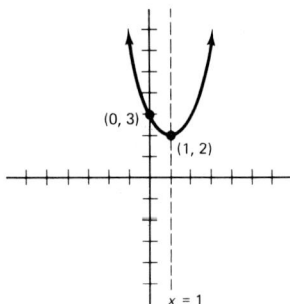

13.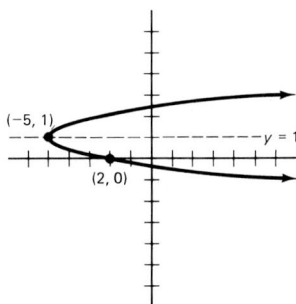

15. $y = (x - 1)^2$; parabola; vertex: $(1, 0)$; symmetry: $x = 1$; intercept: $(0, 1)$; opens up
17. $x = (y + 2)^2 + 2$; parabola; vertex: $(2, -2)$; symmetry: $y = -2$; intercept: $(6, 0)$; opens right
19. $y = -(x - 2)^2 + 5$; parabola; vertex: $(2, 5)$; symmetry: $x = 2$; intercept: $(0, 1)$; opens down
21. $(x - 1)^2 + (y - 5)^2 = 4$ **23.** $x^2 + (y + 4)^2 = 25$ **25.** $(x - 3)^2 + (y - 2)^2 = 13$ **27.** $(x - 2)^2 + (y - 4)^2 = 16$

29.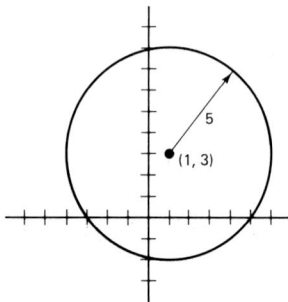

31.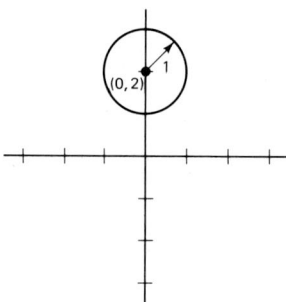

33. Center: $(-1, 0)$; radius: 2

35. Center: $(-2, 3)$; radius: 4 **37.** Center: $\left(\dfrac{-3}{2}, \dfrac{9}{2}\right)$; radius $= \dfrac{\sqrt{94}}{2}$ **39.**

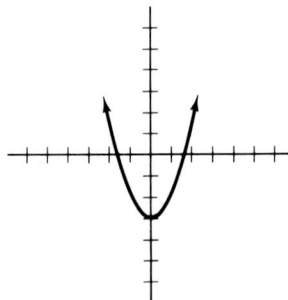

41.

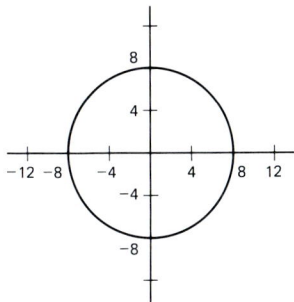

43. Circle; center: $(0, 0)$; radius: 5   **45.** Parabola; vertex: $\left(\frac{13}{4}, \frac{1}{2}\right)$; opens left

47. Circle; center: $(2, 1)$; radius: 4   **49.** Parabola; vertex: $(-10, -3)$; opens right   **51.** $(x - 2)^2 + (y - 4)^2 = 16$

53. $(x - 3)^2 + (y - 5)^2 = 1$ **55.** $y - 4 = -(x - 2)^2$

57. (a)   (b) 104.77   (c) 30

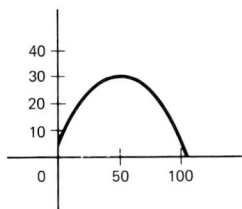

PROBLEM SET 6.3

1.

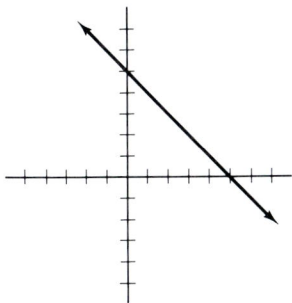

3.

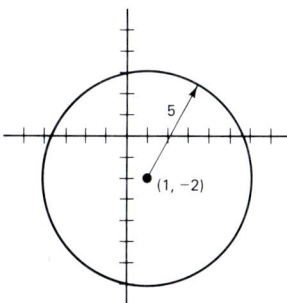

5.

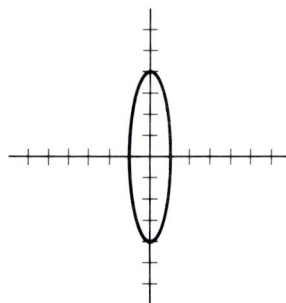

7. Ellipse; vertices: $(0, 4), (0, -4), (2, 0), (-2, 0)$ **9.** Ellipse; vertices: $(-1, 5), (-1, -9), (-5, -2), (3, -2)$

11. Ellipse; vertices: $(2, -1), (6, -1), (4, 0), (4, -2)$ **13.**

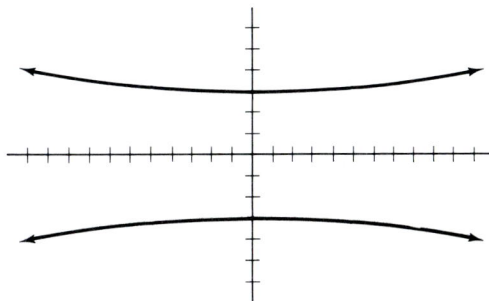

15.

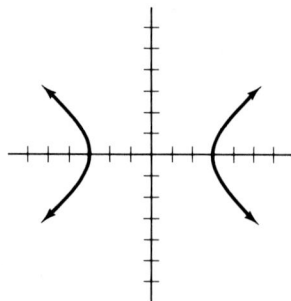

17. Hyperbola; vertices: $(-5, 3), (-5, -5)$, opens up and down

19. Hyperbola; vertices: $(0, 8), (0, -8)$, opens up and down   **21.** Ellipse; vertices: $(0, 5), (0, -5), (2, 0), (-2, 0)$

23. Ellipse; vertices: $(-1, 5), (-1, -3), (1, 1), (-3, 1)$ **25.** Parabola; vertex: $(-1, 1)$, opens right

27. Hyperbola; vertices: $(2, -1), (-6, -1)$, opens right and left

29. $f(x) = \sqrt{4 - x^2}; f^{-1}(x) = \sqrt{4 - x^2}$; domain $= [0, 2]$; range $= [0, 2]$

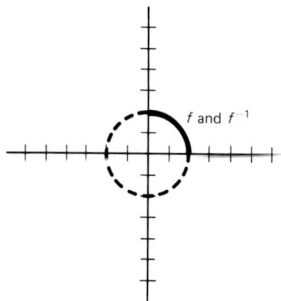

31. $f(x) = \frac{1}{2}\sqrt{x^2 - 4}; f^{-1}(x) = \sqrt{4x^2 + 4}$; domain $= (2, \infty)$; range $= [0, \infty)$

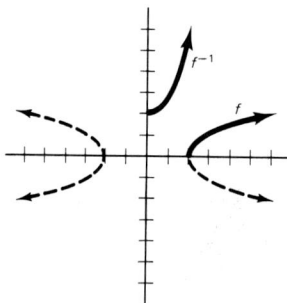

33. $f(x) = \frac{5}{2}\sqrt{4 - x^2}; f^{-1}(x) = \frac{2}{5}\sqrt{25 - x^2}$; domain $= [0, 2]$; range $= [0, 5]$

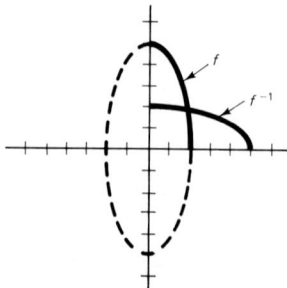

PROBLEM SET 6.4

1.

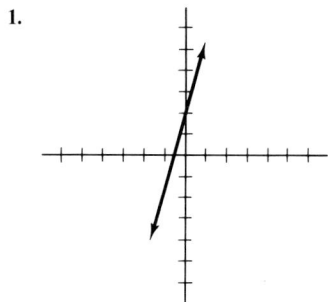

3.

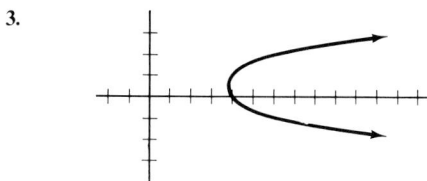

5.

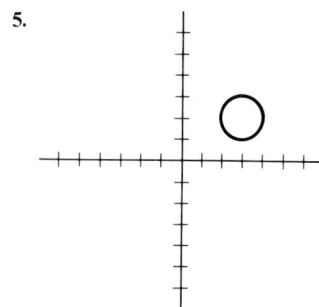

7.

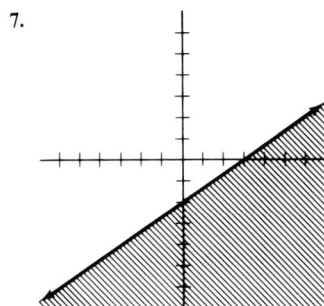

9.

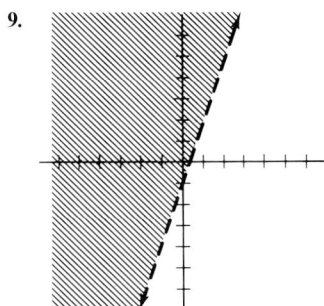

11.

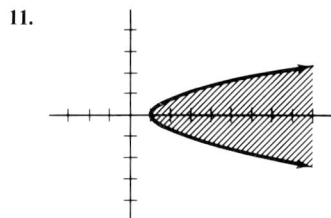

13.

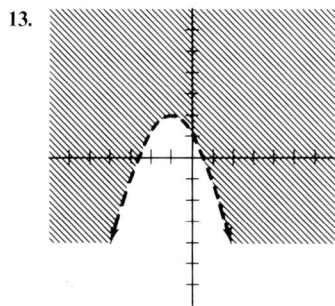

15.

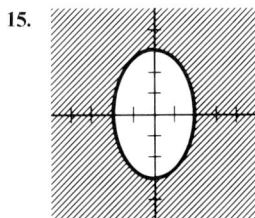

17.

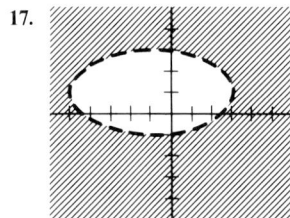

19.

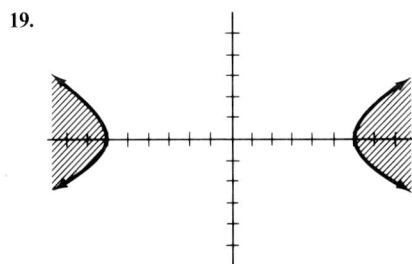

21.

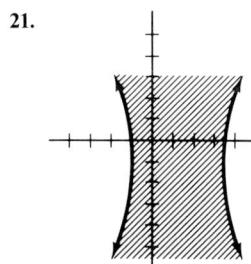

23.

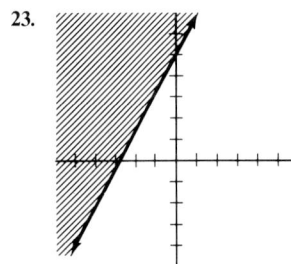

25.

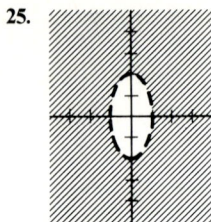

27.

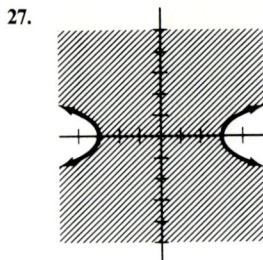

29.

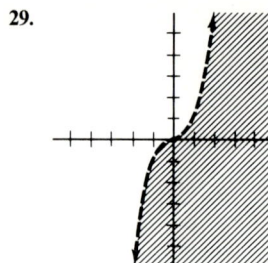

31.

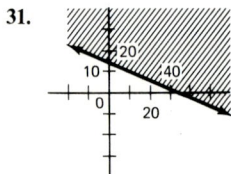

33. $(x - 4)^2 + (y - 10)^2 \leq 49$

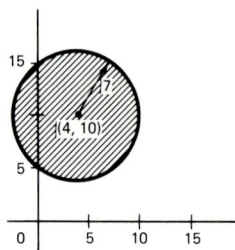

PROBLEM SET 6.5

1. 12   **3.** 20   **5.** 10   **7.** $\dfrac{3ay}{40}$ **9.** $u = \dfrac{k}{v^4}; k = 64; \dfrac{64}{81}$ **11.** $a = \dfrac{kx^3}{y^2\sqrt[4]{z}}; k = 18; \dfrac{243}{2}$

13. $a = \dfrac{k}{\sqrt{t}}; k = 100; 50$ **15.** $p = \dfrac{k\sqrt{u}}{v^3}; k = 40; 240$ **17.** y varies directly with the square of x; $y = 8x^2$

19. 120,000   **21.** 4   **23.** 787   **25.** 3200   **27.** 2.37×10^3 **29.** 19,200

CHAPTER 6 REVIEW EXERCISES

1. $y = (x - 5)^4$ **2.** $(y - 2)^2 = (x + 3)^3$ **3.**

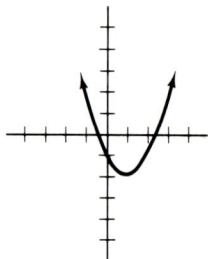

4.

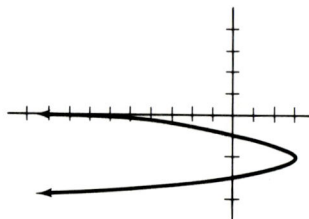

5. Parabola; vertex: $(-3, 2)$; opens right   **6.** Parabola; vertex: $(1, -7)$; opens up   **7.**

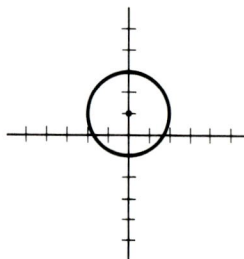

8. Circle; center: $(-2, 3)$; radius: 4   **9.**

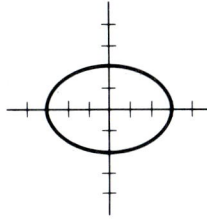

10.

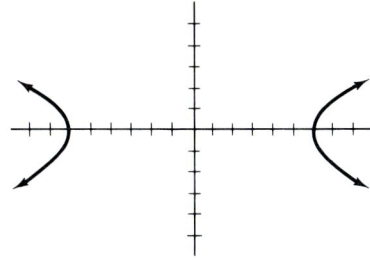

11. Ellipse; vertices: $(0, 5)$, $(0, -5)$, $(4, 0)$, $(-4, 0)$ **12.** Hyperbola; vertices: $(0, 6)$, $(0, -6)$; opens up and down

13. Hyperbola; vertices: $(-1, -1)$, $(3, -1)$; opens left and right   **14.** Ellipse; vertices: $(-2, 8)$, $(-2, -2)$, $(1, 3)$, $(-5, 3)$

15.

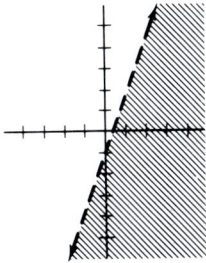

16.

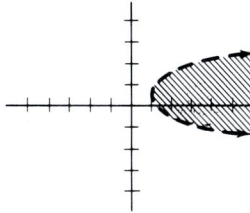

17.

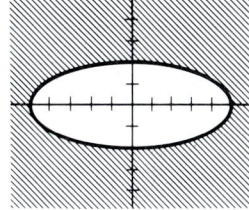

18. $x = k\sqrt{y}$; $k = 5$; 50   **19.** $p = \dfrac{kx}{t^3}$; $k = 64$; 7   **20.** $w = klg^2$; $k = \dfrac{1}{80}$; 6.75

PROBLEM SET 7.1

1. 25   **3.** 1   **5.** $\dfrac{1}{8}$ **7.** 3   **9.** 80   **11.** 11,625.2   **13.**

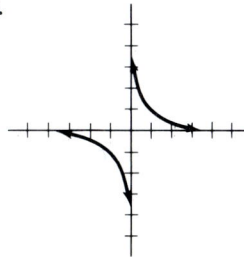

15.

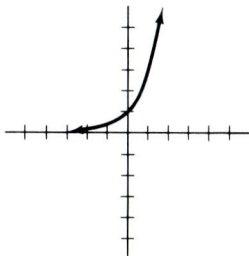

17.

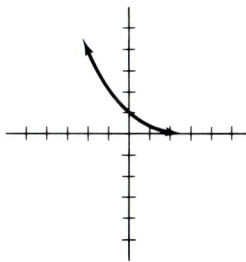

19.

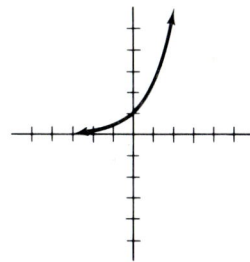

21.

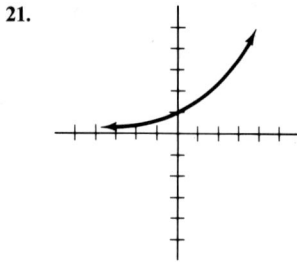

23.

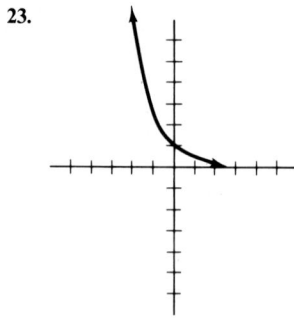

25.

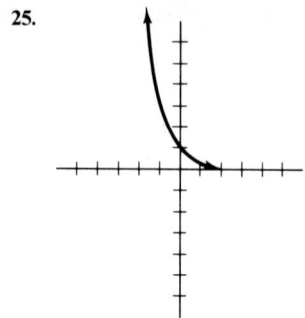

27.

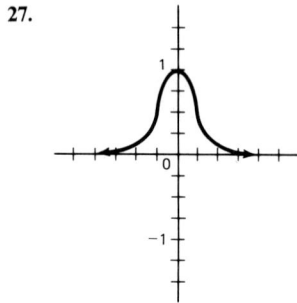

29.

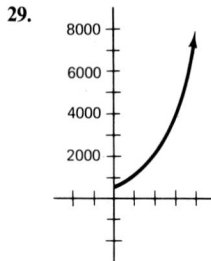

31.

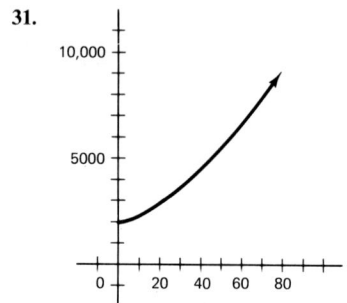

33.

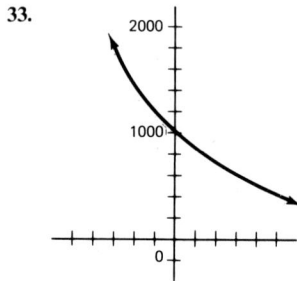

35.

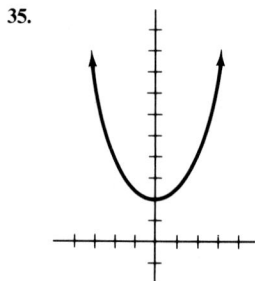

37. -2 **39.** 0 **41.** -3

43. -2 **45.** $\dfrac{3}{2}$ **47.** ± 2 **49.** (a) 10,000; 16,105; 25,937; 41,772; 67,275
(b) 500,000; 921,220; 1,697,292; 3,127,160; 5,761,604 (c) 2000; 2939; 4318; 6344; 9322
(d) 60,000; 120,682; 242,734; 488,226; 981,998 Graph for part (a):

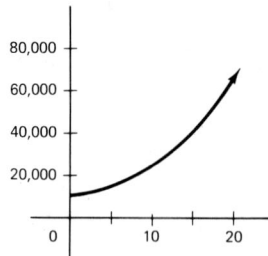

51. 100; 50; 25; 12.5

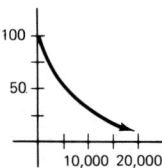

53. 0; 200; 300; 350; 375

PROBLEM SET 7.2

1. 25 **3.** 625 **5.** $\dfrac{1}{5}$ **7.** 3 **9.** 16 **11.** $\dfrac{1}{5}$ **13.** 4 **15.** $\dfrac{5}{3}$ **17.** 49 **19.** 3 **21.** -1 **23.** $-\overline{3}$

25. 3 **27.** 5 **29.** -1 **31.** $\log_2 64 = 6$ **33.** $\log_{81} 9 = \dfrac{1}{2}$ **35.** $\log_4 \dfrac{1}{16} = -2$ **37.** $\log_{25} \dfrac{1}{5} = \dfrac{-1}{2}$

39. $6^2 = 36$ **41.** $16^{1/2} = 4$ **43.** $6^{-1} = \dfrac{1}{6}$ **45.** $1000^{-2/3} = \dfrac{1}{100}$ **47.** $\dfrac{2}{3}$ **49.** 7 **51.** 4 **53.** 3

55. $\dfrac{1}{8}$ **57.**

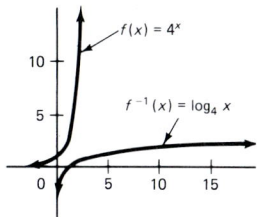

59.

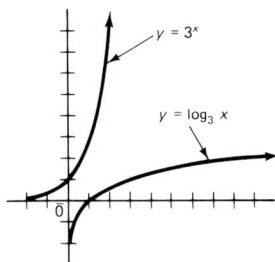

61. $\log_{10} 5 + \log_{10} x + \log_{10} y$

63. $\log_7 6 - \log_7 p$ **65.** $4 \log_{10} x$ **67.** $\dfrac{1}{2} \log_7 z$ **69.** $\dfrac{2}{3} \log_2 p + \log_2 q - \dfrac{1}{3} \log_2 r$ **71.** $\log_3 ab$ **73.** $\log_2 \dfrac{7}{x}$

75. $\log_2 5^p$ **77.** $\log_5 x^2 y^3$ **79.** $\log_{10} \sqrt[6]{\dfrac{k^3}{m^2}}$ **81.** 1.1292 **83.** 1.4248 **85.** 1.277

87. $D = (0, \infty); f^{-1}(x) = 2^x$ **89.** $D = (-1, \infty); f^{-1}(x) = 3^x - 1$ **91.** $D = (1, \infty); f^{-1}(x) = \sqrt[3]{10^x + 1}$
93. $D = $ all reals; $f^{-1}(x) = \log_4 x$ **95.** $D = $ all reals; $f^{-1}(x) = 1 + \log_2 x$ **97.** 30; 60; 80; 110; 120 **99.** 18; 21; 27
101. 1.75

PROBLEM SET 7.3

1. 6×10^4 **3.** 5.21×10^1 **5.** 4×10^{-4} **7.** $\log_{10} 3 + \log_{10} x + \log_{10} y$ **9.** $\log 6 - \log x$ **11.** 1 **13.** 12
15. -5 **17.** 0.8927 **19.** 1.8463 **21.** 5.6990 **23.** $7.5315 - 10 = -2.4685$ **25.** $6.3424 - 10 = -3.6576$
27. $8.8904 - 10 = -1.1096$ **29.** 39.5 **31.** 560,000 **33.** 0.00646 **35.** 0.000325 **37.** 744 **39.** 15.8
41. 3.8513 **43.** $8.4771 - 10 = -1.5229$ **45.** 5 **47.** 120 **49.** 17,000 **51.** 2.5 **53.** 41 **55.** 1.5
57. 7500 **59.** 5.8; 4.1; 3.4; 13.5; 6.4 **61.** 60; 63; 65; 70; 80 **63.** 7.7; 8.3; 8.7

PROBLEM SET 7.4

1. 0.8573 **3.** $6.4150 - 10$ **5.** 2860 **7.** $\log_7 a + \log_7 b + \log_7 c$ **9.** $\dfrac{4}{3} \log t$ **11.** 3.0913 **13.** 5.9005

15. $7.0924 - 10$ **17.** $6.7447 - 10$ **19.** 13,320,000 **21.** 1149 **23.** 0.04534 **25.** 2.79×10^{-7} **27.** 17.17
29. 0.001059 **31.** 78.84 **33.** 279,900 **35.** 399,200 **37.** 40 **39.** 32.31 **41.** 0.1768 **43.** 3.291
45. 5.604 **47.** 13.08 **49.** 0.2763 **51.** 751.6 **53.** 0.2033; 0.3503; 0.1304

PROBLEM SET 7.5

1. 7 **3.** 4 **5.** 7, 2 **7.** $x \log 3$ **9.** $3 \log x$ **11.** $\log_2 5x$ **13.** 1.85 **15.** 8.04 **17.** 11.14
19. 4.894 **21.** 1.072 **23.** 1.5502 **25.** 2000 **27.** 49 **29.** 9 **31.** 7 **33.** 1.086 **35.** 3.00

37. 16 **39.** 2.15 **41.** $\dfrac{\log b}{\log a}$ **43.** 256 **45.** 6 **47.** 5 **49.** 8.04 **51.** 20.8 % **53.** 19
55. $P_{\text{out}} = P_{\text{in}} \cdot 10^{\text{PG}/10}$

PROBLEM SET 7.6

1. 2 **3.** 4.6721 **5.** $8.3010 - 10 = -1.6990$ **7.** 4 **9.** 7 **11.** 2.81 **13.** -0.86 **15.** 1.5
17. 0.732 **19.** -0.672 **21.** 0.693 **23.** -0.916 **25.** A **27.** 0.693 **29.** 9.163 **31.** 0.693

33. 1.833 **35.** 1.209 **37.** 1.065 **39.** 2.708 **41.** 0.357 **43.** 1 **45.** 2.737 **47.** ± 1.646

49. 1.668; 0.332 **51.** 0.0139; $t = \dfrac{-1}{50} \ln\left(1 - \dfrac{i}{2}\right)$ **53.** 3.466; 11.51; $5 \ln\left(\dfrac{P}{10{,}000}\right)$

55. (a)

(b) 23.1 (c) $t = \dfrac{-1}{0.06} \ln \dfrac{T - 20}{40}$

CHAPTER 7 REVIEW EXERCISES

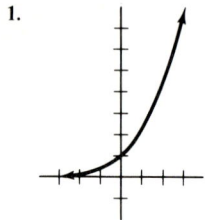

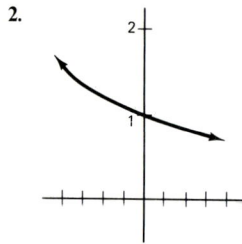

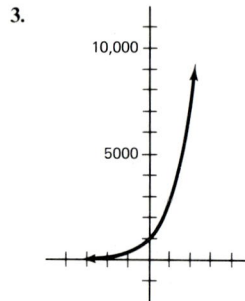

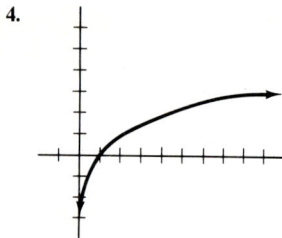

1.

2.

3.

4.

5. -2 **6.** -4 **7.** $2; -2$ **8.** $\log_8 32 = \dfrac{5}{3}$ **9.** $\log_{25} \dfrac{1}{125} = \dfrac{-3}{2}$

10. $10^4 = 10{,}000$ **11.** $8^{-2/3} = \dfrac{1}{4}$ **12.** $\dfrac{1}{2}$ **13.** 4 **14.** $\dfrac{1}{125}$ **15.** $\log_7 6 + \log_7 a + 2 \log_7 b$

16. $4 \log_2 t - 3 \log_2 s$ **17.** $\dfrac{2}{7} \log_{10} x + \dfrac{5}{7} \log_{10} y - \dfrac{6}{7} \log_{10} z$ **18.** $\log_7 \dfrac{ab}{c}$ **19.** $\log \dfrac{x^5 y^{2/3}}{z^{3/5}}$ **20.** 1 **21.** 3.1761

22. $-2.1418 = 7.8582 - 10$ **23.** 1253 **24.** 0.0605 **25.** 516 **26.** 177,800 **27.** 0.6795 **28.** 200.5

29. 1.831 **30.** 1.869 **31.** 1.606 **32.** 1.754 **33.** 4 **34.** $\sqrt{32}$ **35.** -0.683 **36.** 1 **37.** 1.292

38. -0.102 **39.** 1.946 **40.** -0.3567 **41.** 0.7675 **42.** 2.1 **43.** 1.98

PROBLEM SET 8.1

1–5.

7. 200π **9.** 12π **11.** $72°$ **13.** $270°$ **15.** $\dfrac{1}{24}$ **17.** $\dfrac{5}{6}$

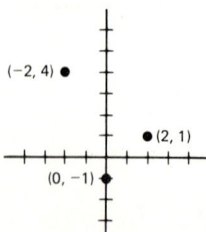

19. $-340°$; $380°$ **21.** $305°$ **23.** $205°$ **25.** $25°$ **27.** $197°$ **29.** $120°$

31-37. Terminal sides are noted.   **39.** $25°$ **41.** $260°$ **43.** 12.2 **45.** 209 **47.** $60°$ **49.** $90°$

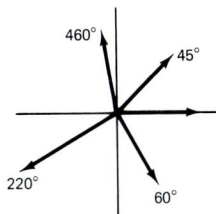

51. $210°$; 12.2; 14.0; 36.7; 62.8 **53.** $90°$; $90°$; $180°$; 25π; 100π **55.** 6; $120°$; $210°$; π; 7π **57.** $\dfrac{1}{4}$; 20π; 5π; $\dfrac{\pi}{2}$

59. $\dfrac{1}{2}$; 60π; 30π; π **61.** $\dfrac{1}{6}$; 20π; $\dfrac{10\pi}{3}$; $\dfrac{\pi}{3}$ **63.** 1; 80π; 80π; 2π **65.** $45°$; 15.7

67. $25,000$; true circumference: $24,881$

PROBLEM SET 8.2

1. 60 **3.** $\dfrac{50}{9}$ **5.** $\sqrt{29}$ **7.** $\sqrt{40}$ **9.** $50°$ **11.** $48°$ **13.** $1.8°$ **15.** $x = 30°$ **17.** $\dfrac{3}{5}$; $\dfrac{4}{5}$; $\dfrac{3}{4}$; $\dfrac{4}{3}$; $\dfrac{5}{4}$; $\dfrac{5}{3}$

19. $\dfrac{48}{52}$; $\dfrac{20}{52}$; $\dfrac{48}{20}$; $\dfrac{20}{48}$; $\dfrac{52}{20}$; $\dfrac{52}{48}$ **21.** $\sqrt{113}$; $\dfrac{7}{\sqrt{113}}$; $\dfrac{8}{\sqrt{113}}$; $\dfrac{8}{\sqrt{113}}$; $\dfrac{7}{\sqrt{113}}$ **23.** $6\sqrt{3}$; $\dfrac{1}{2}$; $\dfrac{\sqrt{3}}{2}$; $\dfrac{\sqrt{3}}{2}$; $\dfrac{1}{2}$

25. $\sqrt{41}$; $\dfrac{4}{\sqrt{41}}$; $\dfrac{5}{\sqrt{41}}$; $\dfrac{5}{\sqrt{41}}$; $\dfrac{4}{\sqrt{41}}$ **27.** $\sqrt{51}$; $\dfrac{7}{10}$; $\dfrac{\sqrt{51}}{10}$; $\dfrac{\sqrt{51}}{10}$; $\dfrac{7}{10}$ **29.** 10; $10\sqrt{3}$; $\dfrac{\sqrt{3}}{2}$; $\dfrac{\sqrt{3}}{2}$; $\dfrac{1}{2}$

31. $6\sqrt{5}$; 18; $\dfrac{\sqrt{5}}{3}$; $\dfrac{2}{3}$; $\dfrac{\sqrt{5}}{3}$ **33.** $\sqrt{65}$; $\dfrac{4}{\sqrt{65}}$; $\dfrac{\sqrt{65}}{4}$; $\dfrac{\sqrt{65}}{4}$; $\dfrac{4}{\sqrt{65}}$ **35.** 13.3; 24.0; 0.67; 1.5; 0.67

37. 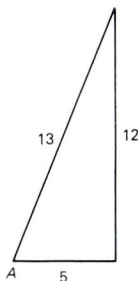 $; \dfrac{4}{5}$ **39.** 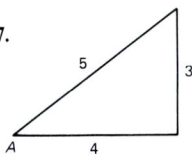 $; \dfrac{12}{5}$ **41.** 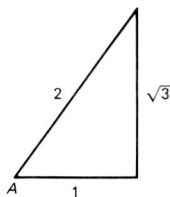 $; \dfrac{\sqrt{3}}{2}$

45. $y = 2x$

PROBLEM SET 8.3

1. $411°$; $-309°$ **3.** $142°$; $-218°$ **5.** $343°$; $703°$ **7.** $c = \sqrt{13}$; $\dfrac{2}{\sqrt{13}}$; $\dfrac{3}{\sqrt{13}}$; $\dfrac{2}{3}$; $\dfrac{3}{2}$; $\dfrac{\sqrt{13}}{3}$; $\dfrac{\sqrt{13}}{2}$ **9.** $\dfrac{5}{13}$, $\dfrac{-12}{13}$, $\dfrac{-5}{12}$

11. $\dfrac{5}{\sqrt{26}}$; $\dfrac{1}{\sqrt{26}}$; 5 **13.** 0; 1; 0 **15.** $\dfrac{-2}{\sqrt{29}}$; $\dfrac{-5}{\sqrt{29}}$; $\dfrac{2}{5}$ **17.** 1 **19.** 0 **21.** $\dfrac{-1}{\sqrt{2}}$ **23.** $\dfrac{1}{\sqrt{3}}$

25. Undefined   **27.** $\dfrac{1}{2}$ **29.** $\dfrac{\sqrt{3}}{2}$ **31.** 1 **33.** -1 **35.** Undefined   **37.** Undefined

39.

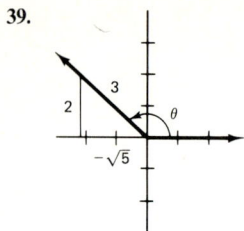

41.

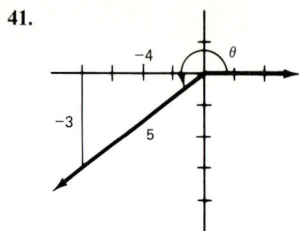

43.

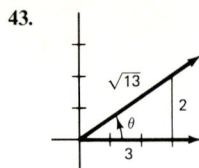

45.

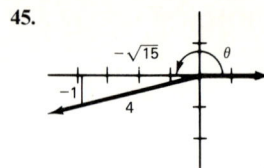

47. $\dfrac{\sqrt{3}}{2}$; $\dfrac{1}{\sqrt{3}}$ **49.** $\dfrac{3}{5}$; $\dfrac{-4}{5}$ **51.** $\dfrac{-\sqrt{2}}{2}$; 1   **53.** $\dfrac{2}{\sqrt{5}}$; $\dfrac{-1}{2}$

55.

θ	$\sin \theta$	$\cos \theta$	$\tan \theta$	$\cot \theta$	$\sec \theta$	$\csc \theta$
0°	0	1	0	undef.	1	undef.
90°	1	0	undef.	0	undef.	1
180°	0	−1	0	undef.	−1	undef.
270°	−1	0	undef.	0	undef.	−1

57.

θ	$\sin \theta$	$\cos \theta$	$\tan \theta$	$\cot \theta$	$\sec \theta$	$\csc \theta$
30°	1/2	$\sqrt{3}/2$	$1/\sqrt{3}$	$\sqrt{3}$	$2/\sqrt{3}$	2
150°	1/2	$-\sqrt{3}/2$	$-1/\sqrt{3}$	$-\sqrt{3}$	$-2/\sqrt{3}$	2
210°	−1/2	$-\sqrt{3}/2$	$1/\sqrt{3}$	$\sqrt{3}$	$-2/\sqrt{3}$	−2
330°	−1/2	$\sqrt{3}/2$	$-1/\sqrt{3}$	$-\sqrt{3}$	$2/\sqrt{3}$	−2

59. (b) 3   (c) 0; 2.1; 3; 2.1; 0; −2.1; −3; −2.1; 0

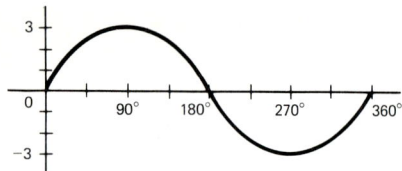

PROBLEM SET 8.4

1. $\dfrac{5}{\sqrt{41}}$; $\dfrac{4}{\sqrt{41}}$; $\dfrac{5}{4}$; $\dfrac{4}{5}$; $\dfrac{\sqrt{41}}{4}$; $\dfrac{\sqrt{41}}{5}$ **3.** $\dfrac{2}{\sqrt{29}}$; $\dfrac{-5}{\sqrt{29}}$; $\dfrac{-2}{5}$; $\dfrac{-5}{2}$; $\dfrac{-\sqrt{29}}{5}$; $\dfrac{\sqrt{29}}{2}$ **5.** $\dfrac{7}{\sqrt{58}}$; $\dfrac{3}{\sqrt{58}}$; $\dfrac{7}{3}$; $\dfrac{3}{7}$; $\dfrac{\sqrt{58}}{3}$; $\dfrac{\sqrt{58}}{7}$

7. $\dfrac{-7}{\sqrt{74}}$; $\dfrac{-5}{\sqrt{74}}$; $\dfrac{7}{5}$; $\dfrac{5}{7}$; $\dfrac{-\sqrt{74}}{5}$; $\dfrac{-\sqrt{74}}{7}$ **9.** 0   **11.** 1   **13.** −1   **15.** $\dfrac{5}{\sqrt{41}}$; $\dfrac{4}{\sqrt{41}}$; $\dfrac{5}{4}$; $\dfrac{4}{5}$; $\dfrac{\sqrt{41}}{4}$; $\dfrac{\sqrt{41}}{5}$

17. 0.7431; 0.6691; 0.9004; 1.1106; 1.4945; 1.3456   **19.** 0.7760; 0.6428; 1.1918; 0.8391; 1.5557; 1.3054   **21.** $\dfrac{6}{5}$; $\dfrac{\sqrt{61}}{6}$; $\dfrac{\sqrt{61}}{5}$

23. 2.4752; 1.0785; 2.6695   **25.** −1.3765; −1.2361; 1.7013   **27.** $\dfrac{1}{5}$; 5   **29.** $\dfrac{-7}{3}$; $\dfrac{-3}{7}$ **31.** 0.5543; 1.8040

33. $\dfrac{\sqrt{3}}{2}$ **35.** $\dfrac{-\sqrt{3}}{2}$ **37.** $\dfrac{-5}{\sqrt{29}}$ **39.** $\dfrac{4}{5}$; $\dfrac{3}{4}$; $\dfrac{4}{3}$; $\dfrac{5}{4}$; $\dfrac{5}{3}$ **41.** $\dfrac{\sqrt{3}}{2}$; $-\sqrt{3}$; $\dfrac{-1}{\sqrt{3}}$; -2; $\dfrac{2}{\sqrt{3}}$

43. $\dfrac{-\sqrt{15}}{4}$; $\dfrac{1}{\sqrt{15}}$; $\sqrt{15}$; $\dfrac{-4}{\sqrt{15}}$; -4 **45.** $\dfrac{-3}{\sqrt{13}}$; $\dfrac{-3}{2}$; $\dfrac{-2}{3}$; $\dfrac{\sqrt{13}}{2}$; $\dfrac{-\sqrt{13}}{3}$ **47.** $\dfrac{4}{5}$; $\dfrac{3}{5}$; $\dfrac{3}{4}$; $\dfrac{5}{3}$; $\dfrac{5}{4}$

49. III; $\dfrac{-2}{\sqrt{5}}$; 2; $\dfrac{-\sqrt{5}}{2}$; $-\sqrt{5}$

PROBLEM SET 8.5

1. $19°$ **3.** $33.3°$ **5.** $\dfrac{1}{\tan 51°}$ **7.** $\dfrac{1}{\sin(-23°)}$ **9.** $40°$ **11.** $245°$ **13.** $350°$ **15.** $+, +, +$

17. $-, -, +$ **19.** $56°$ **21.** $38.2°$ **23.** $50.2°$ **25.** $78°$ **27.** 0.2924 **29.** 0.0017 **31.** 1.4335

33. 573.0 **35.** 0.3584 **37.** 0.7547 **39.** -19.081 **41.** -5.6713 **43.** 0.51504 **45.** -0.17193

47. 6.31375 **49.** 1 **51.** 0.25882 **53.** 5729.58 **55.** $18.1°$ **57.** $51.3°$ **59.** $66.4°$ **61.** $5.6°$

63. $6.8°$ **65.** $5.7°$ **67.** $X = Y = 0.9848$ **69.** $X = Y = 0.1736$ **71.** $X = Y = -0.9397$ **73.** Increase

75. Decrease **77.** Increase **79.** $11.5^{\cup}$ **81.** $126.9°$ **83.** $248.2°$ **85.** $109.3^{\cup}$ **87.** $0.0871557; 0.0871557$

89. $0.8660214; 0.8660254$ **91.** $28.9°; 65.1°; 24.3^{\cup}; 52.6^{\cup}; 1.53; 1.53$ **93.** $0.903; 0.903; 0.902; 0.897$

PROBLEM SET 8.6

1.

Angle	sin	cos	tan	cot	sec	csc
A	$\dfrac{3}{5}$	$\dfrac{4}{5}$	$\dfrac{3}{4}$	$\dfrac{4}{3}$	$\dfrac{5}{4}$	$\dfrac{5}{3}$
B	$\dfrac{4}{5}$	$\dfrac{3}{5}$	$\dfrac{4}{3}$	$\dfrac{3}{4}$	$\dfrac{5}{3}$	$\dfrac{5}{4}$

3. 1.2349 **5.** 0.9806 **7.** 0.8746 **9.** 36 **11.** 5 **13.** 2.5 **15.** 18.4 **17.** 0.17 **19.** 1061

21. $51.3°$ **23.** $33.7°$ **25.** $39°$ **27.** 94.6 **29.** 229 **31.** 149 **33.** $69°$ **35.** 26.4

37. $x = 35.9; y = 12.8; \theta = 12.9°; \phi = 38.7°$

39. $10°: U = 17.4; W = 98.5; 85°: U = 99.6; W = 8.7; 145^{\cup}: U = 57.4; W = -81.9$ (b) Halfway up ($\theta = 90^{\cup}$), where $U = F$

CHAPTER 8   REVIEW EXERCISES

1. $\dfrac{5}{12}$ **2.** $40°$ **3.** $-330°; 390°$ **4.** $240°; 130°; 325°$ **5.**

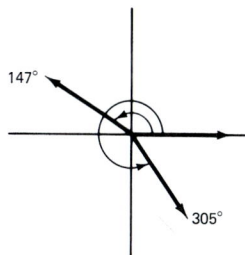

6. $13.1; 192.0$ **7.** $\sqrt{58}; \sqrt{96}$ **8.** $65.6°$

9.

Angle	sin	cos	tan	cot	sec	csc
A	$\dfrac{10}{\sqrt{181}}$	$\dfrac{9}{\sqrt{181}}$	$\dfrac{10}{9}$	$\dfrac{9}{10}$	$\dfrac{\sqrt{181}}{9}$	$\dfrac{\sqrt{181}}{10}$
B	$\dfrac{9}{\sqrt{181}}$	$\dfrac{10}{\sqrt{181}}$	$\dfrac{9}{10}$	$\dfrac{10}{9}$	$\dfrac{\sqrt{181}}{10}$	$\dfrac{\sqrt{181}}{9}$

10. $\dfrac{3}{5}; \dfrac{-4}{5}; \dfrac{-3}{4}, \dfrac{-4}{3}$ **11.** $\dfrac{-3}{\sqrt{73}}; \dfrac{-8}{\sqrt{73}}; \dfrac{3}{8}; \dfrac{8}{3}$ **12.** 0 **13.** $\dfrac{\sqrt{3}}{2}$ **14.** 1 **15.** $\dfrac{\sqrt{3}}{2}$ **16.** -1

17. Undefined **18.** $\dfrac{2}{\sqrt{53}}; \dfrac{7}{\sqrt{53}}; \dfrac{7}{2}; \dfrac{2}{7}; \dfrac{\sqrt{53}}{7}; \dfrac{\sqrt{53}}{.\ 2}$ **19.** $\dfrac{-2}{3}; \dfrac{\sqrt{13}}{2}; \dfrac{-\sqrt{13}}{3}$ **20.** $-1.33; -0.75; 1.67; -1.25$

21. $A: \dfrac{-\sqrt{3}}{2}; \dfrac{-1}{\sqrt{3}}; -\sqrt{3}; \dfrac{-2}{\sqrt{3}}; 2$ $\ B: \text{II}; -0.75; -1.33; -1.25; 1.67$ **22.** $29°; 0.4848; -0.8746; -0.5543; -1.8040$

23. $81°; -0.9877; 0.1564; -6.3138; -0.1584$ **24.** $22°; -0.3746; -0.9272; 0.4040; 2.4751$

25. $26°; 0.4384; 0.8988; 0.4877; 2.0503$ **26.** $19°; -0.3256; 0.9455; -0.3443; -2.9042$ **27.** $35°$ **28.** 123.5

29. 111

PROBLEM SET 9.1

1. 5π **3.** $\dfrac{100\pi}{3}$ **5.** $180°$ **7.** $\dfrac{1}{12}; \dfrac{1}{2}; \dfrac{\sqrt{3}}{2}; \dfrac{1}{\sqrt{3}}$ **9.** $\dfrac{1}{18}; 0.3420; 0.9397; 0.3640$

11. $18°; 0.3090; 0.9511; 0.3249$ **13.** 1 **15.** 3.5 **17.** $180°$ **19.** $45°$ **21.** $72°$ **23.** $720°$ **25.** $114.6°$

27. $5.7°$ **29.** $-57.3°$ **31.** $\dfrac{\pi}{2}$ **33.** π **35.** $\dfrac{\pi}{9}$ **37.** $\dfrac{3\pi}{2}$ **39.** $\dfrac{50\pi}{9}$ **41.** $\dfrac{\pi}{1800}$

43. $200\pi; \dfrac{\pi}{2}; 50\pi$ **45.** $40\pi; 45°; \dfrac{\pi}{4}$ **47.** $60\pi; 180°; 30\pi$

49. $4; 8\pi; 45°$ **51.** $12; \dfrac{\pi}{6}; 2\pi$ **53.** $100; 90°; \dfrac{\pi}{2}$ **55.** $60°; \dfrac{\sqrt{3}}{2}; \dfrac{1}{2}; \sqrt{3}$

57. $180°; 0; -1; 0$ **59.** $\dfrac{\pi}{2}; 1, 0, \text{undefined}$ **61.** $\dfrac{2\pi}{3}, \dfrac{\sqrt{3}}{2}, \dfrac{-1}{2}, -\sqrt{3}$ **63.** π **65.** $\dfrac{\pi}{4}$ **67.** $\dfrac{\pi}{2}$

69. (a) $\pi; 200^g$ (b) $\dfrac{\pi}{4}; 50^g$ (c) $30°; 33\frac{1}{3}^g$ (d) $450°; 500^g$ (e) $135°; \dfrac{3\pi}{4}$ (f) $630°; \dfrac{7\pi}{2}$

71. In BASIC or FORTRAN: (a) $Y = 6 * COS(X)$ (b) $Y = TAN(X)/(5 * X)$ (c) $Y = 1 - 2 * SIN(X) * SIN(X)$
(d) $Y = T * SQRT(SIN(B - C)) + U$ (e) $Y = 1/TAN(X)$ (f) $Y = 1/COS(X)$

PROBLEM SET 9.2

1. $143°; \text{II}; 37°; 0.6018; -0.7986; -0.7536$ **3.** $84°; \text{I}; 84°; 0.9945; 0.1045; 9.1544$

5. $120°; \text{II}; 60°; 0.8660; -0.5000; -1.7321$ **7.** $135°; \text{II}; 45°; \dfrac{\sqrt{2}}{2}; \dfrac{-\sqrt{2}}{2}; -1$ **9.** $\text{I}; 0.10; 0.0998; 0.9950; 0.1003$

11. $\text{I}; 0.56; 0.5312; 0.8473; 0.6269$ **13.** $\text{I}; 1.50; 0.9975; 0.0707; 14.101$ **15.** $\text{III}; 0.60; -0.5633; -0.8262; 0.6818$

17. $\text{IV}; 0.25; -0.2505; 0.9681; -0.2587$ **19.** $\text{IV}; 1.14; -0.9086; 0.4176; -2.1759$ **21.** $\text{I}; \dfrac{\pi}{4}; \dfrac{\sqrt{2}}{2}; \dfrac{\sqrt{2}}{2}; 1$

23. Positive y-axis; $\dfrac{\pi}{2}; 1; 0;$ undefined **25.** Negative x-axis; $0; 0; -1; 0$ **27.** Negative y-axis; $\dfrac{\pi}{2}; -1; 0;$ undefined

29. 0.98 **31.** 1.13 **33.** 0.20 **35.** 1.43 **37.** 0.79 **39.** $0.9602; -0.2794; \text{IV}$ **41.** $0.8660; 1.0472$

43. $0.7042; 2.3603$ **45.** $0.5724; 5.322$ **47.** $5.356; \text{IV}$ **49.** 0.07 **51.** 0.14 **53.** -1.0 **55.** 1.7

PROBLEM SET 9.3

1.

3.

5. $0; 0.2588; 0.5; 0.7071; 0.8660; 0.9659; 1.0000$

7.

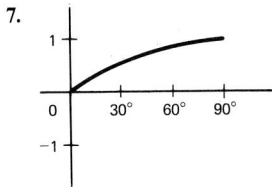

9.

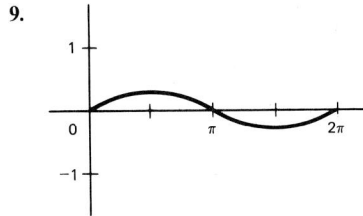

11.

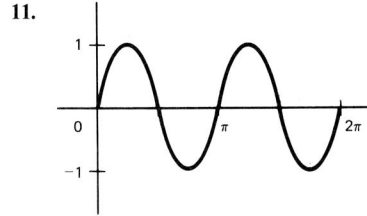

13.

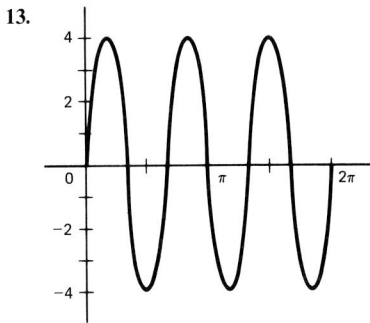

15.

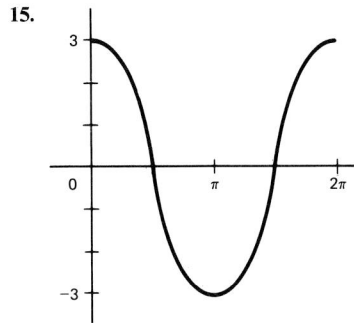

17.

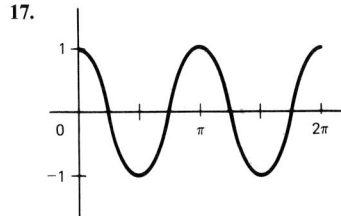

19.

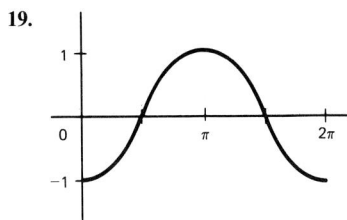

21.

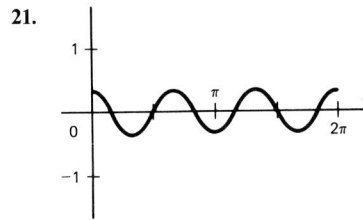

23.

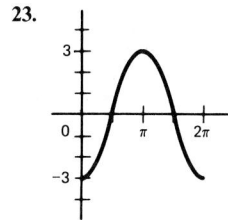

25.

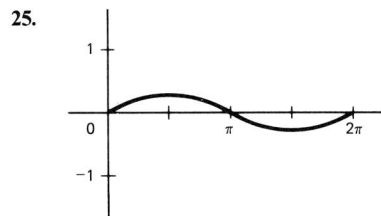

27.

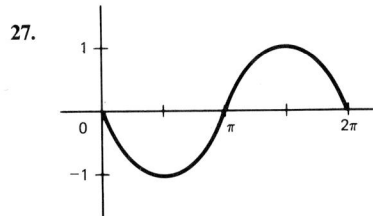

29.

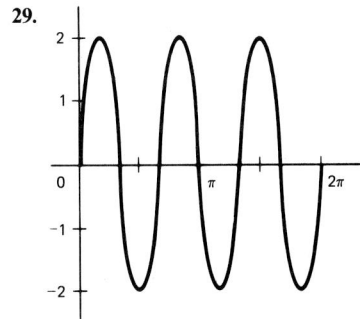

31.

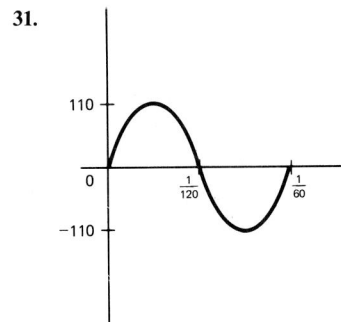

33.

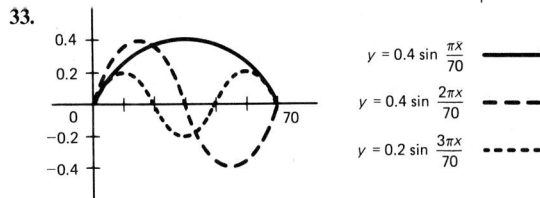

$y = 0.4 \sin \dfrac{\pi x}{70}$ ————

$y = 0.4 \sin \dfrac{2\pi x}{70}$ ― ― ―

$y = 0.2 \sin \dfrac{3\pi x}{70}$ - - - - -

PROBLEM SET 9.4

1. 0.6; 1.33; 0.75; 1.67; 1.25   **3.** −0.44; −2.06; −0.48; −2.29; 1.11   **5.** −0.87; 1.73; 0.58; −2; −1.15

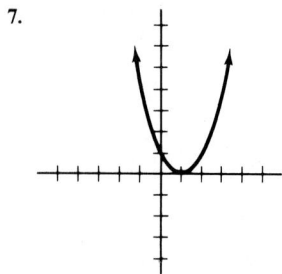

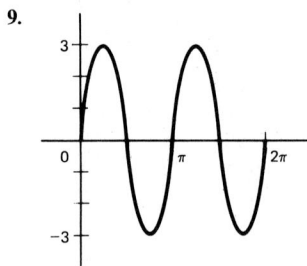

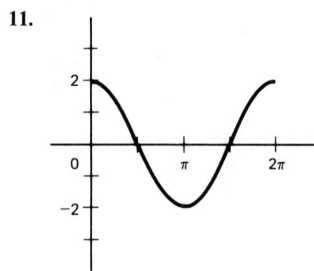

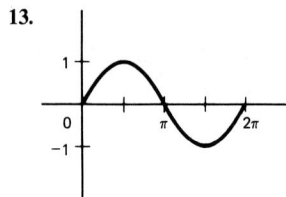

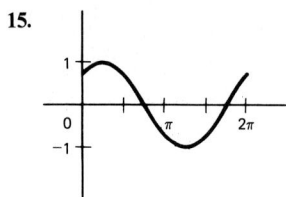

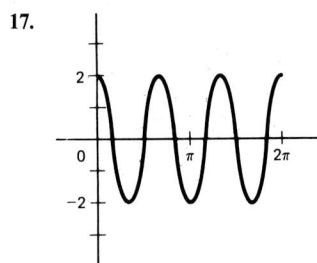

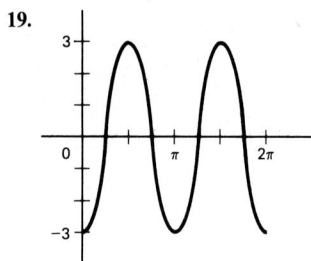

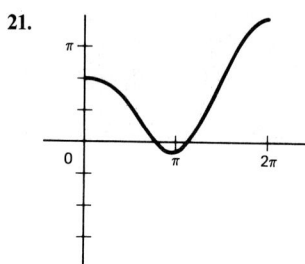

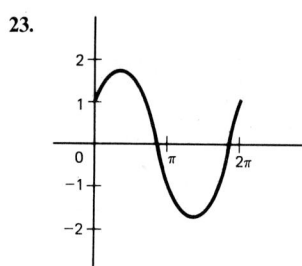

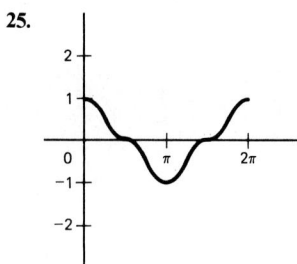

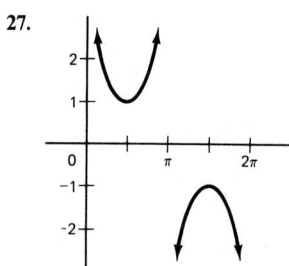

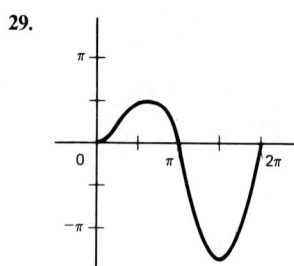

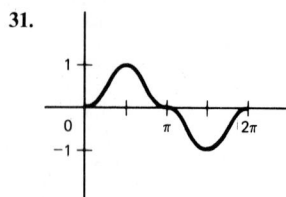

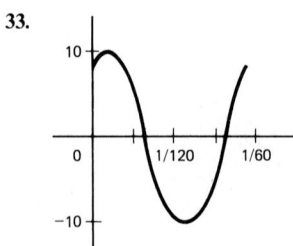

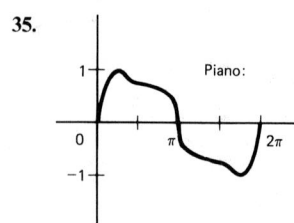

7.

9.

11.

13.

15.

17.

19.

21.

23.

25.

27.

29.

31.

33.

35.

Piano:

PROBLEM SET 9.5

1. $(-0.416, 0.909)$

3. $(-1, 0)$

5. $(-0.654, 0.757)$

7.

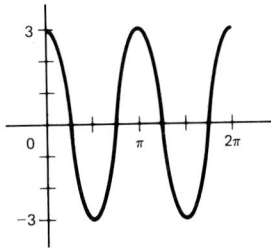

9.

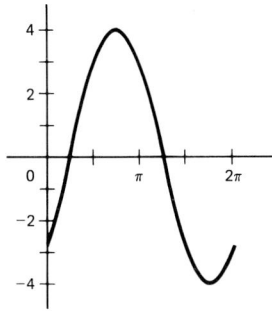

11.

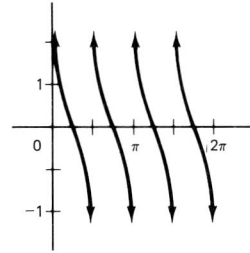

13.

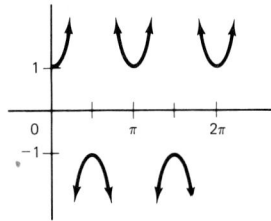

15.

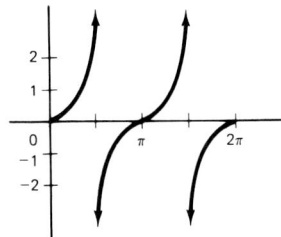

17.

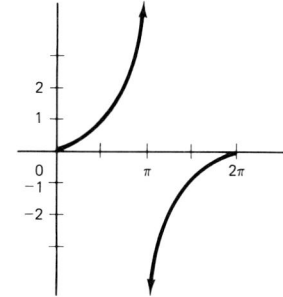

19.

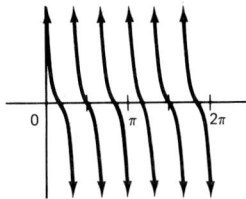

21.

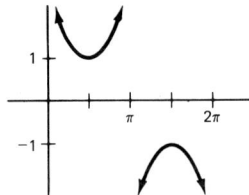

23.

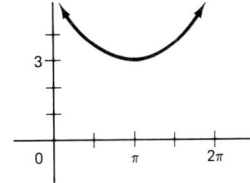

25.

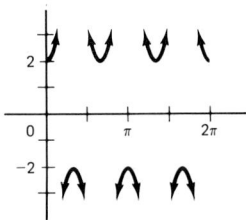

27.

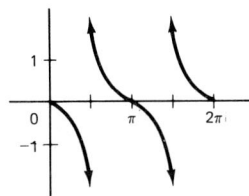

29.

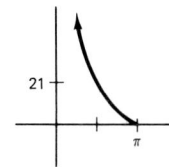

31.

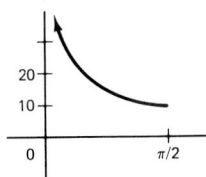

CHAPTER 9 REVIEW EXERCISES

1. $60°; \dfrac{10\pi}{3}; \dfrac{\sqrt{3}}{2}; \dfrac{1}{2}; \sqrt{3}$ 2. $135°; \dfrac{15\pi}{4}; \dfrac{\sqrt{2}}{2}; \dfrac{-\sqrt{2}}{2}; -1$ 3. $\dfrac{\pi}{4}; 25\pi; \dfrac{\sqrt{2}}{2}; \dfrac{\sqrt{2}}{2}; 1$

4. $\pi; 20\pi; 0; -1; 0$ 5. $\dfrac{\pi}{2}; 90°; 1; 0;$ undefined 6. $\dfrac{\pi}{2}; 90°; 25\pi;$ undefined 7. $0.15;$ I; $0.15; 0.1494; 0.9888; 0.1511$

8. $5.02;$ IV; $1.27; -0.9540; 0.2997; -3.1828$ 9. $5.56;$ IV; $0.72; -0.6594; 0.7518; -0.8771$

10. $\dfrac{3\pi}{4};$ II; $\dfrac{\pi}{4}; 0.7071; -0.7071; -1$ 11. $0.59; 0.59; 0.59; 0.83; 0.68$ 12. $2.35; 2.35; 0.79; 0.71; -1.02$

13.
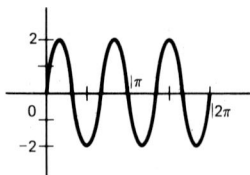

14.
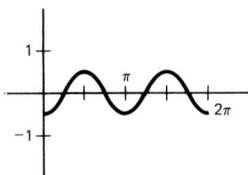

15.
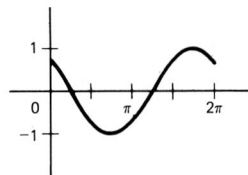

16.
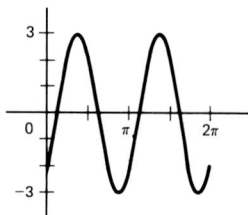

17.
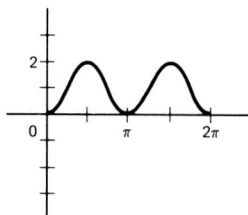

18.
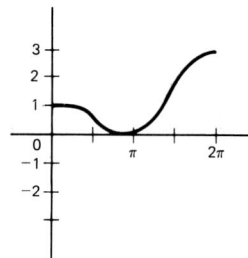

19.
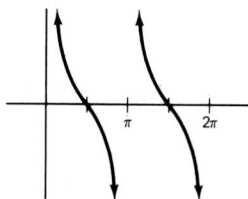

20.
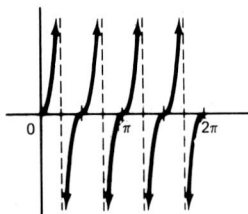

21.
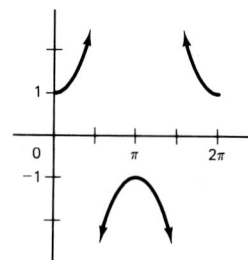

22.
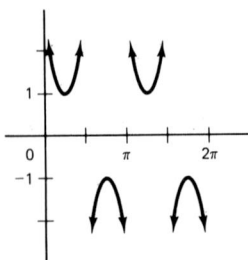

PROBLEM SET 10.1

1. $x^2 - 9$ **3.** $\dfrac{2x + 2\sqrt{3}}{x^2 - 3}$ **5.** $\tan x = \dfrac{\sin x}{\cos x}$; $\cot x = \dfrac{\cos x}{\sin x}$ **7.** $\dfrac{4\cos x + 7\sin x}{\sin x \cos x}$ **9.** $\dfrac{2 - 2\sin x}{\cos^2 x}$ **11.** $\cos x$

63. False **65.** True **67.** False **69.** True **71.** False **73.** True **75.** True **77.** True

PROBLEM SET 10.2

1. 0.9397; 0.8660 **3.** 1.9696; -0.3420 **5.** -1.1133; 0.5 **7.** $30°$; $\dfrac{1}{2}$; $\dfrac{\sqrt{3}}{2}$; $\dfrac{1}{\sqrt{3}}$ **9.** $60°$; $\dfrac{\sqrt{3}}{2}$; $\dfrac{1}{2}$; $\sqrt{3}$

11. $120°$; $\dfrac{\sqrt{3}}{2}$; $\dfrac{-1}{2}$; $-\sqrt{3}$ **13.** $150°$; $\dfrac{1}{2}$; $\dfrac{-\sqrt{3}}{2}$; $\dfrac{-1}{\sqrt{3}}$ **15.** $\dfrac{3}{5}$; $\dfrac{4}{3}$ **17.** -0.92; 2.29 **19.** $\dfrac{\sqrt{2} - \sqrt{6}}{4}$

21. $\dfrac{-\sqrt{2} - \sqrt{6}}{4}$ **23.** $\dfrac{\sqrt{2} + \sqrt{6}}{4}$ **25.** $\dfrac{\sqrt{6} - \sqrt{2}}{4}$ **27.** $\dfrac{\sqrt{3} - 1}{\sqrt{3} + 1}$ **29.** $\dfrac{1 - \sqrt{3}}{1 + \sqrt{3}}$ **31.** $\cos 80°$

33. $\sin \dfrac{11\pi}{28}$ **35.** $\tan 35°$ **37.** $\cos 80°$ **39.** $\dfrac{4}{5}$; $\dfrac{5}{13}$; $\dfrac{63}{65}$; $\dfrac{-16}{65}$; II **41.** $\dfrac{\sqrt{3}}{2}$; $\dfrac{-\sqrt{2}}{2}$; $\dfrac{\sqrt{2} - \sqrt{6}}{4}$; $\dfrac{\sqrt{2} + \sqrt{6}}{4}$; IV

43. -0.9798; -0.7141; 0.5597; -0.8287; II

59. $\cos A \cos B \cos C - \sin A \sin B \cos C - \sin A \cos B \sin C - \cos A \sin B \sin C$ **61.** $8.1°$; $56.3°$; $60.3°$; $15.9°$

PROBLEM SET 10.3

1. $\sin x \cos y + \sin y \cos x$ **3.** $\cos x \cos y - \sin x \sin y$ **5.** $\cos^2 x - \cos^2 y$ **7.** $\dfrac{\tan x + \tan y}{1 - \tan x \tan y}$ **9.** $\dfrac{12}{13}$; $\dfrac{5}{12}$

11. $\dfrac{-3}{5}$; $\dfrac{3}{4}$ **13.** 0.173 **15.** 0.939 **17.** 0.258 **19.** $2\sin\theta\cos\theta$ **21.** $2\sin\theta\cos^3\theta - 2\cos\theta\sin^3\theta$

23. $16(\cos^6 x - \sin^6 x) + 24(\sin^4 x - \cos^4 x) + 9(\cos^2 x - \sin^2 x)$ **25.** $\dfrac{4\tan\theta - 4\tan^3\theta}{1 - 6\tan^2\theta + \tan^4\theta}$ **27.** $\dfrac{3}{5}$; $\dfrac{24}{25}$; $\dfrac{-7}{25}$; II

29. $\dfrac{3}{5}$; $\dfrac{-24}{25}$; $\dfrac{7}{25}$; IV **31.** -0.714; 0.9998; 0.0198; I **33.** -0.866; -0.866; -0.5; III **35.** $\dfrac{1}{2}\sqrt{2 - \sqrt{3}}$

37. $\dfrac{1}{2}\sqrt{2 + \sqrt{2}}$ **39.** $\dfrac{\sqrt{2}}{2}$ **41.** $\dfrac{5}{13}$; $0 < \dfrac{\theta}{2} < 45°$; $\dfrac{2}{\sqrt{13}}$; $\dfrac{3}{\sqrt{13}}$; $\dfrac{2}{3}$ **43.** $\dfrac{12}{13}$; $\dfrac{\pi}{4} < \dfrac{\theta}{2} < \dfrac{\pi}{2}$; $\dfrac{3}{\sqrt{13}}$; $\dfrac{2}{\sqrt{13}}$; $\dfrac{3}{2}$

45. $\dfrac{-\sqrt{8}}{3}$; $90° < \dfrac{\theta}{2} < 135°$; $\sqrt{\dfrac{3 + \sqrt{8}}{6}}$; $-\sqrt{\dfrac{3 - \sqrt{8}}{6}}$; $-\sqrt{\dfrac{3 + \sqrt{8}}{3 - \sqrt{8}}}$

PROBLEM SET 10.4

1. 1 **3.** $\cos A \cos B + \sin A \sin B$ **5.** $2\sin x \cos x$ **7.** $\pm\sqrt{\dfrac{1 + \cos t}{2}}$ **9.** $\sin(p - q)$ **11.** $-\sin x$

13. $44.4°$ **15.** $120.5°$ **17.** $156.4°$ **19.** $\dfrac{1}{2}(\cos 0° - \cos 90°)$ **21.** $\dfrac{1}{2}\left(\cos\dfrac{\pi}{4} + \cos\dfrac{3\pi}{4}\right)$ **23.** $\dfrac{1}{2}\left(\sin\dfrac{7\pi}{12} - \sin\dfrac{\pi}{12}\right)$

25. $\dfrac{1}{2}(\sin 6x - \sin 2x)$ **27.** $2\sin 90° \cos 0°$ **29.** $2\sin\dfrac{5\pi}{24}\cos\dfrac{\pi}{24}$ **31.** $-2\sin\left(x + \dfrac{h}{2}\right)\sin\dfrac{h}{2}$

33. $2\cos 150A \cos 50A$ **35.** $5\sin(x + 53.1°)$ **37.** $13\sin(x + 67.4°)$ **39.** $\sqrt{2}\sin(x + 135°)$

41. $\sqrt{17}\sin(x + 76.0°)$ **43.** $2\sin 5x \cos 3x$ **45.** $\dfrac{1}{2}[\cos 10(x - y) - \cos 10(x + y)]$ **47.** $\sqrt{2}\sin(x - 45°)$

49. $\dfrac{1}{2}[\cos(z - y) + \cos(2x - y - z)]$

59. (a) $2\sin 881\pi t \cos \pi t$ (b) $2\cos 405\pi t \cos 5\pi t$ (c) $2\sin\pi(f_1 + f_2)t \cos\pi(f_1 - f_2)t$
First term is dominant tone; second term is beating.

61. $y = 15\sin(4t + 36.9°)$; amplitude $= 15$; phase angle $= 9.2°$

PROBLEM SET 10.5

1. 4 **3.** 2; 5 **5.** 1 **7.** $\dfrac{\pi}{6}; \dfrac{5\pi}{6}$ **9.** No solution **11.** 0; π **13.** 0.49; 2.65 **15.** 2.82; 3.46

17. 60°; 120°; 240°; 300° **19.** 0°; 45°; 180°; 225° **21.** 90°; 210°; 330° **23.** 30°; 150° **25.** 90°; 120°; 240°; 270°
27. 0°; 180° **29.** 31.6°; 121.6°; 211.6°; 301.6° **31.** 216.1° **33.** 0°; 120°; 240° **35.** 30°; 90°; 150°; 270°
37. 30° **39.** 180°; 240° **41.** 0°; 90°; 180° **43.** 38.2°; 141.8° **45.** 0°; 45°; 135°; 180°; 225°; 315°
47. 56.3°; 236.3° **49.** 208.5° **51.** 38.2°; 141.8° **53.** 46.2°; 73.8°; 166.2°; 193.8°; 286.2°; 313.8°
55. 0°; 45°; 60°; 90° **57.** (a) 6.3°; 83.7° (b) 12.9°; 77.1° (c) 30.3°; 59.7° (d) Impossible (e) 45° produces $x = 91.8$

PROBLEM SET 10.6

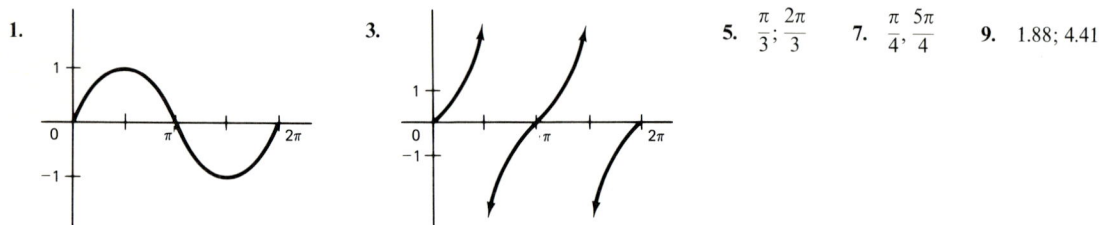

1.

3.

5. $\dfrac{\pi}{3}; \dfrac{2\pi}{3}$ **7.** $\dfrac{\pi}{4}, \dfrac{5\pi}{4}$ **9.** 1.88; 4.41

11. $f^{-1}(x) = x + 2$ **13.** $f^{-1}(x) = \dfrac{x-1}{2}$ **15.** 0 **17.** $\dfrac{-\pi}{2}$ **19.** $\dfrac{\pi}{3}$ **21.** 2.61 **23.** 0.687 **25.** -0.631

27. $\dfrac{\pi}{2}$ **29.** $\dfrac{2\pi}{3}$ **31.** $\dfrac{\pi}{2}$ **33.** $\dfrac{\pi}{2}$

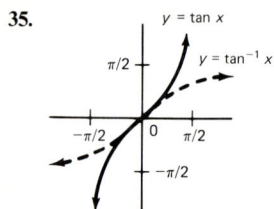

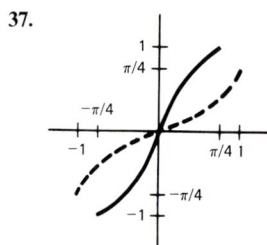

35.

37.

39. $D = [-1, 1]; R = \left[\dfrac{-\pi}{2}, \dfrac{\pi}{2}\right]$

41. $D = \left[-\dfrac{1}{2}, \dfrac{1}{2}\right]; R = [0, \pi]$ **43.** $D = (-\infty, \infty); R = \left(-\dfrac{\pi}{2}, \dfrac{\pi}{2}\right)$ **45.** $D = [1, 3]; R = [-\pi, \pi]$

47. $D = [0, 1]; R = \left[0, \sqrt{\dfrac{\pi}{2}}\right]$

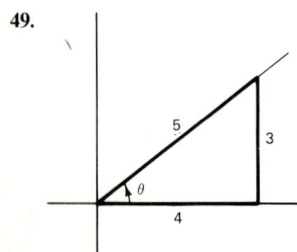

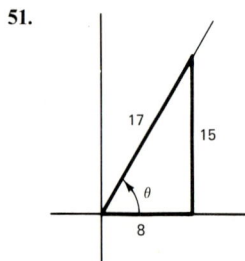

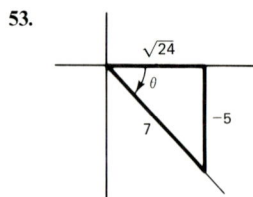

49.

51.

53.

A50 Answers to Selected Exercises

55.

57. $\dfrac{12}{13}$ **59.** Undefined **61.** $\dfrac{4}{5}$ **63.** $\dfrac{\sqrt{1-x^2}}{x}$ **65.** $\sqrt{1-x^2}$

67. $\dfrac{-56}{33}$ **69.** $\dfrac{2}{9}$ **71.** $2x\sqrt{1-x^2}$ **73.** $\sqrt{\sin^{-1} y}$ **75.** $\sqrt{1+\cos^{-1} y}$

77. $\cos\left(\dfrac{y}{4}\right)-1$ **79.** $\sqrt[3]{\sin\left(\dfrac{y}{3}\right)+1}$ **81.** $1-\cos[2(y-3)]$ **83.** $\sqrt{\dfrac{\sqrt[3]{\sin^{-1}\left(\dfrac{y}{3}\right)}-1}{2}}$ **85.** $7+\cos\left(\dfrac{y}{2}\right)$

87. $26.6°; 78.7°; 31.0°; 0°$ **89.** $I=\tan^{-1}(2\cot\theta); \theta=\cot^{-1}\left(\dfrac{1}{2}\tan I\right)$

CHAPTER 10   REVIEW EXERCISES

1. $\dfrac{15}{17}; \dfrac{15}{8}; \dfrac{8}{15}; \dfrac{17}{8}; \dfrac{17}{15}$ **2.** $\dfrac{-4}{5}; \dfrac{-3}{5}; \dfrac{3}{4}; \dfrac{-5}{3}; \dfrac{-5}{4}$ **3.** $\dfrac{\cos x - 1}{\cos x + 1}$ **4.** $\cos x$ **19.** 0.9798 **20.** -0.8

21. 0.2041 **22.** -0.75 **23.** 5 **24.** 1.67 **25.** 1.02 **26.** -1.25 **27.** 0.428 **28.** -0.748
29. -0.904 **30.** -0.664 **31.** -0.474 **32.** 1.126 **33.** 0.392 **34.** -0.960 **35.** 0.280 **36.** 0.920
37. 0.100 **38.** 0.949 **39.** 0.316 **40.** 0.995 **41.** 0.693 **42.** 0.568 **43.** $13\sin(x-67.4°)$

44. $\sqrt{2}\sin(x+225°)$ **45.** $2\cos\left(x+\dfrac{h}{2}\right)\sin\dfrac{h}{2}$ **46.** $2\sin\left(x-\dfrac{h}{2}\right)\sin\dfrac{h}{2}$ **47.** $60°; 300°$ **48.** $0°; 180°$

49. $135°; 315°$ **50.** $65.5°; 294.5°$ **51.** $309.6°; 230.4°$ **52.** $105.7°; 285.7°$ **53.** $0°$ **54.** $0; 3\pi/4; \pi; 7\pi/4$

55. $0; \pi$ **56.** $\dfrac{\pi}{2}; \dfrac{3\pi}{2}$ **57.** $0.506; 2.635; 3.648; 5.777$ **58.** $0; \pi; 7\pi/6; 11\pi/6$ **59.** $\pi/3; \pi$

60. $\pi/6; \pi/3; 2\pi/3; 5\pi/6; 7\pi/6; 4\pi/3; 5\pi/3; 11\pi/6$ **61.** $\dfrac{\pi}{6}$ **62.** $\dfrac{\pi}{2}$ **63.** $\dfrac{-\pi}{4}$ **64.** $\dfrac{\pi}{3}$ **65.** 0.627 **66.** -0.217

67.

68.

69.

70. $\dfrac{4}{3}$ **71.** $\dfrac{1}{2}$

72. $\dfrac{\sqrt{4-x^2}}{2}$ **73.** $\dfrac{5}{\sqrt{26}}$ **74.** 1 **75.** $\dfrac{2x}{1+x^2}$ **76.** $\dfrac{\sin\left(\dfrac{y}{5}\right)-3}{2}$ **77.** $\dfrac{1-\sin^{-1}\sqrt{\dfrac{y}{3}}}{3}$

PROBLEM SET 11.1

1. 0.6820   **3.** -0.8746 **5.** 44.7°   **7.** 5.67   **9.** 9   **11.** 0.25   **13.** 12.9; 58.2°; 32.8°
15. 25; 53.1°; 36.9°   **17.** 83.3°; 52.6°; 44.0°   **19.** 60°; 60°; 60°   **21.** 50°; 25.4; 54.4   **23.** 161°; 4.6; 20.7
25. 92.7°; 87.3°; 94.3   **27.** 42.6°; 137.4°; 143.5   **31.** 39.1   **33.** 63.7   **35.** 87.6°; 92.4°; 93.6°   **37.** 975

PROBLEM SET 11.2

1. 0.3420   **3.** 0.0017   **5.** 0.0017   **7.** 45   **9.** $\dfrac{y \sin A}{\sin B}$ **11.** 13.5; 15.3; 80°   **13.** 143.1; 26.4; 10°
15. 63.3; 47.0; 20°   **17.** 37.9; 29.9; 110.9°   **19.** 17.3; 20; 30°; 60°; 90°   **21.** 17.0; 19.5; 63°   **23.** 29.7; 5.9; 87.7°
25. 95.9; 80.3°; 47.7°   **27.** 51.3°; 59.2°; 69.5°   **29.** 31.8; 24.0; 92°   **33.** 273   **35.** 13.8

PROBLEM SET 11.3

1. 0.9659   **3.** 0.3931   **5.** 45.5°   **7.** Undefined   **9.** 35.2°   **11.** $\dfrac{a}{\sin A} = \dfrac{b}{\sin B} = \dfrac{c}{\sin C}$ **13.** 13.6; 23.7°; 114.3°
15. 10.8; 26.5°; 74.5°   **17.** No such triangle   **19.** 259.8; 60°; 90°   **21.** No such triangle   **23.** No such triangle
25. 48.5; 57.5; 111°   **27.** 41.5°; 62.1°; 76.4   **29.** 83.2; 26.7°; 42.3°   **31.** Infinite number of triangles
33. 13.1; 32.4°; 47.6°   **35.** 121.3 or 29.7 feet from crane; 53.1°

PROBLEM SET 11.4

1. 127.2; 47.6°; 62.4°   **3.** 326.1; 368.6; 70°   **5.** 87.0; 47.6°; 32.4°

7–21.

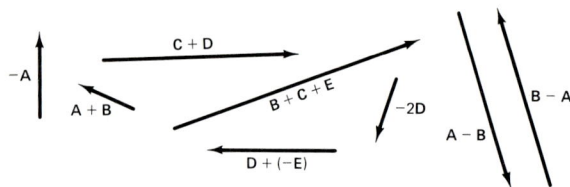

23. 22.4; 63.4°   **25.** 111.8; 153.4°

27. 269.3; 248.2°   **29.** 832.4; 62.1°   **31.** 34.2; 94.0   **33.** 125.8; -81.7 **35.** 18.0; 56.3°   **37.** 53.9; $-21.8°$
39. $|D| = 12.8$; $\theta_D = 141.3°$ **41.** $|D| = 11.2$; $\theta_D = 63.4°$ **43.** $|D| = 3.3$; $\theta_D = 46.4°$ **45.** 180; 40; 184.4; 12.5°
47. -12; 2; 12.17; 170.5°   **49.** 80; -200; 215.4; $-68.2°$ **51.** $A_x = 75.5$; $A_y = 65.6$; $B_x = -10.5$; $B_y = 149.6$; $\mathbf{A} \cdot \mathbf{B} = 9021$
53. $-12{,}287$; -4915; 0   **55.** $|S| = 434$, $\theta_S = 44.9$; $|S| = 306$, $\theta_S = 112.5°$

CHAPTER 11   REVIEW EXERCISES

1. $a = 75.5$; $b = 65.6$ **2.** $a = 182.8$; $d = 189.5$ **3.** $b = 80.8$; $p = 27.2$ **4.** $a = 59.6$; $A = 6.97°$
5. 31.1; 53.5°; 86.5° *and* 7.2; 126.5°; 13.5°   **6.** 94.3; 69.5°; 48.5°   **7.** 29.0°; 46.6°; 104.5°   **8.** 8.4; 9.6; 67°
9. 639.8; 171.1; 139.5°   **10.** 106.3; 22.1°; 87.9°   **11.** 250; 36.9°   **12.** 244.1; 125°   **13.** 189.3; 55.8°
14. 228.2; 79.7°   **15.** 566.4; 824.1   **16.** 640.3; 38.7°

PROBLEM SET 12.1

1. 4   **3.** $2\sqrt{3}$ **5.** $11 - x$ **7.** $4 - 2x$ **9.** $6 - 7x - 5x^2$ **11.** $49 - 25x^2$

13. Real part = 4; imaginary part = -5 **15.** Real part = $\dfrac{5}{2}$; imaginary part = $\dfrac{\sqrt{3}}{2}$

17. Real part $= -7$; imaginary part $= 0$ **19.** $2i$ **21.** $i\sqrt{3}$ **23.** $2i\sqrt{5}$ **25.** $12 - i$ **27.** $-10 + 14i$
29. $11 + 3i$ **31.** $-2 + 2i$ **33.** 2 **35.** $13 - 6i$ **37.** $47 - 35i$ **39.** $-48 + 14i$ **41.** 26
43. $-i$ **45.** $-i$ **47.** $15 - 2i$ **49.** $52 + 23i$ **51.** $39 + 80i$ **53.** $74 + 128i$
55. $1000 + 1500j$; $2500 - 100j$; 500

PROBLEM SET 12.2

1. $10i$ **3.** $-i$ **5.** $12 + i$ **7.** $45 - 15i$ **9.** 41 **11.** $\dfrac{\sqrt{2}}{2}$ **13.** $\dfrac{5 + \sqrt{7}}{3}$

15. $5 - 6i$ **17.** $\dfrac{5 + 3i}{4}$ **19.** $-7i$ **21.** -2 **23.** 6; 25 **25.** 16; 68 **27.** 8; 19 **29.** 0; 100

31. Both $= 9 + i$ **33.** Both $= 22 + 3i$ **35.** Both $= 15 - 8i$ **37.** $\dfrac{1}{10} + \dfrac{4i}{5}$ **39.** $\dfrac{11}{10} - \dfrac{4i}{5}$

41. $\dfrac{3}{17} - \dfrac{12i}{17}$ **43.** $1 - 4i$ **45.** $-i$ **47.** $5i$ **49.** Both $= (a + c) + (-b - d)i$ **51.** Both $= (a^2 - b^2) - 2abi$

53. $3 - 7i$ **55.** $\dfrac{9}{5} + \dfrac{7i}{5}$ **57.** $1000 - 1500j$; $1000 + 1500j$; $3{,}250{,}000$; 1803

PROBLEM SET 12.3

1–4.

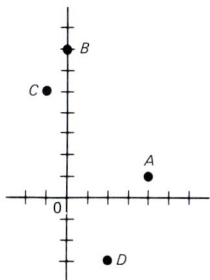

5. 4 **7.** $3 + 4i$ **9.** $-4 + 2i$ **11.** $-3 - 3i$ **13.** $5 - 2i$

15–21.

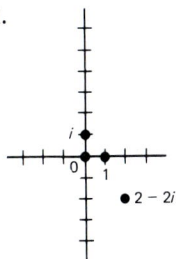

23. $3 + 4i$; parallelogram

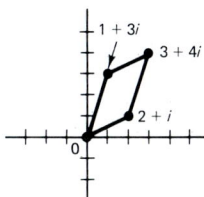

25. $5 - 2i$; parallelogram

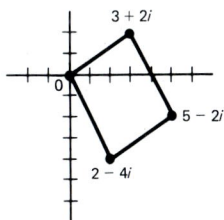

27. 5; $53.1°$ **29.** $\sqrt{40}$; $161.6°$ **31.** $\sqrt{53}$; $254.1°$

33. $5(\cos 53.1° + i \sin 53.1°)$ **35.** $2(\cos 60° + i \sin 60°)$ **37.** $\sqrt{41}(\cos 231.3° + i \sin 231.3°)$

39. $\sqrt{85}[(\cos(-77.5°) + i \sin(-77.5°)]$ **41.** $6(\cos 90° + i \sin 90°)$ **43.** $5(\cos 0° + i \sin 0°)$ **45.** $1 + 1.73i$

47. $7.07 + 7.07i$ **49.** $2.83 + 2.83i$ **51.** $-2.74 + 7.52i$ **53.** $3i$ **55.** $4.92 - 0.87i$

57. $\sqrt{52}[\cos(-33.7°) + i \sin(-33.7°)]$ **59.** $1.96 + 0.39i$ **61.** $10(\cos 270° + i \sin 270°)$ **63.** $-3.54 + 3.54i$

65. $180.3\underline{/56.3°}$; $0.0023 + 0.0019j$; $8163\underline{/31.0°}$; $28.5 + 106.3j$; $0.00022\underline{/-26.6°}$; $1108 - 1418j$

PROBLEM SET 12.4

1. $23 + 14i$ **3.** $24 - 10i$ **5.** $\sqrt{58}(\cos 23.2° + i \sin 23.2°)$ **7.** $8(\cos 0° + i \sin 0°)$ **9.** $3.46 + 2i$ **11.** -3

13. 3 **15.** 10 **17.** $6(\cos 30° + i \sin 30°)$ **19.** $3(\cos 75° + i \sin 75°)$ **21.** $12(\cos 265° + i \sin 265°)$

23. $3(\cos 15° + i \sin 15°)$ **25.** $\frac{1}{32}[\cos(-50°) + i \sin(-50°)]$ **27.** $-278 + 29i$ **29.** $15(\cos 135° + i \sin 135°)$

31. $40(\cos 85° + i \sin 85°)$ **33.** $30(\cos 45° + i \sin 45°)$ **35.** $4(\cos 65° + i \sin 65°)$

37. $58[\cos(-26.6°) + i \sin(-26.6°)]$

39. $\cos 8° + i \sin 8°$; $\cos 80° + i \sin 80°$; $\cos 152° + i \sin 152°$; $\cos 224° + i \sin 224°$; $\cos 296° + i \sin 296°$

41. $3(\cos 50° + i \sin 50°)$; $3(\cos 140° + i \sin 140°)$; $3(\cos 230° + i \sin 230°)$; $3(\cos 320° + i \sin 320°)$

43. $2(\cos 45° + i \sin 45°)$; $2(\cos 225° + i \sin 225°)$ **45.** $3(\cos 90° + i \sin 90°)$; $3(\cos 210° + i \sin 210°)$; $3(\cos 330° + i \sin 330°)$

47. $\sqrt[4]{2}(\cos 22.5° + i \sin 22.5°)$; $\sqrt[4]{2}(\cos 202.5° + i \sin 202.5°)$ **49.** $2i$; $-2i$ **51.** $8(\cos 30° + i \sin 30°)$

53. $2(\cos 20° + i \sin 20°)$; $2(\cos 140° + i \sin 140°)$; $2(\cos 260° + i \sin 260°)$

55. $550,000\underline{/33°}$; $2\underline{/62°}$; $0.02\underline{/-11°}$; $25,000\underline{/-102°}$

PROBLEM SET 12.5

1–5.

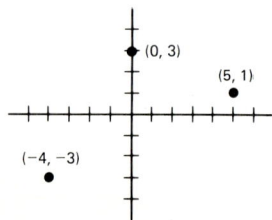

7. $\sqrt{65}(\cos 60.3° + i \sin 60.3°)$ **9.** $\sqrt{34}[\cos 121.0° + i \sin 121.0°]$

11. $2.95 + 0.52i$ **13.** $-2 - 3.46i$ **15–25.**

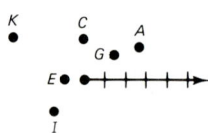

   27. $(2.60, 1.5)$ **29.** $(0, 2)$

31. $(-1.41, -1.41)$ **33.** $(-3.46, 2)$ **35.** $(\sqrt{13}, 56.3°)$ **37.** $(\sqrt{20}, 116.6°)$ **39.** $(5, 216.9°)$ **41.** $(\sqrt{65}, -60.3°)$

43. $r \sin \theta = 7$ **45.** $r(\cos \theta + \sin \theta) = 5$ **47.** $r = 2$ **49.** $r^2 = \dfrac{2}{\sin 2\theta}$ **51.** $r^2(4 \cos^2 \theta + 9 \sin^2 \theta) = 36$

53. $x = 5$ **55.** $x^2 + y^2 = 2x$ **57.** $\sqrt{x^2 + y^2}\cos\sqrt{x^2 + y^2} = x$ **59.** $x^2 + y^2 = 4\sqrt{x^2 + y^2} + 4x$

61. $y^2 = 16 + 8x$ **63.** $y^2 - 3x^2 - 12x = 9$ **65.** $(-2, 210°)$; $(2, -330°)$; $(2, 390°)$

67. $\left(-20, \dfrac{7\pi}{6}\right)$; $\left(20, \dfrac{-11\pi}{6}\right)$; $\left(20, \dfrac{13\pi}{6}\right)$ **69.** $(-5, -30°)$; $(5, -210°)$; $(5, 510°)$

71. 1; 0.7; 0; −0.7; −1; −0.7; 0; 0.7; 1   **73.** 0; 1; 0; −1; 0; 1; 0; −1; 0   **75.** Polar

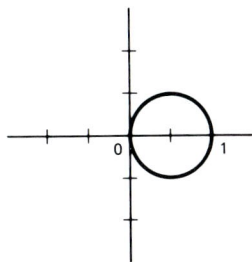

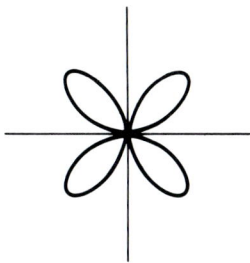

PROBLEM SET 12.6

1–5.

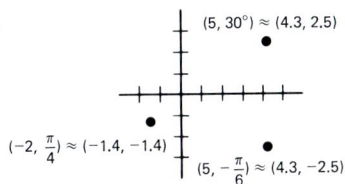

(5, 30°) ≈ (4.3, 2.5)

$(-2, \frac{\pi}{4}) \approx (-1.4, -1.4)$

$(5, -\frac{\pi}{6}) \approx (4.3, -2.5)$

7. $(\sqrt{29}, 68.2°)$ **9.** $(\sqrt{5}, 206.6°)$

11.

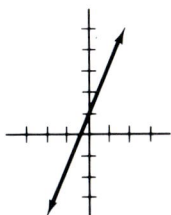

13.

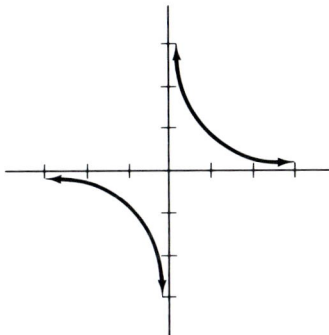

15.

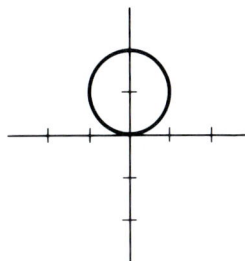

17.

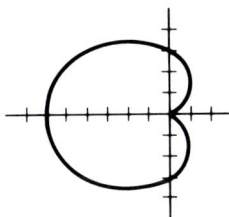

19.

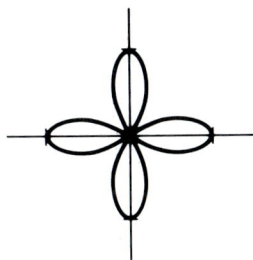

21.

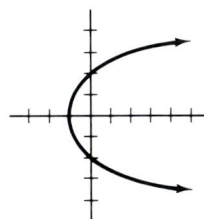

23.

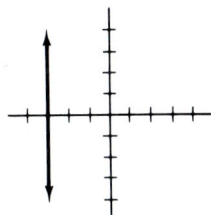

25.

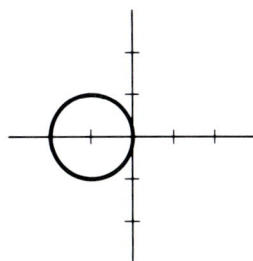

27.

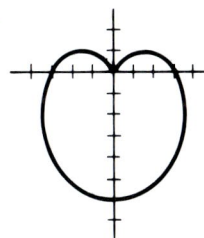

29.

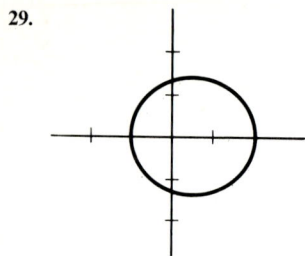

31.

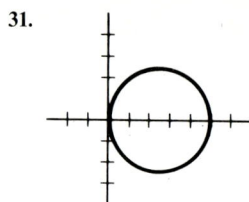

33.

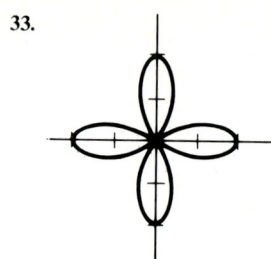

35.

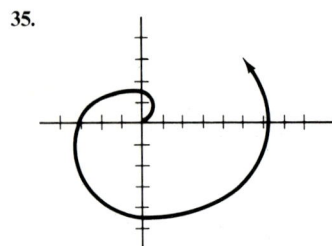

37.

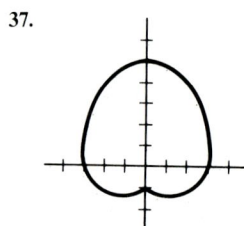

39.

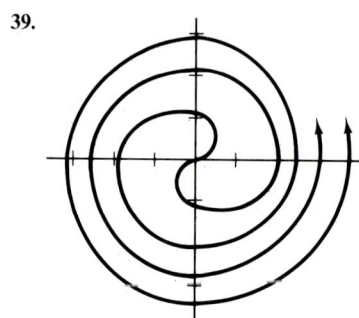

41.

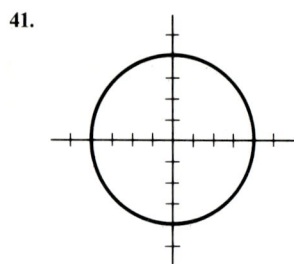

43.

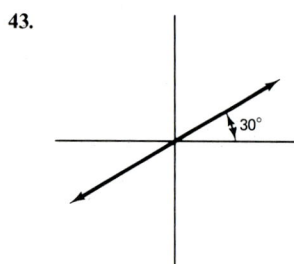

45.

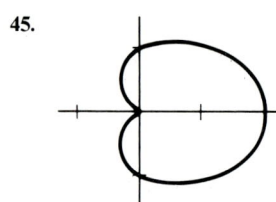

47.

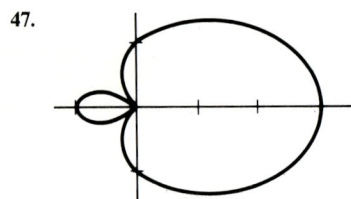

CHAPTER 12   REVIEW EXERCISES

1. Real part $= 7$; imaginary part $= -3$ **2.** Real part $= \dfrac{3}{4}$; imaginary part $= \dfrac{\sqrt{11}}{4}$ **3.** $2i$ **4.** $15 - i$

5. $-3 - 7i$ **6.** $37 - 3i$ **7.** $-21 - 20i$ **8.** i **9.** 58 **10.** 6 **11.** $\dfrac{11}{10} - \dfrac{7i}{10}$

12. Both $= 7 - i$ **13.** Both $= 14 - 2i$ **14.** Both $= 8 + 6i$ **15–18.**

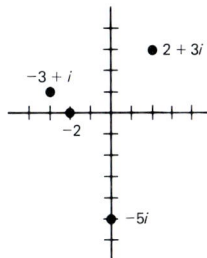

19. $\sqrt{29} \, (\cos 111.8° + i \sin 111.8°)$ **20.** $4(\cos 90° + i \sin 90°)$ **21.** $1.03 + 2.82i$ **22.** $-1.93 + 0.52i$

23. $8(\cos 95° + i \sin 95°)$ **24.** $32(\cos 100° + i \sin 100°)$ **25.** 16

26. $2(\cos 25° + i \sin 25°); 2(\cos 145° + i \sin 145°); 2(\cos 265° + i \sin 265°)$

27. $\sqrt[10]{2} \, (\cos 63° + i \sin 63°); \ \sqrt[10]{2} \, (\cos 135° + i \sin 135°); \ \sqrt[10]{2} \, (\cos 207° + i \sin 207°);$
   $\sqrt[10]{2} \, (\cos 279° + i \sin 279°); \ \sqrt[10]{2} \, (\cos 351° + i \sin 351°)$ **28.** $(\sqrt{58}, -66.8°)$ **28.** $(4, \pi)$ **30.** $(1.55, 5.80)$

31. $(0, -5)$ **32.** $r(\cos \theta + \sin \theta) = 7$ **33.** $(x - 3)^2 + y^2 = 9$

34.

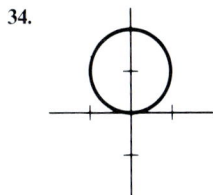

35.

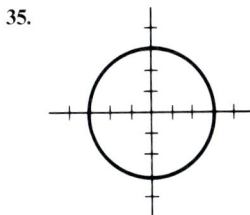

36.

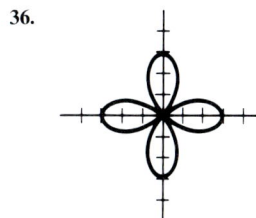

37.

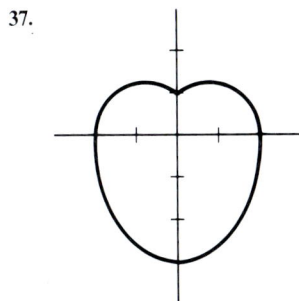

38.

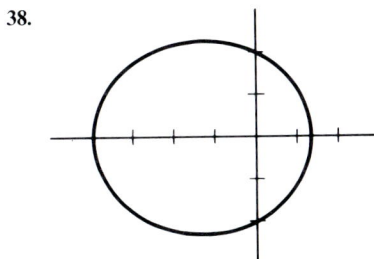

39.

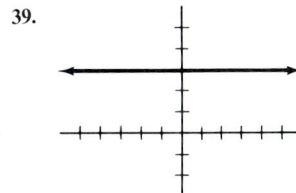

PROBLEM SET 13.1

1. -2 **3.** 50 **5.** -6 **7.** -2 **9.** $\dfrac{20}{13}$ **11.** $\dfrac{c - b}{a}$ **13.** $(-5, -15)$ **15.** $(2, 5)$ **17.** $(7, 0)$

19. No solution   **21.** $(-1, -4)$ **23.** $(100, 10)$ **25.** $(-2, -1)$ **27.** $(4, 7)$ **29.** $(5, 21)$

31. No solution   **33.** $(1, -2)$ **35.** $(4, 17)$ **37.** $(1, 3)$ **39.** $(1, -1)$ **41.** $(2, 7)$ **43.** $(2, 3)$

45. $\left(\dfrac{de - bf}{ad - bc}, \dfrac{af - ce}{ad - bc} \right)$ **47.** $(57, 34)$ **49.** $(20, 3)$ **51.** $(20, 9)$ **53.** $(20{,}000, \$2900)$ **55.** $(50{,}000, 450{,}000)$

57. $(49{,}450, -165)$

PROBLEM SET 13.2

1. $(7, 4)$ **3.** $(3, 1)$ **5.** $(-2, 5)$ **7.** $(4, 3)$ **9.** $(2, 0, 5)$ **11.** $(-1, 2, -3)$ **13.** $(6, -5, 0)$

15. $(1, 2\frac{1}{2}, 3\frac{1}{2})$ **17.** $(-2, 1, 3)$ **19.** $(10, 4, 1)$ **21.** $(1, 2, 3, 4)$ **23.** $(2, 3, -1, 3, 6)$ **25.** $(10, 27, 29)$

27. Nickels: 10; dimes: 7; quarters: 4   **29.** $(2, -3, 1)$

1. 7 **3.** 5; 6 **5.** 5; −4 **7.** (11, 6) **9.** (1, 3) **11.** (2, −7) **13.** (−1, 2); (1, 2) **15.** (1, 7); (−1, 1)

17. (10, 1); (55, 6) **19.** $(1, 3); \left(\dfrac{31}{9}, \dfrac{-17}{9}\right)$ **21.** (1, 2); (3, 10) **23.** (−3, −2); (−3, 2); (3, −2); (3, 2)

25. (2, 5); (−5, −23) **27.** (−2, −5); (5, 2) **29.** (1, 8); (−17, 5) **31.** (1, 1); (1, −1); (−1, 1); (−1, −1)

33. (2, 3); (3, 2) **35.** (0, 0); (1, 1) **37.** 2 and 9 **39.** 7 and 3 **41.** (100, 60,000); (400, 240,000)

PROBLEM SET 13.4

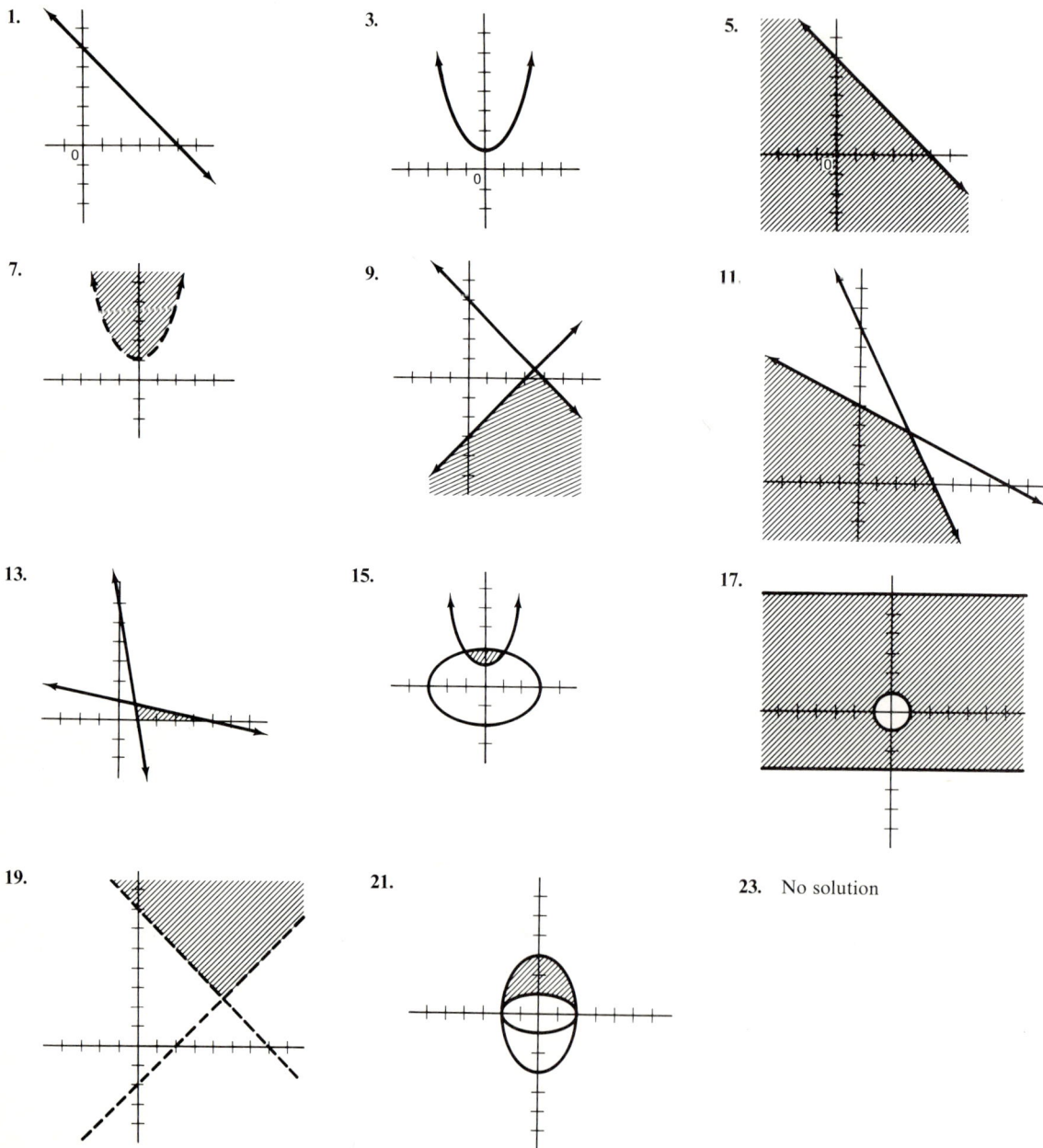

1.

3.

5.

7.

9.

11.

13.

15.

17.

19.

21.

23. No solution

25.

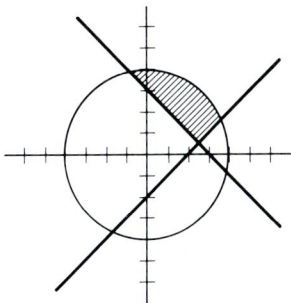

27.

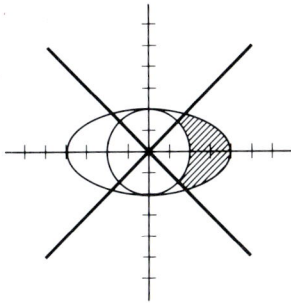

29.

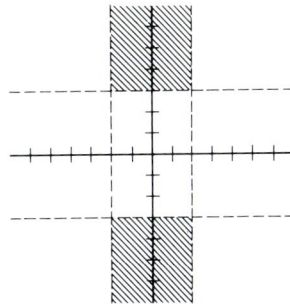

31.

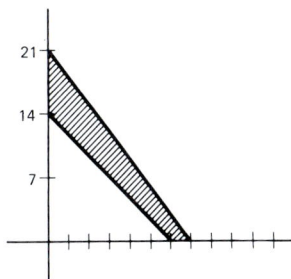

33. (a) $x^2 + y^2 \leq 15^2$
(b) $(x - 20)^2 + (y - 10)^2 \leq 15^2$
(c)

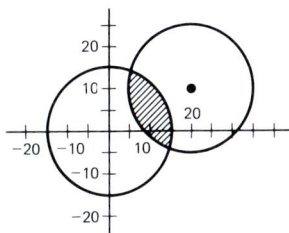

PROBLEM SET 13.5

1. $(11, 4)$　　**3.** $(2, 1)$　　**5.**

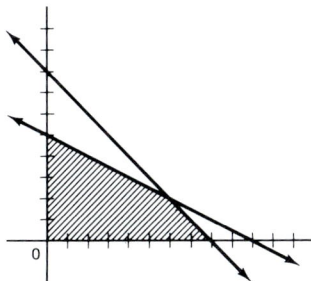

7. $P = 600$ at $(1, 6)$　　**9.** $P = 550$ at $(5, 5)$

11. $C = 1350$ at $(5, 3)$　　**13.** $C = 7300$ at $(50, 30)$　　**15.** $P = 475{,}000$ at $(250, 150)$　　**17.** $C = 220{,}000$ at $(1600, 1200)$
19. $C = 390$ at $(0, 3)$

CHAPTER 13　REVIEW EXERCISES

1. $(12, -3)$　　**2.** $(1, -5)$　　**3.** $(5, 7)$　　**4.** $(7, 39)$　　**5.** Dimes: 9; quarters: 8　　**6.** $(7, 1, -2)$　　**7.** $(1, 0, -5)$

8. $(1, -2, 5)$　　**9.** $(5000, 3000, 2000)$　　**10.** $(1, 1); (-4, 1)$　　**11.** $(5, -2); \left(\dfrac{-63}{13}, \dfrac{38}{13}\right)$

12. $(3, 1); (-3, 1); (3, -1); (-3, -1)$ **13.** $(2, 64)$ **14.**

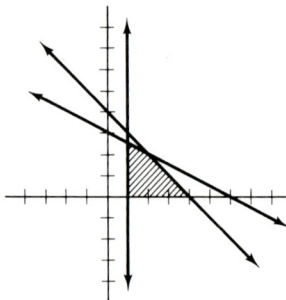

15.

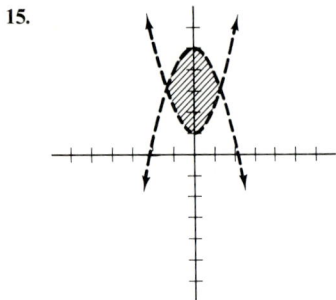

16.

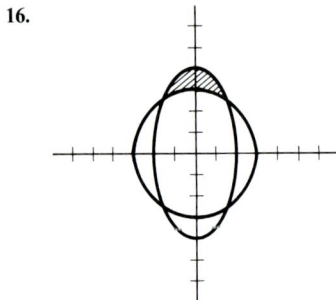

17.

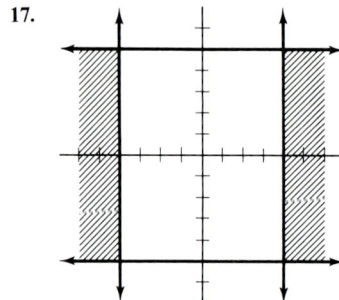

18. $P = 470$ at $(4, 9)$ **19.** $C = 490$ at $(0, 7)$ **20.** $P = 47,500$ at $(2500, 1500)$

PROBLEM SET 14.1

1. -3 **3.** -12 **5.** 8 **7.** 38 **9.** 8 **11.** $\begin{bmatrix} 45 & 52 \\ 39 & 78 \\ 41 & 29 \\ 53 & 59 \\ 27 & 46 \end{bmatrix}; \begin{bmatrix} 49 & 63 \\ 25 & 80 \\ 49 & 50 \\ 50 & 60 \\ 35 & 35 \end{bmatrix}$ **13.** 2 by 3; 7; 9

15. 1 by 5; 3; -4 **17.** Yes; 3 by 6   **19.** Yes; 3 by 4   **21.** No   **23.** $\begin{bmatrix} 13 & -7 & 7 \\ 5 & 0 & 8 \end{bmatrix}$ **25.** Not possible

27. $\begin{bmatrix} 0 & 0 & 0 \\ 0 & 0 & 0 \\ 0 & 0 & 0 \end{bmatrix}$ **29.** $\begin{bmatrix} 14 & 56 \\ 21 & 0 \end{bmatrix}$ **31.** $\begin{bmatrix} -4 & -6 & 14 \\ -6 & 10 & 0 \end{bmatrix}$ **33.** Not possible   **35.** $\begin{bmatrix} 4 & -5 & 6 \\ 2 & 1 & 9 \\ 8 & 10 & -3 \end{bmatrix}$ **37.** $\begin{bmatrix} -15 & 6 \\ -5 & 2 \end{bmatrix}$

39. Not possible   **41.** $\begin{bmatrix} 2 & -5 \\ -6 & 7 \end{bmatrix}$ **43.** $\begin{bmatrix} 5 & -5 & 10 \\ -1 & 7 & -6 \end{bmatrix}$ **45.** $\begin{bmatrix} -5 & 2 \\ 3 & 6 \end{bmatrix}$ **47.** $\begin{bmatrix} 14 & -12 \\ -1 & 21 \end{bmatrix}$

49. $\begin{bmatrix} 94 & 115 \\ 64 & 158 \\ 90 & 79 \\ 103 & 119 \\ 62 & 81 \end{bmatrix}$; two-year totals   **51.** $(8, 5)$ **53.** $(3, 7)$ **55.** $(3, 1)$ **57.** $(3, -1)$

59. (a) $M = \begin{bmatrix} 55 & 60 & 110 \\ 70 & 115 & 225 \\ 45 & 40 & 120 \end{bmatrix}$; $T = \begin{bmatrix} 50 & 65 & 100 \\ 75 & 120 & 195 \\ 45 & 50 & 130 \end{bmatrix}$; (b) 3 by 3; (c) $S = \begin{bmatrix} 105 & 125 & 210 \\ 145 & 235 & 420 \\ 90 & 90 & 250 \end{bmatrix}$; two-day totals;

(d) $D = \begin{bmatrix} -5 & 5 & -10 \\ 5 & 5 & -30 \\ 0 & 10 & 10 \end{bmatrix}$; increase (decrease) from Monday to Tuesday;

(e) $M \cdot C = \begin{bmatrix} 1175 \\ 1990 \\ 1045 \end{bmatrix}$; $T \cdot C = \begin{bmatrix} 1110 \\ 1935 \\ 1125 \end{bmatrix}$; total daily calories for each person

PROBLEM SET 14.2

1. 2 by 2　　**3.** -3　　**5.** $\begin{bmatrix} 1 & 1 \\ 3 & 16 \end{bmatrix}$　　**7.** $\begin{bmatrix} 6 & -23 \\ -14 & 83 \end{bmatrix}$　　**9.** $\begin{bmatrix} -7 & 12 & 5 \\ 9 & -28 & 25 \end{bmatrix}$　　**11.** $(6, 4)$　　**13.** $(4, 0, -2)$

15. $(-2, 5)$　　**17.** $(1, 2, -3)$　　**19.** $\begin{bmatrix} 2 & -3 & 4 & \vdots & 8 \\ 1 & 1 & 3 & \vdots & 1 \\ 5 & -1 & 2 & \vdots & -2 \end{bmatrix}$　　**21.** $\begin{bmatrix} 1 & 1 & 2 & 4 & \vdots & 3 \\ 1 & -1 & 7 & -1 & \vdots & -2 \\ 2 & 1 & -1 & 2 & \vdots & 0 \\ 5 & 4 & 1 & -1 & \vdots & -8 \end{bmatrix}$　　**23.** $\begin{aligned} 2x - y &= 3 \\ 8x + 3y &= -2 \end{aligned}$

25. $\begin{aligned} x + z &= 4 \\ -y + z &= -3 \\ x - 2y &= 2 \end{aligned}$　　**27.** $(-2, -3)$　　**29.** $(1, 7, -5)$　　**31.** $(1, -4)$　　**33.** $(2, 0, -3)$　　**35.** $(-3, -6, 7, 0)$

37. $(4, 3); (4, -3); (-4, 3); (-4, -3)$　　**39.** $(5, -7, 2)$

PROBLEM SET 14.3

1. $\begin{bmatrix} 8 & -6 & 6 \\ 7 & -1 & 1 \end{bmatrix}$　　**3.** $\begin{bmatrix} 3 & -4 & 5 \\ -6 & 8 & -10 \end{bmatrix}$　　**5.** $\begin{bmatrix} 2 \\ -5 \end{bmatrix}$　　**7.** $(6, -1)$　　**9.** $(3, 1, 2)$　　**11.** No　　**13.** No

15. $\begin{bmatrix} -5 & 3 \\ 7 & -4 \end{bmatrix}$　　**17.** $\begin{bmatrix} \frac{9}{2} & -2 \\ -2 & 1 \end{bmatrix}$　　**19.** $\begin{bmatrix} \frac{1}{8} & -\frac{5}{8} \\ \frac{1}{8} & \frac{3}{8} \end{bmatrix}$　　**21.** $\begin{bmatrix} 3 & 0 & 2 \\ -5 & 1 & -5 \\ 1 & 0 & 1 \end{bmatrix}$　　**23.** $\begin{bmatrix} 5 & -1 & 2 \\ 3 & \frac{-1}{2} & \frac{3}{2} \\ -4 & 1 & -2 \end{bmatrix}$

25. $\begin{bmatrix} \frac{1}{2} & 0 \\ 0 & \frac{1}{3} \end{bmatrix}$　　**27.** $\begin{bmatrix} \frac{1}{a} & 0 \\ 0 & \frac{1}{b} \end{bmatrix}$ $a \neq 0, b \neq 0$　　**29.** $\begin{bmatrix} \frac{1}{2} & 0 & 0 \\ 0 & \frac{1}{3} & 0 \\ 0 & 0 & \frac{1}{4} \end{bmatrix}$　　**31.** $\begin{bmatrix} 1 & 0 \\ 0 & 1 \end{bmatrix}$　　**33.** No inverse

35. $\begin{bmatrix} 0 & \frac{1}{2} & 0 \\ 1 & 0 & 0 \\ 0 & 0 & \frac{1}{3} \end{bmatrix}$　　**37.** $\begin{bmatrix} \frac{1}{2} & \frac{1}{2} & \frac{-1}{2} \\ \frac{1}{2} & \frac{-1}{2} & \frac{1}{2} \\ \frac{-1}{2} & \frac{1}{2} & \frac{1}{2} \end{bmatrix}$

PROBLEM SET 14.4

1. $\begin{bmatrix} 4 \\ 6 \end{bmatrix}$ **3.** $\begin{bmatrix} 1 & 0 \\ 0 & 1 \end{bmatrix}$ **5.** $\begin{bmatrix} x + 2y \\ 3x + 4y \end{bmatrix}$ **7.** 7 **9.** -21 **11.** $\begin{bmatrix} -4 & 3 \\ 7 & -5 \end{bmatrix}$ **13.** $\begin{bmatrix} 2 & -3 & 4 \\ 1 & 1 & 3 \\ 5 & -1 & 2 \end{bmatrix} \begin{bmatrix} x \\ y \\ z \end{bmatrix} = \begin{bmatrix} 5 \\ 1 \\ -2 \end{bmatrix}$

15. $\begin{aligned} 2x - y &= 3 \\ 8x + 3y &= -2 \end{aligned}$ **17.** $\begin{bmatrix} 1 & 0 & -4 \\ 1 & 5 & 0 \\ 0 & 1 & 7 \end{bmatrix} \begin{bmatrix} x \\ y \\ z \end{bmatrix} = \begin{bmatrix} 0 \\ -2 \\ 3 \end{bmatrix}$ **19.** $\begin{aligned} x + z &= 4 \\ -y + z &= -3 \\ x - 2y &= 2 \end{aligned}$ **21.** $\begin{bmatrix} 13 \\ -18 \end{bmatrix}$ **23.** $\begin{bmatrix} -37 \\ 17 \end{bmatrix}$

25. $\begin{bmatrix} -7/8 \\ 17/8 \end{bmatrix}$ **27.** $\begin{bmatrix} 9 \\ -20 \\ 4 \end{bmatrix}$ **29.** $\begin{bmatrix} 36 \\ 23.5 \\ -32 \end{bmatrix}$ **31.** $\begin{bmatrix} 3 \\ 4 \end{bmatrix}$ **33.** $\begin{bmatrix} c/a \\ d/b \end{bmatrix}$ **35.** $\begin{bmatrix} 5 \\ 6 \\ 2 \end{bmatrix}$ **37.** $\begin{bmatrix} 3 \\ -2 \end{bmatrix}$

39. No solution **41.** $\begin{bmatrix} 4 \\ 5 \\ -2 \end{bmatrix}$ **43.** $\begin{bmatrix} 1 \\ 2 \\ 3 \end{bmatrix}$ **45.** $\begin{bmatrix} 1 & 1 & 1 \\ 10 & -30 & 0 \\ 0 & -30 & 50 \end{bmatrix} \begin{bmatrix} x \\ y \\ z \end{bmatrix} = \begin{bmatrix} 0 \\ 110 \\ 140 \end{bmatrix}$; $A^{-1} = \begin{bmatrix} \frac{150}{230} & \frac{8}{230} & \frac{-3}{230} \\ \frac{50}{230} & \frac{-5}{230} & \frac{-1}{230} \\ \frac{30}{230} & \frac{-3}{230} & \frac{4}{230} \end{bmatrix}$; $\begin{bmatrix} 2 \\ -3 \\ 1 \end{bmatrix}$

PROBLEM SET 14.5

1. $\begin{bmatrix} 7 & 8 \\ 8 & 7 \end{bmatrix}$ **3.** $\begin{bmatrix} 47 & 18 \\ 34 & 13 \end{bmatrix}$ **5.** $\begin{bmatrix} 37 & 63 \\ 27 & 46 \end{bmatrix}$ **7.** $\begin{bmatrix} 32 & -8 \\ 8 & -32 \end{bmatrix}$ **9.** $\begin{bmatrix} -5 & 7 \\ 3 & -4 \end{bmatrix}$ **11.** $\begin{bmatrix} 37 \\ -21 \end{bmatrix}$ **13.** D1

15. D3 **17.** D5 **19.** D4 **21.** 11 **23.** 1 **25.** -15 **27.** -1 **29.** -1 **31.** 286 **33.** -104

35. -258 **37.** 0 **39.** 1 **41.** abc **43.** $-3i + 6j - 3k$ **47.** 2 or 7

PROBLEM SET 14.6

1. $\begin{bmatrix} 11 & 0 \\ 1 & 7 \end{bmatrix}$ **3.** $\begin{bmatrix} 18 & 4 \\ -14 & 18 \end{bmatrix}$ **5.** $\begin{bmatrix} 0 & -9 & 5 \\ 4 & 10 & -7 \\ -6 & 6 & -3 \end{bmatrix}$ **7.** $\begin{bmatrix} -2 & 23 & 5 \\ 10 & -9 & -21 \\ -7 & -89 & 22 \end{bmatrix}$ **9.** 10 **11.** 165

13. $\left(\frac{43}{23}, \frac{9}{23} \right)$ **15.** $\left(\frac{143}{230}, \frac{125}{230} \right)$ **17.** $\left(\frac{41}{92}, \frac{8}{92} \right)$ **19.** No solution **21.** $\left(\frac{38}{85}, \frac{-71}{85}, \frac{204}{85} \right)$ **23.** $\left(\frac{231}{75}, \frac{309}{75}, \frac{15}{75} \right)$

25. $\left(\frac{149}{62}, \frac{57}{62}, \frac{-46}{62} \right)$ **27.** $(4e - 3f, f - e)$ **29.** $\left(\frac{g - f - e}{-2}, \frac{f - e - g}{-2}, \frac{e - f - g}{-2} \right)$

CHAPTER 14   REVIEW EXERCISES

1. $\begin{bmatrix} 10 & 15 & 13 \\ 21 & 32 & 45 \\ 17 & 25 & 60 \end{bmatrix}$; $\begin{bmatrix} 12 & 14 & 17 \\ 19 & 35 & 50 \\ 20 & 22 & 55 \end{bmatrix}$ **2.** 3 by 3 **3.** $\begin{bmatrix} 22 & 29 & 30 \\ 40 & 67 & 95 \\ 37 & 47 & 115 \end{bmatrix}$ **4.** $\begin{bmatrix} 2 & -1 & 4 \\ -2 & 3 & 5 \\ 3 & -3 & -5 \end{bmatrix}$

5. $\begin{bmatrix} 202 \\ 497 \\ 493 \end{bmatrix}$; $\begin{bmatrix} 232 \\ 511 \\ 488 \end{bmatrix}$ **6.** $\begin{aligned} 2x - 4y + 5z &= 3 \\ 7x + 8y + z &= -8 \\ 9x + 3y &= 10 \end{aligned}$ **7.** $\begin{aligned} x &= -3 \\ y &= 5 \\ z &= 7 \end{aligned}$ **8.** $\begin{bmatrix} 1 & -11 & -6 \\ 3 & 4 & 19 \end{bmatrix}$; $(5, 1)$

9. $\begin{bmatrix} 1 & -3 & -4 & -6 \\ 2 & 1 & 0 & 7 \\ 0 & 1 & -1 & 7 \end{bmatrix}$; $(1, 5, -2)$ **10.** Yes **11.** No **12.** $\begin{bmatrix} 4 & -9 \\ -3 & 7 \end{bmatrix}$ **13.** $\begin{bmatrix} 3 & -2 & -1 \\ 10 & -6 & -5 \\ -8 & 5 & 4 \end{bmatrix}$

14. $\begin{bmatrix} 22 \\ -17 \end{bmatrix}$ **15.** $\begin{bmatrix} 31 \\ 103 \\ -83 \end{bmatrix}$ **16.** 31 **17.** 35 **18.** $\left(\frac{107}{31}, \frac{82}{31} \right)$ **19.** $\left(\frac{37}{35}, \frac{119}{35}, \frac{10}{35} \right)$

PROBLEM SET 15.1

1. $\dfrac{-5}{2}$ **3.** $1; -5$ **5.** $-5x - 8$ **7.** $x^3 + 2x^2 - 34x + 7$ **9.** 6 R 1   **11.** 29 R 9   **13.** 194 R 53

15. $x^2 + 2x + 4$ R (-2) **17.** $x^2 - 5x + 2$ **19.** $2x^2 + 3x - 5$ R $(-2x - 2)$ **21.** $x^3 + x$ R 1

23. $4x^2 + x - 5$ **25.** $5x^2 - 2x + 3$ R (-4) **27.** $x^3 - 2x^2 + 3x - 4$ **29.** $2x^3 - 7x^2 + 5x - 3$ R (-2)

31. $x^3 - 2x^2 + 4x - 8$ R 32   **33.** $x^3 - x^2 + x - 1$ R 1   **35.** $2x^2 + 6x + 1$ R 18   **37.** $2x^2 + 3$

39. $x^2 + ix + (3 + i)$ R $4i$ **41.** $x^2 - 2x + 3$ R 11   **43.** $x - 1 - i$ R i

PROBLEM SET 15.2

1. 3   **3.** 2   **5.** 2   **7.** $x - 4$ R (-8) **9.** $x^2 - 2x + 5$ R 3   **11.** $x^4 - x^3 + x^2 - x + 1$ R (-8)

13. $Q = x - 7; R = 0$ **15.** $Q = 2x - 1; R = -5$ **17.** $Q = 4x^2 + 5x + 17; R = 26$ **19.** $22; -62$ **21.** $179; 7$

23. No   **25.** Yes   **27.** Yes   **29.** Yes   **31.** No   **33.** Yes   **35.** No   **37.** Yes   **39.** No

41. No   **43.** $12; 15; 0; x + 3$ **45.** $0; 0; 0; x, x - 1, x + 1$ **47.** $8a^3; 0; a^3; x + a$

PROBLEM SET 15.3

1. 3   **3.** $2; -2$ **5.** $\dfrac{1 \pm i\sqrt{19}}{2}$ **7.** $0; 1; -1$ **9.** $ax^2 + ax - 6a$ **11.** $ax^4 + 9ax^2$

13. $(x - 2)(x - 3)(x + 4)$ **15.** $(x - 2i)(x + 2i)$ **17.** $(x - 1 - 3i)(x - 1 + 3i)$ **19.** $x(x - 1)(x + 5)$

21. $(x - 1 - i\sqrt{5})(x - 1 + i\sqrt{5})$ **23.** 2, 7 (mult. = 1)   **25.** 0 (mult. = 4); 1 (mult. = 1)

27. 0 (mult. = 3); 1 (mult. = 2); -2 (mult. = 5)   **29.** $-3x^3 + 21x - 18$ **31.** $5x^3 + 5x^2 + 45x + 45$

33. $3x^2 - 12x + 15$ **35.** -4 (mult. = 1); $2(x + 4)$ **37.** 2, 4 (mult. = 1); $(x - 2)(x - 4)$

39. $\dfrac{4}{3}, \dfrac{-1}{2}$ (mult. = 1); $6\left(x - \dfrac{4}{3}\right)\left(x + \dfrac{1}{2}\right)$ **41.** $2i, -2i, 1$ (mult. = 1); $-1(x - 2i)(x + 2i)(x - 1)$

43. 0 (mult. = 3); $\pm 3i, 2 \pm i$ (mult. = 1); $2x^3(x - 3i)(x + 3i)(x - 2 - i)(x - 2 + i)$

PROBLEM SET 15.4

1. $x - 9$ R 15   **3.** $2x^2 + 3x + 5$ R 11   **5.** $x^3 + x^2 + x + 1$ **7.** $x^2 + 7x + 2$ **9.** $x^4 - x^3 + x^2 - x + 1$

11. $1; 1; 12; -1$ **13.** $1; 1; 1; -18$ **15.** 3 or 1; 0; 1; 0   **17.** 2 or 0; 1; 1; -2 **19.** 2 or 0; 2 or 0; 1; -1

21. 2 or 0; 1; 1; -1 **23.** 0; 3 or 1; 0; -2 **25.** 1; 0; 1; 0   **27.** 1; 2 or 0; 2; -1 **29.** 3 or 1; 3 or 1; 2; -1

PROBLEM SET 15.5

1. $\dfrac{-2}{3}(x - 1)(x + 3)$ **3.** $\dfrac{3}{2}(x - 2)(x - i)(x + i)$ **5.** (a) 1, 1; (b) $-1, 8$ **7.** (a) 3 or 1, 0; (b) 0, 1   **9.** $1, -1, -2, -3$

11. $3, -2, \dfrac{-1}{2}$ **13.** $2, \dfrac{-1}{2}, \dfrac{-1}{3}$ **15.** $\dfrac{1}{2}, \dfrac{-4}{3}, -5$ **17.** $2, 2, -1, -1$ **19.** (a) 1; (b) $i, -i$ **21.** (a) 2; (b) $\sqrt{5}, -\sqrt{5}$

23. (a) 2; (b) $1 - \sqrt{5}, 1 + \sqrt{5}$ **25.** (a) $1, -1$; (b) $i, -i$ **27.** (a) $-1, -2$; (b) $\dfrac{3 \pm i\sqrt{19}}{2}$

PROBLEM SET 15.6

1. (a) 0, 1; (b) $-3, 0$; (c) $\dfrac{-5}{2}$ **3.** (a) 2 or 0; 0; (b) 0, 4; (c) 1, 4   **5.** (a) 2 or 0; 1; (b) $-2, 2$; (c) 1

7. (a) 2 or 0; 2 or 0; (b) $-2, 2$; (c) $\pm 1, \pm 2$.   **9.** $(1, 2), (3, 4), (-2, -1)$ **11.** $(1, 2), (4, 5), (-1, 0)$ **13.** $(2, 3)$

15. $(0, 1), (2, 3), (-3, -2) (-1, 0)$ **17.** 0.7 **19.** 1.7 **21.** -1.5 **23.** 1.9 **25.** -0.7 **27.** $1; 1.6; -0.6$

29. $1, \dfrac{-1 \pm i\sqrt{3}}{2}$ **31.** $1; 2; \dfrac{-1 \pm i\sqrt{23}}{2}$ **33.** $-1; i, -i$ **35.** $0.67; 2.7 \pm 0.6i$

CHAPTER 15   REVIEW EXERCISES

1. 107 R 9 **2.** $(x^2 + 2x - 3)$ R 2 **3.** $x^4 + 1$ **4.** $2x^2 + 3x - 4$ R 5 **5.** $4x^3 + 3x^2 - 2x + 1$

6. $x^3 - 3x^2 + 9x - 27$ R 162 **7.** 3 **8.** -12 **9.** Yes   **10.** No   **11.** $(x - 4)(x - i)(x + i)$

12. $\left(x + \dfrac{1}{2} + \dfrac{i\sqrt{11}}{2}\right)\left(x + \dfrac{1}{2} - \dfrac{i\sqrt{11}}{2}\right)$ **13.** 0(mult. = 4); $1, -1$ (mult. = 1)   **14.** 1 (mult. = 1); -2 (mult. = 3); 3 (mult. = 5)

15. $-3x^2 - 9x + 30$ **16.** $3x^2 - 6x + 6$ **17.** (a) 2 or 0; 0; (b) 1; 0   **18.** (a) 1; 2 or 0; (b) 3; -1 **19.** $1, 5, \dfrac{-1}{2}, -2$

20. $1, -3, \dfrac{-1}{2} \pm \dfrac{i\sqrt{19}}{2}$ **21.** 4.5 **22.** 2.2

PROBLEM SET 16.1

1. 9 **3.** 2 **5.** 1 **7.** $2, 5, 8, 11, 14$ **9.** $100, 99, 98, 97, 96$ **11.** $-1, 8, -27, 64, -125$ **13.** $\dfrac{1}{2}, \dfrac{1}{4}, \dfrac{1}{8}, \dfrac{1}{16}, \dfrac{1}{32}$

15. $\dfrac{1}{2}, \dfrac{2}{3}, \dfrac{3}{4}, \dfrac{4}{5}, \dfrac{5}{6}$ **17.** $\dfrac{2}{6}, \dfrac{3}{7}, \dfrac{4}{8}, \dfrac{5}{9}, \dfrac{6}{10}$ **19.** $\dfrac{-1}{2}, \dfrac{1}{3}, \dfrac{-1}{4}, \dfrac{1}{5}, \dfrac{-1}{6}$ **21.** $\dfrac{1}{x}, \dfrac{1}{x^2}, \dfrac{1}{x^3}, \dfrac{1}{x^4}, \dfrac{1}{x^5}$ **23.** n^3 **25.** 2

27. $(-1)^n 3n$ **29.** $\dfrac{1}{n}$ **31.** $\dfrac{n}{n+1}$ **33.** $\dfrac{(-1)^n}{n^2}$ **35.** $0.41, 0.32, 0.27, 0.24$ **37.** $162; 2(3^{n-1})$

39. $0, 0.30, 0.48, 0.60$ **41.** $-6; 14 - 4n$ **43.** $i, -1, -i, 1$ **45.** $\dfrac{(-1)^n 2n}{10^n}$

47. $7{,}000{,}000; 3{,}500{,}000; 2{,}333{,}333; 1{,}750{,}000$ **49.** $5, 8, 13, 21, 34$

PROBLEM SET 16.2

1. $1, 8, 27, 64, 125$ **3.** $-1, 1, -1, 1, -1$ **5.** $\dfrac{0}{2}, \dfrac{1}{3}, \dfrac{2}{4}, \dfrac{3}{5}, \dfrac{4}{6}$ **7.** $3n$ **9.** $(-1)^n 2^n$ **11.** 100 **13.** 38 **15.** $\dfrac{25}{12}$

17. 3 **19.** $\dfrac{1641}{420}$ **21.** $1^x + 2^x + 3^x + 4^x + 5^x$ **23.** $x_1^2 + x_2^2 + x_3^2 + x_4^2 + x_5^2 + x_6^2 + x_7^2$ **25.** $\displaystyle\sum_{n=3}^{7} n^2$ **27.** $\displaystyle\sum_{n=2}^{5} \dfrac{1}{2^n}$

29. $\displaystyle\sum_{n=3}^{7} (-1)^n n$ **31.** $\displaystyle\sum_{n=1}^{4} \dfrac{n}{n+1}$ **33.** $\displaystyle\sum_{n=3}^{7} x^n$ **35.** $\displaystyle\sum_{n=1}^{4} a_n^3$ **37.** $1, 3, 6, 10, 15; \dfrac{n(n+1)}{2}$ **39.** $2, 6, 14, 30, 62; 2^{n+1} - 2$

41. $-1, 0, -1, 0, -1; \dfrac{(-1)^n - 1}{2}$ **43.** $\dfrac{1}{2}, \dfrac{3}{4}, \dfrac{7}{8}, \dfrac{15}{16}, \dfrac{31}{32}; \dfrac{2^n - 1}{2^n}$ **45.** $\displaystyle\sum_{n=1}^{50} (-1)^{n+1} \dfrac{x^n}{n}$ **47.** 1.875

PROBLEM SET 16.3

1. $25; n^2$ **3.** $-32; (-1)^n 2^n$ **5.** $11; 2n + 1$ **7.** $2; 12 - 2n$ **9.** $\displaystyle\sum_{n=1}^{4} n^3$ **11.** $\displaystyle\sum_{n=1}^{4} \dfrac{1}{2^n}$ **13.** $\displaystyle\sum_{n=1}^{4} (3n + 2)$

15. $41; 8$ **17.** $-10; -4$ **19.** $-125; -50$ **21.** $5.9; 0.2$ **23.** $9, 5, 1, -3; 13 - 4n$ **25.** $-6, -2, 2, 6; -10 + 4n$

27. $5, 5.6, 6.2, 6.8; 4.4 + 0.6n$ **29.** 145 **31.** -146 **33.** -2375 **35.** 24.9 **37.** 1387 **39.** -2720

41. $-57{,}500$ **43.** 1500 **45.** $\displaystyle\sum_{n=1}^{1000} n = 500{,}500$ **47.** $\displaystyle\sum_{n=1}^{10} (-2 + 4n) = 200$ **49.** $\displaystyle\sum_{n=1}^{50} (-10 + 20n) = 25{,}000$

51. $\sum\limits_{n=1}^{16}(-1+3n)=392$ **53.** $\sum\limits_{n=1}^{31}(203-3n)=4805$ **55.** $95+5n$; 195; 2950 **57.** $\dfrac{(n-1)n}{2}$

59. 18,000, 18,800, 19,600, 20,400; 25,200; 216,000

PROBLEM SET 16.4

1. 5; 1; 10; 55 **3.** 16; 3; 31; 175 **5.** 4; 0.5; 6.5; 42.5 **7.** -6; -4; -26; -80 **9.** 2592; 6 **11.** $25; \dfrac{1}{2}$

13. 5.2488; 0.9 **15.** $\dfrac{1}{32}; \dfrac{-1}{2}$ **17.** $18, 6, 2, \dfrac{2}{3}; \dfrac{2}{243}$ **19.** 7, -70, 700, -7000; $-70{,}000{,}000$ **21.** $\dfrac{1}{5}, \dfrac{1}{10}, \dfrac{1}{20}, \dfrac{1}{40}; \dfrac{1}{640}$

23. 93,312 **25.** 1.5625 **27.** 4.251528 **29.** $\dfrac{1}{512}$ **31.** 111,974 **33.** 798.4375 **35.** 41.73624 **37.** $\dfrac{513}{1536}$

39. Since $a_1 r^n = a_{n+1}$ **41.** $\sum\limits_{n=1}^{6} 2^n = 126$ **43.** $\sum\limits_{n=1}^{5}\left(\dfrac{1}{3}\right)^n = \dfrac{121}{243}$ **45.** $\sum\limits_{n=1}^{7} 3^{n-1} = 1093$ **47.** $\sum\limits_{n=1}^{7}\dfrac{1}{2^n} = \dfrac{127}{128}$

49. 1.1; 97,436 **51.** $(1+i)$; $\dfrac{P[(1+i)^n-1]}{i}$; \$51,160; \$82,247 **53.** \$43,231.45; \$49,463

PROBLEM SET 16.5

1. (a) Geometric; 2; (b) 16; 2^{n-1}; (c) 255 **3.** (a) Arithmetic; -4; (b) 5; $25-4n$; (c) 56

5. (a) Geometric; -1; (b) 10; $10(-1)^{n-1}$; (c) 0 **7.** (a) Geometric; $\dfrac{1}{3}$; (b) $\dfrac{1}{81}$; $\dfrac{1}{3^{n-1}}$; (c) $\dfrac{3280}{2187}$

9. (a) Neither; $-$; (b) 125; n^3; (c) 1296 **11.** (a) Geometric; $\dfrac{-1}{2}$; (b) $\dfrac{1}{2}$; $8\left(\dfrac{-1}{2}\right)^{n-1}$; (c) $\dfrac{85}{16}$ **13.** 1960; 2305.6; 2484.88; 2500

15. 3.5; 3.875; 3.996; 4 **17.** 1 **19.** $\dfrac{1}{3}$ **21.** $\dfrac{32}{7}$ **23.** $\dfrac{500}{3}$ **25.** No sum **27.** $\dfrac{5}{9}$ **29.** $\dfrac{7}{99}$ **31.** $\dfrac{235}{999}$

33. $\sum\limits_{n=1}^{\infty}\left(\dfrac{1}{4}\right)^n = \dfrac{1}{3}$ **35.** $\sum\limits_{n=1}^{\infty} 1600\left(\dfrac{1}{2}\right)^{n-1} = 3200$ **37.** $\sum\limits_{n=1}^{\infty} 1000\left(\dfrac{-1}{2}\right)^{n-1} = \dfrac{2000}{3}$

39. $\dfrac{1000}{1+r}; \dfrac{1}{1+r}; \dfrac{1000}{r}$ **41.** 20; 16; 36

CHAPTER 16   REVIEW EXERCISES

1. 17, 14, 11, 8, 5 **2.** $\dfrac{-1}{2}, \dfrac{1}{3}, \dfrac{-1}{4}, \dfrac{1}{5}, \dfrac{-1}{6}$ **3.** $2(3^{n-1})$ **4.** $\dfrac{(-1)^n x^n}{10^n}$ **5.** $\dfrac{1}{1}+\dfrac{1}{2}+\dfrac{1}{3}+\dfrac{1}{4}+\dfrac{1}{5}+\dfrac{1}{6}=\dfrac{147}{60}$

6. $\dfrac{-1}{1}+\dfrac{1}{4}-\dfrac{1}{9}+\dfrac{1}{16}=\dfrac{-115}{144}$ **7.** $\sum\limits_{n=1}^{5} n^2$ **8.** $\sum\limits_{n=1}^{5}\dfrac{3n}{x^n}$ **9.** -4 **10.** 14 **11.** 320 **12.** $54-4n$

13. 2 **14.** 2560 **15.** 5115 **16.** $5(2^{n-1})$ **17.** $\dfrac{10{,}000}{7}$ **18.** $\dfrac{5}{3}$ **19.** $\dfrac{29}{99}$

PROBLEM SET 17.1

1. 120 **3.** 5040 **5.** 0 **7.** 30 **9.** 720 **11.** 117,600 **13.** 2 **15.** 120 **17.** 40,320
19. 4; 8; 1024; 1,048,576 **21.** 172,800 **23.** 6840 **25.** 720 **27.** 970,200 **29.** 480; 1080; 56
31. $7! = 5040$ **33.** $2^{32} \approx 4.3 \times 10^9$ **35.** 5^{10}; 2^{20}; $5^{10} \cdot 2^{20} \approx 10^{13}$ **37.** 3.9×10^8

PROBLEM SET 17.2

1. 840 **3.** 2,193,360 **5.** 720 **7.** 13,144 **9.** $9! = 362,880$; $8! = 40,320$ **11.** 35 **13.** 455 **15.** 4950

17. $\binom{250}{2} = 31,125$; $\binom{250}{3} = 2,573,000$; $\binom{250}{5} \approx 7.8 \times 10^9$ **19.** $\binom{15}{10} = 3003$ **21.** $\binom{10}{4} = 210$

23. $\binom{12}{3} = 220$; $\binom{10}{4} = 210$; 46,200 **25.** $\binom{52}{10} = 1.58 \times 10^{10}$ **27.** $\binom{36}{5} = 376,992$ **29.** 1, 5, 10, 10, 5, 1; total $= 32$

31. 56 **33.** 45

PROBLEM SET 17.3

1. 0.3 **3.** $\dfrac{1}{5}$ **5.** 120 **7.** 120 **9.** 55,440 **11.** 220 **13.** 0.45 **15.** $\dfrac{3}{8}$ **17.** $\dfrac{2}{3}$ **19.** $\dfrac{7}{9}$

21. 0.17 **23.** Counting: $\dfrac{1}{13}$; $\dfrac{1}{52}$; $\dfrac{1}{2}$; $\dfrac{3}{26}$ **25.** Counting: $\dfrac{1}{6}$; $\dfrac{1}{3}$; $\dfrac{1}{2}$ **27.** Scientific: 0.99 **29.** Counting: 0.05; 0.02

31. Guesstimate:? **33.** Counting: 0.318; 0.227; 0.091; 0.065 **35.** Scientific: 0.034 **37.** Scientific: 0.947

39. Scientific: 0.998

PROBLEM SET 17.4

1. $a^2 - 2ab + b^2$ **3.** $-x^7$ **5.** $256t^8$ **7.** 210 **9.** 7 **11.** 1225 **13.** $x^4 + 4x^3 + 6x^2 + 4x + 1$

15. $y^3 - 6y^2 + 12y - 8$ **17.** $p^5 - 5p^4q + 10p^3q^2 - 10p^2q^3 + 5pq^4 - q^5$ **19.** $m^3 - 15m^2 + 75m - 125$

21. $x^{14} + 14x^{13}y + 91x^{12}y^2 + 364x^{11}y^3$ **23.** $x^{20} - 60x^{19} + 1710x^{18} - 30,780x^{17}$

25. $512m^9 + 11,520m^8n + 115,200m^7n^2 + 672,000m^6n^3$ **27.** $x^8 - 56x^7 + 1372x^6 - 19,208x^5$

29. $8192u^{13} + 53,248u^{12}v + 159,744u^{11}v^2 + 292,864u^{10}v^3$ **31.** $u^{60} - 20u^{57}v^5 + 190u^{54}v^{10} - 1140u^{51}v^{15}$

33. (a) $\dfrac{1}{256} + \dfrac{12}{256} + \dfrac{54}{256} + \dfrac{108}{256} + \dfrac{81}{256}$

(b) $\dfrac{1}{1024} + \dfrac{15}{1024} + \dfrac{90}{1024} + \dfrac{270}{1024} + \dfrac{405}{1024} + \dfrac{243}{1024}$

(c) $\dfrac{1}{4096} + \dfrac{18}{4096} + \dfrac{135}{4096} + \dfrac{540}{4096} + \dfrac{1215}{4096} + \dfrac{1458}{4096} + \dfrac{729}{4096}$

PROBLEM SET 17.5

1. $(n + 2)(n + 1)$ **3.** $2(2^{n+1} - 1)$ **5.** $\dfrac{n + 1}{n + 2}$

CHAPTER 17 REVIEW EXERCISES

1. 720 **2.** 1 **3.** 90 **4.** 24 **5.** 56 **6.** 21 **7.** 10,080 **8.** $2^{10} = 1024$ **9.** 120 **10.** 210

11. 20,358,520 **12.** 168,960 **13.** $\dfrac{7}{12}$ **14.** 0.13 **15.** $\dfrac{1}{2}$ **16.** $\dfrac{2}{15}$ **17.** 0.0498

18. $x^5 + 10x^4 + 40x^3 + 80x^2 + 80x + 32$ **19.** $a^6 - 6a^5 + 15a^4 - 20a^3 + 15a^2 - 6a + 1$

20. $x^{15} - 30x^{14} + 420x^{13} - 3640x^{12}$ **21.** $z^{10} + 5z^9 + \dfrac{45}{4}z^8 + 15z^7$

INDEX
OF MATHEMATICAL
TERMS

INDEX
OF APPLICATIONS

IMPORTANT GRAPHS OF TRIGONOMETRY

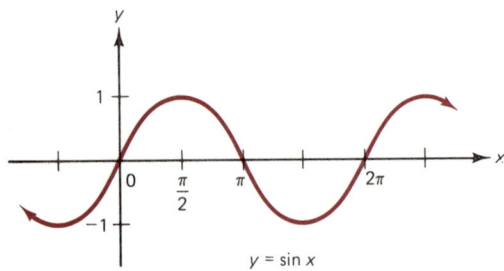

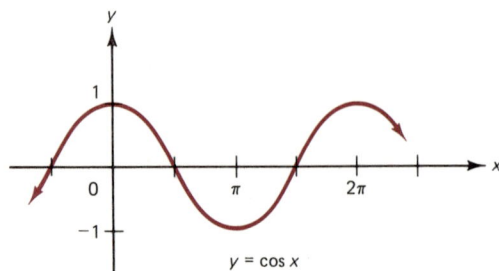

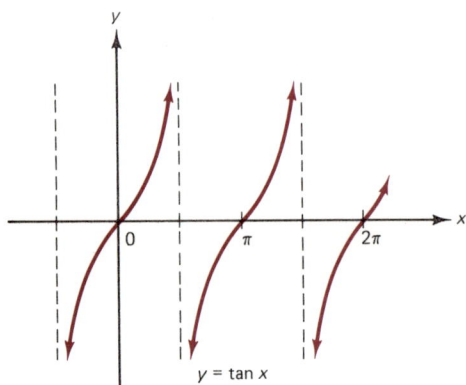

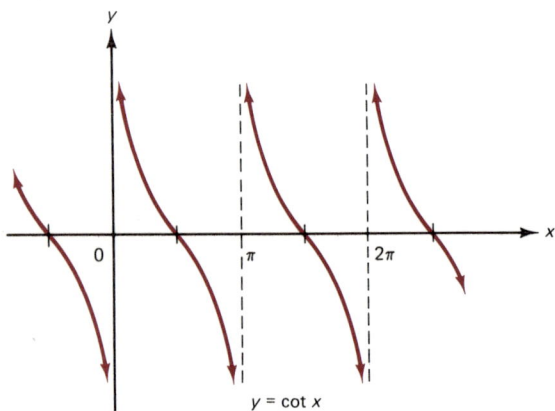

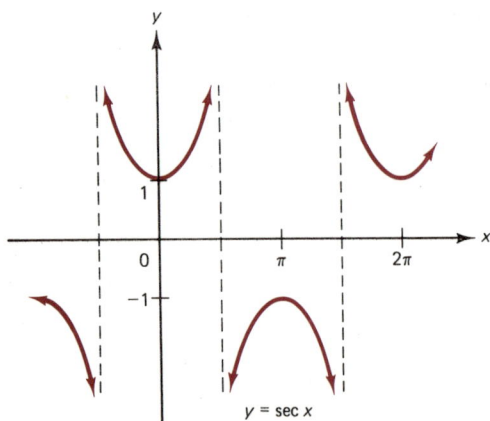

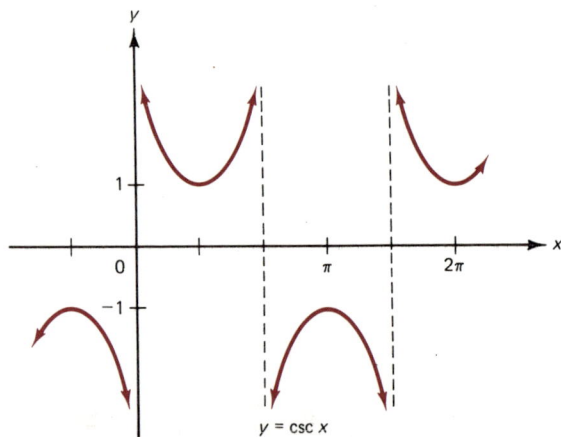

$y = \sin x$

$y = \cos x$

$y = \tan x$

$y = \cot x$

$y = \sec x$

$y = \csc x$